Texts in Applied Mathematics 2

Texts in Applied Mathematics

1. *Sirovich:* **Introduction to Applied Mathematics.**
2. *Wiggins:* **Introduction to Applied Nonlinear Dynamical Systems and Chaos.**
3. *Hale/Koçak:* **Differential Equations: An Introduction to Dynamics and Bifurcations.**
4. *Chorin/Marsden:* **A Mathematical Introduction to Fluid Mechanics, 2nd ed.**

Stephen Wiggins

Introduction to Applied Nonlinear Dynamical Systems and Chaos

With 291 Illustrations

Springer-Verlag
New York Berlin Heidelberg
London Paris Tokyo Hong Kong

Stephen Wiggins
Department of Applied Mechanics
California Institute of Technology
Pasadena, California 91125, USA

Library of Congress Cataloging-in-Publication Data
Wiggins, Stephen.
Introduction to applied nonlinear dynamical systems and chaos /
Stephen Wiggins.
p. cm. — (Texts in applied mathematics)
1. Differentiable dynamical systems. 2. Nonlinear theories.
3. Chaotic behavior in systems. I. Title. II. Series.
QA614.8.W54 1990
003'.85—dc20 89-27832

Mathematics Subject Classification (1980): 58 Fxx, 34Cxx, 70Kxx

Camera-ready copy prepared using a LaTeX file.
Printed and bound by R.R. Donnelley & Sons, Inc., Harrisonburg, Virginia.
Printed in the United States of America.

9 8 7 6 5 4 3 2 1 Printed on acid-free paper.

ISBN 0-387-97003-7 Springer-Verlag New York Berlin Heidelberg
ISBN 3-540-97003-7 Springer-Verlag Berlin Heidelberg New York

To Samantha,

for all the wild nights,

And To Meredith,

for all the calm days.

Series Preface

Mathematics is playing an ever more important role in the physical and biological sciences, provoking a blurring of boundaries between scientific disciplines and a resurgence of interest in the modern as well as the classical techniques of applied mathematics. This renewal of interest, both in research and teaching, has led to the establishment of the series: *Texts in Applied Mathematics (TAM)* .

The development of new courses is a natural consequence of a high level of excitement on the research frontier as newer techniques, such as numerical and symbolic computer systems, dynamical systems, and chaos, mix with and reinforce the traditional methods of applied mathematics. Thus, the purpose of this textbook series is to meet the current and future needs of these advances and encourage the teaching of new courses.

TAM will publish textbooks suitable for use in advanced undergraduate and beginning graduate courses, and will complement the *Applied Mathematical Sciences (AMS)* series, which will focus on advanced textbooks and research level monographs.

Preface

This textbook was developed from material presented in a year-long, graduate-level course in nonlinear dynamics that I taught at Caltech over the past five years. It contains the basic techniques and results I believe to be necessary for graduate students to begin research in the field. The ideal prerequisite for a nonlinear dynamics course would be a thorough knowledge of Arnold's *Ordinary Differential Equations* (Arnold [1973]) or Hirsch and Smale's *Differential Equations, Dynamical Systems, and Linear Algebra* (Hirsch and Smale [1974]). Because only in the rarest instances have I found this prerequisite to be met, I have rapidly reviewed the necessary background material in Section 1.1 of this book.

My main goal in the classroom and in the pages of this book is to provide students with a large arsenal of techniques to increase their chances of success when faced with a nonlinear problem. Inevitably, however, the methods and techniques one has learned often do not quite work for the problems one needs to solve. Consequently, I also try to provide students with a sufficiently strong theoretical base so that they will have the tools and overview they need to develop their own methods and techniques. As a result, this book is long on detail and contains more material than can be covered in a year's worth of lectures. However, because the book *does* contain detailed treatment of its subject, it is possible to cover all the topics presented in an academic year and to merely make reading assignments in topics for which there is no time to lecture in depth.

I would like to make a few remarks concerning content. While in Chapter 3 I spend a fair amount of time on the concept of the codimension of a bifurcation, it is not something that a student first learning the subject needs to worry about in detail. For example, I typically give a one-hour lecture on the subject and make a reading assignment. This is not to say that I believe the idea to be an unimportant one but, rather, that a certain amount of "mathematical maturity" concerning the subject of bifurcation theory is needed before it can be really appreciated. I included such a detailed treatment because it is difficult to find a complete discussion of the subject in the context of *dynamics*. In this regard I have followed the seminal paper of Arnold [1972], from which I learned the subject. All of the above comments notwithstanding, when reading this section the reader should ask him- or herself just how much of it is merely formalized mathematical common sense.

Chapter 4 is concerned with global aspects of dynamics. It has been my experience that most students have had virtually no previous exposure to such ideas. I have thus limited most of the geometrical constructions to two dimensions for maps and three dimensions for vector fields. In this way I hope the student can more easily develop his or her geometrical intuition. All of the results are nonetheless valid in higher dimensions; the interested reader should refer to *Global Bifurcations and Chaos—Analytical Methods* [Wiggins 1988] for a discussion at this level.

Finally, although nonlinear dynamics and chaos have become something of a fad over the past decade, it is still true that an understanding of nonlinear phenomena requires a solid mathematical background and a lot of hard work. Hopefully, those who seek the latter will find this book useful.

At this time I would like to acknowledge all of the help and encouragement I have received in the development of this book. The reader will notice the influence of Philip Holmes throughout the book. Phil was my first teacher in this subject and showed me the beauty of geometry and dynamics. He has influenced my approach to the study of dynamics in many ways—with the exception of my propensity for long and detailed discussions such as those appearing throughout this book, and for which he should not be blamed. Steve Shaw read the entire book, caught many mistakes, and made many useful suggestions. Pat Sethna also offered much good advice concerning content and style. He patiently listened to me present much of the material in the book and often provided me with new insights. Jerry Marsden and Marty Golubitsky also read substantial portions of the manuscript, caught a number of errors, and provided invaluable advice.

The artwork in the book was done by Peggy Firth. Working with Peggy was a real pleasure. She was able to take my roughest sketches and vaguest descriptions and transform them into beautiful and instructive illustrations. Her willingness to cheerfully endure seemingly endless revisions, often on a moment's notice, contributed immensely to the book. I also wish to thank my wife Meredith for copyediting the book. Despite our careful planning, the birth of this book coincided with the birth of our daughter Samantha, and Meredith was forced to juggle copyediting and colic amidst the demands of an often unbearable author. For this sacrifice I will always be grateful. Finally, I wish to thank the National Science Foundation and the Office of Naval Research for the support of my research program.

Contents

Introduction

In this book we will study equations of the following form

$$\dot{x} = f(x, t; \mu) \tag{0.1}$$

and

$$x \mapsto g(x; \mu), \tag{0.2}$$

with $x \in U \subset \mathbb{R}^n$, $t \in \mathbb{R}^1$, and $\mu \in V \subset \mathbb{R}^p$ where U and V are open sets in $\mathbb{R}^n$ and $\mathbb{R}^p$, respectively. The overdot in (0.1) means "$\frac{d}{dt}$," and we view the variables μ as parameters. In the study of dynamical systems the dependent variable is often referred to as "time." We will use this terminology from time to time also. We refer to (0.1) as a *vector field* or *ordinary differential equation* and to (0.2) as a *map* or *difference equation.* Both will be termed *dynamical systems.* Before discussing what we might want to know about (0.1) and (0.2), we need to establish a bit of terminology.

By a solution of (0.1) we mean a map, x, from some interval $I \subset \mathbb{R}^1$ into $\mathbb{R}^n$, which we represent as follows

$$\begin{aligned} x: I &\to \mathbb{R}^n, \\ t &\mapsto x(t), \end{aligned}$$

such that $x(t)$ satisfies (0.1), i.e.,

$$\dot{x}(t) = f\big(x(t), t; \mu\big).$$

The map x has the geometrical interpretation of a curve in $\mathbb{R}^n$, and (0.1) gives the tangent vector at each point of the curve, hence the reason for referring to (0.1) as a vector field. We will refer to the space of dependent variables of (0.1) (i.e., $\mathbb{R}^n$) as the *phase space* of (0.1), and, abstractly, our goal will be to understand the geometry of solution curves in phase space. We remark that in many applications the structure of the phase space may be more general than $\mathbb{R}^n$; frequent examples are cylindrical, spherical, or toroidal phase spaces. We will discuss these situations as they are encountered; for now we incur no loss of generality if we take the phase space of our maps and vector fields to be open sets in $\mathbb{R}^n$.

It will often prove useful to build a little more information into our notation for solutions, which we describe below.

Dependence on Initial Conditions

It may be useful to distinguish a solution curve by a particular point in phase space that it passes through at a specific time, i.e., for a solution $x(t)$ we have $x(t_0) = x_0$. We refer to this as specifying an initial condition. This is often included in the expression for a solution by writing $x(t, t_0, x_0)$. In some situations explicitly displaying the initial condition may be unimportant, in which case we will denote the solution merely as $x(t)$. In still other situations the initial time may be always understood to be a specific value, say $t_0 = 0$; in this case we would denote the solution as $x(t, x_0)$.

Dependence on Parameters

Similarly, it may be useful to explicitly display the parametric dependence of solutions. In this case we would write $x(t, t_0, x_0; \mu)$, or, if we weren't interested in the initial condition, $x(t; \mu)$. If parameters play no role in our arguments we will often omit any specific paramter dependence from the notation.

Some Terminology

1. There are several different terms which are somewhat synonymous with the term *solution* of (0.1). $x(t, t_0, x_0)$ may also be referred to as the *trajectory* or *phase curve* through the point x_0 at $t = t_0$.

2. The graph of $x(t, t_0, x_0)$ over t is referred to as an *integral curve.* More precisely, graph $x(t, t_0, x_0) = \{ (x, t) \in \mathbb{R}^n \times \mathbb{R}^1 \mid x = x(t, t_0, x_0),\ t \in I \}$ where I is the time interval of existence.

3. Let x_0 be a point in the phase space of (0.1). By the *orbit through* x_0, denoted $O(x_0)$, we mean the set of points in phase space that lie on a trajectory passing through x_0. More precisely, for $x_0 \in U \subset \mathbb{R}^n$, the orbit through x_0 is given by $O(x_0) = \{ x \in \mathbb{R}^n \mid x = x(t, t_0, x_0),\ t \in I \}$. Note that for any $T \in I$, it follows that $O(x(T, t_0, x_0)) = O(x_0)$.

Let us now give an example that illustrates the difference between trajectories, integral curves, and orbits.

EXAMPLE 0.1 Consider the equation

$$\begin{aligned} \dot{u} &= v, \\ \dot{v} &= -u, \end{aligned} \qquad (u, v) \in \mathbb{R}^1 \times \mathbb{R}^1. \tag{0.3}$$

The *solution* passing through the point $(u, v) = (1, 0)$ at $t = 0$ is given by $\big(u(t), v(t)\big) = (\cos t, -\sin t)$. The *integral curve* passing through $(u, v) = (1, 0)$ at $t = 0$ is given by $\{ (u, v, t) \in \mathbb{R}^1 \times \mathbb{R}^1 \times \mathbb{R}^1 \mid \big(u(t), v(t)\big) = (\cos t, -\sin t)$, for all $t \in \mathbb{R} \}$. The *orbit* passing through $(u, v) = (1, 0)$ is given by the circle $u^2 + v^2 = 1$. Figure 0.1 gives a geometrical interpretation of these different definitions for this example.

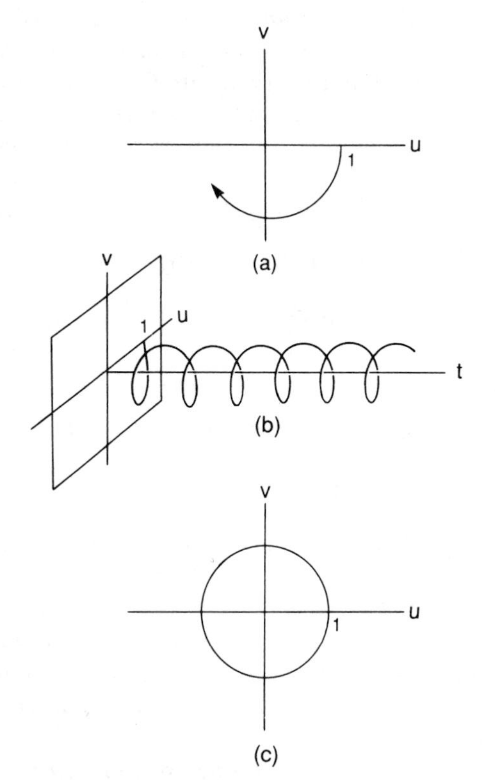

FIGURE 0.1. a) Solution through $(1, 0)$ at $t = 0$. b) Integral curve through $(1, 0)$ at $t = 0$. c) Orbit of $(1, 0)$.

The astute reader will note that we have apparently gotten a bit ahead of ourselves in that we have tacitly assumed that (0.1) has solutions. Of course, this is by no means obvious, and apparently some conditions must be placed on $f(x, t; \mu)$ (as of yet, none have been stated) in order for solutions to exist. Moreover, additional properties of solutions, such as uniqueness and differentiability with respect to initial conditions and parameters, are necessary in applications. When we explicitly consider these questions in Section 1.1, we will see that these properties also are inherited from conditions on $f(x, t; \mu)$. For now, we will merely state without proof that if $f(x, t; \mu)$ is $\mathbf{C}^r$ $(r \geq 1)$ in x, t, and μ then solutions through any $x_0 \in \mathbb{R}^n$ exist and are unique on some time interval. Moreover, the solutions themselves are $\mathbf{C}^r$ functions of t, t_0, x_0, and μ. (Note: recall that a function is said to be $\mathbf{C}^r$ if it is r times differentiable and each derivative is continuous; if $r = 0$ then the function is said to be continuous.)

At this stage we have said nothing about maps, i.e., Equation (0.2). In a broad sense, we will study two types of maps depending on $g(x; \mu)$; *noninvertible* maps if $g(x; \mu)$ as a function of x for fixed μ has no inverse, and *invertible* maps if $g(x;\mu)$ has an inverse. The map will be referred to as a $\mathbf{C}^r$ *diffeomorphism* if $g(x;\mu)$ is invertible, with the inverse denoted $g^{-1}(x;\mu)$, and both $g(x;\mu)$ and $g^{-1}(x;\mu)$ are $\mathbf{C}^r$ maps (recall that a map is invertible if it is one-to-one and onto). Our goal will be to study the orbits of (0.2), i.e., the bi-infinite (if g is invertible) sequences of points

$$\{\cdots, g^{-n}(x_0;\mu), \cdots, g^{-1}(x_0;\mu), x_0, g(x_0;\mu), \cdots g^n(x_0;\mu), \cdots\}, \tag{0.4}$$

where $x_0 \in U$ and g^n is defined inductively by

$$g^n(x_0;\mu) \equiv g\big(g^{n-1}(x_0;\mu)\big), \qquad n \geq 2, \tag{0.5a}$$

$$g^{-n}(x_0;\mu) \equiv g^{-1}\big(g^{-n+1}(x_0;\mu)\big), \qquad n \geq 2, \tag{0.5b}$$

or the infinite (if g is noninvertible) sequences of points

$$\{x_0, g(x_0;\mu), \cdots, g^n(x_0;\mu), \cdots\}, \tag{0.6}$$

where $x_0 \in U$ and g^n is defined inductively by (0.5a). (Note: it should be clear that we must assume $g^{n-1}(x_0;\mu)$, $g^{-n+1}(x_0;\mu) \in U$, $n \geq 2$, for (0.4) to make sense and $g^{n-1}(x_0;\mu) \in U$, $n \geq 2$, for (0.6) to make sense.) Notice that questions of existence and uniqueness of orbits for maps are obvious and that differentiability of orbits with respect to initial conditions and parameters is a consequence of the applicability of the chain rule of elementary calculus.

With these preliminaries out of the way, we can now turn to the main business of this book.

1

The Geometrical Point of View of Dynamical Systems: Background Material, Poincaré Maps, and Examples

Our main goal in the study of dynamical systems is simple. Given a specific dynamical system, give a complete characterization of the geometry of the orbit structure. If the dynamical system depends on parameters, then characterize the change in the orbit structure as the parameters are varied.

Unfortunately, it is not possible to realize this goal for every dynamical system we study. However, in this book we will develop techniques and a point of view which will allow us to make some progress on many problems (more on some, less on others) and to point out where the gaps in our knowledge lie. In order to do this, we will need to bring to bear a wide variety of (seemingly) disparate mathematical techniques on a given problem. Consequently, a fair amount of background material must be introduced. In developing the necessary background material, we will simultaneously focus on a specific dynamical system. In this context we will develop a variety of ideas and techniques which will be used to obtain as much information on our specific dynamical system as possible. We feel that this approach will also best illustrate our strategy for dealing with dynamical systems in applications in that it will show how one takes a variety of ideas and techniques and "puts them all together" in the analysis of a specific dynamical system.

The dynamical system around which we will develop our background material is the damped, forced Duffing oscillator which is given by

$$\begin{aligned}\dot{x} &= y,\\ \dot{y} &= x - x^3 - \delta y + \gamma \cos \omega t,\end{aligned}$$

where δ, γ, and ω are real parameters and the phase space is the plane $\Re^2$. Physically, δ can be regarded as dissipation, γ as the amplitude of the forcing, and ω as the frequency; for this reason we will take $\delta, \gamma, \omega \geq 0$. The damped, forced Duffing oscillator arises in a variety of applications, e.g., see Guckenheimer and Holmes [1983] for specific applications and references.

Vector fields which depend explicitly on time are called *nonautonomous*, and vector fields which are independent of time are called *autonomous*.

We will see that in two dimensions there is a vast difference in the possible dynamics of autonomous versus nonautonomous vector fields. In particular, chaos is possible in the nonautonomous case but not the autonomous case. For this reason we will begin by considering the unforced case.

1.1 Background Material from Dynamical Systems Theory

In Section 1.1 we will develop much of the background material that we will use throughout this book. We will organize much of this material around the example of the unforced, damped Duffing oscillator. The unforced damped, Duffing oscillator is given by

$$\begin{aligned} \dot{x} &= y, \\ \dot{y} &= x - x^3 - \delta y, \end{aligned} \qquad \delta \geq 0. \tag{1.1.1}$$

The easiest way to begin to understand the orbit structure of (1.1.1) is to study the nature of its equilibria.

1.1A Equilibrium Solutions: Linearized Stability

Consider a general autonomous vector field

$$\dot{x} = f(x), \qquad x \in \mathbb{R}^n. \tag{1.1.2}$$

An *equilibrium solution* of (1.1.2) is a point $\bar{x} \in \mathbb{R}^n$ such that

$$f(\bar{x}) = 0,$$

i.e., a solution which does not change in time. Other terms often substituted for the term "equilibrium solution" are "fixed point," "stationary point," "rest point," "singularity," "critical point," or "steady state." In this book we will utilize the terms equilibrium point or fixed point exclusively.

Once we find any solution of (1.1.2) it is natural to try to determine if the solution is stable.

Stability

Let $\bar{x}(t)$ be any solution of (1.1.2). Then, roughly speaking, $\bar{x}(t)$ is *stable* if solutions starting "close" to $\bar{x}(t)$ at a given time remain close to $\bar{x}(t)$ for all later times. It is *asymptotically stable* if nearby solutions actually converge to $\bar{x}(t)$ as $t \to \infty$. Let us formalize these ideas.

Definition 1.1.1 (Liapunov Stability) $\bar{x}(t)$ is said to be *stable* (or *Liapunov stable*) if, given $\varepsilon > 0$, there exists a $\delta = \delta(\varepsilon) > 0$ such that, for any other solution, $y(t)$, of (1.1.2) satisfying $|\bar{x}(t_0) - y(t_0)| < \delta$, then $|\bar{x}(t) - y(t)| < \varepsilon$ for $t > t_0$, $t_0 \in \mathbb{R}$.

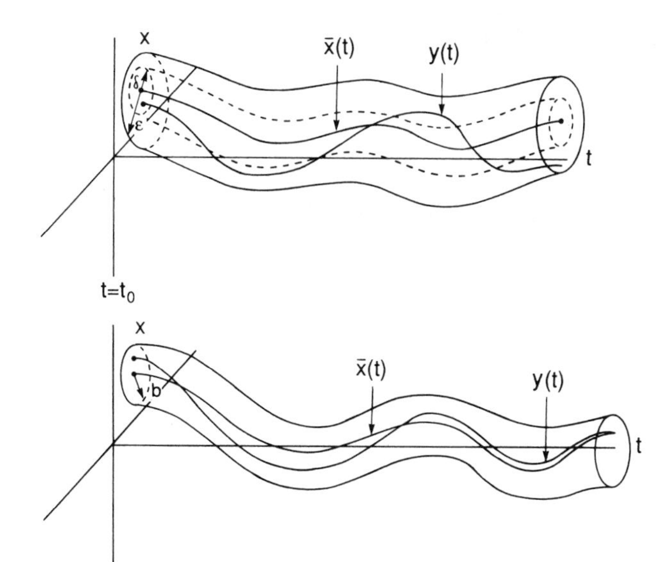

FIGURE 1.1.1. a) Liapunov stability. b) Asymptotic stability.

We remark that a solution which is not stable is said to be *unstable.*

DEFINITION 1.1.2 (ASYMPTOTIC STABILITY) $\bar{x}(t)$ is said to be *asymptotically stable* if it is Liapunov stable and if there exists a constant $b > 0$ such that, if $|\bar{x}(t_0) - y(t_0)| < b$, then $\lim_{t\to\infty} |\bar{x}(t) - y(t)| = 0$.

See Figure 1.1.1 for a geometrical interpretation of these two definitions. Notice that these two definitions imply that we have information on the infinite time existence of solutions. This is obvious for equilibrium solutions but is not necessarily so for nearby solutions. Also, these definitions are for autonomous systems, since in the nonautonomous case it may be that δ and b depend explicitly on t_0 (more about this later).

Definitions 1.1.1 and 1.1.2 are mathematically very tidy; however, they do not provide us with a method for determining whether or not a given solution is stable. We now turn our attention to this question.

Linearization

In order to determine the stability of $\bar{x}(t)$ we must understand the nature of solutions near $\bar{x}(t)$. Let

$$x = \bar{x}(t) + y. \tag{1.1.3}$$

Substituting (1.1.3) into (1.1.2) and Taylor expanding about $\bar{x}(t)$ gives

$$\dot{x} = \dot{\bar{x}}(t) + \dot{y} = f(\bar{x}(t)) + Df(\bar{x}(t))y + \mathcal{O}(|y|^2), \tag{1.1.4}$$

where Df is the derivative of f and $|\cdot|$ denotes a norm on $\mathbb{R}^n$ (note: in order to obtain (1.1.4) f must be at least twice differentiable). Using the fact that $\dot{\bar{x}}(t) = f(\bar{x}(t))$, (1.1.4) becomes

$$\dot{y} = Df(\bar{x}(t))y + \mathcal{O}(|y|^2). \tag{1.1.5}$$

Equation (1.1.5) describes the evolution of orbits near $\bar{x}(t)$. For stability questions we are concerned with the behavior of solutions arbitrarily close to $\bar{x}(t)$, so it seems reasonable that this question could be answered by studying the associated *linear system*

$$\dot{y} = Df(\bar{x}(t))y. \tag{1.1.6}$$

Therefore, the question of stability of $\bar{x}(t)$ involves the following two steps:

1. Determine if the $y = 0$ solution of (1.1.6) is stable.

2. Show that stability (or instability) of the $y = 0$ solution of (1.1.6) implies stability (or instability) of $\bar{x}(t)$.

Step 1 may be equally as difficult as our original problem, since there are no general methods for finding the solution of linear ordinary differential equations with time-dependent coefficients. However, if $\bar{x}(t)$ is an equilibrium solution, i.e., $\bar{x}(t) = \bar{x}$, then $Df(\bar{x}(t)) = Df(\bar{x})$ is a matrix with constant entries, and the solution of (1.1.6) through the point $y_0 \in \mathbb{R}^n$ of $t = 0$ can immediately be written as

$$y(t) = e^{Df(\bar{x})t}y_0. \tag{1.1.7}$$

Thus, $y(t)$ is *asymptotically stable* if all eigenvalues of $Df(\bar{x})$ have negative real parts (cf. Exercise 1.1.22).

The answer to Step 2 can be obtained from the following theorem.

Theorem 1.1.1 *Suppose all of the eigenvalues of $Df(\bar{x})$ have negative real parts. Then the equilibrium solution $x = \bar{x}$ of the nonlinear vector field (1.1.2) is asymptotically stable.*

Proof: We will give the proof of this theorem in Section 1.1.B when we discuss Liapunov functions. □

In Section 1.1B we will give an example in which an equilibrium solution of a nonlinear vector field is stable in the linear approximation, but it is actually unstable. Sometimes the term "linearly stable" is used to describe a solution that is stable in the linear approximation. Thus, linearly stable solutions may be nonlinearly unstable.

In the following sections the reader will see many results that have a similar flavor to Theorem 1.1.1. Namely, if the eigenvalues of the associated linear vector field have nonzero real parts, then the orbit structure *near an equilibrium solution* of the nonlinear vector field is essentially the same as that of the linear vector field. Such equilibrium solutions are given a special name.

DEFINITION 1.1.3 Let $x = \bar{x}$ be a fixed point of $\dot{x} = f(x)$, $x \in \mathbb{R}^n$. Then $\bar{x}$ is called a *hyperbolic* fixed point if none of the eigenvalues of $Df(\bar{x})$ have zero real part.

Maps

Everything discussed thus far applies also for maps; we mention some of the details explicitly.

Consider a $\mathbf{C}^r$ $(r \geq 1)$ map

$$x \mapsto g(x), \qquad x \in \mathbb{R}^n, \tag{1.1.8}$$

and suppose that it has a fixed point at $x = \bar{x}$, i.e., $\bar{x} = g(\bar{x})$. The associated linear map is given by

$$y \mapsto Ay, \qquad y \in \mathbb{R}^n, \tag{1.1.9}$$

where $A \equiv Dg(\bar{x})$.

Definitions of Stability for Maps

The definitions of stability and asymptotic stability for orbits of maps are very similar to the definitions for vector fields. We leave it as an exercise for the reader to formulate these definitions (cf. Exercise 1.1.8).

Stability of Fixed Points of Linear Maps

Choose a point $y_0 \in \mathbb{R}^n$. The orbit of y_0 under the linear map (1.1.9) is given by the bi-infinite sequence (if the map is a $\mathbf{C}^r$, $r \geq 1$, diffeomorphism)

$$\{\cdots, A^{-n}y_0, \cdots, A^{-1}y_0, y_0, Ay_0, \cdots, A^n y_0, \cdots\} \tag{1.1.10}$$

or the infinite sequence (if the map is $\mathbf{C}^r$, $r \geq 1$, but noninvertible)

$$\{y_0, Ay_0, \cdots, A^n y_0, \cdots\}. \tag{1.1.11}$$

From (1.1.10) and (1.1.11) it should be clear the fixed point $y = 0$ of the linear map (1.1.9) is asymptotically stable if all of the eigenvalues of A have moduli strictly less than one (cf. Exercise 1.1.24).

Stability of Fixed Points of Maps Via the Linear Approximation

With the obvious modifications, Theorem 1.1.1 is valid for maps.

Before we apply these ideas to the unforced Duffing oscillator, let us first give some useful terminology.

Terminology

A hyperbolic fixed point of a vector field (resp., map) is called a *saddle* if some, but not all, of the eigenvalues of the associated linearization have real parts greater than zero (resp., moduli greater than one) and the rest of the eigenvalues have real parts less than zero (resp., moduli less than one). If all of the eigenvalues have negative real part (resp., moduli less than one), then the hyperbolic fixed point is called a *stable node* or *sink*, and if all of the eigenvalues have positive real parts (resp., moduli greater than one), then the hyperbolic fixed point is called an *unstable node* or *source*. If the eigenvalues are purely imaginary (resp., have modulus one) and nonzero, the nonhyperbolic fixed point is called a *center*.

Let us now apply our results to the unforced Duffing oscillator.

Application to the Unforced Duffing Oscillator

We recall here Equation (1.1.1)

$$\begin{aligned} \dot{x} &= y, \\ \dot{y} &= x - x^3 - \delta y, \end{aligned} \qquad \delta \geq 0.$$

It is easy to see that this equation has three fixed points given by

$$(x, y) = (0,0), (\pm 1, 0). \tag{1.1.12}$$

The matrix associated with the linearized vector field is given by

$$\begin{pmatrix} 0 & 1 \\ 1 - 3x^2 & -\delta \end{pmatrix}. \tag{1.1.13}$$

Using (1.1.12) and (1.1.13) the eigenvalues λ_1 and λ_2 associated with the fixed point $(0,0)$ are given by $\lambda_{1,2} = -\delta/2 \pm \frac{1}{2}\sqrt{\delta^2 + 4}$, and the eigenvalues associated with the fixed points $(\pm 1, 0)$ are the same for each point and are given by $\lambda_{1,2} = -\delta/2 \pm \frac{1}{2}\sqrt{\delta^2 - 8}$. Hence, for $\delta > 0$, $(0,0)$ is unstable and $(\pm 1, 0)$ are asymptotically stable; for $\delta = 0$, $(\pm 1, 0)$ are stable in the linear approximation.

1.1B Liapunov Functions

The method of Liapunov can often be used to determine the stability of fixed points when the information obtained from linearization is inconclusive (i.e., when the fixed point is nonhyperbolic). Liapunov theory is a large area, and we will examine only an extremely small part of it; for more information, see Lasalle and Lefschetz [1961].

The basic idea of the method is as follows (the method works in n-dimensions and also infinite dimensions, but for the moment we will describe it pictorally in the plane). Suppose you have a vector field in the plane with a fixed point $\bar{x}$, and you want to determine whether or not it is stable. Roughly speaking, according to our previous definitions of stability it would be sufficient to find a neighborhood U of $\bar{x}$ for which orbits starting in U remain in U for all positive times (for the moment we don't distinguish between stability and asymptotic stability). This condition would be satisfied if we could show that the vector field is either tangent to the boundary

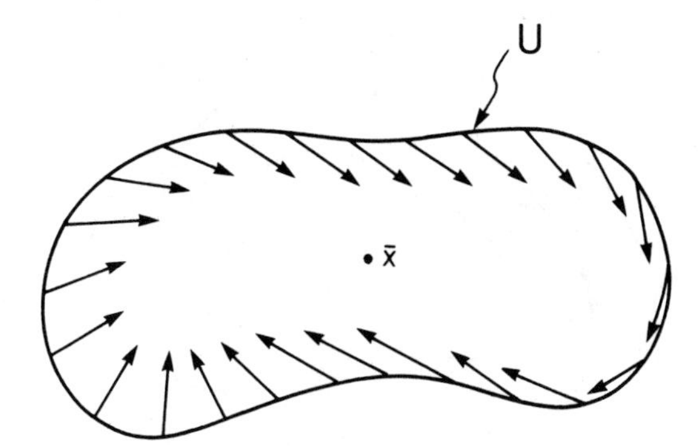

FIGURE 1.1.2. The vector field on the boundary of U.

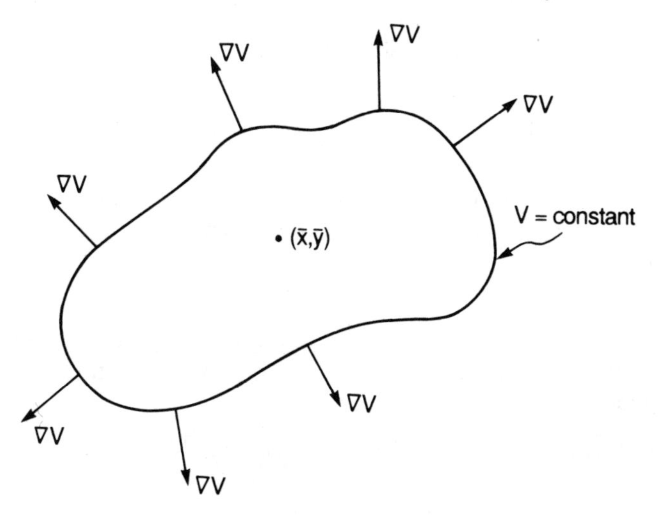

FIGURE 1.1.3. Level set of V and ∇V denoted at various points on the boundary.

of U or pointing inward toward $\bar{x}$ (see Figure 1.1.2). This situation should remain true even as we shrink U down onto $\bar{x}$. Now, Liapunov's method gives us a way of making this precise; we will show this for vector fields in the plane and then generalize our results to $\mathbb{R}^n$.

Suppose we have the vector field

$$\begin{aligned} \dot{x} &= f(x,y), \\ \dot{y} &= g(x,y), \end{aligned} \qquad (x,y) \in \mathbb{R}^2, \tag{1.1.14}$$

which has a fixed point at $(\bar{x}, \bar{y})$ (assume it is stable). We want to show that in any neighborhood of $(\bar{x}, \bar{y})$ the above situation holds. Let $V(x,y)$ be a scalar-valued function on $\mathbb{R}^2$, i.e., $V: \mathbb{R}^2 \to \mathbb{R}^1$ (and at least $\mathbf{C}^1$), with $V(\bar{x}, \bar{y}) = 0$, and such that the locus of points satisfying $V(x,y) = C =$ constant form closed curves for different values of C encircling $(\bar{x}, \bar{y})$ with $V(x,y) > 0$ in a neighborhood of $(\bar{x}, \bar{y})$ (see Figure 1.1.3).

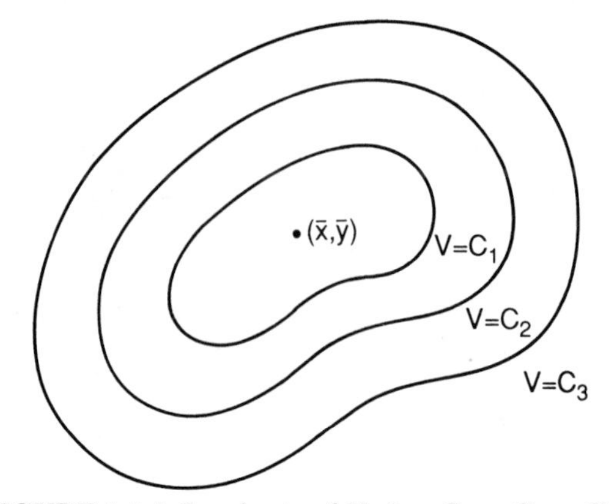

FIGURE 1.1.4. Level sets of V, $0 < C_1 < C_2 < C_3$.

Now recall that the gradient of V, ∇V, is a vector perpendicular to the tangent vector along each curve $V = C$ which points in the direction of increasing V (see Figure 1.1.4). So if the vector field were always to be either tangent or pointing inward for each of these curves surrounding $(\bar{x}, \bar{y})$, we would have

$$\nabla V(x, y) \cdot (\dot{x}, \dot{y}) \leq 0,$$

where the "dot" represents the usual vector scalar product. (This is simply the derivative of V along orbits of (1.1.14).) We now state the general theorem which makes these ideas precise.

Theorem 1.1.2 *Consider the following vector field*

$$\dot{x} = f(x), \qquad x \in \mathbb{R}^n. \tag{1.1.15}$$

Let $\bar{x}$ be a fixed point (1.1.15) and let $V\colon U \to \mathbb{R}$ be a $\mathbf{C}^1$ function defined on some neighborhood U of $\bar{x}$ such that

i) $V(\bar{x}) = 0$ *and* $V(x) > 0$ *if* $x \neq \bar{x}$.

ii) $\dot{V}(x) \leq 0$ *in* $U - \{\bar{x}\}$.

Then $\bar{x}$ is stable. *Moreover, if*

iii) $\dot{V}(x) < 0$ *in* $U - \{\bar{x}\}$

then $\bar{x}$ is asymptotically stable.

Proof: See Exercise 1.1.6. □

We refer to V as a *Liapunov function.* We remark that if U can be chosen to be all of $\mathbb{R}^n$, then $\bar{x}$ is said to be *globally asymptotically stable* if i) and iii) hold.

EXAMPLE 1.1.1 Consider the following vector field

$$\begin{aligned} \dot{x} &= y, \\ \dot{y} &= -x + \varepsilon x^2 y. \end{aligned} \tag{1.1.16}$$

It is easy to verify that (1.1.16) has a nonhyperbolic fixed point at $(x, y) = (0, 0)$. Our goal is to determine if this fixed point is stable.

Let $V(x, y) = (x^2 + y^2)/2$. Clearly $V(0, 0) = 0$ and $V(x, y) > 0$ in any neighborhood of $(0, 0)$. Then

$$\begin{aligned} \dot{V}(x, y) &= \nabla V(x, y) \cdot (\dot{x}, \dot{y}) \\ &= (x, y) \cdot (y, \varepsilon x^2 y - x) \\ &= xy + \varepsilon x^2 y^2 - xy \end{aligned}$$

and hence $\dot{V} = \varepsilon x^2 y^2$. Then, by Theorem 1.1.2, $(0, 0)$ is globally stable for $\varepsilon = 0$ and globally asymptotically stable for $\varepsilon < 0$.

Let us now use Liapunov theory to give an outline of the proof of Theorem 1.1.1. We begin by recalling the set-up of the problem.

Consider the vector field

$$\dot{x} = f(x), \qquad x \in \mathbb{R}^n, \tag{1.1.17}$$

and suppose that (1.1.17) has a fixed point at $x = \bar{x}$, i.e., $f(\bar{x}) = 0$. We translate the fixed point to the origin via the coordinate shift $y = x - \bar{x}$ so that (1.1.17) becomes

$$\dot{y} = f(y + \bar{x}), \qquad y \in \mathbb{R}^n. \tag{1.1.18}$$

Taylor expanding (1.1.18) about $\bar{x}$ gives

$$\dot{y} = Df(\bar{x})y + R(y), \tag{1.1.19}$$

where $R(y) \equiv \mathcal{O}(|y|^2)$.

Now let us introduce the coordinate rescaling

$$y = \varepsilon u, \qquad 0 < \varepsilon < 1. \tag{1.1.20}$$

Thus, taking ε small implies making y small. Under (1.1.20) equation (1.1.19) becomes

$$\dot{u} = Df(\bar{x})u + \bar{R}(u, \varepsilon), \tag{1.1.21}$$

where $\bar{R}(u, \varepsilon) = R(\varepsilon u)/\varepsilon$. It should be clear that $\bar{R}(u, 0) = 0$ since $R(y) = \mathcal{O}(|y|^2)$. We choose as a Liapunov function

$$V(u) = \frac{1}{2}|u|^2.$$

Therefore,

$$\begin{aligned}\dot{V}(u) &= \nabla V(u) \cdot \dot{u} \\ &= \big(u \cdot Df(\bar{x})u\big) + \big(u \cdot \bar{R}(u,\varepsilon)\big).\end{aligned} \tag{1.1.22}$$

From elementary linear algebra the reader should recall that if all eigenvalues of $Df(\bar{x})$ have negative real part, then there exists a real number k such that

$$\big(u \cdot Df(\bar{x})u\big) < k < 0 \tag{1.1.23}$$

for all u (see Arnold [1973] for a proof). Hence, by choosing ε sufficiently small, (1.1.22) is strictly negative, which implies that the fixed point $x = \bar{x}$ is asymptotically stable.

1.1C Invariant Manifolds: Linear and Nonlinear Systems

We will see throughout this book that invariant manifolds, in particular stable, unstable, and center manifolds, play a central role in the analysis of dynamical systems. We will give a simultaneous discussion of these ideas for both vector fields

$$\dot{x} = f(x), \qquad x \in \mathbb{R}^n \tag{1.1.24}$$

and maps

$$x \mapsto g(x), \qquad x \in \mathbb{R}^n. \tag{1.1.25}$$

Definition 1.1.4 Let $S \subset \mathbb{R}^n$ be a set, then

a) (Continuous time) S is said to be *invariant* under the vector field $\dot{x} = f(x)$ if for any $x_0 \in S$ we have $x(t, 0, x_0) \in S$ for all $t \in \mathbb{R}$.

b) (Discrete time) S is said to be *invariant* under the map $x \mapsto g(x)$ if for any $x_0 \in S$ we have $g^n(x_0) \in S$ for all n.

If we restrict ourselves to positive times (i.e., $t \geq 0$, $n \geq 0$) then we refer to S as a *positively invariant set* and, for negative time, as a *negatively invariant set.*

We remark that if g is noninvertible, then only $n \geq 0$ makes sense (although in some instances it may be useful to consider g^{-1} which does have a set theoretic meaning).

Definition 1.1.5 An invariant set $S \subset \mathbb{R}^n$ is said to be a $\mathbf{C}^r$ *($r \geq 1$) invariant manifold* if S has the structure of a $\mathbf{C}^r$ differentiable manifold. Similarly, a positively (resp., negatively) invariant set $S \subset \mathbb{R}^n$ is said to be a $\mathbf{C}^r$ *($r \geq 1$) positively* (resp., *negatively*) *invariant manifold* if S has the structure of a $\mathbf{C}^r$ differentiable manifold.

Evidently, we need to say what we mean by the term "$\mathbf{C}^r$ differentiable manifold." However, this is the subject of a course in itself, so rather than define the concept of a manifold in its full generality, we will describe only that portion of the vast theory that we will need.

Roughly speaking, a manifold is a set which *locally* has the structure of Euclidean space. In applications, manifolds are most often met as m-dimensional surfaces embedded in $\mathbb{R}^n$. If the surface has no singular points, i.e., the derivative of the function representing the surface has maximal rank, then by the implicit function theorem it can locally be represented as a graph. The surface is a $\mathbf{C}^r$ manifold if the (local) graphs representing it are $\mathbf{C}^r$ (note: for a thorough treatment of this particular representation of a manifold see Dubrovin, Fomenko, and Novikov [1985]).

Another example is even more basic. Let $\{s_1, \cdots, s_n\}$ denote the standard basis on $\mathbb{R}^n$. Let $\{s_{i_1}, \cdots, s_{i_j}\}$, $j < n$, denote any j basis vectors from this set; then the span of $\{s_{i_1}, \cdots, s_{i_j}\}$ forms a j-dimensional subspace of $\mathbb{R}^n$ which is trivially a $\mathbf{C}^\infty$ j-dimensional manifold. For a thorough introduction to the theory of manifolds with a view to applications see Abraham, Marsden, and Ratiu [1988].

The main reason for choosing these examples is that, in this book, when the term "manifold" is used, it will be sufficient to think of one of the following two situations:

1. *Linear Settings*: a linear vector subspace of $\mathbb{R}^n$;

2. *Nonlinear Settings*: a surface embedded in $\mathbb{R}^n$ which can be locally represented as a graph (which can be justified via the implicit function theorem).

Let us return to our study of the orbit structure near fixed points to see how some important invariant manifolds arise. We begin with vector fields. Let $\bar{x} \in \mathbb{R}^n$ be a fixed point of

$$\dot{x} = f(x), \qquad x \in \mathbb{R}^n. \tag{1.1.26}$$

Then, by the discussion in Section 1.1A, it is natural to consider the associated linear system

$$\dot{y} = Ay, \qquad y \in \mathbb{R}^n, \tag{1.1.27}$$

where $A \equiv Df(\bar{x})$ is a constant $n \times n$ matrix. The solution of (1.1.27) through the point $y_0 \in \mathbb{R}^n$ at $t = 0$ is given by

$$y(t) = e^{At} y_0, \tag{1.1.28}$$

where

$$e^{At} = \mathrm{id} + At + \frac{1}{2!}A^2t^2 + \frac{1}{3!}A^3t^3 + \cdots \tag{1.1.29}$$

and "id" denotes the $n \times n$ identity matrix. We must assume sufficient background in the theory of linear constant coefficient ordinary differential equations so that (1.1.28) and (1.1.29) make sense to the reader. Excellent references for this theory are Arnold [1973] and Hirsch and Smale [1974]. Our goal here is to extract the necessary ingredients from this theory so as to give a geometrical interpretation to (1.1.28).

Now $\mathbb{R}^n$ can be represented as the direct sum of three subspaces denoted E^s, E^u, and E^c, which are defined as follows:

$$\begin{aligned} E^s &= \text{span}\{e_1, \cdots, e_s\}, \\ E^u &= \text{span}\{e_{s+1}, \cdots, e_{s+u}\}, \qquad\qquad s+u+c=n, \\ E^c &= \text{span}\{e_{s+u+1}, \cdots, e_{s+u+c}\}, \end{aligned} \tag{1.1.30}$$

where $\{e_1, \cdots, e_s\}$ are the (generalized) eigenvectors of A corresponding to the eigenvalues of A having *negative real part*, $\{e_{s+1}, \cdots, e_{s+u}\}$ are the (generalized) eigenvectors of A corresponding to eigenvalues of A having *positive real part*, and $\{e_{s+u+1}, \cdots, e_{s+u+c}\}$ are the (generalized) eigenvectors of A corresponding to the eigenvalues of A having *zero real part* (note: this is proved in great detail in Hirsch and Smale [1974]). E^s, E^u, and E^c are referred to as the stable, unstable, and center subspaces, respectively. They are also examples of invariant subspaces (or manifolds) since solutions of (1.1.27) with initial conditions entirely contained in either E^s, E^u, or E^c must forever remain in that particular subspace for all time (we will motivate this a bit more shortly). Moreover, solutions starting in E^s approach $y = 0$ asymptotically as $t \to +\infty$ and solutions starting in E^u approach $y = 0$ asymptotically as $t \to -\infty$. Let us now illustrate these ideas with three examples where for simplicity and easier visualization we will work in $\mathbb{R}^3$.

EXAMPLE 1.1.2 Suppose the three eigenvalues of A are real and distinct and denoted by λ_1, $\lambda_2 < 0$, $\lambda_3 > 0$. Then A has three linearly independent eigenvectors e_1, e_2, and e_3 corresponding to λ_1, λ_2, and λ_3, respectively. If we form the 3×3 matrix T by taking as columns the eigenvectors e_1, e_2, and e_3, which we write as

$$T \equiv \begin{pmatrix} \vdots & \vdots & \vdots \\ e_1 & e_2 & e_3 \\ \vdots & \vdots & \vdots \end{pmatrix}, \tag{1.1.31}$$

then we have

$$\Lambda \equiv \begin{pmatrix} \lambda_1 & 0 & 0 \\ 0 & \lambda_2 & 0 \\ 0 & 0 & \lambda_3 \end{pmatrix} = T^{-1}AT. \tag{1.1.32}$$

Recall that the solution of (1.1.27) through $y_0 \in \mathbb{R}^3$ at $t = 0$ is given by

$$y(t) = e^{At} y_0 = e^{T\Lambda T^{-1} t} y_0. \tag{1.1.33}$$

Using (1.1.29), it is easy to see that (1.1.33) is the same as

$$\begin{aligned} y(t) &= Te^{\Lambda t}T^{-1}y_0 \\ &= T\begin{pmatrix} e^{\lambda_1 t} & 0 & 0 \\ 0 & e^{\lambda_2 t} & 0 \\ 0 & 0 & e^{\lambda_3 t} \end{pmatrix} T^{-1}y_0 \\ &= \begin{pmatrix} \vdots & \vdots & \vdots \\ e_1 e^{\lambda_1 t} & e_2 e^{\lambda_2 t} & e_3 e^{\lambda_3 t} \\ \vdots & \vdots & \vdots \end{pmatrix} T^{-1}y_0. \end{aligned} \tag{1.1.34}$$

Now we want to give a geometric interpretation to (1.1.34). Recall from (1.1.30) that we have

$$\begin{aligned} E^s &= \operatorname{span}\{e_1, e_2\}, \\ E^u &= \operatorname{span}\{e_3\}. \end{aligned}$$

Invariance

Choose any point $y_0 \in \mathbb{R}^n$. Then T^{-1} is the transformation matrix which changes the coordinates of y_0 with respect to the standard basis on $\mathbb{R}^3$ (i.e., $(1,0,0)$, $(0,1,0)$, $(0,0,1)$) into coordinates with respect to the basis e_1, e_2, and e_3. Thus, for $y_0 \in E^s$, $T^{-1}y_0$ has the form

$$T^{-1}y_0 = \begin{pmatrix} \tilde{y}_{01} \\ \tilde{y}_{02} \\ 0 \end{pmatrix}, \tag{1.1.35}$$

and, for $y_0 \in E^u$, $T^{-1}y_0$ has the form

$$T^{-1}y_0 = \begin{pmatrix} 0 \\ 0 \\ \tilde{y}_{03} \end{pmatrix}. \tag{1.1.36}$$

Therefore, by substituting (1.1.35) (resp., (1.1.36)) into (1.1.34), it is easy to see that $y_0 \in E^s$ (resp., E^u) implies $e^{At}y_0 \in E^s$ (resp., E^u). Thus, E^s and E^u are *invariant* manifolds.

Asymptotic Behavior

Using (1.1.35) and (1.1.34), we can see that, for any $y_0 \in E^s$, we have $e^{At}y_0 \to 0$ as $t \to +\infty$ and, for any $y_0 \in E^u$, we have $e^{At}y_0 \to 0$ as $t \to -\infty$ (hence the reason behind the names stable and unstable manifolds).

See Figure 1.1.5 for an illustration of the geometry of E^s and E^u.

EXAMPLE 1.1.3 Suppose A has two complex conjugate eigenvalues $\rho \pm i\omega$, $\rho < 0$, $\omega \neq 0$ and one real eigenvalue $\lambda > 0$. Then A has three real

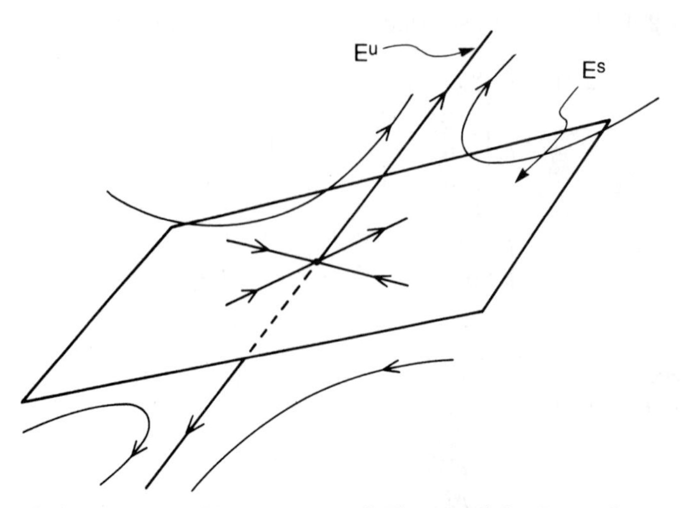

FIGURE 1.1.5. The geometry of E^s and E^u for Example 1.1.2.

generalized eigenvectors e_1, e_2, and e_3, which can be used as the columns of a matrix T in order to transform A as follows

$$\Lambda \equiv \begin{pmatrix} \rho & \omega & 0 \\ -\omega & \rho & 0 \\ 0 & 0 & \lambda \end{pmatrix} = T^{-1}AT. \tag{1.1.37}$$

From Example 1.1.2 it is easy to see that in this example we have

$$\begin{aligned} y(t) &= Te^{\Lambda t}T^{-1}y_0 \\ &= T \begin{pmatrix} e^{\rho t}\cos\omega t & e^{\rho t}\sin\omega t & 0 \\ -e^{\rho t}\sin\omega t & e^{\rho t}\cos\omega t & 0 \\ 0 & 0 & e^{\lambda t} \end{pmatrix} T^{-1}y_0. \end{aligned} \tag{1.1.38}$$

Using the same arguments given in Example 1.1.2 it should be clear that $E^s = \text{span}\{e_1, e_2\}$ is an invariant manifold of solutions that decay exponentially to zero as $t \to +\infty$, and $E^u = \text{span}\{e_3\}$ is an invariant manifold of solutions that decay exponentially to zero as $t \to -\infty$ (see Figure 1.1.6).

EXAMPLE 1.1.4 Suppose A has two real repeated eigenvalues, $\lambda < 0$, and a third distinct eigenvalue $\gamma > 0$ such that there exist generalized eigenvectors e_1, e_2, and e_3 which can be used to form the columns of a matrix T so that A is transformed as follows

$$\Lambda = \begin{pmatrix} \lambda & 1 & 0 \\ 0 & \lambda & 0 \\ 0 & 0 & \gamma \end{pmatrix} = T^{-1}AT. \tag{1.1.39}$$

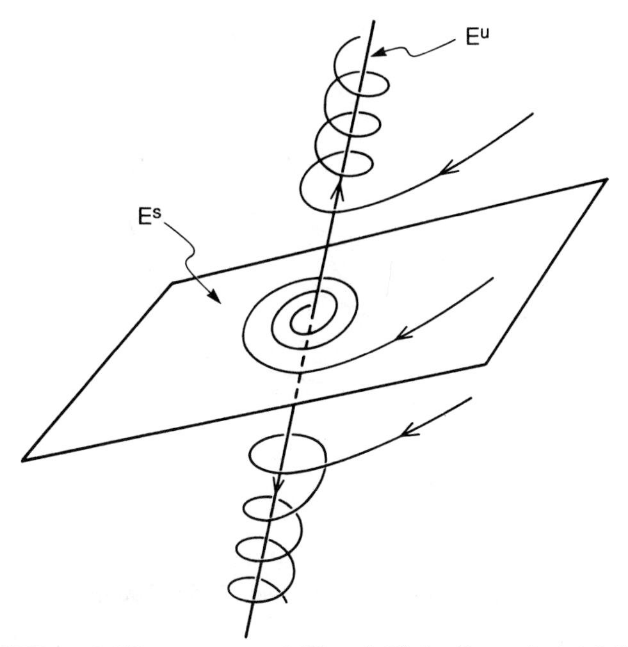

FIGURE 1.1.6. The geometry of E^s and E^u for Example 1.1.3 (for $\omega < 0$).

Following Examples 1.1.2 and 1.1.3, in this example the solution through the point $y_0 \in \mathbb{R}^3$ at $t = 0$ is given by

$$\begin{aligned} y(t) &= Te^{\Lambda t}T^{-1}y_0 \\ &= T\begin{pmatrix} e^{\lambda t} & te^{\lambda t} & 0 \\ 0 & e^{\lambda t} & 0 \\ 0 & 0 & e^{\lambda t} \end{pmatrix} T^{-1}y_0. \end{aligned} \tag{1.1.40}$$

Using the same arguments as in Example 1.1.2, it is easy to see that $E^s = \text{span}\{e_1, e_2\}$ is an invariant manifold of solutions that decay to $y = 0$ as $t \to +\infty$, and $E^u = \text{span}\{e_3\}$ is an invariant manifold of solutions that decay to $y = 0$ as $t \to -\infty$ (see Figure 1.1.7).

The reader should review enough linear algebra so that he or she can justify each step in the arguments given in these examples. We remark that we have not considered an example of a linear vector field having a center subspace. The reader can construct his or her own examples from Example 1.1.3 by setting $\rho = 0$ or from Example 1.1.4 by setting $\lambda = 0$; we leave these as exercises and now turn to the nonlinear system.

The Nonlinear System

Recall that our original motivation for studying the linear system

$$\dot{y} = Ay, \qquad y \in \mathbb{R}^n, \tag{1.1.41}$$

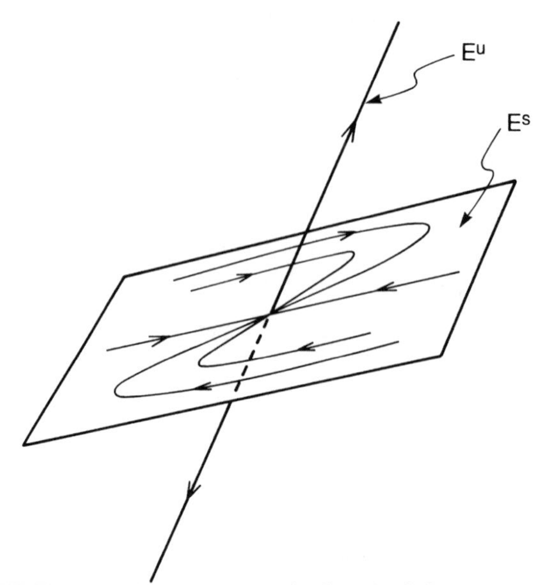

FIGURE 1.1.7. The geometry of E^s and E^u for Example 1.1.4.

where $A = Df(\bar{x})$, was to obtain information about the nature of solutions near the fixed point $x = \bar{x}$ of the nonlinear equation

$$\dot{x} = f(x), \qquad x \in \mathbb{R}^n. \tag{1.1.42}$$

The stable, unstable, and center manifold theorem provides an answer to this question; let us first transform (1.1.42) to a more convenient form.

We first transform the fixed point $x = \bar{x}$ of (1.1.42) to the origin via the translation $y = x - \bar{x}$. In this case (1.1.42) becomes

$$\dot{y} = f(\bar{x} + y), \qquad y \in \mathbb{R}^n. \tag{1.1.43}$$

Taylor expanding $f(\bar{x} + y)$ about $x = \bar{x}$ gives

$$\dot{y} = Df(\bar{x})y + R(y), \qquad y \in \mathbb{R}^n, \tag{1.1.44}$$

where $R(y) = \mathcal{O}(|y|^2)$ and we have used $f(\bar{x}) = 0$. From elementary linear algebra (see Hirsch and Smale [1974]) we can find a linear transformation T which transforms the linear equation (1.1.41) into block diagonal form

$$\begin{pmatrix} \dot{u} \\ \dot{v} \\ \dot{w} \end{pmatrix} = \begin{pmatrix} A_s & 0 & 0 \\ 0 & A_u & 0 \\ 0 & 0 & A_c \end{pmatrix} \begin{pmatrix} u \\ v \\ w \end{pmatrix}, \tag{1.1.45}$$

where $T^{-1}y \equiv (u, v, w) \in \mathbb{R}^s \times \mathbb{R}^u \times \mathbb{R}^c$, $s + u + c = n$, A_s is an $s \times s$ matrix having eigenvalues with negative real part, A_u is an $u \times u$ matrix having eigenvalues with positive real part, and A_c is an $c \times c$ matrix having eigenvalues with zero real part (note: we point out the (hopefully) obvious fact that the "0" in (1.1.45) are not scalar zero's but rather the appropriately sized block consisting of all zero's. This notation will be used throughout the book). Using this same linear transformation to transform the coordinates of the nonlinear vector field (1.1.44) gives the equation

$$\begin{aligned} \dot{u} &= A_s u + R_s(u, v, w), \\ \dot{v} &= A_u v + R_u(u, v, w), \\ \dot{w} &= A_c w + R_c(u, v, w), \end{aligned} \tag{1.1.46}$$

where $R_s(u, v, w)$, $R_u(u, v, w)$, and $R_c(u, v, w)$ are the first s, u, and c components, respectively, of the vector $TR(T^{-1}y)$.

Now consider the linear vector field (1.1.45). From our previous discussion (1.1.45) has an s-dimensional invariant stable manifold, a u-dimensional invariant unstable manifold, and a c-dimensional invariant center manifold all intersecting in the origin. The following theorem shows how this structure changes when the nonlinear vector field (1.1.46) is considered.

Theorem 1.1.3 (Local, Stable, Unstable, and Center Manifolds of Fixed Points) *Suppose (1.1.46) is* $\mathbf{C}^r$, $r \geq 2$. *Then the fixed point* $(u, v, w) = 0$ *of (1.1.46) possesses a* $\mathbf{C}^r$ *s-dimensional local, stable manifold,* $W^s_{\text{loc}}(0)$, *a* $\mathbf{C}^r$ *u-dimensional local, unstable manifold,* $W^u_{\text{loc}}(0)$, *and a* $\mathbf{C}^r$ *c-dimensional local, center manifold,* $W^c_{\text{loc}}(0)$, *all intersecting at* $(u, v, w) = 0$. *These manifolds are all tangent to the respective invariant manifolds of the linear vector field (1.1.45) at the origin and, hence, are locally representable as graphs. In particular, we have*

$$W^s_{\text{loc}}(0) = \{(u, v, w) \in \mathbb{R}^s \times \mathbb{R}^u \times \mathbb{R}^c \mid v = h^s_v(u), w = h^s_w(u);$$
$$Dh^s_v(0) = 0, Dh^s_w(0) = 0; |u| \textit{ sufficiently small}\}$$

$$W^u_{\text{loc}}(0) = \{(u, v, w) \in \mathbb{R}^s \times \mathbb{R}^u \times \mathbb{R}^c \mid u = h^u_u(v), w = h^u_w(v);$$
$$Dh^u_u(0) = 0, Dh^u_w(0) = 0; |v| \textit{ sufficiently small}\}$$

$$W^c_{\text{loc}}(0) = \{(u, v, w) \in \mathbb{R}^s \times \mathbb{R}^u \times \mathbb{R}^c \mid u = h^c_u(w), v = h^c_v(w);$$
$$Dh^c_u(0) = 0, Dh^c_v(0) = 0; |w| \textit{ sufficiently small}\}$$

where $h^s_v(u)$, $h^s_w(u)$, $h^u_u(v)$, $h^u_w(v)$, $h^c_u(w)$, *and* $h^c_v(w)$ *are* $\mathbf{C}^r$ *functions. Moreover,* $W^s_{\text{loc}}(0)$ *and* $W^u_{\text{loc}}(0)$ *have the asymptotic properties of* E^s *and* E^u, *respectively. Namely, solutions of 1.1.46 with initial conditions in* $W^s_{\text{loc}}(0)$ *(resp.,* $W^u_{\text{loc}}(0)$*) approach the origin at an exponential rate asymptotically as* $t \to +\infty$ *(resp.,* $t \to -\infty$*).*

Proof: See Fenichel [1971] or Hirsch, Pugh, and Shub [1977] for details and see Wiggins [1988] for some history and further references on invariant manifolds. □

Some remarks on this important theorem are now in order.

Remark 1. First some terminology. Very often one hears the terms "stable manifold," "unstable manifold," or "center manifold" used alone; however, alone they are not sufficient to describe the dynamical situation. Notice that Theorem 1.1.3 is entitled stable, unstable, and center manifolds *of fixed points.* The phrase "of fixed points" is the key: one must say the stable, unstable, or center manifold *of something* in order to make sense. The "somethings" studied thus far have been fixed points; however, more general invariant sets also have stable, unstable, and center manifolds. See Wiggins [1988] for a discussion.

Remark 2. The conditions $Dh_v^s(0) = 0$, $Dh_w^s(0) = 0$, etc., reflect that the nonlinear manifolds are tangent to the associated linear manifolds at the origin.

Remark 3. Suppose the fixed point is hyperbolic, i.e., $E^c = \emptyset$. In this case an interpretation of the theorem is that solutions of the nonlinear vector field in a sufficiently small neighborhood of the origin behave the same as solutions of the associated linear vector field.

Remark 4. In general, the nature of solutions in $W_{\text{loc}}^c(0)$ cannot be inferred from the nature of solutions in E^c. More refined techniques are needed and are developed in Chapter 2.

Maps

An identical theory can be developed for maps. We summarize the details below. Consider a $\mathbf{C}^r$ diffeomorphism

$$x \mapsto g(x), \qquad x \in \mathbb{R}^n. \tag{1.1.47}$$

Suppose (1.1.47) has a fixed point at $x = \bar{x}$ and we want to know the nature of orbits near this fixed point. Then it is natural to consider the associated linear map

$$y \mapsto Ay, \qquad y \in \mathbb{R}^n, \tag{1.1.48}$$

where $A = Dg(\bar{x})$. The linear map (1.1.48) has invariant manifolds given by

$$\begin{aligned}
E^s &= \operatorname{span}\{e_1, \cdots, e_s\}, \\
E^u &= \operatorname{span}\{e_{s+1}, \cdots, e_{s+u}\}, \\
E^c &= \operatorname{span}\{e_{s+u+1}, \cdots, e_{s+u+c}\},
\end{aligned}$$

where $s + u + c = n$ and $e_1, \cdots, e_s$ are the (generalized) eigenvectors of A corresponding to the eigenvalues of A having *modulus less than one,*

$e_{s+1}, \cdots, e_{s+u}$ are the (generalized) eigenvectors of A corresponding to the eigenvalues of A having *modulus greater than one*, and $e_{s+u+1}, \cdots, e_{s+u+c}$ are the (generalized) eigenvectors of A corresponding to the eigenvalues of A having *modulus equal to one*. The reader should find it easy to prove this by putting A in Jordan canonical form and noting that the orbit of the linear map 1.1.48 through the point $y_0 \in \mathbb{R}^n$ is given by

$$\{\cdots, A^{-n}y_0, \cdots, A^{-1}y_0, y_0, Ay_0, \cdots, A^n y_0, \cdots\}. \tag{1.1.49}$$

Now we address the question of how this structure goes over to the nonlinear map (1.1.47). In the case of maps Theorem 1.1.3 holds identically. Namely, the nonlinear map (1.1.47) has a $\mathbf{C}^r$ invariant s-dimensional stable manifold, a $\mathbf{C}^r$ invariant u-dimensional unstable manifold, and a $\mathbf{C}^r$ invariant c-dimensional center manifold all intersecting in the fixed point. Moreover, these manifolds are all tangent to the respective invariant manifolds of the linear map (1.1.48) at the fixed point.

Essentially, everything about stable, unstable, and center manifolds for fixed points of vector fields holds for fixed points of maps. We will give examples in the exercises at the end of Section 1.1. However, before completing our discussion of invariant manifolds let us apply our results to the unforced Duffing oscillator.

Application to the Unforced Duffing Oscillator

In Section 1.1A we have seen that the equation

$$\begin{aligned} \dot{x} &= y, \\ \dot{y} &= x - x^3 - \delta y, \end{aligned} \qquad \delta > 0,$$

has a saddle-type fixed point of $(x, y) = (0, 0)$, and sinks at $(\pm 1, 0)$ for $\delta > 0$. From Theorem 1.1.3 we now know that $(\pm 1, 0)$ have two-dimensional stable manifolds (this is obvious) and $(0, 0)$ has a one-dimensional stable manifold and a one-dimensional unstable manifold as shown in Figure 1.1.8 (note: we have drawn the figure for $0 < \delta < \sqrt{8}$. The reader should show how the solutions near the sinks are modified for $\delta > \sqrt{8}$). Note that Theorem 1.1.3 also tells us that a good local approximation to the stable and unstable manifolds of $(0, 0)$ is given by the corresponding invariant linear manifolds, which are relatively easy to calculate. The case $\delta = 0$ is treated in great detail in Section 1.1E.

Let us consider a final example from Guckenheimer and Holmes [1983].

EXAMPLE 1.1.5 Consider the planar vector field

$$\begin{aligned} \dot{x} &= x, \\ \dot{y} &= -y + x^2, \end{aligned} \qquad (x, y) \in \mathbb{R}^1 \times \mathbb{R}^1,$$

which has a hyperbolic fixed point at $(x, y) = (0, 0)$. The associated linearized system is given by

$$\begin{aligned} \dot{x} &= x, \\ \dot{y} &= -y, \end{aligned}$$

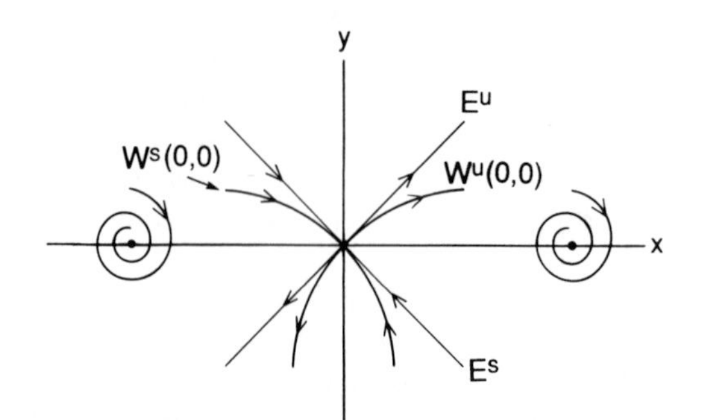

FIGURE 1.1.8. Local invariant manifold structure in the unforced Duffing oscillator, $0 < \delta < \sqrt{8}$.

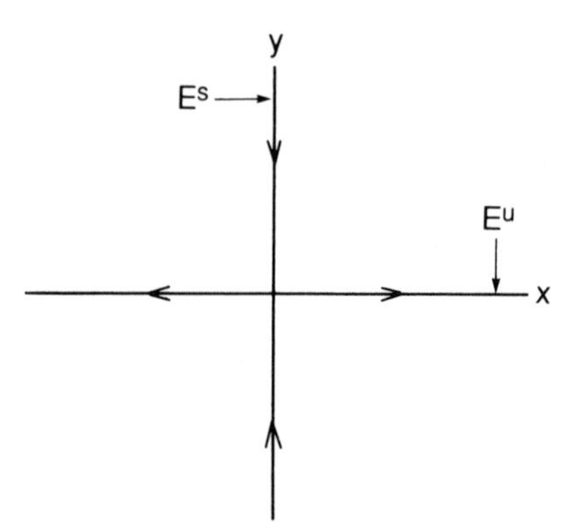

FIGURE 1.1.9. The stable and unstable subspaces in Example 1.1.5.

with stable and unstable subspaces given by

$$
\begin{aligned}
E^s &= \{ (x,y) \in \mathbb{R}^2 \mid x = 0 \}, \\
E^u &= \{ (x,y) \in \mathbb{R}^2 \mid y = 0 \}
\end{aligned}
$$

(see Figure 1.1.9).

Now we turn our attention to the nonlinear vector field for which, in this case, the solution can be obtained explicitly as follows. Eliminating time as the independent variable gives

$$\frac{\dot{y}}{\dot{x}} = \frac{dy}{dx} = \frac{-y}{x} + x,$$

which can be solved to obtain

$$y(x) = \frac{x^2}{3} + \frac{c}{x},$$

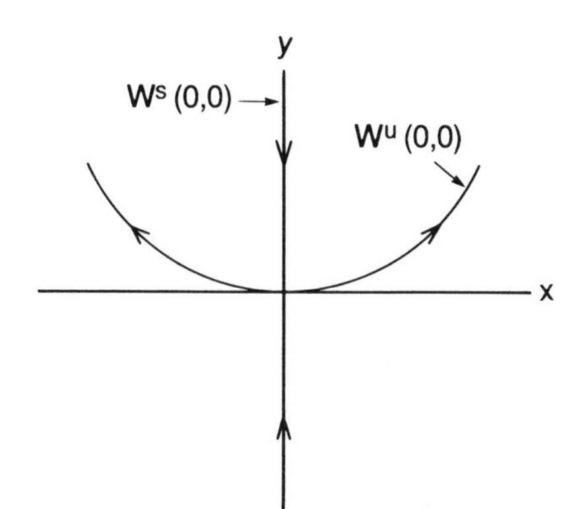

FIGURE 1.1.10. Stable and unstable manifolds of $(x, y) = (0, 0)$ in Example 1.1.5.

where c is some constant. Now $W^u_{\text{loc}}(0,0)$ can be represented by a graph over the x variables, i.e., $y = h(x)$ with $h(0) = h'(0) = 0$. Varying c in the solution above takes us from orbit to orbit; we seek the value of c which corresponds to the unstable manifold—this is $c = 0$. Therefore, we have

$$W^u_{\text{loc}}(0,0) = \left\{ (x,y) \in \mathbb{R}^2 \;\middle|\; y = \frac{x^2}{3} \right\},$$

which is also the global unstable manifold of the origin (see Exercise 1.1.28). Finally, note that if we have initial conditions with the x component equal to zero, i.e., $(0, y)\ \forall y$, then the solution stays on the y axis and approaches $(0, 0)$ as $t \uparrow \infty$; thus, $E^s = W^s(0,0) = \{ (x,y) \mid x = 0 \}$ (see Figure 1.1.10).

1.1D Periodic Solutions

We consider vector fields

$$\dot{x} = f(x), \qquad x \in \mathbb{R}^n, \tag{1.1.50}$$

and maps

$$x \mapsto g(x), \qquad x \in \mathbb{R}^n. \tag{1.1.51}$$

Definition 1.1.6 (Vector Fields) A solution of (1.1.50) through the point x_0 is said to be *periodic of period* T if there exists $T > 0$ such that $x(t, t_0) = x(t + T, x_0)$ for all $t \in \mathbb{R}$. (Maps) The orbit of $x_0 \in \mathbb{R}^n$ is said to be *periodic of period* $k > 0$ if $g^k(x_0) = x_0$.

We remark that if a solution of (1.1.50) is periodic of period T then evidently it is periodic of period nT for any integer $n > 1$. However, by the period of an orbit we mean the smallest possible $T > 0$ such that Definition 1.1.6 holds. A similar statement holds for periodic orbits of maps.

We will discuss stability of periodic orbits of vector fields when we discuss Poincaré maps in Section 1.2; for stability of periodic orbits of maps see Exercise 1.1.5.

Now we will learn a useful and easily applicable trick for establishing the nonexistence of periodic solutions of autonomous vector fields on the plane. We will denote these vector fields by

$$\begin{aligned} \dot{x} &= f(x, y), \\ \dot{y} &= g(x, y), \end{aligned} \qquad (x, y) \in \mathbb{R}^2, \tag{1.1.52}$$

where f and g are at least $\mathbf{C}^1$.

Theorem 1.1.4 (Bendixson's criterion) *If on a simply connected region $D \subset \mathbb{R}^2$ (i.e., D has no holes in it) the expression $\frac{\partial f}{\partial x} + \frac{\partial g}{\partial y}$ is not identically zero and does not change sign, then (1.1.52) has no closed orbits lying entirely in D.*

Proof: This is a simple result of Green's theorem on the plane; see Abraham, Marsden, and Ratiu [1988]. Using (1.1.52) and applying the chain rule we find that on any closed orbit Γ we have

$$\int_{\Gamma} f\,dy - g\,dx = 0. \tag{1.1.53}$$

By Green's theorem this implies

$$\int_{S} \left(\frac{\partial f}{\partial x} + \frac{\partial g}{\partial y} \right) dx\,dy = 0, \tag{1.1.54}$$

where S is the interior bounded by Γ. But if $\frac{\partial f}{\partial x} + \frac{\partial g}{\partial y} \neq 0$ and doesn't change sign, then this obviously can't be true. Therefore, there must be no closed orbits in D. $\square$

A generalization of Bendixson's criterion due to Dulac is the following.

Theorem 1.1.5 *Let $B(x, y)$ be $\mathbf{C}^1$ on a simply connected region $D \subset \mathbb{R}^2$. If $\frac{\partial (Bf)}{\partial x} + \frac{\partial (By)}{\partial y}$ is not identically zero and does not change sign in D, then (1.1.52) has no closed orbits lying entirely in D.*

Proof: The proof is very similar to the previous theorem so we omit it and leave it as an exercise. $\square$

Application to the Unforced Duffing Oscillator

Consider the vector field

$$\begin{aligned} \dot{x} &= y \equiv f(x, y), \\ \dot{y} &= x - x^3 - \delta y \equiv g(x, y), \end{aligned} \qquad \delta \geq 0. \tag{1.1.55}$$

An easy calculation shows that

$$\frac{\partial f}{\partial x} + \frac{\partial g}{\partial y} = -\delta.$$

Thus, for $\delta > 0$, (1.1.55) has no closed orbits. We will answer the question of what happens when $\delta = 0$ in Section 1.1E.

The next example shows how Theorem 1.1.4 allows us to restrict regions in the plane where closed orbits might exist.

EXAMPLE 1.1.6 Consider the following modification of the unforced Duffing oscillator

$$\begin{aligned} \dot{x} &= y \equiv f(x,y), \\ \dot{y} &= x - x^3 - \delta y + x^2 y \equiv g(x,y), \end{aligned} \qquad \delta \geq 0. \tag{1.1.56}$$

This equation has three fixed points at $(x,y) = (0,0)$, $(\pm 1, 0)$ with the eigenvalues, $\lambda_{1,2}$, of the associated linearization about each fixed point given by

$$(0,0) \Rightarrow \lambda_{1,2} = \frac{-\delta}{2} \pm \frac{1}{2}\sqrt{\delta^2 + 4}, \tag{1.1.57a}$$

$$(1,0) \Rightarrow \lambda_{1,2} = \frac{-\delta + 1}{2} \pm \frac{1}{2}\sqrt{(-\delta + 1)^2 - 8}, \tag{1.1.57b}$$

$$(-1,0) \Rightarrow \lambda_{1,2} = \frac{-\delta + 1}{2} \pm \frac{1}{2}\sqrt{(-\delta + 1)^2 - 8}. \tag{1.1.57c}$$

Thus, $(0,0)$ is a saddle, and $(\pm 1, 0)$ are sinks for $\delta > 1$ and sources for $0 \leq \delta < 1$.

A simple calculation gives

$$\frac{\partial f}{\partial x} + \frac{\partial g}{\partial y} = -\delta + x^2. \tag{1.1.58}$$

Thus, (1.1.58) vanishes on the lines $x = \pm\sqrt{\delta}$. These two lines divide the plane into three disjoint regions which we label (from left to right) R_1, R_2, and R_3 as shown in Figure 1.1.11.

Now from Theorem 1.1.4, we can immediately conclude that (1.1.56) can have no closed orbits lying entirely in either region R_1, R_2, or R_3. However, we cannot rule out the existence of closed orbits which overlap these regions as shown in Figure 1.1.12. When we discuss index theory in Section 1.1F we will see how to reduce the number of possibilities even further. We finally remark that it is not a coincidence that the lines $x = \pm\sqrt{\delta}$ fall on the fixed points $(\pm 1, 0)$ when the real parts of the eigenvalues of these fixed points vanish. We will learn what is going on in this case when we study the Poincaré–Andronov–Hopf bifurcation in Chapter 3.

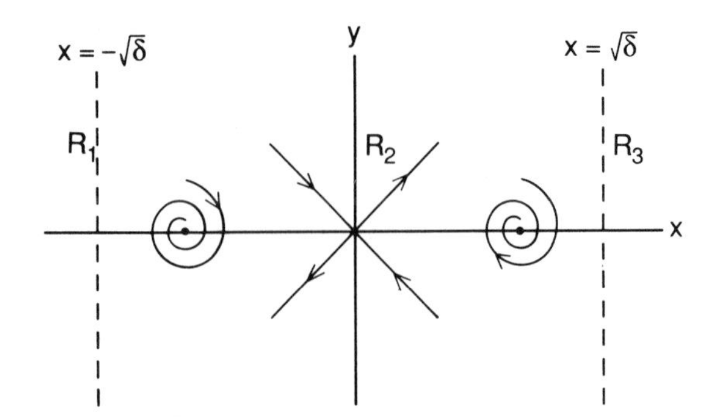

FIGURE 1.1.11. The regions defined by $x = \pm\sqrt{\delta}$ (the figure is drawn for $\delta > 1$).

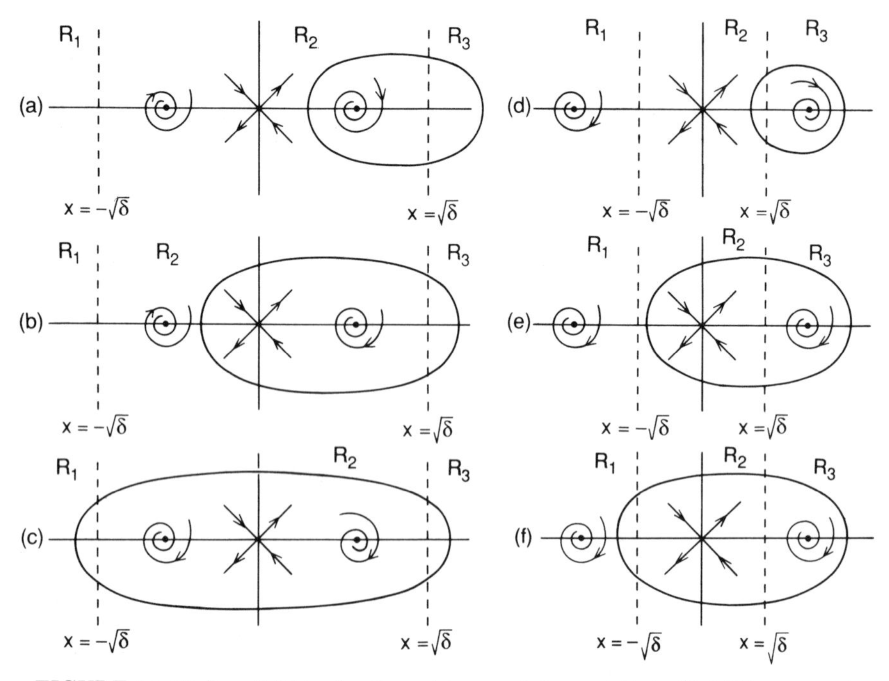

FIGURE 1.1.12. Possibilities for the existence of close orbits in (1.1.55). a–c apply to the case for $\delta > 1$, d–f to $0 < \delta < 1$.

1.1E Integrable Vector Fields on Two-Manifolds

In applications, three types of two-dimensional phase spaces occur frequently; they are (1) the plane, $\mathbb{R}^2 = \mathbb{R}^1 \times \mathbb{R}^1$, (2) the cylinder, $\mathbb{R}^1 \times S^1$, and (3) the two-torus, $T^2 = S^1 \times S^1$. The vector field can be written as

$$\begin{aligned} \dot{x} &= f(x, y), \\ \dot{y} &= g(x, y), \end{aligned} \tag{1.1.59}$$

where f and g are $\mathbf{C}^r$ $(r \geq 1)$, and as $(x, y) \in \mathbb{R}^1 \times \mathbb{R}^1$ for a vector field on the plane, as $(x, y) \in \mathbb{R}^1 \times S^1$ for a vector field on the cylinder, and as $(x, y) \in S^1 \times S^1$ for a vector field on the torus, where S^1 denotes the circle (which is sometimes referred to as a 1-torus, T^1). We now want to give some examples of how these different phase spaces arise and at the same time introduce the idea of an *integrable* vector field. We begin with the unforced Duffing oscillator.

EXAMPLE 1.1.7: THE UNFORCED DUFFING OSCILLATOR We have been slowly discovering the global structure of the phase space of the unforced Duffing oscillator given by

$$\ddot{x} - x + \delta\dot{x} + x^3 = 0, \tag{1.1.60}$$

or, written as a system,

$$\begin{aligned} \dot{x} &= y, \\ \dot{y} &= x - x^3 - \delta y, \end{aligned} \qquad (x, y) \in \mathbb{R}^1 \times \mathbb{R}^1, \quad \delta \geq 0. \tag{1.1.61}$$

Thus far we know the local structure near the three fixed points $(x, y) = (0, 0)$, $(\pm 1, 0)$ and that for $\delta > 0$ there are no closed orbits. The next step is to understand the geometry of the global orbit structure. In general, this is a formidable task. However, for the special parameter value $\delta = 0$, we can understand completely the global geometry, which, we will see, provides a framework for understanding the global geometry for $\delta \neq 0$.

The reason we can do this is that, for $\delta = 0$, the unforced, undamped Duffing oscillator has a *first integral* or a function of the dependent variables whose level curves give the orbits. Alternately, in more physical terms, the unforced, undamped Duffing oscillator is a conservative system having an energy function which is constant on orbits. This can be seen as follows—take the unforced, undamped Duffing oscillator, multiply it by $\dot{x}$, and integrate as below.

$$\dot{x}\ddot{x} - \dot{x}x + \dot{x}x^3 = 0$$

or

$$\frac{d}{dt}\left(\frac{1}{2}\dot{x}^2 - \frac{x^2}{2} + \frac{x^4}{4}\right) = 0; \tag{1.1.62}$$

hence,

$$\frac{1}{2}\dot{x}^2 - \frac{x^2}{2} + \frac{x^4}{4} = h = \text{constant}$$

or

$$h = \frac{y^2}{2} - \frac{x^2}{2} + \frac{x^4}{4}. \tag{1.1.63}$$

This is a first integral for the unforced, undamped Duffing oscillator or, if you think of $y^2/2$ as the kinetic energy (mass has been scaled to be 1) and

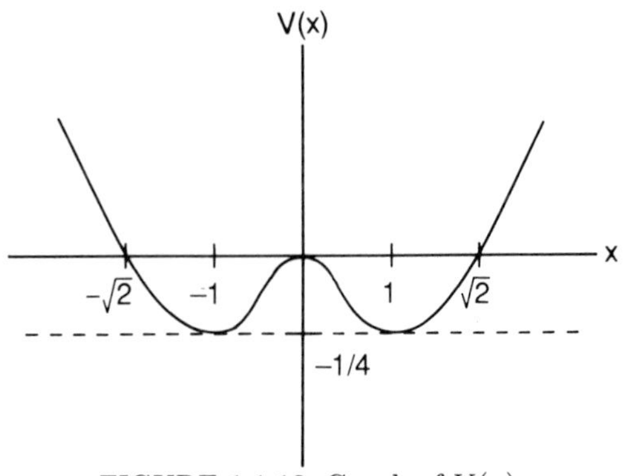

FIGURE 1.1.13. Graph of $V(x)$.

$-x^2/2 + x^4/4 \equiv V(x)$ as potential energy, h can be thought of as the total energy of the system. Therefore, the level curves of this function give the global structure of the phase space.

In general, for one-degree-of-freedom problems (i.e., vector fields on a two-dimensional phase space) that have a first integral that can be viewed as the sum of a kinetic and potential energy, there is an easy, graphical method for drawing the phase space. We will illustrate the method for the unforced, undamped Duffing oscillator. As a preliminary step, we point out the shape of the graph of $V(x)$ in Figure 1.1.13.

Now suppose that the first integral is given by

$$h = \frac{y^2}{2} + V(x);$$

then

$$y = \pm\sqrt{2}\sqrt{h - V(x)}. \tag{1.1.64}$$

Our goal is to draw the level sets of h. Imagine sitting at the point $(0, 0)$, with h fixed. Now move toward the right (i.e., let x increase). A glance at the graph of $V(x)$ shows that V begins to decrease. Then, since $y = +\sqrt{2}\sqrt{h - V(x)}$ (we take the + sign for the moment) and h is fixed, y must increase until the minimum of the potential is reached, and then it decreases until the boundary of the potential is reached (why can't you go farther?) (see Figure 1.1.14). Now $y = +$ or $-\sqrt{2}\sqrt{h - V(x)}$; hence the entire orbit through $(0, 0)$ for fixed h is as in Figure 1.1.15. (Note: why are the arrows drawn in their particular directions in Figure 1.1.15?) By symmetry, there is another *homoclinic orbit* to the left as in Figure 1.1.16, and if you repeat this procedure for different points you can draw the entire phase plane as shown in Figure 1.1.17. The homoclinic orbit is sometimes called a *separatrix* because it is the boundary between two

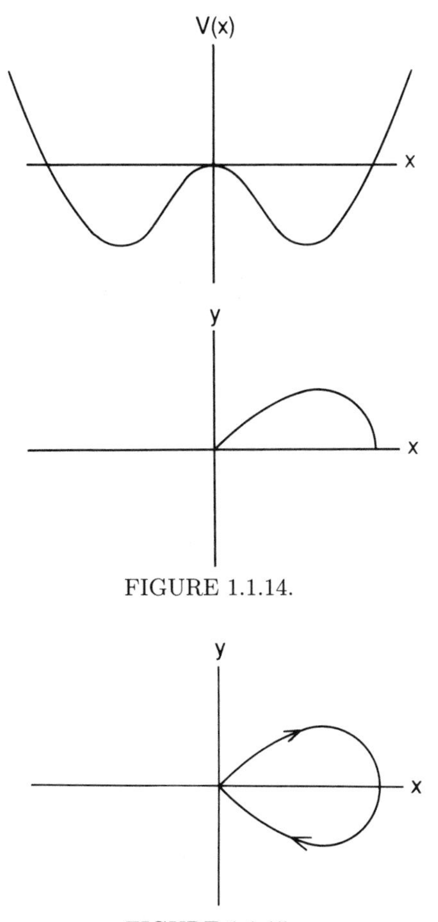

FIGURE 1.1.14.

y

x

FIGURE 1.1.15.

distinctly different types of motions. We will study homoclinic orbits in some detail in Chapter 4.

Denoting the first integral of the unforced, undamped Duffing oscillator by h was meant to be suggestive. The unforced, undamped Duffing oscillator is actually a *Hamiltonian System*, i.e., there exists a function $h = h(x, y)$ such that the vector field is given by

$$\begin{aligned} \dot{x} &= \frac{\partial h}{\partial y}, \\ \dot{y} &= -\frac{\partial h}{\partial x} \end{aligned} \tag{1.1.65}$$

(we will study these in more detail later). Note that all the solutions lie on level curves of h which are topologically the same as S^1 (or T^1). This Hamiltonian system is an *integrable* Hamiltonian system and it has a char-

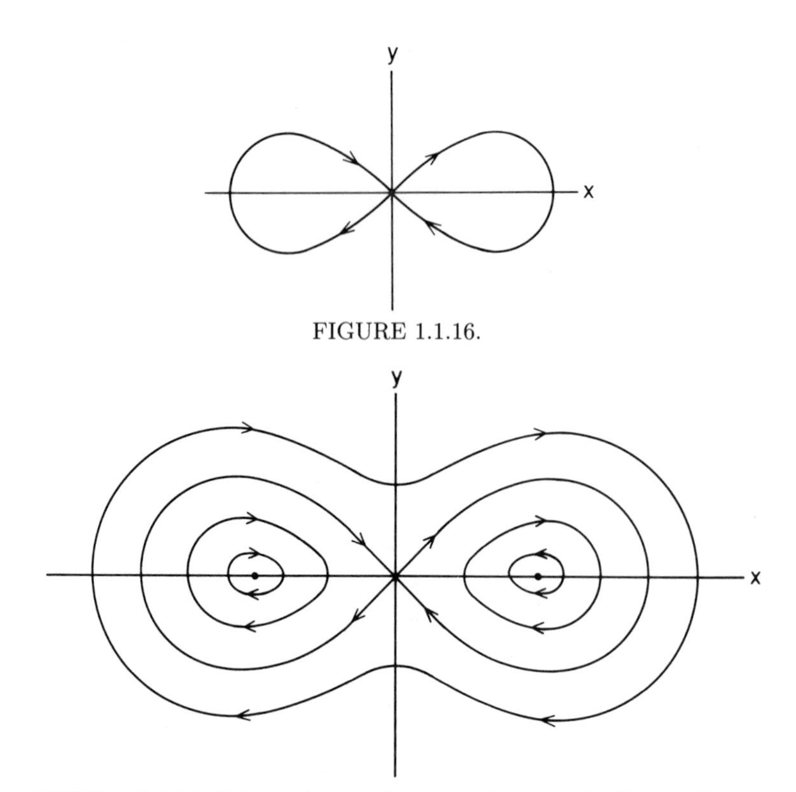

FIGURE 1.1.16.

FIGURE 1.1.17. Orbits of the unforced, undamped Duffing oscillator.

acteristic of all n-degree-of-freedom integrable Hamiltonian systems in that its bounded motions lie on n-dimensional tori or homoclinic and heteroclinic orbits (see Arnold [1978] or Abraham and Marsden [1978]). (Note that all one-degree-of-freedom Hamiltonian systems are integrable.)

EXAMPLE 1.1.8: THE PENDULUM The equation of motion of a simple pendulum (again, all physical constants are scaled out) is given by

$$\ddot{\phi} + \sin\phi = 0 \tag{1.1.66}$$

or, written as a system,

$$\begin{aligned} \dot{\phi} &= v, \\ \dot{v} &= -\sin\phi, \end{aligned} \qquad (\phi, v) \in S^1 \times \mathbb{R}^1. \tag{1.1.67}$$

This equation has fixed points at $(0,0)$, $(\pm\pi, 0)$, and simple calculations show that $(0,0)$ is a *center* (i.e., the eigenvalues are purely imaginary) and $(\pm\pi, 0)$ are saddles, but since the phase space is the cylinder and not the plane, $(\pm\pi, 0)$ are really the same point (see Figure 1.1.18). (Think of the pendulum as a physical object and you will see that this is obvious.)

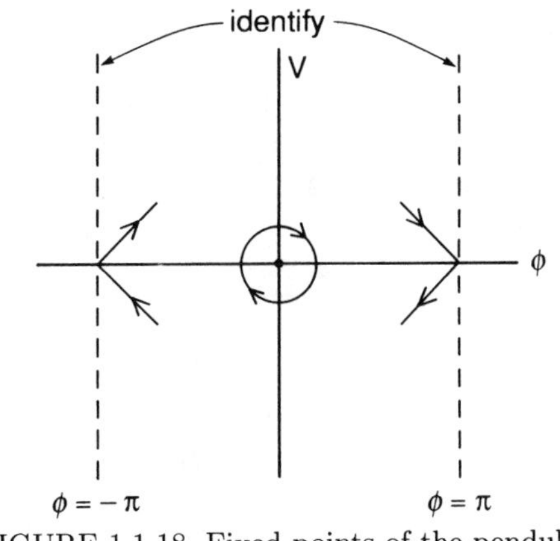

FIGURE 1.1.18. Fixed points of the pendulum.

Now, just as in Example 1.1.7, the pendulum is a Hamiltonian system with a first integral given by

$$h = \frac{v^2}{2} - \cos\phi. \tag{1.1.68}$$

Again, as in Example 1.1.7, this fact allows the global phase portrait for the pendulum to be drawn, as shown in Figure 1.1.19a. Alternatively, by gluing the two lines $\phi = \pm\pi$ together, we obtain the orbits on the cylinder as shown in Figure 1.1.19b.

EXAMPLE 1.1.9: A VECTOR FIELD ON A TWO-TORUS Now we want to consider an example of a vector field on the two-torus. Since this may appear a little bit unnatural, we begin with a simple example to motivate the situation. Suppose we have an undamped two-degree-of-freedom system consisting of two coupled linear oscillators. Under general conditions we can perform a change of variables to canonical coordinates (the "normal modes"), which uncouples the system; we will suppose this has been done so that the vector field takes the form

$$\begin{aligned} \ddot{x} + \omega_1^2 x &= 0, \\ \ddot{y} + \omega_2^2 y &= 0, \end{aligned} \tag{1.1.69}$$

or, written as a system,

$$\begin{aligned} \dot{x}_1 &= x_2, \\ \dot{x}_2 &= -\omega_1^2 x_1, \\ \dot{y}_1 &= y_2, \\ \dot{y}_2 &= -\omega_2^2 y_1, \end{aligned} \qquad (x_1, x_2, y_1, y_2) \in \mathbb{R}^4. \tag{1.1.70}$$

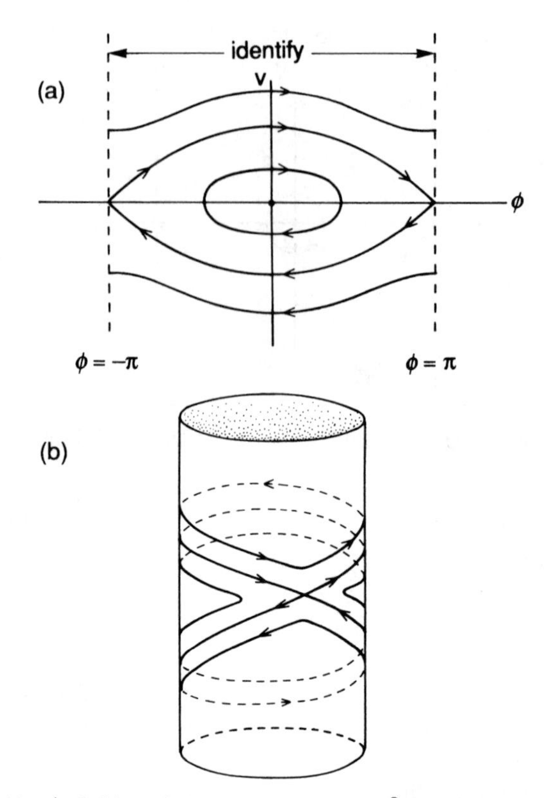

FIGURE 1.1.19. a) Orbits of the pendulum on $\mathbb{R}^2$ with $\phi = \pm\pi$ identified. b) Orbits of the pendulum on the cyliner.

This system is integrable, since we have the two independent functions of the dependent variables given by

$$\begin{aligned} h_1 &= \frac{x_2^2}{2} + \frac{\omega_1^2 x_1^2}{2}, \\ h_2 &= \frac{y_2^2}{2} + \frac{\omega_2^2 y_1^2}{2}. \end{aligned} \tag{1.1.71}$$

The level curves of these functions are compact sets (topological circles); therefore, the orbits in the four-dimensional phase space actually lie on two-tori. This can be made more apparent by making the change of variables

$$\begin{aligned} x_1 &= \sqrt{2I_1/\omega_1}\sin\theta_1, \qquad x_2 = \sqrt{2\omega_1 I_1}\cos\theta_1, \\ y_1 &= \sqrt{2I_2/\omega_2}\sin\theta_2, \qquad y_2 = \sqrt{2\omega_2 I_2}\cos\theta_2, \end{aligned} \tag{1.1.72}$$

which results in the new equations

$$\begin{aligned} \dot{I}_1 &= 0, \qquad \dot{I}_2 = 0, \\ \dot{\theta}_1 &= \omega_1, \qquad \dot{\theta}_2 = \omega_2. \end{aligned} \tag{1.1.73}$$

Hence, I_1 and I_2 are constants, and, therefore, the dynamics are contained in the equations

$$\begin{aligned} \dot{\theta}_1 &= \omega_1, \\ \dot{\theta}_2 &= \omega_2, \end{aligned} \qquad (\theta_1, \theta_2) \in S^1 \times S^1 = T^2. \tag{1.1.74}$$

The flow defined by this vector field will be discussed in more detail in Section 1.2A.

1.1F Index Theory

Before we describe some of the uses of index theory, we will give a heuristic description of the idea.

Suppose we have a vector field in the plane (this is a two-dimensional method only). Let Γ be any closed loop in the plane *which contains no fixed points* of the vector field. You can imagine at each point, p, on the loop Γ that there is an arrow representing the value of the vector field at p (see Figure 1.1.20).

Now as you move around Γ in the counter-clockwise sense (call this the positive direction), the vectors on Γ rotate, and when you get back to the point at which you started, they will have rotated through an angle $2\pi k$, where k is some integer. This integer, k, is called the *index of* Γ.

The index of a closed loop containing no fixed points can be calculated by integrating the angle of the vectors at each point on Γ around Γ (this angle is measured with respect to some chosen coordinate system). For a vector field on the plane given by

$$\begin{aligned} \dot{x} &= f(x, y), \\ \dot{y} &= g(x, y), \end{aligned} \qquad (x, y) \in \mathbb{R}^1 \times \mathbb{R}^1, \tag{1.1.75}$$

the index of Γ, k, is found by computing

$$\begin{aligned} k = \frac{1}{2\pi} \oint_\Gamma d\phi &= \frac{1}{2\pi} \oint_\Gamma d\left(\tan^{-1} \frac{g(x, y)}{f(x, y)}\right) \\ &= \frac{1}{2\pi} \oint_\Gamma \frac{f\,dg - g\,df}{f^2 + g^2}. \end{aligned} \tag{1.1.76}$$

This integral has several properties, one of the most important being that it retains the same value if Γ is smoothly deformed, as long as it is not deformed through some fixed point of the vector field. From the definition of the index given above (if not by just drawing pictures), one can prove the following theorems.

Theorem 1.1.6

i) *The index of a sink, a source, or a center is* $+1$.

ii) *The index of a hyperbolic saddle point is* -1.

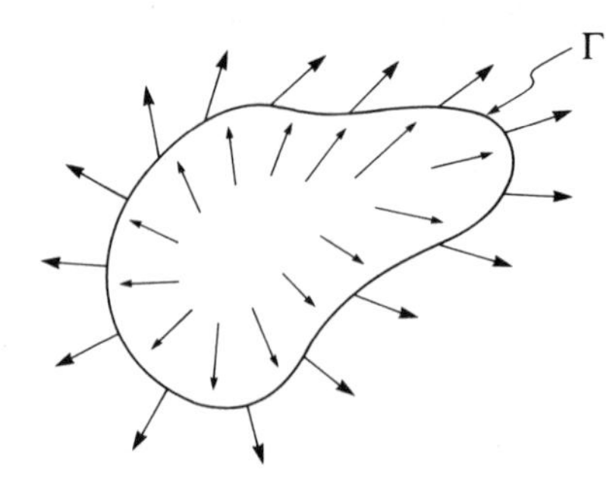

FIGURE 1.1.20. Vector field on the closed curve Γ.

iii) *The index of a closed orbit is* $+1$.

iv) *The index of a closed curve not containing any fixed points is 0.*

v) *The index of a closed curve is equal to the sum of the indices of the fixed points within it.*

An immediate corollary of this is the following.

Corollary 1.1.7 *Inside any closed orbit* γ *there must be at least one fixed point. If there is only one, then it must be a sink, source, or center. If all the fixed points within* γ *are hyperbolic, then there must be an odd number,* $2n+1$, *of which* n *are saddles and* $n+1$ *either sinks, sources, or centers.*

For more information on index theory, see Andronov et al. [1971].

EXAMPLE 1.1.6 REVISITED Using the above results, the reader should be able to verify that the phase portraits shown in Figures 1.1.12b, 1.1.12e, and 1.1.12f cannot occur. This example shows how Bendixson and Dulac's criteria used with index theory can go a long way toward describing the global structure of phase portraits on the plane. We remark that a higher dimensional generalization of index theory is *degree theory.* For an introduction to the use of degree theory in dynamical systems and bifurcation theory we refer the reader to Chow and Hale [1982] or Smoller [1983].

1.1G Some General Properties of Vector Fields: Existence, Uniqueness, Differentiability, and Flows

In this section we want to give some of the basic theorems describing general properties of solutions of vector fields. Since it is just as easy to treat the nonautonomous case we will do so.

Consider the vector field

$$\dot{x} = f(x, t), \tag{1.1.77}$$

where $f(x, t)$ is $\mathbf{C}^r$, $r \geq 1$, on some open set $U \subset \mathbb{R}^n \times \mathbb{R}^1$.

Existence, Uniqueness, Differentiability with Respect to Initial Conditions

Theorem 1.1.8 *Let $(x_0, t_0) \in U$. Then there exists a solution of (1.1.77) through the point x_0 at $t = t_0$, denoted $x(t, t_0, x_0)$ with $x(t_0, t_0, x_0) = x_0$, for $|t - t_0|$ sufficiently small. This solution is unique in the sense that any other solution of (1.1.77) through x_0 at $t = t_0$ must be the same as $x(t, t_0, x_0)$ on their common interval of existence. Moreover, $x(t, t_0, x_0)$ is a $\mathbf{C}^r$ function of t, t_0, and x_0.*

Proof: See Arnold [1973], Hirsch and Smale [1974], or Hale [1980]. □

We remark that it is possible to weaken the assumptions on $f(x, t)$ and still obtain existence and uniqueness. We refer the reader to Hale [1980] for a discussion.

Theorem 1.1.8 only guarantees existence and uniqueness for sufficiently small time intervals. The following result allows us to uniquely extend the time interval of existence.

Continuation of Solutions

Let $C \subset U \subset \mathbb{R}^n \times \mathbb{R}^1$ be a compact set containing (x_0, t_0).

Theorem 1.1.9 *The solution $x(t, t_0, x_0)$ can be uniquely extended backward and forward in t up to the boundary of C.*

Proof: See Hale [1980]. □

Theorem 1.1.9 tells us how solutions fail to exist; namely, they "blow up." Consider the following example.

Example 1.1.10 Consider the equation

$$\dot{x} = x^2, \qquad x \in \mathbb{R}^1. \tag{1.1.78}$$

The solution of (1.1.78) through x_0 at $t = 0$ is given by

$$x(t, 0, x_0) = \frac{-x_0}{x_0 t - 1}. \tag{1.1.79}$$

It should be clear that (1.1.79) does not exist for all time, since it becomes infinite at $t = 1/x_0$. This example also shows that the time interval of existence may depend on x_0.

In practice we often encounter vector fields depending on parameters, and it is often necessary to differentiate the solutions with respect to the parameters. The following result covers this situation.

Differentiability with Respect to Parameters

Consider the vector field

$$\dot{x} = f(x, t; \mu), \tag{1.1.80}$$

where $f(x, t; \mu)$ is $\mathbf{C}^r$ $(r \geq 1)$ on some open set $U \subset \mathbb{R}^n \times \mathbb{R}^1 \times \mathbb{R}^p$.

Theorem 1.1.10 *For $(t_0, x_0, \mu) \in U$ the solution $x(t, t_0, x_0, \mu)$ is a $\mathbf{C}^r$ function of t, t_0, x_0, and μ.*

Proof: See Arnold [1973] or Hale [1980]. □

At this stage we would like to point out some special properties of $\mathbf{C}^r$, $r \geq 1$, autonomous vector fields which will prove useful.

Autonomous Vector Fields

Consider the vector field

$$\dot{x} = f(x), \qquad x \in \mathbb{R}^n, \tag{1.1.81}$$

where $f(x)$ is $\mathbf{C}^r$, $r \geq 1$, on some open set $U \in \mathbb{R}^n$. For simplicity, let us suppose that the solutions exist for all time (we leave it as an exercise to make the necessary modifications when solutions exist only on finite time intervals). The following three results are very useful in applications.

Proposition 1.1.11 *If $x(t)$ is a solution of (1.1.81), then so is $x(t + \tau)$ for any $\tau \in \mathbb{R}$.*

Proof: By definition

$$\frac{dx(t)}{dt} = f(x(t)). \tag{1.1.82}$$

Hence, we have

$$\left.\frac{dx(t+\tau)}{dt}\right|_{t=t_0} = \left.\frac{dx(t)}{dt}\right|_{t=t_0+\tau} = f(x(t_0+\tau)) = f(x(t+\tau))\big|_{t=t_0}$$

or

$$\left.\frac{dx(t+\tau)}{dt}\right|_{t=t_0} = f(x(t+\tau))\big|_{t=t_0}. \tag{1.1.83}$$

Since (1.1.83) is true for any $t_0 \in \mathbb{R}$, the result follows. □

Note that Proposition 1.1.11 does not hold for nonautonomous vector fields. Consider the following example.

EXAMPLE 1.1.11 Consider the nonautonomous vector field

$$\dot{x} = e^t, \qquad x \in \mathbb{R}^1. \tag{1.1.84}$$

The solution of (1.1.84) is given by

$$x(t) = e^t, \tag{1.1.85}$$

and it should be clear that

$$x(t+\tau) = e^{t+\tau} \tag{1.1.86}$$

is not a solution of (1.1.84) for $\tau \neq 0$.

The following proposition lies at the heart of the Poincaré-Bendixson theorem.

Proposition 1.1.12 *For any $x_0 \in \mathbb{R}^n$ there exists only one solution of (1.1.81) passing through this point.*

Proof: We will show that if this proposition weren't true, then uniqueness of solutions would be violated.

Let $x_1(t)$, $x_2(t)$ be solutions of (1.1.81) satisfying

$$\begin{aligned} x_1(t_1) &= x_0, \\ x_2(t_2) &= x_0. \end{aligned}$$

By Proposition 1.1.11

$$\tilde{x}_2(t) \equiv x_2\big(t - (t_1 - t_2)\big)$$

is also a solution of (1.1.81) satisfying

$$\tilde{x}_2(t_1) = x_0.$$

Hence, by Theorem 1.1.8, $x_1(t)$ and $x_2(t)$ must be identical. □

Since for autonomous vector fields time-translated solutions remain solutions (i.e., Proposition 1.1.11 holds), it suffices to choose a fixed initial time, say $t_0 = 0$, which is understood and therefore often omitted from the notation (as we do now).

Proposition 1.1.13

i) $x(t, x_0)$ *is* $\mathbf{C}^r$.

ii) $x(0, x_0) = x_0$.

iii) $x(t+s, x_0) = x\big(t, x(s, x_0)\big)$.

Proof: i) follows from Theorem 1.1.8, ii) is by definition, and iii) follows from Proposition 1.1.12; namely, $\tilde{x}(t, x_0) \equiv x(t+x, x_0)$ and $x\big(t, x(s, x_0)\big)$ are both solutions of 1.1.81 satisfying the same initial conditions at $t = 0$. Hence, by uniqueness, they must coincide. □

Proposition 1.1.13 shows that the solutions of (1.1.81) form a one-parameter family of $\mathbf{C}^r$, $r \geq 1$, diffeomorphisms of the phase space (invertibility comes from iii)). This is referred to as a *phase flow* or just a *flow*. A common notation for flows is $\phi(t,x)$ or $\phi_t(x)$.

Let us comment a bit more on this notation $\phi_t(x)$. The part of Theorem 1.1.8 dealing with differentiability of solutions with respect to x_0 (regarding t and t_0 as fixed) allows us to think differently about the solutions of ordinary differential equations. More precisely, in the solution $x(t,t_0,x_0)$, we can think of t and t_0 as fixed and then study how the map $x(t,t_0,x_0)$ moves sets of points around in phase space. This is the global, geometrical view of the study of dynamical systems. For a set $U \subset \mathbb{R}^n$, we would denote its image under this map by $x(t,t_0,U)$. Since points in phase space are also labeled by the letter x, it is often less confusing to change the notation for the solutions, which is why we use the symbol ϕ. This point of view will become more apparent when we study the construction of Poincaré maps in Section 1.2.

Finally, let us note that in the study of ordinary differential equations one might believe the problem to be finished when the "solution" $x(t,t_0,x_0)$ is found. The rest of the book will show that this is not the case, but, on the contrary, that this is when the story begins to get really interesting.

Nonautonomous Vector Fields

It should be clear that Propositions 1.1.11, 1.1.12, and 1.1.13 do not hold for nonautonomous vector fields. However, we can always make a nonautonomous vector field autonomous by redefining time as a new dependent variable. This is done as follows.

By writing (1.1.77) as

$$\frac{dx}{dt} = \frac{f(x,t)}{1} \tag{1.1.87}$$

and using the chain rule, we can introduce a new independent variable s so that (1.1.87) becomes

$$\begin{aligned} \frac{dx}{ds} &\equiv x' = f(x,t), \\ \frac{dt}{ds} &\equiv t' = 1. \end{aligned} \tag{1.1.88}$$

If we define $y = (x,t)$ and $g(y) = (f(x,t),1)$, we see that (1.1.88) becomes

$$y' = g(y), \qquad y \in \mathbb{R}^n \times \mathbb{R}^1. \tag{1.1.89}$$

Of course, knowledge of the solutions of (1.1.89) implies knowledge of the solutions of (1.1.77) and vice versa. For example, if $x(t)$ is a solution of (1.1.77) passing through x_0 at $t = t_0$, i.e., $x(t_0) = x_0$, then $y(s) = (x(s+t_0), t(s) = s + t_0)$ is a solution of (1.1.89) passing through $y_0 \equiv \big(x(t_0), t_0\big)$ at $s = 0$.

Every vector field can thus be viewed as an autonomous vector field. This apparently trivial trick is a great conceptual aid in the construction of Poincaré maps for time-periodic and quasiperiodic vector fields, as we shall see in Section 1.2. Notice, however, that in redefining time as a *dependent* variable, it may then be introduced in various situations requiring specification of initial positions (i.e., specifying x_0); in particular, the reader should reexamine the definition of stability given in Section 1.1A. For an alternative view of nonautonomous vector fields see Sell [1971].

For the most part in this book we will be considering autonomous vector fields or maps constructed from nonautonomous vector fields (more specifically, maps constructed from time-periodic and quasiperiodic vector fields). Consequently, henceforth we will state definitions in the context of autonomous vector fields and maps.

1.1H Asymptotic Behavior

We now develop a technical apparatus to deal with the notions of "long term" and "observable" behavior for orbits of dynamical systems. We will be concerned with $\mathbf{C}^r$ $(r \geq 1)$ maps and autonomous vector fields on $\mathbb{R}^n$ denoted as follows.

$$\textit{Vector Field:} \qquad \dot{x} = f(x), \qquad x \in \mathbb{R}^n, \tag{1.1.90}$$

$$\textit{Map:} \qquad x \mapsto g(x), \qquad x \in \mathbb{R}^n. \tag{1.1.91}$$

The *flow* generated by (1.1.90) (see Section 1.1G) will be denoted as $\phi(t, x)$.

As we shall see in Section 1.1I, the Poincaré-Bendixson theorem characterizes the nature of the α and ω limit sets of flows on certain two manifolds. We now define α and ω limit sets.

Definition 1.1.7 A point $x_0 \in \mathbb{R}^n$ is called an ω *limit point* of $x \in \mathbb{R}^n$, denoted $\omega(x)$, if there exists a sequence $\{t_i\}$, $t_i \longrightarrow \infty$, such that

$$\phi(t_i, x) \longrightarrow x_0.$$

α *limit points* are defined similarly by taking a sequence $\{t_i\}$, $t_i \longrightarrow -\infty$.

Example 1.1.12 Consider a vector field on the plane with a hyperbolic saddle point, x, as shown in Figure 1.1.21. Then x is the ω limit point of any point on the stable manifold and the α limit point of any point on the unstable manifold.

Example 1.1.13 This example shows why it is necessary to take a subsequence in time, $\{t_i\}$, and not to simply let $t \uparrow \infty$ in the definition of the α and ω limit point. Consider a vector field on the plane with a globally attracting closed orbit, γ, as shown in Figure 1.1.22. Then orbits not starting on γ "wrap onto" γ.

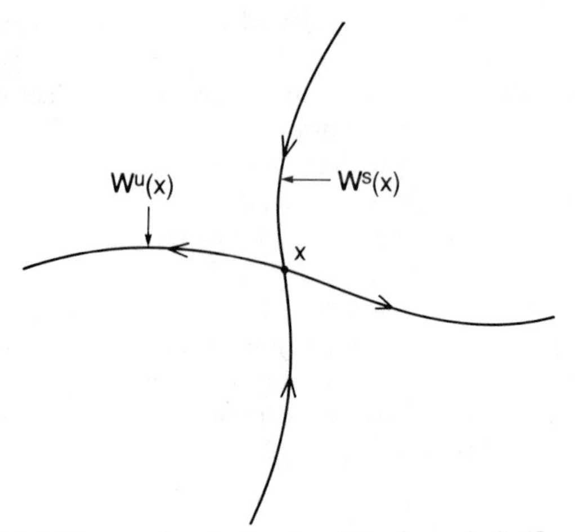

FIGURE 1.1.21. α and ω limit sets of the hyperbolic fixed point x.

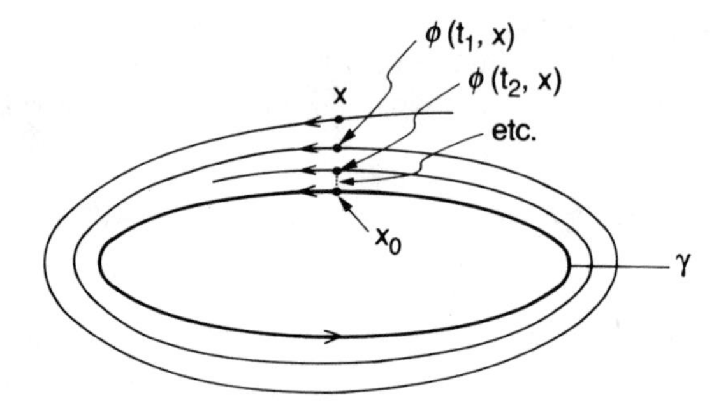

FIGURE 1.1.22. The point $x_0 \in \gamma$ is the ω limit point of x.

Now for each point on γ, we can find a subsequence $\{t_i\}$ such that $\phi(t_i, x)$, $x \in \mathbb{R}^2$, approaches that point as $i \uparrow \infty$. Therefore, γ is the ω limit set of x as you would expect. However, $\lim_{t \to \infty} \phi(t, x) \neq \gamma$.

DEFINITION 1.1.8 The set of all ω limit points of a flow or map is called the ω *limit set*. The α limit set is similarly defined.

We will need the idea of α and ω limit sets in the context of flows only, so we leave it to the reader to modify Definition 1.1.8 for maps as an exercise.

For maps, the notion of a nonwandering point has been more fashionable; however, we will explore the relationship between these two concepts in the exercises.

DEFINITION 1.1.9 A point x_0 is called *nonwandering* if the following holds.

Flows: For *any* neighborhood U of x_0, there exists some $t \neq 0$ such that

$$\phi(t, U) \cap U \neq \emptyset.$$

Maps: For *any* neighborhood U of x_0, there exists some $n \neq 0$ such that

$$g^n(U) \cap U \neq \emptyset.$$

Note that if the map is noninvertible, then we must take $n > 0$.

Fixed points and periodic orbits are nonwandering; we will see more complicated examples in Chapter 4.

DEFINITION 1.1.10 The set of all nonwandering points of a map or flow is called the *nonwandering set* of that particular map or flow.

Definitions 1.1.8 and 1.1.9 do not address the question of stability of those asymptotic motions. For this we want to develop the idea of an attractor.

DEFINITION 1.1.11 A closed invariant set $A \subset \mathbb{R}^n$ is called an *attracting set* if there is some neighborhood U of A such that:

flows: $\forall x \in U, \quad \forall t \geq 0, \qquad \phi(t,x) \in U$ and $\phi(t,x) \underset{t\uparrow\infty}{\longrightarrow} A.$

maps: $\forall x \in U, \quad \forall n \geq 0, \qquad g^n(x) \in U$ and $g^n(x) \underset{n\uparrow\infty}{\longrightarrow} A.$

If we have an attracting set it is natural to ask which points in phase space approach the attracting set asymptotically.

DEFINITION 1.1.12 The *domain or basin of attraction* of A is given by

flows: $\bigcup_{t \leq 0} \phi(t, U),$

maps: $\bigcup_{n \leq 0} g^n(U),$

where U is defined in Definition 1.1.11.

Note that even if g is noninvertible, g^{-1} still makes sense in a set theoretic sense. Namely, $g^{-1}(U)$ is the set of points in $\mathbb{R}^n$ that maps into U under g; g^{-n}, $n > 1$ is then defined inductively.

In practice, a way of locating attracting sets is to first find a trapping region.

DEFINITION 1.1.13 A closed, connected set $\mathcal{M}$ is called a *trapping region* if $\phi(t, \mathcal{M}) \subset \mathcal{M}$, $\forall t \geq 0$ or, equivalently, if the vector field on the boundary of $\mathcal{M}$ (denoted $\partial\mathcal{M}$) is pointing toward the interior of $\mathcal{M}$. Then

$$\bigcap_{t>0} \phi(t, \mathcal{M}) \stackrel{\text{def}}{=} A$$

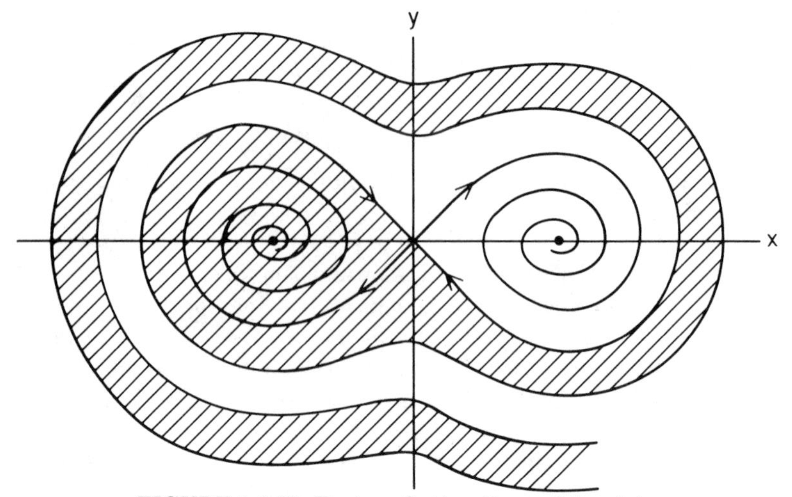

FIGURE 1.1.23. Basins of attractions of the sinks.

is an *attracting set.*

A similar definition can be given for maps. By now the necessary modifications should be obvious, and we leave the details as an exercise for the reader. It should be evident to the reader that finding a Liapunov function is equivalent to finding a trapping region (cf. Section 1.1B). Also, let us mention a technical point; by Theorem 1.1.8 it follows that all solutions starting in a trapping region exist for all positive times. This is useful in noncompact phase spaces such as $\mathbb{R}^2$ for proving existence on semi-infinite time intervals.

Application to the Unforced Duffing Oscillator

As we've seen, the unforced Duffing oscillator has, for $\delta > 0$, two attractors which are fixed points. The boundaries of the domains of attraction for the two attractors are defined by the stable manifold of the saddle of the origin (see Figure 1.1.23).

Now we want to motivate the idea of an *attractor* as opposed to attracting set. We do this with the following example taken from Guckenheimer and Holmes [1983].

EXAMPLE 1.1.14 Consider the planar autonomous vector field

$$\begin{aligned} \dot{x} &= x - x^3, \\ \dot{y} &= -y, \end{aligned} \qquad (x, y) \in \mathbb{R}^1 \times \mathbb{R}^1.$$

This vector field has a saddle at $(0,0)$ and two sinks at $(\pm 1, 0)$. The y-axis is the stable manifold of $(0,0)$. We choose an ellipse, $\mathcal{M}$, containing the three fixed points as shown in Figure 1.1.24. It should be clear that $\mathcal{M}$ is

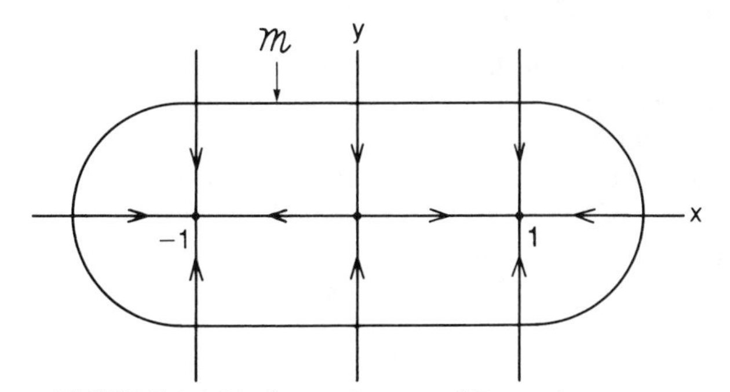

FIGURE 1.1.24. Attracting set of Example 1.1.14.

a trapping region and that the closed interval $[-1, 1] = \bigcap_{t \geq 0} \phi(t, \mathcal{M})$ is an attracting set.

Example 1.1.14 points out what some might regard as a possible deficiency in our Definition 1.1.11 of an attracting set. In this example, almost all points in the plane will *eventually* end up near one of the sinks. Hence, the attracting set, the interval $[-1, 1]$, contains two *attractors*, the sinks $(\pm 1, 0)$. Therefore, if we are interested in describing where most points in phase space ultimately go, the idea of an attracting set is not quite precise enough. Somehow we want to incorporate into the definition of an attracting set the notion that it is not a collection of distinct *attractors*, but rather that all points in the attracting set eventually come arbitrarily close to every other point in the attracting set under the evolution of the flow or map. We now want to make this mathematically precise.

DEFINITION 1.1.14 A close invariant set A is said to be *topologically transitive* if, for any two open sets U, $V \subset A$

flows: $\exists \; t \in \mathbb{R} \ni \phi(t, U) \cap V \neq \emptyset$,

maps: $\exists \; n \in \mathbb{Z} \ni g^n(U) \cap V \neq \emptyset$.

DEFINITION 1.1.15 An *attractor* is a topologically transitive attracting set.

We remark that the study of attractors and their basin boundaries in dynamical systems is rapidly evolving and, consequently, the theory is incomplete. For more information see Conley [1978], Guckenheimer and Holmes [1983], Milnor [1985], and Ruelle [1981].

1.1I The Poincaré-Bendixson Theorem

The Poincaré-Bendixson theorem gives us a complete determination of the asymptotic behavior of a large class of flows on the plane, cylinder, and two-sphere. It is remarkable in that it assumes no detailed information about the vector field, only uniqueness of solutions, properties of ω limit sets, and some properties of the geometry of the underlying phase space. We begin by setting the framework and giving some preliminary definitions.

We will consider $\mathbf{C}^r$, $r \geq 1$, vector fields

$$\begin{aligned} \dot{x} &= f(x,y), \\ \dot{y} &= g(x,y), \end{aligned} \qquad (x,y) \in \mathcal{P},$$

where $\mathcal{P}$ denotes the phase space, which may be the plane, cylinder, or two-sphere. We denote the flow generated by this vector field by

$$\phi_t(\cdot),$$

where the "$\cdot$" in this notation denotes a point $(x,y) \in \mathcal{P}$. The following proposition is fundamental and is *independent* of the dimension of phase space (as long as it is finite).

Proposition 1.1.14 *Let $\phi_t(\cdot)$ be a flow generated by a vector field and let $\mathcal{M}$ be a positively invariant compact set for this flow. Then, for $p \in \mathcal{M}$, we have*

i) $\omega(p) \neq \emptyset$;

ii) $\omega(p)$ *is closed;*

iii) $\omega(p)$ *is invariant under the flow, i.e., $\omega(p)$ is a union of orbits;*

iv) $\omega(p)$ *is connected.*

Proof: i) Choose a sequence $\{t_i\}$, $\lim_{i\to\infty} t_i = \infty$, and let $\{p_i = \phi_{t_i}(p)\}$. Since $\mathcal{M}$ is compact, $\{p_i\}$ has a convergent subsequence whose limit belongs to $\omega(p)$. Thus, $\omega(p) \neq \emptyset$.

ii) It suffices to show that the complement of $\omega(p)$ is open. Choose $q \notin \omega(p)$. Then there must exist some neighborhood of q, $U(q)$, that is disjoint from the set of points $\{\, \phi_t(p) \mid t \geq T \,\}$ for some $T > 0$. Hence, q is contained in some open set that contains no points in $\omega(p)$. Since q is arbitrary, we are done.

iii) Let $q \in \omega(p)$ and $\tilde{q} = \phi_s(q)$. Choose a sequence $t_i \underset{i\uparrow\infty}{\longrightarrow} \infty$ with $\phi_{t_i}(p) \to q$. Then $\phi_{t_i+s}(p) = \phi_s\big(\phi_{t_i}(p)\big)$ (cf. the notation for flows following Proposition 1.1.13) converges to $\tilde{q}$ as $i \to \infty$. Hence, $\tilde{q} \in \omega(p)$, and therefore $\omega(p)$ is invariant. However, there is a slight hole in this argument that needs to be filled; namely, it is not immediately obvious that $\phi_s(\cdot)$ exists for all s.

We begin by arguing that $\phi_s(q)$ exists for $s \in (-\infty, \infty)$ *when* $q \in \omega(p)$. It should be clear that this is true for $s \in (0, \infty)$ since $\mathcal{M}$ is a positively invariant compact set (cf. Theorem 1.1.9). Therefore, it suffices to show that this is true for $s \in (-\infty, 0]$.

Now $q \in \omega(p)$, so by definition we can find a sequence $\{t_i\}$, $t_i \underset{i\uparrow\infty}{\longrightarrow} \infty$, such that $\phi_{t_i}(p) \to q$ as $i \to \infty$. Let us order the sequence so that $t_1 < t_2 < \cdots < t_n < \cdots$. Next consider $\phi_s(\phi_{t_i}(p))$. By Proposition 1.1.13 this is valid for $s \in [-t_i, 0]$. Taking the limit as $i \to \infty$ and using continuity as well as the fact that $\phi_{t_i} \to q$ as $i \to \infty$, we see that $\phi_s(q)$ exists for $s \in (-\infty, 0]$.

iv) The proof is by contradiction. Suppose $\omega(p)$ is not connected. Then we can choose open sets V_1, V_2 such that $\omega(p) \subset V_1 \cup V_2$, $\omega(p) \cap V_1 \neq \emptyset$, $\omega(p) \cap V_2 \neq \emptyset$, and $\bar{V}_1 \cap \bar{V}_2 = \emptyset$. The orbit of p accumulates on points in both V_1 and V_2; hence, given $T > 0$, there exists $t > T$ such that $\phi_t(p) \in \mathcal{M} - (V_1 \cup V_2) = K$. Then we can find a sequence $\{t_n\}$, $t_n \underset{n\uparrow\infty}{\longrightarrow} \infty$, with $\phi_{t_n}(p) \in K$. Passing to a subsequence if necessary we have $\phi_{t_n}(p) \to q$, $q \in K$. But this implies that $q \in \omega(p) \in V_1 \cup V_2$. *Contradiction.* □

The following definition will be useful.

DEFINITION 1.1.16 Let Σ be a continuous, connected arc in $\mathcal{P}$. Then Σ is said to be *transverse* to the vector field on $\mathcal{P}$ if the vector dot product of the unit normal at each point on Σ with the vector field at that point is not zero and does not change sign on Σ. Or equivalently, since the vector field is $\mathbf{C}^r$, $r \geq 1$, the vector field has no fixed points on Σ and is never tangent to Σ.

Now we are in a position to actually prove the Poincaré-Bendixson theorem. We will first prove several lemmas from which the theorem will follow easily. Our presentation follows closely Palis and de Melo [1982]. In all that follows, $\mathcal{M}$ is understood to be a positively invariant compact set in $\mathcal{P}$. For any point $p \in \mathcal{P}$, we will denote the orbit of p under the flow $\phi_t(\cdot)$ *for positive times* $O_+(p)$ (also called the positive semiorbit of p).

Lemma 1.1.15 *Let $\Sigma \subset \mathcal{M}$ be an arc transverse to the vector field. The positive orbit through any point $p \in \mathcal{M}$, $O_+(p)$, intersects Σ in a monotone sequence; that is, if p_i is the ith intersection of $O_+(p)$ with Σ, then $p_i \in [p_{i-1}, p_{i+1}]$.*

Proof: Consider the piece of the orbit $O_+(p)$ from p_{i-1} to p_i along with the segment $[p_{i-1}, p_i] \subset \Sigma$ (see Figure 1.1.25). (Note: of course, if $O_+(p)$ intersects Σ only once then we are done.) This forms a positively invariant region $\mathcal{D}$. Hence, $O_+(p_i) \subset \mathcal{D}$, and therefore we must have p_{i+1} (if it exists) contained in $\mathcal{D}$. Thus we have shown that $p_i \in [p_{i-1}, p_{i+1}]$. □

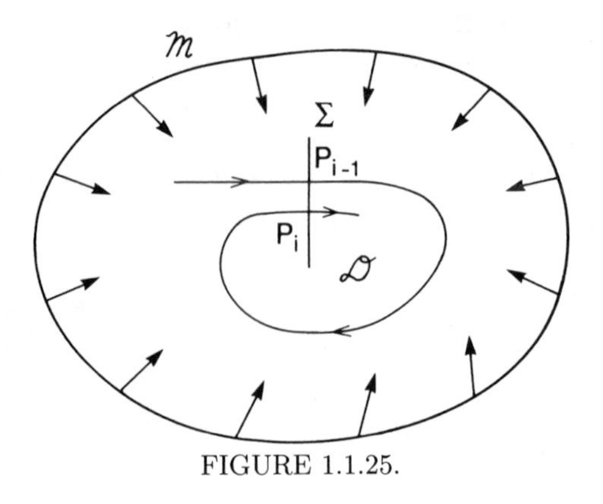

FIGURE 1.1.25.

We remark that Lemma 1.1.15 does not apply immediately to toroidal phase spaces. This is because the piece of the orbit from p_{i-1} to p_i along with the segment $[p_{i-1}, p_i] \subset \Sigma$ divides $\mathcal{M}$ into two "disjoint pieces." This would not be true for orbits completely encircling a torus. However, the lemma would apply to pieces of the torus that behave as $\mathcal{M}$ described above.

Corollary 1.1.16 *The ω-limit set of p $(\omega(p))$ intersects Σ in at most one point.*

Proof: The proof is by contradiction. Suppose $\omega(p)$ intersects Σ in two points, q_1 and q_2. Then by the definition of ω-limit sets, we can find sequences of points along $O_+(p)$, $\{p_n\}$ and $\{\bar{p}_n\}$, which intersect Σ such that $p_n \underset{n\uparrow\infty}{\longrightarrow} q_1$ and $\bar{p}_n \underset{n\uparrow\infty}{\longrightarrow} q_2$. However, if this were true, then it would contradict the previous lemma on monotonicity of the intersections of $O_+(p)$ with Σ. □

Lemma 1.1.17 *If $\omega(p)$ does not contain fixed points, then $\omega(p)$ is a closed orbit.*

Proof: The strategy is to choose a point $q \in \omega(p)$, show that the orbit of q is closed, and then show that $\omega(p)$ is the same as the orbit of q.

Choose $x \in \omega(q)$; then x is not a fixed point, since $\omega(p)$ is connected and closed and is a union of orbits containing no fixed points. Construct an arc transverse to the vector field at x (call it Σ). Now $O_+(q)$ intersects Σ in a monotone sequence, $\{q_n\}$, with $q_n \underset{n\uparrow\infty}{\rightarrow} x$, but since $q_n \in \omega(p)$, by the previous corollary we must have $q_n = x$ for all n. Since $x \in \omega(q)$, the orbit of q must be a closed orbit.

It only remains to show that the orbit of q and $\omega(p)$ are the same thing. Taking a transverse arc, Σ, at q, we see by the previous corollary that $\omega(p)$ intersects Σ only at q. Since $\omega(p)$ is a union of orbits, contains no fixed points, and is connected, we know that $O(q) = \omega(p)$. □

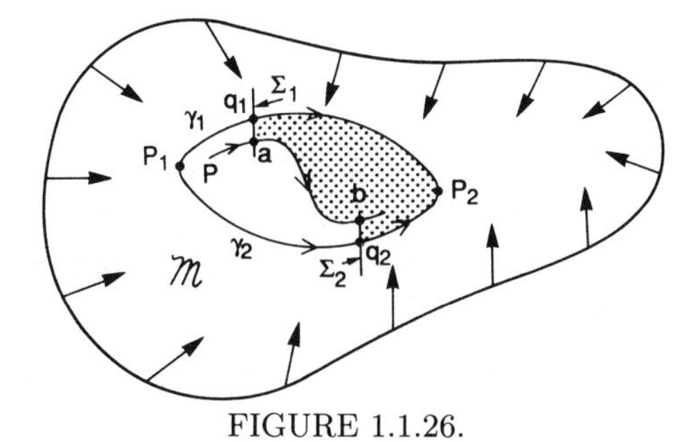

FIGURE 1.1.26.

Lemma 1.1.18 *Let p_1 and p_2 be distinct fixed points of the vector field contained in $\omega(p)$, $p \in \mathcal{M}$. Then there exists at most one orbit $\gamma \subset \omega(p)$ such that $\alpha(\gamma) = p_1$ and $\omega(\gamma) = p_2$. (Note: by $\alpha(\gamma)$ we mean the α limit set of every point on γ; similarly for $\omega(\gamma)$.)*

Proof: The proof is by contradiction. Suppose there exist two orbits γ_1, $\gamma_2 \in \omega(p)$ such that $\alpha(\gamma_i) = p_1$, $\omega(\gamma_i) = p_2$, $i = 1, 2$. Choose points $q_1 \in \gamma_1$ and $q_2 \in \gamma_2$ and construct arcs Σ_1, Σ_2 transverse to the vector field at each of these points (see Figure 1.1.26).

Since γ_1, $\gamma_2 \subset \omega(p)$, $O_+(p)$ intersects Σ_1 in a point a and later intersects Σ_2 in a point b. Hence, the region bounded by the points q_1, a, b, q_2, p_2 is a positively invariant region, but this leads to a contradiction, since γ_1, $\gamma_2 \subset \omega(p)$. □

Now we can finally prove the theorem.

Theorem 1.1.19 (Poincaré-Bendixson) *Let $\mathcal{M}$ be a positively invariant region for the vector field containing a finite number of fixed points. Let $p \in \mathcal{M}$, and consider $\omega(p)$. Then one of the following possibilities holds.*

i) *$\omega(p)$ is a fixed point;*

ii) *$\omega(p)$ is a closed orbit;*

iii) *$\omega(p)$ consists of a finite number of fixed points $p_1, \cdots, p_n$ and orbits γ with $\alpha(\gamma) = p_i$ and $\omega(\gamma) = p_j$.*

Proof: If $\omega(p)$ contains only fixed points, then it must consist of a unique fixed point, since the number of fixed points in $\mathcal{M}$ is finite and $\omega(p)$ is a connected set.

If $\omega(p)$ contains no fixed points, then, by Lemma 1.1.17, it must be a closed orbit. Suppose that $\omega(p)$ contains fixed points and nonfixed points (sometimes called regular points). Let γ be a trajectory in $\omega(p)$ consisting of regular points. Then $\omega(\gamma)$ and $\alpha(\gamma)$ must be fixed points since, if they were not, then, by Lemma 1.1.17, $\omega(\gamma)$ and $\alpha(\gamma)$ would be closed orbits, which is absurd, since $\omega(p)$ is connected and contains fixed points.

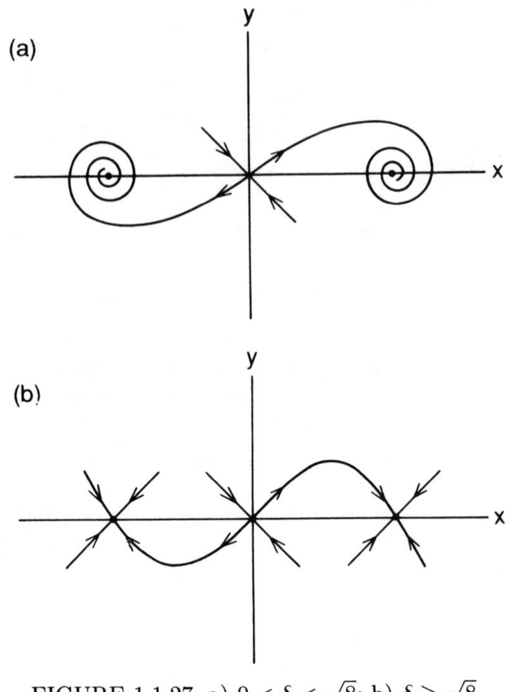

FIGURE 1.1.27. a) $0 < \delta < \sqrt{8}$; b) $\delta \geq \sqrt{8}$.

We have thus shown that every regular point in $\omega(p)$ has a fixed point for an α and ω limit set. This proves iii) and completes the proof of the Poincaré-Bendixson theorem. □

For an example illustrating the necessity of a finite number of fixed points in the hypotheses of Theorem 1.1.19 see Palis and de Melo [1982]. For generalizations of the Poincaré-Bendixson theorem to arbitrary closed two-manifolds see Schwartz [1963].

Application to the Unforced Duffing Oscillator

We now want to apply the Poincaré-Bendixson theorem to the unforced Duffing oscillator which, we recall, is given by

$$\begin{aligned} \dot{x} &= y, \\ \dot{y} &= x - x^3 - \delta y, \end{aligned} \qquad \delta > 0.$$

Using the fact that the level sets of $V(x, y) = y^2/2 - x^2/2 + x^4/4$ form positively invariant sets for $\delta > 0$, we see that the unstable manifold of the saddle must fall into the sinks as shown in Figure 1.1.26; see Exercise 1.1.36. The reader should convince him- or herself that Figure 1.1.27 is rigorously justified based on analytical techniques developed in this chapter. Note

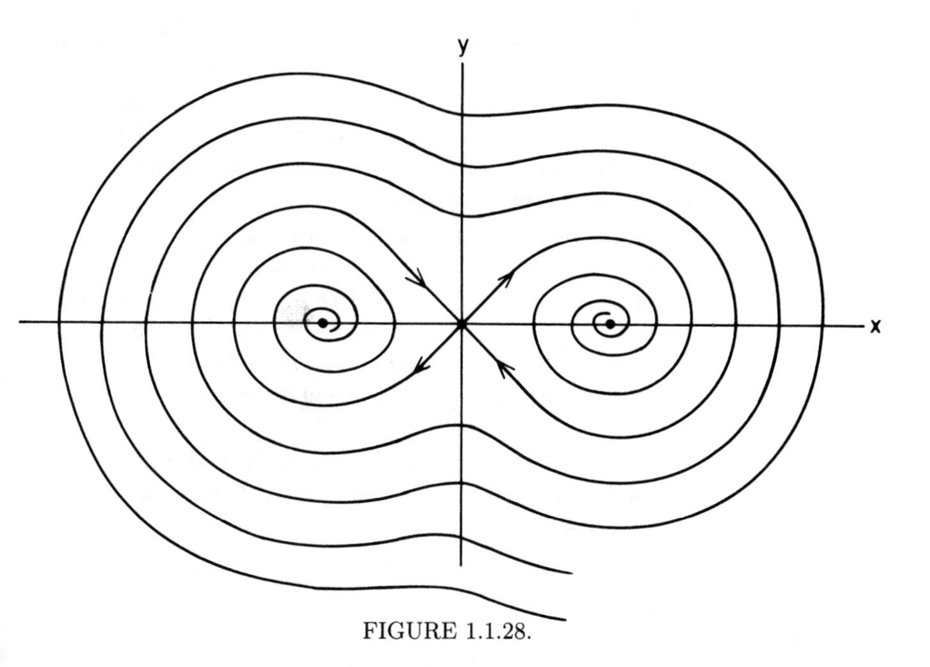

FIGURE 1.1.28.

that we have not proved anything about the global behavior of the stable manifold of the saddle. Qualitatively, it behaves as in Figure 1.1.28, but, we stress, this has not been rigorously justified.

Exercises

1.1.1 Consider the following linear vector fields on $\mathbb{R}^2$.

a) $\begin{pmatrix} \dot{x}_1 \\ \dot{x}_2 \end{pmatrix} = \begin{pmatrix} \lambda & 0 \\ 0 & \mu \end{pmatrix} \begin{pmatrix} x_1 \\ x_2 \end{pmatrix}, \qquad \begin{matrix} \lambda < 0 \\ \mu > 0 \end{matrix}.$

b) $\begin{pmatrix} \dot{x}_1 \\ \dot{x}_2 \end{pmatrix} = \begin{pmatrix} \lambda & 0 \\ 0 & \mu \end{pmatrix} \begin{pmatrix} x_1 \\ x_2 \end{pmatrix}, \qquad \begin{matrix} \lambda < 0 \\ \mu < 0 \end{matrix}.$

c) $\begin{pmatrix} \dot{x}_1 \\ \dot{x}_2 \end{pmatrix} = \begin{pmatrix} \lambda & -\omega \\ \omega & \lambda \end{pmatrix} \begin{pmatrix} x_1 \\ x_2 \end{pmatrix}, \qquad \begin{matrix} \lambda < 0 \\ \omega > 0 \end{matrix}.$

d) $\begin{pmatrix} \dot{x}_1 \\ \dot{x}_2 \end{pmatrix} = \begin{pmatrix} 0 & 0 \\ 0 & \lambda \end{pmatrix} \begin{pmatrix} x_1 \\ x_2 \end{pmatrix}, \qquad \lambda < 0.$

e) $\begin{pmatrix} \dot{x}_1 \\ \dot{x}_2 \end{pmatrix} = \begin{pmatrix} 0 & \lambda \\ 0 & 0 \end{pmatrix} \begin{pmatrix} x_1 \\ x_2 \end{pmatrix}, \qquad \lambda > 0.$

f) $\begin{pmatrix} \dot{x}_1 \\ \dot{x}_2 \end{pmatrix} = \begin{pmatrix} 0 & 0 \\ 0 & 0 \end{pmatrix} \begin{pmatrix} x_1 \\ x_2 \end{pmatrix}.$

1) For each vector field compute all trajectories and illustrate them graphically on the phase plane. Describe the stable, unstable, and center manifolds of the origin.

2) For vector field a), discuss the cases $|\lambda| < \mu$, $|\lambda| = \mu$, and $|\lambda| > \mu$. What are the qualitative and quantitative differences in the dynamics for these three cases? Can the unstable manifold of the origin be considered an attracting set and/or an attractor and do the relative magnitudes of the eigenvalues affect these conclusions?

3) For vector field b), discuss the cases $\lambda < \mu$, $\lambda = \mu$, $\lambda > \mu$. What are the qualitative and quantitative differences in the dynamics for these three cases? Describe all zero- and one-dimensional invariant manifolds for this vector field. Describe the nature of the trajectories at the origin. In particular, which trajectories are tangent to either the x_1 or x_2 axis?

4) In vector field c), describe how the trajectories depend on the relative magnitudes of λ and ω. What happens when $\lambda = 0$? When $\omega = 0$?

5) Describe the effect of linear perturbations on each of the vector fields.

6) Describe the effect *near the origin* of nonlinear perturbations on each of the vector fields. Can you say anything about the effects of nonlinear perturbations on the dynamics outside of a neighborhood of the origin?

We remark that 6) is a difficult problem for the nonhyperbolic fixed points. We will study this situation in great detail in Chapter 3.

1.1.2 Consider the following linear maps on $\mathbb{R}^2$.

a) $$\begin{pmatrix} x_1 \\ x_2 \end{pmatrix} \mapsto \begin{pmatrix} \lambda & 0 \\ 0 & \mu \end{pmatrix} \begin{pmatrix} x_1 \\ x_2 \end{pmatrix}, \qquad \begin{matrix} |\lambda| < 1 \\ |\mu| > 1 \end{matrix}.$$

b) $$\begin{pmatrix} x_1 \\ x_2 \end{pmatrix} \mapsto \begin{pmatrix} \lambda & 0 \\ 0 & \mu \end{pmatrix} \begin{pmatrix} x_1 \\ x_2 \end{pmatrix}, \qquad \begin{matrix} |\lambda| < 1 \\ |\mu| < 1 \end{matrix}.$$

c) $$\begin{pmatrix} x_1 \\ x_2 \end{pmatrix} \mapsto \begin{pmatrix} \lambda & -\omega \\ \omega & \lambda \end{pmatrix} \begin{pmatrix} x_1 \\ x_2 \end{pmatrix}, \qquad \omega > 0.$$

d) $$\begin{pmatrix} x_1 \\ x_2 \end{pmatrix} \mapsto \begin{pmatrix} 1 & 0 \\ 0 & \lambda \end{pmatrix} \begin{pmatrix} x_1 \\ x_2 \end{pmatrix}, \qquad |\lambda| < 1.$$

e) $$\begin{pmatrix} x_1 \\ x_2 \end{pmatrix} \mapsto \begin{pmatrix} 1 & \lambda \\ 0 & 1 \end{pmatrix} \begin{pmatrix} x_1 \\ x_2 \end{pmatrix}, \qquad \lambda > 0.$$

f) $$\begin{pmatrix} x_1 \\ x_2 \end{pmatrix} \mapsto \begin{pmatrix} 1 & 0 \\ 0 & 1 \end{pmatrix} \begin{pmatrix} x_1 \\ x_2 \end{pmatrix}.$$

1) For each map compute all the orbits and illustrate them graphically on the phase plane. Describe the stable, unstable, and center manifolds of the origin.

2) For map a), discuss the cases $\lambda, \mu > 0$; $\lambda = 0$, $\mu > 0$; $\lambda, \mu < 0$; and $\lambda < 0$, $\mu > 0$. What are the qualitative differences in the dynamics for these four cases? Discuss how the orbits depend on the relative magnitudes of the eigenvalues. Discuss the attracting nature of the unstable manifold of the origin and its dependence on the relative magnitudes of the eigenvalues.
3) For map b), discuss the cases $\lambda, \mu > 0$; $\lambda = 0$, $\mu > 0$; $\lambda, \mu < 0$; and $\lambda < 0$, $\mu > 0$. What are the qualitative differences in the dynamics for these four cases? Describe all zero- and one-dimensional invariant manifolds for this map. Do all orbits lie on invariant manifolds?
4) For map c), consider the cases $\lambda^2+\omega^2 < 1$, $\lambda^2+\omega^2 > 1$, and $\lambda + i\omega = e^{i\alpha}$ for α rational and α irrational. Describe the qualitative differences in the dynamics for these four cases.
5) Describe the effect of linear perturbations on each of the maps.
6) Describe the effect *near the origin* of nonlinear perturbations on each of the maps. Can you say anything about the effects of nonlinear perturbations on the dynamics outside of a neighborhood of the origin?

We remark that 6) is very difficult for nonhyperbolic fixed points (more so than the analogous case for vector fields in Exercise 1.1.1) and will be treated in great detail in Chapter 3.

1.1.3 Consider the following vector fields.

a) $$\begin{aligned} \dot{x} &= y, \\ \dot{y} &= -\delta y - \mu x, \end{aligned} \qquad (x, y) \in \mathbb{R}^2.$$

b) $$\begin{aligned} \dot{x} &= y, \\ \dot{y} &= -\delta y - \mu x - x^2, \end{aligned} \qquad (x, y) \in \mathbb{R}^2.$$

c) $$\begin{aligned} \dot{x} &= y, \\ \dot{y} &= -\delta y - \mu x - x^3, \end{aligned} \qquad (x, y) \in \mathbb{R}^2.$$

d) $$\begin{aligned} \dot{x} &= -\delta x - \mu y + xy, \\ \dot{y} &= \mu x - \delta y + \tfrac{1}{2}(x^2 - y^2), \end{aligned} \qquad (x, y) \in \mathbb{R}^2.$$

e) $$\begin{aligned} \dot{x} &= -x + x^3, \\ \dot{y} &= x + y, \end{aligned} \qquad (x, y) \in \mathbb{R}^2.$$

f) $$\begin{aligned} \dot{r} &= r(1 - r^2), \\ \dot{\theta} &= \cos 4\theta, \end{aligned} \qquad (r, \theta) \in \mathbb{R}^+ \times S^1.$$

g) $$\begin{aligned} \dot{r} &= r(\delta + \mu r^2 - r^4), \\ \dot{\theta} &= 1 - r^2, \end{aligned} \qquad (r, \theta) \in \mathbb{R}^+ \times S^1.$$

h) $\begin{aligned} \dot{\theta} &= v, \\ \dot{v} &= -\sin\theta - \delta v + \mu, \end{aligned}$ $\quad (\theta, v) \in S^1 \times \mathbb{R}.$

i) $\begin{aligned} \dot{\theta}_1 &= \omega_1, \\ \dot{\theta}_2 &= \omega_2 + \theta_1^n, \ n \geq 1, \end{aligned}$ $\quad (\theta_1, \theta_2) \in S^1 \times S^1.$

j) $\begin{aligned} \dot{\theta}_1 &= \theta_2 - \sin\theta_1, \\ \dot{\theta}_2 &= -\theta_2, \end{aligned}$ $\quad (\theta_1, \theta_2) \in S^1 \times S^1.$

k) $\begin{aligned} \dot{\theta}_1 &= \theta_1^2, \\ \dot{\theta}_2 &= \omega_2, \end{aligned}$ $\quad (\theta_1, \theta_2) \in S^1 \times S^1.$

Find all fixed points and discuss their stability in the linear approximation. Describe the nature of the stable and unstable manifolds of the fixed points by drawing phase portraits. Can you determine anything about the global behavior of the manifolds? Do there exist periodic, homoclinic, or heteroclinic orbits? (*Hint:* use index theory, Bendixson's criteria, the Poincaré–Bendixson theorem, etc.; you also may want to look ahead to Exercise 1.1.16.)

Describe the attractors for each vector field and discuss what might determine their domains of attraction. In a), b), c), d), g), and h) consider the cases $\delta < 0$, $\delta = 0$, $\delta > 0$, $\mu < 0$, $\mu = 0$, and $\mu > 0$. In i) and k) consider $\omega_1 > 0$ and $\omega_2 > 0$.

1.1.4 Consider the following maps.

a) $\begin{aligned} x &\mapsto x, \\ y &\mapsto x + y, \end{aligned}$ $\quad (x, y) \in \mathbb{R}^2.$

b) $\begin{aligned} x &\mapsto x^2, \\ y &\mapsto x + y, \end{aligned}$ $\quad (x, y) \in \mathbb{R}^2.$

c) $\begin{aligned} \theta_1 &\mapsto \theta_1, \\ \theta_2 &\mapsto \theta_1 + \theta_2, \end{aligned}$ $\quad (\theta_1, \theta_2) \in S^1 \times S^1.$

d) $\begin{aligned} \theta_1 &\mapsto \sin\theta_1, \\ \theta_2 &\mapsto \theta_1, \end{aligned}$ $\quad (\theta_1, \theta_2) \in S^1 \times S^1.$

e) $\begin{aligned} x &\mapsto \tfrac{2xy}{x+y}, \\ y &\mapsto \left(\tfrac{2xy^2}{x+y}\right)^{1/2}, \end{aligned}$ $\quad (x, y) \in \mathbb{R}^2.$

f) $\begin{aligned} x &\mapsto \tfrac{x+y}{2}, \\ y &\mapsto (xy)^{1/2}, \end{aligned}$ $\quad (x, y) \in \mathbb{R}^2.$

g) $\begin{aligned} x &\mapsto \mu - \delta y - x^2, \\ y &\mapsto x, \end{aligned}$ $\quad (x, y) \in \mathbb{R}^2.$

h) $\begin{aligned} \theta &\mapsto \theta + v, \\ v &\mapsto \delta v - \mu\cos(\theta + v), \end{aligned}$ $\quad (\theta, v) \in S^1 \times \mathbb{R}^1.$

Find all the fixed points and discuss their stability in the linear approximation. Describe the nature of the stable and unstable manifolds of the fixed points by drawing phase portraits. Can you determine any higher period orbits or global behavior? Describe the nature of the attractors and what might determine their domains of attraction. In g) and h) consider the cases $\delta < 0$, $\delta = 0$, $\delta > 0$, $\mu < 0$, $\mu = 0$, and $\mu > 0$.

1.1.5 Consider a $\mathbf{C}^r$ $(r \geq 1)$ diffeomorphism

$$x \mapsto f(x), \qquad x \in \mathbb{R}^n.$$

Suppose f has a hyperbolic periodic orbit of period k. Denote the orbit by

$$O(p) = \{p, f(p), f^2(p), \cdots, f^{k-1}(p), f^k(p) = p\}.$$

Show that stability of $O(p)$ is determined by the linear map

$$y \mapsto Df^k(f^j(p))y$$

for any $j = 0, 1, \cdots, k-1$. Does the same result hold for periodic orbits of noninvertible maps?

1.1.6 Prove Liapunov's theorem for flows, i.e., let $\bar{x}$ be a fixed point of

$$\dot{x} = f(x), \qquad x \in \mathbb{R}^n,$$

and $V\colon W \to \mathbb{R}$ be a differentiable function defined on some neighborhood W of $\bar{x}$ such that

i) $V(\bar{x}) = 0$ and $V(x) > 0$ if $x \neq \bar{x}$; and
ii) $\dot{V}(x) \leq 0$ in $W - \{\bar{x}\}$.

Then $\bar{x}$ is stable. Moreover, if

iii) $\dot{V}(x) < 0$ in $W - \{\bar{x}\}$

then $\bar{x}$ is asymptotically stable. Also, show that if $V(x) > 0$, then $x = \bar{x}$ is unstable.

1.1.7 Prove Dirichlet's theorem (Siegel and Moser [1971]). Consider a $\mathbf{C}^r$ vector field $(r \geq 1)$

$$\dot{x} = f(x), \qquad x \in \mathbb{R}^n,$$

which has a fixed point at $x = \bar{x}$. Let $H(x)$ be a first integral of this vector field defined in a neighborhood of $x = \bar{x}$ such that $x = \bar{x}$ is a nondegenerate minimum of $H(x)$. Then $x = \bar{x}$ is stable.

1.1.8 Formulate the definitions of Liapunov stability and asymptotic stability for maps.

1.1.9 Prove Liapunov's theorem for maps, i.e., consider a $\mathbf{C}^r$ diffeomorphism

$$x \mapsto f(x), \qquad x \in \mathbb{R}^n,$$

and suppose that we have a scalar-valued function

$$V: U \to \mathbb{R}^1$$

defined on some open set $U \in \mathbb{R}^n$ satisfying

i) $V(x_0) = 0$;

ii) $V(x) > 0$ for $x \neq x_0$;

iii) $V \circ f(x) \leq V(x)$ with equality if and only if $x = x_0$.

Then $x = x_0$ is a stable fixed point. Moreover, if strict inequality holds in iii), then $x = x_0$ is asymptotically stable. Does the same result hold for noninvertible maps?

1.1.10 Show that hyperbolic fixed points of maps which are asymptotically stable in the linear approximation are nonlinearly asymptotically stable.

1.1.11 Give examples of fixed points of vector fields and maps that are stable in the linear approximation but are nonlinearly unstable.

1.1.12 Show that there is a solid ellipsoid E given by

$$\rho x^2 + \sigma y^2 + \sigma(z - 2\rho)^2 \leq c < \infty$$

such that all solutions of the Lorenz equations,

$$\begin{aligned} \dot{x} &= \sigma(y - x), \\ \dot{y} &= \rho x - y - xz, \qquad \sigma, \beta, \rho \geq 0, \\ \dot{z} &= -\beta z + xy, \end{aligned}$$

enter E within finite time and thereafter remain in E.

1.1.13 Let $V: \mathbb{R}^n \to \mathbb{R}$ be a $\mathbf{C}^r$ map. Then the vector field

$$\dot{x} = -\nabla V(x)$$

is called a gradient vector field. Show that the nonwandering set of a gradient vector field on $\mathbb{R}^2$ contains only fixed points and that no periodic or homoclinic orbits are possible. (*Hint:* use $V(x)$ like a Liapunov function, but see Hirsch and Smale [1974] if you need help.)

1.1.14 Find fixed points and some low period points of the map

$$x \mapsto \mu x(1-x), \qquad \mu \in [0,4].$$

Can you find *bifurcation values* of which new periodic points appear? Discuss the results graphically.

1.1.15 Consider the vector field

$$\begin{aligned} \dot{x} &= x, \\ \dot{y} &= -y, \end{aligned} \qquad (x,y) \in \mathbb{R}^2.$$

The origin is a hyperbolic fixed point with stable and unstable manifolds given by

$$W^s(0,0) = \{(x,y) \,|\, x = 0\}, \quad W^u(0,0) = \{(x,y) \,|\, y = 0\}.$$

Let

$$\begin{aligned} U_s &= \{(x,y) \,|\, |y - y_0| \le \varepsilon, 0 \le x \le \varepsilon, \text{for some } \varepsilon > 0\}, \\ U_u &= \{(x,y) \,|\, |x - x_0| \le \bar{\varepsilon}, 0 \le y \le \bar{\varepsilon}, \text{for some } \bar{\varepsilon} > 0\}; \end{aligned}$$

see Figure E1.1.1.

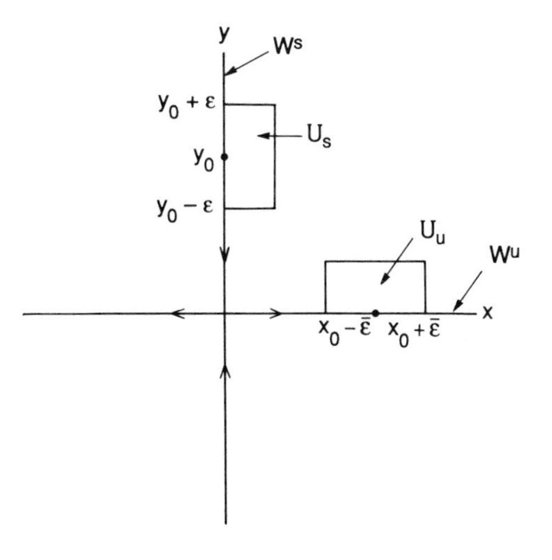

FIGURE E1.1.1

Show that you can find smaller closed sets $\tilde{U}_s \subset U_s$, $\tilde{U}_u \subset U_u$, such that $\tilde{U}_s$ maps onto $\tilde{U}_u$ under the time T flow map (T must be chosen carefully; it depends on the size of the $\tilde{U}_s$, $\tilde{U}_u$) *and* such that horizontal and vertical boundaries of $\tilde{U}_s$ correspond to horizontal and vertical boundaries of $\tilde{U}_u$. How would this problem be formulated and solved for maps?

(Note: This seemingly silly exercise is important for the understanding of chaotic invariant sets. We will use it later when we study the orbit structure near homoclinic orbits in Chapter 4.)

1.1.16 Consider the following vector fields.

a) $\ddot{x} + \mu x = 0, \qquad x \in \mathbb{R}^1.$

b) $\ddot{x} + \mu x + x^2 = 0, \qquad x \in \mathbb{R}^1.$

c) $\ddot{x} + \mu x + x^3 = 0, \qquad x \in \mathbb{R}^1.$

d) $\begin{aligned} \dot{x} &= -\mu y + xy, \\ \dot{y} &= \mu x + \tfrac{1}{2}(x^2 - y^2), \end{aligned} \qquad (x, y) \in \mathbb{R}^2.$

1) Write a), b) and c) as systems.
2) Find and determine the nature of the stability of the fixed points.
3) Find the first integrals and draw all phase curves for $\mu < 0$, $\mu = 0$, and $\mu > 0$.

1.1.17 Euler's equations of motion for a free rigid body are

$$\begin{aligned} \dot{m}_1 &= \frac{I_2 - I_3}{I_2 I_3} m_2 m_3, \\ \dot{m}_2 &= \frac{I_3 - I_1}{I_1 I_3} m_1, m_3, \qquad (m_1, m_2, m_3) \in \mathbb{R}^3, \\ \dot{m}_3 &= \frac{I_1 - I_2}{I_1 I_2} m_1 m_2, \end{aligned}$$

where $m_i = I_i \omega_i, \; i = 1, 2, 3, \; I_1 > I_2 > I_3.$

1) Find and determine the nature of the stability of the fixed points.
2) Show that the functions

$$\begin{aligned} H(m_1, m_2, m_3) &= \frac{1}{2}\left[\frac{m_1^2}{I_1} + \frac{m_2^2}{I_2} + \frac{m_3^2}{I_3}\right], \\ L(m_1, m_2, m_3) &= m_1^2 + m_2^2 + m_3^2 \end{aligned}$$

are constants on orbits.

3) For fixed L, draw all phase curves.

1.1.18 Show that the following vector field on the cylinder

$$\begin{aligned} \dot{v} &= -v, \\ \dot{\theta} &= 1, \end{aligned} \qquad (v, \theta) \in \mathbb{R}^1 \times S^1,$$

has a periodic orbit. Explain why Bendixson's criterion does not hold.

1.1.19 Use the Poincaré–Bendixson theorem to show that the vector field

$$\begin{aligned} \dot{x} &= \mu x - y - x(x^2 + y^2), \\ \dot{y} &= x + \mu y - y(x^2 + y^2), \end{aligned} \qquad (x, y) \in \mathbb{R}^2,$$

has a closed orbit for $\mu > 0$. (*Hint:* transform to polar coordinates.)

1.1.20 There are six phase portraits of vector fields on the plane shown in Figure E1.1.2. Using various phase plane techniques (e.g., index theory, the Poincaré–Bendixson theorem) determine which phase portraits are correct and which are incorrect. Modify the incorrect phase portraits to make them correct, *not by deleting any orbits shown* but by changing the stability types of existing orbits or adding new orbits.

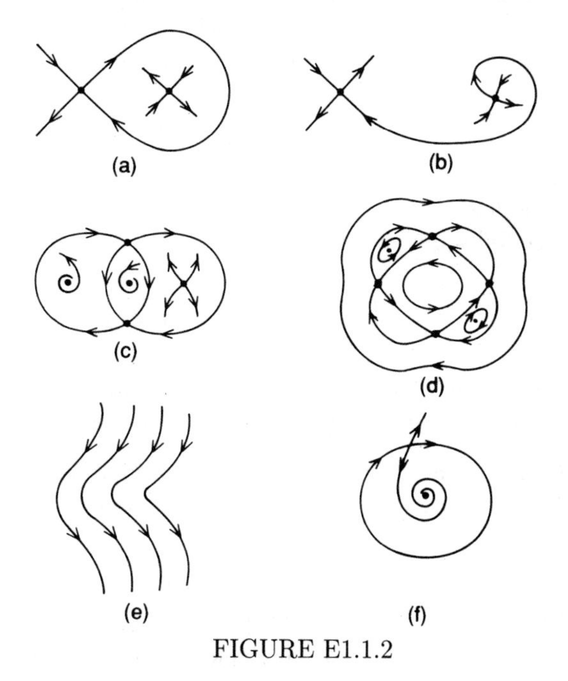

FIGURE E1.1.2

1.1.21 Construct a vector field in $\mathbb{R}^3$ with negative divergence that possesses a periodic orbit. Construct a vector field in $\mathbb{R}^3$ with negative divergence that contains a continuous family of periodic orbits.

1.1.22 Consider the linear vector field

$$\dot{x} = Ax, \qquad x \in \mathbb{R}^n,$$

where A is an $n \times n$ constant matrix. Suppose all the eigenvalues of A have negative real parts. Then prove that $x = 0$ is an asymptotically stable fixed point for this linear vector field. (*Hint:* utilize a linear transformation of the coordinates which transforms A into Jordan canonical form.)

1.1.23 Suppose that the matrix A in Exercise 1.1.22 has some eigenvalues with zero real parts (and the rest have negative real parts). Does it follow that $x = 0$ is stable? Answer this question by considering the following example.

$$\begin{pmatrix} \dot{x}_1 \\ \dot{x}_2 \end{pmatrix} = \begin{pmatrix} 0 & 1 \\ 0 & 0 \end{pmatrix} \begin{pmatrix} x_1 \\ x_2 \end{pmatrix}.$$

1.1.24 Consider the linear map

$$x \mapsto Ax, \qquad x \in \mathbb{R}^n,$$

where A is an $n \times n$ constant matrix. Suppose all of the eigenvalues of A have modulus less than one. Then prove that $x = 0$ is an asymptotically stable fixed point for this linear map (use the same hint given for Exercise 1.1.22).

1.1.25 Suppose that the matrix A in Exercise 1.1.24 has some eigenvalues having modulus one (with the rest having modulus less than one). Does it follow that $x = 0$ is stable? Answer this question by considering the following example.

$$\begin{pmatrix} x_1 \\ x_2 \end{pmatrix} \mapsto \begin{pmatrix} 1 & 1 \\ 0 & 1 \end{pmatrix} \begin{pmatrix} x_1 \\ x_2 \end{pmatrix}.$$

1.1.26 Consider the stable and unstable manifolds of a hyperbolic fixed point of saddle-type of a $\mathbf{C}^r$ $(r \geq 1)$ vector field.

1) Can the stable (resp., unstable) manifold intersect itself?
2) Can the stable (resp., unstable) manifold intersect the stable (resp., unstable) manifold of another fixed point?
3) Can the stable manifold intersect the unstable manifold? If so, can the intersection consist of a discrete set of points?
4) Can the stable (resp., unstable) manifold intersect a periodic orbit?

These questions are independent of the dimension of the vector field (as long as it is finite); however, justify each of your answers with a geometrical argument for vector fields on $\mathbb{R}^2$. (*Hint:* the key to this problem is uniqueness of solutions.)

1.1.27 Consider the stable and unstable manifolds of a hyperbolic fixed point of saddle-type of a $\mathbf{C}^r$ $(r \geq 1)$ diffeomorphism.

1) Can the stable (resp., unstable) manifold intersect itself?
2) Can the stable (resp., unstable) manifold intersect the stable (resp., unstable) manifold of another fixed point?

3) Can the stable manifold intersect the unstable manifold? If so, can the intersection consist of a discrete set of points?

These questions are independent of the dimension of the diffeomorphism (as long as it is finite); however, justify each of your answers with a geometrical argument for diffeomorphisms on $\mathbb{R}^2$. Are the arguments the same as for vector fields?

1.1.28 Consider the $\mathbf{C}^r$ $(r \geq 1)$ vector field

$$\dot{x} = f(x), \qquad x \in \mathbb{R}^n.$$

Let $\phi_t(x)$ denote the flow generated by this vector field, which we assume exists for all $t \in \mathbb{R}$, $x \in \mathbb{R}^n$. Suppose that the vector field has a hyperbolic fixed point at $x = \overline{x}$ having an s-dimensional stable manifold, $W^s(\overline{x})$, and a u-dimensional unstable manifold, $W^u(\bar{x})$ $(s + u = n)$. The typical way of proving their existence (see, e.g., Palis and deMelo [1982] or Fenichel [1971]) is to prove the existence of the local manifolds $W^s_{\text{loc}}(\bar{x})$ and $W^u_{\text{loc}}(\bar{x})$ via a contraction mapping type of argument. Then the global manifolds are defined by

$$W^s(\bar{x}) = \bigcup_{t \leq 0} \phi_t(W^s_{\text{loc}}(\bar{x})),$$

$$W^u(\bar{x}) = \bigcup_{t \geq 0} \phi_t(W^u_{\text{loc}}(\bar{x})).$$

1) Show that $W^s(\bar{x})$ and $W^u(\bar{x})$ defined in this way are invariant for *all* $t \in \mathbb{R}$.

2) If $W^s_{\text{loc}}(\bar{x})$ and $W^u_{\text{loc}}(\bar{x})$ are $\mathbf{C}^r$, does it follow by this definition that $W^s(\bar{x})$ and $W^u(\bar{x})$ are $\mathbf{C}^r$?

3) Discuss this definition of the stable and unstable manifolds in the context of how one might compute the manifolds numerically.

1.1.29 Consider the situation described in Exercise 1.1.28 in the context of $\mathbf{C}^r$ diffeomorphisms. Existence of stable and unstable manifolds of a hyperbolic fixed point is proved similarly (i.e., local manifolds are shown to exist via a contraction mapping argument), and the global manifolds are defined by

$$W^s(\bar{x}) = \bigcup_{n \leq 0} g^n(W^s_{\text{loc}}(\bar{x})),$$

$$W^u(\bar{x}) = \bigcup_{n \geq 0} g^n(W^u_{\text{loc}}(\bar{x})),$$

where g denotes the diffeomorphism and $\bar{x}$ the hyperbolic fixed point. Answer 1), 2), and 3) from Exercise 1.1.28 in the context of $\mathbf{C}^r$ diffeomorphisms.

1.1.30 Consider a hyperbolic fixed point of a $\mathbf{C}^r$ ($r \geq 1$) vector field on $\mathbb{R}^2$ whose stable and unstable manifolds intersect along a homoclinic orbit, as shown in Figure E1.1.3. Show that any point on the homoclinic orbit cannot reach the fixed point in finite time.

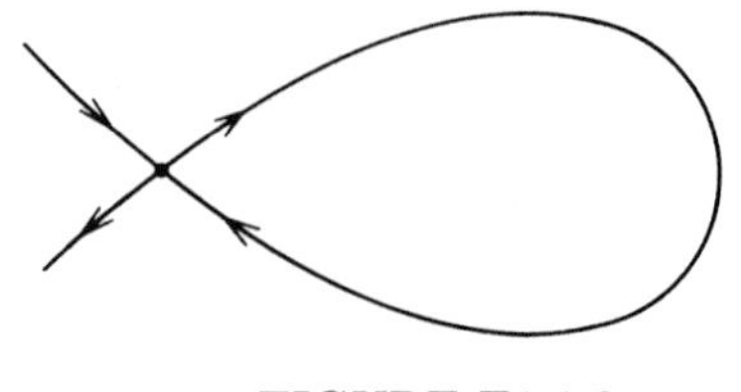

FIGURE E1.1.3

1.1.31 Consider a periodic orbit (of either a $\mathbf{C}^r$ ($r \geq 1$) vector field or map) that is contained in a compact region of phase space. Can the period of the orbit be infinite?

1.1.32 Let $\phi_t(x)$ denote a flow generated by a $\mathbf{C}^r$ ($r \geq 1$) vector field on $\mathbb{R}^n$ that exists for all $x \in \mathbb{R}^n$, $t \in \mathbb{R}$.

1) Show that the α and ω limit sets of the flow are contained in the nonwandering set of the flow.
2) Is the nonwandering set contained in the union of the α and ω limit set?

1.1.33 Let $\phi_t(x)$ denote a flow generated by a $\mathbf{C}^r$ ($r \geq 1$) vector field on $\mathbb{R}^n$ that exists for all $x \in \mathbb{R}^n$, $t \in \mathbb{R}$. Suppose A is an attracting set, and let U be a neighborhood of A that is attracted to A. Then is it true that

$$A = \bigcap_{t>0} \phi_t(U)?$$

1.1.34 Suppose A is an attracting set (of either a vector field or map), and suppose that $\bar{x} \in A$ is a hyperbolic fixed point of saddle-type. Must the following be true

1) $W^s(\bar{x}) \subset A$,
2) $W^u(\bar{x}) \subset A$?

1.1.35 Consider the union of the homoclinic orbit and the hyperbolic fixed point that it connects (shown in Figure E1.1.3). Can this set be an attracting set?

1.1.36 Prove that for $\delta > 0$ the unstable manifold of the saddle-type fixed point of the unforced Duffing oscillator falls into the sinks as shown in Figure 1.1.27.

1.1.37 Consider the $\mathbf{C}^r$ $(r \geq 1)$ vector field

$$\dot{x} = f(x), \qquad x \in \mathbb{R}^n,$$

with flow $\phi(t, x)$ defined for all $t \in \mathbb{R}$, $x \in \mathbb{R}^n$. Prove that if $\operatorname{tr} Df(x) = 0 \; \forall \, x \in \mathbb{R}^n$, then the flow $\phi(t, x)$ preserves volume.

1.1.38 Consider the $\mathbf{C}^r$ $(r \geq 1)$ diffeomorphism

$$x \mapsto g(x), \qquad x \in \mathbb{R}^n.$$

Suppose that $\det Dg(x) = 1 \; \forall \, x \in \mathbb{R}^n$. Prove that the diffeomorphism preserves volume.

1.1.39 Discuss the relationship between Exercises 1.1.37 and 1.1.38.

1.1.40 Consider the following vector field on $\mathbb{R}^2$.

$$\begin{pmatrix} \dot{x}_1 \\ \dot{x}_2 \end{pmatrix} = \begin{pmatrix} -\lambda & 0 \\ 0 & \lambda \end{pmatrix} \begin{pmatrix} x_1 \\ x_2 \end{pmatrix}, \qquad \lambda > 0.$$

The stable manifold of the origin is given by

$$W^s(0,0) = \left\{ (x_1, x_2) \in \mathbb{R}^2 \,|\, x_2 = 0 \right\}.$$

Consider a line segment contained in $W^s(0,0)$. Under the evolution of the flow generated by the vector field the length of the line segment shrinks to zero as $t \to \infty$. Does this violate the result of Exercise 1.1.37? Why or why not?

1.1.41 Consider the $\mathbf{C}^r$ $(r \geq 1)$ diffeomorphism

$$x \mapsto g(x), \qquad x \in \mathbb{R}^n.$$

Suppose $x = \bar{x}$ is a nonwandering point, i.e., for any neighborhood U of $\bar{x}$, there exists an $n \neq 0$ such that $g^n(U) \cap U \neq \emptyset$ (cf. Definition 1.1.9). Is it possible that there may exist only one such n, or, if there exists one n, must there be a countable (infinity?) of such n? Does the same result hold for flows?

1.1.42 Define what is meant by the term "perturbation" in the context of dynamical systems.

1.2 Poincaré Maps: Theory, Construction, and Examples

In the second half of Chapter 1 we will develop more quantitative, global techniques for the analysis of dynamical systems. In particular, we will emphasize the idea of a Poincaré map, especially the intuitive, geometrical, and computational aspects. We will apply our methods to a study of the periodically forced, damped Duffing oscillator given by

$$\begin{aligned} \dot{x} &= y, \\ \dot{y} &= x - x^3 - \delta y + \gamma \cos \omega t, \end{aligned} \qquad (x, y) \in \mathbb{R}^2, \tag{1.2.1}$$

where $\delta \geq 0$ and $\gamma, \omega > 0$. We will see that the dynamics of this equation are much more complicated than the unforced (nonautonomous) case. Indeed, we know that orbits of nonautonomous systems may intersect and that the Poincaré-Bendixson theorem does not hold. This leads to the possibility of much more exotic dynamics than we have seen thus far. In Chapter 4 we will see that (1.2.1) may exhibit deterministic chaos.

1.2A Poincaré Maps: Examples

The idea of reducing the study of continuous time systems (flows) to the study of an associated discrete time system (map) is due to Poincaré [1899], who first utilized it in his studies of the three body problem in celestial mechanics. Nowadays virtually any discrete time system that is associated with an ordinary differential equation is referred to as a *Poincaré map*. This technique offers several advantages in the study of ordinary differential equations, including the following:

1. *Dimensional Reduction.* Construction of the Poincaré map involves the elimination of *at least* one of the variables of the problem resulting in the study of a lower dimensional problem.

2. *Global Dynamics.* In lower dimensional problems (say, dimension ≤ 4) numerically computed Poincaré maps provide an insightful and striking display of the global dynamics of a system; see Guckenheimer and Holmes [1983] and Lichtenberg and Lieberman [1982] for examples of numerically computed Poincaré maps.

3. *Conceptual Clarity.* Many concepts that are somewhat cumbersome to state for ordinary differential equations may often be succinctly stated for the associated Poincaré map. An example would be the notion of orbital stability of a periodic orbit of an ordinary differential equation (see Hale [1980]). In terms of the Poincaré map, this problem would reduce to the problem of the stability of a fixed point of the map, which is simply characterized in terms of the eigenvalues of the

map linearized about the fixed point (see Case 1 to follow in this section).

It would be useful to give methods for constructing the Poincaré map associated with an ordinary differential equation. Unfortunately, there exist no general methods applicable to arbitrary ordinary differential equations, since construction of the Poincaré map requires some knowledge of the geometrical structure of the phase space of the ordinary differential equation. Thus, construction of a Poincaré map requires ingenuity specific to the problem at hand; however, in three cases that come up frequently, the construction of a specific type of Poincaré map can in some sense be said to be canonical. The three cases are:

1. In the study of the orbit structure near a periodic orbit of an ordinary differential equation.

2. In the case where the phase space of an ordinary differential equation is periodic, such as in periodically forced oscillators.

3. In the study of the orbit structure near a homoclinic or heteroclinic orbit.

We begin by considering Case 1.

Case 1: Poincaré Map Near a Periodic Orbit

Consider the following ordinary differential equation

$$\dot{x} = f(x), \qquad x \in \mathbb{R}^n, \tag{1.2.2}$$

where $f: U \to \mathbb{R}^n$ is $\mathbf{C}^r$ on some open set $U \subset \mathbb{R}^n$. Let $\phi(t, \cdot)$ denote the flow generated by (1.2.2). Suppose that (1.2.2) has a periodic solution of period T which we denote by $\phi(t, x_0)$, where $x_0 \in \mathbb{R}^n$ is any point through which this periodic solution passes (i.e., $\phi(t+T, x_0) = \phi(t, x_0)$). Let Σ be an $n-1$ dimensional surface transverse to the vector field at x_0 (note: "transverse" means that $f(x) \cdot n(x) \neq 0$ where "$\cdot$" denotes the vector dot product and $n(x)$ is the normal to Σ at x); we refer to Σ as a cross-section to the vector field (1.2.2). Now in Theorem 1.1.8 we proved that $\phi(t, x)$ is $\mathbf{C}^r$ if $f(x)$ is $\mathbf{C}^r$; thus, we can find an open set $V \subset \Sigma$ such that the trajectories starting in V return to Σ in a time close to T. The map that associates points in V with their points of first return to Σ is called the *Poincaré map,* which we denote by P. To be more precise,

$$\begin{aligned} P: V &\to \Sigma, \\ x &\mapsto \phi\big(\tau(x), x\big), \end{aligned} \tag{1.2.3}$$

where $\tau(x)$ is the time of first return of the point x to Σ. Note that, by construction, we have $\tau(x_0) = T$ and $P(x_0) = x_0$.

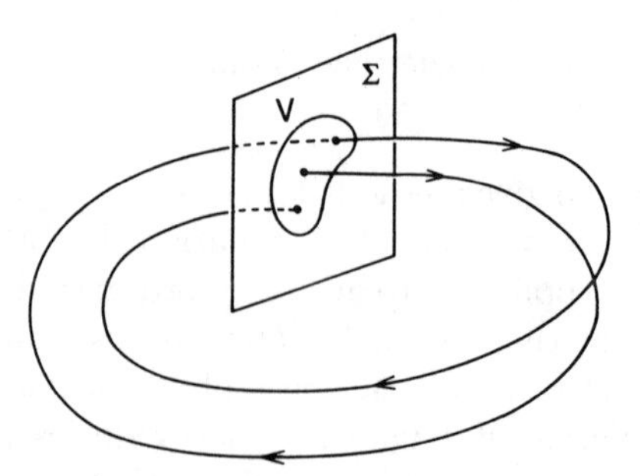

FIGURE 1.2.1. The geometry of the Poincaré map for a periodic orbit.

Therefore, a fixed point of P corresponds to a periodic orbit of (1.2.2), and a period k point of P (i.e., a point $x \in V$ such that $P^k(x) = x$ provided $P^i(x) \in V$, $i = 1, \cdots, k$) corresponds to a periodic orbit of (1.2.2) that pierces Σ k times before closing; see Figure 1.2.1. In applying this technique to specific examples, the following questions immediately arise.

1. How is Σ chosen?

2. How does P change as Σ is changed?

Question 1 cannot be answered in a general way, since in any given problem there will be many possible choices of Σ. This fact makes the answer to Question 2 even more important. However, for now we will postpone answering this question in order to consider a specific example.

EXAMPLE 1.2.1 Consider the following vector field on $\mathbb{R}^2$

$$\begin{aligned} \dot{x} &= \mu x - y - x(x^2 + y^2), \\ \dot{y} &= x + \mu y - y(x^2 + y^2), \end{aligned} \qquad (x, y) \in \mathbb{R}^2, \tag{1.2.4}$$

where $\mu \in \mathbb{R}^1$ is a parameter (note: we will meet (1.2.4) later when we study the Poincaré–Andronov–Hopf bifurcation). Our goal is to study (1.2.4) by constructing an associated one-dimensional Poincaré map and studying the dynamics of the map. According to our previous discussion, we need to find a periodic orbit of (1.2.4), construct a cross-section to the orbit, and then study how points on the cross-section return to the cross-section under the flow generated by (1.2.4). Considering (1.2.4) and thinking about how to carry out these steps should bring home the point stated at the beginning of this section — constructing a Poincaré map requires some knowledge of the geometry of the flow generated by (1.2.4). In this example the procedure is greatly facilitated by considering the vector field in a "more appropriate" coordinate system; in this case, polar coordinates.

Let

$$\begin{aligned} x &= r \cos\theta, \\ y &= r \sin\theta; \end{aligned} \tag{1.2.5}$$

then (1.2.4) becomes

$$\begin{aligned} \dot{r} &= \mu r - r^3, \\ \dot{\theta} &= 1. \end{aligned} \tag{1.2.6}$$

We will require $\mu > 0$, in which case the flow generated by (1.2.6) is given by

$$\phi_t(r_0, \theta_0) = \left(\left(\frac{1}{\mu} + \left(\frac{1}{r_0^2} - \frac{1}{\mu} \right) e^{-2\mu t} \right)^{-1/2}, t + \theta_0 \right). \tag{1.2.7}$$

It should be clear that (1.2.6) has a periodic orbit given by $\phi_t(\sqrt{\mu}, \theta_0)$. We now construct a Poincaré map near this periodic orbit.

We define a cross-section Σ to the vector field (1.2.6) by

$$\Sigma = \{ (r, \theta) \in \mathbb{R} \times S^1 \mid r > 0, \ \theta = \theta_0 \}. \tag{1.2.8}$$

The reader should verify that Σ is indeed a cross-section. From (1.2.6) we see that the "time of flight" for orbits starting on Σ to return to Σ is given by $t = 2\pi$. Using this information, the Poincaré map is given by

$$\begin{aligned} &P\colon \Sigma \to \Sigma \\ &(r_0, \theta_0) \mapsto \phi_{2\pi}(r_0, \theta_0) = \left(\left(\tfrac{1}{\mu} + \left(\tfrac{1}{r_0^2} - \tfrac{1}{\mu} \right) e^{-4\pi\mu} \right)^{-1/2}, \theta_0 + 2\pi \right), \end{aligned} \tag{1.2.9}$$

or simply

$$r \mapsto \left(\frac{1}{\mu} + \left(\frac{1}{r^2} - \frac{1}{\mu} \right) e^{-4\pi\mu} \right)^{-1/2}, \tag{1.2.10}$$

where we have dropped the subscript '0' on r for notational convenience. The Poincaré map has a fixed point at $r = \sqrt{\mu}$. We can compute the stability of the fixed point by computing the eigenvalue (which is just the derivative for a one-dimensional map) of $DP(\sqrt{\mu})$. A simple calculation gives

$$DP(\sqrt{\mu}) = e^{-4\pi\mu}. \tag{1.2.11}$$

Therefore, the fixed point $r = \sqrt{\mu}$ is asymptotically stable.

Before leaving this example there are several points to make.

1. Viewing (1.2.4) in the correct coordinate system was the key to this problem. This made the choice of a cross-section virtually obvious and provided "nice" coordinates on the cross-section (i.e., r and θ "decoupled" as well). Later we will learn a general technique called *normal form theory* which can be used to transform vector fields into the "nicest possible" coordinate systems.

2. We know that the fixed point of P corresponds to a periodic orbit of (1.2.6) and that the fixed point of P is asymptotically stable. Does this imply that the corresponding periodic orbit of (1.2.6) is also asymptotically stable? It does, but we have not proved it yet (note:

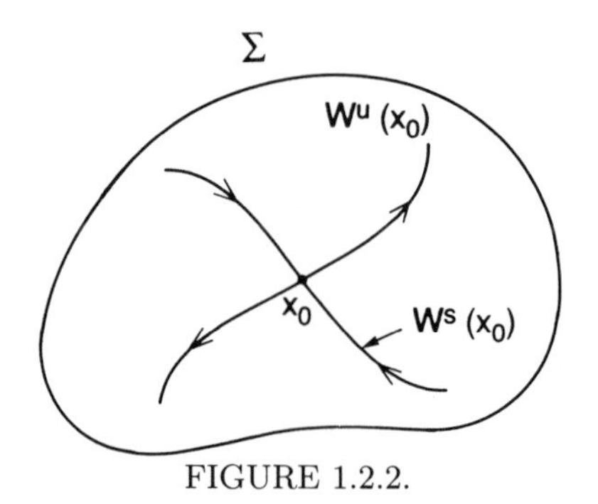

FIGURE 1.2.2.

the reader should think about this in the context of this example until it feels "obvious"). We will consider this point when we consider how the Poincaré map changes when the cross-section is varied.

Before leaving Case 1, let us illustrate how the study of Poincaré maps near periodic orbits may simplify the geometry.

Consider a vector field in $\mathbb{R}^3$ generating a flow given by $\phi_t(x)$, $x \in \mathbb{R}^3$. Suppose also that it has a periodic orbit, γ, of period $T > 0$ passing through the point $x_0 \in \mathbb{R}^3$, i.e.,

$$\phi_t(x_0) = \phi_{t+T}(x_0).$$

We construct in the usual way a Poincaré map, P, near this periodic orbit by constructing a cross-section, Σ, to the vector field through x_0 and considering the return of points to Σ under the flow generated by the vector field; see Figure 1.2.1.

Now consider the Poincaré map P. The map has a fixed point at x_0. Suppose that the fixed point is of saddle type having a one-dimensional stable manifold, $W^s(x_0)$, and a one-dimensional unstable manifold $W^u(x_0)$; see Figure 1.2.2. We now want to show how these manifolds are manifested in the flow and how they are related to γ. Very simply, using them as initial conditions, they generate the two-dimensional stable and unstable manifolds of γ. Mathematically, this is represented as follows

$$W^s(\gamma) = \bigcup_{t \leq 0} \phi_t(W^s_{\text{loc}}(x_0)),$$

$$W^u(\gamma) = \bigcup_{t \geq 0} \phi_t(W^u_{\text{loc}}(x_0)).$$

It should be clear that $W^s(\gamma)$ (resp. $W^u(\gamma)$) is just as differentiable as $W^s_{\text{loc}}(x_0)$ (resp. $W^u_{\text{loc}}(x_0)$), since $\phi_t(x)$ is differentiable with respect to x; see Figure 1.2.3 for an illustration of the geometry. Hence, in $\mathbb{R}^3$, $W^s(\gamma)$ and $W^u(\gamma)$ are two two-dimensional surfaces which intersect in the closed

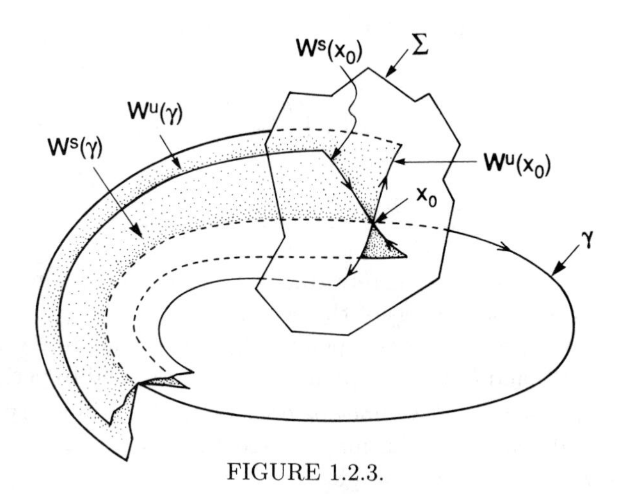

FIGURE 1.2.3.

curve. This should serve to show that it is somewhat simpler geometrically to study periodic orbits and their associated stable and unstable manifolds by studying the associated Poincaré map.

We now turn to Case 2.

Case 2: Poincaré Map of a Time-Periodic Ordinary Differential Equation

Consider the following ordinary differential equation

$$\dot{x} = f(x,t), \qquad x \in \mathbb{R}^n, \tag{1.2.12}$$

where $f: U \to \mathbb{R}^n$ is $\mathbf{C}^r$ on some open set $U \subset \mathbb{R}^n \times \mathbb{R}^1$. Suppose the time dependence of (1.2.12) is periodic with fixed period $T = 2\pi/\omega > 0$, i.e., $f(x,t) = f(x, t+T)$. We rewrite (1.2.12) in the form of an autonomous equation in $n+1$ dimensions (see Section 1.1G) by defining the function

$$\begin{aligned} \theta: \mathbb{R}^1 &\to S^1, \\ t &\mapsto \theta(t) = \omega t, \quad \mod 2\pi. \end{aligned} \tag{1.2.13}$$

Using (1.2.13) equation (1.2.12) becomes

$$\begin{aligned} \dot{x} &= f(x,\theta), \\ \dot{\theta} &= \omega, \end{aligned} \qquad (x,\theta) \in \mathbb{R}^n \times S^1. \tag{1.2.14}$$

We denote the flow generated by (1.2.14) by $\phi(t) = \big(x(t), \theta(t) = \omega t + \theta_0 \pmod{2\pi}\big)$. We define a cross-section $\Sigma^{\bar{\theta}_0}$ to the vector field (1.2.14) by

$$\Sigma^{\bar{\theta}_0} = \{\, (x,\theta) \in \mathbb{R}^n \times S^1 \mid \theta = \bar{\theta}_0 \in (0, 2\pi] \,\}. \tag{1.2.15}$$

The unit normal to $\Sigma^{\bar{\theta}_0}$ in $\mathbb{R}^n \times S^1$ is given by the vector $(0,1)$, and it is clear that $\Sigma^{\bar{\theta}_0}$ is transverse to the vector field (1.2.14) for all $x \in \mathbb{R}^n$, since $\big(f(x,\theta),\omega\big) \cdot (0,1) = \omega \neq 0$. In this case $\Sigma^{\bar{\theta}_0}$ is called a *global cross-section*.

We define the Poincaré map of $\Sigma^{\bar{\theta}_0}$ as follows:

$$P_{\bar{\theta}_0}: \Sigma^{\bar{\theta}_0} \to \Sigma^{\bar{\theta}_0},$$
$$\left(x\left(\frac{\bar{\theta}_0 - \theta_0}{\omega}\right), \bar{\theta}_0\right) \mapsto \left(x\left(\frac{\bar{\theta}_0 - \theta_0 + 2\pi}{\omega}\right), \bar{\theta}_0 + 2\pi \equiv \bar{\theta}_0\right),$$

or

$$x\left(\frac{\bar{\theta}_0 - \theta_0}{\omega}\right) \mapsto x\left(\frac{\bar{\theta}_0 - \theta_0 + 2\pi}{\omega}\right). \tag{1.2.16}$$

Thus, the Poincaré map merely tracks initial conditions in x at a fixed phase after successive periods of the vector field.

It should be clear that fixed points $P_{\bar{\theta}_0}$ correspond to $2\pi/\omega$-periodic orbits of (1.2.12) and k-periodic points of $P_{\bar{\theta}_0}$ correspond to periodic orbits of (1.2.12) that pierce $\Sigma^{\bar{\theta}_0}$ k times before closing. We will worry about the effect on the dynamics of the map caused by changing the cross-section later. Now we consider an example.

EXAMPLE 1.2.2: PERIODICALLY FORCED LINEAR OSCILLATORS Consider the following ordinary differential equation

$$\ddot{x} + \delta\dot{x} + \omega_0^2 x = \gamma \cos \omega t. \tag{1.2.17}$$

This is an equation which most students learn to solve in elementary calculus courses. Our goal here is to study the nature of solutions of (1.2.17) from our more geometrical setting in the context of Poincaré maps. This will enable the reader to obtain a new point of view on something relatively familiar and, we hope, to see the value of this new point of view.

We begin by first obtaining the solution of (1.2.17). Recall (see, e.g., Arnold [1973] or Hirsch and Smale [1974]) that the general solution of (1.2.17) is the sum of the solution of the homogeneous equation (i.e., the solution for $\gamma = 0$), sometimes called the free oscillation, and a particular solution, sometimes called the forced oscillation. For $\delta > 0$ there are several possibilities for the homogeneous solution, which we state below.

$\delta > 0$: *The Homogeneous Solution, $x_h(t)$.* There are three cases depending on the sign of the quantity $\delta^2 - 4\omega_0^2$.

$$\delta^2 - 4\omega_0^2 > 0 \Rightarrow x_h(t) = C_1 e^{r_1 t} + C_2 e^{r_2 t}, \tag{1.2.18}$$

where

$$r_{1,2} = -\delta/2 \pm (1/2)\sqrt{\delta^2 - 4\omega_0^2},$$
$$\delta^2 - 4\omega_0^2 = 0 \Rightarrow x_h(t) = (C_1 + C_2 t)e^{-(\delta/2)t},$$
$$\delta^2 - 4\omega_0^2 < 0 \Rightarrow x_h(t) = e^{-(\delta/2)t}(C_1 \cos \bar{\omega} t + C_2 \sin \bar{\omega} t),$$

and where $\bar{\omega} = (1/2)\sqrt{4\omega_0^2 - \delta^2}$. In all three cases C_1 and C_2 are unknown constants which are fixed when initial conditions are specified. Also, notice that in all three cases $\lim_{t\to\infty} x_h(t) = 0$. We now turn to the particular solution.

The Particular Solution, $x_p(t)$.

$$x_p(t) = A\cos\omega t + B\sin\omega t, \tag{1.2.19}$$

where

$$A \equiv \frac{(\omega_0^2 - \omega^2)\gamma}{(\omega_0^2 - \omega^2)^2 + (\delta\omega)^2}, \qquad B \equiv \frac{\delta\gamma\omega}{(\omega_0^2 - \omega^2)^2 + (\delta\omega)^2}.$$

Next we turn to the construction of the Poincaré map. For this we will consider only the case $\delta^2 - 4\omega_0^2 < 0$. The other two cases are similar, and we leave them as exercises for the reader.

The Poincaré Map: $\delta^2 - 4\omega_0^2 < 0$. Rewriting (1.2.17) as a system, we obtain

$$\begin{aligned} \dot{x} &= y, \\ \dot{y} &= -\omega_0^2 x - \delta y + \gamma\cos\omega t. \end{aligned} \tag{1.2.20}$$

By rewriting (1.2.20) as an autonomous system, as was described at the beginning of our discussion of Case 2, we obtain

$$\begin{aligned} \dot{x} &= y, \\ \dot{y} &= -\omega_0^2 x - \delta y + \gamma\cos\theta, \qquad (x, y, \theta) \in \mathbb{R}^1 \times \mathbb{R}^1 \times S^1. \\ \dot{\theta} &= \omega, \end{aligned} \tag{1.2.21}$$

The flow generated by (1.2.21) is given by

$$\phi_t(x_0, y_0, \theta_0) = \big(x(t), y(t), \omega t + \theta_0\big), \tag{1.2.22}$$

where, using (1.2.18c) and (1.2.19), $x(t)$ is given by

$$x(t) = e^{-(\delta/2)t}(C_1\cos\bar{\omega}t + C_2\sin\bar{\omega}t) + A\cos\omega t + B\sin\omega t$$

with

$$y(t) = \dot{x}(t). \tag{1.2.23}$$

The constants C_1 and C_2 are obtained by requiring

$$\begin{aligned} x(0) &= x_0, \\ y(0) &= y_0, \end{aligned}$$

which yield

$$\begin{aligned} C_1 &= x_0 - A, \\ C_2 &= \frac{1}{\bar{\omega}}\left(\frac{\delta}{2}x_0 + y_0 - \frac{\delta}{2}A - \omega B\right). \end{aligned} \tag{1.2.24}$$

Notice from (1.2.20) that we can set $\theta_0 = 0$ in (1.2.22) (cf. (1.2.16)).

We construct a cross-section at $\bar{\theta}_0 = 0$ (note: this is why we specified the initial conditions at $t = 0$) as follows

$$\Sigma^0 \equiv \Sigma = \{\, (x, y, \theta) \in \mathbb{R}^1 \times \mathbb{R}^1 \times S^1 \mid \theta = 0 \in [0, 2\pi)\,\}, \tag{1.2.25}$$

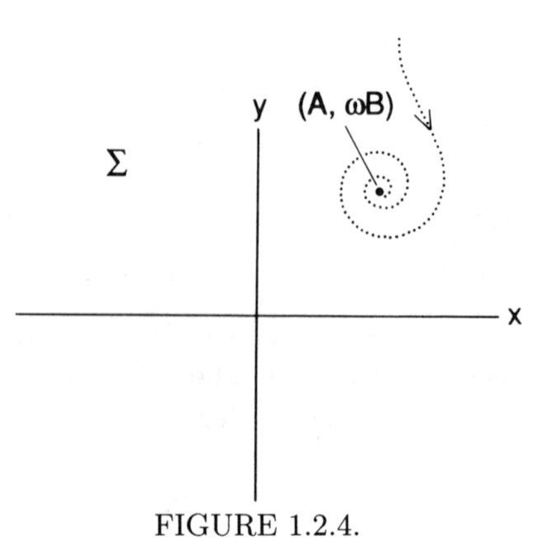

FIGURE 1.2.4.

where we have dropped the subscript "0" on x, y, and θ for notational convenience. Using (1.2.23), the Poincaré map is given by

$$
\begin{aligned}
&P\colon \Sigma \to \Sigma,\\
&\begin{pmatrix} x \\ y \end{pmatrix} \mapsto e^{-\delta\pi/\omega} \begin{pmatrix} \mathcal{C} + \frac{\delta}{2\bar{\omega}}\mathcal{S} & \frac{1}{\bar{\omega}}\mathcal{S} \\ -\frac{\omega_0^2}{\bar{\omega}}\mathcal{S} & \mathcal{C} - \frac{\delta}{2\bar{\omega}}\mathcal{S} \end{pmatrix} \begin{pmatrix} x \\ y \end{pmatrix} \\
&+ \begin{pmatrix} e^{-\delta\pi/\omega}\left[-A\mathcal{C} + \left(-\frac{\delta}{2\bar{\omega}}A - \frac{\omega}{\bar{\omega}}B\right)\mathcal{S}\right] + A \\ e^{-\delta\pi/\omega}\left[-\omega B\mathcal{C} + \left(\frac{\omega_0^2}{\bar{\omega}}A + \frac{\delta\omega}{2\bar{\omega}}B\right)\mathcal{S}\right] + \omega B \end{pmatrix},
\end{aligned}
\tag{1.2.26}
$$

where

$$\mathcal{C} \equiv \cos 2\pi \frac{\bar{\omega}}{\omega},$$

$$\mathcal{S} \equiv \sin 2\pi \frac{\bar{\omega}}{\omega}.$$

Equation (1.2.26) is an example of an *affine map*, i.e., it is a linear map plus a translation.

The Poincaré map has a single fixed point given by

$$(x, y) = (A, \omega B) \tag{1.2.27}$$

(note: this should not be surprising). The next question is whether or not the fixed point is stable. A simple calculation shows that the eigenvalues of $DP(A, \omega B)$ are given by

$$\lambda_{1,2} = e^{-\delta\pi/\omega \pm i2\pi\bar{\omega}/\omega}. \tag{1.2.28}$$

Thus the fixed point is asymptotically stable with nearby orbits appearing as in Figure 1.2.4. (Note: the "spiraling" of orbits near the fixed point is

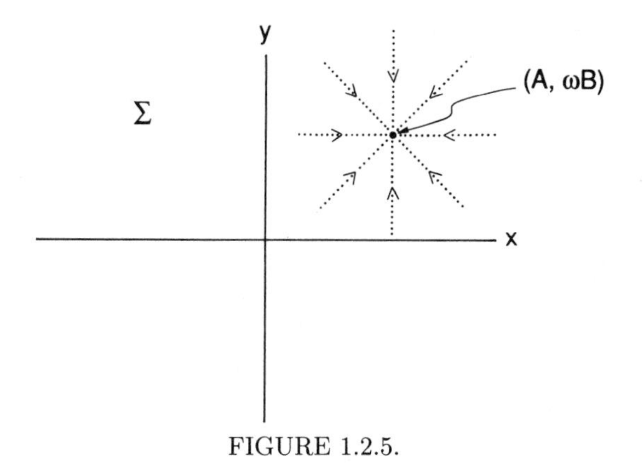

FIGURE 1.2.5.

due to the imaginary part of the eigenvalues.) Figure 1.2.4 is drawn for $A > 0$; see (1.2.19).

The Case of Resonance: $\bar{\omega} = \omega$. We now consider the situation where the driving frequency is equal to the frequency of the free oscillation. In this case the the Poincaré map becomes

$$P: \Sigma \to \Sigma,$$
$$\begin{pmatrix} x \\ y \end{pmatrix} \mapsto e^{-\delta\pi/\omega} \begin{pmatrix} 1 & 0 \\ 0 & 1 \end{pmatrix} \begin{pmatrix} x \\ y \end{pmatrix} + \begin{pmatrix} A(1 - e^{-\delta\pi/\omega}) \\ \omega B(1 - e^{-\delta\pi/\omega}) \end{pmatrix}. \tag{1.2.29}$$

This map has a unique fixed point at

$$(x, y) = (A, \omega B). \tag{1.2.30}$$

The eigenvalues of $DP(A, \omega B)$ are identical and are equal to

$$\lambda = e^{-\delta\pi/\omega}. \tag{1.2.31}$$

Thus, the fixed point is asymptotically stable with nearby orbits appearing as in Figure 1.2.5. (Note: in this case orbits do not spiral near the fixed point since the eigenvalues are purely real.)

For $\delta > 0$, in all cases the free oscillation dies out and we are left with the forced oscillation of frequency ω which is represented as an attracting fixed point of the Poincaré map. We will now examine what happens for $\delta = 0$.

$\delta = 0$: *Subharmonics, Ultraharmonics, and Ultrasubharmonics.* In this case the equation becomes

$$\begin{aligned} \dot{x} &= y, \\ \dot{y} &= -\omega_0^2 x + \gamma \cos\theta, \\ \dot{\theta} &= \omega. \end{aligned} \tag{1.2.32}$$

Using (1.2.18c) and (1.2.19), we see that the general solution of (1.2.32) is given by

$$\begin{aligned} x(t) &= C_1 \cos \omega_0 t + C_2 \sin \omega_0 t + \bar{A} \cos \omega t, \\ y(t) &= \dot{x}(t), \end{aligned} \tag{1.2.33}$$

where

$$\bar{A} \equiv \frac{\gamma}{\omega_0^2 - \omega^2}, \tag{1.2.34}$$

and C_1 and C_2 are found by solving

$$\begin{aligned} x(0) &\equiv x_0 = C_1 + \bar{A}, \\ y(0) &\equiv y_0 = C_2 \omega_0. \end{aligned} \tag{1.2.35}$$

It should be evident that, for now, we must require $\omega \neq \omega_0$ in order for (1.2.33) to be valid.

Before writing down the Poincaré map, there is an important distinction to draw between the cases $\delta > 0$ and $\delta = 0$. As mentioned above, for $\delta > 0$, the free oscillation eventually dies out leaving only the forced oscillation of frequency ω. This corresponds to the associated Poincaré map having a single asymptotically stable fixed point. In the case $\delta = 0$, by examining (1.2.33), we see that this does not happen. In general, for $\delta = 0$, it should be clear that the solution is a superposition of solutions of frequencies ω and ω_0. The situation breaks down into several cases depending on the relationship of ω to ω_0. We will first write down the Poincaré map and then consider each case individually.

The Poincaré map is given by

$$\begin{aligned} P\colon \Sigma &\to \Sigma, \\ \begin{pmatrix} x \\ y \end{pmatrix} &\mapsto \begin{pmatrix} \cos 2\pi \frac{\omega_0}{\omega} & \frac{1}{\omega_0} \sin 2\pi \frac{\omega_0}{\omega} \\ -\omega_0 \sin 2\pi \frac{\omega_0}{\omega} & \cos 2\pi \frac{\omega_0}{\omega} \end{pmatrix} \begin{pmatrix} x \\ y \end{pmatrix} \\ &\quad + \begin{pmatrix} \bar{A}\left(1 - \cos 2\pi \frac{\omega_0}{\omega}\right) \\ \omega_0 \bar{A} \sin 2\pi \frac{\omega_0}{\omega} \end{pmatrix}. \end{aligned} \tag{1.2.36}$$

Our goal is to study the orbits of P. As mentioned above, this will depend on the relationship of ω and ω_0. We begin with the simplest case.

1) Harmonic Response. Consider the point

$$(x, y) = (\bar{A}, 0). \tag{1.2.37}$$

It is easy to verify that this is a fixed point of P corresponding to a solution of (1.2.32) having frequency ω.

We now want to describe a somewhat more geometrical way of viewing this solution which will be useful later on. Using (1.2.37), (1.2.35) and (1.2.33), the fixed point (1.2.37) corresponds to the solution

$$\begin{aligned} x(t) &= \bar{A} \cos \omega t, \\ y(t) &= -\bar{A}\omega \sin \omega t. \end{aligned} \tag{1.2.38}$$

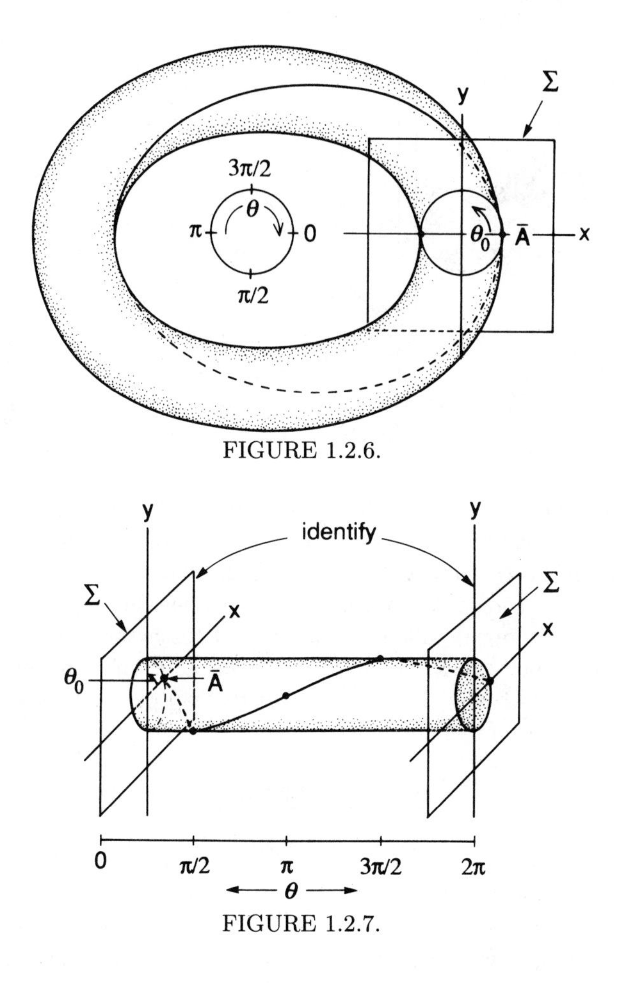

FIGURE 1.2.6.

FIGURE 1.2.7.

If we view this solution in the x–y plane, it traces out a circle which closes after time $2\pi/\omega$. If we view this solution in the x-y-θ phase space, it traces out a spiral which can be viewed as lying on the surface of a cylinder. The cylinder can be thought of as an extension of the circle traced out by (1.2.38) in the x-y plane into the θ-direction. Since θ is periodic, the ends of the cylinder are joined to become a torus, and the trajectory traces out a curve on the surface of the torus which makes *one* complete revolution on the torus before closing. The torus can be parameterized by two angles; the angle θ is the longitudinal angle. We will call the latitudinal angle θ_0, which is the angle through which the circular trajectory turns in the x-y plane. This situation is depicted geometrically in Figure 1.2.6. Trajectories which wind many times around the torus may be somewhat difficult to draw, as in Figure 1.2.6; we now want to show an easier way to represent the same information. First, we cut open the torus and identify the two ends as shown in Figure 1.2.7. Then we cut along the longitudinal angle θ and

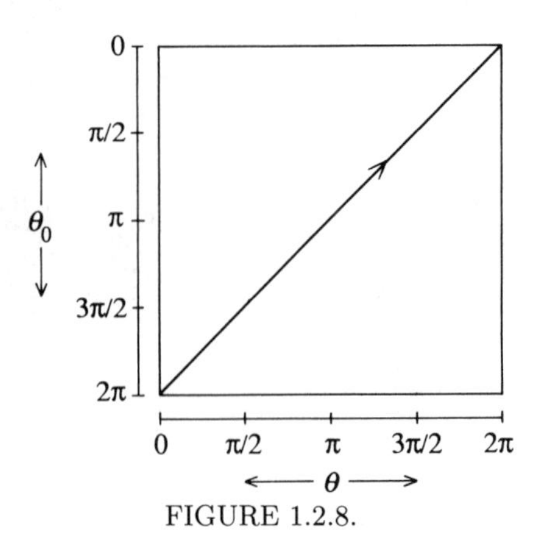

FIGURE 1.2.8.

flatten it out into a square as shown in Figure 1.2.8. This square is really a torus if we identify the two vertical sides and the two horizontal sides. This means that a trajectory that runs off the top of the square reappears at the bottom of the square at the same θ value where it intersected the top edge. For a more detailed description of trajectories on a torus, see Abraham and Shaw [1984]. We stress that this construction works because *all* trajectories of (1.2.32) lie on circles in the x-y plane. Motion on tori is a characteristic of multifrequency systems.

2) Subharmonic Response of Order m. Suppose we have

$$\omega = m\omega_0, \qquad m > 1, \tag{1.2.39}$$

where m is an integer. Consider all points on Σ *except* $(x, y) = (\bar{A}, 0)$ (we already know about this point). Using (1.2.33) and the expression for the Poincaré map given in (1.2.36), it is easy to see that all points *except* $(x, y) = (\bar{A}, 0)$ are *period* m *points*, i.e., they are fixed points of the mth iterate of the Poincaré map (note: this statement assumes that by the phrase "period of a point" we mean the smallest possible period). Let us now see what they correspond to in terms of motion on the torus.

Using (1.2.39) and (1.2.33) it should be clear that $x(t)$ and $y(t)$ have frequency ω/m. Thus, after a time $t = 2\pi/\omega$, the solution has turned through an angle $2\pi/m$ in the x-y plane, i.e., θ_0 has changed by $2\pi/m$. Therefore, the solution makes m longitudinal circuits and one latitudinal circuit around the torus before closing up. The m distinct points of intersection that the trajectory makes with $\theta = 0$ are all period m points of P, or equivalently, fixed points of the mth iterate of P. Such solutions are called *subharmonics of order* m. In Figure 1.2.9 we show examples for $m = 2$ and $m = 3$.

3) Ultraharmonic Response of Order n. Suppose we have

$$n\omega = \omega_0, \qquad n > 1, \tag{1.2.40}$$

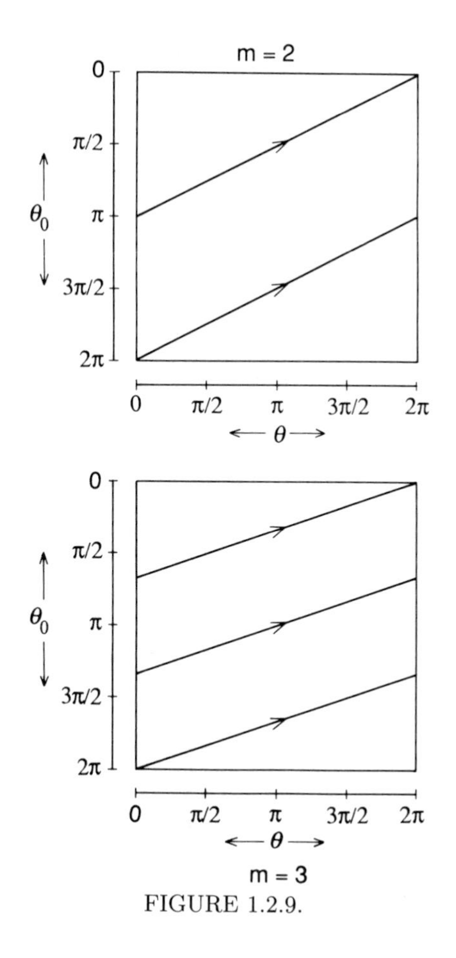

FIGURE 1.2.9.

where n is an integer. Consider all points on Σ *except* $(x, y) = (\bar{A}, 0)$. Using (1.2.40) and (1.2.33) it is easy to see that every point is a fixed point of the Poincaré map. Let us see what this corresponds to in terms of motion on the torus.

Using (1.2.33) and (1.2.40), we see that $x(t)$ and $y(t)$ have frequency $n\omega$. This means that after a time $t = 2\pi/\omega$, the solution has turned through an angle $2\pi n$ in the x-y plane before closing up. Since $2\pi n = 2\pi \pmod{2\pi}$, this explains the nature of these fixed points of P: they correspond to solutions which make n latitudinal and one longitudinal circuits around the torus before closing up. We illustrate the situation geometrically for $n = 2$ and $n = 3$ in Figure 1.2.10. Such solutions are referred to as *ultraharmonics of order n*.

4) Ultrasubharmonic Response of Order m, n. Suppose we have

$$m\omega = m\omega_0, \qquad m,\ n > 1, \tag{1.2.41}$$

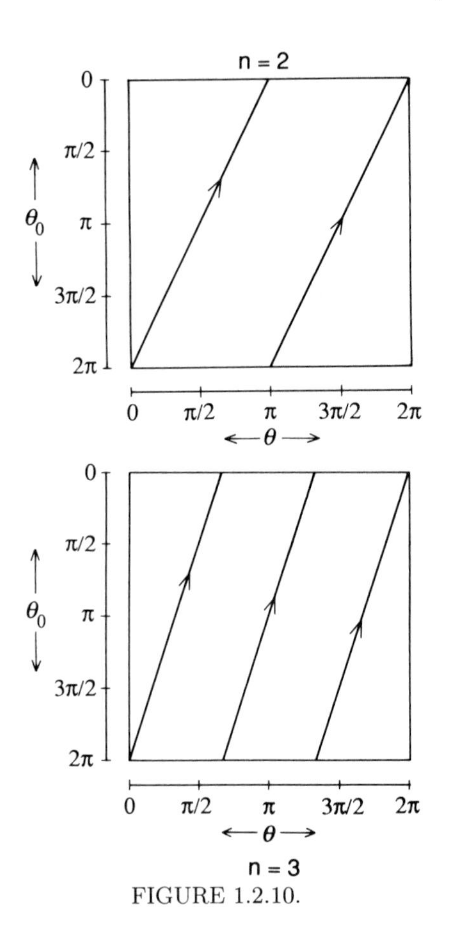

FIGURE 1.2.10.

where m and n are *relatively prime integers*, which means that all common factors of n/m have been divided out. Using exactly the same arguments as those given above, it is easy to show that all points in Σ *except* $(x,y) = (\bar{A}, 0)$ are period m points which correspond to trajectories making m longitudinal and n latitudinal circuits around the torus before closing up. These solutions are referred to as *ultrasubharmonics of order* m, n. We illustrate the situation for $(n,m) = (2,3)$ and $(n,m) = (3,2)$ in Figure 1.2.11.

5) Quasiperiodic Response. For the final case, suppose we have

$$\frac{\omega}{\omega_0} = \text{irrational number}. \tag{1.2.42}$$

Then for all points in Σ *except* $(x,y) = (\bar{A}, 0)$, the orbit of the point densely fills out a circle on Σ which corresponds to an invariant two-torus in x-y-θ space. We will prove this rigorously in Example 1.2.3.

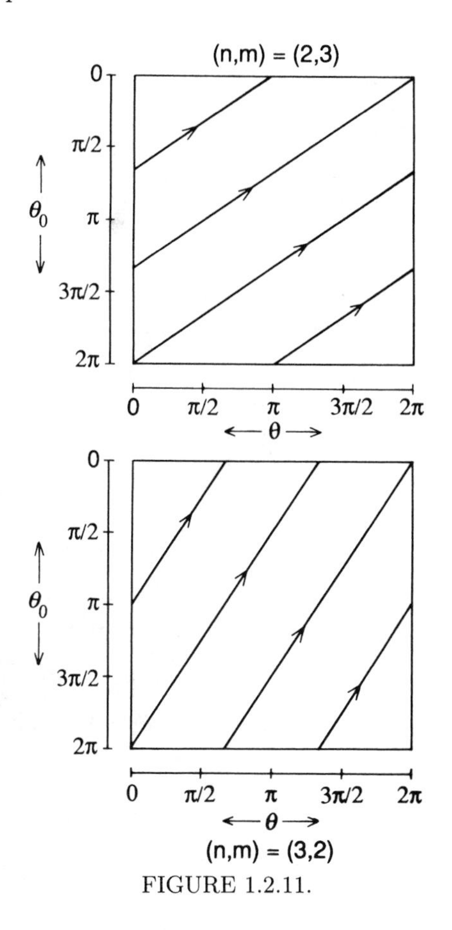

FIGURE 1.2.11.

EXAMPLE 1.2.3: THE STUDY OF COUPLED OSCILLATORS VIA CIRCLE MAPS In Example 1.1.9 we saw that (for I_1, $I_2 \neq 0$) the study of two linearly coupled, linear undamped oscillators in a four-dimensional phase space could be reduced to the study of the following two-dimensional vector field

$$\begin{aligned} \dot{\theta}_1 &= \omega_1, \\ \dot{\theta}_2 &= \omega_2, \end{aligned} \qquad (\theta_1, \theta_2) \in S^1 \times S^2. \tag{1.2.43}$$

The flow generated by (1.2.43) is defined on the two-torus, $S^1 \times S^1 \equiv T^2$, and θ_1 and θ_2 are called the longitude and latitude; see Figure 1.2.12. As in Example 1.2.2, it is often easier to visualize flows on tori by cutting open the torus, flattening it out, and identifying horizontal and vertical sides of the resulting square as shown in Figure 1.2.13. The flow generated by (1.2.43) is simple to compute and is given by

$$\begin{aligned} \theta_1(t) &= \omega_1 t + \theta_{10}, \\ \theta_2(t) &= \omega_2 t + \theta_{20}, \end{aligned} \qquad (\text{mod}\, 2\pi). \tag{1.2.44}$$

However, orbits under this flow will depend on how ω_1 and ω_2 are related.

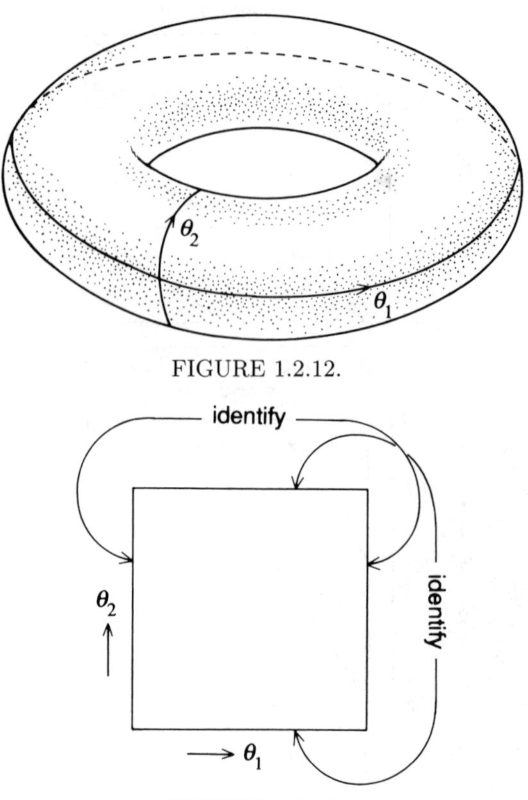

FIGURE 1.2.12.

FIGURE 1.2.13.

DEFINITION 1.2.1 ω_1 and ω_2 are said to be *incommensurate* if the equation

$$m\omega_1 + n\omega_2 = 0$$

has no solutions consisting of n, $m \in \mathbb{Z}$ (integers). Otherwise, ω_1 and ω_2 are *commensurate.*

Theorem 1.2.1 *If ω_1 and ω_2 are commensurate, then every phase curve of (1.2.43) is closed. However, if ω_1 and ω_2 are incommensurate, then every phase curve of (1.2.43) is everywhere dense on the torus.*

To prove this theorem, we need the following lemma.

Lemma 1.2.2 *Suppose the circle S^1 is rotated through an angle α, and α is incommensurate with 2π. Then the sequence*

$$S = \{\theta, \theta + \alpha, \theta + 2\alpha, \cdots, \theta + n\alpha, \cdots, (\text{mod}\, 2\pi)\}$$

is everywhere dense on the circle (note: n is an integer).

Proof:

$$\theta + m\alpha (\text{mod}\, 2\pi) = \begin{cases} \theta + m\alpha & \text{if} \quad m\alpha - 2\pi < 0, \\ \theta + (m\alpha - 2\pi k) & \text{if} \quad m\alpha - 2\pi k > 0, k > 1 \\ & \quad \text{and } m\alpha - 2\pi(k+1) < 0, \end{cases}$$

so, in particular, since α and 2π are incommensurate, the sequence S is infinite and never repeats.

We will use the "pigeonhole principle," i.e., if you have n holes and $n+1$ pigeons, then one hole must contain at least two pigeons.

Divide the circle into k half-open intervals of equal length $2\pi/k$. Then, among the first $k+1$ elements of the sequence S, at least two must be in the same half-open interval; call these points $\theta + p\alpha$, $\theta + q\alpha (\text{mod}\, 2\pi)$ with $p > q$. Thus, $(p-q)\alpha \equiv s\alpha < 2\pi/k (\text{mod}\, 2\pi)$. Any two consecutive points of the sequence $\bar{S}$ given by

$$\bar{S} = \{\theta, \theta + s\alpha, \theta + 2s\alpha, \cdots, \theta + ns\alpha, \cdots, (\text{mod}\, 2\pi)\}$$

are therefore the same distance d apart, where $d < 2\pi/k$ (note that $\bar{S} \subset S$).

Now choose any point on S^1 and construct an ε-neighborhood around it. If k is chosen such that $2\pi/k < \varepsilon$, then at least one of the elements of $\bar{S}$ will lie in the ε-neighborhood. This proves the lemma. □

Now we prove Theorem 1.2.1.

Proof: First, suppose ω_1 and ω_2 are commensurate, i.e., $\exists\, n, m \in \mathbb{Z}$ such that $\omega_1 = (n/m)\omega_2$. We construct a Poincaré map as follows. Let the cross-section Σ be defined as

$$\Sigma^{\theta_{10}} = \{ (\theta_1, \theta_2) \mid \theta_1 = \theta_{10} \}. \tag{1.2.45}$$

Then, using (1.2.45), we have

$$\begin{aligned} P_{\theta_{10}} \colon \Sigma^{\theta_{10}} &\to \Sigma^{\theta_{10}}, \\ \theta_2 &\mapsto \theta_2 + \omega_2 \frac{2\pi}{\omega_1}. \end{aligned} \tag{1.2.46}$$

However, $\omega_2/\omega_1 = m/n$; hence, we have

$$\theta_2 \mapsto \theta_2 + 2\pi \frac{m}{n} (\text{mod}\, 2\pi). \tag{1.2.47}$$

This is a map of the circle onto itself (called a *circle map*); the number ω_2/ω_1 is called the *rotation number*. (Rotation numbers are also defined for nonlinear circle maps, as we shall see later.)

It is clear that the nth iterate of this map is given by

$$\theta_2 \mapsto \theta_2 + 2\pi m (\text{mod}\, 2\pi) = \theta_2. \tag{1.2.48}$$

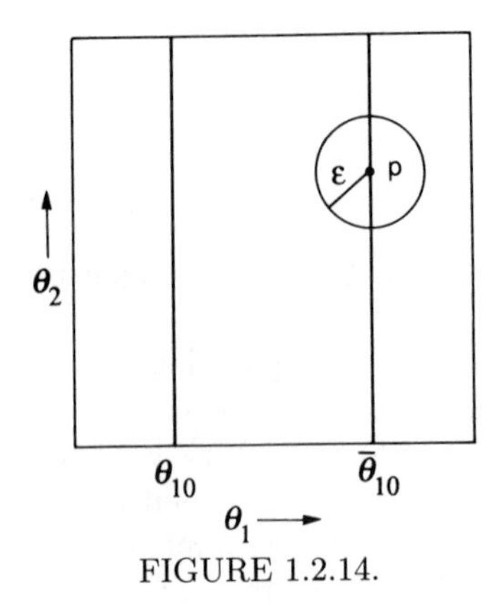

FIGURE 1.2.14.

Thus, every θ_2 is a periodic point; hence the flow consists entirely of closed orbits. This proves the first part of the theorem.

Now suppose ω_1 and ω_2 are incommensurate; then $\omega_2/\omega_1 = \alpha$, where α is irrational. The Poincaré map is then given by

$$\theta_2 \mapsto \theta_2 + 2\pi\alpha (\text{mod}\, 2\pi); \tag{1.2.49}$$

thus, by Lemma 1.2.2, the orbit of any point θ_2 is dense in the circle.

Next choose any point p on T^2 and construct an ε-neighborhood of p. To finish the proof of Theorem 1.2.1 we need to show that, given any orbit on T^2, it eventually passes through this ε-neighborhood of p. This is done as follows.

First, we are able to construct a new cross-section $\Sigma^{\bar{\theta}_{10}}$ which passes through the ε-neighborhood of p; see Figure 1.2.14. We have seen that the orbits of $P_{\theta_{10}}\colon \Sigma^{\theta_{10}} \to \Sigma^{\theta_{10}}$ are all dense on $\Sigma^{\theta_{10}}$ *for any* θ_{10}. Therefore, we can take any point on $\Sigma^{\theta_{10}}$ and look at its first intersection point with $\Sigma^{\bar{\theta}_{10}}$ under the flow (1.2.44). From this it follows that the iterates of this point under $P_{\bar{\theta}_{10}}$ are dense in $\Sigma^{\bar{\theta}_{10}}$. This completes the proof. □

Let us make a final remark before leaving this example. In our introductory motivational remarks we stated that Poincaré maps allow a dimensional reduction of the problem by *at least one*. In this example, we have seen how the study of a four-dimensional system can be reduced to the study of a one-dimensional system. This was possible because of our understanding of the geometry of the phase space; i.e., the phase space was made up of families of two-tori. It will be a common theme throughout this book that a good qualitative feel for the geometry of the phase space will put us in the best position for quantitative analysis.

Finally, we note that these results for linear vector fields on T^2 actually remain true for nonlinear differentiable vector fields on T^2, namely, that the ω limit sets for vector fields with no singular points are either closed orbits or the entire torus; see Hale [1980].

Case 3: Poincaré Map Near a Homoclinic Orbit

We now want to give an example of the construction of a Poincaré map in the neighborhood of a homoclinic orbit. The general analysis for orbits homoclinic to hyperbolic fixed points of autonomous ordinary differential equations in arbitrary dimensions is rather involved and can be found in Wiggins [1988]. Rather than getting entangled in technical details, we will concentrate on a specific example in two dimensions which illustrates the main ideas. Section 4.8 contains more examples.

Consider the ordinary differential equation

$$\begin{aligned} \dot{x} &= \alpha x + f_1(x, y; \mu), \\ \dot{y} &= \beta y + f_2(x, y; \mu), \end{aligned} \qquad (x, y, \mu) \in \mathbb{R}^1 \times \mathbb{R}^1 \times \mathbb{R}^1, \tag{1.2.50}$$

with $f_1, f_2 = \mathcal{O}(|x|^2 + |y|^2)$ and $\mathbf{C}^r$, $r \geq 2$ and where μ is regarded as a parameter. We make the following hypotheses on (1.2.50).

Hypothesis 1. $\alpha < 0$, $\beta > 0$, and $\alpha + \beta \neq 0$.

Hypothesis 2. At $\mu = 0$ (1.2.50) possesses a homoclinic orbit connecting the hyperbolic fixed point $(x, y) = (0, 0)$ to itself, and on both sides of $\mu = 0$ the homoclinic orbit is broken. Furthermore, the homoclinic orbit breaks in a transverse manner in the sense that the stable and unstable manifolds have different orientations on different sides of $\mu = 0$. For definiteness, we will assume that, for $\mu < 0$, the stable manifold lies outside the unstable manifold, for $\mu > 0$, the stable manifold lies inside the stable manifold and, for $\mu = 0$, they coincide; see Figure 1.2.15.

Hypothesis 1 is of a local nature, since it concerns the nature of the eigenvalues of the vector field linearized about the fixed point. Hypothesis 2 is global in nature, since it supposes the existence of a homoclinic orbit and describes the nature of the parameter dependence of the homoclinic orbit.

Now an obvious question is why this scenario? Why not stable inside for $\mu > 0$ and unstable inside for $\mu < 0$? Certainly this could happen; however, this is not important for us to consider at the moment. We need to know only that, on one side of $\mu = 0$, the stable manifold lies inside the unstable manifold, and on the other side of $\mu = 0$, the unstable manifold lies inside the stable manifold. Of course, in applications, we would want to determine which case actually occurs, and in Chapter 4, we will learn a method for doing this (Melnikov's method); however, now we will simply study the consequences of a homoclinic orbit to a hyperbolic fixed point of a planar vector field breaking in the manner described above.

Let us remark that it is certainly possible for the eigenvalues α and β to depend on the parameter μ. However, this will be of no consequence provided that Hypothesis 1 is satisfied for each parameter value and that this is true for μ sufficiently close to zero.

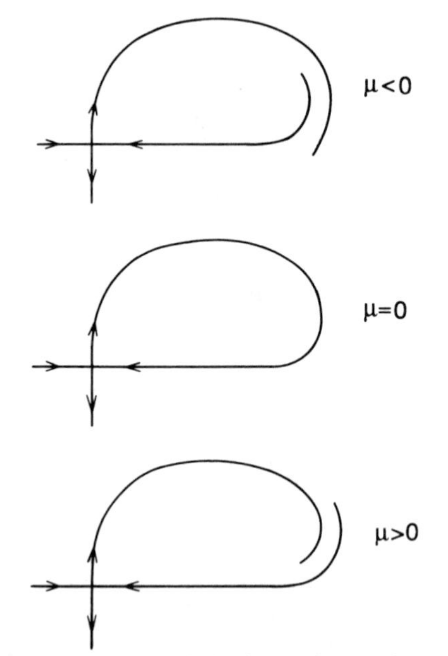

FIGURE 1.2.15. Behavior of the homoclinic orbit as μ is varied.

The question we ask is the following: *What is the nature of the orbit structure near the homoclinic orbit for μ near $\mu = 0$?* We will answer this question by computing a Poincaré map near the homoclinic orbit and studying the orbit structure of the Poincaré map. The Poincaré map that we construct will be very different from those we constructed in Cases 1 and 2 in that it will be the composition of two maps. One of the maps, P_0, will be constructed from the flow near the origin (which we will take to be the flow generated by the linearization of (1.2.50) about the origin). The other map, P_1, will be constructed from the flow outside of a neighborhood of the fixed point, which, if we remain close enough to the homoclinic orbit, can be made to be as close to a rigid motion as we like. The resulting Poincaré map, P, will then be given by $P \equiv P_1 \circ P_0$. Evidently, with these approximations, our Poincaré map will be valid (meaning that its dynamics reflect the dynamics of (1.2.50)) only when it is defined sufficiently close to the (broken) homoclinic orbit. We will discuss the validity of our approximations later on, but for now we begin our analysis.

The analysis will proceed in several steps.

Step 1. Set up the domain for the Poincaré map.

Step 2. Compute P_0.

Step 3. Compute P_1.

Step 4. Examine the dynamics of $P = P_1 \circ P_0$.

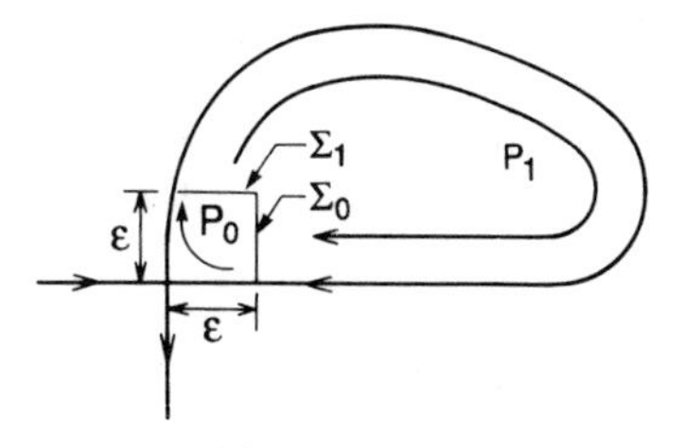

FIGURE 1.2.16.

Step 1: *Set Up the Domain for the Poincaré Map.* For the domain of P_0 we choose

$$\Sigma_1 = \{ (x, y) \in \mathbb{R}^2 \mid x = \varepsilon > 0, \ y > 0 \}, \tag{1.2.51}$$

and for the domain P_1 we choose

$$\Sigma_0 = \{ (x, y) \in \mathbb{R}^2 \mid x > 0, \ y = \varepsilon > 0 \}. \tag{1.2.52}$$

We will take ε small; the need for this will become apparent later on. See Figure 1.2.16 for an illustration of the geometry of Σ_0 and Σ_1.

Step 2: *Compute* P_0. We will use the flow generated by the linear vector field

$$\begin{aligned} \dot{x} &= \alpha x, \\ \dot{y} &= \beta y, \end{aligned} \tag{1.2.53}$$

in order to compute the map, P_0, of points on Σ_0 to Σ_1. For this to be a good approximation, it should be clear that we must take ε and y small. We will discuss the validity of this approximation later.

The flow generated by (1.2.53) is given by

$$\begin{aligned} x(t) &= x_0 e^{\alpha t}, \\ y(t) &= y_0 e^{\beta t}. \end{aligned} \tag{1.2.54}$$

The time of flight, T, needed for a point $(\varepsilon, y_0) \in \Sigma_0$ to reach Σ_1 under the action of (1.2.54) is given by solving

$$\varepsilon = y_0 e^{\beta T} \tag{1.2.55}$$

to obtain

$$T = \frac{1}{\beta} \log \frac{\varepsilon}{y_0}. \tag{1.2.56}$$

From (1.2.56) it is clear that we must require $y_0 \leq \varepsilon$.

$$\begin{aligned} P_0 \colon \Sigma_0 &\to \Sigma_1, \\ (\varepsilon, y_0) &\mapsto \left(\varepsilon \left(\frac{\varepsilon}{y_0} \right)^{\alpha/\beta}, \varepsilon \right). \end{aligned} \tag{1.2.57}$$

Step 3: *Compute* P_1. Using Theorem 1.1.8, by smoothness of the flow with respect to initial conditions and the fact that it only takes a finite time to flow from Σ_1 to Σ_0 along the homoclinic orbit, we can find a neighborhood $U \subset \Sigma_1$ which is mapped onto Σ_0 under the flow generated by (1.2.53). We denote this map by

$$P_1(x, \varepsilon; \mu) = \big(P_{11}(x, \varepsilon; \mu), P_{12}(x, \varepsilon; \mu)\big): U \subset \Sigma_1 \to \Sigma_0, \tag{1.2.58}$$

where $P_1(0, \varepsilon; 0) = (\varepsilon, 0)$. Taylor expanding (1.2.58) about $(x, \varepsilon; \mu) = (0, \varepsilon; 0)$ gives

$$P_1(x, \varepsilon; \mu) = (\varepsilon, ax + b\mu) + \mathcal{O}(2). \tag{1.2.59}$$

The expression "$\mathcal{O}(2)$" in (1.2.59) represents higher order nonlinear terms which can be made small by taking ε, x, and μ small. For now, we will neglect these terms and take as our map

$$\begin{aligned} P_1 &: U \subset \Sigma_1 \to \Sigma_0, \\ (x, \varepsilon) &\mapsto (\varepsilon, ax + b\mu), \end{aligned} \tag{1.2.60}$$

where $a > 0$ and $b > 0$. The reader should study Figure 1.2.15 to determine why we must have a, $b > 0$.

Step 4: *Examine the Dynamics of* $P = P_1 \circ P_0$. We have

$$\begin{aligned} P = P_1 \circ P_0 &: V \subset \Sigma_0 \to \Sigma_0, \\ (\varepsilon, y_0) &\mapsto \left(\varepsilon, a\varepsilon \left(\frac{\varepsilon}{y_0}\right)^{\alpha/\beta} + b\mu\right), \end{aligned} \tag{1.2.61}$$

where $V = (P_0)^{-1}(U)$, or

$$P(y; \mu): y \to Ay^{|\alpha/\beta|} + b\mu, \tag{1.2.62}$$

where $A \equiv a\varepsilon^{1+(\alpha/\beta)} > 0$ (we have left the subscript "0" off the y_0 for the sake of a less cumbersome notation). (Note: of course, we are assuming also that U is sufficiently small so that $(P_0)^{-1}(U) \subset \Sigma_0$.)

Let $\delta = |\alpha/\beta|$; then $\alpha + \beta \neq 0$ implies $\delta \neq 1$. We will seek fixed points of the Poincaré map, i.e., $y \in V$ such that

$$P(y; \mu) = Ay^\delta + b\mu = y. \tag{1.2.63}$$

The fixed points can be displayed graphically as the intersection of the graph of $P(y; \mu)$ with the line $y = P(y; \mu)$ for fixed μ.

There are two distinct cases.

Case 1 $|\alpha| > |\beta|$ or $\delta > 1$. For this case $D_y P(0; 0) = 0$, and the graph of P appears as in Figure 1.2.17 for $\mu > 0$, $\mu = 0$, and $\mu < 0$. Thus, for $\mu > 0$ and small μ, (1.2.62) has a fixed point. The fixed point is stable and hyperbolic,

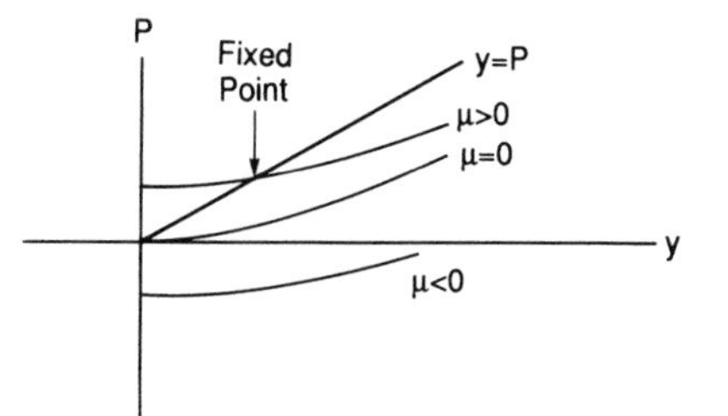

FIGURE 1.2.17. Graph of P for $\mu > 0$, $\mu = 0$, and $\mu < 0$ with $\delta > 1$.

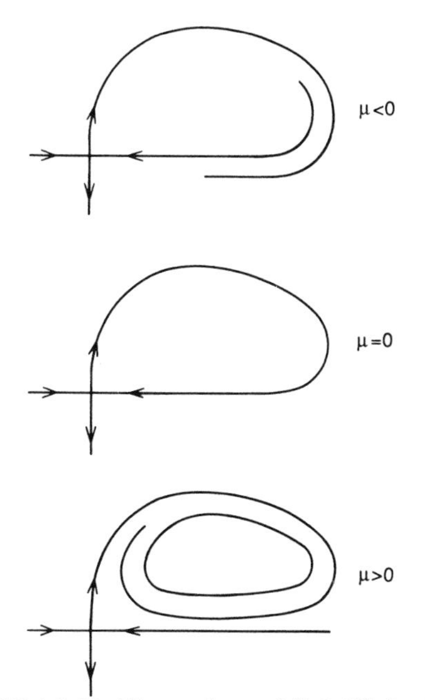

FIGURE 1.2.18. Phase plane of (1.2.50) for $\delta > 1$.

since $0 < D_y P < 1$ for μ sufficiently small. By construction we therefore see that this fixed point corresponds to an attracting periodic orbit of (1.2.50) (provided that we can justify our approximations); see Figure 1.2.18. We remark that if the homoclinic orbit were to break in the manner opposite to that shown in Figure 1.2.15, then the fixed point of (1.2.62) would occur for $\mu < 0$.

Case 2 $|\alpha| < |\beta|$ or $\delta < 1$. For this case, $D_y P(0; 0) = \infty$, and the graph of P appears as in Figure 1.2.19. Thus, for $\mu < 0$, (1.2.62) has a repelling fixed point. By construction we can therefore conclude that this corresponds to a repelling periodic orbit for (1.2.50); see Figure 1.2.20. We remark that if

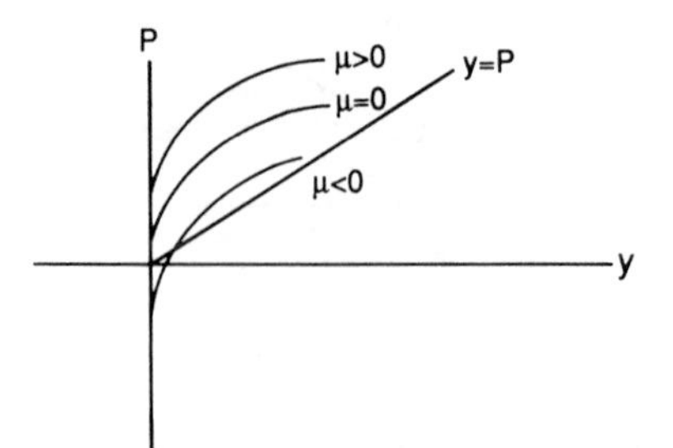

FIGURE 1.2.19. Graph of P for $\mu > 0$, $\mu = 0$, and $\mu < 0$ with $\delta < 1$.

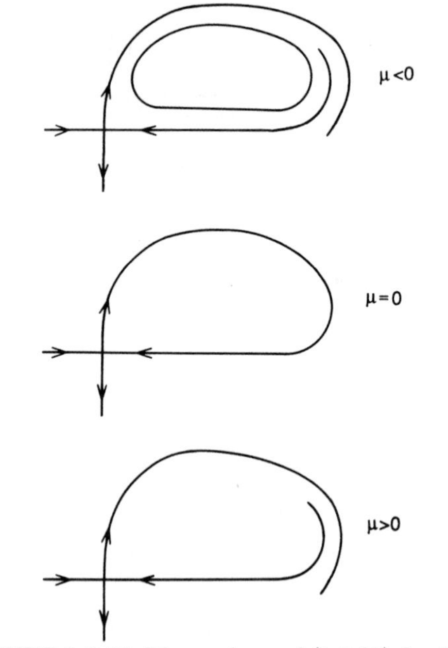

FIGURE 1.2.20. Phase plane of (1.2.50) for $\delta < 1$.

the homoclinic orbit were to break in the manner opposite to that shown in Figure 1.2.15, then the fixed point of (1.2.62) would occur for $\mu > 0$.

We summarize our results in the following theorem.

Theorem 1.2.3 *Consider a system where Hypothesis 1 and Hypothesis 2 hold. Then we have, for μ sufficiently small:* i) *If $\alpha + \beta < 0$, there exists a unique stable periodic orbit on one side of $\mu = 0$; on the opposite side of μ there are no periodic orbits.* ii) *If $\alpha + \beta > 0$, the same conclusion holds as in* i), *except that the periodic orbit is unstable.*

We remark that if the homoclinic orbit breaks in the manner opposite that shown in Figure 1.2.15, then Theorem 1.2.3 still holds except that the periodic orbits occur for μ values having the opposite sign as those given in Theorem 1.2.3. Theorem 1.2.3 is a classical result which can be found in Andronov et al. [1971]. Additional proofs can be found in Guckenheimer and Holmes [1983] and Chow and Hale [1982].

Before leaving this example we must address an important point, which is that we have not rigorously proven Theorem 1.2.3, since the Poincaré map we computed was only an approximation. We must therefore show that the dynamics of the exact Poincaré map are contained in the dynamics of the approximate Poincaré map. Because our main goal is to demonstrate how to construct a Poincaré map near a homoclinic orbit, we refer the reader to Wiggins [1988] for the proof of this fact under the condition that we remain sufficiently close to the (broken) homoclinic orbit, i.e., for ϵ and μ sufficiently small.

1.2B Varying the Cross-Section: Conjugacies of Maps

We now turn to answering the question of how the choice of cross-section affects the Poincaré map. The point of view that we develop will be that Poincaré maps defined on different cross-sections are related by a (in general, nonlinear) coordinate transformation. The importance of coordinate transformations in the study of dynamical systems cannot be overestimated. For example, in the study of systems of linear constant coefficient ordinary differential equations, coordinate transformations allow one to decouple the system and hence reduce the system to a set of decoupled linear first-order equations which are easily solved. In the study of completely integrable Hamiltonian systems, the transformations to action-angle coordinates results in a trivially solvable system (see Arnold [1978]), and these coordinates are also useful in the study of near integrable systems. If we consider general properties of dynamical systems, coordinate transformations provide us with a way of classifying dynamical systems according to properties which remain unchanged after a coordinate transformation. In Section 1.2C we will see that the notion of structural stability is based on such a classification scheme.

Before considering Poincaré maps, we want to discuss coordinate transformations, or, to use the more general mathematical term, *conjugacies,* giving some results that describe properties which must be retained by a map or vector field after a coordinate transformation of a specific differentiability class. Let us begin with an example which should be familiar to the reader.

Example 1.2.4 We want to motivate how coordinate transformations affect the orbits of maps.

Consider two linear, invertible maps

$$x \mapsto Ax, \qquad x \in \mathbb{R}^n \tag{1.2.64a}$$

$$y \mapsto By, \qquad y \in \mathbb{R}^n. \tag{1.2.64b}$$

For $x_0 \in \mathbb{R}^n$, we denote the orbit of x_0 under A by

$$O_A(x_0) = \{\cdots, A^{-n}x_0, \cdots, A^{-1}x_0, x_0, Ax_0, \cdots, A^n x_0, \cdots\}, \tag{1.2.65a}$$

and, for $y_0 \in \mathbb{R}^n$, we denote the orbit of y_0 under B by

$$O_B(y_0) = \{\cdots, B^{-n}y_0, \cdots, B^{-1}y_0, y_0, By_0, \cdots, B^n y_0, \cdots\}. \tag{1.2.65b}$$

Now suppose A and B are related by a similarity transformation, i.e., there is an invertible matrix T such that

$$B = TAT^{-1}. \tag{1.2.66}$$

We could think of T as transforming A into B, and, hence, since it does no harm in the linear setting to confuse the map with the matrix that generates it, T transforms (1.2.64a) into (1.2.64b). We represent this in the following diagram

$$\begin{array}{ccc} \mathbb{R}^n & \xrightarrow{A} & \mathbb{R}^n \\ \Big\downarrow T & & \Big\downarrow T. \\ \mathbb{R}^n & \xrightarrow{B} & \mathbb{R}^n \end{array} \tag{1.2.67}$$

The question we want to answer is this: when (1.2.64a) is transformed into (1.2.64b) via (1.2.66), how are orbits of A related to orbits of B? To answer this question, note that from (1.2.66) we have

$$B^n = TA^nT^{-1} \qquad \text{for all } n. \tag{1.2.68}$$

Hence, using (1.2.68) and comparing (1.2.64a) and (1.2.66), we see that orbits of A are mapped to orbits of B under the transformation $y = Tx$. Moreover, we know that since similar matrices have the same eigenvalues, the stability types of these orbits coincide under the transformation T.

Now we want to consider coordinate transformation in a more general, nonlinear setting. However, the reader will see that the essence of the ideas is contained in this example.

Let us consider two $\mathbf{C}^r$ diffeomorphisms $f: \mathbb{R}^n \to \mathbb{R}^n$ and $g: \mathbb{R}^n \to \mathbb{R}^n$, and a $\mathbf{C}^k$ diffeomorphism $h: \mathbb{R}^n \to \mathbb{R}^n$.

DEFINITION 1.2.2 f and g are said to be $\mathbf{C}^k$ *conjugate* ($k \leq r$) if there exists a $\mathbf{C}^k$ diffeomorphism $h: \mathbb{R}^n \to \mathbb{R}^n$ such that $g \circ h = h \circ f$. If $k = 0$, f and g are said to be *topologically conjugate*.

The conjugacy of two diffeomorphisms is often represented by the following diagram.

$$\begin{array}{ccc} \mathbb{R}^n & \xrightarrow{f} & \mathbb{R}^n \\ \Big\downarrow h & & \Big\downarrow h. \\ \mathbb{R}^n & \xrightarrow{g} & \mathbb{R}^n \end{array} \tag{1.2.69}$$

The diagram is said to *commute* if the relation $g \circ h = h \circ f$ holds, meaning that you can start at a point in the upper left-hand corner of the diagram and reach the same point in the lower right-hand corner of the diagram by either of the two possible routes. We note that h need not be defined on all of $\mathbb{R}^n$ but possibly only locally about a given point. In such cases, f and g are said to be *locally* $\mathbf{C}^k$ *conjugate.*

If f and g are $\mathbf{C}^k$ conjugate, then we have the following results.

Proposition 1.2.4 *If f and g are $\mathbf{C}^k$ conjugate, then orbits of f map to orbits of g under h.*

Proof: Let $x_0 \in \mathbb{R}^n$; then the orbit of x_0 under f is given by

$$O(x_0) = \{\cdots, f^{-n}(x_0), \cdots, f^{-1}(x_0), x_0, f(x_0), \cdots, f^n(x_0), \cdots\}. \tag{1.2.70}$$

From Definition 1.2.2, we have that $f = h^{-1} \circ g \circ h$, so for a given $n > 0$ we have

$$\begin{aligned} f^n(x_0) &= \underbrace{(h^{-1} \circ g \circ h) \circ (h^{-1} \circ g \circ h) \circ \cdots \circ (h^{-1} \circ g \circ h)}_{n \text{ factors}}(x_0) \\ &= h^{-1} \circ g^n \circ h(x_0) \end{aligned} \tag{1.2.71}$$

or

$$h \circ f^n(x_0) = g^n \circ h(x_0). \tag{1.2.72}$$

Also from Definition 1.2.2, we have that $f^{-1} = h^{-1} \circ g^{-1} \circ h$, so by the same argument, for $n > 0$ we obtain

$$h \circ f^{-n}(x_0) = g^{-n} \circ h(x_0). \tag{1.2.73}$$

Therefore, from (1.2.71) and (1.2.73) we see that the orbit of x_0 under f is mapped by h to the orbit of $h(x_0)$ under g. □

Proposition 1.2.5 *If f and g are $\mathbf{C}^k$ conjugate, $k \geq 1$, and x_0 is a fixed point of f, then the eigenvalues of $Df(x_0)$ are equal to the eigenvalues of $Dg\big(h(x_0)\big)$.*

Proof: From Definition 1.2.2, $f(x) = h^{-1} \circ g \circ h(x)$. Note that since x_0 is a fixed point then $h^{-1} \circ g \circ h(x_0) = x_0$. Also, by the inverse function theorem, we have $Dh^{-1} = (Dh)^{-1}$. Using this and the fact that h is differentiable, we have

$$Df\big|_{x_0} = Dh^{-1}\big|_{x_0} Dg\big|_{h(x_0)} Dh\big|_{x_0}. \tag{1.2.74}$$

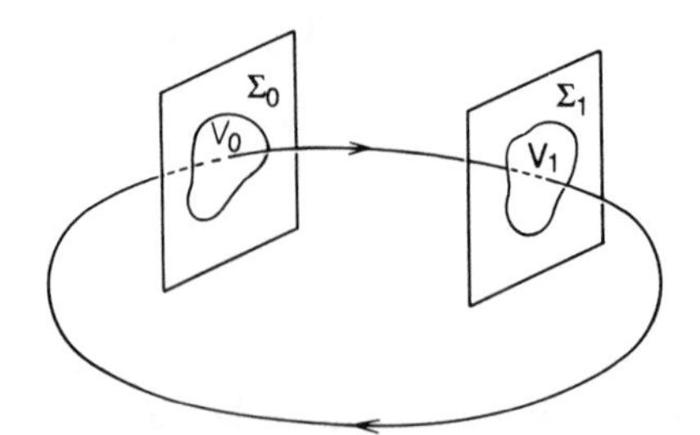

FIGURE 1.2.21. The cross-sections Σ_0 and Σ_1.

Therefore, recalling that similar matrices have equal eigenvalues gives the result. □

Now we will return to the specific question of what happens to the Poincaré map when the cross-section is changed. We begin with Case 1, a Poincaré map defined near a periodic orbit.

Case 1: Variation of the Cross-Section. Let x_0 and x_1 be two points on the periodic solution of (1.2.2), and let Σ_0 and Σ_1 be two $(n-1)$-dimensional surfaces at x_0 and x_1, respectively, which are transverse to the vector field, and suppose that Σ_1 is chosen such that it is the image of Σ_0 under flow generated by (1.2.2); see Figure 1.2.21. By Theorem 1.1.8, this defines a $\mathbf{C}^r$ diffeomorphism

$$h\colon \Sigma_0 \to \Sigma_1. \tag{1.2.75}$$

We define Poincaré maps P_0 and P_1 as in the previous construction.

$$\begin{aligned} P_0\colon V_0 &\to \Sigma_0, \\ x_0 &\mapsto \phi(\tau(\bar{x}_0), \bar{x}_0), \qquad \bar{x}_0 \in V_0 \subset \Sigma_0, \end{aligned} \tag{1.2.76}$$

$$\begin{aligned} P_1\colon V_1 &\to \Sigma_1, \\ x_1 &\mapsto \phi(\tau(\bar{x}_1), \bar{x}_1), \qquad \bar{x}_1 \in V_1 \subset \Sigma_1. \end{aligned} \tag{1.2.77}$$

Then we have the following result.

Proposition 1.2.6 *P_0 and P_1 are locally $\mathbf{C}^r$ conjugate.*

Proof: We need to show that

$$P_1 \circ h = h \circ P_0,$$

from which the result follows immediately since h is a $\mathbf{C}^r$ diffeomorphism. However, we need to worry a bit about the domains of the maps. We have

$$\begin{aligned} h(\Sigma_0) &= \Sigma_1, \\ P_0(V_0) &\subset \Sigma_0, \\ P_1(V_1) &\subset \Sigma_1. \end{aligned} \tag{1.2.78}$$

Thus, $h \circ P_0 \colon V_0 \to \Sigma_1$ is well defined but $P_1 \circ h$ need not be defined, since P_1 is not defined on all of Σ_1; however, this problem is solved if we choose Σ_1 such that $V_1 = h(V_0)$ and take V_0 sufficiently small. □

Case 2: Variation of the Cross-Section. Consider the Poincaré map $P_{\bar{\theta}_0}$ defined on the cross-section $\Sigma^{\bar{\theta}_0}$ as defined in (1.2.16). Suppose we construct a different Poincaré map, $P_{\bar{\theta}_1}$, in the same manner but on the cross-section

$$\Sigma^{\bar{\theta}_1} = \left\{ (x, \theta) \in \mathbb{R}^n \times S^1 \mid \theta = \bar{\theta}_1 \in (0, 2\pi] \right\}. \tag{1.2.79}$$

Then we have the following result.

Proposition 1.2.7 *$P_{\bar{\theta}_0}$ and $P_{\bar{\theta}_1}$ are $\mathbf{C}^r$ conjugate.*

Proof: The proof follows a construction similar to that given in Proposition 1.2.6. We construct a $\mathbf{C}^r$ diffeomorphism, h, of $\Sigma^{\bar{\theta}_0}$ into $\Sigma^{\bar{\theta}_1}$ by mapping points on $\Sigma^{\bar{\theta}_0}$ into $\Sigma^{\bar{\theta}_1}$ under the action of the flow generated by (1.2.14). Points starting on $\Sigma^{\bar{\theta}_0}$ have initial time $t_0 = (\bar{\theta}_0 - \theta_0)/\omega$, and they reach $\Sigma^{\bar{\theta}_1}$ after time

$$t = \frac{\bar{\theta}_1 - \bar{\theta}_0}{\omega};$$

thus we have

$$\begin{aligned} h \colon \Sigma^{\bar{\theta}_0} &\to \Sigma^{\bar{\theta}_1}, \\ \left(x\left(\frac{\bar{\theta}_0 - \theta_0}{\omega} \right), \bar{\theta}_0 \right) &\mapsto \left(x\left(\frac{\bar{\theta}_1 - \theta_0}{\omega} \right), \bar{\theta}_1 \right). \end{aligned} \tag{1.2.80}$$

Using (1.2.80) and the expression for the Poincaré maps defined on the different cross-sections, we obtain

$$\begin{aligned} h \circ P_{\bar{\theta}_0} \colon \Sigma^{\bar{\theta}_0} &\to \Sigma^{\bar{\theta}_1}, \\ \left(x\left(\frac{\bar{\theta}_0 - \theta_0}{\omega} \right), \bar{\theta}_0 \right) &\mapsto \left(x\left(\frac{\bar{\theta}_1 - \theta_0 + 2\pi}{\omega} \right), \bar{\theta}_1 + 2\pi \equiv \bar{\theta}_1 \right), \end{aligned} \tag{1.2.81}$$

and

$$\begin{aligned} P_{\bar{\theta}_1} \circ h \colon \Sigma^{\bar{\theta}_0} &\to \Sigma^{\bar{\theta}_1}, \\ \left(x\left(\frac{\bar{\theta}_0 - \theta_0}{\omega} \right), \bar{\theta}_0 \right) &\mapsto \left(x\left(\frac{\bar{\theta}_1 - \theta_0 + 2\pi}{\omega} \right), \bar{\theta}_1 + 2\pi \equiv \bar{\theta}_1 \right). \end{aligned} \tag{1.2.82}$$

Thus, from (1.2.81) and (1.2.82), we have that

$$h \circ P_{\bar{\theta}_0} = P_{\bar{\theta}_1} \circ h. \quad \square \tag{1.2.83}$$

Therefore, Propositions 1.2.4 and 1.2.5 imply that, as long as we remain sufficiently close to the periodic orbit, changing the cross-section does not have any dynamical effect in the sense that we will still have the same orbits with the same stability type. However, geometrically there may be an apparent difference in the sense that the locations of the orbits as well as their stable and unstable manifolds may "move around" under a change in cross-section. It may also be possible that an intelligent choice of the cross-section could result in a "more symmetric" Poincaré map which could facilitate the analysis. We will see an example of this later.

We note that Case 3, the Poincaré map near a homoclinic orbit, can be treated in the same way with the same results. We leave this as an exercise for the reader.

We remark that it should be clear from these results that a Poincaré map constructed according to Case 2 (i.e., the global cross-section) has information concerning all possible dynamics of the vector field. When only a local cross-section can be constructed (e.g., as in Cases 1 and 3), then the Poincaré map will not, in general, contain information on all possible dynamics of the vector field. Different Poincaré maps defined on different cross-sections may not have the same dynamics.

1.2c Structural Stability, Genericity, and Transversality

The mathematical models we devise to make sense of the world around and within us can only be approximations. Therefore, it seems reasonable that if they are to accurately reflect reality, the models themselves must be somewhat insensitive to perturbations. The attempts to give mathematical substance to these rather vague ideas have led to the concept of *structural stability*. Before giving a definition of structural stability, let us consider a specific example which illustrates many of the issues that need to be addressed.

Example 1.2.5 Consider the simple harmonic oscillator

$$\begin{aligned} \dot{x} &= y, \\ \dot{y} &= -\omega_0^2 x, \end{aligned} \qquad (x,y) \in \mathbb{R}^2. \tag{1.2.84}$$

We know everything about this system. It has a nonhyperbolic fixed point of $(x, y) = (0, 0)$ surrounded by a one-parameter family of periodic orbits, each having frequency ω_0. The phase portrait of (1.2.84) is shown in Figure 1.2.22 (note: strictly speaking, the phase curves are circles for $\omega_0 = 1$ and ellipses otherwise). Is (1.2.84) stable with respect to perturbations (note: this is a new concept of stability, as opposed to the idea of stability of specific solutions discussed in Section 1.1A)? Let us try a few perturbations and see what happens.

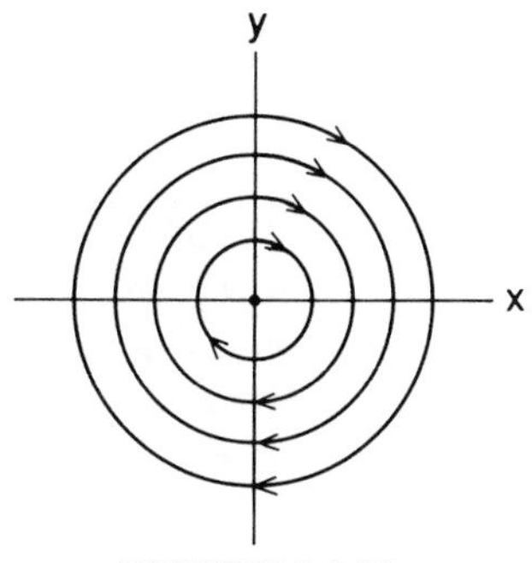

FIGURE 1.2.22.

Linear, Dissipative Perturbation

Consider the perturbed system

$$\begin{aligned}\dot{x} &= y,\\ \dot{y} &= -\omega_0^2 x - \varepsilon y.\end{aligned} \tag{1.2.85}$$

It is easy to see that the origin is a hyperbolic fixed point, a sink for $\varepsilon > 0$ and a source for $\varepsilon < 0$. However, all the periodic orbits are destroyed (use Bendixson's criteria). Thus, this perturbation radically alters the structure of the phase space of (1.2.84); see Figure 1.2.23.

Nonlinear Perturbation

Consider the perturbed system

$$\begin{aligned}\dot{x} &= y,\\ \dot{y} &= -\omega_0^2 x + \varepsilon x^2.\end{aligned} \tag{1.2.86}$$

The perturbed system now has two fixed points given by

$$\begin{aligned}(x, y) &= (0, 0),\\ (x, y) &= (\omega_0^2/\varepsilon, 0).\end{aligned} \tag{1.2.87}$$

The origin is still a center (i.e., unchanged by the perturbation), and the new fixed point is a saddle and far away for ε small.

This particular perturbation has the property of preserving a first integral. In particular, (1.2.86) has a first integral given by

$$h(x, y) = \frac{y^2}{2} + \frac{\omega_0^2 x^2}{2} - \varepsilon\frac{x^3}{3}. \tag{1.2.88}$$

This enables us to draw all phase curves for (1.2.86), which are shown in Figure 1.2.24. From Figure 1.2.24, we make the following observations.

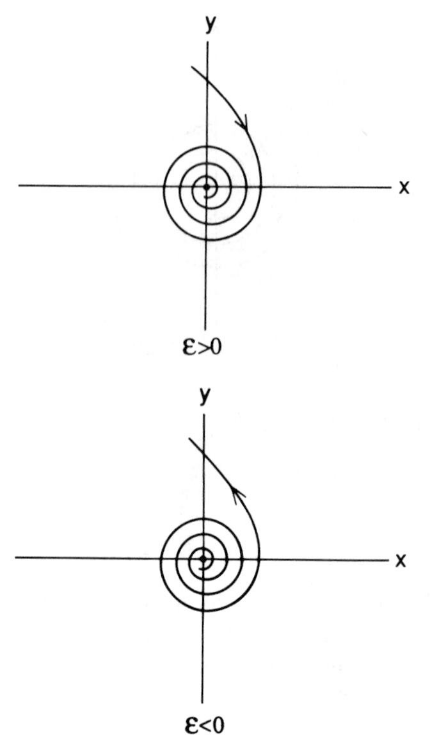

FIGURE 1.2.23.

1. This particular perturbation preserves the symmetry of (1.2.86) implied by the existence of a first integral. Therefore, sufficiently close to $(x, y) = (0, 0)$ the phase portraits of (1.2.84) and (1.2.86) look the same. However, for (1.2.86), it is important to note that the frequency of the periodic orbits changes with distance from the origin, as opposed to (1.2.84).

2. The phase space of (1.2.84) is unbounded. Therefore, no matter how small we take ε, far enough away from the origin the perturbation is no longer a perturbation. This is evidenced in Figure 1.2.24 by the saddle point and the homoclinic orbit connecting it to itself. Thus, there is a problem in discusing perturbations of vector fields on unbounded phase spaces.

Time-Dependent Perturbation

Consider the system

$$\begin{aligned} \dot{x} &= y, \\ \dot{y} &= -\omega_0^2 x + \varepsilon x \cos t. \end{aligned} \tag{1.2.89}$$

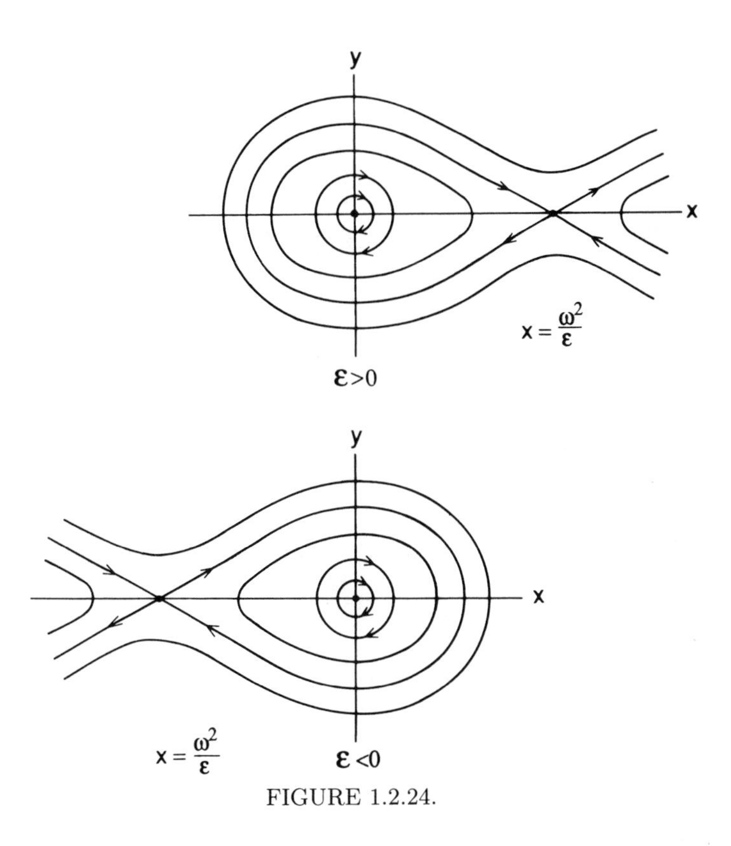

FIGURE 1.2.24.

This perturbation is of a very different character than the previous two. Writing (1.2.89) as an autonomous system (see Section 1.1G)

$$\begin{aligned} \dot{x} &= y, \\ \dot{y} &= -\omega_0^2 x + \varepsilon x \cos\theta, \\ \dot{\theta} &= 1, \end{aligned} \tag{1.2.90}$$

we see that the time-dependent perturbation has the effect of enlarging the dimension of the system. However, in any case, $(x, y) = (0, 0)$ is still a fixed point of (1.2.89), although it is interpreted as a periodic orbit of (1.2.90). We now ask what the nature of the flow is near $(x, y) = (0, 0)$, which is a difficult question to answer due to the time dependence. Equation (1.2.89) is known as the Mathieu equation, and for $\omega_0 = n/2$, n an integer, it is possible for the system to exhibit parametric resonance resulting in a solution starting near the origin that grows without bound. Thus, the flow of (1.2.90) near the origin differs very much from the flow of (1.2.84) near the origin. For more information on the Mathieu equation see Nayfeh and Mook [1979].

This simple example illustrates several points that need to be considered when discussing whether or not a system is stable under perturbations.

1. It is important to specify the type of perturbations that are allowed. For example, if the system has a symmetry, then one might want to consider only perturbations which preserve the symmetry. The idea of structural stability thus depends on the type of dynamical system under consideration.

2. In discussing the idea of a perturbation of a dynamical system, it is necessary to specify what it means for two vector fields or maps to be "close." In our example we used an ε and required ε to be small. However, we saw that this did not work well when the phase space was unbounded.

3. It is necessary to quantify the statement "two dynamical systems have qualitatively the same dynamics." This must be specified if one is to decide when a system is structurally stable.

Up to this point our discussion has been very heuristic. Indeed, our main purpose has been to get the reader to worry about whether the systems they are studying are stable under perturbations. We will see throughout this book, especially when we study bifurcation theory, that a consideration of this question often reveals much about the underlying dynamics of dynamical systems. However, now we want to say a little about the mathematical formulation of the notion of structural stability.

Definitions of Structural Stability and Genericity

The concept of structural stability was introduced by Andronov and Pontryagin [1931] and has played a central role in the development of dynamical systems theory. Roughly speaking, a dynamical system (vector field or map) is said to be struturally stable if nearby systems have qualitatively the same dynamics. Therefore, in defining structural stability one must provide a recipe for determining when two systems are "close," and then one must specify what is meant by saying that, qualitatively, two systems have the same dynamics. We will discuss each question separately.

Let $\mathbf{C}^r(\mathbb{R}^n, \mathbb{R}^n)$ denote the space of $\mathbf{C}^r$ maps of $\mathbb{R}^n$ into $\mathbb{R}^n$. In terms of dynamical systems, we can think of the elements of $\mathbf{C}^r(\mathbb{R}^n, \mathbb{R}^n)$ as being vector fields. We denote the subset of $\mathbf{C}^r(\mathbb{R}^n, \mathbb{R}^n)$ consisting of the $\mathbf{C}^r$ diffeomorphisms by $\text{Diff}^r(\mathbb{R}^n, \mathbb{R}^n)$. We remark that if one is studying dynamical systems that have certain symmetries, then additional constraints must be put on these spaces.

Two elements of $\mathbf{C}^r(\mathbb{R}^n, \mathbb{R}^n)$ are said to be $\mathbf{C}^r$ ε-close ($k \leq r$), or just $\mathbf{C}^k$ close, if they, along with their first k derivatives, are within ε as measured in some norm. There is a problem with this definition; namely, $\mathbb{R}^n$ is unbounded, and the behavior at infinity needs to be brought under control (note: this explains why most of dynamical systems theory has been developed using compact phase spaces; however, in applications this is not sufficient and appropriate modifications must be made).

There are several ways of handling this difficulty. For the purpose of our discussion we will choose the usual way and assume that our maps act on compact, boundaryless n-dimensional differentiable manifolds, M, rather than all of $\mathbb{R}^n$. The topology induced on $\mathbf{C}^r(M, M)$ by this measure of distance between two elements of $\mathbf{C}^r(M, M)$ is called the $\mathbf{C}^k$ *topology*, and we refer the reader to Palis and de Melo [1982] or Hirsch [1976] for a more thorough discussion.

The question of what is meant by saying that two dynamical systems are close is usually answered in terms of conjugacies. Specifically, $\mathbf{C}^0$ conjugate maps have qualitatively the same orbit structure in the sense of the propositions given in Section 1.2B. For vector fields there is a similar notion to $\mathbf{C}^k$ conjugacies for maps called a $\mathbf{C}^k$ equivalence. We will discuss this in more detail in Chapter 3 when we study bifurcation theory (note: in some sense the study of bifurcation theory will be the study of structural instability). In this section we will state the definitions for maps along with vector fields; the reader should refer back to these definitions when we study the related ideas for vector fields.

We are now at the point where we can formally define structural stability.

DEFINITION 1.2.3 Consider a map $f \in \text{Diff}^r(M, M)$ (resp. a $\mathbf{C}^r$ vector field in $\mathbf{C}^r(M, M)$); then f is said to be *structurally stable* if there exists a neighborhood $\mathcal{N}$ of f in the $\mathbf{C}^k$ topology such that f is $\mathbf{C}^0$ conjugate (resp. $\mathbf{C}^0$ equivalent) to every map (resp. vector field) in $\mathcal{N}$.

Now that we have defined structural stability, it would be nice if we could determine the characteristics of a specific system which result in that system being structurally stable. From the point of view of the applied scientist, this would be useful, since one might presume that a dynamical system used to model phenomena occurring in nature should possess the property of structural stability. Unfortunately, such a characterization does not exist, although some partial results are known, which we will describe shortly. One approach to the characterization of structural stability has been through the identification of typical or generic properties of dynamical systems, and we now discuss this idea.

Naively, one might expect a typical or generic property of a dynamical system to be one that is common to a dense set of dynamical systems in $\mathbf{C}^r(M, M)$. This is not quite adequate, since it is possible for a set and its complement to both be dense. For example, the set of rational numbers is dense in the real line, and so is its complement, the set of irrational numbers. However, there are many more irrational numbers than rational numbers, and one might expect the irrationals to be more typical than the rationals in some sense. The proper sense in which this is true is captured by the idea of a *residual set*.

DEFINITION 1.2.4 Let X be a topological space, and let U be a subset of X. U is called a *residual set* if it is the intersection of a countable number of sets each of which are open and dense in X. If a residual set in X is itself dense in X, then X is called a *Baire space.*

We remark that $\mathbf{C}^r(M, M)$ equipped with the $\mathbf{C}^r$ topology $(k \le r)$ is a Baire space (see Palis and deMelo [1982]). We now give the definition of a generic property.

DEFINITION 1.2.5 A property of a map (resp. vector field) is said to be $\mathbf{C}^k$ *generic* if the set of maps (resp. vector fields) possessing that property contains a residual subset in the $\mathbf{C}^k$ topology.

EXAMPLE 1.2.6 In the class of dynamical systems having fixed points, hyperbolic fixed points are structurally stable and generic.

In utilizing the idea of a generic property to characterize the structurally stable systems, one first identifies some generic property. Then, since a structurally stable system is $\mathbf{C}^0$ conjugate (resp. equivalent for vector fields) to all nearby systems, structurally stable systems must have this property if the property is one that is preserved under $\mathbf{C}^0$ conjugacy (resp. equivalence for vector fields). One would like to go the other way with this argument; namely, it would be nice to show that structurally stable systems are generic. For two-dimensional vector fields on compact manifolds, we have the following result due to Peixoto [1962].

Theorem 1.2.8 *A $\mathbf{C}^r$ vector field on a compact boundaryless two-dimensional manifold M is structurally stable if and only if*

i) *the number of fixed points and periodic orbits is finite and each is hyperbolic;*

ii) *there are no orbits connecting saddle points;*

iii) *the nonwandering set consists of fixed points and periodic orbits.*

Moreover, if M is orientable, then the set of such vector fields is open and dense in $\mathbf{C}^r(M, M)$ (note: this is stronger than generic).

This theorem is useful because it spells out precise conditions under which the dynamics of a vector field on a compact boundaryless two manifold are structurally stable. Unfortunately, we do not have a similar theorem in higher dimensions. This is in part due to the presence of complicated recurrent motions (e.g., the Smale horseshoe; see Chapter 4) which are not possible for two-dimensional vector fields. Even more disappointing is the fact that structural stability is not a generic property for n-dimensional diffeomorphisms $(n \ge 2)$ or n-dimensional vector fields $(n \ge 3)$. This fact was first demonstrated by Smale [1966].

At this point we will conclude our brief discussion of the ideas of structural stability and genericity. For more information, we refer the reader to Chillingworth [1976], Hirsch [1976], Arnold [1982], Nitecki [1971], Smale [1967], and Shub [1987]. However, before ending this section, we want to comment on the relevance of these ideas to the applied scientist, i.e., someone who must discover what types of dynamics are present in a specific dynamical system.

Genericity and structural stability as defined above have been guiding forces behind much of the development of dynamical systems theory. The approach often taken has been to postulate some "reasonable" form of dynamics for a certain class of dynamical systems and then to prove that this form of dynamics is structurally stable and/or generic within this class. If one is persistent with this approach one is occasionally successful and eventually a significant catalogue of generic and structurally stable dynamical properties is obtained. This catalogue is useful to the applied scientist in that it gives some idea of what dynamics to expect in a specific dynamical system. However, this is hardly adequate. Given a specific dynamical system, is it structurally stable and/or generic?

We would like to give computable conditions under which a specific dynamical system is structurally stable and/or generic. For certain special types of motions such as periodic orbits and fixed points, this can be done in terms of the eigenvalues of the linearized system. However, for more general, global motions such as homoclinic orbits and quasiperiodic orbits, this cannot be done so easily, since the nearby orbit structure may be exceedingly complicated and defy any local description (see Chapter 4). What this boils down to is that to determine whether or not a specific dynamical system is structurally stable, one needs a fairly complete understanding of its orbit structure, or to put it more cynically, one needs to know the answer before asking the question. It might therefore seem that these ideas are of little use to the applied scientist; however, this is not exactly true, since the theorems describing structural stability and generic properties do give one a good idea of what to *expect*, although they cannot tell what is precisely happening in a specific system. Also, the reader should always ask him or herself whether or not the dynamics are stable and/or typical in some sense. Probably the best way of mathematically quantifying these two notions for the applied scientist has yet to be determined.

Transversality

Before leaving this section let us introduce the idea of *transversality*, which will play a central role in many of our geometrical arguments.

Transversality is a geometric notion which deals with the intersection of surfaces or manifolds. Let M and N be differentiable (at least $\mathbf{C}^1$) manifolds in $\mathbb{R}^n$.

DEFINITION 1.2.6 Let p be a point in $\mathbb{R}^n$; then M and N are said to be *transversal at* p if $p \notin M \cap N$; or, if $p \in M \cap N$, then $T_pM + T_pN = \mathbb{R}^n$, where T_pM and T_pN denote the tangent spaces of M and N, respectively,

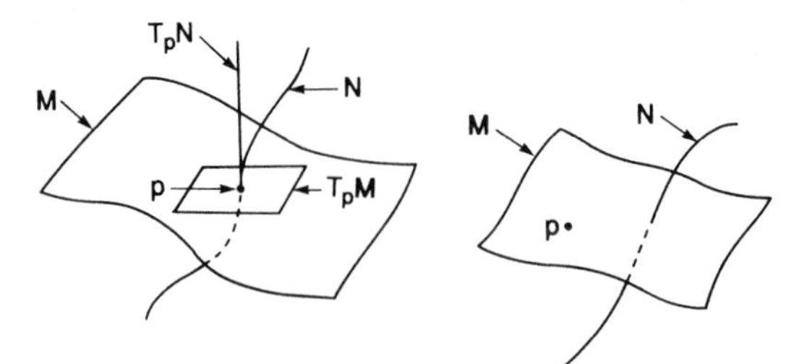

FIGURE 1.2.25. M and N transversal at p.

at the point p. M and N are said to be *transversal* if they are transversal at every point $p \in \mathbb{R}^n$; see Figure 1.2.25.

Whether or not the intersection is transversal can be determined by knowing the dimension of the intersection of M and N. This can be seen as follows. Using the formula for the dimension of the intersection of two vector subspaces we have

$$\dim(T_pM + T_pN) = \dim T_pM + \dim T_pN - \dim(T_pM \cap T_pN). \quad (1.2.91)$$

From Definition 1.2.6, if M and N intersect transversely at p, then we have

$$n = \dim T_pM + \dim T_pN - \dim(T_pM \cap T_pN). \quad (1.2.92)$$

Since the dimensions of M and N are known, then knowing the dimension of their intersection allows us to determine whether or not the intersection is transversal.

Note that transversality of two manifolds at a point requires more than just the two manifolds geometrically piercing each other at the point. Consider the following example.

EXAMPLE 1.2.7 Let M be the x axis in $\mathbb{R}^2$, and let N be the graph of the function $f(x) = x^3$; see Figure 1.2.26. Then M and N intersect at the origin in $\mathbb{R}^2$, but they are not transversal at the origin, since the tangent space of M is just the x axis and the tangent space of N is the span of the vector $(1, 0)$; thus, $T_{(0,0)}N = T_{(0,0)}M$ and, therefore, $T_{(0,0)}N + T_{(0,0)}M \neq \mathbb{R}^2$.

The most important characteristic of transversality is that it persists under sufficiently small perturbations. This fact will play a useful role in many of our geometric arguments; we remark that a term often used synonymously for transversal is *general position*, i.e., two or more manifolds which are transversal are said to be in general position.

Let us end this section by giving a few "dynamical" examples of transversality.

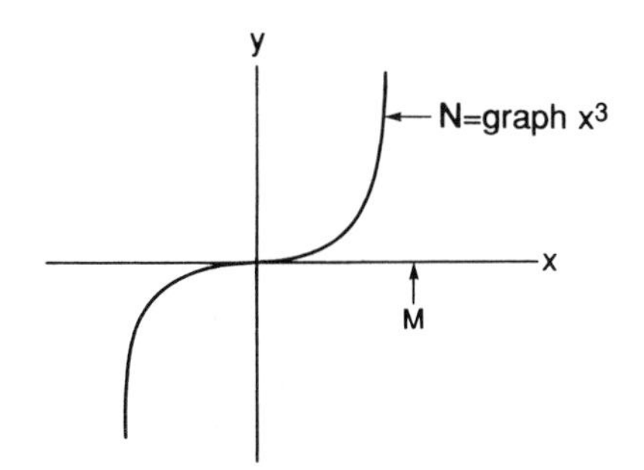

FIGURE 1.2.26. Nontransversal manifolds.

EXAMPLE 1.2.8 Consider a hyperbolic fixed point of a $\mathbf{C}^r$, $r \geq 1$, vector field on $\mathbb{R}^n$. Suppose the matrix associated with the linearization of the vector field about the fixed point has $n-k$ eigenvalues with positive real part and k eigenvalues with negative real part. Thus this fixed point has an $(n-k)$-dimensional unstable manifold and a k-dimensional stable manifold. If these two manifolds intersect in a point, then by uniqueness of solutions and invariance of the manifolds, they must intersect along a (at least) one-dimensional orbit. Hence, by (1.2.92), the intersection *cannot* be transverse.

EXAMPLE 1.2.9 Suppose the vector field of Example 1.2.8 is Hamiltonian so that all orbits are restricted to lie in $(n-1)$-dimensional "energy" surfaces given by the level sets of the Hamiltonian. Then it is possible for the stable and unstable manifolds of the hyperbolic fixed point to intersect transversely *in the $(n-1)$-dimensional energy surface.*

EXAMPLE 1.2.10 Consider a hyperbolic periodic orbit of a $\mathbf{C}^r$, $r \geq 1$, vector field on $\mathbb{R}^n$. Suppose that the Poincaré map associated with the periodic orbit linearized about the fixed point has $n-k-1$ eigenvalues with modulus greater than one and k eigenvalues with modulus less than one. Then the periodic orbit has an $(n-k)$-dimensional unstable manifold and a $(k+1)$-dimensional stable manifold. Therefore, by (1.2.92), if these manifolds intersect transversely, the dimension of the intersection must be one. This is possible without violating uniqueness of solutions and invariance of the manifolds.

1.2D Construction of the Poincaré Map

We will now develop two methods for constructing Poincaré maps for nonlinear systems. They will be perturbation methods and will apply to systems of the following form.

$$\dot{x} = f(x) + \varepsilon g(x, t, \epsilon), \qquad x \in \mathbb{R}^n, \tag{1.2.93}$$

where

$$f: U \to \mathbb{R}^n,$$
$$g: U \times \mathbb{R}^1 \times [0, \varepsilon_0) \to \mathbb{R}^n,$$

are $\mathbf{C}^r$ $(r \geq 1)$ functions on their respective domains of definition with U an open set in $\mathbb{R}^n$. We think of ε as being small and fixed. Equation (1.2.93) may also depend on parameters; however, we will omit any explicit parametric dependence when it does not affect our arguments.

Setting $\varepsilon = 0$, we obtain the following equation

$$\dot{x} = f(x), \qquad x \in \mathbb{R}^n, \tag{1.2.94}$$

which we refer to as the *unperturbed equation.* As with most perturbation methods, the idea is to use knowledge of the solutions of (1.2.94) to infer results concerning the solutions of (1.2.93). Our approach will be more geometrical in that we will use knowledge about the geometrical structure of the phase space of (1.2.94) to infer results concerning the geometrical structure of the phase space of (1.2.93). This will be a common theme throughout this book.

Before discussing the specific methods for constructing Poincaré maps, we need a general result giving us an estimate of how close trajectories of (1.2.93) and (1.2.94) remain during their evolution in time. More specifically, let $x_\varepsilon(t)$ and $x_0(t)$ denote solutions of (1.2.93) and (1.2.94), respectively. Suppose $x_\varepsilon(t)$ and $x_0(t)$ are "close." Then for how long do $x_\varepsilon(t)$ and $x_0(t)$ remain "close"? In order to answer this question, the following lemma will be useful.

Lemma 1.2.9 (Gronwall's Inequality) *Suppose the functions $u(s)$ and $v(s)$ are continuous and nonnegative on the interval $[t_0, t]$, and the function $c(s)$ is $\mathbf{C}^1$ and nonnegative on the interval $[t_0, t]$ with*

$$v(t) \leq c(t) + \int_{t_0}^{t} u(s)v(s)\, ds;$$

then

$$v(t) \leq c(t_0) \exp\left(\int_{t_0}^{t} u(s)\, ds\right) + \int_{t_0}^{t} \dot{c}(s)\left(\exp \int_{s}^{t} u(\tau)\, d\tau\right) ds.$$

Proof: See Guckenheimer and Holmes [1983] or Hale [1980]. □

Gronwall's inequality is the basic tool for estimating the difference between solutions of (1.2.93) and (1.2.94) on *finite* time intervals. We see this in the following proposition.

Proposition 1.2.10 *Suppose $|x_\varepsilon(t_0) - x_0(t_0)| = \mathcal{O}(\varepsilon)$; then $|x_\varepsilon(t) - x_0(t)| = \mathcal{O}(\varepsilon)$ for $|t - t_0| = \mathcal{O}(1)$.*

Proof: Subtract (1.2.94) from (1.2.93) to obtain

$$\dot{x}_\varepsilon - \dot{x}_0 = f(x_\varepsilon) - f(x_0) + \varepsilon g(x_\varepsilon, t, \varepsilon). \tag{1.2.95}$$

By integrating (1.2.95) and considering absolute values, we obtain the following estimate

$$|x_\varepsilon(t) - x_0(t)| \le |x_\varepsilon(t_0) - x_0(t_0)| + \int_{t_0}^{t} \big|f\big(x_\varepsilon(s)\big) - f\big(x_0(s)\big)\big|\, ds$$
$$+ \varepsilon \int_{t_0}^{t} \big|g\big(x_\varepsilon(s), s, \varepsilon\big)\big|\, ds. \tag{1.2.96}$$

Now, since g is $\mathbf{C}^r$ $(r \ge 1)$, there exists a constant $M \ge 0$ such that

$$|g(x, t, \varepsilon)| \le M \qquad \text{on } U \times I \times [0, \varepsilon_0), \tag{1.2.97}$$

where I is a compact interval in $\mathbb{R}^1$. By the mean value theorem, we have

$$\big|f\big(x_\varepsilon(s)\big) - f\big(x_0(s)\big)\big| \le L|x_\varepsilon(s) - x_0(s)| \tag{1.2.98}$$

for some $L \ge 0$. Substituting (1.2.97) and (1.2.98) into (1.2.96) gives

$$|x_\varepsilon(t) - x_0(t)| \le |x_\varepsilon(t_0) - x_0(t_0)| + L \int_{t_0}^{t} |x_\varepsilon(s) - x_0(s)|\, ds + \varepsilon M(t - t_0). \tag{1.2.99}$$

Now, applying Gronwall's inequality to (1.2.99) gives (with $c(t) = \varepsilon M(t - t_0) + |x_\varepsilon(t_0) - x_0(t_0)|$, $u(s) = L$, and $v(s) = |x_\varepsilon(s) - x_0(s)|$)

$$|x_\varepsilon(t) - x_0(t)| \le |x_\varepsilon(t_0) - x_0(t_0)| \exp\left(\int_{t_0}^{t} L\, ds\right)$$
$$+ \int_{t_0}^{t} \varepsilon M \left(\exp \int_{s}^{t} L\, d\tau\right) ds$$
$$\le \left(|x_\varepsilon(t_0) - x_0(t_0)| + \varepsilon \frac{M}{L}\right) \exp L(t - t_0). \tag{1.2.100}$$

Therefore. since $|x_\varepsilon(t_0) - x_0(t_0)| = \mathcal{O}(\varepsilon)$, from (1.2.100) it follows that $|x_\varepsilon(t) - x_0(t)| = \mathcal{O}(\varepsilon)$ for $0 < L(t - t_0) < N$ where N is some constant independent of ε. In other words,

$$|x_\varepsilon(t) - x_0(t)| = \mathcal{O}(\varepsilon) \qquad \text{for } t_0 \le t \le t_0 + \frac{N}{L}. \quad \square$$

Now we are ready to develop our first method for constructing Poincaré maps.

i) THE METHOD OF AVERAGING

This is a method which has appeared in the applied mathematics and engineering literature in various forms over (approximately) the last two hundred years; see Arnold [1982], Lochak and Meunier [1988], and Sanders and Verhulst [1985] for a historical survey and additional references. We will consider only a very small portion of the general theory here.

The method of averaging is concerned with equations of the following form.

$$\dot{x} = \varepsilon f(x,t) + \varepsilon^2 g(x,t,\varepsilon) \qquad \text{for } x \in \mathbb{R}^n, \tag{1.2.101}$$

where

$$\begin{aligned} &f: U \times \mathbb{R}^1 \to \mathbb{R}^n, \\ &g: U \times \mathbb{R}^1 \times [0, \varepsilon_0) \to \mathbb{R}^n \end{aligned}$$

are $\mathbf{C}^r$ $(r \geq 1)$ on their respective domains of definition with U some open set in $\mathbb{R}^n$. We will make the additional assumption that f and g are periodic in t with the same period $T > 0$ (note: time periodicity is *not* a prerequisite for this method; see Hale [1980] or Sanders and Verhulst [1985]). The associated averaged equation is given by

$$\dot{y} = \varepsilon \bar{f}(y), \qquad y \in \mathbb{R}^n, \tag{1.2.102}$$

where

$$\bar{f}(y) = \frac{1}{T} \int_0^T f(y,t)\, dt.$$

The idea behind the method is simple. Presumably (1.2.102) is easier to study than (1.2.101), so can we infer properties of the dynamics of (1.2.101) based on an understanding of the dynamics of (1.2.102)? Obviously, much more specific questions arise, such as the following.

1. Suppose (1.2.102) has a fixed point or periodic orbit. What do these special solutions correspond to in (1.2.101)?

2. Are stable and unstable manifolds of special solutions of (1.2.102) related to stable and unstable manifolds of some other special solutions of (1.2.101)?

The averaging theorem will be our starting point for relating the dynamics of (1.2.102) to the dynamics of (1.2.101) and in the process we will answer these two questions. However, before stating and proving this theorem, two important points must be addressed.

1. At the beginning of this section we stated that the systems of interest are of the form (1.2.93). Evidently (1.2.101) is not of this form. How then can (1.2.93) be transformed into the form of (1.2.101)? (Ultimately, we will answer why it is necessary to have the small parameter, ε, multiplying the vector field when employing the method of averaging.)

2. The title of this section is "Construction of the Poincaré Map." How does employing the method of averaging facilitate the construction of the Poincaré map?

We will begin by answering the first question and, in the process, we will shed some light on the second.

Transformation of (1.2.93) to the Form (1.2.101)

In (1.2.93) let us now suppose that g is periodic in t with period $T = 2\pi/\omega$. Let $x(t, x_0)$ be a *periodic* solution of the *unperturbed* equation (1.2.94) having frequency ω_0 (note: the necessity for requiring $x(t, x_0)$ to be periodic will be discussed shortly). Our goal will be to derive an ordinary differential equation which governs the time evolution of the initial condition x_0 in the solution $x(t, x_0)$. From the discussion of Poincaré maps in Section 1.2A, this procedure should immediately call to mind the spirit of the Poincaré map. We will elaborate on this as we go along.

Let us suppose that the initial condition, x_0, is a function of time so that the resulting function

$$y(t) \equiv x\big(t, x_0(t)\big) \tag{1.2.103}$$

is a solution of the perturbed equation (1.2.93). If this is true, then $x_0(t)$ must have a certain form which can be found by differentiating (1.2.103) and substituting the result into the perturbed equation (1.2.93). Carrying out this procedure gives

$$\dot{y} = \dot{x} + (D_{x_0}x)\dot{x}_0 = f\big(x(t, x_0)\big) + \varepsilon g\big(x(t, x_0), t, \varepsilon\big)$$

or

$$\dot{x}_0 = (D_{x_0}x)^{-1}\left(f\big(x(t, x_0)\big) - \dot{x} + \varepsilon g\big(x(t, x_0), t, \varepsilon\big)\right). \tag{1.2.104}$$

Now, if $x(t, x_0)$ is a solution of the unperturbed equation (1.2.94), then $\dot{x} = f\big(x(t, x_0)\big)$, so that (1.2.104) reduces to

$$\dot{x}_0 = \varepsilon(D_{x_0}x)^{-1}g\big(x(t, x_0), t, \varepsilon\big). \tag{1.2.105}$$

This vector field is now apparently of the same form as equation (1.2.101). However, a problem has arisen; namely, that (1.2.101) is periodic with period T but (1.2.105) is quasiperiodic with frequencies ω_0 and ω, since $x(t, x_0)$ has frequency ω_0 and $g(x, t, \varepsilon)$ has frequency ω in t. Since our goal is to have the time dependence of (1.2.105) periodic (so that it has the same form as (1.2.101)), we must show how this can be done. It can be done in one of two ways.

1. Restrict the application of the method to situations where the frequencies ω and ω_0 are equal. In this case (1.2.105) is now a time-periodic vector field and the method of averaging can be applied. Another way to obtain a time-periodic vector field which contains the above as a special case is as follows.

2. Suppose that $n\omega$ is "close" to $m\omega_0$ where m and n are integers. (Note: we assume that m and n are *relatively prime* in the sense that all common factors of m and n in the ratio n/m have been cancelled.) Assume that the frequency of $x(t, x_0)$ is $(n/m)\omega$. In this case $x(t, x_0)$ is no longer a solution of the unperturbed problem (1.2.94) so that (1.2.104) does not reduce to (1.2.105). In order that the method of averaging would apply, however, in some sense we would expect $f\big(x(t, x_0)\big)$ to be close $(\mathcal{O}(\varepsilon))$ to $\dot{x}$ so that (1.2.105) would be periodic in t. This admittedly sounds rather strange and unmotivated, but the following example should clear this up.

EXAMPLE 1.2.11 Consider the following vector field

$$\begin{aligned} \dot{u} &= v, \\ \dot{v} &= -\omega_0^2 u + \varepsilon h(u, v, t, \varepsilon), \end{aligned} \qquad (u, v) \in \mathbb{R}^2, \tag{1.2.106}$$

where h is $\mathbf{C}^r$ $(r \geq 1)$ and periodic in t with period $T = 2\pi/\omega$. Our goal is to transform (1.2.106) into the general form (1.2.101).

We begin by considering the unperturbed vector field

$$\begin{aligned} \dot{u} &= v, \\ \dot{v} &= -\omega_0^2 u. \end{aligned} \tag{1.2.107}$$

From (1.2.33), the solution of (1.2.107) is given by

$$\begin{aligned} u(t) &= u_0 \cos \omega_0 t + \frac{v_0}{\omega_0} \sin \omega_0 t, \\ v(t) &= \dot{u}(t) = -u_0 \omega_0 \sin \omega_0 t + v_0 \cos \omega_0 t. \end{aligned} \tag{1.2.108}$$

Now we assume that

$$n\omega = m\omega_0; \tag{1.2.109}$$

we then substitute $(n/m)\omega$ for ω_0 in (1.2.108) and go through the procedure outlined above. For this example we have

$$x(t, x_0) \equiv \big(u(t, u_0, v_0), v(t, u_0, v_0)\big), \tag{1.2.110}$$

so that

$$D_{x_0} x = \begin{pmatrix} \cos \frac{\omega}{k} t & \frac{k}{\omega} \sin \frac{\omega}{k} t \\ -\frac{\omega}{k} \sin \frac{\omega}{k} t & \cos \frac{\omega}{k} t \end{pmatrix}, \tag{1.2.111}$$

with

$$(D_{x_0} x)^{-1} = \begin{pmatrix} \cos \frac{\omega}{k} t & -\frac{k}{\omega} \sin \frac{\omega}{k} t \\ \frac{\omega}{k} \sin \frac{\omega}{k} t & \cos \frac{\omega}{k} t \end{pmatrix} \tag{1.2.112}$$

and

$$f\big(x(t, x_0)\big) = \begin{pmatrix} -u_0 \frac{\omega}{k} \sin \frac{\omega}{k} t + v_0 \cos \frac{\omega}{k} t \\ -u_0 \omega_0^2 \cos \frac{\omega}{k} t - v_0 \frac{k \omega_0^2}{\omega} \sin \frac{\omega}{k} t \end{pmatrix}, \tag{1.2.113}$$

$$g\big(x(t, x_0), t, \varepsilon\big) = \begin{pmatrix} 0 \\ h\big(u(t, u_0, v_0), v(t, u_0, v_0), t, \epsilon\big) \end{pmatrix}, \tag{1.2.114}$$

where we have set

$$k = \frac{m}{n}.$$

Substituting (1.2.110), (1.2.112), (1.2.113), and (1.2.114) into (1.2.104) gives

$$\begin{pmatrix} \dot{u}_0 \\ \dot{v}_0 \end{pmatrix} = \frac{\omega^2 - k^2\omega_0^2}{k^2} \begin{pmatrix} -\frac{k}{\omega}(u_0 \cos\frac{\omega}{k}t + \frac{kv_0}{\omega}\sin\frac{\omega}{k}t)\sin\frac{\omega}{k}t \\ (u_0\cos\frac{\omega}{k}t + \frac{kv_0}{\omega}\sin\frac{\omega}{k}t)\cos\frac{\omega}{k}t \end{pmatrix}$$
$$+\varepsilon \begin{pmatrix} -\frac{k}{\omega}h\big(u(t,u_0,v_0), v(t,u_0,t_0), t, \varepsilon\big)\sin\frac{\omega}{k}t \\ h\big(u(t,u_0,v_0), v(t,u_0,t_0), t, \varepsilon\big)\cos\frac{\omega}{k}t) \end{pmatrix}. \quad (1.2.115)$$

Now let

$$k^2\omega_0^2 - \omega^2 \equiv \varepsilon\rho, \quad (1.2.116)$$

where ρ is referred to as the "detuning" parameter. In this case the vector field (1.2.115) is periodic in t and has the form of (1.2.101). Finally, let us remark that if we let

$$v_0 \to -\frac{\omega}{k}v_0, \quad (1.2.117)$$

then (1.2.115) takes the more "symmetric" form

$$\begin{pmatrix} \dot{u}_0 \\ \dot{v}_0 \end{pmatrix} = \frac{\varepsilon\rho}{k\omega} \begin{pmatrix} (u_0\cos\frac{\omega}{k}t - v_0\sin\frac{\omega}{k}t)\sin\frac{\omega}{k}t \\ (u_0\cos\frac{\omega}{k}t - v_0\sin\frac{\omega}{k}t)\cos\frac{\omega}{k}t \end{pmatrix}$$
$$+\varepsilon \begin{pmatrix} h\big(u(t,u_0,v_0), v(t,u_0,v_0), t, \epsilon\big)\sin\frac{\omega}{k}t \\ h\big(u(t,u_0,v_0), v(t,u_0,v_0), t, \varepsilon\big)\cos\frac{\omega}{k}t \end{pmatrix}. \quad (1.2.118)$$

With the rescaled variables (1.2.117), the transformation (1.2.110) has the form of the familiar "van der Pol transformation" from the theory of nonlinear oscillations; see Sanders and Verhulst [1985].

At this stage we have said little about the interpretation of solutions of the averaged equations. We will postpone this discussion until we have given the averaging theorem. However, before this, let us see how we might transform the forced Duffing oscillator into a form suitable for application of the method of averaging.

EXAMPLE 1.2.12 Consider the equation

$$\begin{aligned} \dot{x} &= y, \\ \dot{y} &= x - x^3 - \varepsilon\delta y + \varepsilon\gamma\cos\omega t, \end{aligned} \qquad (x,y) \in \mathbb{R}^2. \quad (1.2.119)$$

We want to study (1.2.119) near a specific solution $(\hat{x}(t), \hat{y}(t))$ using the method of averaging. First we must transform (1.2.119) into the standard form for this method. Let

$$\begin{aligned} x(t) &= \hat{x}(t) + \mu u(t), \\ y(t) &= \hat{y}(t) + \mu v(t), \end{aligned} \quad (1.2.120)$$

where $\mu = \mu(\varepsilon)$ is a small parameter that requires some care in its specification; we will worry about this shortly. Substituting (1.2.120) into (1.2.119) gives

$$\begin{aligned} \dot{u} &= v, \\ \dot{v} &= -(3\hat{x}^2(t) - 1)u - 3\mu\hat{x}(t)u^2 - \varepsilon\delta v - \mu^2 u^3. \end{aligned} \tag{1.2.121}$$

Suppose we are interested in the dynamics of (1.2.119) near small amplitude resonant periodic solutions close to $(x, y) = (1, 0)$ (note: by symmetry the same results will be valid near $(x, y) = (-1, 0)$). The points $(x, y) = (\pm 1, 0)$ are center type fixed points of (1.2.119) for $\varepsilon = 0$. We must therefore solve for $(\hat{x}(t), \hat{y}(t))$ using perturbation theory. There are two cases.

Case 1: $\omega = \sqrt{2}$ – 1:1 Resonance. In this case regular perturbation methods cannot be used to approximate $(\hat{x}(t), \hat{y}(t))$ due to the presence of secular terms. Using either Lindstedt's method or two-timing (see Kevorkian and Cole [1981]) a periodic solution of frequency $\omega = \sqrt{2}$ can be computed and is found to have the form

$$\begin{aligned} \hat{x}(t) &= 1 + \mathcal{O}(\varepsilon^{1/3}), \\ \hat{y}(t) &= \mathcal{O}(\varepsilon^{1/3}). \end{aligned} \tag{1.2.122}$$

Equation (1.2.122) is then substituted into (1.2.121), which becomes

$$\begin{aligned} \dot{u} &= v, \\ \dot{v} &= -2u + \mathcal{O}(\varepsilon^{1/3}) + \mathcal{O}(\varepsilon^{2/3}) + \mathcal{O}(\mu) + \mathcal{O}(\mu\varepsilon^{1/3}) + \mathcal{O}(\varepsilon) + \mathcal{O}(\mu^2). \end{aligned} \tag{1.2.123}$$

We then define

$$\mu = \varepsilon^{1/3}$$

so that (1.2.123) becomes

$$\begin{aligned} \dot{u} &= v, \\ \dot{v} &= -2u + \mathcal{O}(\varepsilon^{1/3}). \end{aligned} \tag{1.2.124}$$

Now the van der Pol transformation from Example 1.2.11 can be used to transform (1.2.124) into the standard form for application of the method of averaging (where $\varepsilon^{1/3}$ is regarded as the small parameter. The details will be left to Exercise 1.2.18.

Case 2: $\omega = m\sqrt{2}$, $m > 1$ – 1:m Resonance. In this case regular perturbation theory can be used to approximate a periodic solution of (1.2.119) of frequency $\omega = m\sqrt{2}$, $m > 1$. The solution has the form

$$\begin{aligned} \dot{\hat{x}}(t) &= 1 + \mathcal{O}(\varepsilon), \\ \dot{\hat{y}}(t) &= \mathcal{O}(\varepsilon). \end{aligned} \tag{1.2.125}$$

Equation (1.2.125) is then substituted into (1.2.121), which becomes

$$\begin{aligned} \dot{u} &= v, \\ \dot{v} &= -2u + \mathcal{O}(\varepsilon) + \mathcal{O}(\mu) + \mathcal{O}(\mu\varepsilon) + \mathcal{O}(\mu^2). \end{aligned} \tag{1.2.126}$$

We then define

$$\mu = \varepsilon$$

so that (1.2.126) becomes

$$\begin{aligned} \dot{u} &= v, \\ \dot{v} &= -2u + \mathcal{O}(\varepsilon). \end{aligned} \tag{1.2.127}$$

The van der Pol transformation from Example 1.2.11 can now be used to transform (1.2.127) into the standard form for application of the method of averaging. We leave the details to Exercise 1.2.18. There are two points to be made in light of this example.

1. The results obtained from the method of averaging will be valid only near $(x, y) = (1, 0)$. Thus, the method of averaging gives us information concerning dynamics that are local in phase space.

2. The transformations in Example 1.2.12 imply that the dynamics of the averaged equations should give us information only near resonant solutions close to $(x, y) = (1, 0)$. Thus, the method may not detect other types of solutions that may exist near $(x, y) = (1, 0)$.

Now we will state and prove the averaging theorem.

Theorem 1.2.11 *There exists a* $\mathbf{C}^r$ *change of coordinates* $x = y + \varepsilon\omega(y, t)$ *under which (1.2.101) becomes*

$$\dot{y} = \varepsilon\bar{f}(y) + \varepsilon^2 f_1(y, t, \varepsilon), \tag{1.2.128}$$

where f_1 *is of period* T *in* t. *Moreover,*

i) *if* $x(t)$ *and* $y(t)$ *are solutions of (1.2.101) and (1.2.102), respectively, with* $x(t_0) = x_0$, $y(t_0) = y_0$, *and* $|x_0 - y_0| = \mathcal{O}(\varepsilon)$, *then* $|x(t) - y(t)| = \mathcal{O}(\varepsilon)$ *on a time scale* $\mathcal{O}(1/\varepsilon)$ *provided* $y(t) \in U$ *on a time scale* $\mathcal{O}(1/\varepsilon)$;

ii) *if* p_0 *is a hyperbolic fixed point of (1.2.102), then there exists* $\varepsilon_0 > 0$ *such that, for all* $0 < \varepsilon \leq \varepsilon_0$, *(1.2.101) possesses an isolated hyperbolic periodic orbit* $\gamma_\varepsilon(t) = p_0 + \mathcal{O}(\varepsilon)$ *of the same stability type as* p_0;

iii) *if* $x^s(t) \in W^s(\gamma_\varepsilon)$ *is a solution of (1.2.101) lying in the stable manifold of the hyperbolic periodic orbit* $\gamma_\varepsilon(t) = p_0 + \mathcal{O}(\varepsilon)$, $y^s(t) \in W^s(p_0)$ *is a solution of (1.2.102) lying in the stable manifold of the hyperbolic fixed point* p_0, *and if* $|x(0) - y(0)| = \mathcal{O}(\varepsilon)$, *then* $|x^s(t) - y^s(t)| = \mathcal{O}(\varepsilon)$ *for* $t \in [0, \infty)$. *A similar statement holds for solutions lying in the unstable manifold on the time interval* $(-\infty, 0]$.

Proof: We begin by constructing the change of coordinates which transforms (1.2.101) into (1.2.128). The effect of the coordinate change is to eliminate the explicit time dependence of $\mathcal{O}(\varepsilon)$ by "moving it up" to $\mathcal{O}(\varepsilon^2)$.

We first decompose $f(x,t)$ in (1.2.101) into its oscillating and mean parts as follows

$$f(x,t) = \bar{f}(x) + \tilde{f}(x,t), \tag{1.2.129}$$

where

$$\bar{f}(x) = \frac{1}{T}\int_0^T f(x,t)\,dt, \qquad \tilde{f}(x,t) = f(x,t) - \frac{1}{T}\int_0^T f(x,t)\,dt.$$

Using (1.2.129), we rewrite (1.2.101) as follows

$$\dot{x} = \varepsilon\bar{f}(x) + \varepsilon\tilde{f}(x,t) + \varepsilon^2 g(x,t,\varepsilon). \tag{1.2.130}$$

Now we make the change of coordinates

$$x = y + \varepsilon w(y,t),$$

where w will be defined shortly, to obtain

$$\begin{aligned}\dot{x} = \dot{y} + \varepsilon D_y w\dot{y} + \varepsilon\frac{\partial w}{\partial t} &= \varepsilon\bar{f}(y+\varepsilon w) + \varepsilon\tilde{f}(y+\varepsilon w,t)\\ &+\varepsilon^2 g(y+\varepsilon w,t,\varepsilon).\end{aligned} \tag{1.2.131}$$

Next, we expand the right-hand side of (1.2.131) in powers of ε

$$\begin{aligned}\dot{y} + \varepsilon D_y w\dot{y} + \varepsilon\frac{\partial w}{\partial t} = \varepsilon&\left(\bar{f}(y) + \tilde{f}(y,t)\right)\\ &+\varepsilon^2\left(D_y\bar{f}(y)w + D_y\tilde{f}(y,t)w + g(y,t,0)\right)\\ &+\mathcal{O}(\varepsilon^3).\end{aligned} \tag{1.2.132}$$

We remark that it is possible to compute the $\mathcal{O}(\varepsilon^3)$ term explicitly in (1.2.132) if needed. We rewrite (1.2.132) as

$$\begin{aligned}(I + \varepsilon D_y w)\dot{y} = \varepsilon&\left(\bar{f}(y) + \tilde{f}(y,t) - \frac{\partial w}{\partial t}\right)\\ &+\varepsilon^2\left(D_y\bar{f}(y)\omega + D_y\tilde{f}(y,t)w + g(y,t,0)\right) + \mathcal{O}(\varepsilon^3)\end{aligned}$$

or

$$\begin{aligned}\dot{y} = (I + \varepsilon D_y w)^{-1}&\left\{\varepsilon\left(\bar{f}(y) + \tilde{f}(y,t) - \frac{\partial w}{\partial t}\right)\right.\\ &\left.+\,\varepsilon^2\left(D_y\bar{f}(y)\omega + D_y\tilde{f}(y,t)w + g(y,t,0)\right) + \mathcal{O}(\varepsilon^3)\right\},\end{aligned} \tag{1.2.133}$$

where "I" denotes the $n \times n$ identity matrix. Now, for ε small, $(I + \varepsilon D_y w)^{-1} = I - \varepsilon D_y w + \mathcal{O}(\varepsilon^2)$, so (1.2.133) becomes

$$\begin{aligned}\dot{y} = \varepsilon&\left(\bar{f}(y) + \tilde{f}(y,t) - \frac{\partial w}{\partial t}\right)\\ &+\varepsilon^2\left(D_y\bar{f}(y)w + D_y\tilde{f}(y,t)w + g(y,t,0)\right.\\ &\left.-\,D_y w\bar{f}(y) - D_y w\tilde{f}(y,t) + D_y w\frac{\partial w}{\partial t}\right) + \mathcal{O}(\varepsilon^3).\end{aligned} \tag{1.2.134}$$

We then define

$$f_1(y,t,\varepsilon) \equiv D_y\bar{f}(y)w + D_y\tilde{f}(y,t)w + g(y,t,0)$$
$$-D_yW\bar{f}(y) - D_yw\tilde{f}(y,t) + D_yw\frac{\partial w}{\partial t} + \mathcal{O}(\varepsilon)$$

so that (1.2.134) takes the form

$$\dot{y} = \varepsilon\left(\bar{f}(y) + \tilde{f}(y,t) - \frac{\partial w}{\partial t}\right) + \varepsilon^2 f_1(y,t,\varepsilon). \tag{1.2.135}$$

Next, we choose w so that

$$\frac{\partial w}{\partial t} = \tilde{f}(y,t); \tag{1.2.136}$$

then (1.2.135) becomes

$$\dot{y} = \varepsilon\bar{f}(y) + \varepsilon^2 f_1(y,t,\varepsilon), \tag{1.2.137}$$

where, tracing through the above steps, it should be clear that f_1 is T-periodic in t.

Now we prove i).

i) We begin by comparing solutions of (1.2.102) and (1.2.128), which we rewrite below as

$$\begin{aligned} \dot{y} &= \varepsilon\bar{f}(y) + \varepsilon^2 f_1(y,t,\varepsilon), \qquad & y(t_0) &= y_0,\\ \dot{x} &= \varepsilon\bar{f}(x), & x(t_0) &= x_0. \end{aligned}$$

We integrate and subtract (1.2.128) and (1.2.102) to obtain

$$y(t) - x(t) = y_0 - x_0 + \varepsilon\int_{t_0}^{t}\left(\bar{f}(y(s)) - \bar{f}(x(s))\right)ds + \varepsilon^2\int_{t_0}^{t} f_1(y(s),s,\varepsilon)\,ds,$$

from which follows

$$|y(t) - x(t)| \le |y_0 - x_0| + \varepsilon\int_{t_0}^{t}\left|\bar{f}(y(s)) - \bar{f}(x(s))\right|ds$$
$$+\,\varepsilon^2\int_{t_0}^{t}|f_1(y(s),s,\varepsilon)|\,ds. \tag{1.2.138}$$

Using the fact that the vector field is $\mathbf{C}^r$ ($r \ge 1$), on U we have

$$|\bar{f}(y) - \bar{f}(x)| \le L|y - x| \tag{1.2.139}$$

for some constant $L > 0$, and on $U \times \mathbb{R} \times (0,\varepsilon_0]$ we have

$$|f_1(y,t,\varepsilon)| \le M \tag{1.2.140}$$

for some constant $M > 0$ (note: f_1 is periodic in t). Now let $v(t) \equiv |x(t) - y(t)|$. Then, using (1.2.139) and (1.2.140), (1.2.138) becomes

$$v(t) \leq v(t_0) + \varepsilon^2 M(t - t_0) + \varepsilon L \int_{t_0}^{t} v(s)\, ds. \tag{1.2.141}$$

Next, we apply Gronwall's inequality to (1.2.141) to obtain

$$\begin{aligned} v(t) &\leq v(t_0) \exp \int_{t_0}^{t} \varepsilon L\, ds + \int_{t_0}^{t} \varepsilon^2 M \left(\exp \int_{s}^{t} \varepsilon L\, d\tau \right) ds \\ &= v(t_0) \exp \varepsilon L(t - t_0) + \varepsilon^2 M \exp(\varepsilon L t) \int_{t_0}^{t} \exp(-\varepsilon L s)\, ds \\ &= v(t_0) \exp \varepsilon L(t - t_0) \\ &\qquad + \varepsilon^2 M \exp(\varepsilon L t) \left(\frac{1}{\varepsilon L} (\exp(-\varepsilon L t_0) - \exp(-\varepsilon L t)) \right) \\ &\leq v(t_0) \exp \varepsilon L(t - t_0) + \frac{\varepsilon M}{L} \exp \varepsilon L(t - t_0) \end{aligned} \tag{1.2.142}$$

or

$$|x(t) - y(t)| \leq \left(|x_0 - y_0| + \frac{\varepsilon M}{L} \right) \exp \varepsilon L(t - t_0).$$

Therefore, for $|x_0 - y_0| = \mathcal{O}(\varepsilon)$, $|x(t) - y(t)| = \mathcal{O}(\varepsilon)$ on the time scale $0 \leq \varepsilon L(t - t_0) \leq N$, or equivalently, for $t_0 \leq t \leq t_0 + N/(\varepsilon L)$, where N is some constant independent of ε. Now we must relate this result to (1.2.101), which we rewrite below as

$$\dot{x} = \varepsilon f(x, t) + \varepsilon^2 g(x, t, \varepsilon).$$

Solutions of (1.2.101) can be written as

$$x_\varepsilon(t) = y(t) + \varepsilon w(y, t) \qquad \text{on } t_0 \leq t \leq t_0 + \frac{N}{\varepsilon L},$$

and thus,

$$\begin{aligned} |x_\varepsilon(t) - x(t)| &= |y(t) + \varepsilon w(y, t) - x(t)| \\ &\leq |y(t) - x(t)| + \varepsilon |w(y, t)| = \mathcal{O}(\varepsilon) \end{aligned}$$

on $t_0 \leq t \leq t_0 + N/(\varepsilon L)$, where the bound of $|w(y, t)|$ uses the fact that $y(t) \in U$ on a time scale $\mathcal{O}(1/\varepsilon)$. This completes the proof of i).

ii) We will explicitly construct a Poincaré map using regular perturbation theory.

Using the fact that solutions of (1.2.128) depend on ε in a $\mathbf{C}^r$ manner, we may Taylor expand the solution of (1.2.128) to obtain

$$y(t, \varepsilon) = y(t, 0) + \varepsilon y_1(t) + \varepsilon^2 y_2(t) + \mathcal{O}(\varepsilon^3), \tag{1.2.143}$$

where $y(t,0)$ is the solution of

$$\dot{y}(t,0) = 0, \tag{1.2.144}$$

$y_1(t)$ is the solution of

$$\dot{y}_1 = \bar{f}\big(y(t,0)\big), \tag{1.2.145}$$

and $y_2(t)$ is the solution of

$$\dot{y}_2 = D_y\bar{f}\big(y(t,0)\big) + f_1\big(y(t,0),t,0\big). \tag{1.2.146}$$

Equations (1.2.145) and (1.2.146) are referred to as the first and second variational equations, respectively, and are obtained by differentiating (1.2.128) with respect to ε; see Hale [1980] for more details.

We construct a standard (see Section 1.2A, Case 2) time $2\pi/\omega$ Poincaré map as follows

$$y(0,\varepsilon) \mapsto y(2\pi/\omega,\varepsilon). \tag{1.2.147}$$

Using (1.2.143) and choosing initial conditions such that

$$y(0,\varepsilon) = y(0,0) = y_0 \tag{1.2.148}$$

by replacing $y_i(0) = 0$ for $i \geq 1$, (1.2.147) becomes

$$y_0 \mapsto y_0 + \varepsilon y_1\left(\frac{2\pi}{\omega}\right) + \varepsilon^2 y_2\left(\frac{2\pi}{\omega}\right) + \mathcal{O}(\varepsilon^3). \tag{1.2.149}$$

We will also consider the map obtained by truncating (1.2.149) at $\mathcal{O}(\varepsilon^2)$

$$y_0 \mapsto y_0 + \varepsilon y_1\left(\frac{2\pi}{\omega}\right). \tag{1.2.150}$$

Now, from (1.2.145) and (1.2.148) we obtain

$$y_1\left(\frac{2\pi}{\omega}\right) = \frac{2\pi}{\omega}\bar{f}(y_0) \tag{1.2.151}$$

so that (1.2.150) becomes

$$y_0 \mapsto y_0 + \varepsilon\frac{2\pi}{\omega}\bar{f}(y_0). \tag{1.2.152}$$

From (1.2.152) we see that a fixed point of the averaged equations corresponds to a fixed point of (1.2.152). Also, if $D\bar{f}(y_0)$ is hyperbolic, then it follows (see Kato [1980]) that, for ε sufficiently small,

$$\mathrm{id} + \varepsilon\frac{2\pi}{\omega}D\bar{f}(y_0) \tag{1.2.153}$$

is likewise hyperbolic and, moreover, (1.2.153) will have the same number of eigenvalues inside (resp. outside) the unit circle as $D\bar{f}(y_0)$ has eigenvalues in the left (resp. right) half plane; see Exercise 1.2.24. These facts will be useful shortly.

Now let us consider the full Poincaré map (1.2.149). The condition for (1.2.149) to have a fixed point is

$$y_0 = y_0 + \varepsilon \frac{2\pi}{\omega} \bar{f}(y_0) + \mathcal{O}(\varepsilon^2)$$

or

$$\left(\frac{2\pi}{\omega} \bar{f}(y_0) + \mathcal{O}(\varepsilon) \right) \equiv g(y_0, \varepsilon) = 0. \tag{1.2.154}$$

We are now in a position to complete the proof. Suppose the averaged equation (1.2.102) has a hyperbolic fixed point at $y = \bar{y}_0$. This then corresponds to a hyperbolic fixed point of the $\mathcal{O}(\varepsilon^2)$ truncated Poincaré map (1.2.152) having the same stability type. We want to show that these conditions are sufficient for this fixed point to persist in the full Poincaré map (1.2.149) without change of stability and, hence, for the hyperbolic fixed point of the averaged equations to correspond to a hyperbolic periodic orbit of the full equations having the same stability type. We do this by applying the implicit function theorem to (1.2.154). We know

$$g(\bar{y}_0, 0) = 0,$$

and this matrix

$$D_{y_0} g(\bar{y}_0, 0) = \frac{2\pi}{\omega} D\bar{f}(\bar{y}_0) \tag{1.2.155}$$

is hyperbolic. Hence, for ε sufficiently small, there exists a $\mathbf{C}^r$ function of ε, $y_0(\varepsilon)$ such that

$$g\big(y_0(\varepsilon), \varepsilon\big) = 0.$$

iii) This result follows from the persistence theory for normally hyperbolic invariant manifolds and is beyond the scope of this book. See Fenichel [1971], [1974], [1977], [1979]; Hirsch, Pugh, and Shub [1978]; Schecter [1988]; and Murdock and Robinson [1980]. □

Interpretation of the Dynamics of the Averaged Equations

We now want to discuss the dynamics of the averaged equations and the relationship with the fully time-dependent equations. We will do this in the context of Example 1.2.11.

Recall that in Example 1.2.11 the following equation was considered.

$$\begin{aligned} \dot{u} &= v, \\ \dot{v} &= -\omega_0^2 u + \varepsilon h(u, v, t, \varepsilon). \end{aligned} \tag{1.2.156}$$

The solution to the unperturbed equation is given by $\big(u(t, u_0, v_0), v(t, u_0, v_0)\big)$ and is periodic in t with frequency ω_0. We used this solution to transform (1.2.156) into a form where the method of averaging could be applied. This was accomplished in two steps.

Step 1. Assume $n\omega$ is "close" to $m\omega_0$ and substitute $(n/m)\omega$ for ω_0 in the frequency of $\big(u(t,u_0,v_0), v(t,u_0,v_0)\big)$.

Step 2. Assume the initial conditions (u_0, v_0) are functions of time and use the "almost" solution of the unperturbed equation to derive a vector field for $\big(u_0(t), v_0(t)\big)$ such that $(u(t,u_0(t),v_0(t)),\ v(t,u_0(t),v_0(t)))$ is a solution of (1.2.156).

This vector field for $\big(u_0(t), v_0(t)\big)$ is in the standard form for the method of averaging and is subsequently averaged over the period $T = 2\pi/\omega$. From the averaging theorem we know that hyperbolic fixed points of the averaged equations correspond to periodic orbits of period $2\pi/\omega$.

Let us suppose we found a hyperbolic fixed point of the averaged equations. Then from (1.2.108) and the averaging theorem, we have

$$\begin{aligned} u(t) &= u_0(t)\cos\frac{n}{m}\omega t + v_0(t)\frac{m}{n\omega}\sin\frac{n}{m}\omega t,\\ v(t) &= -u_0(t)\frac{n}{m}\omega\sin\frac{n}{m}\omega t + v_0(t)\cos\frac{n}{m}\omega t, \end{aligned} \tag{1.2.157}$$

where $u_0(t)$ and $v_0(t)$ are periodic with period $2\pi/\omega$. Because (1.2.157) is periodic in t with period $2\pi m/\omega$, it is natural to think of (1.2.157) in the context of a Poincaré map where initial conditions are mapped under the flow generated by (1.2.156) to their image under the flow at time $2\pi/\omega$ (see Section 1.2A). We consider three cases.

1. $n = 1$. In this case the solution (1.2.157) pierces the Poincaré cross-section m times before returning to its starting point. Thus, the fixed point of the averaged equation corresponds to a period m point of the Poincaré map, or equivalently, a subharmonic of order m.

2. $m = 1$. In this case the solution (1.2.157) returns to its starting point on the Poincaré maps. Thus, the fixed point of the averaged equation corresponds to a fixed point of the Poincaré map or, equivalently, an ultraharmonic of order n.

3. $n > 1$, $m > 1$. Following the above arguments, the reader should be able to verify that, in this case, hyperbolic fixed points of the averaged equations correspond to period m points of the Poincaré map or, equivalently, ultrasubharmonics of order m, n.

ii) The Subharmonic Melnikov Theory

We will now develop a method for constructing Poincaré maps which is similar in spirit to averaging yet much more geometrical. However, it will be necessary for us to restrict (1.2.93) somewhat. Specifically, we will be considering two-dimensional systems that are periodic in t, which we write in component form as follows

$$\begin{aligned} \dot{x} &= f_1(x,y) + \varepsilon g_1(x,y,t,\varepsilon),\\ \dot{y} &= f_2(x,y) + \varepsilon g_2(x,y,t,\varepsilon), \end{aligned} \qquad (x,y)\in\mathbb{R}^2. \tag{1.2.158}$$

When it will prove notationally convenient we may write (1.2.158) in the following vector form

$$\dot{q} = f(q) + \varepsilon g(q, t, \varepsilon), \tag{1.2.159}$$

where $q \equiv (x, y)$, $f \equiv (f_1, f_2)$, and $g \equiv (g_1, g_2)$. We have the same differentiability assumptions on f and g as stated for (1.2.93). However, now we will assume that g is periodic in t with period $T = 2\pi/\omega$. Our system may depend on parameters; however, for now we will not consider this possibility.

Our strategy will be to use the global geometric structure as a framework or skeleton on which to construct our analytical techniques for studying the perturbed system. In order to do this we will have to introduce some assumptions regarding the geometrical structure of the unperturbed system. Our first step in doing this will be to assume that the unperturbed vector field is *Hamiltonian.* By this we mean that there is a $\mathbf{C}^{r+1}$ scalar valued function, $H(x, y)$, such that

$$\begin{aligned} f_1(x, y) &= \frac{\partial H(x, y)}{\partial y}, \\ f_2(x, y) &= -\frac{\partial H(x, y)}{\partial x}. \end{aligned} \tag{1.2.160}$$

(Note: we are assuming that f and H have the same domains.) Thus (1.2.158) takes the form

$$\begin{aligned} \dot{x} &= \frac{\partial H(x, y)}{\partial y} + \varepsilon g_1(x, y, t, \varepsilon), \\ \dot{y} &= -\frac{\partial H(x, y)}{\partial x} + \varepsilon g_2(x, y, t, \varepsilon), \end{aligned} \tag{1.2.161}$$

or, in vector form

$$\dot{q} = JDH(q) + \varepsilon g(q, t, \varepsilon), \tag{1.2.162}$$

where

$$DH \equiv \begin{pmatrix} \frac{\partial H}{\partial x} \\ \\ \frac{\partial H}{\partial y} \end{pmatrix}$$

and

$$J \equiv \begin{pmatrix} 0 & 1 \\ -1 & 0 \end{pmatrix}.$$

We remark that the perturbation (i.e., g) need not be Hamiltonian although, as we will see later, in the case where the perturbation is Hamiltonian versus the case where it is not, the dynamics are very different.

Up to this point everything has been very general; now we want to make the following assumptions on the structure of the unperturbed system.

Assumption 1. The unperturbed system possesses a hyperbolic fixed point, p_0, connected to itself by a homoclinic orbit $q_0(t) \equiv \big(x^0(t), y^0(t)\big)$.

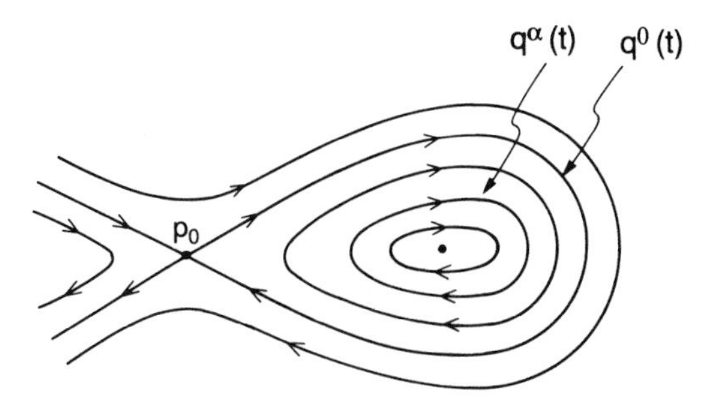

FIGURE 1.2.27.

Assumption 2. Let $\Gamma_{p_0} = \{ q \in \mathbb{R}^2 \mid q = q_0(t), t \in \mathbb{R} \} \cup \{p_0\} = W^s(p_0) \cap W^u(p_0) \cup \{p_0\}$. The interior of Γ_{p_0} is filled with a continuous family of periodic orbits $q^\alpha(t)$, $\alpha \in (-1, 0)$, with period T^α. We assume that $\lim_{\alpha \to 0} q^\alpha(t) = q_0(t)$ and $\lim_{\alpha \to 0} T^\alpha = \infty$.

See Figure 1.2.27 for an illustration of the geometry of the unperturbed phase space. At this point we wish to make the following remarks.

Remark 1. α is simply a parameter that indexes the periodic orbits inside Γ_{p_0}. In specific problems α may have the interpretation of energy, action, elliptic modulus, etc. The fact that it lies in the interval $(-1, 0)$ is not important (as we shall see); the interval could be arbitrary. We chose this particular interval for notational convenience only.

Remark 2. If more than one hyperbolic fixed point with associated homoclinic orbit and family of periodic orbits appears in a specific equation, then the method can be applied to each separately.

For now, we will concentrate on the dynamics of the one-parameter family of periodic orbits under perturbation. In Chapter 4 we will develop techniques for studying the behavior of Γ_{p_0} under perturbation and learn that it contains many delightfully complicated surprises.

The following lemma will be very useful.

Lemma 1.2.12 *Let $q^\alpha(t - t_0)$ be a periodic orbit of the unperturbed system with period T^α. Then there exists a perturbed orbit, not necessarily periodic, which can be expressed as*

$$q_\varepsilon^\alpha(t, t_0) = q^\alpha(t - t_0) + \varepsilon q_1^\alpha(t - t_0) + \mathcal{O}(\varepsilon^2),$$

uniformly in $t \in [t_0, t_0 + T^\alpha]$ for ϵ sufficiently small and all $\alpha \in (-1, 0)$.

Proof: Suppose we choose any $\alpha = \hat{\alpha} < 0$ and restrict ourselves to consideration of $q^\alpha(t - t_0)$, $\alpha \in (-1, \hat{\alpha}]$. In this case the period, T^α, is bounded from above and the approximation problem is only on a finite time interval. In this case Proposition 1.2.10 immediately applies.

The aspect that makes this lemma significant is the fact that $\lim_{\alpha\to 0} T^\alpha = \infty$. In this case crude estimates such as those used in Proposition 1.2.10 do not work. The way to salvage this situation is to use the geometrical structure associated with the stable and unstable manifolds of the hyperbolic fixed point. This is discussed in depth in Guckenheimer and Holmes [1983], and we refer the reader there for the details. □

Let us make the following remarks which highlight some of the important implications of Lemma 1.2.12.

Remark 1. As mentioned above, this lemma is nontrivial. It says that we can approximate *uniformly* (meaning one ε is sufficient for all the q^α, $\alpha \in (-1,0)$) perturbed orbits by unperturbed orbits as their periods go to ∞ (i.e., as they limit on Γ_{p_0}). We are able to do this because of the control we have on the geometric structure of the stable and unstable manifolds of the hyperbolic fixed point. We will see this more explicitly in Chapter 4 when we study the global behavior of Γ_{p_0} under perturbation.

Remark 2. $q_1^\alpha(t,t_0)$ is a solution of the linear *first variational equation*, i.e., we have

$$\dot{q}_1^\alpha = JD^2H\big(q^\alpha(t-t_0)\big)q_1^\alpha + g\big(q^\alpha(t-t_0),t,0\big) \tag{1.2.163}$$

for $t \in [t_0, t_0 + T^\alpha]$ (cf. Exercise 1.2.11).

Remark 3. We are only able to approximate perturbed orbits by unperturbed orbits uniformly as they limit on Γ_{p_0} for *one* unperturbed period (even though this period goes to ∞). Geometrically, this means that approximate perturbed orbits are only allowed to pass once through a neighborhood of the hyperbolic fixed point, which, as we will see in Section 4.5, becomes a hyperbolic periodic orbit for the perturbed vector field. This is because as orbits get closer and closer to Γ_{p_0} they spend longer and longer near the hyperbolic periodic orbit and, consequently, any small error may be magnified by an arbitrarily large amount. We can control this error for one passage through a neighborhood of the hyperbolic periodic orbit simply by choosing the initial conditions correctly; however, the second time through we have no control. If we stay bounded away from Γ_{p_0} so that α remains bounded, then we can approximate perturbed orbits by unperturbed orbits for nT^α, $n > 1$, integer, although in this case $\varepsilon = \varepsilon(n)$ and $\varepsilon(n) \to 0$ as $n \uparrow \infty$, i.e., we lose uniformity.

Remark 4. Remark 3 is of interest because we will be interested in *resonant* periodic orbits, i.e., orbits whose periods are related to the period of the external perturbation by a relation of the following form:

$$nT^\alpha = mT,$$

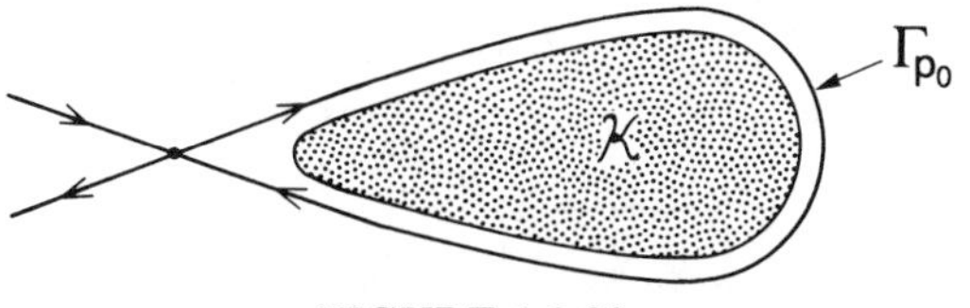

FIGURE 1.2.28.

where $T = 2\pi/\omega$ is the period of the perturbation and m and n are relatively prime integers.

We now begin the analysis. For the moment we restrict ourselves to a region inside and bounded away from Γ_{p_0}; call it $\mathcal{K}$; see Figure 1.2.28. In $\mathcal{K}$, the periods of the unperturbed orbits are uniformly bounded above, say by a constant C. It is well known that in such a situation a Hamiltonian system can be transformed to a new coordinate system, the so-called action-angle coordinate system (see, e.g., Arnold [1978], Goldstein [1980], or Percival and Richards [1982]). This coordinate change is represented by the functions

$$I = I(x, y), \qquad \theta = \theta(x, y),$$

with inverse

$$x = x(I, \theta), \quad y = y(I, \theta).$$

In such a coordinate system the unperturbed vector field takes the form

$$\begin{aligned} \dot{I} &= 0, \\ \dot{\theta} &= \Omega(I); \end{aligned}$$

hence, I is a constant on unperturbed orbits and the Hamiltonian takes the form

$$H = H(I)$$

and also, $\Omega(I) = \frac{\partial H}{\partial I}$, where now in the action-angle coordinate system I plays the role of the more general parameter α introduced earlier; i.e., specifying I specifies a periodic orbit. If we transform the coordinates of the *perturbed* vector field using the action-angle transformation for the unperturbed Hamiltonian vector field, we obtain

$$\begin{aligned} \dot{I} &= \varepsilon \left(\frac{\partial I}{\partial x} g_1 + \frac{\partial I}{\partial y} g_2 \right) \equiv \varepsilon F(I, \theta, t, \varepsilon), \\ \dot{\theta} &= \Omega(I) + \varepsilon \left(\frac{\partial \theta}{\partial x} g_1 + \frac{\partial \theta}{\partial y} g_2 \right) \equiv \Omega(I) + \varepsilon G(I, \theta, t, \varepsilon). \end{aligned}$$

(Note: an obvious question is that if we are given a system in Cartesian coordinates, how do we transform to action-angle variables? We will see that it is often unnecessary to do this. The action-angle variables are merely a convenience that make the geometric interpretation clearer and, in the

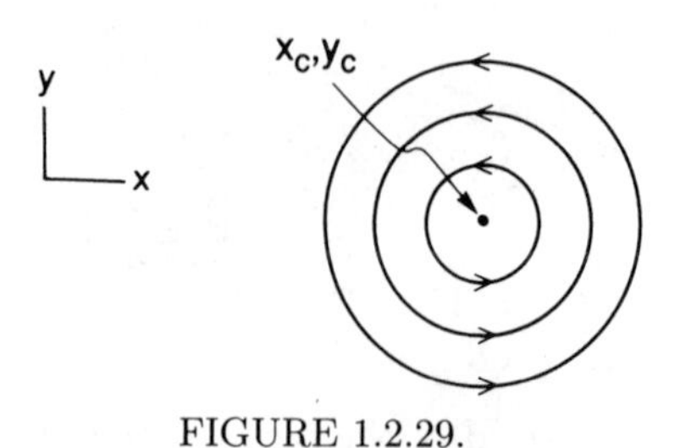

FIGURE 1.2.29.

end, computations can be made without explicitly transforming the vector field into the action-angle coordinate system.)

At this stage the introduction of action-angle variables probably appears to be somewhat unmotivated. However, we would like to argue that they are the most geometrically natural and revealing coordinates for our system. In order to do this we must make a slight digression and give a derivation of action-angle variables for our system that is sensitive to the underlying geometry.

Digression: Action-Angle Variables

The "traditional" development of the idea of action-angle variables utilizes the familiar idea of *generating functions* from classical mechanics (see, e.g., Goldstein [1980] or Landau and Lifshitz [1976]). For an excellent derivation of action-angle variables from this point of view we refer the reader to Percival and Richards [1982]. We will take a somewhat different approach (loosely inspired by Melnikov [1963]) which more directly utilizes the underlying geometry of the vector field.

We consider a $\mathbf{C}^r$ $(r \geq 1)$ Hamiltonian vector field on the plane given as follows

$$\begin{aligned} \dot{x} &= \frac{\partial H}{\partial y}(x, y), \\ \dot{y} &= -\frac{\partial H}{\partial x}(x, y), \end{aligned} \qquad (x, y) \in \mathbb{R}^2, \tag{1.2.164}$$

with the following structural assumption.

Assumption 1. In some open set in $\mathbb{R}^2$ there exists a fixed point of center type, (x_c, y_c), surrounded by a one-parameter family of periodic orbits, i.e., $H(x, y) = H =$ constant consists of closed, non-self-intersecting curves surrounding (x_c, y_c) in this open set; see Figure 1.2.29.

Our goal is to find a new set of coordinates in which the differential equation has the simplest possible structure, i.e., you can simply look at it and write down the solution. Let us consider the following motivational example.

EXAMPLE 1.2.13 Consider the simple harmonic oscillator

$$\begin{aligned} \dot{x} &= y, \\ \dot{y} &= -x. \end{aligned} \tag{1.2.165}$$

It is easy to verify that (1.2.165) has a fixed point at $(x, y) = (0, 0)$ having purely imaginary eigenvalues (i.e., it is a center) and is surrounded by a one-parameter family of periodic orbits given by $H = y^2/2 + x^2/2$. If we transform (1.2.165) into polar coordinates using

$$\begin{aligned} x &= r \sin\theta, \\ y &= r \cos\theta, \end{aligned}$$

then the vector field becomes

$$\begin{aligned} \dot{r} &= 0, \\ \dot{\theta} &= 1, \end{aligned}$$

which has the obvious solution

$$\begin{aligned} r &= \text{constant}, \\ \theta &= t + \theta_0. \end{aligned}$$

For this example polar coordinates work very nicely because (1.2.165) is linear and, therefore, all of the periodic orbits have the same period.

For the general nonlinear vector field (1.2.164) we will seek a coordinate transformation that has the same effect. Namely, we will seek a coordinate transformation

$$(x, y) \mapsto \big(\theta(x, y), I(x, y)\big)$$

with inverse

$$(\theta, I) \mapsto \big(x(I, \theta), y(I, \theta)\big)$$

such that the vector field (1.2.164) in the (θ, I) coordinate satisfies the following conditions:

1. $\dot{I} = 0$;
2. θ changes linearly in time on the closed orbits.

We might then think of θ and I heuristically as "nonlinear polar coordinates."

In the construction of the action-angle variables the following steps will be carried out.

1. *Definition of the Transformation.* This will rely exclusively on the geometry of the phase space of (1.2.164).
2. *Write the Vector Field in the New Coordinates.* In this step we will show that in the (θ, I) coordinates (1.2.164) takes the form

$$\begin{aligned} \dot{I} &= 0, \\ \dot{\theta} &= \Omega(I). \end{aligned} \tag{1.2.166}$$

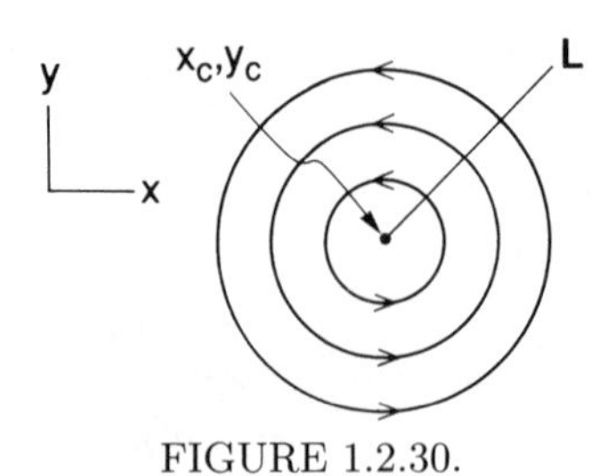

FIGURE 1.2.30.

3. *Show that the Transformation Preserves the Hamiltonian Structure.* This step implies the following. Note that (1.2.166) is Hamiltonian for some function $K(I)$, where $\Omega(I) = \frac{\partial K}{\partial I}$. If the transformation preserves the Hamiltonian structure, then we must have

$$K(I) = H\big(x(I,\theta), y(I,\theta)\big).$$

We begin with Step 1.

Step 1. We first define $\theta(x,y)$. Let L denote a curve which emanates from (x_c, y_c) and intersects each periodic orbit only once; see Figure 1.2.30. We denote L in parametric form as follows

$$L = \left\{ \big(x_0(s), y_0(s)\big) \in \mathbb{R}^2 \mid s \in \text{some interval in } \mathbb{R} \right\}.$$

We denote solutions of (1.2.164) starting on L by $\big(x(t,s), y(t,s)\big)$ where $x(0,s) = x_0(s)$ and $y(0,s) = y_0(s)$. Let (x,y) be a point on the orbit of $\big(x(t,s), y(t,s)\big)$ and let $t = t(x,y)$ be the time taken for the solution starting at $(\big(x_0(s), y_0(s)\big)$ to reach (x,y).

(Note: using the implicit function theorem it can be shown that t has the same degree of differentiability in x and y as the vector field.) We denote the period of each periodic orbit defined by $H(x,y) = H = \text{constant}$ by $T(H)$. We now define the angle variable, $\theta(x,y)$, by

$$\theta(x,y) = \frac{2\pi}{T(H)} t(x,y), \tag{1.2.167}$$

where $(x,y) \in H = \text{constant}$; see Figure 1.2.31. Note that $\theta(x,y)$ is multi-valued, i.e., there is a multiple of 2π ambiguity in its definition. However, $\frac{\partial\theta}{\partial x}$ and $\frac{\partial\theta}{\partial y}$ are single-valued functions.

Next we define the action variable $I(x,y)$. The area enclosed by any closed curve is constant in time; this area is called the *action* and is defined as

$$I = \frac{1}{2\pi} \oint_H y\, dx, \tag{1.2.168}$$

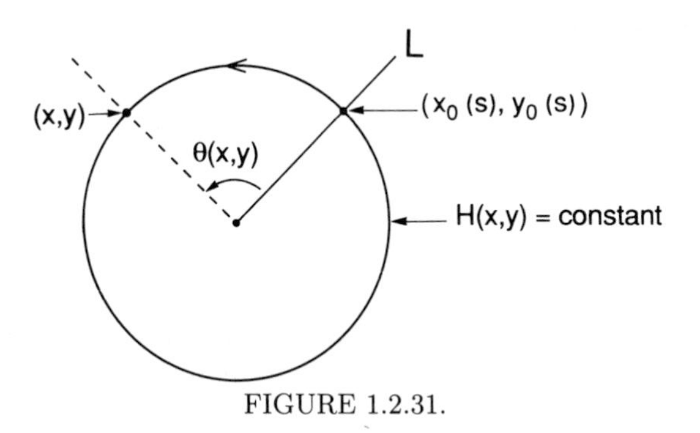

FIGURE 1.2.31.

where H denotes the periodic orbit defined by $H(x, y) = H =$ constant. The normalization factor of $1/(2\pi)$ is traditional for the following reason; differentiating (1.2.167) with respect to H gives

$$\frac{\partial I}{\partial H} = \frac{1}{2\pi} \int_H \frac{\partial y}{\partial H} \, dx, \tag{1.2.169}$$

and, from (1.2.164) we obtain

$$\frac{\partial H}{\partial y} = \frac{dx}{dt}.$$

Using the chain rule, it is easy to see that

$$\frac{\partial y}{\partial H} \, dx = dt, \tag{1.2.170}$$

so that (1.2.169) becomes

$$\frac{\partial I}{\partial H} = \frac{1}{2\pi} T(H). \tag{1.2.171}$$

Thus, $\frac{\partial H}{\partial I}$ can be interpreted as the frequency of oscillation. Also, from (1.2.168) it should be clear that we have

$$I = I(H). \tag{1.2.172}$$

Now, using (1.2.171), we can invert (1.2.172) to obtain

$$H = H(I), \tag{1.2.173}$$

which can be substituted into (1.2.171) to obtain

$$T = T(I) \tag{1.2.174}$$

so that

$$\frac{\partial H}{\partial I} = \frac{2\pi}{T(I)} \equiv \Omega(I). \tag{1.2.175}$$

Step 2. By definition (1.2.168) it should be evident that

$$\dot{I} = 0,$$

and by using (1.2.167) and (1.2.175), it follows that

$$\dot{\theta} = \frac{2\pi}{T(I)} = \Omega(I).$$

Before proceeding to Step 3, we want to derive two relationships amongst the partial derivatives of the transformation functions which will prove useful later on. Differentiating (1.2.168) with respect to t along orbits gives

$$\dot{I} = \frac{\partial I}{\partial x}\dot{x} + \frac{\partial I}{\partial y}\dot{y} = 0, \tag{1.2.176}$$

and substituting (1.2.164) into (1.2.176) gives

$$\frac{\partial I}{\partial x}\frac{\partial H}{\partial y} - \frac{\partial I}{\partial y}\frac{\partial H}{\partial x} = 0. \tag{1.2.177}$$

Using the same idea, differentiating (1.2.167) along orbits gives

$$\frac{\partial \theta}{\partial x}\frac{\partial H}{\partial y} - \frac{\partial \theta}{\partial y}\frac{\partial H}{\partial x} = \Omega(I) = \frac{\partial H}{\partial I}. \tag{1.2.178}$$

Next, using (1.2.173) and the fact that $I = I(x, y)$, differentiating (1.2.173) gives

$$\begin{aligned} \frac{\partial H}{\partial y} &= \frac{\partial H}{\partial I}\frac{\partial I}{\partial y}, \\ \frac{\partial H}{\partial x} &= \frac{\partial H}{\partial I}\frac{\partial I}{\partial x}. \end{aligned} \tag{1.2.179}$$

Now, on the periodic orbits, $\frac{\partial H}{\partial x}$ and $\frac{\partial H}{\partial y}$ cannot simultaneously vanish, so (1.2.179) implies $\frac{\partial H}{\partial I} \neq 0$ (which also follows from (1.2.171)). Therefore, substituting (1.2.179) into (1.2.178) gives

$$\frac{\partial \theta}{\partial x}\frac{\partial I}{\partial y} - \frac{\partial \theta}{\partial y}\frac{\partial I}{\partial x} = 1. \tag{1.2.180}$$

The reader may also recognize (1.2.180) as the Jacobian of the transformation to action-angle variables. The fact that the Jacobian is one implies that area is preserved under the action-angle transformation.

Step 3. We have shown that in action-angle variables (1.2.164) takes the form

$$\begin{aligned} \dot{I} &= 0, \\ \dot{\theta} &= \Omega(I). \end{aligned} \tag{1.2.181}$$

As mentioned earlier, it is evident that (1.2.181) is Hamiltonian with Hamiltonian function $K(I)$. We now want to explicitly show that the action-angle transformation preserves the Hamiltonian structure in the sense that

$$H\big(x(I,\theta), y(I,\theta)\big) = K(I). \tag{1.2.182}$$

Therefore, thinking of H as a function of θ and I, we must show that

$$\frac{\partial H}{\partial I} = \frac{\partial K}{\partial I},$$
$$\frac{\partial H}{\partial \theta} = 0.$$

We proceed as follows. We have

$$\begin{aligned}
\frac{\partial K}{\partial I} = \dot{\theta} &= \frac{\partial \theta}{\partial x}\dot{x} + \frac{\partial \theta}{\partial y}\dot{y} \\
&= \frac{\partial \theta}{\partial x}\frac{\partial H}{\partial y} - \frac{\partial \theta}{\partial y}\frac{\partial H}{\partial x} \\
&= \frac{\partial \theta}{\partial x}\left(\frac{\partial H}{\partial \theta}\frac{\partial \theta}{\partial y} + \frac{\partial H}{\partial I}\frac{\partial I}{\partial y}\right) - \frac{\partial \theta}{\partial y}\left(\frac{\partial H}{\partial \theta}\frac{\partial \theta}{\partial x} + \frac{\partial H}{\partial I}\frac{\partial I}{\partial x}\right) \\
&= \frac{\partial H}{\partial I}\left(\frac{\partial \theta}{\partial x}\frac{\partial I}{\partial y} - \frac{\partial \theta}{\partial y}\frac{\partial I}{\partial x}\right) \\
&= \frac{\partial H}{\partial I} \qquad \text{(using (1.2.180))}.
\end{aligned}$$

A similar calculation shows that

$$-\frac{\partial K}{\partial \theta} = \dot{I} = -\frac{\partial H}{\partial \theta} = 0.$$

Now let us consider two useful examples.

EXAMPLE 1.2.14 Consider a Hamiltonian vector field given by the Hamiltonian

$$H(x,y) = \frac{y^2}{2} + V(x) \tag{1.2.183}$$

where $V(x)$ is shown is Figure 1.2.32a and the phase curves of the corresponding vector field are shown in Figure 1.2.32b. We will consider orbits interior to the region bounded by the homoclinic orbit in Figure 1.2.32b. (Note: for the more physically minded, (1.2.183) has the form of a kinetic energy term plus a potential energy term.)

From (1.2.183) we obtain

$$y = \pm\sqrt{2}\sqrt{H - V(x)}. \tag{1.2.184}$$

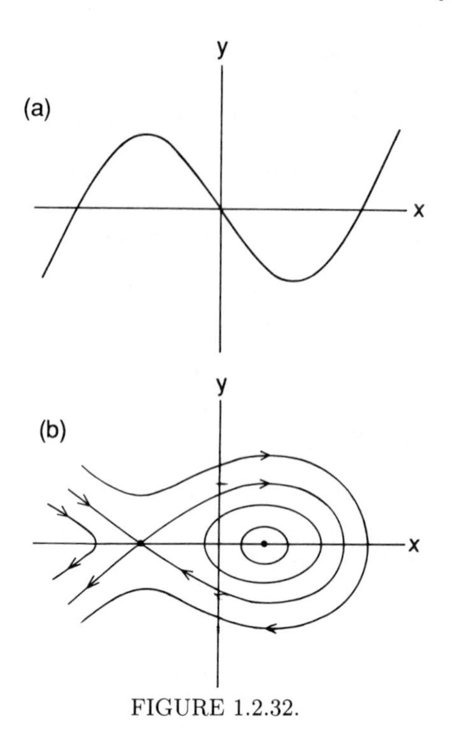

FIGURE 1.2.32.

Using (1.2.168) and (1.2.184), we obtain

$$I = \frac{\sqrt{2}}{\pi} \int_{x_{\min}}^{x_{\max}} \sqrt{H - V(x)}\, dx \tag{1.2.185}$$

where $x_{\min}$ is the left-most intersection of H = constant with the x-axis and $x_{\max}$ is the right-most intersection. (Note: the reader should figure out what became of the choice of plus or minus in (1.2.185).)

Using (1.2.164), (1.2.167), and (1.2.184) yields

$$\theta(x, y) = \frac{2\pi}{T(H)} \int_{x_{\min}}^{x} \frac{dx}{\sqrt{2}\sqrt{H - V(x)}}, \tag{1.2.186}$$

from which we obtain

$$T(H) = 2 \int_{x_{\min}}^{x_{\max}} \frac{dx}{\sqrt{2}\sqrt{H - V(x)}}. \tag{1.2.187}$$

By direct computation, using (1.2.171), the reader may verify directly that

$$\frac{\partial I}{\partial H} = \frac{T(H)}{2\pi}. \tag{1.2.188}$$

EXAMPLE 1.2.15 Recall that the perturbed equations (1.2.161) are given by

$$\begin{aligned} \dot{x} &= \frac{\partial H}{\partial y}(x, y) + \varepsilon g_1(x, y, t, \varepsilon), \\ \dot{y} &= -\frac{\partial H}{\partial x}(x, y) + \varepsilon g_2(x, y, t, \varepsilon). \end{aligned} \tag{1.2.189}$$

We want to transform this vector field using the action-angle transformation for the unperturbed vector. Differentiating this transformation gives

$$\begin{aligned} \dot{I} &= \frac{\partial I}{\partial x}\dot{x} + \frac{\partial I}{\partial y}\dot{y}, \\ \dot{\theta} &= \frac{\partial \theta}{\partial x}\dot{x} + \frac{\partial \theta}{\partial y}\dot{y}. \end{aligned} \tag{1.2.190}$$

Substituting (1.2.189) into (1.2.190) gives

$$\begin{aligned} \dot{I} &= \left(\frac{\partial I}{\partial x}\frac{\partial H}{\partial y} - \frac{\partial I}{\partial y}\frac{\partial H}{\partial x}\right) + \varepsilon\left(\frac{\partial I}{\partial x}g_1 + \frac{\partial I}{\partial y}g_2\right), \\ \dot{\theta} &= \left(\frac{\partial \theta}{\partial x}\frac{\partial H}{\partial y} - \frac{\partial \theta}{\partial y}\frac{\partial H}{\partial x}\right) + \varepsilon\left(\frac{\partial \theta}{\partial x}g_1 + \frac{\partial \theta}{\partial y}g_2\right). \end{aligned} \tag{1.2.191}$$

Using (1.2.177) and (1.2.180), (1.2.191) reduces to

$$\begin{aligned} \dot{I} &= \varepsilon F(I, \theta, t, \varepsilon), \\ \dot{\theta} &= \Omega(I) + \varepsilon G(I, \theta, t, \varepsilon), \end{aligned} \tag{1.2.192}$$

where

$$\begin{aligned} F(I, \theta, t, \varepsilon) = {} & \frac{\partial I}{\partial x}\big(x(I,\theta), y(I,\theta)\big)g_1\big(x(I,\theta), y(I,\theta), t, \varepsilon\big) \\ & + \frac{\partial I}{\partial y}\big(x(I,\theta), y(I,\theta)\big)g_2\big(x(I,\theta), y(I,\theta), t, \varepsilon\big) \\ G(I, \theta, t, \varepsilon) = {} & \frac{\partial \theta}{\partial x}\big(x(I,\theta), y(I,\theta)\big)g_1\big(x(I,\theta), y(I,\theta), t, \varepsilon\big) \\ & + \frac{\partial \theta}{\partial y}\big(x(I,\theta), y(I,\theta)\big)g_2\big(x(I,\theta), y(I,\theta), t, \varepsilon\big). \end{aligned}$$

It should be evident that F and G are 2π periodic in θ and $T = 2\pi/\omega$ periodic in t.

Our treatment of action-angle variables has been strictly two-dimensional. For higher dimensional generalizations we refer the reader to Arnold [1978], Goldstein [1980], or Nehorošev [1972].

Now let us return to the construction of the Poincaré map for (1.2.161). Writing (1.2.192) as an autonomous system gives

$$\begin{aligned} \dot{I} &= \varepsilon F(I, \theta, \phi, \varepsilon), \\ \dot{\theta} &= \Omega(I) + \epsilon G(I, \theta, \phi, \varepsilon), \qquad (I, \theta, \phi) \in \mathbb{R}^+ \times S^1 \times S^1, \\ \dot{\phi} &= \omega, \end{aligned} \tag{1.2.193}$$

(cf. Section 1.2A). We construct a global cross-section, Σ, to this vector field defined as follows

$$\Sigma^{\phi_0} = \{ (I, \theta, \phi) \mid \phi = \phi_0 \}. \tag{1.2.194}$$

(Note: for this definition of Σ in the context of Section 1.2A, we have chosen $\bar{\phi}_0 = 0$.) If we denote the (I, θ) components of solutions of (1.2.193) by $\big(I_\varepsilon(t), \theta_\varepsilon(t)\big)$ and the (I, θ) components of solutions of (1.2.193) for $\varepsilon = 0$ by $\big(I_0, \Omega(I_0)t + \theta_0\big)$, then the (perturbed) Poincaré map is given by

$$\begin{aligned} P_\varepsilon \colon \Sigma^{\phi_0} &\to \Sigma^{\phi_0}, \\ \big(I_\varepsilon(0), \theta_\varepsilon(0)\big) &\mapsto \big(I_\varepsilon(T), \theta_\varepsilon(T)\big) \end{aligned} \tag{1.2.195}$$

and the mth *iterate* of the Poincaré map is given by

$$\begin{aligned} P_\varepsilon^m \colon \Sigma^{\phi_0} &\to \Sigma^{\phi_0}, \\ \big(I_\varepsilon(0), \theta_\varepsilon(0)\big) &\mapsto \big(I_\varepsilon(mT), \theta_\varepsilon(mT)\big). \end{aligned}$$

Now we can approximate the solutions to the perturbed problem by using Lemma 1.2.12.

$$\begin{aligned} I_\varepsilon(t) &= I_0 + \varepsilon I_1(t) + \mathcal{O}(\varepsilon^2), \\ \theta_\varepsilon(t) &= \theta_0 + \Omega(I_0)t + \varepsilon\theta_1(t) + \mathcal{O}(\varepsilon^2), \end{aligned}$$

where I_0 is the unperturbed (constant) action value and I_1, θ_1 can be found by solving the first variational equation

$$\dot{q}_1^\alpha = Df(q^\alpha)q_1^\alpha + g(q^\alpha, \phi(t), 0),$$

where $\phi(t) = \omega t + \phi_0$, or written in the action-angle variables,

$$\begin{pmatrix} \dot{I}_1 \\ \dot{\theta}_1 \end{pmatrix} = \begin{pmatrix} 0 & 0 \\ \frac{\partial \Omega}{\partial I}(I_0) & 0 \end{pmatrix} \begin{pmatrix} I_1 \\ \theta_1 \end{pmatrix} + \begin{pmatrix} F(I_0, \Omega(I_0)t + \theta_0, \phi(t), 0) \\ G(I_0, \Omega(I_0)t + \theta_0, \phi(t), 0) \end{pmatrix}.$$

Now we can see the tremendous advantage we have gained in using the action-angle coordinates; namely, the solution of this equation is trivial because the matrix is constant coefficient (normally it would be time dependent in an arbitrary coordinate system, and we know that there are no general methods for solving linear equations with *nonconstant* coefficients). Thus we have that

$$\begin{aligned} P_\varepsilon^m \colon \Sigma^{\phi_0} &\to \Sigma^{\phi_0}, \\ \big(I_\varepsilon(0), \theta_\varepsilon(0)\big) &\mapsto \big(I_\varepsilon(mT), \theta_\varepsilon(mT)\big) \\ = (I_0, \theta_0) &\mapsto \big(I_0 + \varepsilon I_1(mT), \theta_0 + mT\Omega(I_0) + \varepsilon\theta_1(mT)\big) + \mathcal{O}(\varepsilon^2), \end{aligned}$$

where we have chosen

$$\begin{aligned} I_\varepsilon(0) &= I_0, \\ \theta_\varepsilon(0) &= \theta_0. \end{aligned}$$

From the first variational equation, $I_1(mT)$ and $\theta_1(mT)$ are given by

$$I_1(mT) = \int_0^{mT} F(I_0, \Omega(I_0)t + \theta_0, \omega t + \phi_0, 0)\, dt \equiv M_1^{m/n}(I_0, \theta_0; \phi_0)$$

$$\theta_1(mT) = \left.\frac{\partial \Omega}{\partial I}\right|_{I=I_0} \int_0^{mT} \int_0^t F\big(I_0, \Omega(I_0)\xi + \theta_0, \omega\xi + \phi_0, 0\big)\, d\xi\, dt$$
$$+ \int_0^{mT} G\big(I_0, \Omega(I_0)t + \theta_0, \omega t + \phi_0, 0\big)\, dt \equiv M_2^{m/n}(I_0, \theta_0; \phi_0).$$

The Poincaré map therefore takes the form

$$\begin{aligned} P_\varepsilon^m &: \Sigma^{\phi_0} \to \Sigma^{\phi_0}, \\ &(I, \theta) \mapsto \big(I, \theta + mT\Omega(I)\big) + \varepsilon\left(M_1^{m/n}(I, \theta; \phi_0), M_2^{m/n}(I, \theta; \phi_0)\right) \\ &\quad + \mathcal{O}(\varepsilon^2), \end{aligned} \tag{1.2.196}$$

where we have dropped the subscripts 0's on I and θ for notational convenience and where ϕ_0 denotes the dependence of the Poincaré map on the cross-section Σ^{ϕ_0}.

We define

$$M^{m/n}(I, \theta; \phi_0) \equiv \left(M_1^{m/n}(I, \theta; \phi_0), M_2^{m/n}(I, \theta; \phi_0)\right) \tag{1.2.197}$$

to be the *subharmonic Melnikov vector* named in honor of V. K. Melnikov. We make the following remarks.

Remark 1. We superscripted the Melnikov vector with m/n to denote our search for periodic orbits which satisfy the resonance relation

$$nT(I) = mT.$$

This relation will enter into the computation of the Melnikov vector as we explain in Remark 4 following the proof of Theorem 1.2.13. Thus, the superscript m/n implies that the value of I in the Melnikov vector satisfies the resonance relation. It will also aid us in avoiding confusion with the homoclinic Melnikov function in Section 4.5.

Remark 2. The subharmonic Melnikov *function* defined in previous expositions of the Melnikov theory for periodic orbits (cf. Guckenheimer and Holmes [1983]) is, up to a normalization factor that is constant on orbits, the first component of our Melnikov vector. We will discuss this in more detail following the proof of Theorem 1.2.13.

Remark 3. Let us make a few comments regarding our potpourri of notation above.

a) I and α play the same role.

b) From the resonance relation

$$nT(I) = mT,$$

we see that n, m, and I are functionally related. Thus, our practice of tagging n, m, and I (or α) on the subharmonic Melnikov vector is a bit redundant; however, it has become traditional.

Occasionally, either I (or α) or m/n may not be explicitly shown, i.e., we may write $M(I, \theta; \phi_0)$, $M(\alpha, t_0; \phi_0)$, $M^{m/n}(\theta; \phi_0)$, or $M^{m/n}(t_0; \phi_0)$. However, the reader should keep in mind that the resonance relation $nT(I) = mT$ (or $nT^\alpha = mT$) is the analytical and geometrical assumption.

Now we state the main theorem concerning the existence of subharmonic periodic orbits.

Theorem 1.2.13 *Suppose there exists a point $(\bar{I}, \bar{\theta})$ at which $T(\bar{I}) = (m/n)T$, and one of the following conditions is satisfied:*

FP1) $$M_1^{m/n}(\bar{I}, \bar{\theta}; \phi_0) = 0 \quad \textit{and} \quad \left. \left(\frac{\partial \Omega}{\partial I} \frac{\partial M_1^{m/n}}{\partial \theta} \right) \right|_{(\bar{I}, \bar{\theta})} \neq 0;$$

FP2) $$M^{m/n}(\bar{I}, \bar{\theta}; \phi_0) = 0, \qquad \left. \frac{\partial \Omega}{\partial I} \right|_{\bar{I}} = 0, \qquad \textit{and}$$

$$\left. \left(\frac{\partial M_1^{m/n}}{\partial I} \frac{\partial M_2^{m/n}}{\partial \theta} - \frac{\partial M_2^{m/n}}{\partial I} \frac{\partial M_1^{m/n}}{\partial \theta} \right) \right|_{(\bar{I}, \bar{\theta})} \neq 0.$$

Then, for $0 < \varepsilon \leq \varepsilon(n)$, the Poincaré map P_ε^m has a fixed point of period m. If $n = 1$, the result is uniformly valid in $0 < \varepsilon \leq \varepsilon(1)$.

Proof: Case FP1)

$$M_1(\bar{I}, \bar{\theta}; \phi_0) = 0, \qquad \left. \frac{\partial \Omega}{\partial I} \frac{\partial M_1}{\partial \theta} \right|_{(\bar{I}, \bar{\theta})} \neq 0.$$

(Note: we have dropped the superscript m/n for notational convenience.) Then we have

$$P_\varepsilon^m(\bar{I}, \bar{\theta}) - (\bar{I}, \bar{\theta}) = \big(0, mT\Omega(\bar{I})\big) + \varepsilon\big(0, M_2(\bar{I}, \bar{\theta}; \phi_0)\big) + \mathcal{O}(\varepsilon^2).$$

Let us perturb the action by an amount ΔI, let $\hat{I} = \bar{I} + \Delta I$, and expand the right-hand side about $\bar{I}$ to obtain

$$\begin{aligned} P_\varepsilon^m(\hat{I}, \bar{\theta}) - (\hat{I}, \bar{\theta}) = {} & \left(0, mT\Omega(\bar{I}) + mT \left. \frac{\partial \Omega}{\partial I} \right|_{\bar{I}} \Delta I + \mathcal{O}((\Delta I)^2) \right) \\ & + \quad \varepsilon\big(0, M_2(\bar{I}, \bar{\theta}; \phi_0)\big) + \mathcal{O}(\varepsilon \Delta I) + \mathcal{O}(\varepsilon^2). \end{aligned}$$

We recall from the resonance relation that

$$mT\Omega(\bar{I}) = 2\pi n = 0(\mathrm{mod}\, 2\pi),$$

and that if we choose ΔI such that

$$\Delta I = -\varepsilon \frac{M_2(\bar{I}, \bar{\theta}; \phi_0)}{mT \frac{\partial \Omega}{\partial I}\big|_{\bar{I}}},$$

we obtain

$$P_\varepsilon^m(\hat{I}, \bar{\theta}) - (\hat{I}, \bar{\theta}) = \mathcal{O}(\varepsilon^2)$$

since $\Delta I = \mathcal{O}(\varepsilon)$. We have therefore shown that the mth iterate of the Poincaré map has a fixed point up to an error of $\mathcal{O}(\varepsilon^2)$. The fact that the map has an exact fixed point follows immediately from the implicit function theorem provided that

$$\det\left((DP_\epsilon^m - \mathrm{id})\Big|_{(\hat{I}, \bar{\theta})} \right) \neq 0,$$

where "id" denotes the 2×2 identity matrix. An easy calculation shows that this is equivalent to

$$\left(\frac{\partial \Omega}{\partial I} \frac{\partial M_1}{\partial \theta} \right)_{(\bar{I}, \bar{\theta})} \neq 0.$$

Case FP2) The argument is very similar to that given for FP1. We leave the details to the reader. □

Let us now make a few remarks concerning the consequences and implications of Theorem 1.2.13.

Remark 1. Earlier we remarked that the action-angle variables were merely a convenience to allow us to view the problem in its intrinsic geometric context. We saw that their use allowed us to approximate the Poincaré map to $\mathcal{O}(\varepsilon^2)$ by rendering the first variational equation (1.2.163) trivially solvable. In any other coordinate system (1.2.163) might be analytically intractable. However, in order to use Theorem 1.2.13 to find ultrasubharmonics we must compute the subharmonic Melnikov vector. To do this, it would appear that our original perturbed equation (1.2.161) must first be transformed into (1.2.193) via the action-angle transformation. We want to show that in Case FP1, this is unnecessary.

Let us now assume that

$$\frac{\partial \Omega}{\partial I} \neq 0.$$

In this case, by Theorem 1.2.13, to determine the existence of ultrasubharmonics we need information only on $M_1^{m/n}(I,\theta;\phi_0)$. Recall from (1.2.196) that $M_1^{m/n}(I,\theta;\phi_0)$ is given by

$$M_1^{m/n}(I,\theta) = \int_0^{mT} F\big(I, \Omega(I)t+\theta, \omega t+\phi_0, 0\big)\, dt, \tag{1.2.198}$$

where

$$F = \frac{\partial I}{\partial x} g_1 + \frac{\partial I}{\partial y} g_2.$$

We now have $\frac{\partial H}{\partial I} = \Omega(I) \neq 0$; hence we can invert this and write $I = I\big(H(x,y)\big)$. Using the chain rule we obtain

$$\begin{aligned} \frac{\partial I}{\partial x} &= \frac{\partial I}{\partial H}\frac{\partial H}{\partial x} = \frac{1}{\Omega(I)}\frac{\partial H}{\partial x}, \\ \frac{\partial I}{\partial y} &= \frac{\partial I}{\partial H}\frac{\partial H}{\partial y} = \frac{1}{\Omega(I)}\frac{\partial H}{\partial y}, \end{aligned}$$

so that F becomes

$$F = \frac{1}{\Omega(I)}\left(\frac{\partial H}{\partial x} g_1 + \frac{\partial H}{\partial y} g_2\right),$$

and we have

$$M_1^{m/n} = \frac{1}{\Omega(I)} \int_0^{mT} (DH \cdot g)(\text{unperturbed orbit})\, dt, \tag{1.2.199}$$

where "$\cdot$" represents the usual dot product for vectors.

In integrating the expression $DH \cdot g$ around an unperturbed periodic orbit it does not matter whether we express the unperturbed orbit in action-angle variables or in Cartesian coordinates, since the Jacobian of the transformation between the two coordinate systems is identically one. Thus, for the the "unperturbed orbit" in (1.2.199), we may substitute the expression

$$q^\alpha\left(\frac{\theta-\theta_0}{\Omega(I)}\right), \tag{1.2.200}$$

where the $q^\alpha(\cdot)$ denotes an unperturbed periodic orbit in the coordinates of (1.2.161). The argument for (1.2.200) is found by solving $\theta = \Omega(I)t + \theta_0$ for t. We can let $\theta/\Omega(I) = t$ and $\theta_0/\Omega(I) = t_0$ so that (1.2.199) becomes

$$M_1^{m/n}(\alpha, t_0; \phi_0) = \frac{1}{\Omega(I)} \int_0^{mT} (DH \cdot g)\big(q^\alpha(t-t_0), \omega t+\phi_0, 0\big)\, dt. \tag{1.2.201}$$

Then, letting $t \to t+t_0$, and using periodicity in t of the vector field, (1.2.201) becomes

$$M_1^{m/n}(\alpha, t_0; \phi_0) = \frac{1}{\Omega(I)} \int_0^{mT} (DH \cdot g)\big(q^\alpha(t), \omega t+\omega t_0+\phi_0, 0\big)\, dt. \tag{1.2.202}$$

If we define

$$\bar{M}_1^{m/n}(\alpha, t_0; \phi_0) = \int_0^{mT} (DH \cdot g)\big(q^\alpha(t), \omega t + \omega t_0 + \phi_0, 0\big)\, dt, \quad (1.2.203)$$

it is easy to see that $\bar{M}_1^{m/n}(\alpha, t_0; \phi_0)$ has a zero at which $\frac{\partial \bar{M}_1^{m/n}(\alpha,t_0;\phi_0)}{\partial t_0} \neq 0$ if and only if $M_1^{m/n}(\alpha, t_0; \phi_0)$ has a zero at which $\frac{\partial M_1^{m/n}(\alpha,t_0;\phi_0)}{\partial t_0} \neq 0$. Thus, the computation of (1.2.203) in the original coordinate system is sufficient for verifying the hypotheses of Theorem 1.2.13. Note that $\bar{M}_1^{m/n}(\alpha, t_0; \phi_0)$ is the standard subharmonic Melnikov function that can be found in Guckenheimer and Holmes [1983].

In the Case FP2, where $\frac{\partial \Omega}{\partial I} = 0$, we have not found analogously simple transformations which transform the subharmonic Melnikov vector derived in action-angle coordinates into the original coordinates of (1.2.161). It appears that in this case the action-angle coordinate transformations must be explicitly computed. A situation in which FP2 would arise would be in perturbations of linear systems.

Remark 2. By construction it is easy to see that $M^{m/n}(I, \theta; \phi_0)$ is 2π-periodic in θ. We will explore the geometrical implications of this for Case FP1 and leave the details of Case FP2 for the reader.

Suppose we locate a point $(\bar{I}, \bar{\theta})$ satisfying the hypotheses of Theorem 1.2.13 for Case FP1. Then the point $(\bar{I}, \bar{\theta})$ is $\mathcal{O}(\varepsilon)$ close to a period m point of P_ε. Now, since $M_1^{m/n}(\bar{I}, \theta; \phi_0)$ is periodic in θ with period 2π, this implies that $M_1^{m/n}(\bar{I}, \theta; \phi_0)$ has at least m zero's for $\theta \in [0, 2\pi)$. These zero's are simply the orbit of the period m point under P_ε, or equivalently, the m fixed points of P_ε^m. We can go even further. First, we note that since $M_1^{m/n}(\bar{I}, \theta; \phi_0)$ is periodic in θ, $\frac{\partial M_1^{m/n}}{\partial \theta}(\bar{I}, \theta; \phi_0)$ is also. Therefore, $\frac{\partial M_1^{m/n}}{\partial \theta}(\bar{I}, \theta; \phi_0)$ is identical and nonzero at each of these fixed points for P_ε^m. Hence, by the mean value theorem, between any two of these fixed points of P_ε^m there must be at least one more point where $M_1^{m/n}(\bar{I}, \theta; \phi_0) = 0$ with $\frac{\partial M_1^{m/n}}{\partial \theta}(\bar{I}, \theta; \phi_0) \neq 0$. Using the same argument as above, we conclude there must be m such points.

To summarize, a point $(\bar{I}, \bar{\theta})$ satisfying the hypotheses of Theorem 1.2.13 for Case FP1 implies the existence of $2m$ fixed points for P_ε^m, or equivalently, two period m orbits for P_ε. We will see later that these two orbits have different stability characteristics.

Remark 3. As we have mentioned previously, the global geometry of the unperturbed phase space provides us with a framework for developing our analysis of the perturbed orbit structure. With this in mind, we now discuss the geometry of the unperturbed Poincaré map, which we rewrite below as

$$(I, \theta) \mapsto \big(I, \theta + mT\Omega(I)\big).$$

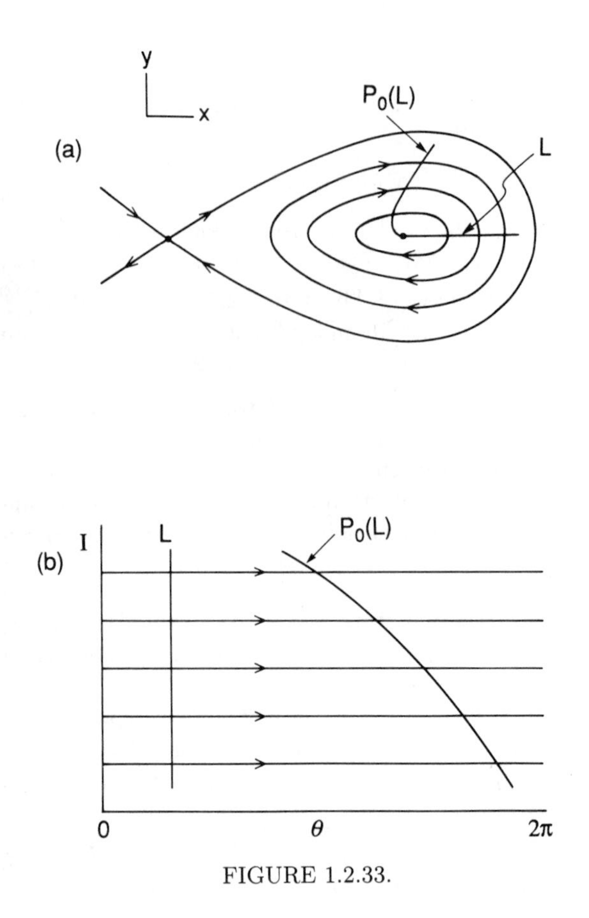

FIGURE 1.2.33.

This is a very simple map; all of the orbits lie on closed curves (as they should from the structure of the unperturbed problem), and $mT\Omega(I)$ tells us how much the points move around the curve on each iteration. Let us assume that $\frac{\partial\Omega}{\partial I} < 0$ inside the homoclinic orbit. If $\frac{\partial\Omega}{\partial I}$ should vanish at isolated values of I, then we can simply apply our arguments to a subset inside the homoclinic orbit (i.e., a range of I values) where $\frac{\partial\Omega}{\partial I}$ does not change sign. With this assumption, the frequency, $\Omega(I)$, decreases monotonically to zero as the homoclinic orbit is approached. Therefore, if we examine an image of a radial line under P_0 it appears as in Figure 1.2.33.

Thus, points closer to the homoclinic orbit do not move as far as points closer to the center. Maps of this type are called *twist maps*. Clearly, since the twist is an $\mathcal{O}(1)$ property, the perturbed map is still a twist map. Now, examining the Case FP1, we can see what the twist condition does for us. Since we are dealing with a two-dimensional map, to determine fixed points normally we would have to satisfy two conditions; however, in our search for resonant periodic points we can see from the proof for Case FP1 that the twist condition (i.e., $\frac{\partial\Omega}{\partial I} \neq 0$) guarantees that we return to the Poincaré section at the correct θ value. Thus, we only need to check whether or not the radial (i.e., I) coordinates of the image and preimage match up, and this is measured by $M_1^{m/n}$.

In Case FP2, where the twist is zero, it is clear that two conditions must be satisfied for a fixed point of P_ε^m to exist; namely, $(M_1^{m/n}, M_2^{m/n}) = (0, 0)$.

Remark 4. What about the m and n? As these may appear to be rather mysterious, we want to go into some detail about what they mean. We are looking for periodic orbits in the perturbed system that are in resonance with the external forcing and that satisfy the *resonance relation*

$$nT(I) = mT, \qquad n,\, m \text{ relatively prime}$$

(note: this can be taken as the definition of resonant orbits) or $T(I) = (m/n)T$. Now our map is constructed from the ordinary differential equation by taking points (initial conditions) on Σ^{ϕ_0} and letting them evolve in time until they return to Σ^{ϕ_0} (by definition of Σ^{ϕ_0}, this occurs after time T). Thus, if a point returns to where it started on Σ^{ϕ_0} after m iterations of P_ε, we call it a period m point for the map P_ε. Next we ask how this period m point for the map is related to periodic orbits in the ordinary differential equation. The reader should recall that computation of the Melnikov vector involved integrating around an unperturbed periodic orbit whose period we denoted $T(I)$. If we find a zero of the Melnikov function corresponding to a period m point of the map at, say, $(I, \theta) = (\bar{I}, \bar{\theta})$, then the period of the unperturbed orbit on which this particular Melnikov vector is computed is given by $T(\bar{I})$. In order for this orbit to return to Σ^{ϕ_0} and be a period m fixed point for the map, it is necessary that $nT(I) = mT$. We can therefore see that the n enters into the Melnikov vector through the fact that we calculate the Melnikov vector on a specific periodic orbit and we substitute mT/n for $T(I)$ in the expression for the periodic orbit. Thus, we speak of a period $(m/n)T$ ultrasubharmonic being preserved in the perturbed vector field.

We also urge the reader to review the geometrical description of Poincaré maps given in Section 1.2A.

Remark 5. The sign of $M_1^{m/n}(I, \theta; \phi_0)$ together with the twist condition $\frac{\partial\Omega}{\partial I} \neq 0$ tells us a great deal about the orbit structure near the fixed points on a resonance band, i.e., a neighborhood of $I = \bar{I}$ satisfying the resonance relation $nT(I) = mT$. We now want to illustrate this idea.

Suppose, for definiteness, that we have

$$\frac{\partial\Omega}{\partial I} < 0,$$

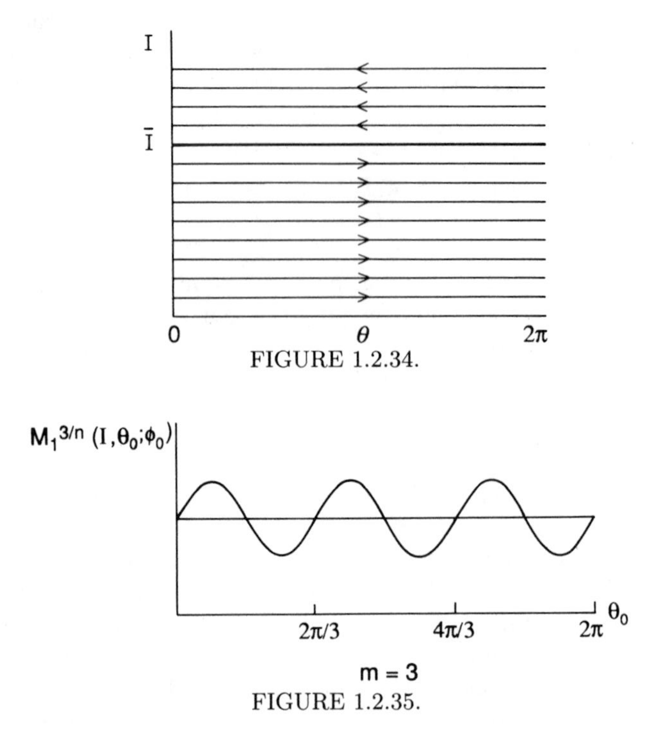

FIGURE 1.2.34.

FIGURE 1.2.35.

and we have found $I = \bar{I}$ such that

$$T(\bar{I}) = \frac{3T}{n}.$$

In Figure 1.2.34 we denote the invariant circles in action-angle coordinates. The arrows on the invariant circles represent the action of P_0^3 on the circles. In this case the circle labeled by $I = \bar{I}$ is a circle of fixed points and, due to $\frac{\partial \Omega}{\partial I} < 0$, points on the circles above $I = \bar{I}$ move to the left, and points on the circle below $I = \bar{I}$ move to the right.

Furthermore, suppose we have found a $\bar{\theta}$ such that

$$\begin{aligned} &M_1^{m/n}(\bar{I}, \bar{\theta}; \phi_0) = 0, \\ &\frac{\partial \Omega}{\partial I} \frac{\partial M_1^{m/n}}{\partial \theta}(\bar{I}, \bar{\theta}; \phi_0) \neq 0; \end{aligned}$$

then, by Theorem 1.2.13 and Remark 2 following the proof of the theorem, we know that P_ε^3 has six fixed points near $(\mathcal{O}(\varepsilon))$ $I = \bar{I}$. In Figure 1.2.35 we have graphed $M_1^{3/n}(\bar{I}, \theta; \phi_0)$ on top of the unperturbed circle of fixed points given by $I = \bar{I}$. The intersection of the graph of $M_1^{3/n}(\bar{I}, \theta; \phi_0)$ represents the fixed points of P_ε^3.

By construction, $M_1^{3/n}(\bar{I}, \theta; \phi_0)$ measures the "push" in the direction normal to the unperturbed closed orbit defined by $I = \bar{I}$ due to the per-

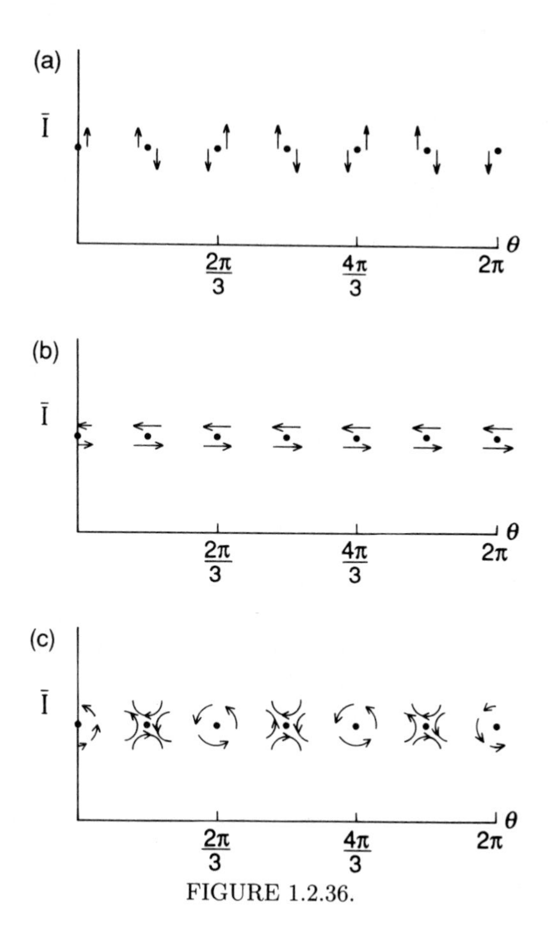

FIGURE 1.2.36.

turbation. Thus, if $M_1^{3/n}(\bar{I},\theta;\phi_0) > 0$, then points starting near $(\bar{I},\theta)$ are pushed above $I = \bar{I}$ and, if $M_1^{3/n}(\bar{I},\theta;\phi_0) < 0$, then points starting near $(\bar{I},\theta)$ are pushed below $I = \bar{I}$. Using this, along with Figure 1.2.35, we represent by arrows in Figure 1.2.36a the direction in which points will be pushed due to the I component of the perturbation near the fixed points. In Figure 1.2.36b, using the fact that the twist ($\frac{\partial\Omega}{\partial I} \neq 0$) is an order one effect, we represent by arrows the direction in which points will be pushed due to the θ motion. In Figure 1.2.36c we superimpose the I and θ dynamics near the fixed points, as shown in Figure 1.2.36a and 1.2.36b, and obtain that, near the fixed points satisfying $\frac{\partial\Omega}{\partial I}\frac{\partial M_1^{3/n}}{\partial\theta} > 0$, points move in a hyperbolic manner and, near the fixed points satisfying $\frac{\partial\Omega}{\partial I}\frac{\partial M_1^{3/n}}{\partial\theta} < 0$, nearby points seem to circulate around the fixed points. (Note: what would happen if we had originally assumed $\frac{\partial\Omega}{\partial I} > 0$?)

We stress that this is a heuristic argument only; we will address stability directly in the next section. However, much of what we have said is true in a more general sense. On an order m/n resonance band we have $2m$ fixed points. These fixed points will (generally) alternate in stability type as one moves around the resonance band with m of the fixed points being saddles. The stability of the remaining fixed points is a more delicate matter (note: in the above argument we deliberately used the vague statement that nearby points tended to "circulate" around the fixed points). For dissipative perturbations these fixed points will be sinks and for Hamiltonian perturbation they will be elliptic fixed points (i.e., their eigenvalues will both have modulus one).

Remark 6. For Case FP1, we recall from (1.2.203) that the first component of the subharmonic Melnikov vector in the $x-y$ coordinate system (neglecting the nonzero normalization factor $1/\Omega(I)$) is given by

$$\bar{M}_1^{m/n}(\alpha, t_0; \phi_0) = \int_0^{mT} (DH \cdot g)(q^\alpha(t), \omega t + \omega t_0 + \phi_0, 0)dt.$$

We can thus see that, due to the periodicity of the vector field, varying t_0 and varying ϕ_0 have the same effect on $\bar{M}_1^{m/n}(\alpha, t_0; \phi_0)$. Therefore, varying the cross-section Σ^{ϕ_0} on which the Poincaré map is defined corresponds to shifting the phase of the periodic orbits.

iia) *Stability*

One of the major advantages of working with the Poincaré map is that stability questions follow easily. Recall that stability of the fixed points of P_ε^m may often be determined by linearizing the map about the fixed point and computing the eigenvalues. Cases FP1 and FP2 are different; therefore, we treat each separately. Recall also the mth iterate of the Poincaré map is given by

$$P_\varepsilon^m(I,\theta) = \big(I, \theta + mT\Omega(I)\big) + \varepsilon\big(M_1^{m/n}(I,\theta;\phi_0), M_2^{m/n}(I,\theta;\phi_0)\big) + \mathcal{O}(\varepsilon^2);$$

hence, the linearization is

$$DP_\varepsilon^m = \begin{pmatrix} 1+\varepsilon M_{1,I}^{m/n} & \varepsilon M_{1,\theta}^{m/n} \\ mT\Omega_I + \varepsilon M_{2,I}^{m/n} & 1+\varepsilon M_{2,\theta}^{m/n} \end{pmatrix} + \mathcal{O}(\varepsilon^2),$$

where, for notational convenience, we denote partial derivatives by

$$\frac{\partial M_1^{m/n}}{\partial \theta} \equiv M_{1,\theta}^{m/n},$$
$$\frac{\partial M_1^{m/n}}{\partial I} \equiv M_{1,I}^{m/n},$$
$$\frac{\partial \Omega}{\partial I} \equiv \Omega_I,$$

and similarly for $M_2^{m/n}$.

Case FP1. Suppose we have found a point $(\bar{I}, \bar{\theta})$ such that $nT(\bar{I}) = mT$ and

$$M_1^{m/n}(\bar{I}, \bar{\theta}) = 0, \qquad \left(\frac{\partial \Omega}{\partial I} \frac{\partial M_1^{m/n}}{\partial \theta}\right)\bigg|_{(\bar{I},\bar{\theta})} \neq 0.$$

Then, by Theorem 1.2.13, we know there exists a point $(\hat{I}, \hat{\theta}) = (\bar{I}, \bar{\theta}) + \mathcal{O}(\varepsilon)$ such that $(\hat{I}, \hat{\theta})$ is a period m point for P_ε.

We want to compute the stability of this fixed point. The eigenvalues of DP_ε^m are given by

$$\lambda_{1,2} = \frac{\mathrm{tr} DP_\varepsilon^m}{2} \pm \frac{1}{2}\sqrt{(\mathrm{tr}(DP_\varepsilon^m))^2 - 4\det(DP_\varepsilon^m)}$$

and (we drop the superscript m/n for easier notation)

$$\begin{aligned}
\mathrm{tr}\, DP_\varepsilon^m &= 2 + \varepsilon(M_{1,I} + M_{2,\theta}) + \mathcal{O}(\varepsilon^2), \\
\det DP_\varepsilon^m &= 1 - \varepsilon mT\Omega_I M_{1,\theta} + \varepsilon(M_{1,I} + M_{2,\theta}) + \mathcal{O}(\varepsilon^2), \\
(\mathrm{tr}\, DP_\varepsilon^m)^2 - 4\det DP_\varepsilon^m &= 4 + 4\varepsilon(M_{1,I} + M_{2,\theta}) + \mathcal{O}(\varepsilon^2) \\
&\quad - 4 + 4\varepsilon mT\Omega_I M_{1,\theta} - 4\varepsilon(M_{1,I} + M_{2,\theta}) + \mathcal{O}(\varepsilon^2), \\
&= 4\varepsilon mT\Omega_I M_{1,\theta} + \mathcal{O}(\varepsilon^2).
\end{aligned}$$

At this point the question arises as to where we evaluate the partial derivatives since we don't know the exact fixed point, only an $\mathcal{O}(\varepsilon)$ approximation which is a zero of M_1. However, we can see from the above expressions for $\mathrm{tr}\, DP_\varepsilon^m$ and $\det DP_\varepsilon^m$ that by Taylor expanding these expressions about $(\bar{I}, \bar{\theta})$ and substituting in $(\hat{I}, \hat{\theta}) = (\bar{I}, \bar{\theta}) + \mathcal{O}(\varepsilon)$, we incur an error of only $\mathcal{O}(\varepsilon^2)$. *Therefore, we evaluate all partial derivatives at* $(\bar{I}, \bar{\theta})$, i.e., the zero for M_1 of which $nT(I) = mT$ is satisfied. We thus obtain

$$\lambda_{1,2} = 1 + \frac{\varepsilon}{2}(M_{1,I} + M_{2,\theta}) \pm \sqrt{\varepsilon mT\Omega_I M_{1,\theta} + \mathcal{O}(\varepsilon^2)} + \mathcal{O}(\varepsilon^2).$$

Expanding the "square-root" part of the expression in a Taylor series gives

$$\lambda_{1,2} = 1 \pm \sqrt{\varepsilon}\sqrt{mT\Omega_I M_{1,\theta}} + \frac{\varepsilon}{2}(M_{1,I} + M_{2,\theta}) + \mathcal{O}(\varepsilon^{3/2}). \tag{1.2.204}$$

From this expression we may determine stability provided both the $\mathcal{O}(\sqrt{\varepsilon})$ and $\mathcal{O}(\varepsilon)$ terms are nonzero. Note that, for $mT\Omega_I M_{1,\theta} > 0$, for ε sufficiently small the $\mathcal{O}(\sqrt{\varepsilon})$ term in (1.2.204) is sufficient for determining stability; cf. Remark 5 following the proof of Theorem 1.2.13.

Case FP2. Suppose that we have found a point $(\bar{I}, \bar{\theta})$ such that $nT(\bar{I}) = mT$ and

$$M(\bar{I}, \bar{\theta}) = 0, \qquad \frac{\partial \Omega}{\partial I}\bigg|_{\bar{I}} = 0, \qquad \left(\frac{\partial M_1}{\partial I}\frac{\partial M_2}{\partial \theta} - \frac{\partial M_2}{\partial I}\frac{\partial M_1}{\partial \theta}\right)\bigg|_{(\bar{I},\bar{\theta})} \neq 0.$$

Then, from Theorem 1.2.13, we have a period m point for P_ε at $(\hat{I}, \hat{\theta}) = (\bar{I}, \bar{\theta}) + \mathcal{O}(\varepsilon)$. We compute the eigenvalues of the linearized map at the fixed point in a similar manner as in Case FP1. In this case we have to be a bit more careful with the $\mathcal{O}(\varepsilon^2)$ terms, so we include them explicitly in the expression for DP_ε^m as follows

$$DP_\varepsilon^m = \begin{pmatrix} 1 + \varepsilon M_{1,I} + \varepsilon^2 A & \varepsilon M_{1,\theta} + \varepsilon^2 B \\ \varepsilon M_{2,I} + \varepsilon^2 C & 1 + \varepsilon M_{2,\theta} + \varepsilon^2 D \end{pmatrix},$$

where, of course, A, B, C, and D are unknown. The eigenvalues of DP_ε^m are given by

$$\lambda_{1,2} = \frac{(\operatorname{tr} DP_\varepsilon^m)}{2} \pm \frac{1}{2}\sqrt{(\operatorname{tr} DP_\varepsilon^m)^2 - 4 \det DP_\varepsilon^m},$$

where

$$\begin{aligned}
\operatorname{tr} DP_\varepsilon^m &= 2 + \varepsilon(M_{1,I} + M_{2,\theta}) + \varepsilon^2(A + D), \\
\det DP_\varepsilon^m &= 1 + \varepsilon(M_{1,I} + M_{2,\theta}) + \varepsilon^2(M_{1,I}M_{2,\theta} - M_{1,\theta}M_{2,I}) \\
&\quad + \varepsilon^2(A + D) + \mathcal{O}(\varepsilon^3), \\
(\operatorname{tr} DP_\varepsilon^m)^2 - 4 \det DP_\varepsilon^m &= 4 + 4\varepsilon(M_{1,I} + M_{2,\theta}) + \varepsilon^2(M_{1,I} + M_{2,\theta})^2 \\
&\quad + 4\varepsilon^2(A + D) + \mathcal{O}(\varepsilon^3) - 4 - 4\varepsilon(M_{1,I} + M_{2,\theta}) \\
&\quad - 4\varepsilon^2(M_{1,I}M_{2,\theta} - M_{2,I}M_{1,\theta}) \\
&\quad - 4\varepsilon^2(A + D) + \mathcal{O}(\varepsilon^3) \\
&= \varepsilon^2(M_{1,I} + M_{2,\theta})^2 - 4\varepsilon^2(M_{1,I}M_{2,\theta} - M_{2,I}M_{1,\theta}) \\
&\quad + \mathcal{O}(\varepsilon^3).
\end{aligned}$$

(Note the fortuitous cancelling of the unknown $\mathcal{O}(\varepsilon^2)$ terms above.) Thus, we obtain

$$\begin{aligned}
\lambda_{1,2} = 1 &+ \frac{\varepsilon}{2}(M_{1,I} + M_{2,\theta}) \\
&\pm \frac{\varepsilon}{2}\sqrt{(M_{1,I} + M_{2,\theta})^2 - 4(M_{1,I}M_{2,\theta} - M_{1,\theta}M_{2,I})} \\
&+ \mathcal{O}(\varepsilon^2)
\end{aligned} \tag{1.2.205}$$

and, similarly to Case FP1, all partial derivatives are evaluated at the zeros of the Melnikov vector $M = (M_1, M_2)$ at which the resonance relation $nT(I) = mT$ is also satisfied. The above expression can thus be used to determine stability in Case FP2.

We inject a word of caution at this point regarding the use of these expressions for the eigenvalues of P_ε^m linearized about the period m point. The only way in which they are useful for stability considerations is if, for ε sufficiently small, the known part of the expression dominates the unknown part in the sense that inclusion of the higher order unknown terms does not cause the eigenvalue(s) to move across the unit circle in the complex plane. Let us consider a specific example.

EXAMPLE 1.2.16 Consider the number

$$\lambda(\varepsilon) = 1 + i\sqrt{\varepsilon}a + \varepsilon b + \mathcal{O}(\varepsilon^{3/2}), \tag{1.2.206}$$

where a and b are real (note: this would correspond to Case FP1 where $\frac{\partial \Omega}{\partial I}\frac{\partial M_1 m/n}{\partial \theta} < 0$). A simple calculation gives

$$|\lambda(\varepsilon)| = \sqrt{1 + \varepsilon(a^2 + 2b) + \mathcal{O}(\varepsilon^2)}. \tag{1.2.207}$$

Thus, from (1.2.207) it is easy to see that both the $\mathcal{O}(\sqrt{\varepsilon})$ and $\mathcal{O}(\varepsilon)$ terms of (1.2.206) are important for determining whether $|\lambda(\varepsilon)|$ is greater than or less than one for ε sufficiently small.

We refer the reader to Murdock and Robinson [1980] for more information concerning such issues. Finally, we remark that if the perturbation is Hamiltonian also, then many stability considerations are, in a certain sense, "beyond all orders" for perturbation theory. In this case a different approach must be taken. We will comment on these issues as they arise.

iib) *Structure of the Resonance Bands*

We will refer to the region on the Poincaré cross-section near an action value satisfying the resonance relation as a *resonance band.* The techniques developed thus far enable us to determine the existence of ultrasubharmonics and possibly even their stability. Now we would like to develop a technique for studying the global dynamics near a resonance band. The idea will be to derive an ordinary differential equation that describes the dynamics near a specific resonance band and then to use the method of averaging. The original idea appears to be due to Melnikov [1963] and can also be found in Guckenheimer and Holmes [1983] and Greenspan and Holmes [1983].

We begin by recalling the form of the perturbed system in action-angle coordinates, which we rewrite below as

$$\begin{aligned} \dot{I} &= \varepsilon\left(\frac{\partial I}{\partial x}g_1 + \frac{\partial I}{\partial y}g_2\right) \equiv F(I,\theta,t,\varepsilon), \\ \dot{\theta} &= \Omega(I) + \varepsilon\left(\frac{\partial \theta}{\partial x}g_1 + \frac{\partial \theta}{\partial y}g_2\right) \equiv \Omega(I) + \varepsilon G(I,\theta,t,\varepsilon). \end{aligned} \tag{1.2.208}$$

Recall that the variable I in this case is the parameter which labels the unperturbed periodic orbits and that we are interested in the behavior of *resonant* periodic orbits under the perturbation, i.e., orbits labeled by I which satisfy the relation

$$\text{(resonance relation)} \qquad nT(I) = mT,$$

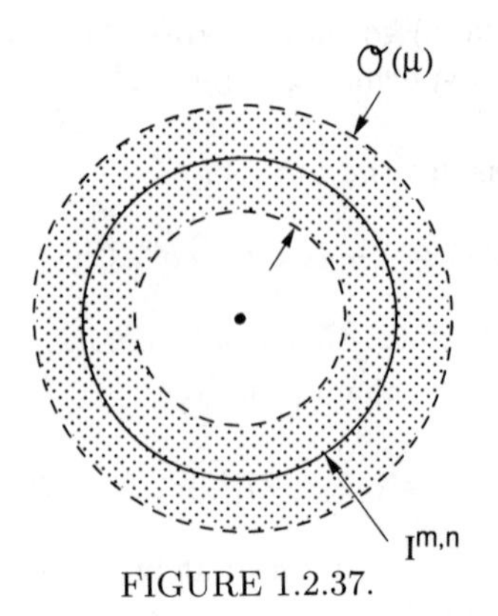

FIGURE 1.2.37.

where T is the period of the perturbation. We label such I as $I^{m,n}$. We now introduce the following transformation valid in the neighborhood of a *fixed* resonance band.

$$\begin{aligned} I &= I^{m,n} + \mu h, \\ \theta &= \Omega(I^{m,n})t + \phi = \left(\frac{2\pi n}{mT}\right) t + \phi, \end{aligned} \tag{1.2.209}$$

where μ is a small parameter which has yet to be determined.

Figure 1.2.37 illustrates that our transformation is valid in the shaded region of width $\mathcal{O}(\mu)$. Later we will determine μ in terms of ε.

Substituting this transformation into equation (1.2.208) gives

$$\begin{aligned} \mu \dot{h} &= \varepsilon F\big(I^{m,n} + \mu h, \Omega(I^{m,n})t + \phi, t, \varepsilon\big), \\ \Omega(I^{m,n}) + \dot{\phi} &= \Omega(I^{m,n} + \mu h) + \varepsilon G\big(I^{m,n} + \mu h, \Omega(I^{m,n})t + \phi, t, \varepsilon\big). \end{aligned} \tag{1.2.210}$$

Expanding the right-hand side of (1.2.210) in powers of μ and ε gives

$$\begin{aligned} \mu \dot{h} = {} & \varepsilon F\big(I^{m,n}, \Omega(I^{m,n})t + \phi, t, 0\big) + \varepsilon\mu \frac{\partial F}{\partial I}\big(I^{m,n}, \Omega(I^{m,n})t + \phi, t, 0\big) h \\ & + \mathcal{O}(\varepsilon\mu^2) + \mathcal{O}(\varepsilon^2), \end{aligned}$$

$$\begin{aligned} \Omega(I^{m,n}) + \dot{\phi} = {} & \Omega(I^{m,n}) + \mu \frac{\partial \Omega}{\partial I}(I^{m,n}) h + \mu^2 \frac{1}{2} \frac{\partial^2 \Omega}{\partial I^2}(I^{m,n}) h^2 \\ & + \varepsilon G\big(I^{m,n}, \Omega(I^{m,n})t + \phi, t, 0\big) + \mathcal{O}(\mu^3) + \mathcal{O}(\varepsilon\mu) + \mathcal{O}(\varepsilon^2), \end{aligned}$$

or (where we neglect the arguments of the functions for the sake of a simpler notation)

$$\begin{aligned} \dot{h} &= \frac{\varepsilon}{\mu} F + \varepsilon \frac{\partial F}{\partial I} h + \mathcal{O}(\varepsilon\mu) + \mathcal{O}\left(\frac{\varepsilon^2}{\mu}\right), \\ \dot{\phi} &= \mu \frac{\partial \Omega}{\partial I} h + \varepsilon G + \mu^2 \frac{1}{2} \frac{\partial^2 \Omega}{\partial I^2} h^2 + \mathcal{O}(\varepsilon\mu) + \mathcal{O}(\mu^3) + \mathcal{O}(\varepsilon^2). \end{aligned} \tag{1.2.211}$$

Now we address the question of how μ should be related to ε.

Our goal is to apply the method of averaging to (1.2.211). In order to do this we must have a small parameter multiplying the vector field; hence we will choose μ such that this situation holds. Therefore, requiring

$$\frac{\varepsilon}{\mu} = \mu,$$

we obtain

$$\mu = \sqrt{\varepsilon}.$$

Equation (1.2.211) thus becomes

$$\begin{aligned}
\dot{h} &= \sqrt{\varepsilon} F\big(I^{m,n}, \Omega(I^{m,n})t + \phi, t, 0\big) + \varepsilon \frac{\partial F}{\partial I}\big(I^{m,n}, \Omega(I^{m,n})t + \phi, t, 0\big)h \\
&\quad + \mathcal{O}(\varepsilon^{3/2}), \\
\dot{\phi} &= \sqrt{\varepsilon}\frac{\partial \Omega}{\partial I}(I^{m,n})h + \varepsilon\left(G\big(I^{m,n}, \Omega(I^{m,n})t + \phi, t, 0\big) + \frac{1}{2}\frac{\partial^2 \Omega}{\partial I^2}(I^{m,n})h^2\right) \\
&\quad + \mathcal{O}(\varepsilon^{3/2}).
\end{aligned} \tag{1.2.212}$$

We make the very important remark that the choice of $\mu = \sqrt{\varepsilon}$ depends on the fact that $\frac{\partial \Omega}{\partial I}(I^{m,n}) \neq 0$; otherwise, a different fractional power of ε is required. This issue is discussed in Morozov and Silnikov [1984]; see also Exercises 1.2.36 and 1.2.37. Henceforth we will assume that $\frac{\partial \Omega}{\partial I}(I^{m,n}) \neq 0$.

It should be clear that (1.2.212) is in the form in which the averaging theorem can be applied. However, first we will simplify the first term of the $\dot{h}$ component of (1.2.212) in a way that is computationally beneficial and makes an explicit connection with the subharmonic Melnikov theory. Using the expressions for the action-angle transformations, the chain rule, and implicit differentiation we obtain

$$\begin{aligned}
\frac{\partial I}{\partial x} &= \frac{\partial I}{\partial H}\frac{\partial H}{\partial x} = \frac{1}{\Omega(I)}\frac{\partial H}{\partial x}, \\
\frac{\partial I}{\partial y} &= \frac{\partial I}{\partial H}\frac{\partial H}{\partial y} = \frac{1}{\Omega(I)}\frac{\partial H}{\partial y};
\end{aligned}$$

thus,

$$F\big(I^{m,n}, \Omega(I^{m,n})t + \phi, t, 0\big) = \frac{1}{\Omega(I^{m,n})}(DH \cdot g)\big(I^{m,n}, \Omega(I^{m,n})t + \phi, t, 0\big),$$

where $DH \cdot g \equiv \frac{\partial H}{\partial x}g_1 + \frac{\partial H}{\partial y}g_2$. Therefore, to first order in $\sqrt{\varepsilon}$, (1.2.212) becomes

$$\begin{aligned}
\dot{h} &= \frac{\sqrt{\varepsilon}}{\Omega(I^{m,n})}(DH \cdot g)\big(I^{m,n}, \Omega(I^{m,n})t + \phi, t, 0\big), \\
\dot{\phi} &= \sqrt{\varepsilon}\frac{\partial \Omega}{\partial I}(I^{m,n})h.
\end{aligned} \tag{1.2.213}$$

Now we apply the averaging theorem to (1.2.213). Recall that since the explicit time dependence of the right-hand side is periodic, we merely average over one period of the perturbation and the averaging theorem allows us to draw conclusions about the full equations from the averaged equations. This we now do and obtain

$$\begin{aligned}\dot{\bar{h}} &= \frac{\sqrt{\varepsilon}}{\Omega(I^{m,n})}\frac{1}{T}\int_0^T (DH\cdot g)\big(I^{m,n},\Omega(I^{m,n})t+\bar{\phi},t,0\big)\,dt,\\ \dot{\bar{\phi}} &= \sqrt{\varepsilon}\frac{\partial\Omega}{\partial I}(I^{m,n})\bar{h}.\end{aligned} \tag{1.2.214}$$

To simplify (1.2.214), we use the following facts.

1. From the resonance relation, $T = \dfrac{2\pi n}{m\Omega(I^{m,n})}$.

2. From (1.2.180), the transformation
$$\big(x(I,\theta),y(I,\theta)\big)\leftrightarrow\big(I(x,y),\theta(x,y)\big)$$
has Jacobian equal to one, so we can transform the coordinates on which the integrand is evaluated back to our original coordinates, resulting in no change in the equations.

Combining these two remarks, we arrive at the averaged equations

$$\begin{aligned}\dot{\bar{h}} &= \frac{\sqrt{\varepsilon}}{\Omega(I^{m,n})mT}\int_0^{mT}(DH\cdot g)\left(q^{I^{m,n}}\left(t-\frac{\bar{\phi}}{\Omega(I^{m,n})}\right),t,0\right)dt,\\ \dot{\bar{\phi}} &= \sqrt{\varepsilon}\frac{\partial\Omega}{\partial I}(I^{m,n})\bar{h},\end{aligned} \tag{1.2.215}$$

where $q^{I^{m,n}}\left(t-\frac{\bar{\phi}}{\Omega(I^{m,n})}\right)$ represents the unperturbed periodic orbit with its period satisfying

$$nT(I,m,n) = mT.$$

(For a discussion regarding the nature of the argument of $q^{I^{m,n}}(\cdot)$, see the first remark following the proof of Theorem 1.2.13.)

Note that (unsurprisingly) the first term of the $\dot{h}$ equation is just the normalized first component of the subharmonic Melnikov vector. The averaged equations describing the dynamics near a resonance band are thus given by

$$\begin{aligned}\dot{\bar{h}} &= \sqrt{\varepsilon}\frac{1}{2\pi n}\bar{M}_1^{m/n}\left(\frac{\bar{\phi}}{\Omega(I^{m,n})}\right),\\ \dot{\bar{\phi}} &= \sqrt{\varepsilon}\frac{\partial\Omega}{\partial I}(I^{m,n})\bar{h},\end{aligned} \tag{1.2.216}$$

where $\bar{M}_1^{m/n}$ is defined in Remark 1 following the proof of Theorem 1.2.13. Therefore, the conditions

$$\begin{aligned}\bar{M}_1^{m/n}\left(\frac{\bar{\phi}}{\Omega(I^{m,n})}\right) &= 0,\\ \bar{h} &= 0,\end{aligned}$$

correspond to subharmonic periodic orbits for the original equation (just as we might have guessed, and in fact have already shown using the subharmonic Melnikov theory). However, we can obtain more information about the structure of the resonance band simply by examining (1.2.216), since the averaging theorem tells us that stable and unstable manifolds of hyperbolic fixed points of the averaged equations are close to corresponding structures in the full equations.

For this particular equation, however, there is a problem. Notice that the first-order averaged equations are a (structurally unstable with respect to arbitrary perturbations) Hamiltonian system with the Hamiltonian given by

$$H = \sqrt{\varepsilon}\left(\frac{\partial \Omega}{\partial I}(I^{m,n})\frac{\bar{h}^2}{2} - V(\bar{\phi})\right), \tag{1.2.217}$$

where

$$V(\bar{\phi}) = \frac{1}{2\pi n}\int \bar{M}_1^{m/n}\left(\frac{\bar{\phi}}{\Omega(I^{m,n})}\right) d\bar{\phi}.$$

Thus, if the perturbation is not Hamiltonian and autonomous (i.e., $g_1(x,y,\varepsilon) = \frac{\partial \tilde{H}}{\partial y}(x,y,\varepsilon)$, $g_2(x,y,\varepsilon) = -\frac{\partial \tilde{H}}{\partial y}(x,y,\varepsilon)$ for some $\mathbf{C}^{r+1}$ function $\tilde{H}$, then the first-order averaged equations cannot possibly capture the correct qualitative dynamics near the resonance band. Therefore, we will have to carry out the averaging procedure at least to second-order in $\sqrt{\varepsilon}$.

Now let us go back to the full, original equations

$$\begin{aligned}
\dot{h} &= \frac{\sqrt{\varepsilon}}{2\pi n}(DH\cdot g)(I^{m,n},\Omega(I^{m,n})t+\phi,t,0) \\
&\qquad + \varepsilon\frac{\partial F}{\partial I}(I^{m,n},\Omega(I^{m,n})t+\phi,t,0)h + \mathcal{O}(\varepsilon^{3/2}), \\
\dot{\phi} &= \sqrt{\varepsilon}\frac{\partial \Omega}{\partial I}(I^{m,n})h + \varepsilon\Big(G(I^{m,n},\Omega(I^{m,n})t+\phi,t,0) \\
&\qquad + \frac{1}{2}\frac{\partial^2 \Omega}{\partial I^2}(I^{m,n})h^2\Big) + \mathcal{O}(\varepsilon^{3/2}).
\end{aligned}$$

We recall from the proof of the averaging theorem that the method is effected by choosing a coordinate change which annihilates the time dependence at the highest order, and the nontrivial part of the coordinate change is chosen to be the antiderivative (with repect to time) of the oscillating part of the highest order term. The oscillating part of the $\mathcal{O}(\sqrt{\varepsilon})$ part of the $\dot{h}$ component of the vector field is given by

$$\begin{aligned}
\tilde{F}(I^{m,n},\Omega(I^{m,n})t+\phi,t,0) &= \frac{1}{\Omega(I^{m,n})}(DH\cdot g)(I^{m,n},\Omega(I^{m,n})t+\phi,t,0) \\
&\qquad - \frac{1}{2\pi n}\bar{M}_1^{m/n}\left(\frac{\phi}{\Omega(I^{m,n})}\right);
\end{aligned}$$

since the $\mathcal{O}(\sqrt{\varepsilon})$ part of the $\dot{\phi}$ component of the vector field is constant, we choose as the averaging transformation

$$\begin{aligned} h &\to \bar{h} + \sqrt{\varepsilon} \int \tilde{F}(I^{m,n}, \Omega(I^{m,n})t + \phi, t, 0) \\ \phi &\to \bar{\phi}, \end{aligned}$$

and note that

$$\begin{aligned} \frac{d}{dt} \int \tilde{F}(I^{m,n}, \Omega(I^{m,n})t + \phi, t, 0) = \tilde{F}(I^{m,n}, \Omega(I^{m,n})t + \phi, t, 0) \\ + \frac{\partial}{\partial \phi} \int \tilde{F}(I^{m,n}, \Omega(I^{m,n})t + \phi, t, 0)\dot{\phi}. \end{aligned}$$

Substituting this into the equation and doing some algebra (as in the proof of the averaging theorem) we arrive at

$$\begin{aligned} \dot{\bar{h}} &= \frac{\sqrt{\varepsilon}}{2\pi n} \bar{M}_1^{m/n} \left(\frac{\bar{\phi}}{\Omega(I^{m,n})} \right) + \varepsilon \left(\frac{\partial F}{\partial I}(I^{m,n}, \Omega(I^{m,n})t + \bar{\phi}, t, 0)\bar{h} \right. \\ & \left. - \frac{\partial}{\partial \phi} \int \frac{\partial \Omega}{\partial I}(I^{m,n}) \bar{h} \tilde{F}(I^{m,n}, \Omega(I^{m,n})t + \bar{\phi}, t, 0) \right), + \mathcal{O}(\varepsilon^{3/2}) \\ \dot{\bar{\phi}} &= \sqrt{\varepsilon} \frac{\partial \Omega}{\partial I}(I^{m,n})\bar{h} + \varepsilon \left(\frac{1}{2} \frac{\partial^2 \Omega}{\partial I^2}(I^{m,n})\bar{h}^2 \right. \\ & + G(I^{m,n}, \Omega(I^{m,n})t + \bar{\phi}, t, 0) \\ & \left. + \frac{\partial \Omega}{\partial I}(I^{m,n}) \int \tilde{F}(I^{m,n}, \Omega(I^{m,n})t + \bar{\phi}, t, 0) \right) + \mathcal{O}(\varepsilon^{3/2}). \quad (1.2.218) \end{aligned}$$

Averaging (1.2.218) at second order in $\sqrt{\varepsilon}$ and using the fact that $\tilde{F}$ has zero average (and, hence, so do $\int \tilde{F}$ and $\frac{\partial}{\partial \phi} \int \tilde{F}$) gives

$$\begin{aligned} \dot{\bar{h}} &= \frac{\sqrt{\epsilon}}{2\pi n} \bar{M}_1^{m/n} \left(\frac{\bar{\phi}}{\Omega(I^{m,n})} \right) + \varepsilon \overline{\frac{\partial F}{\partial I}}(\bar{\phi})\bar{h}, \\ \dot{\bar{\phi}} &= \sqrt{\varepsilon} \frac{\partial \Omega}{\partial I}(I^{m,n})\bar{h} + \varepsilon \left(\frac{1}{2} \frac{\partial^2 \Omega}{\partial I^2}(I^{m,n})\bar{h}^2 + \bar{G}(\bar{\phi}) \right), \end{aligned} \quad (1.2.219)$$

where we define

$$\begin{aligned} \overline{\frac{\partial F}{\partial I}}(\bar{\phi}) &\equiv \frac{1}{mT} \int_0^{mT} \frac{\partial F}{\partial I}(I^{m,n}, \Omega(I^{m,n})t + \bar{\phi}, t, 0)dt, \\ \overline{G}(\bar{\phi}) &= \frac{1}{mT} \int_0^{mT} G(I^{m,n}, \Omega(I^{m,n})t + \bar{\phi}, t, 0)dt. \end{aligned}$$

In the case where the perturbation is not Hamiltonian, equations (1.2.219) will often be sufficient to enable us to determine much of the dynamics near a particular resonance band. When the perturbation is Hamiltonian, special problems arise which we shall discuss as they are encountered.

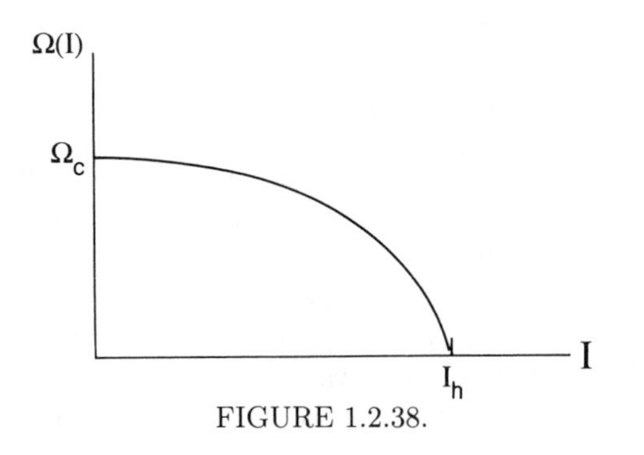

FIGURE 1.2.38.

iic) *Nonresonance*

Up to this point the discussion has been entirely concerned with resonances, i.e., the orbit structure in the neighborhood of an I value satisfying

$$nT(I) = mT.$$

Let us consider the nature of all the periodic orbits inside Γ_{p_0}. Consider Figure 1.2.38. In this figure we graph $\Omega(I)$ versus I, where Ω_c represents the frequency of the vector field linearized about the center type fixed point and I_h represents the value of action on the homoclinic orbit at which $\Omega(I_h) = 0$ (note: the action is defined on the homoclinic orbit but the angle variable is not defined on the homoclinic orbit). So, for $I \in [0, I_h]$, Ω takes all values between Ω_c and 0 (note: in Figure 1.2.38 we have illustrated the case where $\frac{\partial \Omega}{\partial I} < 0$; however, this presents no loss of generality). Therefore, for T fixed, for every m and n satisfying

$$n\frac{2\pi}{\Omega_c} \geq mT,$$

we can find a unique I value such that

$$n\frac{2\pi}{\Omega(I)} = mT.$$

Hence, there are a countable infinity of such resonant I levels or resonance bands. (Note: it should be clear to the reader that a possible effect from having $\frac{\partial \Omega}{\partial I} = 0$ might be to have more than one I value satisfying the resonance relation for a given n and m.) P_0^m fixes each resonant I level as a circle of fixed points. Clearly, this is a structurally unstable situation, and we would expect arbitrary perturbations to break up this circle of fixed points into a finite number of fixed points (cf. the remarks following the proof of Theorem 1.2.13).

It should be evident that there are an uncountable infinity of I values contained in the interval $[0, I_h]$ at which the resonance relation is not satisfied, i.e., I values for which

$$\frac{T(I)}{T} = \text{irrational number.} \tag{1.2.220}$$

It is natural to ask what is the nature of the orbit structure near these I values for the perturbed Poincaré map. As opposed to the resonance case, no iterate of P_ε^m will result in a circle of fixed points but, rather, orbits on the nonresonant I levels remain on these *invariant circles* and densely fill the circle under iteration by P_ε^m (cf. Section 1.2, Example 1.2.3). In this case on might expect that (some) invariant circles may be preserved under the perturbation. It turns out that this depends very much on whether or not the perturbation is Hamiltonian.

Non-Hamiltonian Perturbations

In this case there are no general theorems. However, a quantity that can give us much information is the determinant of the linearized Poincaré map denoted

$$\det DP_\varepsilon. \tag{1.2.221}$$

Recall that (1.2.221) is a local measure of the contraction or expansion of area. Thus, it follows that if (1.2.221) is constant everywhere (in general, DP_ε varies from point to point) and bounded away from one, then DP_ε cannot have any invariant circles; see Exercise 1.2.26. In this case one would expect the nonresonant invariant circles of the unperturbed Poincaré map to be destroyed. If (1.2.221) is not constant, then a more careful analysis is required.

Hamiltonian Perturbations

In this case, (1.2.221) is identically one and we are in the situation dealt with by the famous KAM (for Kolmogorov, Arnold, and Moser) theorem and the Moser twist theorem; see Moser [1973]. Roughly speaking, these theorems tell us that the nonresonant invariant circles having the property that (1.2.220) is poorly approximated by rational numbers in a number-theoretic sense are preserved as invariant circles filled with quasiperiodic orbits for the perturbed Poincaré map.

We now state the Moser twist theorem in a more general setting. Consider the unperturbed integrable map

$$\begin{aligned} I &\mapsto I, \\ \theta &\mapsto \theta + \alpha(I), \end{aligned} \tag{1.2.222}$$

defined on the annulus

$$A = \left\{(I,\theta) \in \mathbb{R}^+ \times S^1 \mid I \in [I_1, I_2]\right\}$$

and the perturbed map

$$\begin{aligned} I &\mapsto I + f(I,\theta), \\ \theta &\mapsto \theta + \alpha(I) + g(I,\theta), \end{aligned} \tag{1.2.223}$$

with f and g also defined on A (we will worry about differentiability shortly.) In order for (1.2.223) to be regarded as a perturbation of (1.2.222), f and g must be "small." Let $\mathbf{C}^r(A)$ denote the class of $\mathbf{C}^r$ functions defined on A. Then a norm on $\mathbf{C}^r(A)$, denoted $|\cdot|_r$, is defined as follows

$$h \in \mathbf{C}^r(A) \Rightarrow |h|_r = \sup_{\substack{i+j\leq r \\ A}} \left| \frac{\partial^{i+j} h}{\partial I^i \partial \theta^j} \right|.$$

We are now able to state the Moser twist theorem.

Theorem 1.2.14 (Moser [1973]) *Let $\varepsilon > 0$ be a positive number with $\alpha(I) \in \mathbf{C}^r$, $r \geq 5$, and $|\frac{\partial \alpha}{\partial I}| \geq \nu > 0$ in A. There then exists a δ depending on ε, r, and $\alpha(r)$ such that (1.2.223) with f and g, $\mathbf{C}^r$, $r \geq 5$, on A and*

$$|f(I,\theta) - I|_r + |g(I,\theta) - \alpha(I)|_r < \nu\delta$$

possesses an invariant circle in A with the parametric representation

$$I = \bar{I} + u(t), \quad \theta = t + v(t), \quad t \in [0, 2\pi),$$

where u and v are $\mathbf{C}^1$ with period 2π and satisfy

$$|u|_1 + |v|_1 < \varepsilon$$

with $\bar{I} \in [I_1, I_2]$. Moreover, the map restricted to this invariant circle is given by

$$t \to t + \omega, \qquad t \in [0, 2\pi),$$

where ω is incommensurate with 2π and satisfies the infinitely many conditions

$$\left| \frac{\omega}{2\pi} - \frac{p}{q} \right| \geq \gamma q^{-\tau} \tag{1.2.224}$$

for some $\gamma, \tau > 0$ and all integers $p, q > 0$. In fact, each choice of $\omega \in [\Omega(I_1), \Omega(I_2)]$ satisfying (1.2.224) gives rise to such an invariant circle.

Proof: See Moser [1973]. □

Several remarks are now in order.

Remark 1. Equation (1.2.224) indicates that the irrational number $\omega/2\pi$ is poorly approximated by rational numbers. Certainly there exist irrational numbers that do not satisfy this condition. The theorem says nothing about these.

Remark 2. Recall the mth iterate of the Poincaré map from (1.2.196),

$$\begin{aligned} I &\mapsto I + \varepsilon M_1(I,\theta) + \mathcal{O}(\varepsilon^2), \\ \theta &\mapsto \theta + mT\Omega(I) + \varepsilon M_2(I,\theta) + \mathcal{O}(\varepsilon^2). \end{aligned} \tag{1.2.225}$$

Note that we have left off the superscripts m/n on M_1 and M_2 since we are now interested in nonresonant dynamics.

a) The size of the perturbation of (1.2.223) from the integrable case is controlled by ε. This presents no problems, since M_1 and M_2, along with their first r derivatives, are bounded on bounded subsets of $\mathbb{R}^+ \times S^1$.

b) The quantity $mT\Omega(I)$ in (1.2.225) plays the role of $\alpha(I)$ in Theorem 1.2.14. Thus, $\frac{\partial \alpha}{\partial I} \neq 0$ if and only if $\frac{\partial \Omega}{\partial I} \neq 0$. We can therefore conclude from Theorem 1.2.14 that, for each $I = \bar{I} \in (0, I_h)$ such that

$$\left| \frac{mT\Omega(\bar{I})}{2\pi} - \frac{p}{q} \right| \geq \gamma q^{-\tau},$$

(1.2.223) possesses an invariant circle close to $I = \bar{I}$.

Remark 3. The existence of invariant circles can prove very useful for stability arguments. This is because the region of phase space enclosed by an invariant circle is an invariant set; see Exercises 1.2.25 and 1.2.27.

Remark 4. Theorem 1.2.14 gives rise to a natural question: Do all quasiperiodic orbits of (1.2.223) lie on invariant circles or, more precisely, is the closure of a quasiperiodic orbit of (1.2.223) an invariant circle? The answer to this question is no — there may be quasiperiodic orbits whose closure is a Cantor set. This structure is often called a *Cantorus.* For more information on Cantori the reader is referred to Aubry [1983a], [1983b], Mather [1982], and Percival [1979].

Remark 5. We have not directly defined the notion of a quasiperiodic orbit for a two-dimensional map. One definition might be an orbit whose closure is an invariant circle. If one thinks of the map as arising as a Poincaré map of a time-periodic ordinary differential equation, then the quasiperiodic orbit is just a two-frequency solution of the ordinary differential equation, where the two frequencies are incommensurate. The discovery of Cantori indicates that one might want to generalize the notion of a quasiperiodic orbit; see Mather [1982] for a discussion. However, at present, there is only a general existence theory for Cantori in two-dimensional, area-preserving twist maps (but see Katok and Bernstein [1987]).

Remark 6. Theorem 1.2.14 is a perturbation theorem, i.e., it asserts the existence of invariant circles only for a perturbation of unknown size. An interesting (and practical) problem would be to locate a quasiperiodic invariant circle in the unperturbed map and study what becomes of it as the strength of the perturbation increases. Recently, there has been much analytical and numerical work along these lines, and we refer the reader to Celletti and Chierchia [1988], Herman [1988], de la Llave and Rana [1988], MacKay [1988], MacKay, Meiss, and Stark [1989], MacKay and Percival [1985], Mather [1984, 1986], and Stark [1988].

Remark 7. The KAM theorem is more general than Theorem 1.2.14 in that it is concerned with the preservation of n-frequency quasiperiodic motions in perturbations of completely integrable, n-degree-of-freedom Hami systems. We refer the reader to Arnold [1978], Moser [1973], Sie Moser [1971], and Bost [1986] for the precise statement and some ... izations of the theorem.

1.2E Application to the Dynamics of the Damped, Forced Duffing Oscillator

We now want to apply this mass of theory to the damped, forced Duffing oscillator. Recall that this equation is given by

$$\begin{aligned} \dot{x} &= y, \\ \dot{y} &= x - x^3 + \varepsilon(\gamma \cos \omega t - \delta y), \end{aligned} \tag{1.2.226}$$

where ε is assumed small (note: this is so that we can rigorously apply our theory) and γ, δ, ω are positive parameters. The unperturbed system is given by

$$\begin{aligned} \dot{x} &= y, \\ \dot{y} &= x - x^3, \end{aligned} \tag{1.2.227}$$

and is Hamiltonian with Hamiltonian function

$$H(x, y) = \frac{y^2}{2} - \frac{x^2}{2} + \frac{x^4}{4}. \tag{1.2.228}$$

The first thing to do is to obtain a complete understanding of the geometry of the unperturbed phase space which we illustrate in Figure 1.2.39. As stated in Section 1.1E, all orbits are given by the level sets of the Hamiltonian (1.2.228). In the following we will gave analytic expressions for these orbits, leaving the details of the derivation of the expressions to the dedicated reader.

There are three equilibrium points at the following coordinates with the following stability types

$$\begin{aligned} (x, y) &= (\pm 1, 0) \text{ — centers,} \\ (x, y) &= (0, 0) \text{ — saddle.} \end{aligned}$$

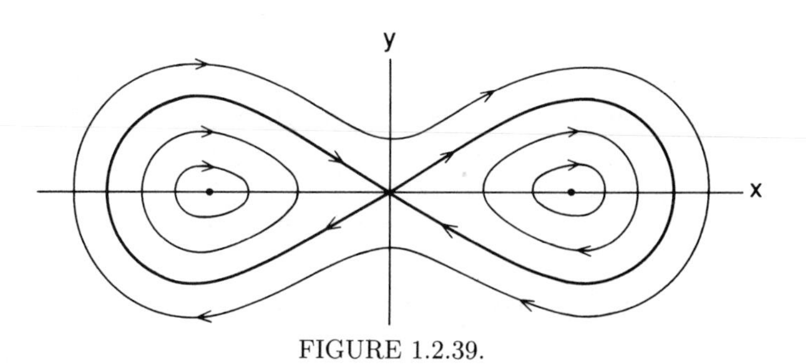

FIGURE 1.2.39.

The saddle point is connected to itself by a two homoclinic orbits given by

$$q_+^0(t) = (\sqrt{2}\,\mathrm{sech}\,t, -\sqrt{2}\,\mathrm{sech}\,t\tanh t),$$
$$q_-^0(t) = -q_+^0(t).$$

There are two families of periodic orbits inside the corresponding homoclinic orbits given by

$$q_+^k(t) = \left(\frac{\sqrt{2}}{\sqrt{2-k^2}}\,\mathrm{dn}\left(\frac{t}{\sqrt{2-k^2}}, k\right),\right.$$
$$\left.\frac{-\sqrt{2}k^2}{2-k^2}\,\mathrm{sn}\left(\frac{t}{\sqrt{2-k^2}}, k\right)\mathrm{cn}\left(\frac{t}{\sqrt{2-k^2}}, k\right)\right),$$
$$q_-^k(t) = -q_+^k(t), \qquad k \in (0,1),$$

where k is the elliptic modulus and $\mathrm{sn}(\cdot)$, $\mathrm{cn}(\cdot)$, and $\mathrm{dn}(\cdot)$ are elliptic functions (see Byrd and Friedman [1971]). Substituting the above expressions for the periodic orbits into the expression for the Hamiltonian gives the following relationship between the Hamiltonian and the elliptic modulus.

$$H\left(q_\pm^k(t)\right) \equiv H(k) = \frac{k^2-1}{(2-k^2)^2} \qquad \text{(constant on orbits)}.$$

Elementary properties of elliptic functions give the period of the above orbits as

$$T(k) = 2K(k)\sqrt{2-k^2},$$

where $K(k)$ is the complete elliptic integral of the first kind.

Also, there exists a family of periodic orbits outside the homoclinic orbits given by

$$q^k(t) = \left(\sqrt{\frac{2k^2}{2k^2-1}}\,\mathrm{cn}\left(\frac{t}{\sqrt{2k^2-1}}, k\right),\right.$$
$$\left.\frac{-\sqrt{2}k}{2k^2-1}\,\mathrm{sn}\left(\frac{t}{\sqrt{2k^2-1}}, k\right)\mathrm{dn}\left(\frac{t}{\sqrt{2k^2-1}}, k\right)\right), \quad k \in (1, 1/\sqrt{2}).$$

The periods of these orbits are given by

$$T(k) = 4K(k)\sqrt{2k^2-1}.$$

It is a simple matter to check that, in the limit as $k \to 1$, both families of periodic orbits converge to the homoclinic orbits; see Exercise 1.2.29 for justification of these statements.

Now we set the stage for our study of the perturbed system (1.2.226). Rewriting (1.2.226) as a third-order autonomous system gives

$$\begin{aligned}
\dot{x} &= y, \\
\dot{y} &= x - x^3 + \varepsilon(\gamma\cos\phi - \delta y), \qquad (x, y, \phi) \in \mathbb{R}^2 \times S^1, \\
\dot{\phi} &= \omega,
\end{aligned}$$

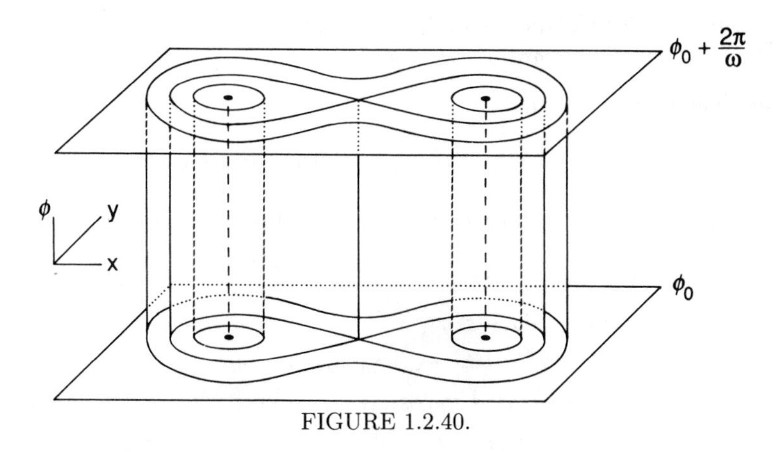

FIGURE 1.2.40.

where S^1 is the circle of length $2\pi/\omega$ and $\phi(t) = \omega t + \phi_0$. The *unperturbed* suspended phase space appears as in Figure 1.2.40. We form the global cross-section to the flow

$$\Sigma^{\phi_0} = \left\{ (x, y, \phi) \;\middle|\; \phi = \phi_0 \in \left[0, \frac{2\pi}{\omega}\right) \right\}$$

and the associated Poincaré map is given by

$$\begin{aligned} P\colon \Sigma^{\phi_0} &\to \Sigma^{\phi_0}, \\ \big(x(0), y(0)\big) &\mapsto \big(x(2\pi/\omega), y(2\pi/\omega)\big). \end{aligned}$$

Let us begin our study of the perturbed dynamics with two preliminary lemmas. We first want to show that there exists a closed convex set in $\mathbb{R}^2$, called D, such that the vector field is pointing strictly inward on the boundary of D (∂D) for all times. Thus we will establish the existence of a trapping region which will be of some use later on.

Consider the following scalar-valued function

$$L(x, y) = \frac{y^2}{2} + \nu xy - \frac{x^2}{2} + \frac{x^4}{4}, \qquad 0 < \nu < \varepsilon\delta.$$

Consider the level sets of this function

$$L(x, y) = C.$$

For C large the level sets are essentially ellipses (note that L is only a slight modification of the Hamiltonian); moreover, any line $y = \alpha x$ intersects any given level set (for C sufficiently large) in exactly two diametrically opposite points.

Let

$$\begin{aligned} D &= \{ (x, y) \mid L(x, y) \leq C,\ C \text{ large} \}, \\ \partial D &= \{ (x, y) \mid L(x, y) = C \}. \end{aligned}$$

Lemma 1.2.15 (Holmes and Whitley [1984]) *The perturbed vector field points strictly inward on ∂D for all times.*

Proof: The lemma is true if

$$\nabla L \cdot (\dot{x}, \dot{y}) < 0 \qquad \text{on } \partial D \text{ for all time,}$$

where $(\dot{x}, \dot{y})$ is the perturbed vector field. Calculating this expression gives

$$\begin{aligned} \nabla L \cdot (\dot{x}, \dot{y}) &= (\nu y - x + x^3, y + \nu x) \cdot \big(y, x - x^3 + \varepsilon(\gamma \cos \omega t - \delta y)\big) \\ &= \nu y^2 - xy + x^3 y + xy - x^3 y + \varepsilon(\gamma y \cos \omega t - \delta y^2) \\ &\quad + \nu x^2 - \nu x^4 + \varepsilon(\nu \gamma x \cos \omega t) - \nu \delta x y) \\ &= (\nu - \varepsilon\delta) y^2 + \varepsilon\gamma \cos \omega t (y + \nu x) - \nu x^2 (x^2 - 1) - \varepsilon \nu \delta x y \\ &\leq -(\varepsilon\delta - \nu) y^2 - \nu x^2 (x^2 - 1) + \varepsilon \nu \delta |xy| + \varepsilon\gamma |y + \nu x|. \end{aligned}$$

Recall that we require $\nu < \varepsilon\delta$. On the y-axis this expression becomes

$$\nabla L \cdot (\dot{x}, \dot{y}) \leq -(\varepsilon\delta - \nu) y^2 + \varepsilon\gamma |y|;$$

hence, for y large, this is strictly negative. Also, on any line $y = \alpha x$, the expression becomes

$$\nabla L \cdot (\dot{x}, \dot{y}) \leq -(\varepsilon\delta - \nu)\alpha^2 x^2 - \nu x^2 (x^2 - 1) + \varepsilon \nu \delta |\alpha| x^2 + \varepsilon\gamma |x||1 + \nu\alpha|.$$

As α varies between $(-\infty, \infty)$, all points are swept out in D, so we see that, for x sufficiently large, on any line $y = \alpha x$, $\alpha \in (-\infty, \infty)$, the expression is strictly negative. This proves the lemma. $\square$

Remark 1. The above lemma depends entirely on the damping (δ) term; without that term it is not true.

Lemma 1.2.16

$$\det DP = e^{-2\pi\varepsilon\delta/\omega} < 1, \qquad \delta > 0.$$

Proof: We give a general proof for the determinant of any Poincaré map. Consider a general t-periodic ordinary differential equation

$$\dot{x} = f(x, t), \qquad \text{with } f(x, t) = f(x, t + T).$$

Suppose $\bar{x}(t)$ is a solution. Linearize about this solution with the following variational equation

$$\dot{\xi} = D_x f(\bar{x}(t), t))\xi.$$

This equation has the fundamental solution matrix

$$X(t),$$

so the general solution of the linearized equation is $\xi(t) = X(t)\xi_0$. Therefore, the linearized Poincaré map of the original equation is given by

$$x_0 \mapsto X(T)x_0;$$

the Jacobian is thus

$$DP = X(T).$$

By Liouville's formula (Arnold [1973]), we have that

$$\det X(T) = \det DP = \exp \int_0^T \operatorname{tr} D_x f\big(\bar{x}(t), t\big)\, dt.$$

For our system (1.2.226), we thus have

$$f = \begin{pmatrix} y \\ x - x^3 - \varepsilon\delta y + \varepsilon\gamma\cos\omega t \end{pmatrix},$$

$$Df = \begin{pmatrix} 0 & 1 \\ 1 - 3x^2 & -\varepsilon\delta \end{pmatrix},$$

so that $\operatorname{tr} Df = -\varepsilon\delta =$ constant and, therefore, $\det DP = e^{-2\pi\varepsilon\delta/\omega}$; see Exercise 1.2.39. □

A consequence of Lemma 1.2.16 is that, for $\delta > 0$, the Poincaré map is area-contracting. Therefore, it cannot possess any invariant circles.

To obtain more detailed information concerning the dynamics of the perturbed Poincaré map, we compute the Melnikov function $\bar{M}_1^{m/n}(t_0)$, where we omit denoting the explicit dependence on the cross-section ϕ_0 and α (whose role will be played by the elliptic modulus k), since we will be interested in resonant periodic orbits, which will be labeled by m/n through the resonance relation. The reason that we compute only one component of the subharmonic Melnikov vector is that

$$\frac{\partial\Omega}{\partial I} \neq 0. \tag{1.2.229}$$

We will leave it to the reader to verify (1.2.229); however, it should become apparent from various expressions for the frequency which we will derive during the course of our analysis.

The Melnikov function for the two families of periodic orbits inside the homoclinic orbits is given by

$$\bar{M}_1^{m/n}(t_0; \gamma, \delta, \omega) = -\delta J_1(m, n) \pm \gamma J_2(m, n, \omega) \sin\omega t_0, \tag{1.2.230}$$

where m, n are relatively prime positive integers satisfying the resonance relation

$$2K(k)\sqrt{2 - k^2} = \frac{2\pi m}{\omega n},$$

$+$ and $-$ refer to the right-hand and left-hand families of periodic orbits, respectively (note: by monotonicity of the period this equation has a solution for each choice of m, n with $2\pi m/(\omega n) > \sqrt{2}\pi$), and $J_1(m,n)$, $J_2(m,n,\omega)$ are complicated positive expressions involving elliptic functions which are given as follows (note: $K(k)$ and $E(k)$ denote the complete elliptic integrals of the first and second kind, respectively)

$$J_1(m,n) = \frac{4}{3}\left((2-k^2)E(k) - 2k'^2K(k)\right)/(2-k^2)^{3/2},$$

where $k' = \sqrt{1-k^2}$, and

$$J_2(m,n,\omega) = \begin{cases} 0 & \text{for } n \neq 1 \\ \sqrt{2}\pi\omega \operatorname{sech} \dfrac{\pi m K(k')}{K(k)} & \text{for } n = 1 \end{cases}.$$

Details of these calculations can be found in Greenspan and Holmes [1984]; see also Exercise 1.2.30. Thus we see that no ultrasubharmonics inside the homoclinic orbits are excited by the perturbation. Therefore, hereafter we will write

$$\bar{M}_1^{m/1}(t_0;\gamma,\delta,\omega) = -\delta J_1(m,1) \pm \gamma J_2(m,1,\omega)\sin\omega t_0. \tag{1.2.231}$$

If we define

$$R^m(\omega) = \frac{J_1(m,1)}{J_2(m,1,\omega)},$$

it is easy to see that the condition for the existence of zero's of the Melnikov function becomes

$$-\delta R^m(\omega) \pm \gamma \sin\omega t_0 = 0$$

or

$$\left|\frac{\delta R^m(\omega)}{\gamma}\right| \leq 1$$

or

$$\frac{\gamma}{\delta} \geq R^m(\omega). \tag{1.2.232}$$

We now want to carefully explain geometrically just what this condition means.

Consider the unperturbed Poincaré map on the cross-section Σ and the unperturbed periodic orbit satisfying the resonance relation

$$T(k) = 2K(k)\sqrt{2-k^2} = \frac{2\pi m}{\omega}.$$

We illustrate the solution in Figure 1.2.41. (Note: since, inside the homoclinic orbit, the identical situation occurs in both the right and left half-plane (by the symmetry), hereafter we will draw only the right side and not the homoclinic orbit.)

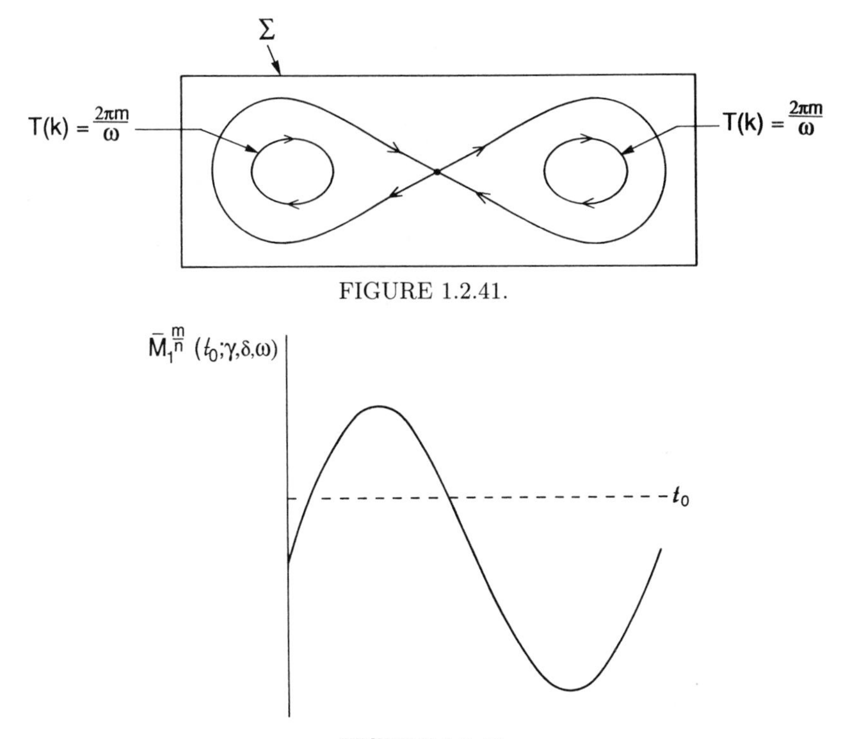

FIGURE 1.2.41.

FIGURE 1.2.42.

Therefore, for the unperturbed map, the above circles are all period m points or, equivalently, fixed points of the mth iterate of the map. We now ask if any of these period m points are preserved for the perturbed map, and the answer is provided by the Melnikov function.

1. For $\gamma/\delta < R^m(\omega)$, the Melnikov function has no zero's; therefore, the perturbed Poincaré map has no period m points for this particular m.

2. For $\gamma/\delta > R^m(\omega)$, the Melnikov function has zero's, and we want to count how many.

Because the Melnikov function is periodic with period $2\pi/\omega$, in one period the Melnikov function appears as in Figure 1.2.42. Thus, during one period of the perturbation, the Melnikov function has two zero's. From our earlier work concerning stability, we know that one zero corresponds to a period m saddle. With additional work, we will show that the other zero corresponds to a period m sink for $\delta > 0$.

The length of time it takes to get around the unperturbed orbit (i.e., its period) is $2\pi m/\omega$; hence, we see that during this time the Melnikov function passes through m periods. Therefore, we have that on this particular resonant level, the perturbed Poincaré map has $2m$ periodic points (or equivalently, the mth iterate has $2m$ fixed points), m of which are saddles and, as we will see later, m of which are sinks.

Now let us examine condition (1.2.232) a bit more closely. For m fixed, with $\gamma/\delta < R^m(\omega)$, there are no period m points on the resonance band and, for $\gamma/\delta > R^m(\omega)$, there are $2m$ period m points on the resonance band. Thus, the parameter values $\gamma/\delta = R^m(\omega)$ are "critical" in some sense. In Chapter 3 we will see that this is an example of a *bifurcation*.

We could also consider the limit of $R^m(\omega)$ as $m \to \infty$, i.e.,

$$\lim_{m\to\infty} R^m(\omega) \equiv R^0(\omega). \tag{1.2.233}$$

It is easy to verify that this limit exists, but what does it mean? Interpreting it in the same manner as for the periodic orbits, we would conclude that for

$$\frac{\gamma}{\delta} > R^0(\omega),$$

we would have *infinitely* many periodic orbits. When we study the breakup of Γ_{p_0} under the perturbation in Chapter 4 we will see that this is indeed the case (note: in the limit $m \to \infty$, for resonance $T(k) \to \infty$, which implies that we are approaching the homoclinic orbit). Moreover, this phenomenon will lie at the heart of what we call *chaos*.

Equations (1.2.232) and (1.2.233) have implications for the global structure of the Poincaré map. For example, for $\omega = 1$ it can be verified that $R^m(1)$ approaches $R^0(1)$ monotonically. Thus, if we pick any M, inner (referring to inside the homoclinic orbit) subharmonics of order M exist if $\gamma > R^M(1)\delta$. Also, since $R^m(1) < R^M(1)$ $\forall m < M$, we have that

$$\gamma > R^M(1)\delta > R^m(1)\delta \qquad \forall m < M.$$

Thus, all inner subharmonics of order m, $m \leq M$ are also excited. We remark that for $\omega \neq 1$, the sequence may not be monotonic; see Exercise 1.2.30.

For the orbits outside the homoclinic orbit we can carry out the same computation of the Melnikov function (see Exercise 1.2.30) and obtain

$$\bar{M}_1^{m/1}(t_0;\gamma,\delta,\omega) = -\delta\hat{J}_1(m,1) - \gamma\hat{J}_2(m,1,\omega)\sin\omega t_0,$$

where $\hat{J}_1$ and $\hat{J}_2$ are complicated positive expressions involving elliptic functions (note: again we have $n = 1$) where m must be odd. Letting $\hat{R}^m(\omega) = \hat{J}_1(m,1)/\hat{J}_2(m,1,\omega)$, the condition (in terms of the parameters) for the existence of subharmonics of order m outside the homoclinc orbits is given by

$$\frac{\gamma}{\delta} > \hat{R}^m(\omega), \qquad m \text{ odd}.$$

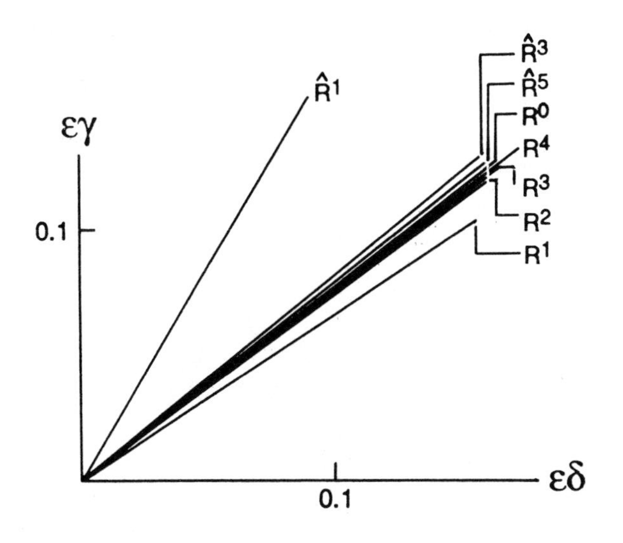

FIGURE 1.2.43.

As for the inner subharmonics, we can verify

$$\lim_{m \to \infty} \hat{R}^m(\omega) \equiv R^0(\omega),$$

and for $\omega = 1$, this $R^0(\omega)$ is approached monotonically. In Figure 1.2.43 we show the subharmonic bifurcation curves (this will be justified in Chapter 3) for $\omega = 1$.

i) THE RESONANCE BANDS IN DUFFING'S EQUATION

We now apply the previously developed theory to a study of the resonance bands in Duffing's equation; we follow Greenspan and Holmes [1983, 1984] and Morozov [1976]. We will discuss only the inner right-hand subharmonics because essentially the same behavior is exhibited by the other subharmonics (with the exception of m being *odd* for the outer subharmonics). Recall that we are studying periodic orbits in resonance with the external time-periodic forcing which satisfy the resonance relation

$$nT(I) = mT.$$

In terms of the expressions for the unperturbed periodic orbits, this can be written as

$$2K(k)\sqrt{2 - K^2} = \frac{2\pi m}{n\omega},$$

where the relation between I and K is given by

$$I(k) = \frac{2(2 - k^2)E(k) - 4k'^2 K(k)}{3\pi(2 - k^2)^{3/2}}, \qquad k'^2 = 1 - k^2$$

(note: it is not hard to verify that the relation between I and k is monotonic).

Previously we had the monotonic relation between H and k given by

$$H(k) = \frac{k^2 - 1}{(2 - k^2)^2}.$$

Thus, using these two relations we can compute

$$\begin{aligned}
\frac{\partial \Omega}{\partial I} = \frac{\partial}{\partial I}\left(\frac{2\pi}{T(I)}\right) &= -\frac{2\pi}{T(k)^2}\frac{\partial}{\partial k}(T(k)) \Big/ \frac{\partial}{\partial k}(I(k)) \\
&= -\frac{\pi^2(2-k^2)\big((2-k^2)E(k) - 2k'^2K(k)\big)}{2k^4k'^2K(k)^3} \equiv \Omega'(m),
\end{aligned}$$

where we denote the argument of Ω' by m to indicate the unique m selected for each k by the resonance relation.

Now, using the transformation valid in the neighborhood of a fixed resonance level, recall that to first order in $\sqrt{\varepsilon}$ our system becomes the following Hamiltonian system

$$\begin{aligned}
\dot{\bar{h}} &= \sqrt{\varepsilon}\frac{1}{2\pi n}\bar{M}_1^{m/n}\left(\frac{\bar{\phi}}{\Omega(m)}\right), \\
\dot{\bar{\phi}} &= \sqrt{\varepsilon}\Omega'(m)\bar{h}
\end{aligned}$$

with Hamiltonian function

$$H = \sqrt{\varepsilon}\left(\frac{\Omega'(m)\bar{h}^2}{2} - V(\bar{\phi})\right).$$

Using our previous calculations of the Melnikov function, we can immediately put our system in this form and write down the Hamiltonian as follows

$$\begin{aligned}
\dot{\bar{h}} &= \frac{\sqrt{\varepsilon}}{2\pi}(-\delta J_1(m) + \gamma J_2(m,\omega)\sin(m\bar{\phi})), \\
\dot{\bar{\phi}} &= \sqrt{\varepsilon}\Omega'(m)\bar{h},
\end{aligned}$$

$$H = \sqrt{\varepsilon}\left(\frac{\Omega'(m)}{2}h^2 + \frac{1}{2\pi}\left(\delta J_1(m)\bar{\phi} + \frac{\gamma J_2(m,\omega)}{m}\cos m\bar{\phi}\right)\right),$$

where we have written $J_1(m,1) \equiv J_1(m)$, $J_2(m,1,\omega) \equiv J_2(m,\omega)$, since we found that, for $n \neq 1$, there were no fixed points for P_ε^m. Now we examine the structure of the mth-order resonance level given by the $\mathcal{O}(\sqrt{\varepsilon})$ truncation of our full system.

Fixed points of our system are given by

$$\begin{aligned}
&-\delta J_1(m) + \gamma J_2(m,\omega)\sin(m\bar{\phi}) = 0, \\
&\bar{h} = 0.
\end{aligned}$$

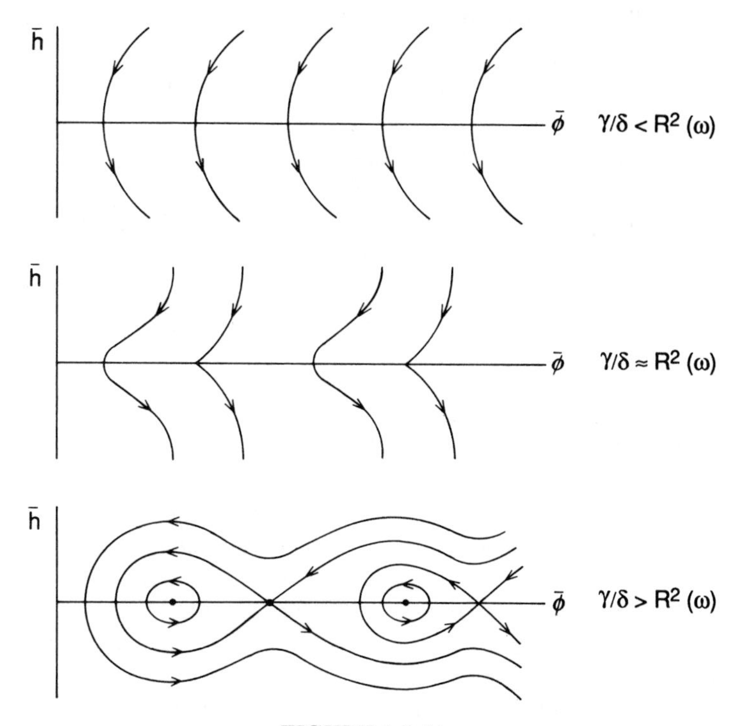

FIGURE 1.2.44.

We can compute the stability type of the fixed points by linearizing about the appropriate fixed point. The linearization of the vector field is given by

$$\begin{pmatrix} 0 & \frac{\sqrt{\varepsilon}\gamma m J_2(m,\omega)}{2\pi}\cos m\bar{\phi} \\ \sqrt{\varepsilon}\Omega'(m) & 0 \end{pmatrix}.$$

When γ and δ are appropriately chosen (assuming that ω is fixed) as $\bar{\phi}$ varies from 0 to 2π (i.e., once around the unperturbed periodic orbit), the Melnikov function passes through m periods; hence it can have at most $2m$ zero's (a result which we have already established using a slightly different argument). It is easy to see from the above matrix that these alternate in stabilty type from saddles to centers (we should also have concluded this from the Hamiltonian structure). For the $\mathcal{O}(\sqrt{\varepsilon})$ system we draw the $m = 2$ resonance band in Figure 1.2.44, where $R^m(\omega) = J_1(m)/J_2(m,\omega)$ (note: we obtain the correct directions for the arrows by examining the averaged vector field and noting that $\Omega'(m) < 0$).

Because we cannot conclude anything about the full system from this structurally unstable Hamiltonian system, we must therefore include second-order ($\mathcal{O}(\varepsilon)$) terms in our equation. Without doing the explicit calculations and using the notation and formalism outlined in the previous section, tedious but routine calculations give

$$\overline{\frac{\partial F}{\partial I}} = -\gamma K_2(m, w) \sin m\bar{\phi} - \delta K_1(m),$$
$$\bar{G} = -\gamma K_2(m, \omega) \frac{\cos m\bar{\phi}}{m},$$

where K_1 and K_2 are positive constants involving elliptic integrals.

Therefore, the averaged system to $\mathcal{O}(\varepsilon)$ is given by

$$\begin{aligned}\dot{\bar{h}} &= \frac{\sqrt{\varepsilon}}{2\pi} \left(-\delta J_1(m) + \gamma J_2(m, \omega) \sin m\bar{\phi}\right) \\ &\quad -\varepsilon\big(\delta K_1(m) + \gamma K_2(m, \omega) \sin m\bar{\phi}\big)\bar{h}, \\ \dot{\bar{\phi}} &= \sqrt{\varepsilon}\frac{\partial \Omega}{\partial I}(m)\bar{h} + \varepsilon \left(\frac{1}{2}\frac{\partial^2 \Omega}{\partial I^2}(m)\bar{h}^2 - \gamma K_2(m, \omega)\frac{\cos m\bar{\phi}}{m}\right).\end{aligned}$$

An easy computation of the trace of the linearized system gives $-\epsilon\delta K_1(m) < 0$, so that the centers are actually sinks. Also, by Bendixson's criterion, we know that there are no closed orbits in a resonance band. Therefore, by appealing to the averaging theorem, we can conclude that the Poincaré map in the neighborhood of a resonance band is diffeomorphic to Figure 1.2.45 (we draw the $m = 2$ resonance band for definiteness).

Next, we want to obtain an estimate of the width of the resonance band and the size of the domains of attraction of the sinks. Recall that, at $\mathcal{O}(\sqrt{\varepsilon})$, the system in a neighborhood of a resonance level is Hamiltonian with Hamiltonian function given by

$$H = \sqrt{\varepsilon} \left(\frac{\Omega'(m)}{2}\bar{h}^2 + \frac{1}{2\pi}\left(\delta J_1(m)\bar{\phi} + \frac{\gamma J_2(m, \omega)}{m} \cos m\bar{\phi}\right)\right),$$

and that, for $\gamma > R^m(\omega)\delta$, the phase portrait of this Hamiltonian system is as shown in Figure 1.2.46, where we draw the $m = 2$ case for definiteness. Recall also that a measure of the width of the energy level would be the length of the vertical lines (labeled $\Delta\bar{h}$) passing through the centers, as shown in Figure 1.2.46. To make the picture clearer, we will draw just one saddle-center pair in Figure 1.2.47, where we have labeled the angular values of the center and saddle by $\bar{\phi}_1$ and $\bar{\phi}_2$, respectively. The value of the Hamiltonian on the homoclinic orbit at the same angular value as the center is labeled as H_1 and the value of the Hamiltonian at the saddle as H_2. Clearly, we must have $H_1 = H_2$. Now, from Figure 1.2.47, the width of the resonance band in the original coordinates is given by

$$\Delta I = \sqrt{\varepsilon}\Delta\bar{h} + \mathcal{O}(\varepsilon),$$

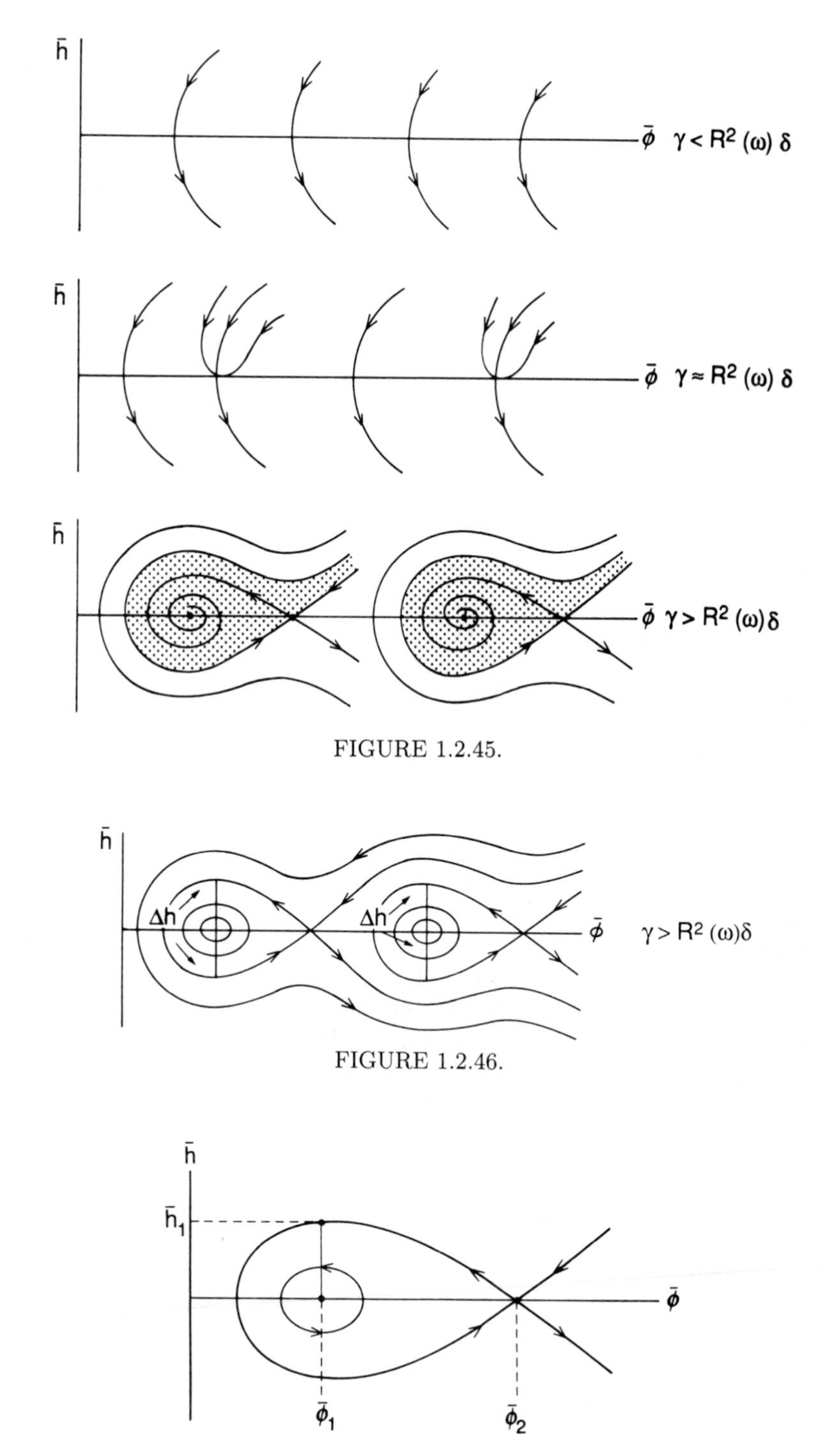

FIGURE 1.2.45.

FIGURE 1.2.46.

FIGURE 1.2.47.

where $\Delta\bar{h} = 2\bar{h}_1$. We can calculate $\bar{h}_1$ as follows

$$H_1 = \sqrt{\varepsilon}\left(\frac{\Omega'}{2}\bar{h}_1^2 + V(\bar{\phi}_1)\right),$$

where

$$V(\bar{\phi}_1) = \frac{1}{2\pi}\left(\delta J_1(m)\bar{\phi}_1 + \frac{\gamma J_2(m,\omega)}{m}\cos m\bar{\phi}_1\right).$$

H_2 can thus be calculated as

$$H_2 = \sqrt{\varepsilon}V(\bar{\phi}_2);$$

we therefore obtain

$$V(\bar{\phi}_2) = \frac{\Omega'}{2}\bar{h}_1^2 + V(\bar{\phi}_1)$$

or

$$\bar{h}_1 = \sqrt{\frac{2}{\Omega'}(V(\bar{\phi}_2) - V(\bar{\phi}_1))}.$$

The width of the resonance band is thus given by

$$\Delta I = 2\sqrt{\varepsilon}\sqrt{\frac{2}{\Omega'}(V(\bar{\phi}_2) - V(\bar{\phi}_1))} + \mathcal{O}(\varepsilon),$$

where $\bar{\phi}_2$, $\bar{\phi}_1$ are zero's of the Melnikov function at an adjacent saddle and center, respectively.

Recall that $V(\bar{\phi})$ is the indefinite integral of the Melnikov function; this is another example of information contained in the Melnikov function.

The case $\delta = 0$. In this case the perturbation is Hamiltonian, and on the order m resonance level we have m saddles and m elliptic fixed points. Using standard perturbation methods, we would find that the stable and unstable manifolds of the saddle coincide as shown in Figure 1.2.48a. However (we will justify this in Chapter 4), we typically expect the stable and unstable manifolds of the saddle to intersect infinitely many times, forming a so-called "stochastic layer" as illustrated in Figure 1.2.48b. The width of such stochastic layers is exponentially small in ε and therefore "beyond all orders" in standard perturbation theory. In Chapter 4 we will discuss how one might verify such behavior with more modern global perturbation methods.

Validity of the Approximation

Now we want to examine the validity of the results we obtained through the method of averaging. Recall that to $\mathcal{O}(\sqrt{\varepsilon})$ the averaged equations were given by

$$\begin{aligned}\dot{\bar{h}} &= \frac{\sqrt{\varepsilon}}{2\pi}(-\delta J_1(m) + \gamma J_2(m,\omega)\sin m\bar{\phi}),\\ \dot{\bar{\phi}} &= \sqrt{\varepsilon}\Omega'(m)\bar{h}.\end{aligned}$$

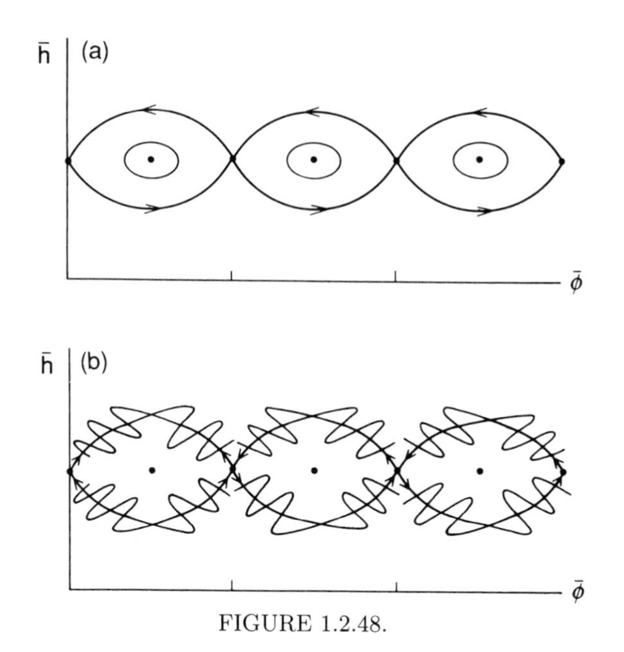

FIGURE 1.2.48.

From the expression for the frequency given in terms of elliptic functions, it is not hard to show that, as $m \uparrow \infty$ (or equivalently, as $k \to 1$ or $k' \to 0$), we have

$$\Omega'(m) \approx \frac{-1}{k'^2\big(\log(4/k')\big)^3} \approx -\frac{\omega^3 \exp(2\pi m/\omega)}{m^3};$$

thus, $\Omega'(m) \to -\infty$ exponentially fast; see Exercise 1.2.31. Therefore, by examining the equations given above, one can see (from the $\dot{\bar{\phi}}$ term) that our stability results have become valid for smaller and smaller ε; we can thus make no claims (other than existence) about subharmonics in an arbitrarily small neighborhood of the homoclinic orbit. (Note: studying the structure of the Poincaré map we derived previously is not a way out of this difficulty, since we used action-angle variables in that derivation, and the action-angle transformation becomes singular on the homoclinic orbit.)

In order to understand what is happening arbitrarily close to the homoclinic orbit, we will have to use a different form of analysis, which we will present in Section 4.7.

One other result which we will state without proof is the following. Recall that we gave an expression for the leading order term of the width of the resonance bands. Using the asymptotic forms for the bahavior of the elliptic integrals involved in that expression, we find that

$$\Delta I \approx \sqrt{\varepsilon \frac{m^3}{\omega^3} \exp\left(\frac{-2\pi m}{\omega}\right)} + \mathcal{O}(\varepsilon) \qquad \text{as } m \uparrow \infty.$$

Thus, the widths of the resonance bands shrink exponentially fast as the homoclinic orbit is approached. This explains why in practice only low order ($m \leq 5$) subharmonics are usually observed.

ii) Interactions Between Resonance Bands

Now that we have some understanding of the dynamics on any given resonance band, we want to gain some understanding as to how adjacent resonance bands can interact; the discussion here will be particular to the Duffing equation; however, the same type of behavior also occurs in other applications.

Recall that the subharmonic Melnikov function for the Duffing equation is given by

$$\bar{M}_1^{m/n}(t_0; \gamma, \delta, \omega) = -\delta J_1(m,n) + \gamma J_2(m,n,\omega) \sin \omega t_0,$$

where

$$J_1(m,n) = \frac{2}{3}((2-k^2)2E(k) - 4k'^2 K(k))/(2-k^2)^{3/2} > 0,$$
$$J_2(m,n,\omega) = \begin{cases} 0 & \text{for } n \neq 1 \\ \sqrt{2}\pi\omega \operatorname{sech} \frac{\pi m K(k')}{K(k)} & \text{for } n = 1 \end{cases}.$$

We thus see that, in the Duffing equation, resonance bands exist only for $n = 1$. Recall also that the Melnikov function is a measure of the "push" in the radial direction (i.e., away from the resonance bands) due to the perturbation. Now let us consider adjacent resonance bands inside the homoclinic orbit (exactly the same argument applies to resonance levels outside the homoclinic orbit, except that the ordering by rational numbers is reversed) of order $(m+1)/1$ and $m/1$, respectively. Between these two resonance bands is a dense set of resonance levels of order $\bar{m}/n$, $n > 1$. However, an examination of the form of the Melnikov function shows that it has no zero's on these resonance levels (and hence there are no periodic orbits between the $(m+1)/1$ resonance band and the m resonance band) and that on each $\bar{m}/n$, $n > 1$, band the Melnikov function is strictly negative. Therefore, we conclude that orbits starting in a neighborhood of the $(m+1)/1$ resonance band, not in the domain of attraction of the sink, make their way to the $m/1$ resonance band. More specifically, one component of the unstable manifold of the saddle on the $(m+1)/1$ resonance level can be found in a neighborhood of the $m/1$ resonance level, and one component of the stable manifold of the saddle on the $m/1$ resonance level can be found in a neighborhood of the $(m+1)/1$ resonance level. We can actually prove more.

Theorem 1.2.17 *The unstable manifold of the saddle-point on the* $(m+1)/1$ *resonance band intersects topologically transversely both components of the stable manifold of the saddle on the* $m/1$ *resonance band.*

Of course, this assumes that the $(m+1)/1$ resonance band exists. The proof of this theorem was originally given by Morozov [1976]. We will merely outline the main ideas and refer the reader to the original paper for all the details. However, first we want to make some general remarks.

Remark 1. By the term "topologically transverse," we mean that the intersection of the manifolds may be of odd order.

Remark 2. The theorem as stated applies to the inner subharmonics; however, the same conclusions hold for the outer subharmonics, but the order of the integers labeling the resonance bands is reversed. The idea is that unstable manifolds of larger amplitude subharmonics intersect the stable manifolds of lower amplitude subharmonics.

Proof: The proof will consist of considering the possible cases. For notation we label the saddles and sinks on the $(m+1)/1$ resonance band by Sa^{m+1}, Si^{m+1}, respectively, and their respective stable and unstable manifolds by $W^s(Sa^{m+1})$, $W^s(Si^{m+1})$, $W^u(Sa^{m+1})$, and $W^u(Si^{m+1})$. The obvious similar notation holds for the saddles and sinks on the $m/1$ resonance band.

Before enumerating the different possible cases we make the following trivial, yet very important, remark. Recall that the *unperturbed* Poincaré map was as given below

$$\begin{aligned} I &\mapsto I, \\ \theta &\mapsto \theta + T\Omega(I), \end{aligned}$$

and the phase space of the map appears as in Figure 1.2.49. Consider the pie-shaped region in Figure 1.2.49 labeled π and its image under one iterate labeled $P(\pi)$. You can see that radial lines are twisted, yet the orientation of π is preserved and, consequently, the ordering of radial lines eminating from the center and foliating π (ordering as measured by an angular coordinate along any circle intersecting the foliation transversely) is also preserved under P. Clearly this property is also preserved for the perturbed map.

Now we give the possible cases.

1. *Case a.* $W^u(Sa^{m+1}) \cap W^s(Sa^m) = \emptyset$.
2. *Case b.* $W^u(Sa^{m+1}) \cap W^s(Si^m) \neq \emptyset$.
3. *Case c.* $W^u(Sa^{m+1})$ intersects *only 1 component* of $W^s(Sa^m)$.
4. *Case d.* $W^u(Sa^{m+1})$ intersects $W^s(Sa^m)$ nontopologically transversely.

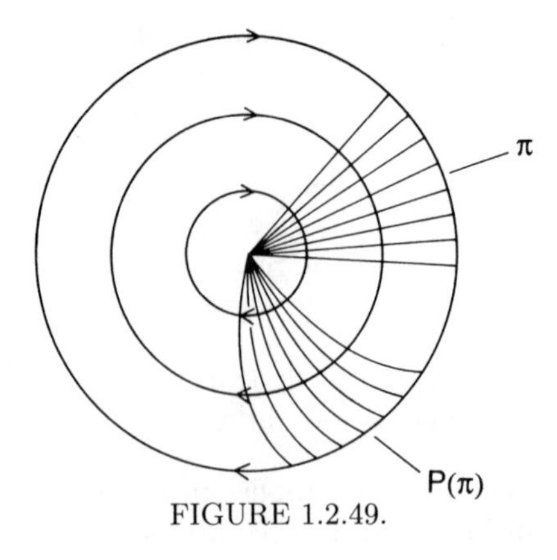

FIGURE 1.2.49.

We will show that Cases a–d cannot occur and thus that the only possibility is for $W^u(Sa^{m+1})$ to intersect topologically transversely both components of $W^u(Sa^m)$. Our conclusions are valid for any integer $m > 0$ for which the $(m+1)/1$ resonance band exists; however, we will draw pictures only for $m = 2$. It is an easy matter to generalize our arguments to the case of arbitrary m. Also, we will draw the unperturbed resonance levels superimposed on the picture for reference; these form an annulus into which $W^u(Sa^{m+1})$ enters on one side and $W^s(Sa^m)$ enters on the other (this will be made clear shortly). We now consider Case a.

Case a. In Figure 1.2.50 we draw the phase plane of the map showing only the third- and second-order resonance bands and assuming that $W^u(Sa^3) \cap W^s(Sa^2) = \emptyset$. We label the outer and inner circles defining the annulus by C_1 and C_2, respectively (note: this annulus is *not* invariant under the perturbed Poincaré map). Since we have assumed $W^u(Sa^3) \cap W^s(Sa^2) = \emptyset$, by the fact that the Melnikov function is negative between C_1 and C_2 we can assume that the stable manifolds of the second-order saddles intersect C_2 as shown in Figure 1.2.50. Thus, $W^s(Sa^2)$ (picking the appropriate components) divides the annulus into two components, with two third-order saddle-sink pairs in one component and the remaining third-order saddle-sink pair in the other component.

We label the outer saddle-sink pairs as O_1, O_2, O_3 and the inner saddle-sink pairs as I_1, I_2; note the relative orientations of the O_i, I_i. There are two important aspects to focus on.

1. The invariant manifolds of the saddles and sinks are fixed (i.e., stationary) in $\mathbb{R}^2$.

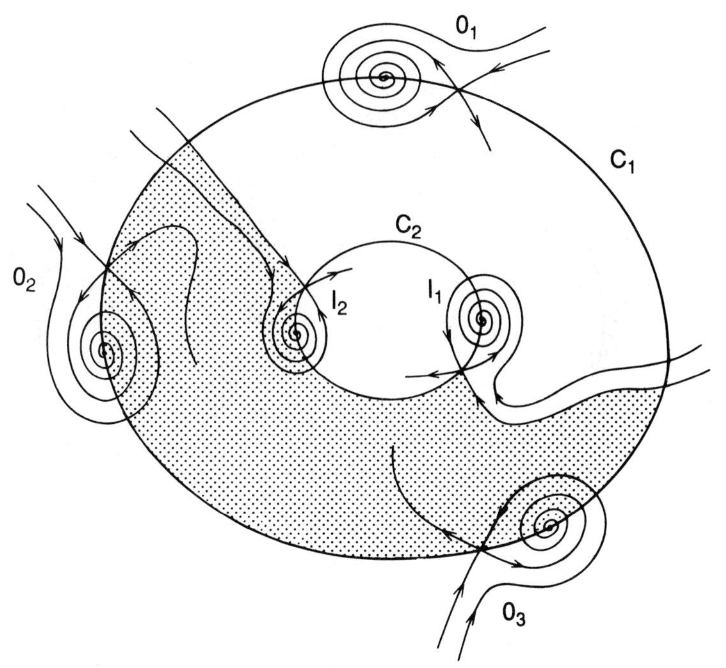

FIGURE 1.2.50.

2. These periodic points, along with points on their stable and unstable manifolds, are not fixed. For example, under iteration by the Poincaré map we have

$$O_1 \to O_2 \to O_3 \to O_1,$$

with

$$W^u(Sa_1^3) \to W^u(Sa_2^3) \to W^u(Sa_3^3) \to W^u(Sa_1^3),$$

where Sa_i^3 denotes the saddle-type periodic orbit in the saddle-sink pair O_i.

Let us now iterate this picture and assume that everything rotates *counter-clockwise*. Then under one iteration we have $O_1 \to O_2$, $O_2 \to O_3$, $O_3 \to O_1$ and $I_1 \to I_2$, $I_2 \to I_1$, with the phase space appearing as in Figure 1.2.51. Thus, we see that for this particular disposition of $W^{u,s}(Sa^3)$, $W^{u,s}(Sa^2)$, the $W^s(Sa^2)$ has jumped ahead of O_2 on C_2 under one iterate. This violates the orientation-perserving property of the map, so we conclude that Case a cannot occur.

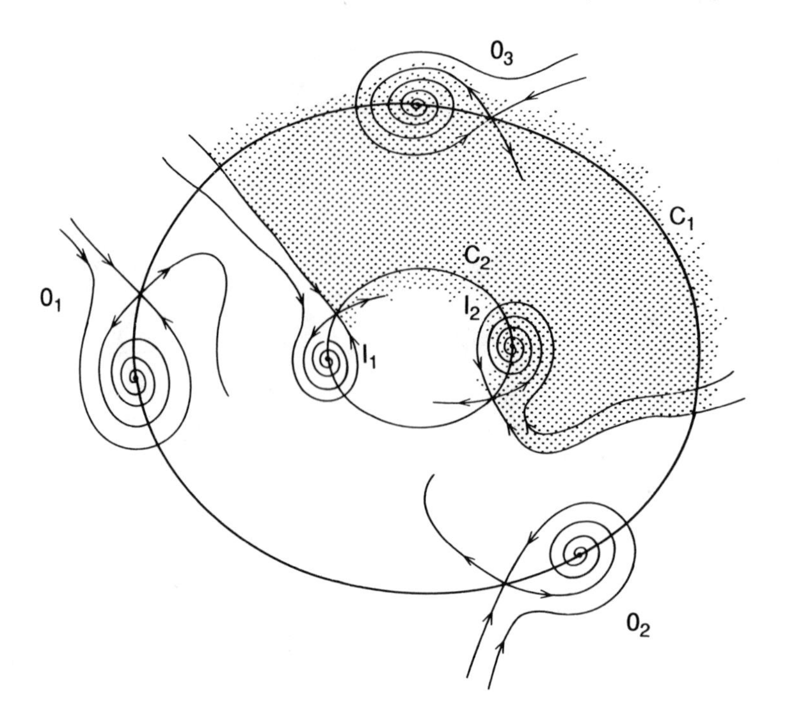

FIGURE 1.2.51.

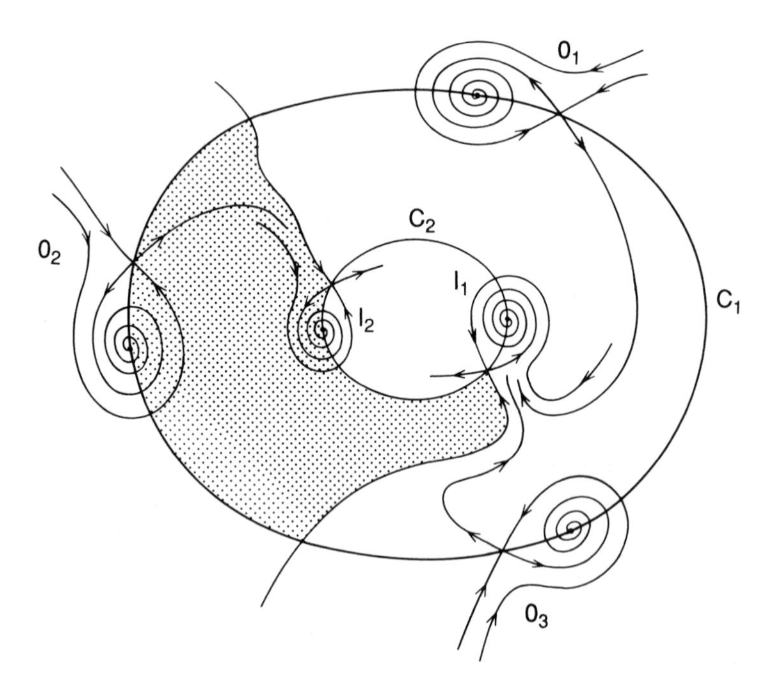

FIGURE 1.2.52.

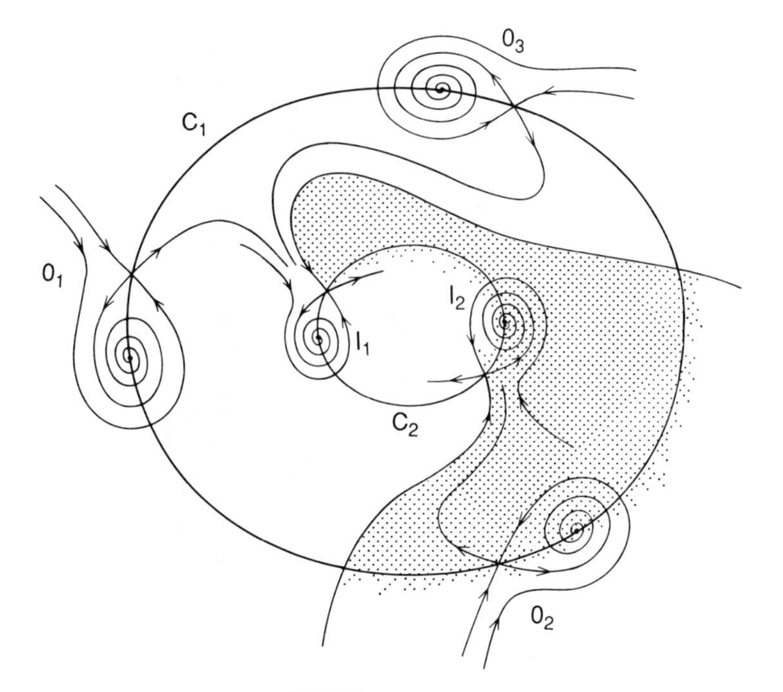

FIGURE 1.2.53.

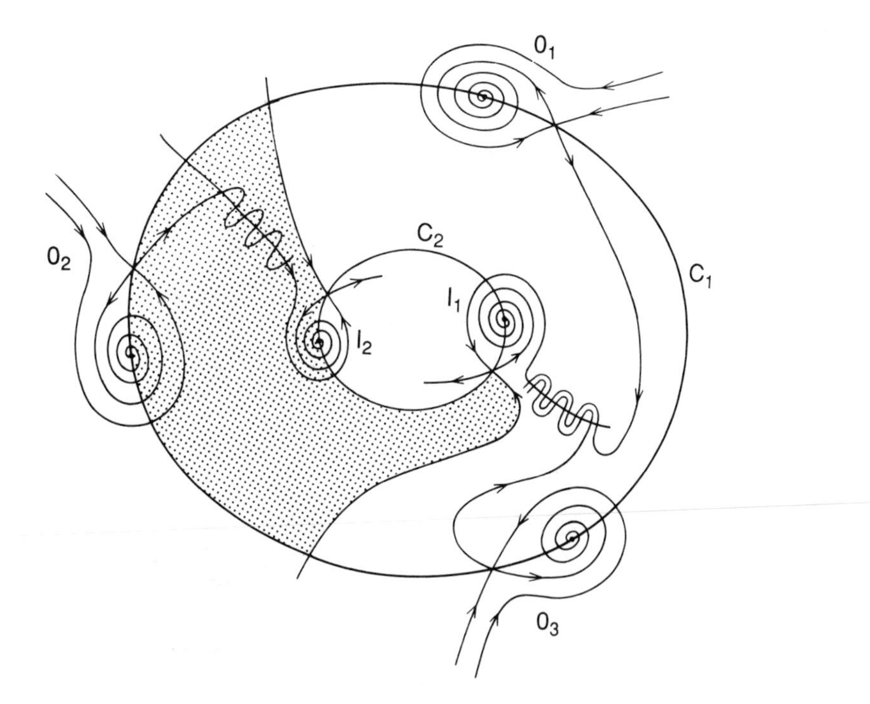

FIGURE 1.2.54.

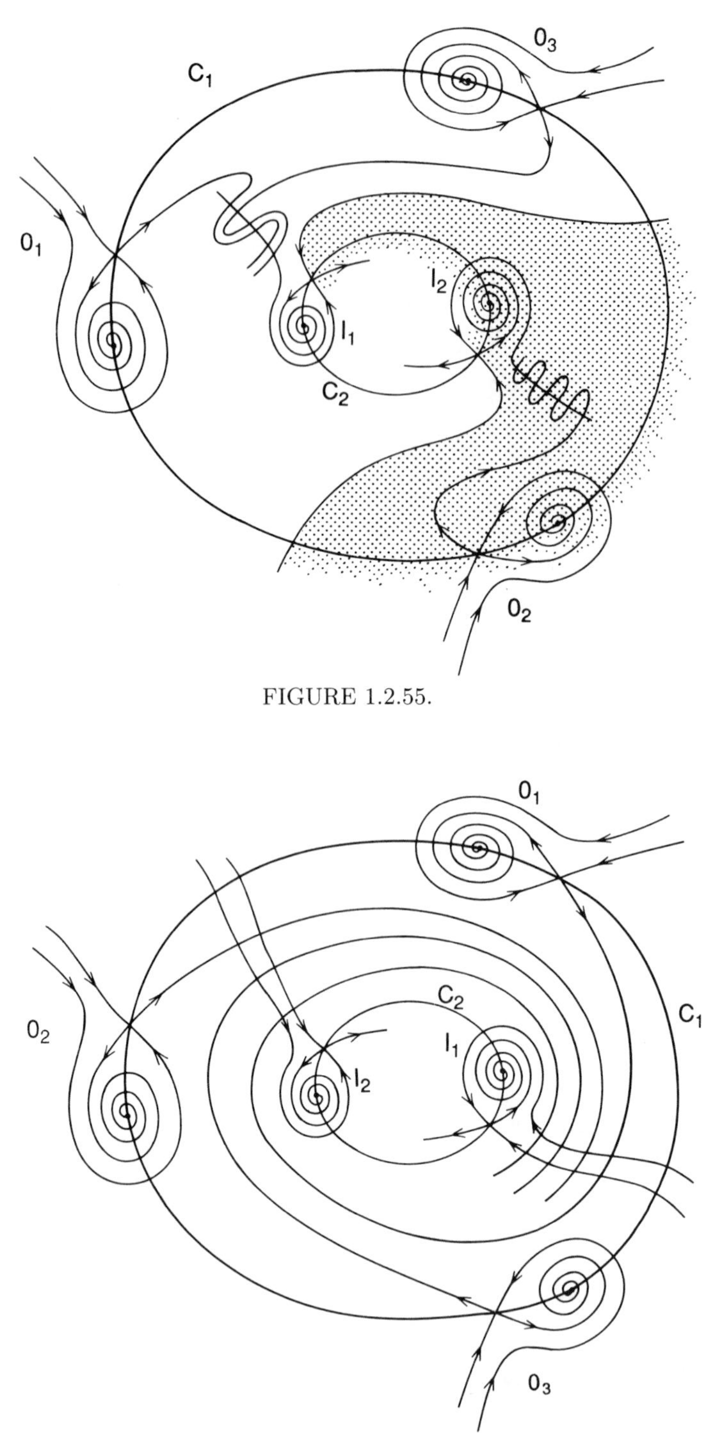

FIGURE 1.2.55.

FIGURE 1.2.56.

Case b. In this case we assume that $W^u(Sa^3)$ falls into the domain of attraction of Si^2 as pictured in Figure 1.2.52. Iterating Figure 1.2.52 once, we have $O_1 \to O_2$, $O_2 \to O_3$, $O_3 \to O_1$ and $I_1 \to I_2$, $I_2 \to I_1$, with the picture appearing as in Figure 1.2.53, since the iterate of orbits attracted to I_1 (resp. I_2) should be attracted to I_2 (resp. I_1). However, Figure 1.2.53 violates the fact that the stable and unstable manifolds of the saddle-sink pairs are fixed in $\mathbb{R}^2$. Therefore, Case b cannot occur.

Case c. There are a variety of possibilities for illustrating Case c. We iterate Figure 1.2.54 and show the result in Figure 1.2.55. Using the same argument as in Case b we see that Figure 1.2.54 violates the fact that the invariant manifolds have fixed locations in $\mathbb{R}^2$. We thus conclude that Case c cannot occur.

Case d. Using exactly the same argument as in Cases b and c, we can conclude that Case d does not occur.

We thus conclude that the unstable manifold of the saddle in a given resonance band intersects the stable manifold of the saddle in the adjacent lower resonance band. We illustrate this in Figure 1.2.56. Exercise 1.2.34 will explore some of the consequences of this result.

Exercises

Exercises 1.2.1–1.2.4 *are concerned with the properties of the following Poincaré map.*

Consider the Poincaré map given in (1.2.26)

$$\begin{pmatrix} x \\ y \end{pmatrix} \mapsto e^{-\delta\pi/\omega} \begin{pmatrix} \mathcal{C} + \frac{\delta}{2\bar{\omega}}\mathcal{S} & \frac{1}{\bar{\omega}}\mathcal{S} \\ -\frac{\omega_0^2}{\bar{\omega}}\mathcal{S} & \mathcal{C} - \frac{\delta}{2\bar{\omega}}\mathcal{S} \end{pmatrix} \begin{pmatrix} x \\ y \end{pmatrix} + \begin{pmatrix} e^{-\delta\pi/\omega}[-A\mathcal{C} + (\frac{-\delta}{2\bar{\omega}}A - \frac{\omega}{\bar{\omega}}B)\mathcal{S}] + A \\ e^{-\delta\pi/\omega}[-\omega B\mathcal{C} + (\frac{\omega_0^2}{\bar{\omega}}A + \frac{\delta\omega}{2\bar{\omega}}B)\mathcal{S}] + \omega B \end{pmatrix},$$

where

$$\mathcal{C} = \cos 2\pi\frac{\bar{\omega}}{\omega}, \qquad \mathcal{S} = \sin 2\pi\frac{\bar{\omega}}{\omega},$$

and

$$\bar{\omega} = \frac{1}{2}\sqrt{4\omega_0^2 - \delta^2}, \quad A = \frac{(\omega_0^2 - \omega^2)\gamma}{(\omega_0^2 - \omega^2)^2 + (\delta\omega)^2}, \quad B = \frac{\delta\gamma\omega}{(\omega_0^2 - \omega^2)^2 + (\delta\omega)^2}.$$

1.2.1 Show that $(x, y) = (A, \omega B)$ is the only fixed point for this map and, hence, argue that it is a global attractor.

1.2.2 Discuss the nature of the orbit structure near $(x, y) = (A, \omega B)$ for different values of $\bar{\omega}/\omega$.

1.2.3 Show how the fixed point of the Poincaré map changes as the cross-section (i.e., phase angle) is varied.

1.2.4 Construct and study "the" Poincaré map for

$$\begin{aligned}\dot{x} &= y,\\ \dot{y} &= -\omega_0^2 x + \gamma \cos\theta,\\ \dot{\theta} &= \omega_0.\end{aligned}$$

Exercises 1.2.5–1.2.9 *are concerned with the properties of the following Poincaré map.*

Consider the Poincaré map ($\omega \neq \omega_0$) given in (1.2.36)

$$\begin{pmatrix} x \\ y \end{pmatrix} \mapsto \begin{pmatrix} \cos 2\pi\frac{\omega_0}{\omega} & \frac{1}{\omega_0}\sin 2\pi\frac{\omega_0}{\omega} \\ -\omega_0 \sin 2\pi\frac{\omega_0}{\omega} & \cos 2\pi\frac{\omega_0}{\omega} \end{pmatrix} \begin{pmatrix} x \\ y \end{pmatrix} + \begin{pmatrix} \bar{A}(1 - \cos 2\pi\frac{\omega_0}{\omega}) \\ \omega_0 \bar{A} \sin 2\pi\frac{\omega_0}{\omega} \end{pmatrix}.$$

1.2.5 For $\omega = m\omega_0$, $m > 1$, show that all points except $(x, y) = (\bar{A}, 0)$ are period m points.

1.2.6 For $n\omega = \omega_0$, $n > 1$, show that all points are fixed points.

1.2.7 Discuss the orbit structure of the Poincaré map when ω/ω_0 is an irrational number.

1.2.8 Discuss stability of the harmonics, subharmonics, ultraharmonics, and ultrasubharmonics.

1.2.9 Discuss the idea of structural stability in the context of Questions 1.2.5, 1.2.6, 1.2.7, and 1.2.8. Discuss the effects of various perturbations on the different cases.

1.2.10 Discuss the idea of structural stability of the following vector fields and maps.

a) $\begin{aligned}\dot{\theta}_1 &= \omega_1,\\ \dot{\theta}_2 &= \omega_2,\end{aligned} \qquad (\theta_1, \theta_2) \in S^1 \times S^1.$

b) $\begin{aligned}\dot{x} &= 1,\\ \dot{y} &= 2,\end{aligned} \qquad (x, y) \in \mathbb{R}^2.$

c) $\begin{aligned}\dot{x} &= y,\\ \dot{y} &= x - x^3,\end{aligned} \qquad (x, y) \in \mathbb{R}^2.$

d) $\begin{aligned}\dot{x} &= y,\\ \dot{y} &= x - x^3 - y,\end{aligned} \qquad (x, y) \in \mathbb{R}^2.$

e) $\theta \mapsto \theta + \omega, \qquad \theta \in S^1$.

f) $\theta \mapsto \theta + \omega + \varepsilon \sin \theta, \qquad \varepsilon$ small, $\qquad \theta \in S^1$.

1.2.11 Consider the equation

$$\dot{x} = \varepsilon x, \qquad x \in \mathbb{R}^1.$$

Solve this equation and discuss the time interval on which this solution is valid. Compute $\left.\frac{\partial x(t,\varepsilon)}{\partial \varepsilon}\right|_{\varepsilon=0}$ and $\left.\frac{\partial^2 x(t,\varepsilon)}{\partial \varepsilon^2}\right|_{\varepsilon=0}$, and compare their behavior in t with $x(t, \varepsilon)$.

1.2.12 We denote the solution of (1.2.93) by $x(t, \varepsilon)$ and assume that it is valid on some time interval I. Derive differential equations for $x(t, 0)$, $\left.\frac{\partial x(t,\varepsilon)}{\partial \varepsilon}\right|_{\varepsilon=0}$, and $\left.\frac{\partial^2 x(t,\varepsilon)}{\partial \varepsilon^2}\right|_{\varepsilon=0}$. Discuss the nature of the time interval on which the solutions of these equations are valid. The equations for $\left.\frac{\partial x(t,\varepsilon)}{\partial \varepsilon}\right|_{\varepsilon=0}$ and $\left.\frac{\partial^2 x(t,\varepsilon)}{\partial \varepsilon^2}\right|_{\varepsilon=0}$ are known as the first and second variational equations, respectively.

1.2.13 *Second-order Averaging.* Construct a change of coordinates

$$y = z + \varepsilon^2 h(z, t)$$

such that (1.2.125) becomes

$$\dot{z} = \varepsilon \bar{f}(z) + \varepsilon^2 \bar{f}_1(z) + \mathcal{O}(\varepsilon^3),$$

where $\overline{f_1(z)} \equiv \frac{1}{T} \int_0^T f_1(z, t, 0) dt$. State and prove a *second-order averaging theorem* that has the same ingredients as Theorem 1.2.11.

1.2.14 Consider the system

$$\begin{aligned} \dot{x} &= \varepsilon f(x, \theta), \\ \dot{\theta} &= \Omega(x) + \theta(\varepsilon), \end{aligned} \qquad (x, \theta) \in \mathbb{R}^n \times S^1,$$

where f and Ω are $\mathbf{C}^r$ $(r \geq 1)$ functions on some open sets in $\mathbb{R}^n \times S^1$ and $\mathbb{R}^n$, respectively. Using the idea of a "near-identity" coordinate transformation to transform the θ-dependence of $f(x, \theta)$ to higher order in ε, state and prove a theorem concerning the system that has all the ingredients of Theorem 1.2.11. What conditions must be put on $\Omega(x)$? *Hint:* use the transformation $x = y + \varepsilon h(y, \theta)$, $\theta = \theta$ and choose h appropriately. What would happen if $\Omega(\bar{x}) = 0$ for some $\bar{x} \in \mathbb{R}^n$?

1.2.15 Consider the system

$$\begin{aligned} \dot{x} &= \varepsilon f(x, \theta_1, \theta_2), \\ \dot{\theta}_1 &= \Omega_1(x) + \mathcal{O}(\varepsilon), \\ \dot{\theta}_2 &= \Omega_2(x) + \mathcal{O}(\varepsilon), \end{aligned} \qquad (x, \theta_1, \theta_2) \in \mathbb{R}^n \times S^1 \times S^1,$$

where f, Ω_1, and Ω_2 are $\mathbf{C}^r$ $(r \geq 2)$ functions on sufficiently large open sets in $\mathbb{R}^n \times S^1 \times S^1$ and $\mathbb{R}^n$, respectively. Introduce the coordinate transformation

$$x = y + \varepsilon h(y, \theta_1, \theta_2).$$

Show that $h(y, \theta_1, \theta_2)$ can be chosen such that the vector field becomes

$$\begin{aligned}\dot{y} &= \varepsilon \bar{f}(y) + \mathcal{O}(\varepsilon^2),\\ \dot{\theta}_1 &= \Omega_1(y) + \mathcal{O}(\varepsilon),\\ \dot{\theta}_2 &= \Omega_2(y) + \mathcal{O}(\varepsilon),\end{aligned}$$

where

$$\bar{f}(y) = \frac{1}{(2\pi)^2} \int_0^{2\pi} \int_0^{2\pi} f(y, \theta_1, \theta_2) d\theta_1 d\theta_2$$

provided that we require

$$n_1 \Omega_1(y) + n_2 \Omega_2(y) \neq 0$$

for integers n_1 and n_2. How might one interpret geometrically in $\mathbb{R}^n \times S^1 \times S^1$ hyperbolic fixed points of

$$\dot{y} = \varepsilon \bar{f}(y)?$$

Hint: split f into its oscillating and mean parts as follows

$$f(y, \theta_1, \theta_2) = \bar{f}(y) + \tilde{f}(y, \theta_1, \theta_2),$$

where

$$\tilde{f}(y, \theta_1, \theta_2) \equiv f(y, \theta_1, \theta_2) - \frac{1}{(2\pi)^2} \int_0^{2\pi} \int_0^{2\pi} f(y, \theta_1, \theta_2) d\theta_1 d\theta_2.$$

Next show that in order to eliminate $\tilde{f}$ from the vector field, h must be chosen so that

$$\frac{\partial h}{\partial \theta_1} \Omega_1 + \frac{\partial h}{\partial \theta_2} \Omega_2 = \tilde{f}.$$

To solve this equation, expand $\tilde{f}$ and $\tilde{h}$ in Fourier series

$$\tilde{f}(y, \theta_1, \theta_2) = \sum_{\|n\| \geq 0} f_n(y) e^{i(n_1\theta_1 + n_2\theta_2)},$$

$$h(y, \theta_1, \theta_2) = \sum_{\|n\| \geq 0} h_n(y) e^{i(n_1\theta_1 + n_2\theta_2)},$$

where $\|n\| = |n_1| + |n_2|$, and solve for the Fourier components of h.

We remark that this example shows that "small divisor" problems arise when attempting to average over more than one frequency. This subject is very broad and beyond the scope of this book. For more information on multifrequency averaging, we refer the reader to Grebenikov and Ryabov [1983] and Lochak and Meunier [1988].

1.2.16 Consider the van der Pol equation

$$\ddot{x} + \varepsilon(x^2 - 1)\dot{x} + x = 0.$$

Using the method of averaging, show that, for $\varepsilon > 0$ sufficiently small, the equation has an attracting periodic orbit having amplitude $2 + \mathcal{O}(\varepsilon^2)$ and phase $\theta_0 + \mathcal{O}(\varepsilon^2)$.

1.2.17 Consider the forced van der Pol equation

$$\ddot{x} + \frac{\varepsilon}{\omega}(x^2 - 1)\dot{x} + x = \varepsilon F \cos \omega t.$$

a) Use the van der Pol transformation discussed in Example 1.2.11 to put this equation into the standard form for application of the method of averaging.

b) Average the equations and obtain

$$\dot{u} = \frac{\varepsilon}{2\omega}\left[u - \sigma v - \frac{u}{4}(u^2 + v^2)\right],$$
$$\dot{v} = \frac{\varepsilon}{2\omega}\left[\sigma u + v - \frac{v}{4}(u^2 + v^2) - F\right],$$

where $\varepsilon\sigma = 1 - \omega^2$.

c) Rescale the averaged equation as follows

$$t \to \frac{2\omega}{\varepsilon}t,$$
$$v \to 2v,$$
$$u \to 2u,$$

and let $\gamma = F/2$ to obtain

$$\dot{u} = u - \sigma v - u(u^2 + v^2),$$
$$\dot{v} = \sigma u + v - v(u^2 + v^2) - \gamma.$$

We will return to a detailed study of the dynamics of this equation in the exercises following Chapter 3.

1.2.18 Consider the equation

$$\begin{aligned} \dot{x} &= y, \\ \dot{y} &= x - x^3 + \varepsilon(\gamma \cos \omega t - \delta y), \end{aligned} \qquad \delta, \gamma, \omega > 0.$$

We are interested in studying the dynamics of small-amplitude periodic orbits near $(x, y) = (1, 0)$; cf. Example 1.2.12.

Harmonic Response — 1:1 Resonance

a) Show that a periodic solution of frequency $\omega = \sqrt{2}$ exists near $(x, y) = (1, 0)$ and has the form

$$\hat{x}(t) = 1 + \mathcal{O}(\varepsilon^{1/3}),$$
$$\hat{y}(t) = \mathcal{O}(\varepsilon^{1/3}).$$

(Note: you will need to use either Lindstedt's method or two-timing; see Kevorkian and Cole [1981]).

b) Substitute this solution into equation (1.2.121) and apply the van der Pol transformation (see Example 1.2.11) to transform (1.2.121) into the standard form for application of the method of averaging.

c) Compute the averaged equations.

Subharmonic Response — 1:3 Resonance

d) Show that a periodic solution of frequency $\omega = 3\sqrt{2}$ exists near $(x, y) = (1, 0)$ and has the form

$$\hat{x}(t) = 1 + \mathcal{O}(\varepsilon),$$
$$\hat{y}(t) = \mathcal{O}(\varepsilon).$$

(Note: a regular perturbation expansion should suffice for finding this periodic solution.)

e) Substitute this equation into equation (1.2.121) and apply the van der Pol transformation to transform (1.2.121) into the standard form for application of the method of averaging.

f) Compute the averaged equations.

We will return to this example and analyze the averaged equations in detail for both cases after we have developed some bifurcation theory in Chapter 3.

1.2.19 The equation of motion describing the librational motion of an arbitrarily shaped satellite in a planar, elliptical orbit is

$$(1 + \varepsilon\mu\cos\theta)\psi'' - 2\varepsilon\mu\sin\theta(\psi' + 1) + 3K_i\sin\psi\cos\psi = 0$$

where $\psi' \equiv \frac{\partial\psi}{\partial\theta}$, $K_i = \frac{I_{xx} - I_{zz}}{I_{yy}}$, ε is small, and $\varepsilon\mu$ is the eccentricity of the orbit; see Modi and Brereton [1969]. The geometry is illustrated in Figure E1.2.1. For ε small, this equation can be written in the form (using the fact that $\frac{1}{1+\varepsilon\mu\cos\theta} = 1 - \varepsilon\mu\cos\theta + \mathcal{O}(\varepsilon^2)$)

$$\psi'' + 3K_i\sin\psi\cos\psi = \varepsilon[2\mu\sin\theta(\psi' + 1) + 3\mu K_i\sin\psi\cos\psi\cos\theta] + \mathcal{O}(\varepsilon^2).$$

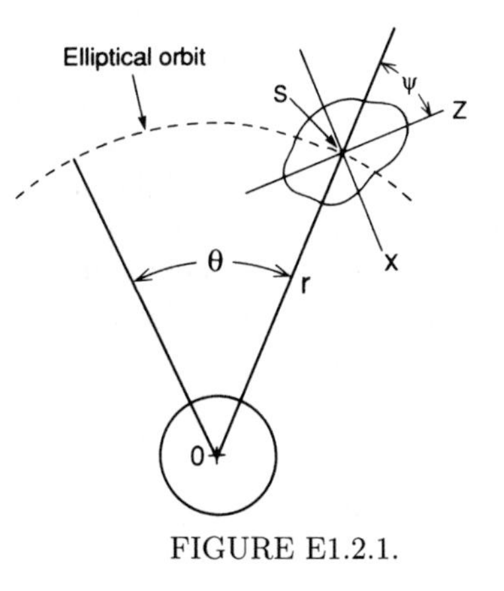

FIGURE E1.2.1.

Describe as much of the dynamics of this equation as possible for different μ, K_i values using Melnikov's method for the subharmonics. Describe physically as well as mathematically the motions that you find.

1.2.20 The driven *Morse oscillator* is an equation frequently used in theoretical chemistry to describe the photodissociation of molecules (see, e.g., Goggin and Milonni [1988]). The equation is given by

$$\begin{aligned} \dot{x} &= y, \\ \dot{y} &= -\mu\left(e^{-x} - e^{-2x}\right) + \varepsilon\gamma\cos\omega t, \end{aligned}$$

with μ, γ, $\omega > 0$.

For $\varepsilon = 0$, the equation is Hamiltonian with Hamiltonian function

$$H(x, y) = \frac{y^2}{2} + \mu\left(-e^{-x} + \frac{1}{2}e^{-2x}\right).$$

a) For $\varepsilon = 0$, use the Hamiltonian function to determine all the orbits. Represent the orbits graphically in the phase plane.

b) Describe the global dynamics for ε small. How do the dynamics depend on μ, γ, and ω?

c) Describe the nature of the dynamics at ∞ for $\varepsilon = 0$ and for $0 < \varepsilon << 1$; in particular, describe the dynamics for the point $(x, y) = (\infty, 0)$.

We will examine the possibility of chaos in this system in Chapter 4.

1.2.21 *Fluid Transport and the Dynamical Systems Point of View.* Consider the Navier–Stokes equations for three-dimensional viscous, incompressible fluid

$$\frac{\partial v}{\partial t} + (v \cdot \nabla)v = -\nabla p + \frac{1}{R}\nabla^2 v,$$

where p denotes the pressure and R the Reynolds number. Additionally, boundary and initial conditions may be specified. The solution of this highly nonlinear partial differential equation gives a velocity field, $v(x,t)$. Suppose we are interested in the transport of infinitesimal fluid elements (referred to as fluid particles) in this flow. The fluid particles move under the influence of two processes: convection (or advection) due to the velocity field and molecular diffusion (since the fluid is not really a continuum). The motion of fluid particles due to convection is determined by

$$\dot{x} = v(x,t), \qquad x \in \mathbb{R}^3.$$

This is simply a finite-dimensional dynamical system where the phase space is actually the physical space occupied by the fluid.

If we consider two-dimensional incompressible inviscid fluid flow, the velocity field can be determined from a stream function, $\psi(x_1, x_2; t)$, where

$$v(x_1, x_2, t) = \left(\frac{\partial \psi}{\partial x_2}, -\frac{\partial \psi}{\partial x_1} \right);$$

see Chorin and Marsden [1979] for background on these statements. The equations for fluid particle motions in this case become

$$\begin{aligned} \dot{x}_1 &= \frac{\partial \psi}{\partial x_2}(x_1, x_2, t), \\ \dot{x}_2 &= -\frac{\partial \psi}{\partial x_1}(x_1, x_2, t). \end{aligned}$$

The reader should note that this is simply a Hamiltonian dynamical system where the stream function plays the role of the Hamiltonian. The study of the transport and mixing of fluids along these lines using the framework of dynamical systems theory is a topic of much current interest; the reader should consult Ottino [1989] for a good introduction and Rom-Kedar, Leonard, and Wiggins [1989] for a specific example.

We will consider the situation of fluid particle transport in modulated traveling waves in a binary fluid mixture heated from below; see Weiss and Knobloch [1989] and Moses and Steinberg [1988]. The stream function (in a moving frame) near an instability leading to time-dependent oscillations is given by

$$\psi(x_1, x_2, t) = \psi_0(x_1, x_2) + \varepsilon \psi_1(x_1, x_2, t),$$

where

$$\psi_0(x_1, x_2) = -x_2 + R \cos x_1 \sin x_2,$$

$$\psi_1(x_1, x_2, t) = \frac{\gamma}{2}\left[\left(1 - \frac{2}{\omega}\right)\cos(x_1 + \omega t + \theta) + \left(1 + \frac{2}{\omega}\right)\cos(x_1 - \omega t - \theta)\right]\sin x_2.$$

In the above $\omega > 0$, θ is a phase, and R, γ, and ε are parameters (amplitude depending on the temperature) with $0 < \varepsilon << 1$; see Weiss and Knobloch [1989] for a detailed discussion. The equations for fluid particle motions are given by

$$\dot{x}_1 = \frac{\partial \psi_0}{\partial x_2}(x_1, x_2) + \varepsilon \frac{\partial \psi_1}{\partial x_2}(x_1, x_2, t),$$
$$\dot{x}_2 = -\frac{\partial \psi_0}{\partial x_1}(x_1, x_2) - \varepsilon \frac{\partial \psi_1}{\partial x_1}(x_1, x_2, t).$$

a) Describe the fluid particle paths for $\varepsilon = 0$. How does the topology of the flow change as R is varied?

b) Describe the fluid particle paths using the subharmonic Melnikov theory for $\varepsilon \neq 0$. How does the topology of the fluid flow change as a function of R, γ, ω, and θ?

c) Compare your results with the experimental results of Moses and Steinberg [1988].

1.2.22 Consider the Poincaré map in the neighborhood of a periodic orbit as discussed in Section 1.1A, Case 1. If the vector field is $\mathbf{C}^r$, does it follow that the first return time $\tau(x)$ is a $\mathbf{C}^r$ function of x? Give a proof or counterexample.

1.2.23 For a planar, autonomous Hamiltonian vector field, describe the action of a homoclinic orbit connecting a saddle-type fixed point. What is the nature of the angle transformation on the homoclinic orbit? Also, discuss the action-angle transformation in the limit as the center-type fixed point is approached (think of polar coordinates and the harmonic oscillator). Compare these two cases.

1.2.24 Suppose the $n \times n$ matrix A has k eigenvalues with negative real part and $n - k$ eigenvalues with positive real parts. Then prove that, for ε sufficiently small, the matrix $id + \varepsilon A$ ("id" denotes the $n \times n$ identity matrix) has k eigenvalues with modulus less than one and $n - k$ eigenvalues with modulus greater than one.

1.2.25 Consider a $\mathbf{C}^r$ $(r \geq 1)$ area-preserving diffeomorphism of the plane, i.e.,

$$f: \mathbb{R}^2 \rightarrow \mathbb{R}^2,$$

with $\det Df(x) = 1 \; \forall x \in \mathbb{R}^2$. Suppose that f has an invariant circle. Then show that the region enclosed by the invariant circle is an invariant set.

1.2.26 Consider a $\mathbf{C}^r$ $(r \geq 1)$ diffeomorphism

$$f: \mathbb{R}^2 \to \mathbb{R}^2,$$

with $\det Df(x) \neq 1 \ \forall x \in \mathbb{R}^2$. Show that f cannot possess any invariant circles.

1.2.27 Consider a $\mathbf{C}^r$ $(r \geq 1)$ diffeomorphism

$$f: \mathbb{R}^2 \to \mathbb{R}^2,$$

with $\det Df(x) = 1 \ \forall x \in \mathbb{R}^2$. Suppose that f has an elliptic fixed point at $x = \bar{x}$, i.e., both eigenvalues of $Df(\bar{x})$ have unit modulus. Discuss the *nonlinear* stability of this fixed point. (*Hint:* this problem needs more work in order to be set up properly; there are several cases to consider. The problem can be solved when the Moser twist theorem (or KAM theorem) along with Exercise 1.2.25 can be applied. Under what conditions might the hypotheses of the Moser twist theorem be satisfied in a neighborhood of the elliptic fixed point?) This problem has a long history; see Siegel and Moser [1971].

1.2.28 Consider the Mathieu equation

$$\ddot{\phi} + (\alpha^2 + \beta \cos t)\phi = 0,$$

where α and $\beta \geq 0$ are parameters. Can the Moser twist theorem or KAM theorem be applied to study the stability of the fixed point $(x, y) = (0, 0)$? Consider the following nonlinear Mathieu equation

$$\ddot{\phi} + (\alpha^2 + \beta \cos t) \sin \phi = 0.$$

Can the Moser twist theorem or KAM theorem be applied to study the stability of the fixed point $(x, y) = (0, 0)$? Compare the nonlinear and linear problems. Nonlinear problems are automatically assumed to be more difficult than linear problems. Comment on this statement in the context of this exercise.

Exercises 1.2.29–1.2.31 *are concerned with carrying out some of the computations in Section 1.2E.*

1.2.29 Consider the unforced, undamped Duffing equation

$$\dot{x} = \frac{\partial H}{\partial y}(x, y) = y,$$

$$\dot{y} = -\frac{\partial H}{\partial x}(x, y) = x - x^3,$$

where $H(x, y) = \frac{y^2}{2} - \frac{x^2}{2} + \frac{x^4}{4}$.

a) Show that the solutions $(x(t), y(t))$ can be found from the integral

$$\int dt = \frac{1}{\sqrt{2}} \int \frac{dx}{\sqrt{H + \frac{x^2}{2} - \frac{x^4}{4}}},$$

where $H = \frac{y^2}{2} - \frac{x^2}{2} + \frac{x^4}{4} = \text{constant}$.

b) Compute this integral explicitly (i.e., use Byrd and Friedman [1971]) to obtain the expressions for the inner periodic orbits, the homoclinic orbits, and the outer periodic orbits given in Section 1.2E.

c) Using the results from b), compute $H(k)$ where k is the elliptic modulus. What is the range of k for the inner and outer periodic orbits?

d) Using the properties of the elliptic functions compute the periods of the inner and outer periodic orbits.

e) Compute the action, $I(k)$, for the inner and outer periodic orbits (see Example 1.2.14).

f) Compute $H(I)$, $\frac{\partial H}{\partial I}(I)$, $\frac{\partial H}{\partial k}(k)$, and $\frac{\partial I}{\partial k}(k)$ for the inner and outer periodic orbits. Describe the nature of the limits as the homoclinic orbit is approached and as the center-type fixed point is approached. Are the limits continuous? Differentiable (cf. Exercise 1.2.23)?

g) Show that the twist condition (i.e., $\frac{\partial \Omega}{\partial I} \neq 0$) is always satisfied for the inner periodic orbits. Is it also satisfied for the outer periodic orbits? Is the twist condition equivalent to any of the following:

1) $\dfrac{\partial \Omega}{\partial K} \neq 0,$

2) $\dfrac{\partial \Omega}{\partial H} \neq 0;$

3) $\dfrac{\partial T}{\partial H} \neq 0;$

4) $\dfrac{\partial T}{\partial K} \neq 0;$

5) $\dfrac{\partial T}{\partial I} \neq 0?$

Justify your answers.

1.2.30 Consider the forced, damped Duffing equation

$$\begin{aligned}
\dot{x} &= y, \\
\dot{y} &= x - x^3 + \varepsilon(\gamma \cos \omega t - \delta y).
\end{aligned}$$

a) Show that the subharmonic Melnikov functions for the inner periodic orbits are given by

$$\bar{M}_1^{m/n}(t_0;\gamma,\delta,\omega) = -\delta J_1(m,n) \pm \gamma J_2(m,n,\omega)\sin\omega t_0,$$

where

$$J_1(m,n) = \frac{4}{3}\big[(2-k^2)E(k) - 2k'^2K(k)\big]/(2-k^2)^{3/2},$$

$$J_2(m,n,\omega) = \begin{cases} 0 & \text{for } n \neq 1 \\ \sqrt{2}\pi\omega\,\mathrm{sech}\frac{\pi m K(k')}{K(k)} & \text{for } n = 1 \end{cases},$$

with $k' = \sqrt{1-k^2}$.

b) Show that the subharmonic Melnikov function for the outer periodic orbits has the form

$$\bar{M}_1^{m/n}(t_0;\gamma,\delta,\omega) = -\delta\hat{J}_1(m,n) - \gamma\hat{J}_2(m,n,\omega)\sin\omega t_0$$

and give explicit expressions for $\hat{J}_1(m,n)$ and $\hat{J}_2(m,n,\omega)$. In particular, show that m must be odd. Compute and discuss the following limits:

$$\lim_{m\to\infty} J_1(m,1), \qquad \lim_{m\to\infty} \hat{J}_1(m,1),$$

and

$$\lim_{m\to\infty} J_2(m,1,\omega), \qquad \lim_{m\to\infty} \hat{J}_2(m,1,\omega).$$

c) Let

$$R^m(\omega) = \frac{J_1(m,1)}{J_2(m,1,\omega)},$$

$$\hat{R}^m(\omega) = \frac{\hat{J}_1(m,1)}{\hat{J}_2(m,1,\omega)}.$$

1) Show that $R^m(\omega)$ and $\hat{R}^m(\omega)$ are positive for all $\omega > 0$.
2) Describe the sequences

$$\{R^1(\omega), R^2(\omega), \cdots, R^m(\omega), \cdots\}$$

$$\{\hat{R}^1(\omega), \hat{R}^3(\omega), \cdots, \hat{R}^m(\omega), \cdots\}$$

as a function of ω. In particular, are the sequences monotone for some or all values of ω? What are the dynamical consequences of monotonicity or nonmonotonicity?

1.2.31 Show that the width of the inner resonance bands in the damped, forced Duffing equation has the following asymptotic behavior

$$\Delta I \approx \sqrt{\varepsilon \frac{m^3}{\omega^3} \exp\left(\frac{-2\pi m}{\omega}\right)} + \mathcal{O}(\varepsilon) \qquad \text{as } m \uparrow \infty.$$

Hints:

1) Using asymptotic properties of $K(k)$ and $E(k)$ (see Byrd and Friedman [1971]), show that

$$\Omega'(m) \approx \frac{-1}{k'^2(\log 4/k')^3} \qquad \text{as } k' \to 0.$$

2) Use the resonance relation $2K(k)\sqrt{2-k^2} = \frac{2\pi m}{\omega}$ to show

$$\frac{-1}{k'^2 \log 4/k'} \sim -\frac{\omega^3 e^{2\pi m/\omega}}{m^3} \qquad \text{as } m \uparrow \infty,\ k' \to 0.$$

3) Show that $J_1(m,1)$ and $J_2(m,1,\omega)$ are $\mathcal{O}(1)$ as $m \uparrow \infty$.
4) Apply 1), 2), and 3) to the expression for ΔI given in Section 1.2.E,i).

1.2.32 The subharmonic Melnikov theory developed in Section 1.2D,ii) applies to time periodically perturbed, single-degree-of-freedom Hamiltonian systems. In this exercise we will show how the method can be extended to a class of two-degree-of-freedom Hamiltonian systems by using the *method of reduction* (see Holmes and Marsden [1982]).

Consider the following two-degree-of-freedom Hamiltonian system

$$\begin{aligned}
\dot{x} &= \frac{\partial F}{\partial y}(x,y) + \varepsilon \frac{\partial H^1}{\partial y}(x,y,I,\theta),\\
\dot{y} &= -\frac{\partial F}{\partial x}(x,y) - \varepsilon \frac{\partial H^1}{\partial x}(x,y,I,\theta),\\
\dot{I} &= -\varepsilon \frac{\partial H^1}{\partial \theta}(x,y,I,\theta),\\
\dot{\theta} &= \frac{\partial G}{\partial I}(I) + \varepsilon \frac{\partial H^1}{\partial I}(x,y,I,\theta),
\end{aligned} \qquad (x,y,I,\theta) \in \mathbb{R} \times \mathbb{R} \times \mathbb{R}^+ \times S^1, \tag{E1.2.1}$$

with the Hamiltonian function

$$H^\varepsilon(x,y,I,\theta) = F(x,y) + G(I) + \varepsilon H^1(x,y,I,\theta).$$

Note that, for $\varepsilon = 0$, the $x-y$ components of the vector field decouple from the $I-\theta$ components.

We make the following assumptions on the unperturbed system (i.e., the vector field with $\varepsilon = 0$). The $x-y$ component of the vector field satisfies the following two assumptions.

Assumption 1. The $x-y$ component of the unperturbed system possesses a hyperbolic fixed point, $p_0 \equiv (x_0, y_0)$, connected to itself by a homoclinic orbit $q^0(t) = (x^0(t), y^0(t))$.

Assumption 2. Let $\Gamma_{p_0} = \{(x,y) \in \mathbb{R}^2 \mid (x,y) = (x^0(t), y^0(t)), t \in \mathbb{R}\} \cup \{p_0\} = W^s(p_0) \cap W^u(p_0) \cup \{p_0\}$. The interior of Γ_{p_0} is filled with a continuous family of periodic orbits, $q^\alpha(t) = (x^\alpha(t), y^\alpha(t))$, $\alpha \in (-1, 0)$, with period T^α. We assume that $\lim_{\alpha\to 0} q^\alpha(t) = q^0(t)$ and $\lim_{\alpha\to 0} T^\alpha = \infty$.

The $I-\theta$ component of the vector field satisfies the following assumption.

Assumption 3. $\dfrac{\partial G}{\partial I}(I) \neq 0$.

Thus, the phase space of the unperturbed system appears as in Figure E1.2.2. The reader should note that Assumptions 1 and 2 above are identical to Assumptions 1 and 2 given in Section 1.2D,ii) for the unperturbed structure of the single-degree-of-freedom, time periodically perturbed systems studied in that section.

a) Show that the three-dimensional surface

$$H^\varepsilon(x,y,I,\theta) = F(x,y) + G(I) + \varepsilon H^1(x,y,I,\theta) \equiv h = \text{constant} \tag{E1.2.2}$$

is invariant under the flow generated by the vector field. *Hint:* invariance implies that the vector field is tangent to the surface.

b) Using Assumption 3, a), and the implicit function theorem show that, for ε sufficiently small, I can be represented as a function of x, y, θ, and h, which we denote as follows

$$I = \mathcal{L}^\varepsilon(x,y,\theta;h) = \mathcal{L}^0(x,y;h) + \varepsilon\mathcal{L}^1(x,y,\theta;h) + \mathcal{O}(\varepsilon^2). \tag{E1.2.3}$$

Are there any restrictions on h? What about x, y, and θ? Is $\mathcal{L}^\varepsilon$ 2π-periodic in θ?

c) Show that

$$\mathcal{L}^0(x,y;h) = G^{-1}(h - F(x,y)) \tag{E1.2.4}$$

and

$$\mathcal{L}^1(x,y,\theta;h) = \frac{-H^1(x,y,\mathcal{L}^0(x,y;h),\theta)}{\Omega(\mathcal{L}^0(x,y;h))}, \tag{E1.2.5}$$

where $\Omega \equiv \frac{\partial G}{\partial I}$.

Hint: substitute (E1.2.3) into (E1.2.2), expand the result in powers of ε, and equate like powers of ε.

Is $\mathcal{L}^1$ 2π-periodic in θ?

d) Derive the following one-parameter family of time periodically perturbed, one-degree-of-freedom Hamiltonian systems

$$\frac{dx}{d\theta} = -\frac{\partial \mathcal{L}^\varepsilon}{\partial y}(x, y, \theta; h) = -\frac{\partial \mathcal{L}^0}{\partial y}(x, y; h)$$
$$- \varepsilon \frac{\partial \mathcal{L}^1}{\partial y}(x, y, \theta; h) + \mathcal{O}(\varepsilon^2), \tag{E1.2.6}$$
$$\frac{dy}{d\theta} = \frac{\partial \mathcal{L}^\varepsilon}{\partial x}(x, y, \theta; h) = \frac{\partial \mathcal{L}^0}{\partial x}(x, y; h)$$
$$+ \varepsilon \frac{\partial \mathcal{L}^1}{\partial x}(x, y, \theta; h) + \mathcal{O}(\varepsilon^2).$$

Hints:

1) $\frac{dx}{d\theta} = \left(\frac{dx}{dt} / \left(\frac{d\theta}{dt}\right)\right)$; similarly for $\frac{dy}{d\theta}$.
2) $\left(\frac{dx}{dt}\right)/\left(\frac{d\theta}{dt}\right) = \left(\frac{\partial H^\varepsilon}{\partial y}\right)/\left(\frac{\partial H^\varepsilon}{\partial I}\right)$; similarly for $\left(\frac{dy}{dt}\right)/\left(\frac{d\theta}{dt}\right)$.
3) $\frac{dH^\varepsilon}{dy} = 0 = \frac{\partial H^\varepsilon}{\partial y} + \frac{\partial H^\varepsilon}{\partial I}\frac{\partial I}{\partial y}$; similarly for $\frac{dH^\varepsilon}{dx}$.
4) Hint 3) implies $\left(\frac{\partial H^\varepsilon}{\partial y}\right)/\left(\frac{\partial H^\varepsilon}{\partial I}\right) = -\frac{\partial I}{\partial y} = -\frac{\partial \mathcal{L}^\varepsilon}{\partial y}$; similarly for $\frac{\partial H^\varepsilon}{\partial x} / \frac{\partial H^\varepsilon}{\partial I}$.

Justify all uses of the chain rule.

e) (E1.2.6) is called the *reduced system.* Show that the *unperturbed reduced system*

$$\frac{dx}{d\theta} = -\frac{\partial \mathcal{L}^0}{\partial y}(x, y; h),$$
$$\frac{dy}{d\theta} = \frac{\partial \mathcal{L}^0}{\partial x}(x, y; h), \tag{E1.2.7}$$

satisfies Assumptions 1 and 2. How does the structure of the phase space depend on the parameter h? How is the one-parameter family of periodic orbits and the homoclinic orbit of the $x - y$ component of the unperturbed two-degree-of-freedom system described in Assumptions 1 and 2 related to the orbits of (E1.2.7)?

f) The subharmonic Melnikov theory can now be applied to the reduced system (E1.2.6). Show that the reduced system satisfies the nonzero twist condition in the region enclosed by the homoclinic orbit. Show that

$$\bar{M}_1^{m/n}(\theta_0; h)$$
$$= \int_0^{2\pi m} \left\{\mathcal{L}^0(x^\alpha(\theta), y^\alpha(\theta); h), \mathcal{L}^1(x^\alpha(\theta), y^\alpha(\theta), \theta + \theta_0; h)\right\} d\theta,$$

where

$$\{\mathcal{L}^0, \mathcal{L}^1\} \equiv \frac{\partial \mathcal{L}^0}{\partial x}\frac{\partial \mathcal{L}^1}{\partial y} - \frac{\partial \mathcal{L}^0}{\partial y}\frac{\partial \mathcal{L}^1}{\partial x}$$

is the *Poisson bracket* of $\mathcal{L}^0$ with $\mathcal{L}^1$. Show that

$$\{\mathcal{L}^0, \mathcal{L}^1\} = \frac{1}{\Omega^2}\{F, H^1\}$$

where, recall, $\Omega \equiv \frac{\partial G}{\partial I}$.

1.2.33 Consider the following two-degree-of-freedom Hamiltonian system

$$\begin{aligned}
\dot{\phi} &= v = \frac{\partial H^\varepsilon}{\partial v}, \\
\dot{v} &= \sin\phi + \varepsilon(x-\phi) = -\frac{\partial H^\varepsilon}{\partial \phi}, \\
\dot{x} &= y = \frac{\partial H^\varepsilon}{\partial y}, \\
\dot{y} &= -\omega^2 x - \varepsilon(x-\phi) = -\frac{\partial H^\varepsilon}{\partial x},
\end{aligned} \qquad (\phi, v, x, y) \in S^1 \times \mathbb{R} \times \mathbb{R} \times \mathbb{R},$$

with

$$H^\varepsilon(\phi, v, x, y) = \frac{v^2}{2} - \cos\phi + \frac{y^2}{2} + \frac{\omega^2 x}{2} + \frac{\varepsilon}{2}(x-\phi)^2.$$

Use the method of reduction discussed in Exercise 1.2.32 to study the dynamics of this system for ε small. *Hint:* you need to transform the x and y components of the unperturbed vector field into action-angle variables so that the vector field takes the form of (E1.2.1).

1.2.34 What are the different possibilities for the ultimate behavior of orbits in the resonance bands of the periodically forced, damped Duffing oscillator described in Section 1.2E?

1.2.35 Recall that a $\mathbf{C}^1$ map, f, preserves orientation when $\det Df > 0$. Show that the three types of Poincaré maps discussed in Section 1.2A preserve orientation. What would be the consequences if they did not?

1.2.36 Recall the study of the resonance bands in Section 1.2D,iib). In this exercise we want to study the scaling with respect to ε of the width of the resonance bands. Suppose that we have

$$\frac{\partial^r \Omega}{\partial I^r}(I^{m,n}) = 0, \qquad r = 1, \cdots, k,$$

and

$$\frac{\partial^{r+1}\Omega}{\partial I^{r+1}}(I^{m,n}) \neq 0.$$

Then show that the small parameter μ in the transformation (1.2.209) is given by

$$\mu = \varepsilon^{1/(r+2)}.$$

(Note: the paper of Morozov and Silnikov [1984] gives a very detailed discussion of this situation.)

1.2.37 Recall Exercise 1.2.18; we were interested in studying the orbit structure near $(x, y) = (1, 0)$ of

$$\begin{aligned} \dot{x} &= y, \\ \dot{y} &= x - x^3 + \varepsilon(\gamma \cos \omega t - \delta y), \end{aligned} \qquad \delta, \gamma, \omega > 0.$$

In this exercise we will use the subharmonic Melnikov theory to study the same questions.

a) Using the techniques described in Section 1.2D,ii) study the orbit structure near the 1:1 resonance, i.e., $\omega = \sqrt{2}$. Does Exercise 1.2.36 explain the $\mathcal{O}(\varepsilon^{1/3})$ amplitude of the periodic response? *Hint:* You will have to consider $\frac{\partial \Omega}{\partial I}(0)$; cf. Exercise 1.2.29, parts f) and g).

b) Do the same analysis for the 1:3 resonance, i.e., $\omega = 3\sqrt{2}$.

c) In Exercise 1.2.18, the van der Pol transformation (cf. Example 1.2.11) introduced the detuning parameter, ρ, which enabled us to study the *passage through resonance* as ρ varied from zero. (Note: we will consider this in detail in Chapter 3 when we study bifurcation theory.) This parameter was necessary because the unperturbed equation was linear; hence, the frequency was constant. However, the class of unperturbed problems studied in Section 1.2D,ii) were nonlinear. Show how the idea of detuning or passage through resonance arises in this case by comparing the structure of the equations derived in Exercise 1.2.18 with the structure of those derived in this exercise. Are the two methods equivalent in the sense that they yield exactly the same dynamics?

d) Compare the computational effort required in Exercise 1.2.18 versus that required in this exercise to obtain similar results.

1.2.38 *Ultrasubharmonics in the Damped, Forced Duffing Oscillator.* We have seen in Section 1.2E that the equation

$$\begin{aligned} \dot{x} &= y, \\ \dot{y} &= x - x^3 + \varepsilon(\gamma \cos \omega t - \delta y), \end{aligned} \qquad \delta, \gamma, \omega > 0, \quad \text{(E1.2.8)}$$

possesses no ultrasubharmonic solutions. However, consider the equation

$$\begin{aligned} \dot{x} &= y, \\ \dot{y} &= x - x^3 + \varepsilon \gamma \cos \omega t - \varepsilon^2 \delta y, \end{aligned} \qquad \delta, \gamma, \omega > 0. \quad \text{(E1.2.9)}$$

Show that (E1.2.9) has ultrasubharmonic solutions. (*Hint:* using the formalism developed in Section 1.2D,ii), compute the Poincaré map through $\mathcal{O}(\varepsilon^2)$, i.e., develop a higher order Melnikov theory.) Compare the dynamics of (E1.2.8) and (E1.2.9).

1.2.39 Justify all steps in the proof of Lemma 1.2.16. In particular, does the fact that, in general, the linear variational equation varies from orbit to orbit cause difficulties?

2

Methods for Simplifying Dynamical Systems

When one thinks of simplifying dynamical systems, two approaches come to mind: one, reduce the dimensionality of the system and two, eliminate the nonlinearity. Two rigorous mathematical techniques that allow substantial progress along both lines of approach are center manifold theory and the method of normal forms. These techniques are the most important, generally applicable methods available in the local theory of dynamical systems, and they will form the foundation of our development of bifurcation theory in Chapter 3.

The center manifold theorem in finite dimensions can be traced to the work of Pliss [1964], Šošitaĭšvili [1975], and Kelley [1967]. Additional valuable references are Guckenheimer and Holmes [1983], Hassard, Kazarinoff, and Wan [1980], Marsden and McCracken [1976], Carr [1981], Henry [1981], and Sijbrand [1985].

The method of normal forms can be traced to the Ph.D thesis of Poincaré [1929]. The books by van der Meer [1985] and Bryuno [1989] give valuable historical background.

2.1 Center Manifolds

Let us begin our discussion with some motivation. Consider the linear systems

$$\dot{x} = Ax, \tag{2.1.1a}$$

$$x \longmapsto Ax, \qquad x \in \mathbb{R}^n, \tag{2.1.1b}$$

where A is an $n \times n$ matrix. Recall from Section 1.1C that each system has invariant subspaces E^s, E^u, E^c, corresponding to the span of the generalized eigenvectors, which in turn correspond to eigenvalues having

Flows: negative real part, positive real part, and zero real part, respectively.

Maps: modulus < 1, modulus > 1, and modulus $= 1$, respectively.

The subspaces were so named because orbits starting in E^s decayed to zero as t (resp. n for maps) $\uparrow \infty$, orbits starting in E^u became unbounded as t (resp. n for maps) $\uparrow \infty$, and orbits starting in E^c neither grew nor decayed exponentially as t (resp. n for maps) $\uparrow \infty$.

If we suppose that $E^u = \emptyset$, then we find that any orbit will rapidly decay to E^c. Thus, if we are interested in long-time behavior (i.e., stability) we need only to investigate the system restricted to E^c.

It would be nice if a similar type of "reduction principle" applied to the study of the stability of nonhyperbolic fixed points of nonlinear vector fields and maps, namely, that there were an invariant *center manifold* passing through the fixed point to which the system could be restricted in order to study its asymptotic behavior in the neighborhood of the fixed point. That this is the case is the content of the center manifold theory.

2.1A Center Manifolds for Vector Fields

We will begin by considering center manifolds for vector fields. The set-up is as follows. We consider vector fields of the following form

$$\begin{aligned} \dot{x} &= Ax + f(x,y), \\ \dot{y} &= By + g(x,y), \qquad (x,y) \in \mathbb{R}^c \times \mathbb{R}^s, \end{aligned} \tag{2.1.2}$$

where

$$\begin{aligned} f(0,0) &= 0, \qquad Df(0,0) = 0, \\ g(0,0) &= 0, \qquad Dg(0,0) = 0. \end{aligned} \tag{2.1.3}$$

(See Section 1.1C for a discussion of how a general vector field is transformed to the form of (2.1.2) in the neighborhood of a fixed point.)

In the above, A is a $c \times c$ matrix having eigenvalues with zero real parts, B is an $s \times s$ matrix having eigenvalues with negative real parts, and f and g are $\mathbf{C}^r$ functions ($r \geq 2$).

Definition 2.1.1 An invariant manifold will be called a center manifold for (2.1.2) if it can locally be represented as follows

$$W^c(0) = \{\,(x,y) \in \mathbb{R}^c \times \mathbb{R}^s \mid y = h(x), |x| < \delta, h(0) = 0, Dh(0) = 0\,\}$$

for δ sufficiently small.

We remark that the conditions $h(0) = 0$ and $Dh(0) = 0$ imply that $W^c(0)$ is tangent to E^c at $(x,y) = (0,0)$. The following three theorems are taken from the excellent book by Carr [1981].

The first result on center manifolds is an existence theorem.

Theorem 2.1.1 *There exists a* $\mathbf{C}^r$ *center manifold for (2.1.2). The dynamics of (2.1.2) restricted to the center manifold is, for u sufficiently small, given by the following c-dimensional vector field*

$$\dot{u} = Au + f(u, h(u)), \qquad u \in \mathbb{R}^c. \tag{2.1.4}$$

Proof: See Carr [1981]. □

The next result implies that the dynamics of (2.1.4) near $u = 0$ determine the dynamics of (2.1.2) near $(x, y) = (0, 0)$.

Theorem 2.1.2 i) *Suppose the zero solution of (2.1.4) is stable (asymptotically stable) (unstable); then the zero solution of (2.1.2) is also stable (asymptotically stable) (unstable).* ii) *Suppose the zero solution of (2.1.4) is stable. Then if* $\big(x(t), y(t)\big)$ *is a solution of (2.1.2) with* $\big(x(0), y(0)\big)$ *sufficiently small, there is a solution* $u(t)$ *of (2.1.4) such that as* $t \to \infty$

$$\begin{aligned} x(t) &= u(t) + \mathcal{O}\big(e^{-\gamma t}\big), \\ y(t) &= h(u(t)) + \mathcal{O}\big(e^{-\gamma t}\big), \end{aligned}$$

where $\gamma > 0$ *is a constant.*

Proof: See Carr [1981]. □

The obvious question now is how do we compute the center manifold so that we can reap the benefits of Theorem 2.1.2? To answer this question we will derive an equation that $h(x)$ must satisfy in order for its graph to be a center manifold for (2.1.2).

Suppose we have a center manifold

$$W^c(0) = \{(x, y) \in \mathbb{R}^c \times \mathbb{R}^s \mid y = h(x), |x| < \delta, h(0) = 0,\ Dh(0) = 0\}, \tag{2.1.5}$$

with δ sufficiently small.

Using invariance of $W^c(0)$ under the dynamics of (2.1.2), we derive a quasilinear partial differential equation that $h(x)$ must satisfy. This is done as follows:

1. The (x, y) coordinates of any point on $W^c(0)$ must satisfy

$$y = h(x). \tag{2.1.6}$$

2. Differentiating (2.1.6) with respect to time implies that the $(\dot{x}, \dot{y})$ coordinates of any point on $W^c(0)$ must satisfy

$$\dot{y} = Dh(x)\dot{x}. \tag{2.1.7}$$

3. Any point on $W^c(0)$ obeys the dynamics generated by (2.1.2). Therefore, substituting

$$\dot{x} = Ax + f\big(x, h(x)\big), \tag{2.1.8a}$$
$$\dot{y} = Bh(x) + g\big(x, h(x)\big) \tag{2.1.8b}$$

into (2.1.7) gives

$$Dh(x)\big[Ax + f\big(x, h(x)\big)\big] = Bh(x) + g\big(x, h(x)\big) \tag{2.1.9}$$

or

$$\mathcal{N}\big(h(x)\big) \equiv Dh(x)\big[Ax + f\big(x, h(x)\big)\big] - Bh(x) - g\big(x, h(x)\big) = 0. \tag{2.1.10}$$

Equation (2.1.10) is a quasilinear partial differential equation that $h(x)$ must satisfy in order for its graph to be an invariant center manifold; see Exercise 2.14 for further discussion of the notion of invariance of a manifold. To find a center manifold, all we need do is solve (2.1.10).

Unfortunately, it is probably more difficult to solve (2.1.10) than our original problem; however, the following theorem gives us a method for computing an approximate solution of (2.1.10) to any desired degree of accuracy.

Theorem 2.1.3 *Let $\phi : \mathbb{R}^c \to \mathbb{R}^s$ be a $\mathbf{C}^1$ mapping with $\phi(0) = D\phi(0) = 0$ such that $\mathcal{N}\big(\phi(x)\big) = \mathcal{O}(|x|^q)$ as $x \to 0$ for some $q > 1$. Then*

$$|h(x) - \phi(x)| = \mathcal{O}(|x|^q) \qquad \text{as } x \to 0.$$

Proof: See Carr [1981]. □

This theorem allows us to compute the center manifold to any desired degree of accuracy by solving (2.1.10) to the same degree of accuracy. For this task, power series expansions will work nicely. Let us consider a concrete example.

EXAMPLE 2.1.1 Consider the vector field

$$\begin{aligned} \dot{x} &= x^2 y - x^5, \\ \dot{y} &= -y + x^2, \qquad (x, y) \in \mathbb{R}^2. \end{aligned} \tag{2.1.11}$$

The origin is obviously a fixed point for (2.1.11), and the question we ask is whether or not it is stable. The eigenvalues of (2.1.11) linearized about $(x, y) = (0, 0)$ are 0 and -1. Thus, since the fixed point is not hyperbolic, we cannot make any conclusions concerning the stability or instability of $(x, y) = (0, 0)$ based on linearization (note: in the linear approximation the origin is stable but not asymptotically stable). We will answer the question of stability using center manifold theory.

From Theorem 2.1.1, there exists a center manifold for (2.1.11) which can locally be represented as follows

$$W^c(0) = \left\{ (x, y) \in \Re^2 \mid y = h(x), |x| < \delta,\ h(0) = Dh(0) = 0 \right\} \quad (2.1.12)$$

for δ sufficiently small. We now want to compute $W^c(0)$. We assume that $h(x)$ has the form

$$h(x) = ax^2 + bx^3 + \mathcal{O}(x^4), \quad (2.1.13)$$

and we substitute (2.1.13) into equation (2.1.10), which $h(x)$ must satisfy to be a center manifold. We then equate equal powers of x, and in that way we can compute $h(x)$ to any desired order of accuracy. In practice, computing only a few terms is usually sufficient to answer questions of stability.

We recall from (2.1.10) that the equation for the center manifold is given by

$$\mathcal{N}\left(h(x)\right) = Dh(x)\left[Ax + f\left(x, h(x)\right)\right] - Bh(x) - g\left(x, h(x)\right) = 0, \quad (2.1.14)$$

where, in this example, we have $(x, y) \in \mathbb{R}^2$,

$$\begin{aligned} A &= 0, \\ B &= -1, \\ f(x, y) &= x^2 y - x^5, \\ g(x, y) &= x^2. \end{aligned} \quad (2.1.15)$$

Substituting (2.1.13) into (2.1.14) and using (2.1.15) gives

$$\begin{aligned} \mathcal{N}\left(h(x)\right) = (2ax + 3bx^2 + \cdots)(ax^4 + bx^5 - x^5 + \cdots) \\ + ax^2 + bx^3 - x^2 + \cdots = 0. \end{aligned} \quad (2.1.16)$$

In order for (2.1.16) to hold, the coefficients of each power of x must be zero; see Exercise 2.20. Thus, equating coefficients on each power of x to zero gives

$$\begin{aligned} x^2 &: a - 1 = 0 \Rightarrow a = 1, \\ x^3 &: b = 0, \\ \vdots &\quad \vdots \end{aligned} \quad (2.1.17)$$

and we therefore have

$$h(x) = x^2 + \mathcal{O}(x^4). \quad (2.1.18)$$

Using (2.1.18) along with Theorem 2.1.1, the vector field restricted to the center manifold is given by

$$\dot{x} = x^4 + \mathcal{O}(x^5). \quad (2.1.19)$$

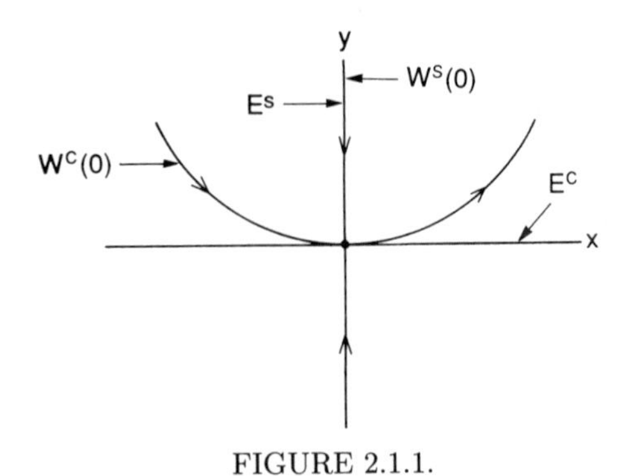

FIGURE 2.1.1.

For x sufficiently small, $x = 0$ is thus unstable in (2.1.19). Hence, by Theorem 2.1.1, $(x, y) = (0, 0)$ is unstable in (2.1.11); see Figure 2.1.1 for an illustration of the geometry of the flow near $(x, y) = (0, 0)$.

This example illustrates an important phenomenon, which we now describe.

The Failure of the Tangent Space Approximation

The idea is as follows. Consider (2.1.11). One might expect that the y components of orbits starting near $(x, y) = (0, 0)$ should decay to zero exponentially fast. Therefore, the question of stability of the origin should reduce to a study of the x component of orbits starting near the origin. One might thus be very tempted to set $y = 0$ in (2.1.11) and study the reduced equation

$$\dot{x} = -x^5. \tag{2.1.20}$$

This corresponds to approximating $W^c(0)$ by E^c. However, $x = 0$ is stable for (2.1.20) and, therefore, we would arrive at the *wrong* conclusion that $(x, y) = (0, 0)$ is stable for (2.1.20). The tangent space approximation might sometimes work, but, as this example shows, it does not always do so.

2.1B Center Manifolds Depending on Parameters

Suppose (2.1.2) depends on a vector of parameters, say $\varepsilon \in \mathbb{R}^p$. In this case we write (2.1.2) in the form

$$\begin{aligned} \dot{x} &= Ax + f(x, y, \varepsilon), \\ \dot{y} &= By + g(x, y, \varepsilon), \end{aligned} \qquad (x, y, \varepsilon) \in \mathbb{R}^c \times \mathbb{R}^s \times \mathbb{R}^p, \tag{2.1.21}$$

where

$$\begin{aligned} f(0,0,0) &= 0, \qquad & Df(0,0,0) &= 0, \\ g(0,0,0) &= 0, \qquad & Dg(0,0,0) &= 0, \end{aligned}$$

and we have the same assumptions on A and B as in (2.1.2), with f and g also being $\mathbf{C}^r$ $(r \geq 2)$ functions in some neighborhood of $(x, y, \varepsilon) = (0, 0, 0)$. An obvious question is why do we not allow the matrices A and B to depend on ε? This will be answered shortly.

The way in which we will handle parametrized systems is to include the parameter ε as a new *dependent variable* as follows

$$\begin{aligned} \dot{x} &= Ax + f(x, y, \varepsilon), \\ \dot{\varepsilon} &= 0, \\ \dot{y} &= By + g(x, y, \varepsilon), \end{aligned} \qquad (x, \varepsilon, y) \in \mathbb{R}^c \times \mathbb{R}^s \times \mathbb{R}^p. \tag{2.1.22}$$

At first glance it might appear that nothing is really gained from this action, but we will argue otherwise.

Let us suppose we are considering (2.1.22) afresh. It obviously has a fixed point at $(x, \varepsilon, y) = (0, 0, 0)$. The matrix associated with the linearization of (2.1.22) about this fixed point has $c + p$ eigenvalues with zero real part and s eigenvalues with negative real part. Now let us apply center manifold theory. Modifying Definition 2.1.1, a center manifold will be represented as a graph over the x *and* ε *variables*, i.e., the graph of $h(x, \varepsilon)$ for x and ε sufficiently small. Theorem 2.1.1 still applies, with the vector field reduced to the center manifold given by

$$\begin{aligned} \dot{u} &= Au + f(u, h(u, \varepsilon), \varepsilon), \\ \dot{\varepsilon} &= 0, \end{aligned} \qquad (u, \varepsilon) \in \mathbb{R}^c \times \mathbb{R}^p. \tag{2.1.23}$$

Theorems 2.1.2 and 2.1.3 also follow (we will worry about any modifications to computing the center manifold shortly). Thus, adding the parameter as a new dependent variable merely acts to augment the matrix A in (2.1.2) by adding p new center directions that have no dynamics, and the theory goes through just the same. However, there is a new concept which will be important when we study *bifurcation theory*; namely, the center manifold exists for all ε in a sufficiently small neighborhood of $\varepsilon = 0$. We will learn in Chapter 3 that it is possible for solutions to be created or destroyed by perturbing nonhyperbolic fixed points. Thus, since the invariant center manifold exists in a sufficiently small neighborhood in both x and ε of $(x, \varepsilon) = (0, 0)$, all bifurcating solutions will be contained in the lower dimensional center manifold.

Let us now worry about computing the center manifold. From the existence theorem for center manifolds, locally we have

$$\begin{aligned} W^c_{\text{loc}}(0) = \{ (x, \varepsilon, y) \in \mathbb{R}^c \times \mathbb{R}^p \times \mathbb{R}^s \mid y = h(x, \varepsilon), |x| < \delta, \\ |\varepsilon| < \bar{\delta},\ h(0, 0) = 0,\ Dh(0, 0) = 0 \} \end{aligned} \tag{2.1.24}$$

for δ and $\bar{\delta}$ sufficiently small. Using invariance of the graph of $h(x, \varepsilon)$ under the dynamics generated by (2.1.22) we have

$$\dot{y} = D_x h(x, \varepsilon)\dot{x} + D_\varepsilon h(x, \varepsilon)\dot{\varepsilon} = Bh(x, \varepsilon) + g\left(x, h(x, \varepsilon), \varepsilon\right). \tag{2.1.25}$$

However,

$$\begin{aligned}\dot{x} &= Ax + f\left(x, h(x,\varepsilon), \varepsilon\right),\\ \dot{\varepsilon} &= 0;\end{aligned} \tag{2.1.26}$$

hence substituting (2.1.26) into (2.1.25) results in the following quasilinear partial differential equation that $h(x,\varepsilon)$ must satisfy in order for its graph to be a center manifold.

$$\begin{aligned}\mathcal{N}\left(h(x,\varepsilon)\right) = D_x h(x,\varepsilon)\left[Ax + f\left(x, h(x,\varepsilon), \varepsilon\right)\right]\\ - Bh(x,\varepsilon) - g\left(x, h(x,\varepsilon), \varepsilon\right) = 0.\end{aligned} \tag{2.1.27}$$

Thus, we see that (2.1.27) is very similar to (2.1.10).

Before considering a specific example we want to point out an important fact. By considering ε as a new dependent variable, terms such as

$$x_i \varepsilon_j, \qquad 1 \le i \le c, \quad 1 \le j \le p,$$

or

$$y_i \varepsilon_j, \qquad 1 \le i \le s, \quad 1 \le j \le p,$$

become *nonlinear terms.* In this case, returning to a question asked at the beginning of this section, the parts of the matrices A and B depending on ε are now viewed as nonlinear terms and are included in the f and g terms of (2.1.22), respectively. We remark that in applying center manifold theory to a given system, it must first be transformed into the standard form (either (2.1.2) or (2.1.22)).

EXAMPLE 2.1.2 Consider the Lorenz equations

$$\begin{aligned}\dot{x} &= \sigma(y - x),\\ \dot{y} &= \bar{\rho}x + x - y - xz, \qquad (x,y,z) \in \mathbb{R}^3,\\ \dot{z} &= -\beta z + xy,\end{aligned} \tag{2.1.28}$$

where σ and β are viewed as fixed positive constants and $\bar{\rho}$ is a parameter (note: in the standard version of the Lorenz equations it is traditional to put $\bar{\rho} = \rho - 1$). It should be clear that $(x,y,z) = (0,0,0)$ is a fixed point of (2.1.28). Linearizing (2.1.28) about this fixed point, we obtain the associated matrix

$$\begin{pmatrix} -\sigma & \sigma & 0 \\ 1 & -1 & 0 \\ 0 & 0 & -\beta \end{pmatrix}. \tag{2.1.29}$$

(Note: recall, $\bar{\rho}x$ is a nonlinear term.)

Since (2.1.29) is in block form, the eigenvalues are particularly easy to compute and are given by

$$0, -\sigma - 1, -\beta, \tag{2.1.30}$$

with eigenvectors

$$\begin{pmatrix} 1 \\ 1 \\ 0 \end{pmatrix}, \begin{pmatrix} \sigma \\ -1 \\ 0 \end{pmatrix}, \begin{pmatrix} 0 \\ 0 \\ 1 \end{pmatrix}. \tag{2.1.31}$$

Our goal is to determine the nature of the stability of $(x, y, z) = (0, 0, 0)$ for $\bar{\rho}$ near zero. First, we must put (2.1.29) into the standard form (2.1.22). Using the eigenbasis (2.1.31), we obtain the transformation

$$\begin{pmatrix} x \\ y \\ z \end{pmatrix} = \begin{pmatrix} 1 & \sigma & 0 \\ 1 & -1 & 0 \\ 0 & 0 & 1 \end{pmatrix} \begin{pmatrix} u \\ v \\ w \end{pmatrix} \tag{2.1.32}$$

with inverse

$$\begin{pmatrix} u \\ v \\ w \end{pmatrix} = \frac{1}{1+\sigma} \begin{pmatrix} 1 & \sigma & 0 \\ 1 & -1 & 0 \\ 0 & 0 & 1+\sigma \end{pmatrix} \begin{pmatrix} x \\ y \\ z \end{pmatrix}, \tag{2.1.33}$$

which transforms (2.1.28) into

$$\begin{aligned} \begin{pmatrix} \dot{u} \\ \dot{v} \\ \dot{w} \end{pmatrix} &= \begin{pmatrix} 0 & 0 & 0 \\ 0 & -(1+\sigma) & 0 \\ 0 & 0 & -\beta \end{pmatrix} \begin{pmatrix} u \\ v \\ w \end{pmatrix} \\ &\quad + \frac{1}{1+\sigma} \begin{pmatrix} \sigma\bar{\rho}(u+\sigma v) - \sigma w(u+\sigma v) \\ -\bar{\rho}(u+\sigma v) + w(u+\sigma v) \\ (1+\sigma)(u+\sigma v)(u-v) \end{pmatrix}, \\ \dot{\bar{\rho}} &= 0. \end{aligned} \tag{2.1.34}$$

Thus, from center manifold theory, the stability of $(x, y, z) = (0, 0, 0)$ near $\bar{\rho} = 0$ can be determined by studying a one-parameter family of first-order ordinary differential equations on a center manifold, which can be represented as a graph over the u and $\bar{\rho}$ variables, i.e.,

$$\begin{aligned} W^c(0) = \{ (u, v, w, \bar{\rho}) \in \mathbb{R}^4 \mid v = h_1(u, \bar{\rho}), w = h_2(u, \bar{\rho}), \\ h_i(0, 0) = 0, Dh_i(0, 0) = 0, i = 1, 2\} \end{aligned} \tag{2.1.35}$$

for u and $\bar{\rho}$ sufficiently small.

We now want to compute the center manifold and derive the vector field on the center manifold. Using Theorem 2.1.3, we assume

$$\begin{aligned} h_1(u, \bar{\rho}) &= a_1 u^2 + a_2 u\bar{\rho} + a_3 \bar{\rho}^2 + \cdots, \\ h_2(u, \bar{\rho}) &= b_1 u^2 + b_2 u\bar{\rho} + b_3 \bar{\rho}^2 + \cdots. \end{aligned} \tag{2.1.36}$$

Recall from (2.1.27) that the center manifold must satisfy

$$\begin{aligned} \mathcal{N}(h(x, \varepsilon)) = D_x h(x, \varepsilon) [Ax + f(x, h(x, \varepsilon), \varepsilon)] \\ - Bh(x, \varepsilon) - g(x, h(x, \varepsilon), \varepsilon) = 0, \end{aligned} \tag{2.1.37}$$

where, in this example,

$$
\begin{aligned}
&x \equiv u, \qquad y \equiv (v, w), \qquad \varepsilon \equiv \bar{\rho}, \qquad h = (h_1, h_2),\\
&A = 0,\\
&B = \begin{pmatrix} -(1+\sigma) & 0 \\ 0 & -\beta \end{pmatrix},\\
f(x, y, \varepsilon) &= \frac{1}{1+\sigma}[\sigma\bar{\rho}(u+\sigma v) - \sigma w(u+\sigma v)],\\
g(x, y, \varepsilon) &= \frac{1}{1+\sigma}\begin{pmatrix} -\bar{\rho}(u+\sigma v) + w(u+\sigma v) \\ (1+\sigma)(u+\sigma v)(u-v) \end{pmatrix}.
\end{aligned}
\tag{2.1.38}
$$

Substituting (2.1.36) into (2.1.37) and using (2.1.38) gives the two components of the equation for the center manifold.

$$
\begin{aligned}
&(2a_1 u + a_2\bar{\rho} + \cdots)\left[\frac{\sigma}{1+\sigma}\left(\bar{\rho}(u+\sigma h_1) - h_2(u+\sigma h_1)\right)\right]\\
&\quad + (1+\sigma)h_1 + \frac{\bar{\rho}}{1+\sigma}(u+\sigma h_1) - \frac{h_2}{1+\sigma}(u+\sigma h_1) = 0,\\
&(2b_1 u + b_2\bar{\rho} + \cdots)\left[\frac{\sigma}{1+\sigma}\left(\bar{\rho}(u+\sigma h_1) - h_2(u+\sigma h_1)\right)\right]\\
&\quad + \beta h_2 - (u+\sigma h_1)(u-h_1) = 0.
\end{aligned}
\tag{2.1.39}
$$

Equating terms of like powers to zero gives

$$
\begin{aligned}
u^2 &: a_1(1+\sigma) = 0 \Rightarrow a_1 = 0,\\
&\quad \beta b_1 - 1 = 0 \Rightarrow b_1 = \frac{1}{\beta},\\
u\bar{\rho} &: (1+\sigma)a_2 + \frac{1}{1+\sigma} = 0 \Rightarrow a_2 = \frac{-1}{(1+\sigma)^2},\\
&\quad \beta b_2 = 0 \Rightarrow b_2 = 0.
\end{aligned}
\tag{2.1.40}
$$

Then, using (2.1.40) and (2.1.36), we obtain

$$
\begin{aligned}
h_1(u, \bar{\rho}) &= -\frac{1}{(1+\sigma)^2}u\bar{\rho} + \cdots,\\
h_2(u, \bar{\rho}) &= \frac{1}{\beta}u^2 + \cdots.
\end{aligned}
\tag{2.1.41}
$$

Finally, substituting (2.1.41) into (2.1.34) we obtain the vector field reduced to the center manifold

$$
\begin{aligned}
\dot{u} &= u\left(\frac{\sigma}{(1+\sigma)}\bar{\rho} - \frac{\sigma}{\beta}u^2 + \cdots\right),\\
\dot{\bar{\rho}} &= 0.
\end{aligned}
\tag{2.1.42}
$$

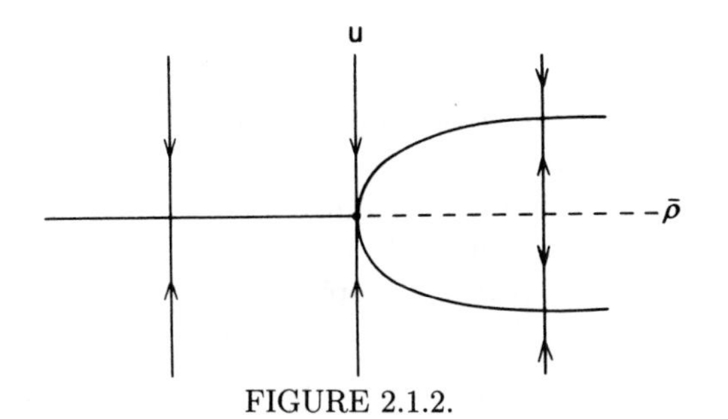

FIGURE 2.1.2.

In Figure 2.1.2 we plot the fixed points of (2.1.42) *neglecting* higher order terms such as $\mathcal{O}(\bar{\rho}^2)$, $\mathcal{O}(u\bar{\rho}^2)$, $\mathcal{O}(u^3)$, etc. It should be clear that $u = 0$ is always a fixed point and is stable for $\bar{\rho} < 0$ and unstable for $\bar{\rho} > 0$. At the point of exchange of stability (i.e., $\bar{\rho} = 0$) two new stable fixed points are created and are given by

$$\frac{1}{1+\sigma}\bar{\rho} = \frac{1}{\beta}u^2. \tag{2.1.43}$$

A simple calculation shows that these fixed points are stable. In Chapter 3 we will see that this is an example of a *pitchfork bifurcation.*

Before leaving this example two comments are in order.

1. Figure 2.1.2 shows the advantage of introducing the parameter as a new dependent variable. In a full neighborhood in parameter space new solutions are "captured" on the center manifold. In Figure 2.1.2, for each fixed $\bar{\rho}$ we have a flow in the u direction; this is represented by the vertical lines with arrows.

2. We have not considered the effects of the higher order terms in (2.1.42) on Figure 2.1.2. In Chapter 3 we will show that they do not qualitatively change the figure (i.e., they do not create, destroy, or change the stability of any of the fixed points) near the origin.

2.1C The Inclusion of Linearly Unstable Directions

Suppose we consider the system

$$\begin{aligned} \dot{x} &= Ax + f(x,y,z), \\ \dot{y} &= By + g(x,y,z), \qquad (x,y,z) \in \mathbb{R}^c \times \mathbb{R}^s \times \mathbb{R}^u, \\ \dot{z} &= Cz + h(x,y,z), \end{aligned} \tag{2.1.44}$$

where

$$\begin{aligned} f(0,0,0) &= 0, \qquad Df(0,0,0) = 0, \\ g(0,0,0) &= 0, \qquad Dg(0,0,0) = 0, \\ h(0,0,0) &= 0, \qquad Dh(0,0,0) = 0, \end{aligned}$$

and f, g, and h are $\mathbf{C}^r (r \geq 2)$ in some neighborhood of the origin, A is a $c \times c$ matrix having eigenvalues with zero real parts, B is an $s \times s$ matrix having eigenvalues with negative real parts, and C is a $u \times u$ matrix having eigenvalues with positive real parts.

In this case $(x, y, z) = (0, 0, 0)$ is unstable due to the existence of a u-dimensional unstable manifold. However, much of the center manifold theory still applies, in particular Theorem 2.1.1 concerning existence, with the center manifold being locally represented by

$$W^c(0) = \big\{(x, y, z) \in \mathbb{R}^c \times \mathbb{R}^s \times \mathbb{R}^u \mid y = h_1(x), z = h_2(x), \\ h_i(0) = 0, Dh_i(0) = 0, i = 1, 2\big\} \tag{2.1.45}$$

for x sufficiently small. The vector field restricted to the center manifold is given by

$$\dot{u} = Au + f\big(u, h_1(u), h_2(u)\big), \qquad u \in \mathbb{R}^c. \tag{2.1.46}$$

Using the fact that the center manifold is invariant under the dynamics generated by (2.1.44), we obtain

$$\begin{aligned} \dot{x} &= Ax + f\big(x, h_1(x), h_2(x)\big), \\ \dot{y} &= Dh_1(x)\dot{x} = Bh_1(x) + g\big(x, h_1(x), h_2(x)\big), \\ \dot{z} &= Dh_2(x)\dot{x} = Ch_2(x) + h\big(x, h_1(x), h_2(x)\big), \end{aligned} \tag{2.1.47}$$

which yields the following quasilinear partial differential equation for $h_1(x)$ and $h_2(x)$

$$\begin{aligned} &Dh_1(x)\big[Ax + f\big(x, h_1(x), h_2(x)\big)\big] \\ &\qquad - Bh_1(x) - g\big(x, h_1(x), h_2(x)\big) = 0, \\ &Dh_2(x)\left[Ax + f\big(x, h_1(x), h_2(x)\big)\right] \\ &\qquad - Ch_2(x) - h\big(x, h_1(x), h_2(x)\big) = 0. \end{aligned} \tag{2.1.48}$$

Theorem 2.1.3 also holds in order that we may justify solving (2.1.48) approximately via power series expansions. We can also include parameters in exactly the same way as in Section 2.1B.

2.1D Center Manifolds for Maps

The center manifold theory can be modified so that it applies to maps with only a slight difference in the method by which the center manifold is calculated. We outline the theory below.

Suppose we have the map

$$\begin{aligned} x &\longmapsto Ax + f(x, y), \\ y &\longmapsto By + g(x, y), \end{aligned} \qquad (x, y) \in \mathbb{R}^c \times \mathbb{R}^s, \tag{2.1.49}$$

or

$$\begin{aligned} x_{n+1} &= Ax_n + f(x_n, y_n), \\ y_{n+1} &= By_n + g(x_n, y_n), \end{aligned}$$

where

$$\begin{aligned} f(0,0) &= 0, \qquad & Df(0,0) &= 0, \\ g(0,0) &= 0, \qquad & Dg(0,0) &= 0, \end{aligned}$$

and f and g are $\mathbf{C}^r (r \geq 2)$ in some neighborhood of the orgin, A is a $c \times c$ matrix with eigenvalues of modulus one, and B is an $s \times s$ matrix with eigenvalues of modulus less than one.

Evidently $(x, y) = (0,0)$ is a fixed point of (2.1.49), and the linear approximation is not sufficient for determining its stability. We have the following theorems, which are completely analogous to Theorems 2.1.1, 2.1.2, and 2.1.3.

Theorem 2.1.4 *There exists a $\mathbf{C}^r$ center manifold for (2.1.49) which can be locally represented as a graph as follows*

$$W^c(0) = \{(x, y) \in \mathbb{R}^c \times \mathbb{R}^s \mid y = h(x), |x| < \delta, h(0) = 0, Dh(0) = 0\} \tag{2.1.50}$$

for δ sufficiently small. Moreover, the dynamics of (2.1.49) restricted to the center manifold is, for u sufficiently small, given by the c-dimensional map

$$u \longmapsto Au + f\big(u, h(u)\big), \qquad u \in \mathbb{R}^c. \tag{2.1.51}$$

Proof: See Carr [1981]. □

The next theorem allows us to conclude that $(x, y) = (0,0)$ is stable or unstable based on whether or not $u = 0$ is stable or unstable in (2.1.51).

Theorem 2.1.5 i) *Suppose the zero solution of (2.1.51) is stable (asymptotically stable) (unstable). Then the zero solution of (2.1.49) is stable (asymptotically stable) (unstable).* ii) *Suppose that the zero solution of (2.1.51) is stable. Let (x_n, y_n) be a solution of (2.1.49) with (x_0, y_0) sufficiently small. Then there is a solution u_n of (2.1.51) such that $|x_n - u_n| \leq k\beta^n$ and $|y_n - h(u_n)| \leq k\beta^n$ for all n where k and β are positive constants with $\beta < 1$.*

Proof: See Carr [1981]. □

Next we want to compute the center manifold so that we can derive (2.1.51). This is done in exactly the same way as for vector fields, i.e., by deriving a nonlinear functional equation that the graph of $h(x)$ must satisfy in order for it to be invariant under the dynamics generated by (2.1.49). In this case we have

$$\begin{aligned} x_{n+1} &= Ax_n + f\big(x_n, h(x_n)\big), \\ y_{n+1} &= h(x_{n+1}) = Bh(x_n) + g\big(x_n, h(x_n)\big), \end{aligned} \tag{2.1.52}$$

or

$$\mathcal{N}\big(h(x)\big) = h\big(Ax + f\big(x, h(x)\big)\big) - Bh(x) - g\big(x, h(x)\big) = 0. \tag{2.1.53}$$

(Note: the reader should compare (2.1.53) with (2.1.10).) The next theorem justifies the approximate solution of (2.1.53) via power series expansions.

Theorem 2.1.6 *Let $\phi : \mathbb{R}^c \to \mathbb{R}^s$ be a $\mathbf{C}^1$ map with $\phi(0) = 0, \phi'(0) = 0$, and $\mathcal{N}\big(\phi(x)\big) = \mathcal{O}(|x|^q)$ as $x \to 0$ for some $q > 1$. Then*

$$h(x) = \phi(x) + \mathcal{O}(|x|^q) \qquad \text{as } x \to 0.$$

Proof: See Carr [1981]. □

We now give an example.

EXAMPLE 2.1.3 Consider the map

$$\begin{pmatrix} u \\ v \\ w \end{pmatrix} \mapsto \begin{pmatrix} -1 & 0 & 0 \\ 0 & -\frac{1}{2} & 0 \\ 0 & 0 & \frac{1}{2} \end{pmatrix} \begin{pmatrix} u \\ v \\ w \end{pmatrix} + \begin{pmatrix} vw \\ u^2 \\ -uv \end{pmatrix}, \qquad (u, v, w) \in \mathbb{R}^3. \tag{2.1.54}$$

It should be clear that $(u, v, w) = (0, 0, 0)$ is a fixed point of (2.1.54), and the eigenvalues associated with the map linearized about this fixed point are -1, $-\frac{1}{2}$, $\frac{1}{2}$. Thus, the linear approximation does not suffice to determine the stability or instability. We will apply center manifold theory to this problem.

The center manifold can locally be represented as follows

$$W^c(0) = \big\{(u, v, w) \in \mathbb{R}^3 \mid v = h_1(u), w = h_2(u), h_i(0) = 0, \\ Dh_i(0) = 0, i = 1, 2\big\} \tag{2.1.55}$$

for u sufficiently small. Recall that the center manifold must satisfy the following equation

$$\mathcal{N}\big(h(x)\big) = h\Big(Ax + f\big(x, h(x)\big)\Big) - Bh(x) - g\big(x, h(x)\big) = 0, \tag{2.1.56}$$

where, in this example,

$$\begin{aligned} x &= u, \quad y \equiv (v, w), \quad h = (h_1, h_2), \\ A &= -1, \\ B &= \begin{pmatrix} -\frac{1}{2} & 0 \\ 0 & \frac{1}{2} \end{pmatrix}, \\ f(u, v, w) &= vw, \\ g(u, v, w) &= \begin{pmatrix} u^2 \\ -uv \end{pmatrix}. \end{aligned} \tag{2.1.57}$$

We assume a center manifold of the form

$$h(u) = \begin{pmatrix} h_1(u) \\ h_2(u) \end{pmatrix} = \begin{pmatrix} a_1u^2 + b_1u^3 + \mathcal{O}(u^4) \\ a_2u^2 + b_2u^3 + \mathcal{O}(u^4) \end{pmatrix}. \tag{2.1.58}$$

Substituting (2.1.58) into (2.1.56) and using (2.1.57) yields

$$\begin{aligned} \mathcal{N}(h(u)) &= \begin{pmatrix} a_1u^2 - b_1u^3 + \mathcal{O}(u^5) \\ a_2u^2 - b_2u^3 + \mathcal{O}(u^5) \end{pmatrix} \\ &\quad - \begin{pmatrix} -1/2 & 0 \\ 0 & 1/2 \end{pmatrix} \begin{pmatrix} a_1u^2 + b_1u^3 + \cdots \\ a_2u^2 + b_2u^3 + \cdots \end{pmatrix} - \begin{pmatrix} u^2 \\ -uh_1(u) \end{pmatrix} \\ &= \begin{pmatrix} 0 \\ 0 \end{pmatrix}. \end{aligned} \tag{2.1.59}$$

Balancing powers of coefficients for each component gives

$$u^2 : \begin{pmatrix} a_1 + \frac{1}{2}a_1 - 1 \\ a_2 - \frac{1}{2}a_2 \end{pmatrix} = \begin{pmatrix} 0 \\ 0 \end{pmatrix} \Rightarrow \begin{array}{ccc} a_1 & = & \frac{2}{3} \\ a_2 & = & 0 \end{array}, \tag{2.1.60}$$

$$u^3 : \begin{pmatrix} -b_1 + \frac{1}{2}b_1 \\ -b_2 - \frac{1}{2}b_2 + a_1 \end{pmatrix} = \begin{pmatrix} 0 \\ 0 \end{pmatrix} \Rightarrow \begin{array}{ccccc} b_1 & = & 0 & & \\ b_2 & = & a_1\frac{2}{3} & = & \frac{4}{9} \end{array};$$

hence, the center manifold is given by the graph of $(h_1(u), h_2(u))$, where

$$\begin{aligned} h_1(u) &= \frac{2}{3}u^2 + \mathcal{O}(u^4), \\ h_2(u) &= \frac{4}{9}u^3 + \mathcal{O}(u^4). \end{aligned} \tag{2.1.61}$$

The map on the center manifold is given by

$$u \longmapsto -u + \frac{8}{27}u^5 + \mathcal{O}(u^6); \tag{2.1.62}$$

thus, the origin is attracting; see Figure 2.1.3.

EXAMPLE 2.1.4 Consider the map

$$\begin{pmatrix} x \\ y \end{pmatrix} \mapsto \begin{pmatrix} 0 & 1 \\ -\frac{1}{2} & \frac{3}{2} \end{pmatrix} \begin{pmatrix} x \\ y \end{pmatrix} + \begin{pmatrix} 0 \\ -y^3 \end{pmatrix}, \qquad (x,y) \in \mathbb{R}^2. \tag{2.1.63}$$

The origin is a fixed point of the map. Computing the eigenvalues of the map linearized about the origin gives

$$\lambda_{1,2} = 1, \frac{1}{2}.$$

Therefore, there is a one-dimensional center manifold and a one-dimensional stable manifold with the orbit structure in a neighborhood of $(0,0)$ determined by the orbit structure on the center manifold.

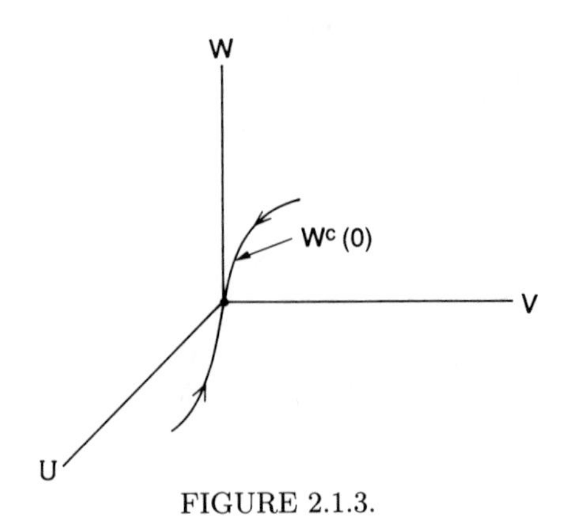

FIGURE 2.1.3.

We wish to compute the center manifold, but first we must put the linear part in block diagonal form as given in (2.1.49). The matrix associated with the linear transformation has columns consisting of the eigenvectors of the linearized map and is easily calculated. It is given by

$$T = \begin{pmatrix} 1 & 2 \\ 1 & 1 \end{pmatrix} \quad \text{with } T^{-1} = \begin{pmatrix} -1 & 2 \\ 1 & -1 \end{pmatrix}. \tag{2.1.64}$$

Thus, letting

$$\begin{pmatrix} x \\ y \end{pmatrix} = T \begin{pmatrix} u \\ v \end{pmatrix},$$

our map becomes

$$\begin{pmatrix} u \\ v \end{pmatrix} \mapsto \begin{pmatrix} 1 & 0 \\ 0 & \frac{1}{2} \end{pmatrix} \begin{pmatrix} u \\ v \end{pmatrix} + \begin{pmatrix} -2(u+v)^3 \\ (u+v)^3 \end{pmatrix}. \tag{2.1.65}$$

We seek a center manifold

$$W^c(0) = \{ (u,v) \mid v = h(u); h(0) = Dh(0) = 0 \} \tag{2.1.66}$$

for u sufficiently small. The next step is to assume $h(u)$ of the form

$$h(u) = au^2 + bu^3 + \mathcal{O}(u^4) \tag{2.1.67}$$

and substitute (2.1.67) into the center manifold equation

$$\mathcal{N}(h(u)) = h\Big(Au + f\big(u, h(u)\big)\Big) - Bh(u) - g\big(u, h(u)\big) = 0, \tag{2.1.68}$$

where, in this example, we have

$$\begin{aligned} A &= 1, \\ B &= \frac{1}{2}, \\ f(u,v) &= -2(u+v)^3, \\ g(u,v) &= (u+v)^3, \end{aligned} \tag{2.1.69}$$

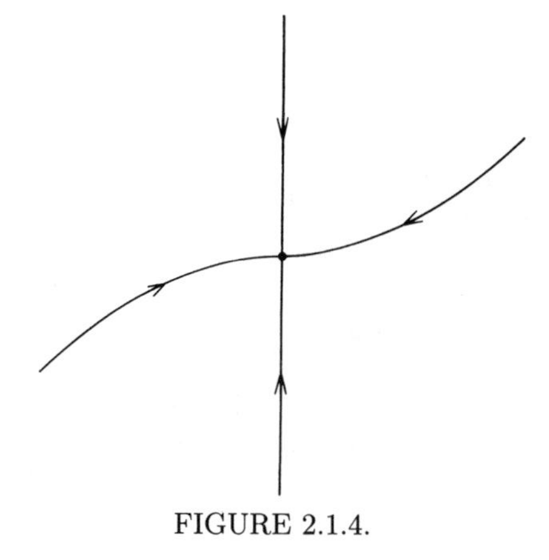

FIGURE 2.1.4.

and (2.1.68) becomes

$$a\Big(u - 2\big(u + au^2 + bu^3 + \mathcal{O}(u^4)\big)^3\Big)^2 + b\Big(u - 2\big(u + au^2 + bu^3 + \mathcal{O}(u^4)\big)^3\Big)^3$$
$$+ \cdots - \frac{1}{2}\big(au^2 + bu^3 + \mathcal{O}(u^4)\big) - \big(u + au^2 + bu^3 + \mathcal{O}(u^4)\big)^3 = 0. \tag{2.1.70}$$

or

$$au^2 + bu^3 - \frac{1}{2}au^2 - \frac{1}{2}bu^3 - u^3 + \mathcal{O}(u^4) = 0. \tag{2.1.71}$$

Equating coefficients of like powers to zero gives

$$\begin{aligned} u^2 &: a - \frac{1}{2}a = 0 \Rightarrow a = 0, \\ u^3 &: b - \frac{1}{2}b - 1 = 0 \Rightarrow b = 2. \end{aligned} \tag{2.1.72}$$

Thus, the center manifold is given by the graph of

$$h(u) = 2u^3 + \mathcal{O}(u^4), \tag{2.1.73}$$

and the map restricted to the center manifold is given by

$$u \mapsto u - 2\big(u + 2u^3 + \mathcal{O}(u^4)\big)^3 \tag{2.1.74}$$

or

$$u \mapsto u - 2u^3 + \mathcal{O}(u^4). \tag{2.1.75}$$

Therefore, the orbit structure in the neighborhood of $(0,0)$ appears as in Figure 2.1.4 and $(0,0)$ is stable.

Some remarks are now in order.

Remark 1. Parametrized Families of Maps. Parameters can be included as new dependent variables for maps in exactly the same way as for vector fields in Section 2.1B.

Remark 2. Inclusion of Linearly Unstable Directions. The case where the origin has an unstable manifold can be treated in exactly the same way as for vector fields in Section 2.1C.

2.1E Properties of Center Manifolds

In this brief section we would like to discuss a few properties of center manifolds. More information can be obtained from Carr [1981] or Sijbrand [1985].

Uniqueness

Although center manifolds exist, they need not be unique. This can be seen from the following example. Consider the vector field

$$\begin{aligned} \dot{x} &= x^2, \\ \dot{y} &= -y, \end{aligned} \qquad (x,y) \in \mathbb{R}^2. \tag{2.1.76}$$

Clearly, $(x,y) = (0,0)$ is a fixed point with stable manifold given by $x = 0$. It should also be clear that $y = 0$ is an invariant center manifold, but there are other center manifolds.

Eliminating t as the independent variable in (2.1.76), we obtain

$$\frac{dy}{dx} = \frac{-y}{x^2}. \tag{2.1.77}$$

The solution of (2.1.77) (for $x \neq 0$) is given by

$$y(x) = \alpha e^{1/x} \tag{2.1.78}$$

for any real constant α. Thus, the curves given by

$$W_\alpha^c(0) = \left\{(x,y) \in \mathbb{R}^2 \mid y = \alpha e^{1/x} \text{ for } x < 0, y = 0 \text{ for } x \geq 0\right\} \tag{2.1.79}$$

are a one-parameter (parametrized by α) family of center manifolds of $(x,y) = (0,0)$; see Figure 2.1.5.

This example immediately brings up two questions.

1. In approximating the center manifold via power series expansions according to Theorem 2.1.3, which center manifold is actually being approximated?

2. Is the dynamical behavior the "same" on all of the center manifolds of a given fixed point?

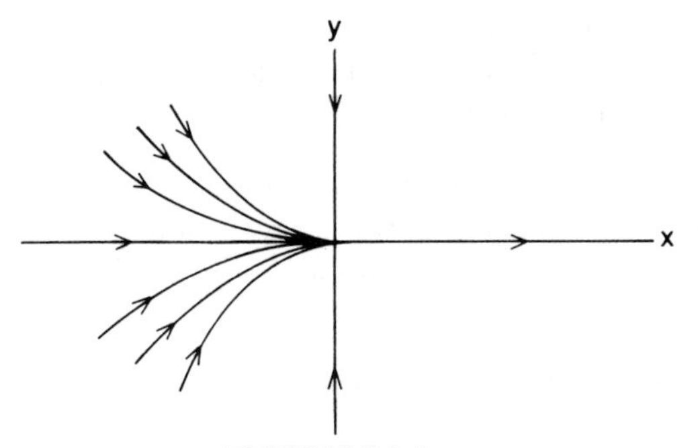

FIGURE 2.1.5.

Regarding Question 1, it can be proven (see Carr [1981] or Sijbrand [1985]) that any two center manifolds of a given fixed point differ by (at most) transcendentally small terms (cf. (2.1.79)). Thus, the Taylor series expansions of any two center manifolds agree to all orders.

This fact emphasizes the importance of Question 2 from a practical point of view. However, it can be shown that due to the attractive nature of the center manifold, certain orbits that remain close to the origin for all time must be on *every* center manifold of a given fixed point, for example, fixed points, periodic orbits, homoclinic orbits, and heteroclinic orbits.

Differentiability

From Theorem 2.1.1 we have that if the vector field is $\mathbf{C}^r$, then the center manifold is also $\mathbf{C}^r$. However, if the vector field is analytic, then the vector field need not be analytic; see Sijbrand [1985].

Preservation of Symmetry

Suppose that the vector field (2.1.2) possesses certain symmetries (e.g., it is Hamiltonian). Does the vector field restricted to the center manifold possess the same symmetries? See Ruelle [1973] for a discussion of these issues.

2.2 Normal Forms

The method of normal forms provides a way of finding a coordinate system in which the dynamical system takes the "simplest" form, where the term "simplest" will be defined as we go along. As we develop the method, three important characteristics should become apparent.

1. The method is local in the sense that the coordinate transformations are generated in a neighborhood of a known solution. For our purposes, the known solution will be a fixed point. However, when we

develop the theory for maps, the results will have immediate applications to periodic orbits of vector fields by considering the associated Poincaré map (cf. Section 1.2A).

2. In general, the coordinate transformations will be nonlinear functions of the dependent variables. However, the important point is that these coordinate transformations are found by solving a sequence of *linear* problems.

3. The structure of the normal form is determined entirely by the nature of the linear part of the vector field.

We now begin the development of the method.

2.2A Normal Forms for Vector Fields

Consider the vector field

$$\dot{w} = G(w), \qquad w \in \mathbb{R}^n, \tag{2.2.1}$$

where G is $\mathbf{C}^r$, with r to be specified as we go along (note: in practice we will need $r \geq 4$). Suppose (2.2.1) has a fixed point at $w = w_0$. We first want to perform a few simple (linear) coordinate transformations that will put (2.2.1) into a form which is easier to work with.

1. First we transform the fixed point to the origin by the translation

$$v = w - w_0, \qquad v \in \mathbb{R}^n,$$

under which (2.2.1) becomes

$$\dot{v} = G(v + w_0) \equiv H(v). \tag{2.2.2}$$

2. We next "split off" the linear part of the vector field and write (2.2.2) as follows

$$\dot{v} = DH(0)v + \bar{H}(v), \tag{2.2.3}$$

where $\bar{H}(v) \equiv H(v) - DH(0)v$. It should be clear that $\bar{H}(v) = \mathcal{O}(|v|^2)$.

3. Finally, let T be the matrix that transforms the matrix $DH(0)$ into (real) Jordan canonical form. Then, under the transformation

$$v = Tx, \tag{2.2.4}$$

(2.2.3) becomes

$$\dot{x} = T^{-1}DH(0)Tx + T^{-1}\bar{H}(Tx). \tag{2.2.5}$$

Denoting the (real) Jordan canonical form of $DH(0)$ by J, we have

$$J \equiv T^{-1} DH(0) T, \tag{2.2.6}$$

and we define

$$F(x) \equiv T^{-1} \bar{H}(Tx)$$

so that (2.2.4) is alternately written as

$$\dot{x} = Jx + F(x), \qquad x \in \mathbb{R}^n. \tag{2.2.7}$$

We remark that the transformation (2.2.4) has simplified the linear part of (2.2.3) as much as possible. We now begin the task of simplifying the nonlinear part, $F(x)$.

First, we Taylor expand $F(x)$ so that (2.2.7) becomes

$$\dot{x} = Jx + F_2(x) + F_3(x) + \cdots + F_{r-1}(x) + \mathcal{O}(|x|^r), \tag{2.2.8}$$

where $F_i(x)$ represent the order i terms in the Taylor expansion of $F(x)$. We next introduce the coordinate transformation

$$x = y + h_2(y), \tag{2.2.9}$$

where $h_2(y)$ is second order in y. Substituting (2.2.9) into (2.2.8) gives

$$\begin{aligned} \dot{x} = \big(\mathrm{id} + Dh_2(y)\big)\dot{y} &= Jy + Jh_2(y) + F_2\big(y + h_2(y)\big) \\ &+ F_3\big(y + h_2(y)\big) + \cdots + F_{r-1}\big(y + h_2(y)\big) + \mathcal{O}(|y|^r), \end{aligned} \tag{2.2.10}$$

where "id" denotes the $n \times n$ identity matrix. Note that each term

$$F_k\big(y + h_2(y)\big), \qquad 2 \le k \le r-1, \tag{2.2.11}$$

can be written as

$$F_k(y) + \mathcal{O}(|y|^{k+1}) + \cdots + \mathcal{O}(|y|^{2k}), \tag{2.2.12}$$

so that (2.2.10) becomes

$$\begin{aligned} \big(\mathrm{id} + Dh_2(y)\big)\dot{y} = Jy + Jh_2(y) + F_2(y) + \tilde{F}_3(y) \\ + \cdots + \tilde{F}_{r-1}(y) + \mathcal{O}(|y|^r), \end{aligned} \tag{2.2.13}$$

where the terms $\tilde{F}_k(y)$ represent the $\mathcal{O}(|y|^k)$ terms which have been modified due to the coordinate transformation.

Now, for y sufficiently small,

$$\big(\mathrm{id} + Dh_2(y)\big)^{-1} \tag{2.2.14}$$

exists and can be represented in a series expansion as follows (see Exercise 2.7)

$$\left(\mathrm{id} + Dh_2(y)\right)^{-1} = \mathrm{id} - Dh_2(y) + \mathcal{O}(|y|^2). \tag{2.2.15}$$

Substituting (2.2.15) into (2.2.13) gives

$$\begin{aligned}\dot{y} = Jy + Jh_2(y) - Dh_2(y)Jy + F_2(y) + \tilde{F}_3(y) \\ + \cdots + \tilde{F}_{r-1}(y) + \mathcal{O}(|y|^r).\end{aligned} \tag{2.2.16}$$

Up to this point $h_2(y)$ has been completely arbitrary. However, now we will choose a specific form for $h_2(y)$ so as to simplify the $\mathcal{O}(|y|^2)$ terms as much as possible. Ideally, this would mean choosing $h_2(y)$ such that

$$Dh_2(y)Jy - Jh_2(y) = F_2(y), \tag{2.2.17}$$

which would eliminate $F_2(y)$ from (2.2.16). Equation (2.2.17) can be viewed as an equation for the unknown $h_2(y)$. We want to motivate the fact that, when viewed in the correct way, it is in fact a linear equation acting on a linear vector space. This will be accomplished by 1) defining the appropriate linear vector space; 2) defining the linear operator on the vector space; and 3) describing the equation to be solved in this linear vector space (which will turn out to be (2.2.17)). We begin with Step 1.

Step 1. The Space of Vector-Valued Monomials of Degree k, H_k. Let $\{s_1, \cdots, s_n\}$ denote a basis of $\mathbb{R}^n$, and let $y = (y_1, \cdots, y_n)$ be coordinates with respect to this basis. Now consider those basis elements with coefficients consisting of monomials of degree k, i.e.,

$$(y_1^{m_1} y_2^{m_2} \cdots y_n^{m_n})s_i, \quad \sum_{j=1}^{n} m_j = k, \tag{2.2.18}$$

where $m_j \geq 0$ are integers. We refer to these objects as *vector-valued monomials of degree k*. The set of all vector-valued monomials of degree k forms a linear vector space, which we denote by H_k. An obvious basis for H_k consists of elements formed by considering all possible monomials of degree k that multiply each s_i. The reader should verify these statements. Let us consider a specific example.

EXAMPLE 2.2.1 We consider the standard basis

$$\begin{pmatrix} 1 \\ 0 \end{pmatrix}, \begin{pmatrix} 0 \\ 1 \end{pmatrix} \tag{2.2.19}$$

on $\mathbb{R}^2$ and denote the coordinates with respect to this basis by x and y, respectively. Then we have

$$H_2 = \mathrm{span}\left\{ \begin{pmatrix} x^2 \\ 0 \end{pmatrix}, \begin{pmatrix} xy \\ 0 \end{pmatrix}, \begin{pmatrix} y^2 \\ 0 \end{pmatrix}, \begin{pmatrix} 0 \\ x^2 \end{pmatrix}, \begin{pmatrix} 0 \\ xy \end{pmatrix}, \begin{pmatrix} 0 \\ y^2 \end{pmatrix} \right\}. \tag{2.2.20}$$

Step 2. The Linear Map on H_k. Now let us reconsider equation (2.2.17). It should be clear that $h_2(y)$ can be viewed as an element of H_2. The reader should easily be able to verify that the map

$$h_2(y) \longmapsto Dh_2(y)Jy - Jh_2(y) \tag{2.2.21}$$

is a linear map of H_2 into H_2. Indeed, for any element $h_k(y) \in H_k$, it similarly follows that

$$h_k(y) \longmapsto Dh_k(y)Jy - Jh_k(y) \tag{2.2.22}$$

is a linear map of H_k into H_k.

Let us mention some terminology associated with Equation (2.2.17) that has become traditional. Due to its presence in Lie algebra theory (see, e.g., Olver [1986]) this map is often denoted as

$$L_J\big(h_k(y)\big) \equiv -\big(Dh_k(y)Jy - Jh_k(y)\big) \tag{2.2.23}$$

or

$$-\ \big(Dh_k(y)Jy - Jh_k(y)\big) \equiv [h_k(y), Jy], \tag{2.2.24}$$

where $[\cdot,\cdot]$ denotes the Lie bracket operation on the vector fields $h_k(y)$ and Jy.

Step 3. The Solution of (2.2.17). We now return to the problem of solving (2.2.17). It should be clear that $F_2(y)$ can be viewed as an element of H_2. From elementary linear algebra, we know that H_2 can be (nonuniquely) represented as follows

$$H_2 = L_J(H_2) \oplus G_2, \tag{2.2.25}$$

where G_2 represents a space complementary to $L_J(H_2)$. Solving (2.2.17) is like solving the equation $Ax = b$ from linear algebra. If $F_2(y)$ is in the range of $L_J(\cdot)$, then all $\mathcal{O}(|y|^2)$ terms can be eliminated from (2.2.17). In any case, we can choose $h_2(y)$ so that only $\mathcal{O}(|y|^2)$ terms that are in G_2 remain. We denote these terms by

$$F_2^r(y) \in G_2 \tag{2.2.26}$$

(note: the superscript r in (2.2.26) denotes the term "resonance," which will be explained shortly).

Thus, (2.2.16) can be simplified to

$$\dot{y} = Jy + F_2^r(y) + \tilde{F}_3(y) + \cdots + \tilde{F}_{r-1}(y) + \mathcal{O}(|y|^r). \tag{2.2.27}$$

At this point the meaning of the phrase "simplify the second-order terms" should be clear. It means the introduction of a coordinate change such that, in the new coordinate system, the only second-order terms are in a space complementary to $L_J(H_2)$. If $L_J(H_2) = H_2$, then all second-order terms can be eliminated.

Next let us simplify the $\mathcal{O}(|y|^3)$ terms. Introducing the coordinate change

$$y \longmapsto y + h_3(y), \tag{2.2.28}$$

where $h_3(y) = \mathcal{O}(|y|^3)$ (note: we will retain the same variables y in our equation), and performing the same algebraic manipulations as in dealing with the second-order terms, (2.2.27) becomes

$$\begin{aligned}\dot{y} = Jy + F_2^r(y) + Jh_3(y) - Dh_3(y)Jy + \tilde{F}_3(y) + \tilde{\tilde{F}}_4(y) \\ + \cdots + \tilde{\tilde{F}}_{r-1}(y) + \mathcal{O}(|y|^r),\end{aligned} \tag{2.2.29}$$

where the terms $\tilde{\tilde{F}}_k(y)$, $4 \leq k \leq r-1$, indicate, as before, that the coordinate transformation has modified the terms of order higher than three. Now, simplifying the third-order terms involves solving

$$Dh_3(y)Jy - Jh_3(y) = \tilde{F}_3(y). \tag{2.2.30}$$

The same comments as for second-order terms apply here. The map

$$h_3(y) \longmapsto Dh_3(y)Jy - Jh_3(y) \equiv -L_J(h_3(y)) \tag{2.2.31}$$

is a linear map of H_3 into H_3. Thus, we can write

$$H_3 = L_J(H_3) \oplus G_3, \tag{2.2.32}$$

where G_3 is some space complementary to $L_J(H_3)$. Thus, the third-order terms can be simplified to

$$F_3^r(y) \in G_3. \tag{2.2.33}$$

If $L_J(H_3) = H_3$, then the third-order terms can be eliminated.

Clearly, this procedure can be iterated so that we obtain the following *normal form theorem.*

Theorem 2.2.1 (Normal Form Theorem) *By a sequence of analytic coordinate changes (2.2.8) can be transformed into*

$$\dot{y} = Jy + F_2^r(y) + \cdots + F_{r-1}^r(y) + \mathcal{O}(|y|^r), \tag{2.2.34}$$

where $F_k^r(y) \in G_k$, $2 \leq k \leq r-1$, and G_k is a space complementary to $L_J(H_k)$. Equation (2.2.34) is said to be in normal form.

Several comments are now in order.

1. The terms $F_k^r(y)$, $2 \leq k \leq r-1$, are referred to as *resonance terms* (hence the superscript r). We will explain what this means in Section 2.2D,i).

2. The structure of the nonlinear terms in (2.2.34) is determined entirely by the linear part of the vector field (i.e., J).

3. It should be clear that simplifying the terms at order k does not modify any lower order terms. However, terms of order higher than k are modified. This happens at each step of the application of the method. If one wanted to actually calculate the coefficients on each term of the normal form in terms of the original vector field, it would be necessary to keep track of how the higher order terms are modified by the successive coordinate transformations.

EXAMPLE 2.2.2 We want to compute the normal form for a vector field on $\mathbb{R}^2$ in the neighborhood of a fixed point where the linear part is given by

$$J = \begin{pmatrix} 0 & 1 \\ 0 & 0 \end{pmatrix}. \tag{2.2.35}$$

Second-Order Terms

We have

$$H_2 = \text{span}\left\{ \begin{pmatrix} x^2 \\ 0 \end{pmatrix}, \begin{pmatrix} xy \\ 0 \end{pmatrix}, \begin{pmatrix} y^2 \\ 0 \end{pmatrix}, \begin{pmatrix} 0 \\ x^2 \end{pmatrix}, \begin{pmatrix} 0 \\ xy \end{pmatrix}, \begin{pmatrix} 0 \\ y^2 \end{pmatrix} \right\}. \tag{2.2.36}$$

We want to compute $L_J(H_2)$. We do this by computing the action of $L_J(\cdot)$ on each basis element on H_2

$$\begin{aligned}
L_J \begin{pmatrix} x^2 \\ 0 \end{pmatrix} &= \begin{pmatrix} 0 & 1 \\ 0 & 0 \end{pmatrix}\begin{pmatrix} x^2 \\ 0 \end{pmatrix} - \begin{pmatrix} 2x & 0 \\ 0 & 0 \end{pmatrix}\begin{pmatrix} y \\ 0 \end{pmatrix} = \begin{pmatrix} -2xy \\ 0 \end{pmatrix} = -2\begin{pmatrix} xy \\ 0 \end{pmatrix}, \\
L_J \begin{pmatrix} xy \\ 0 \end{pmatrix} &= \begin{pmatrix} 0 & 1 \\ 0 & 0 \end{pmatrix}\begin{pmatrix} xy \\ 0 \end{pmatrix} - \begin{pmatrix} y & x \\ 0 & 0 \end{pmatrix}\begin{pmatrix} y \\ 0 \end{pmatrix} = \begin{pmatrix} -y^2 \\ 0 \end{pmatrix} = -1\begin{pmatrix} y^2 \\ 0 \end{pmatrix}, \\
L_J \begin{pmatrix} y^2 \\ 0 \end{pmatrix} &= \begin{pmatrix} 0 & 1 \\ 0 & 0 \end{pmatrix}\begin{pmatrix} y^2 \\ 0 \end{pmatrix} - \begin{pmatrix} 0 & 2y \\ 0 & 0 \end{pmatrix}\begin{pmatrix} y \\ 0 \end{pmatrix} = \begin{pmatrix} 0 \\ 0 \end{pmatrix}, \\
L_J \begin{pmatrix} 0 \\ x^2 \end{pmatrix} &= \begin{pmatrix} 0 & 1 \\ 0 & 0 \end{pmatrix}\begin{pmatrix} 0 \\ x^2 \end{pmatrix} - \begin{pmatrix} 0 & 0 \\ 2x & 0 \end{pmatrix}\begin{pmatrix} y \\ 0 \end{pmatrix} = \begin{pmatrix} x^2 \\ -2xy \end{pmatrix} \\
&= \begin{pmatrix} x^2 \\ 0 \end{pmatrix} - 2\begin{pmatrix} 0 \\ xy \end{pmatrix}, \\
L_J \begin{pmatrix} 0 \\ xy \end{pmatrix} &= \begin{pmatrix} 0 & 1 \\ 0 & 0 \end{pmatrix}\begin{pmatrix} 0 \\ xy \end{pmatrix} - \begin{pmatrix} 0 & 0 \\ y & x \end{pmatrix}\begin{pmatrix} y \\ 0 \end{pmatrix} = \begin{pmatrix} xy \\ -y^2 \end{pmatrix} \\
&= \begin{pmatrix} xy \\ 0 \end{pmatrix} - \begin{pmatrix} 0 \\ y^2 \end{pmatrix}, \\
L_J \begin{pmatrix} 0 \\ y^2 \end{pmatrix} &= \begin{pmatrix} 0 & 1 \\ 0 & 0 \end{pmatrix}\begin{pmatrix} 0 \\ y^2 \end{pmatrix} - \begin{pmatrix} 0 & 0 \\ 0 & 2y \end{pmatrix}\begin{pmatrix} y \\ 0 \end{pmatrix} = \begin{pmatrix} y^2 \\ 0 \end{pmatrix}.
\end{aligned} \tag{2.2.37}$$

From (2.2.37) we have

$$L_J(H_2) = \text{span}\left\{ \begin{pmatrix} 2xy \\ 0 \end{pmatrix}, \begin{pmatrix} -y^2 \\ 0 \end{pmatrix}, \begin{pmatrix} 0 \\ 0 \end{pmatrix}, \begin{pmatrix} x^2 \\ -2xy \end{pmatrix}, \begin{pmatrix} xy \\ -y^2 \end{pmatrix}, \begin{pmatrix} y^2 \\ 0 \end{pmatrix} \right\}. \tag{2.2.38}$$

Clearly, from this set, the vectors

$$\begin{pmatrix} -2xy \\ 0 \end{pmatrix}, \begin{pmatrix} y^2 \\ 0 \end{pmatrix}, \begin{pmatrix} x^2 \\ -2xy \end{pmatrix}, \begin{pmatrix} xy \\ -y^2 \end{pmatrix} \tag{2.2.39}$$

are linearly independent and, hence, second-order terms that are linear combinations of these four vectors can be eliminated. To determine the nature of the second-order terms that cannot be eliminated (i.e., $F_2^r(y)$), we must compute a space complementary to $L_J(H_2)$. This space, denoted G_2, will be two dimensional.

In computing G_2 it will be useful to first obtain a matrix representation for the linear operator $L_J(\cdot)$. This is done with respect to the basis given in (2.2.36) by constructing the columns of the matrix from the coefficients multiplying each basis element that are obtained when $L_J(\cdot)$ acts individually on each basis element of H_2 given in (2.2.36). Using (2.2.37), the matrix representation of $L_J(\cdot)$ is given by

$$\begin{pmatrix} 0 & 0 & 0 & 1 & 0 & 0 \\ -2 & 0 & 0 & 0 & 1 & 0 \\ 0 & -1 & 0 & 0 & 0 & 1 \\ 0 & 0 & 0 & 0 & 0 & 0 \\ 0 & 0 & 0 & -2 & 0 & 0 \\ 0 & 0 & 0 & 0 & -1 & 0 \end{pmatrix}. \tag{2.2.40}$$

One way of finding a complementary space G_2 would be to find two "6-vectors" that are linearly independent and orthogonal (using the standard inner product in $\mathbb{R}^6$) to each column of the matrix (2.2.40), or, in other words, two linearly independent left eigenvectors of zero for (2.2.40). Due to the fact that most entries of (2.2.40) are zero, this is an easy calculation, and two such vectors are found to be

$$\begin{pmatrix} 1 \\ 0 \\ 0 \\ 0 \\ \frac{1}{2} \\ 0 \end{pmatrix}, \begin{pmatrix} 0 \\ 0 \\ 0 \\ 1 \\ 0 \\ 0 \end{pmatrix}. \tag{2.2.41}$$

Hence, the vectors

$$\begin{pmatrix} x^2 \\ \frac{1}{2}xy \end{pmatrix}, \begin{pmatrix} 0 \\ x^2 \end{pmatrix} \tag{2.2.42}$$

span a two-dimensional subspace of H_2 that is complementary to $L_J(H_2)$. This implies that the normal form through second-order is given by

$$\begin{aligned} \dot{x} &= y + a_1 x^2 + \mathcal{O}(3), \\ \dot{y} &= a_2 xy + a_3 x^2 + \mathcal{O}(3), \end{aligned} \tag{2.2.43}$$

where a_1, a_2, and a_3 represent constants.

Now our choice of G_2 is certainly not unique. Another choice might be

$$G_2 = \text{span}\left\{ \begin{pmatrix} x^2 \\ 0 \end{pmatrix}, \begin{pmatrix} 0 \\ x^2 \end{pmatrix} \right\}. \tag{2.2.44}$$

This complementary space can be obtained by taking the vector

$$\begin{pmatrix} x^2 \\ \frac{1}{2}xy \end{pmatrix} \tag{2.2.45}$$

given in (2.2.42) and subtracting from it the vector

$$\begin{pmatrix} 0 \\ -\frac{1}{2}xy \end{pmatrix} \tag{2.2.46}$$

contained in $L_J(H_2)$. This gives the vector

$$\begin{pmatrix} x^2 \\ 0 \end{pmatrix}. \tag{2.2.47}$$

For the other basis element of the complementary space, we simply retain the vector

$$\begin{pmatrix} 0 \\ x^2 \end{pmatrix} \tag{2.2.48}$$

given in (2.2.42). With this choice of G_2 the normal form becomes

$$\begin{aligned} \dot{x} &= y + a_1 x^2 + \mathcal{O}(3), \\ \dot{y} &= a_2 x^2 + \mathcal{O}(3). \end{aligned} \tag{2.2.49}$$

This normal form near a fixed point of a planar vector field with linear part given by (2.2.35) was first studied by Takens [1974].

Another possibility for G_2 is given by

$$G_2 = \text{span}\left\{ \begin{pmatrix} 0 \\ x^2 \end{pmatrix}, \begin{pmatrix} 0 \\ xy \end{pmatrix} \right\}, \tag{2.2.50}$$

where these two vectors are obtained by adding the appropriate linear combinations of vectors in $L_J(H_2)$ to the vectors given in (2.2.38). With this choice of G_2 the normal form becomes

$$\begin{aligned} \dot{x} &= y + \mathcal{O}(3), \\ \dot{y} &= a_1 x^2 + b_2 xy + \mathcal{O}(3); \end{aligned} \tag{2.2.51}$$

this is the normal form for a vector field on $\mathbb{R}^2$ near a fixed point with linear part given by (2.2.35) that was first studied by Bogdanov [1975].

2.2B Normal Forms for Vector Fields with Parameters

Now we want to extend the normal form techniques to systems with parameters. Consider the vector field

$$\dot{x} = f(x, \mu), \qquad x \in \mathbb{R}^n, \quad \mu \in I \subset \mathbb{R}^p, \tag{2.2.52}$$

where I is some open set in $\mathbb{R}^p$ and f is $\mathbf{C}^r$ in each variable. Suppose that

$$f(0, 0) = 0 \tag{2.2.53}$$

(note: the reader should recall from the beginning of Section 2.2A that there is no loss of generality in assuming that the fixed point is located at $(x, \mu) = (0, 0)$). The goal is to transform (2.2.52) into normal form near the fixed point in both phase space and parameter space. The most straightforward way to put (2.2.52) into normal form would be to follow the same procedure as for systems with no parameters except to allow the coefficients of the transformation to depend on the parameters. Rather than develop the general theory along these lines, we illustrate the idea with a specific example which will be of much use later on.

Example 2.2.3 Suppose $x \in \mathbb{R}^2$ and $Df(0, 0)$ has two pure imaginary eigenvalues $\lambda(0) = \pm i\omega(0)$. Then we can find a linear transformation which puts $D_x f(0, \mu)$ in the following form

$$D_x f(0, \mu) = \begin{pmatrix} \text{Re } \lambda(\mu) & -\text{Im } \lambda(\mu) \\ \text{Im } \lambda(\mu) & \text{Re } \lambda(\mu) \end{pmatrix} \tag{2.2.54}$$

for μ sufficiently small. Also, by the implicit function theorem, the fixed point varies in a $\mathbf{C}^r$ manner with μ (for μ sufficiently small) such that, if necessary, we can introduce a parameter-dependent coordinate transformation so that $x = 0$ is a fixed point for all μ sufficiently small; see Exercise 2.13. We will assume that this has been done.

Letting

$$\begin{aligned} \text{Re } \lambda(\mu) &= |\lambda(\mu)| \cos(2\pi\theta(\mu)), \\ \text{Im } \lambda(\mu) &= |\lambda(\mu)| \sin(2\pi\theta(\mu)), \end{aligned} \tag{2.2.55}$$

it is easy to see that (2.2.54) can be put in the form

$$D_x f(0, \mu) = |\lambda(\mu)| \begin{pmatrix} \cos 2\pi\theta(\mu) & -\sin 2\pi\theta(\mu) \\ \sin 2\pi\theta(\mu) & \cos 2\pi\theta(\mu) \end{pmatrix}. \tag{2.2.56}$$

Now we want to put the following equation into normal form

$$\begin{pmatrix} \dot{x} \\ \dot{y} \end{pmatrix} = |\lambda(\mu)| \begin{pmatrix} \cos 2\pi\theta(\mu) & -\sin 2\pi\theta(\mu) \\ \sin 2\pi\theta(\mu) & \cos 2\pi\theta(\mu) \end{pmatrix} \begin{pmatrix} x \\ y \end{pmatrix} + \begin{pmatrix} f^1(x, y; \mu) \\ f^2(x, y; \mu) \end{pmatrix}, \qquad (x, y) \in \mathbb{R}^2, \tag{2.2.57}$$

where the f^i are nonlinear in x and y.

We remark that we will frequently omit the explicit parameter dependence of λ, θ, and possibly other quantities from time to time for the sake of a less cumbersome notation.

In dealing with linear parts of vector fields having complex eigenvalues, it is often easier to calculate the normal form using complex coordinates. We will illustrate this procedure for this example.

We make the following linear transformation

$$\begin{pmatrix} x \\ y \end{pmatrix} = \frac{1}{2}\begin{pmatrix} 1 & 1 \\ -i & i \end{pmatrix}\begin{pmatrix} z \\ \bar{z} \end{pmatrix}; \qquad \begin{pmatrix} z \\ \bar{z} \end{pmatrix} = \begin{pmatrix} 1 & i \\ 1 & -i \end{pmatrix}\begin{pmatrix} x \\ y \end{pmatrix} \tag{2.2.58}$$

to obtain

$$\begin{pmatrix} \dot{z} \\ \dot{\bar{z}} \end{pmatrix} = |\lambda| \begin{pmatrix} e^{2\pi i\theta} & 0 \\ 0 & e^{-2\pi i\theta} \end{pmatrix}\begin{pmatrix} z \\ \bar{z} \end{pmatrix} + \begin{pmatrix} F^1(z, \bar{z}; \mu) \\ F^2(z, \bar{z}; \mu) \end{pmatrix}, \tag{2.2.59}$$

where

$$F^1(z, \bar{z}; \mu) = f^1(x(z, \bar{z}), y(z, \bar{z}); \mu) + if^2(x(z, \bar{z}), y(z, \bar{z}); \mu),$$

$$F^2(z, \bar{z}; \mu) = f^1(x(z, \bar{z}), y(z, \bar{z}); \mu) - if^2(x(z, \bar{z}), y(z, \bar{z}); \mu).$$

Therefore, all we really need to study is

$$\dot{z} = |\lambda|\, e^{2\pi i\theta} z + F^1(z, \bar{z}; \mu), \tag{2.2.60}$$

since the second component of (2.2.59) is simply the complex conjugate of the first component. We will therefore put (2.2.60) in normal form and then transform back to the x, y variables.

Expanding (2.2.60) in a Taylor series gives

$$\dot{z} = |\lambda|\, e^{2\pi i\theta} z + F_2 + F_3 + \cdots + F_{k-1} + \mathcal{O}(|z|^r, |\bar{z}|^r), \tag{2.2.61}$$

where the F_j are polynomials in $z, \bar{z}$ of order j *whose coefficients depend on* μ.

First Simplify Second-Order Terms

We make the transformation

$$z \longmapsto z + h_2(z, \bar{z}), \tag{2.2.62}$$

where $h_2(z, \bar{z})$ is second-order in z and $\bar{z}$ *with coefficients depending on* μ. We neglect displaying the explicit μ dependence.

Under (2.2.62), (2.2.61) becomes

$$\dot{z}\left(1 + \frac{\partial h_2}{\partial z}\right) + \frac{\partial h_2}{\partial \bar{z}}\dot{\bar{z}} = \lambda z + \lambda h_2 + F_2(z, \bar{z}) + \mathcal{O}(3) \tag{2.2.63}$$

or

$$\dot{z} = \left(1 + \frac{\partial h_2}{\partial z}\right)^{-1} \left[\lambda z + \lambda h_2 - \frac{\partial h_2}{\partial \bar{z}}\dot{\bar{z}} + F_2 + \mathcal{O}(3)\right].$$

Note that we have

$$\dot{\bar{z}} = \bar{\lambda}\bar{z} + \bar{F}_2 + \mathcal{O}(3) \tag{2.2.64}$$

and, for $z, \bar{z}$ sufficiently small

$$\left(1 + \frac{\partial h_2}{\partial z}\right)^{-1} = 1 - \frac{\partial h_2}{\partial z} + \mathcal{O}(2). \tag{2.2.65}$$

Thus, using (2.2.64) and (2.2.65), (2.2.63) becomes

$$\dot{z} = \lambda z - \lambda\frac{\partial h_2}{\partial z}z - \bar{\lambda}\frac{\partial h_2}{\partial \bar{z}}\bar{z} + \lambda h_2 + F_2 + \mathcal{O}(3), \tag{2.2.66}$$

so that we can eliminate all second-order terms if

$$\lambda h_2 - \left(\lambda\frac{\partial h_2}{\partial z}z + \bar{\lambda}\frac{\partial h_2}{\partial \bar{z}}\bar{z}\right) + F_2 = 0. \tag{2.2.67}$$

Equation (2.2.67) is very similar to equation (2.2.17) derived earlier. The map

$$h_2 \longmapsto \lambda h_2 - \left(\lambda\frac{\partial h_2}{\partial z}\bar{z} + \bar{\lambda}\frac{\partial h_2}{\partial \bar{z}}\bar{z}\right) \tag{2.2.68}$$

is a linear map of the space of monomials in z and $\bar{z}$ into itself. We denote this space by H_2. F_2 can also be viewed as an element in this space. Thus, solving (2.2.67) is a problem from linear algebra.

Now we have

$$H_2 = \text{span}\left\{z^2, z\bar{z}, \bar{z}^2\right\}. \tag{2.2.69}$$

Computing the action of the linear map (2.2.68) on each of these basis elements gives

$$\begin{aligned}
\lambda z^2 - \left[\lambda\left(\frac{\partial}{\partial z}z^2\right)z + \bar{\lambda}\left(\frac{\partial}{\partial \bar{z}}z^2\right)\bar{z}\right] &= -\lambda z^2, \\
\lambda z\bar{z} - \left[\lambda\left(\frac{\partial}{\partial z}z\bar{z}\right)z + \bar{\lambda}\left(\frac{\partial}{\partial \bar{z}}z\bar{z}\right)\bar{z}\right] &= -\bar{\lambda}z\bar{z}, \\
\lambda \bar{z}^2 - \left[\lambda\left(\frac{\partial}{\partial z}\bar{z}^2\right)z + \bar{\lambda}\left(\frac{\partial}{\partial \bar{z}}\bar{z}^2\right)\bar{z}\right] &= (\lambda - 2\bar{\lambda})\bar{z}^2.
\end{aligned}$$

Thus, (2.2.68) is diagonal in this basis with a matrix representation given by

$$\begin{pmatrix} -\lambda(\mu) & 0 & 0 \\ 0 & -\bar{\lambda}(\mu) & 0 \\ 0 & 0 & \lambda(\mu) - 2\bar{\lambda}(\mu) \end{pmatrix}. \tag{2.2.70}$$

For $\mu = 0$, it should be clear that $\lambda(0) \neq 0$ and $\lambda(0) = -\bar{\lambda}(0)$; hence, for μ sufficiently small, $\lambda(\mu) \neq 0$ and $\lambda(\mu) - 2\bar{\lambda}(\mu) \neq 0$. Therefore, for μ sufficiently small, all second-order terms can be eliminated from (2.2.61) .

Next Simplify Third-Order Terms

We have

$$\dot{z} = \lambda z + F_3 + \mathcal{O}(4). \tag{2.2.71}$$

Let $z \longmapsto z + h_3(z, \bar{z})$; then we obtain

$$\begin{aligned}\dot{z} &= \left(1 + \frac{\partial h_3}{\partial z}\right)^{-1} \left[\lambda z - \frac{\partial h_3}{\partial \bar{z}}\dot{\bar{z}} + \lambda h_3 + F_3(z, \bar{z}) + \mathcal{O}(4)\right] \\ &= \lambda z - \lambda \frac{\partial h_3}{\partial z} z - \bar{\lambda}\frac{\partial h_3}{\partial \bar{z}}\bar{z} + \lambda h_3 + F_3 + \mathcal{O}(4).\end{aligned}$$

We want to solve

$$\lambda h_3 - \lambda \frac{\partial h_3}{\partial z} z - \bar{\lambda}\frac{\partial h_3}{\partial \bar{z}}\bar{z} + F_3 = 0. \tag{2.2.72}$$

Note that we have

$$H_3 = \operatorname{span}\left\{z^3, z^2\bar{z}, z\bar{z}^2, \bar{z}^3\right\}. \tag{2.2.73}$$

We compute the action of the linear map

$$h_3 \longmapsto \lambda h_3 - \left[\lambda \frac{\partial h_3}{\partial z} z + \bar{\lambda}\frac{\partial h_3}{\partial \bar{z}}\bar{z}\right] \tag{2.2.74}$$

on each basis element of H_3 and obtain

$$\begin{aligned}\lambda z^3 - \left[\lambda\left(\frac{\partial}{\partial z} z^3\right) z + \bar{\lambda}\left(\frac{\partial}{\partial \bar{z}} z^3\right)\bar{z}\right] &= -2\lambda z^3, \\ \lambda z^2\bar{z} - \left[\lambda\left(\frac{\partial}{\partial z} z^2\bar{z}\right) z + \bar{\lambda}\left(\frac{\partial}{\partial \bar{z}} z^2\bar{z}\right)\bar{z}\right] &= -\left(\lambda + \bar{\lambda}\right) z^2\bar{z}, \\ \lambda z\bar{z}^2 - \left[\lambda\left(\frac{\partial}{\partial z} z\bar{z}^2\right) z + \bar{\lambda}\left(\frac{\partial}{\partial \bar{z}} z\bar{z}^2\right)\bar{z}\right] &= -2\bar{\lambda} z\bar{z}^2, \\ \lambda \bar{z}^3 - \left[\lambda\left(\frac{\partial}{\partial z} \bar{z}^3\right) z + \bar{\lambda}\left(\frac{\partial}{\partial \bar{z}} \bar{z}^3\right)\bar{z}\right] &= \left(\lambda - 3\bar{\lambda}\right)\bar{z}^3.\end{aligned} \tag{2.2.75}$$

Therefore, a matrix representation for (2.2.74) is given by

$$\begin{pmatrix} -2\lambda(\mu) & 0 & 0 & 0 \\ 0 & -(\lambda(\mu) + \bar{\lambda}(\mu)) & 0 & 0 \\ 0 & 0 & -2\bar{\lambda}(\mu) & 0 \\ 0 & 0 & 0 & \lambda(\mu) - 3\bar{\lambda}(\mu) \end{pmatrix}. \tag{2.2.76}$$

Now, at $\mu = 0$,

$$\lambda(0) + \bar{\lambda}(0) = 0; \tag{2.2.77}$$

however, none of the remaining columns in (2.2.76) are identically zero at $\mu = 0$. Therefore, for μ sufficiently small, third-order terms that are not of the form

$$z^2\bar{z} \tag{2.2.78}$$

can be eliminated.

Thus, the normal form through third-order is given by

$$\dot{z} = \lambda z + c(\mu) z^2 \bar{z} + \mathcal{O}(4), \tag{2.2.79}$$

where $c(\mu)$ is a constant depending on μ.

Next we simplify the fourth-order terms. However, notice that, at each order, simplification depends on whether

$$\lambda h - \left(\lambda z \frac{\partial h}{\partial z} + \bar{\lambda} \bar{z} \frac{\partial h}{\partial \bar{z}} \right) = 0 \tag{2.2.80}$$

for some $h = z^n \bar{z}^m$, where $m + n$ is the order of the term that we want to simplify. Substituting this into (2.2.80) gives

$$\begin{aligned} \lambda z^n \bar{z}^m - \left(n\lambda z^n \bar{z}^m + m\bar{\lambda} z^n \bar{z}^m \right) &= 0, \\ \left(\lambda - n\lambda - m\bar{\lambda} \right) z^n \bar{z}^m &= 0. \end{aligned} \tag{2.2.81}$$

At $\mu = 0$, $\lambda = -\bar{\lambda}$; hence we must not have

$$1 + m - n = 0. \tag{2.2.82}$$

It is easily seen that this can never happen if m and n are even numbers. Therefore, all even-order terms can be removed, and the normal form is given by

$$\dot{z} = \lambda z + c(\mu) z^2 \bar{z} + \mathcal{O}(5) \tag{2.2.83}$$

for μ in some neighborhood of $\mu = 0$.

We can write this in Cartesian coordinates as follows. Let $\lambda(\mu) = \alpha(\mu) + i\omega(\mu)$, and $c(\mu) = a(\mu) + ib(\mu)$. Then

$$\begin{aligned} \dot{x} &= \alpha x - \omega y + (ax - by)(x^2 + y^2) + \mathcal{O}(5), \\ \dot{y} &= \omega x + \alpha y + (bx + ay)(x^2 + y^2) + \mathcal{O}(5). \end{aligned} \tag{2.2.84}$$

In polar coordinates, it can be expressed as

$$\begin{aligned} \dot{r} &= \alpha r + a r^3 + \cdots, \\ \dot{\theta} &= \omega + b r^2 + \cdots. \end{aligned} \tag{2.2.85}$$

We will study the dynamics associated with this normal form in great detail in Chapter 3 when we study the Poincaré–Andronov–Hopf bifurcation.

Differentiability

We make the important remark that in order to obtain the normal form (2.2.83) the vector field must be at least $\mathbf{C}^5$.

2.2C Normal Forms for Maps

We now want to develop the method of normal forms for maps. We will see that it is very much the same as for vector fields with only a slight modification.

Suppose we have a $\mathbf{C}^r$ map which has a fixed point at the origin and is written as follows

$$x \longmapsto Jx + F_2(x) + \cdots + F_{r-1}(x) + \mathcal{O}(|x|^r)$$

or

$$x_{n+1} = Jx_n + F_2(x_n) + \cdots + F_{r-1}(x_n) + \mathcal{O}(|x_n|^r), \tag{2.2.86}$$

where $x \in \mathbb{R}^n$, and the F_j are vector-valued monomials of degree j. We introduce the change of coordinates

$$x \longmapsto y + h_2(y), \qquad h_2 = \mathcal{O}(|y|^2), \tag{2.2.87}$$

under which (2.2.86) becomes

$$x_{n+1} = y_{n+1} + h_2(y_{n+1}) = Jy_n + Jh_2(y_n) + F_2(y_n) + \mathcal{O}(3)$$

or

$$(\mathrm{id} + h_2)(y_{n+1}) = Jy_n + Jh_2(y_n) + F_2(y_n) + \mathcal{O}(3). \tag{2.2.88}$$

Now, for y sufficiently small, the function $(\mathrm{id} + h_2)(\cdot)$ is invertible so that (2.2.88) can be written as

$$y_{n+1} = (\mathrm{id} - h_2)^{-1}\big(Jy_n + Jh_2(y_n) + F_2(y_n) + \mathcal{O}(3)\big). \tag{2.2.89}$$

For y sufficiently small, $(\mathrm{id} - h_2)^{-1}(\cdot)$ can be expressed as follows (see Exercise 2.7)

$$(\mathrm{id} - h_2)^{-1}(\cdot) = (\mathrm{id} - h_2 + \mathcal{O}(4))(\cdot), \tag{2.2.90}$$

so that (2.2.89) becomes

$$y_{n+1} = Jy_n + Jh_2(y_n) - h_2(Jy_n) + F_2(y_n) + \mathcal{O}(3). \tag{2.2.91}$$

Thus, we can eliminate the second-order terms if

$$Jh_2(y) - h_2(Jy) + F_2(y) = 0. \tag{2.2.92}$$

(Compare this with the situation for vector fields.)

This process can be repeated, but it should be clear that the ability to eliminate terms of order j depends upon the operator

$$h_j(y) \longmapsto Jh_j(y) - h_j(Jy) \equiv M_J(h_j(y)), \tag{2.2.93}$$

which (the reader should verify) is a linear map of H_j into H_j, where H_j is the linear vector space of vector-valued monomials of degree j. The analysis proceeds as in the case for vector fields except the equation to solve is slightly different (it has a term involving a composition rather than a matrix multiplication). Let us consider an example which can be viewed as the discrete time analog of Example 2.2.3.

EXAMPLE 2.2.4 Suppose we have a $\mathbf{C}^r$ map of the plane

$$x \longmapsto f(x,\mu), \qquad x \in \mathbb{R}^2, \quad \mu \in I \in \mathbb{R}^p, \tag{2.2.94}$$

where I is some open set in $\mathbb{R}^p$. Suppose also that (2.2.94) has a fixed point at $x = 0$ for μ sufficiently small (cf. Example 2.2.3) and that the eigenvalues of $Df(0,\mu)$, μ small, are given by

$$\lambda_1 = |\lambda(\mu)|\, e^{2\pi i\theta(\mu)}, \qquad \lambda_2 = |\lambda(\mu)|\, e^{-2\pi i\theta(\mu)}, \tag{2.2.95}$$

i.e., $\bar{\lambda}_1 = \lambda_2$. Furthermore, we assume that at $\mu = 0$ the two eigenvalues lie on the unit circle, i.e., $|\lambda(0)| = 1$. As in Example 2.2.3, with a linear change of coordinates we can put the map in the form

$$\begin{pmatrix} x \\ y \end{pmatrix} \longmapsto |\lambda| \begin{pmatrix} \cos 2\pi\theta & -\sin 2\pi\theta \\ \sin 2\pi\theta & \cos 2\pi\theta \end{pmatrix} \begin{pmatrix} x \\ y \end{pmatrix} + \begin{pmatrix} f^1(x,y;\mu) \\ f^2(x,y;\mu) \end{pmatrix}, \tag{2.2.96}$$

where $f^i(x,y;\mu)$ are nonlinear in x and y.

Utilizing the same complex linear transformation as in Example 2.2.3, we reduce the study of the two-dimensional map to the study of the one-dimensional complex map

$$z \longmapsto \lambda(\mu)z + F^1(z,\bar{z};\mu), \tag{2.2.97}$$

where $F^1 = f^1 + if^2$ and $\lambda(\mu) = |\lambda(\mu)|\, e^{2\pi i\theta(\mu)}$.

We want to put this complex map into normal form. As a preliminary transformation, we expand $F^1(z,\bar{z};\mu)$ in a Taylor expansion in z and $\bar{z}$ with *coefficients depending on* μ so that (2.2.97) becomes

$$z_{n+1} = \lambda(\mu)z_n + F_2 + \cdots + F_{r-1} + \mathcal{O}(r), \tag{2.2.98}$$

where F_j is a polynomial of order j in z and $\bar{z}$.

First Simplify Second-Order Terms

Introducing the transformation

$$z \longmapsto z + h_2(z,\bar{z}), \tag{2.2.99}$$

where $h_2(z,\bar{z})$ is a second-order polynomial in z and $\bar{z}$ with *coefficients depending on* μ, (2.2.98) becomes

$$z_{n+1} + h_2(z_{n+1},\bar{z}_{n+1}) = \lambda z_n + \lambda h_2(z_n,\bar{z}_n) + F_2(z_n,\bar{z}_n) + \mathcal{O}(3)$$

or

$$z_{n+1} = \lambda z_n + \lambda h_2(z_n, \bar{z}_n) - h_2(z_{n+1}, \bar{z}_{n+1}) + F_2(z_n, \bar{z}_n) + \mathcal{O}(3). \quad (2.2.100)$$

Let us simplify further the term

$$h_2(z_{n+1}, \bar{z}_{n+1}) \quad (2.2.101)$$

in the right-hand side of (2.2.100). Clearly we have

$$\begin{aligned} z_{n+1} &= \lambda z_n + \mathcal{O}(2), \\ \bar{z}_{n+1} &= \bar{\lambda} \bar{z}_n + \mathcal{O}(2), \end{aligned} \quad (2.2.102)$$

so that

$$h_2(z_{n+1}, \bar{z}_{n+1}) = h_2(\lambda z_n, \bar{\lambda} \bar{z}_n) + \mathcal{O}(3). \quad (2.2.103)$$

Substituting (2.2.103) into (2.2.100) gives

$$z_{n+1} = \lambda z_n + \lambda h_2(z_n, \bar{z}_n) - h_2(\lambda z_n, \bar{\lambda} \bar{z}_n) + F_2 + \mathcal{O}(3). \quad (2.2.104)$$

Therefore, we can eliminate all second-order terms provided we can find $h_2(z, \bar{z})$ so that

$$\lambda h_2(z, \bar{z}) - h_2(\lambda z, \bar{\lambda} \bar{z}) + F_2 = 0. \quad (2.2.105)$$

As in all other situations involving normal forms that we have encountered thus far, this involves a problem from elementary linear algebra. This is because the map

$$h_2(z, \bar{z}) \longmapsto \lambda h_2(z, \bar{z}) - h_2(\lambda z, \bar{\lambda} \bar{z}) \quad (2.2.106)$$

is a linear map of H_2 into H_2 where

$$H_2 = \operatorname{span}\left\{z^2, z\bar{z}, \bar{z}^2\right\}. \quad (2.2.107)$$

In order to compute a matrix representation for (2.2.106) we need to compute the action of (2.2.106) on each basis element in (2.2.107). This is given as follows

$$\begin{aligned} \lambda z^2 - \lambda^2 z^2 &= \lambda(1 - \lambda) z^2, \\ \lambda z\bar{z} - \lambda\bar{\lambda} z\bar{z} &= \lambda(1 - \bar{\lambda}) z\bar{z}, \\ \lambda \bar{z}^2 - \bar{\lambda}^2 \bar{z}^2 &= (\lambda - \bar{\lambda}^2) \bar{z}^2. \end{aligned} \quad (2.2.108)$$

Using (2.2.108), the matrix representation for (2.2.106) with respect to the basis (2.2.107) is

$$\begin{pmatrix} \lambda(\mu)(1 - \lambda(\mu)) & 0 & 0 \\ 0 & \lambda(\mu)(1 - \bar{\lambda}(\mu)) & 0 \\ 0 & 0 & \lambda(\mu) - \bar{\lambda}(\mu)^2 \end{pmatrix}. \quad (2.2.109)$$

Now, by assumption, we have

$$|\lambda(0)| = 1 \qquad \text{and} \qquad \bar{\lambda}(0) = \frac{1}{\lambda(0)}. \tag{2.2.110}$$

Therefore, (2.2.109) is invertible at $\mu = 0$ provided

$$\begin{aligned} &\lambda(0) \neq 1, \\ &\lambda(0) \neq \frac{1}{\lambda(0)^2} \Rightarrow \lambda(0)^3 \neq 1. \end{aligned} \tag{2.2.111}$$

If (2.2.111) are satisfied at $\mu = 0$, then they are also satisfied in a sufficiently small neighborhood of $\mu = 0$. Therefore, if (2.2.111) are satisfied, then all second-order terms can be eliminated from the normal form for μ sufficiently small.

Next Eliminate Third-Order Terms

Using an argument exactly like that given above, third-order terms can be eliminated provided

$$\lambda h_3(z, \bar{z}) - h_3(\lambda z, \bar{\lambda}\bar{z}) + F_3 = 0. \tag{2.2.112}$$

The map

$$h_3(z, \bar{z}) \longmapsto \lambda h_3(z, \bar{z}) - h_2(\lambda z, \bar{\lambda}\bar{z}) \tag{2.2.113}$$

is a linear map of H_3 into H_3 where

$$H_3 = \text{span}\left\{z^3, z^2\bar{z}, z\bar{z}^2, \bar{z}^3\right\}. \tag{2.2.114}$$

The action of (2.2.113) on each element of (2.2.114) is given by

$$\begin{aligned} \lambda z^3 - \lambda^3 z^3 &= \lambda(1 - \lambda^2) z^3, \\ \lambda z^2\bar{z} - \lambda^2\bar{\lambda} z^2\bar{z} &= \lambda(1 - \lambda\bar{\lambda}) z^2\bar{z}, \\ \lambda z\bar{z}^2 - \bar{\lambda}^2\lambda z\bar{z}^2 &= \lambda(1 - \bar{\lambda}^2) z\bar{z}^2, \\ \lambda\bar{z}^3 - \bar{\lambda}^3\bar{z}^3 &= (\lambda - \bar{\lambda}^3)\bar{z}^3. \end{aligned} \tag{2.2.115}$$

Thus, a matrix representation of (2.2.113) with respect to the basis (2.2.114) is given by

$$\begin{pmatrix} \lambda(\mu)(1 - \lambda(\mu)^2) & 0 & 0 & 0 \\ 0 & \lambda(\mu)(1 - \lambda(\mu)\bar{\lambda}(\mu)) & 0 & 0 \\ 0 & 0 & \lambda(\mu)(1 - \bar{\lambda}(\mu)^2) & 0 \\ 0 & 0 & 0 & \lambda(\mu) - \bar{\lambda}(\mu)^3 \end{pmatrix}. \tag{2.2.116}$$

Recall that at $\mu = 0$ we have

$$|\lambda(0)| = 1, \qquad \bar{\lambda}(0) = \frac{1}{\lambda(0)}, \tag{2.2.117}$$

so that at $\mu = 0$ the second column of (2.2.116) is all zero's. The reader can easily check that the remaining columns are all linearly independent at $\mu = 0$ provided

$$\lambda^2(0) \neq 1, \qquad \lambda^4(0) \neq 1. \tag{2.2.118}$$

This situation will also hold for μ sufficiently small. Therefore, the normal form is as follows

$$z \longmapsto \lambda(\mu)z + c(\mu)z^2\bar{z} + \mathcal{O}(4), \tag{2.2.119}$$

where $c(\mu)$ is a constant, provided

$$\lambda^n(0) \neq 1, \qquad n = 1, 2, 3, 4$$

for μ sufficiently small.

More generally, simplification at order k depends on how the linear operator

$$h_k(z, \bar{z}) \longmapsto \lambda h_k(z, \bar{z}) - h_k(\lambda z, \bar{\lambda}\bar{z}) \tag{2.2.120}$$

acts on elements like $h = z^n \bar{z}^m$ where $m + n$ is the order of the term one wishes to simplify. Substituting this into the above equation gives

$$\lambda z^n \bar{z}^m - \lambda^n \bar{\lambda}^m z^n \bar{z}^m = \lambda(1 - \lambda^{n-1}\bar{\lambda}^m) z^n z^m. \tag{2.2.121}$$

At $\mu = 0$, $\bar{\lambda} = 1/\lambda$; hence, we cannot have

$$\lambda^{n-m-1}(0) = 1. \tag{2.2.122}$$

We leave it to the reader to work out general conditions for the elimination of higher order terms based on (2.2.122).

Differentiability

We make the important remark that in order to obtain the normal form (2.2.119) the map must be at least $\mathbf{C}^4$.

2.2D Conjugacies and Equivalences of Vector Fields

In Section 1.2B we discussed the idea of a $\mathbf{C}^r$ conjugacy or coordinate transformation for maps. This was motivated by the question of how the dynamics of a Poincaré map were affected as the cross-section on which the map was defined was changed. The change in cross-section could be viewed as a change of coordinates for the map. A similar question arises here in the context of the method of normal forms; namely, the normal form of a vector field or map is obtained through a coordinate transformation. How does this coordinate transformation modify the dynamics? For maps, the discussion in Section 1.2B applies; however, we did not discuss the notion of conjugacies of vector fields. To address that topic, we begin with a definition.

Let

$$\dot{x} = f(x), \qquad x \in \mathbb{R}^n, \tag{2.2.123a}$$

$$\dot{y} = g(y), \qquad y \in \mathbb{R}^n, \tag{2.2.123b}$$

be two $\mathbf{C}^r$ $(r \geq 1)$ vector fields defined on $\mathbb{R}^n$ (or sufficiently large open sets of $\mathbb{R}^n$).

DEFINITION 2.2.1 The dynamics generated by the vector fields f and g are said to be $\mathbf{C}^k$ *equivalent* $(k \leq r)$ if there exists a $\mathbf{C}^k$ diffeomorphism h which takes orbits of the flow generated by f, $\phi(t, x)$, to orbits of the flow generated by g, $\psi(t, y)$, preserving orientation but not necessarily parameterization by time. If h does preserve parameterization by time, then the dynamics generated by f and g are said to be $\mathbf{C}^k$ *conjugate.*

We remark that, as for maps, the conjugacies do not need to be defined on all of $\mathbb{R}^n$ but, rather, on appropriately chosen open sets in $\mathbb{R}^n$. In this case f and g are said to be *locally* $\mathbf{C}^k$ equivalent or *locally* $\mathbf{C}^k$ conjugate.

Definition 2.2.1 is slightly different than the analogous Definition 1.2.2 for maps. Indeed, in Definition 2.2.1 we have introduced an additional concept: the idea of $\mathbf{C}^k$ *equivalence.* These differences all stem from the fact that in vector fields the independent variable (time) is continuous and in maps time is discrete. Let us illustrate these differences with an example. However, we exhort the reader to keep the following in mind.

The purpose of Definition 2.2.1 *is to provide us with a way of characterizing when two vector fields have qualitatively the same dynamics.*

EXAMPLE 2.2.5 Consider the two vector fields

$$\begin{aligned} \dot{x}_1 &= x_2, \\ \dot{x}_2 &= -x_1, \end{aligned} \qquad (x_1, x_2) \in \mathbb{R}^2, \tag{2.2.124}$$

and

$$\begin{aligned} \dot{y}_1 &= y_2, \\ \dot{y}_2 &= -y_1 - y_1^3, \end{aligned} \qquad (y_1, y_2) \in \mathbb{R}^2. \tag{2.2.125}$$

The phase portraits of each vector field are shown in Figure 2.2.1. Each vector field possesses only a single fixed point, a center at the origin. In each case the fixed point is surrounded by a one-parameter family of periodic orbits. Note also from Figure 2.2.1 that in each case the direction of motion along the periodic orbits is in the same sense; therefore, qualitatively these two vector fields have the same dynamics—the phase portraits appear identical. However, let us see if this is reflected by the idea of $\mathbf{C}^k$ *conjugacy.*

Let us denote the flow generated by (2.2.124) as

$$\phi(t, x), \qquad x \equiv (x_1, x_2),$$

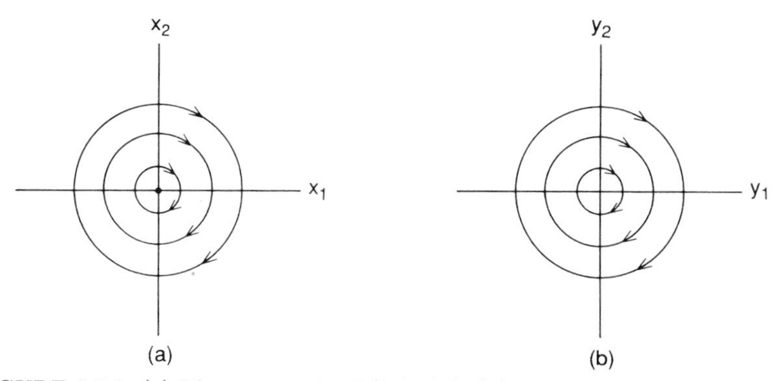

FIGURE 2.2.1. (a) Phase portrait of (2.2.124). (b) Phase portrait of (2.2.125).

and the flow generated by (2.2.125) as

$$\psi(t,y), \qquad y \equiv (y_1, y_2).$$

Suppose we have found a $\mathbf{C}^k$ *diffeomorphism*, h, taking orbits of the flow generated by (2.2.124) into orbits of the flow generated by (2.2.125). Then we have

$$h \circ \phi(t,x) = \psi\big(t, h(x)\big). \tag{2.2.126}$$

Equation (2.2.126) reveals an immediate problem; namely, if $\phi(t,x)$ and $\psi(t,y)$ are periodic in t, then $h \circ \phi(t,x)$ and $\psi\big(t, h(x)\big)$ must have the same period in order for (2.2.126) to hold. However, in general, (2.2.126) cannot be satisfied. Consider (2.2.124) and (2.2.125). The vector field (2.2.124) is linear; therefore, all the periodic orbits have the same period. The vector field (2.2.125) is nonlinear; therefore, the period of the periodic orbits varies with the distance from the fixed point. Thus, (2.2.124) and (2.2.125) are not $\mathbf{C}^k$ *conjugate* (note: one can actually compute the period of the orbits of (2.2.125) since it is a Hamiltonian vector field. It is a tedious exercise involving elliptic integrals).

It is precisely this situation from which the idea of $\mathbf{C}^k$ *equivalence* can rescue us, for rather than only having a $\mathbf{C}^k$ *diffeomorphism* that maps orbits to orbits, we at the same time allow a reparametrization of time along the orbit. We make this idea more quantitative as follows. Let $\alpha(x,t)$ be an increasing function of t along orbits (note: it must be increasing in order to preserve orientations of orbits). Then (2.2.124) and (2.2.125) are $\mathbf{C}^k$ *equivalent* if the following holds

$$h \circ \phi(t,x) = \psi\big(\alpha(x,t), h(x)\big). \tag{2.2.127}$$

Equation (2.2.127) shows that orbits of the flow generated by (2.2.124) are mapped to orbits of the flow generated by (2.2.125); however, the time dependence of the image of an orbit under h may be reparametrized in an orbitally dependent manner. Finally, we remark that the term "preserving orientation" in Definition 2.2.1 refers to the fact that the direction of motion along an orbit is unchanged under $\mathbf{C}^k$ equivalence.

Let us now consider some of the dynamical consequences of Definition 2.2.1.

Proposition 2.2.2 *Suppose f and g are* $\mathbf{C}^k$ conjugate. *Then*

i) *fixed points of f are mapped to fixed points of g;*

ii) *T-periodic orbits of f map to T-periodic orbits of g.*

Proof: That f and g are $\mathbf{C}^k$ conjugate under h implies the following

$$h \circ \phi(t, x) = \psi\big(t, h(x)\big), \tag{2.2.128}$$

$$Dh\dot{\phi} = \dot{\psi}. \tag{2.2.129}$$

The proof of i) follows from (2.2.129) and the proof of ii) follows from (2.2.128). $\square$

Proposition 2.2.3 *Suppose f and g are* $\mathbf{C}^k$ conjugate ($k \geq 1$) *and $f(x_0) = 0$; then $Df(x_0)$ has the same eigenvalues as $Dg\big(h(x_0)\big)$.*

Proof: We have the two vector fields $\dot{x} = f(x)$, $\dot{y} = g(y)$. By differentiating (2.2.128) with respect to t we have

$$Dh|_x\, f(x) = g\big(h(x)\big). \tag{2.2.130}$$

Differentiating (2.2.130) gives

$$D^2h\big|_x\, f(x) + Dh|_x\, Df|_x = Dg|_{h(x)}\, Dh|_x\,. \tag{2.2.131}$$

Evaluating (2.2.131) at x_0 gives

$$Dh|_{x_0}\, Df|_{x_0} = Dg|_{h(x_0)}\, Dh|_{x_0} \tag{2.2.132}$$

or

$$Df|_{x_0} = (Dh)^{-1}\big|_{x_0}\, Dg|_{h(x_0)}\, Dh|_{x_0}\,, \tag{2.2.133}$$

and, since similar matrices have equal eigenvalues, the proof is complete. $\square$

The previous two propositions dealt with $\mathbf{C}^k$ *conjugacies.* We next examine the consequences of $\mathbf{C}^k$ *equivalence* under the assumption that the change in parameterization by time along orbits is $\mathbf{C}^1$. The validity of this assumption must be verified in any specific application.

Proposition 2.2.4 *Suppose f and g are* $\mathbf{C}^k$ equivalent; *then*

i) *fixed points of f are mapped to fixed points of g;*

ii) *periodic orbits of f are mapped to periodic orbits of g, but the periods need not be equal.*

Proof: If f and g are $\mathbf{C}^k$ *equivalent*, then

$$h \circ \phi(t, x) = \psi\big(\alpha(x,t), h(x)\big), \tag{2.2.134}$$

where α is an increasing function of time along orbits (note: α must be increasing in order to preserve orientations of orbits).

Differentiating (2.2.134) gives

$$Dh\dot{\phi} = \frac{\partial \alpha}{\partial t}\frac{\partial \psi}{\partial \alpha}. \tag{2.2.135}$$

Therefore, (2.2.135) implies i), and ii) follows automatically since $\mathbf{C}^k$ *diffeomorphisms* map closed curves to closed curves. (If this were not true, then the inverse would not be continuous.) □

Proposition 2.2.5 *Suppose f and g are* $\mathbf{C}^k$ equivalent ($k \geq 1$) *and $f(x_0) = 0$; then the eigenvalues of $Df(x_0)$ and the eigenvalues of $Dg\big(h(x_0)\big)$ differ by a positive multiplicative constant.*

Proof: Proceeding as in the proof of Proposition 2.2.3, we have

$$Dh\big|_x f(x) = \frac{\partial \alpha}{\partial t} g\big(h(x)\big). \tag{2.2.136}$$

Differentiating (2.2.136) gives

$$D^2h\big|_x f(x) + Dh\big|_x Df|x = \frac{\partial \alpha}{\partial t} Dg|_{h(x)} Dh\big|_x + \left.\frac{\partial^2 \alpha}{\partial x \partial t}\right|_x g\big(h(x)\big). \tag{2.2.137}$$

Evaluating at x_0 gives

$$Dh\big|_{x_0} Df\big|_{x_0} = \frac{\partial \alpha}{\partial t} Dg\big|_{h(x_0)} Dh\big|_{x_0}; \tag{2.2.138}$$

thus, $Df\big|_{x_0}$ and $Dg\big|_{h(x_0)}$ are similar up to the multiplicative constant $\partial\alpha/\partial t$, which is positive, since α is increasing on orbits. □

EXAMPLE 2.2.6 Consider the vector fields

$$\begin{aligned} \dot{x}_1 &= x_1, \\ \dot{x}_2 &= x_2, \end{aligned} \qquad (x_1, x_2) \in \mathbb{R}^2,$$

and

$$\begin{aligned} \dot{y}_1 &= y_1, \\ \dot{y}_2 &= 2y_2, \end{aligned} \qquad (y_1, y_2) \in \mathbb{R}^2.$$

Qualitatively these two vector fields have the same dynamics. However, by Proposition 2.2.3 they are not $\mathbf{C}^k$ *equivalent*, $k \geq 1$.

i) AN APPLICATION: THE HARTMAN-GROBMAN THEOREM

An underlying theme throughout the first chapter of this book was that the orbit structure near a hyperbolic fixed point was qualitatively the same as the orbit structure given by the associated linearized dynamical system. A theorem proved independently by Hartman [1960] and Grobman [1959] makes this precise. We will describe the situation for vector fields.

Consider a $\mathbf{C}^r$ $(r \geq 1)$ vector field

$$\dot{x} = f(x), \qquad x \in \mathbb{R}^n, \tag{2.2.139}$$

where f is defined on a sufficiently large open set of $\mathbb{R}^n$. Suppose that (2.2.139) has a *hyperbolic* fixed point at $x = x_0$, i.e.,

$$f(x_0) = 0,$$

and $Df(x_0)$ has no eigenvalues on the imaginary axis. Consider the associated linear vector field

$$\dot{\xi} = Df(x_0)\xi, \qquad \xi \in \mathbb{R}^n. \tag{2.2.140}$$

Then we have the following theorem.

Theorem 2.2.6 (Hartman and Grobman) *The flow generated by (2.2.139) is* $\mathbf{C}^0$ conjugate *to the flow generated by (2.2.140) in a neighborhood of the fixed point* $x = x_0$.

Proof: See Arnold [1973] or Palis and deMelo [1982]. □

We remark that the theorem can be modified so that it applies to hyperbolic fixed points of maps, and we leave it to the reader to reformulate the theorem along these lines.

A point to note concerning Theorem 2.2.6 is that the conjugacy transforming the nonlinear flow into the linear flow near the hyperbolic fixed point is not differentiable; rather, it is a homeomorphism. This makes the generation of the transformation via, for example, normal form theory, not possible since the coordinate transformations constructed via that theory were power series expansions and, hence, differentiable. However, a closer look at normal form theory will reveal the heart of the problem with "differentiable linearization." Let us expand on this with a brief discussion.

Recall equation (2.2.8)

$$\dot{x} = Jx + F_2(x) + \cdots + F_{r-1}(x) + \mathcal{O}(|x|^r), \qquad x \in \mathbb{R}^n. \tag{2.2.141}$$

A sufficient condition for eliminating the $\mathcal{O}(|x|^k)$ terms $(2 \leq k \leq r-1)$ from (2.2.141) is that the linear operator $L_J(\cdot)$ is invertible on H_k. We want to explore why $L_J(\cdot)$ is noninvertible.

Recall

$$L_J\big(h_k(x)\big) \equiv Jh_k(x) - Dh_k(x)Jx, \tag{2.2.142}$$

with $h_k(x) \in H_k$, where H_k is the linear vector space of vector-valued monomials of degree k. Let us choose a basis for H_k. Suppose J is diagonal with eigenvalues $\lambda_1, \cdots, \lambda_n$ (note: if J is not diagonalizable, then the following argument is still valid, but with slight modifications; see Arnold [1982] or Bryuno [1989]). Let e_i, $1 \le i \le n$, be the standard basis of $\mathbb{R}^n$, i.e., e_i is an n vector with a 1 in the ith component and zeros in the remaining components. Then we have

$$Je_i = \lambda_i e_i. \tag{2.2.143}$$

As a basis for H_k we take the set of elements

$$x_1^{m_1} \cdots x_n^{m_n} e_i, \qquad \sum_{j=1}^{n} m_j = k, \quad m_j \ge 0, \tag{2.2.144}$$

where we consider all possible terms $x_1^{m_1} \cdots x_n^{m_n}$ of degree k multiplying each e_i, $1 \le i \le n$.

Next we consider the action of $L_J(\cdot)$ on each of these basis elements of H_k. Let

$$h_k(x) = x_1^{m_1} \cdots x_n^{m_n} e_i, \qquad \sum_{j=1}^{n} m_j = k, \qquad m_j \ge 0; \tag{2.2.145}$$

then a simple calculation shows that

$$L_J\big(h_k(x)\big) = Jh_k(x) - Dh_k(x)Jx = \left[\lambda_i - \sum_{j=1}^{n} m_j \lambda_j\right] h_k(x). \tag{2.2.146}$$

Thus, the linear operator $L_J(\cdot)$ is diagonal in this basis, with eigenvalues given by

$$\lambda_i - \sum_{j=1}^{n} m_j \lambda_j. \tag{2.2.147}$$

Now we can see the problem. The linear operator $L_J(\cdot)$ will fail to be invertible if it has a zero eigenvalue, which in this case means

$$\lambda_i = \sum_{j=1}^{n} m_j \lambda_j. \tag{2.2.148}$$

Equation 2.2.148 is called a *resonance* and is the origin of the name "resonance terms" for the unremovable nonlinear terms in the normal form described in Theorem 2.2.1. The integer

$$\sum_{j=1}^{n} m_j$$

is called the *order of the resonance.* Thus, the difficulty in finding a differentiable coordinate change that will linearize a vector field in the neighborhood of a hyperbolic fixed point lies in the fact that an eigenvalue of the linearized part may be equal to a linear combination over the nonnegative integers of elements from the set of eigenvalues of the linearized part. Much work has been done on the geometry of the resonances in the complex plane and on differentiable linearizations in the situations where the resonances are avoided. For more information we refer the reader to the fundamental papers by Sternberg [1957], [1958] and also Arnold [1982] and Bryuno [1989].

Let us consider the following example due to Sternberg (see also Meyer [1986]). Consider the vector field

$$\begin{aligned} \dot{x} &= 2x + y^2, \\ \dot{y} &= y, \end{aligned} \qquad (x, y) \in \mathbb{R}^2. \tag{2.2.149}$$

This vector field clearly has a hyperbolic fixed point at the origin. The vector field linearized about the origin is given by

$$\begin{aligned} \dot{x} &= 2x, \\ \dot{y} &= y. \end{aligned} \tag{2.2.150}$$

Eliminating t as the independent variable, (2.2.150) can be written as

$$\frac{dx}{dy} = \frac{2x}{y}, \qquad y \neq 0. \tag{2.2.151}$$

Solving (2.2.151), the orbits of (2.2.150) are given by

$$x = cy^2, \tag{2.2.152}$$

where c is a constant. Clearly (2.2.152) are analytic curves at the origin.

Now consider the nonlinear vector field (2.2.149). Eliminating t as the independent variable, (2.2.149) can be written as

$$\frac{dx}{dy} = \frac{2x}{y} + y, \qquad y \neq 0. \tag{2.2.153}$$

Equation (2.2.153) is a standard first-order linear equation which can be solved via elementary methods (see, e.g., Boyce and DiPrima [1977]). The solution of (2.2.153) is given by

$$x = y^2 \left[k + \log|y|\right], \tag{2.2.154}$$

where k is a constant. Clearly (2.2.154) are $\mathbf{C}^1$ but not $\mathbf{C}^2$ at the origin. Since the property of lying on $\mathbf{C}^2$ curves must be preserved under a $\mathbf{C}^2$ change of coordinates (the chain rule), we conclude that (2.2.149) and (2.2.150) are $\mathbf{C}^1$ but not $\mathbf{C}^2$ conjugate.

In terms of resonances, the problem involves a second-order resonance (hence the problem with $\mathbf{C}^2$ linearization). This can be seen as follows. Let

$$\lambda_1 = 1,$$
$$\lambda_2 = 2,$$

be the eigenvalues of the linearization. Then we have

$$\lambda_2 = m_1\lambda_1 + m_2\lambda_2,$$

with $m_1 = 2$, $m_2 = 0$, and $\sum_{j=1}^{2} m_j = 2$.

ii) An Application: Dynamics Near a Fixed Point

Consider the parameter-dependent vector field

$$\begin{aligned} \dot{x} &= Ax + f(x,y,z,\varepsilon), \\ \dot{y} &= By + g(x,y,z,\varepsilon), \qquad (x,y,z,\varepsilon) \in \mathbb{R}^c \times \mathbb{R}^s \times \mathbb{R}^u \times \mathbb{R}^p, \\ \dot{z} &= Cz + h(x,y,z,\varepsilon), \end{aligned} \tag{2.2.155}$$

where

$$\begin{aligned} f(0,0,0,0) &= 0, & Df(0,0,0,0) &= 0, \\ g(0,0,0,0) &= 0, & Dg(0,0,0,0) &= 0, \\ h(0,0,0,0) &= 0, & Dh(0,0,0,0) &= 0, \end{aligned}$$

and f, g, and h are $\mathbf{C}^r$ $(r \geq 2)$ in some neighborhood of the orgin, A is a $c \times c$ matrix having eigenvalues with zero real parts, B is a $s \times s$ matrix having eigenvalues with negative real parts, and C is a $u \times u$ matrix having eigenvalues with positive real parts.

The center manifold theorem tells us that near the orgin in $\mathbb{R}^c \times \mathbb{R}^s \times \mathbb{R}^u \times \mathbb{R}^p$, the flow generated by (2.2.155) is $\mathbf{C}^0$ *conjugate* to the flow generated by the following vector field

$$\begin{aligned} \dot{x} &= w(x,\varepsilon), \\ \dot{y} &= -y, \qquad (x,y,z,\varepsilon) \in \mathbb{R}^c \times \mathbb{R}^s \times \mathbb{R}^u \times \mathbb{R}^p, \\ \dot{z} &= z, \end{aligned} \tag{2.2.156}$$

where $w(x,\varepsilon)$ represents the $\mathbf{C}^r$ vector field on the center manifold.

2.3 Final Remarks

Remark 1. Nonuniqueness of Normal Forms. It should be clear from our discussion that normal forms need not be unique. However, it may happen that certain properties of a vector field (e.g., symmetries) must be possessed by any normal form. Such questions are explored in Kummer [1971], Bryuno [1989], van der Meer [1985], Baider and Churchill [1988], and Baider [1989].

Remark 2. Divergence of the Normalizing Transformations. In general, normal forms are divergent. This is discussed in detail in Siegel [1941] and Bryuno [1989]. However, this does not affect considerations of local stability.

Remark 3. Computation of Normal Forms. Elphick et al. [1987] describe a very efficient method for constructing normal forms. They also give a nice interpretation and characterization of the resonant terms; see also Cushman and Sanders [1986]. The book of Rand and Armbruster [1987] describes how one may implement the computation of normal forms using computer algebra.

Remark 4. The Method of Amplitude Expansions. This method has been used for many years in studies of hydrodynamic stability and has much in common with the center manifold reduction. This situation has been clarified by Coullet and Spiegel [1983].

Remark 5. The Homological Equation. We want to remark on some terminology. In computing the normal form of vector fields near a fixed point, the equation

$$Dh_k(y)Jy - Jh_k(y) = F_k(y)$$

must be solved in order to simplify the order k terms in the Taylor expansion of the vector field. This equation is called the *homological equation* (see Arnold [1982]). The analogous equation for maps is also called the homological equation.

Remark 6. Nonautonomous Systems. Consider the situation of $\mathbf{C}^r$ (with r as large as necessary) vector fields

$$\dot{x} = f(x), \qquad x \in \mathbb{R}^n. \tag{2.3.1}$$

The method of normal forms as developed in this chapter can be viewed as a method for simplifying the vector field in the neighborhood of a *fixed point.* However, suppose $x(t) = \bar{x}(t)$ is a trajectory of this vector field. Can the method of normal forms then be used to simplify the vector field in the neighborhood of a general (time-dependent) solution? The answer is "sometimes," but there are associated difficulties. Let

$$x = \bar{x}(t) + y;$$

then (2.3.1) becomes

$$\dot{y} = A(t)y + \mathcal{O}(|y|^2), \tag{2.3.2}$$

where

$$A(t) \equiv Df(\bar{x}(t)).$$

In applying the method of normal forms to (2.3.2), the fact that $A(t)$ is time dependent causes problems. If $\bar{x}(t)$ is periodic in t, then $A(t)$ is periodic.

Hence, Floquet theory can be used to transform (2.3.2) to a vector field where the linear part is constant (this is described in Arnold [1982]). In this case the method of normal forms as developed in this chapter can then be applied. Recently, Floquet theory has been generalized to the quasiperiodic case by Johnson [1986, 1987]; using these ideas, the normal form theory can be applied in this case also.

Concerning center manifold theory, Sell [1978] has proved existence theorems for stable, unstable, and center manifolds in nonautonomous systems.

Exercises

2.1 Study the dynamics near the origin for each of the following vector fields. Draw phase portraits. Compute the center manifolds and describe the dynamics on the center manifold. Discuss the stability or instability of the origin.

a) $$\begin{aligned}\dot{\theta} &= -\theta + v^2,\\ \dot{v} &= -\sin\theta,\end{aligned} \qquad (\theta, v) \in S^1 \times \mathbb{R}^1.$$

b) $$\begin{aligned}\dot{x} &= \frac{1}{2}x + y + x^2 y,\\ \dot{y} &= x + 2y + y^2,\end{aligned} \qquad (x, y) \in \mathbb{R}^2.$$

c) $$\begin{aligned}\dot{x} &= x - 2y,\\ \dot{y} &= 3x - y - x^2,\end{aligned} \qquad (x, y) \in \mathbb{R}^2.$$

d) $$\begin{aligned}\dot{x} &= 2x + 2y,\\ \dot{y} &= x + y + x^4,\end{aligned} \qquad (x, y) \in \mathbb{R}^2.$$

e) $$\begin{aligned}\dot{x} &= -y - y^3,\\ \dot{y} &= 2x,\end{aligned} \qquad (x, y) \in \mathbb{R}^2.$$

f) $$\begin{aligned}\dot{x} &= -2x + 3y + y^3,\\ \dot{y} &= 2x - 3y + x^3,\end{aligned} \qquad (x, y) \in \mathbb{R}^2.$$

g) $$\begin{aligned}\dot{x} &= -x - y - xy,\\ \dot{y} &= 2x + y + 2xy,\end{aligned} \qquad (x, y) \in \mathbb{R}^2.$$

h) $$\begin{aligned}\dot{x} &= -x + y,\\ \dot{y} &= -e^x + e^{-x} + 2x,\end{aligned} \qquad (x, y) \in \mathbb{R}^2.$$

i) $$\begin{aligned}\dot{x} &= -2x + y + z + y^2 z,\\ \dot{y} &= x - 2y + z + xz^2,\\ \dot{z} &= x + y - 2z + x^2 y,\end{aligned} \qquad (x, y, z) \in \mathbb{R}^3.$$

j) $$\begin{aligned}\dot{x} &= -x - y + z^2,\\ \dot{y} &= 2x + y - z^2,\\ \dot{z} &= x + 2y - z,\end{aligned} \qquad (x, y, z) \in \mathbb{R}^3.$$

k) $$\begin{aligned}\dot{x} &= -x - y - z - yz,\\ \dot{y} &= -x - y - z - xz,\\ \dot{z} &= -x - y - z - xy,\end{aligned} \qquad (x, y, z) \in \mathbb{R}^3.$$

k) $\begin{aligned} \dot{x} &= -x - y - z - yz, \\ \dot{y} &= -x - y - z - xz, \\ \dot{z} &= -x - y - z - xy, \end{aligned} \qquad (x, y, z) \in \mathbb{R}^2.$

l) $\begin{aligned} \dot{x} &= y + x^2, \\ \dot{y} &= -y - x^2, \end{aligned} \qquad (x, y) \in \mathbb{R}^2.$

m) $\begin{aligned} \dot{x} &= x^2, \\ \dot{y} &= -y - x^2, \end{aligned} \qquad (x, y) \in \mathbb{R}^2.$

n) $\begin{aligned} \dot{x} &= -x + 2y + x^2 y + x^4 y^5, \\ \dot{y} &= y - x^4 y^6 + x^8 y^9, \end{aligned} \qquad (x, y) \in \mathbb{R}^2.$

2.2 Consider the following parametrized families of vector fields with parameter $\varepsilon \in \mathbb{R}^1$. For $\varepsilon = 0$, the origin is a fixed point of each vector field. Study the dynamics near the origin for ε small. Draw phase portraits. Compute the one-parameter family of center manifolds and describe the dynamics on the center manifolds. How do the dynamics depend on ε? Note that, for $\varepsilon = 0$, e.g., a) and a$'$) reduce to a) in Exercise 2.1. Discuss the role played by a parameter by comparing these cases. In, for example, a) and a$'$), the parameter ε multiplies a linear and nonlinear term, respectively. Discuss the differences in these two cases in the most general setting possible.

a) $\begin{aligned} \dot{\theta} &= -\theta + \varepsilon v + v^2, \\ \dot{v} &= -\sin\theta, \end{aligned} \qquad (\theta, v) \in S^1 \times \mathbb{R}^1.$

a$'$) $\begin{aligned} \dot{\theta} &= -\theta + v^2 + \varepsilon v^2, \\ \dot{v} &= -\sin\theta \end{aligned}$

b) $\begin{aligned} \dot{x} &= \frac{1}{2}x + y + x^2 y, \\ \dot{y} &= x + 2y + \varepsilon y + y^2, \end{aligned} \qquad (x, y) \in \mathbb{R}^2.$

b$'$) $\begin{aligned} \dot{x} &= \frac{1}{2}x + y + x^2 y, \\ \dot{y} &= x + 2y + y^2 + \varepsilon y^2, \end{aligned}$

c) $\begin{aligned} \dot{x} &= x - 2y + \varepsilon x, \\ \dot{y} &= 3x - y - x^2, \end{aligned} \qquad (x, y) \in \mathbb{R}^2.$

c$'$) $\begin{aligned} \dot{x} &= x - 2y + \varepsilon x^2, \\ \dot{y} &= 3x - y - x^2, \end{aligned}$

d) $\begin{aligned} \dot{x} &= 2x + 2y + \varepsilon y, \\ \dot{y} &= x + y + x^4, \end{aligned} \qquad (x, y) \in \mathbb{R}^2.$

d$'$) $\begin{aligned} \dot{x} &= 2x + 2y, \\ \dot{y} &= x + y + x^4 + \varepsilon y^2, \end{aligned}$

e) $\begin{aligned} \dot{x} &= -y - \varepsilon x - y^3, \\ \dot{y} &= 2x, \end{aligned} \qquad (x, y) \in \mathbb{R}^2.$

e′) $\begin{aligned}\dot{x} &= -y - y^3,\\ \dot{y} &= 2x + \varepsilon x^2,\end{aligned}$

f) $\begin{aligned}\dot{x} &= -2x + 3y + \varepsilon x + y^3,\\ \dot{y} &= 2x - 3y + x^3,\end{aligned}$ $\qquad (x, y) \in \mathbb{R}^2.$

f′) $\begin{aligned}\dot{x} &= -2x + 3y + y^3 + \varepsilon x^2,\\ \dot{y} &= 2x - 3y + x^3,\end{aligned}$

g) $\begin{aligned}\dot{x} &= -x - y + \varepsilon x - xy,\\ \dot{y} &= 2x + y + 2xy,\end{aligned}$ $\qquad (x, y) \in \mathbb{R}^2.$

g′) $\begin{aligned}\dot{x} &= -x - y - xy + \varepsilon x^2,\\ \dot{y} &= 2x + y + 2xy,\end{aligned}$

h) $\begin{aligned}\dot{x} &= -x + y,\\ \dot{y} &= -e^x + e^{-x} + 2x + \varepsilon y,\end{aligned}$ $\qquad (x, y) \in \mathbb{R}^2.$

h′) $\begin{aligned}\dot{x} &= -x + y + \varepsilon x^2,\\ \dot{y} &= -e^x + e^{-x} + 2x,\end{aligned}$

i) $\begin{aligned}\dot{x} &= -2x + y + z + \varepsilon x - y^2 z,\\ \dot{y} &= x - 2y + z + \varepsilon x + xz^2,\\ \dot{z} &= x + y - 2z + \varepsilon x + x^2 y,\end{aligned}$ $\qquad (x, y, z) \in \mathbb{R}^3.$

i′) $\begin{aligned}\dot{x} &= -2x + y + z + \varepsilon x^2 + y^2 z,\\ \dot{y} &= x - 2y + z + \varepsilon xy + xz^2,\\ \dot{z} &= x + y - 2z + x^2 y.\end{aligned}$

j) $\begin{aligned}\dot{x} &= -x - y + z^2,\\ \dot{y} &= 2x + y + \varepsilon y - z^2,\\ \dot{z} &= x + 2y - z,\end{aligned}$ $\qquad (x, y, z) \in \mathbb{R}^3.$

j′) $\begin{aligned}\dot{x} &= -x - y + \varepsilon x^2 + z^2,\\ \dot{y} &= 2x + y - z^2 + \varepsilon y^2,\\ \dot{z} &= x + 2y - z.\end{aligned}$

k) $\begin{aligned}\dot{x} &= -x - y - z + \varepsilon x - yz,\\ \dot{y} &= -x - y - z - xz,\\ \dot{z} &= -x - y - z - yz,\end{aligned}$ $\qquad (x, y, z) \in \mathbb{R}^3.$

k′) $\begin{aligned}\dot{x} &= -x - y - z - yz + \varepsilon x^2,\\ \dot{y} &= -x - y - z - xz,\\ \dot{z} &= -x - y - z - xy.\end{aligned}$

l) $\begin{aligned}\dot{x} &= y + x^2 + \varepsilon y,\\ \dot{y} &= -y - x^2,\end{aligned}$ $\qquad (x, y) \in \mathbb{R}^2.$

l′) $\begin{aligned}\dot{x} &= y + x^2 + \varepsilon y^2,\\ \dot{y} &= -y - x^2,\end{aligned}$

m) $\begin{aligned}\dot{x} &= x^2 + \varepsilon y,\\ \dot{y} &= -y - x^2,\end{aligned}$ $\qquad (x, y) \in \mathbb{R}^2.$

m′) $\begin{aligned} \dot{x} &= x^2 + \varepsilon y^2, \\ \dot{y} &= -y - x^2. \end{aligned}$

2.3 Study the dynamics near the origin for each of the following maps. Draw phase portraits. Compute the center manifold and describe the dynamics on the center manifold. Discuss the stability or instability of the origin.

a) $\begin{aligned} x &\mapsto -\frac{1}{2}x - y - xy^2, \\ y &\mapsto -\frac{1}{2}x + x^2, \end{aligned} \qquad (x, y) \in \mathbb{R}^2.$

b) $\begin{aligned} x &\mapsto x + 2y + x^3, \\ y &\mapsto 2x + y, \end{aligned} \qquad (x, y) \in \mathbb{R}^2.$

c) $\begin{aligned} x &\mapsto -x + y - xy^2, \\ y &\mapsto y + x^2 y, \end{aligned} \qquad (x, y) \in \mathbb{R}^2.$

d) $\begin{aligned} x &\mapsto 2x + y, \\ y &\mapsto 2x + 3y + x^4, \end{aligned} \qquad (x, y) \in \mathbb{R}^2.$

e) $\begin{aligned} x &\mapsto x, \\ y &\mapsto x + 2y + y^2, \end{aligned} \qquad (x, y) \in \mathbb{R}^2.$

f) $\begin{aligned} x &\mapsto 2x + 3y, \\ y &\mapsto x + x^2 + xy^2, \end{aligned} \qquad (x, y) \in \mathbb{R}^2.$

g) $\begin{aligned} x &\mapsto x - z^3, \\ y &\mapsto 2x - y, \\ z &\mapsto x + \frac{1}{2}z + x^3, \end{aligned} \qquad (x, y, z) \in \mathbb{R}^3.$

h) $\begin{aligned} x &\mapsto x + z^4, \\ y &\mapsto -x - 2y - x^3, \\ z &\mapsto y - \frac{1}{2}z + y^2, \end{aligned} \qquad (x, y, z) \in \mathbb{R}^3.$

i) $\begin{aligned} x &\mapsto y + x^2, \\ y &\mapsto y + xy, \end{aligned} \qquad (x, y) \in \mathbb{R}^2.$

j) $\begin{aligned} x &\mapsto x^2, \\ y &\mapsto y + xy, \end{aligned} \qquad (x, y) \in \mathbb{R}^2.$

2.4 Consider the following parametrized families of maps with parameter $\varepsilon \in \mathbb{R}^1$. For $\varepsilon = 0$, the origin is a fixed point of each vector field. Study the dynamics near the origin for ε small. Draw phase portraits. Compute the one-parameter family of center manifolds and describe the dynamics on the center manifolds. How do the dynamics depend on ε? Note that, for $\varepsilon = 0$, e.g., a) and a′) reduce to a) in Exercise 2.3. Discuss the role played by a parameter by comparing these cases. In, e.g., a) and a′), the parameter ε multiplies a linear and nonlinear term, respectively. Discuss the differences in these two cases in the most general possible setting.

a) $$\begin{aligned} x &\mapsto -\frac{1}{2}x - y - xy^2, \\ y &\mapsto -\frac{1}{2}x + \varepsilon y + x^2, \end{aligned} \qquad (x, y) \in \mathbb{R}^2.$$

a′) $$\begin{aligned} x &\mapsto -\frac{1}{2}x - y - xy^2, \\ y &\mapsto -\frac{1}{2}y + \varepsilon y^2 + x^2. \end{aligned}$$

b) $$\begin{aligned} x &\mapsto x + 2y + x^3, \\ y &\mapsto 2x + y + \varepsilon y, \end{aligned} \qquad (x, y) \in \mathbb{R}^2.$$

b′) $$\begin{aligned} x &\mapsto x + 2y + x^3, \\ y &\mapsto 2x + y + \varepsilon y^2. \end{aligned}$$

c) $$\begin{aligned} x &\mapsto -x + y - xy^2, \\ y &\mapsto y + \varepsilon y + x^2 y, \end{aligned} \qquad (x, y) \in \mathbb{R}^2.$$

c′) $$\begin{aligned} x &\mapsto -x + y - xy^2, \\ y &\mapsto y + \varepsilon y^2 + x^2 y. \end{aligned}$$

d) $$\begin{aligned} x &\mapsto 2x + y, \\ y &\mapsto 2x + 3y + \varepsilon x + x^4, \end{aligned} \qquad (x, y) \in \mathbb{R}^2.$$

d′) $$\begin{aligned} x &\mapsto 2x + y + \varepsilon x^2, \\ y &\mapsto 2x + 3y + x^4. \end{aligned}$$

e) $$\begin{aligned} x &\mapsto x + \varepsilon y, \\ y &\mapsto x + 2y + y^2, \end{aligned} \qquad (x, y) \in \mathbb{R}^2.$$

e′) $$\begin{aligned} x &\mapsto x + \varepsilon y^2, \\ y &\mapsto x + 2y + y^2. \end{aligned}$$

f) $$\begin{aligned} x &\mapsto 2x + 3y, \\ y &\mapsto x + \varepsilon y + x^2 + xy^2, \end{aligned} \qquad (x, y) \in \mathbb{R}^2.$$

f′) $$\begin{aligned} x &\mapsto 2x + 3y, \\ y &\mapsto x + x^2 + \varepsilon y^2 + xy^2. \end{aligned}$$

g) $$\begin{aligned} x &\mapsto x - z^3, \\ y &\mapsto 2x - y + \varepsilon y, \\ z &\mapsto x + \frac{1}{2}z + x^3, \end{aligned} \qquad (x, y, z) \in \mathbb{R}^3.$$

g′) $$\begin{aligned} x &\mapsto x - z^3, \\ y &\mapsto 2x - y + \varepsilon y^2, \\ z &\mapsto x + \frac{1}{2}z + x^3. \end{aligned}$$

h)
$$\begin{aligned} x &\mapsto x + \varepsilon z^4, \\ y &\mapsto -x - 2y - x^3, \\ z &\mapsto y - \frac{1}{2} z + y^2, \end{aligned} \qquad (x, y, z) \in \mathbb{R}^3.$$

h′)
$$\begin{aligned} x &\mapsto x + \varepsilon x + z^4, \\ y &\mapsto -x - 2y - x^3, \\ z &\mapsto y - \frac{1}{2} z + y^2. \end{aligned}$$

i)
$$\begin{aligned} x &\mapsto y + \varepsilon x + x^2, \\ y &\mapsto y + xy, \end{aligned} \qquad (x, y) \in \mathbb{R}^2.$$

i′)
$$\begin{aligned} x &\mapsto y + x^2, \\ y &\mapsto y + xy + \varepsilon x^2. \end{aligned}$$

j)
$$\begin{aligned} x &\mapsto \varepsilon x + x^2, \\ y &\mapsto y + xy, \end{aligned} \qquad (x, y) \in \mathbb{R}^2.$$

j′)
$$\begin{aligned} x &\mapsto x^2 + \varepsilon y, \\ y &\mapsto y + xy. \end{aligned}$$

2.5 Prove that H_k is a linear vector space.

2.6 Suppose $h_k(x) \in H_k$ ($x \in \mathbb{R}^n$), and J is an $n \times n$ matrix of real numbers. Then prove that the maps

a) $h_k(x) \mapsto J h_k(x) - D h_k(x) J x \equiv L_J(h_k(x))$,
b) $h_k(x) \mapsto J h_k(x) - h_k(Jx) \equiv M_J(h_k(x))$,

are linear maps of H_k into H_k.

2.7 Argue that, for $y \in \mathbb{R}^n$ sufficiently small,

a) $(\mathrm{id} + D h_k(y))^{-1}$ exists

and

b) $(\mathrm{id} + D h_k(y))^{-1} = \mathrm{id} - D h_k(y) + \cdots$

for $h_k(y) \in H_k$. Similarly, show that for $y \in \mathbb{R}^n$ sufficiently small

c) $(\mathrm{id} + h_k)^{-1}(y)$ exists

and

d) $(\mathrm{id} + h_k)^{-1}(y) = (\mathrm{id} - h_k + \cdots)(y)$.

2.8 Compute a normal form for a map in the neighborhood of a fixed point having the linear part

$$\begin{pmatrix} 1 & 1 \\ 0 & 1 \end{pmatrix}$$

through second-order terms.

Compare the resulting normal form with the normal form of a vector field near a fixed point having linear part

$$\begin{pmatrix} 0 & 1 \\ 0 & 0 \end{pmatrix}$$

(see Example 2.2.2). Explain the results.

2.9 Consider a third-order autonomous vector field near a fixed point having linear part

$$\begin{pmatrix} 0 & -\omega & 0 \\ \omega & 0 & 0 \\ 0 & 0 & 0 \end{pmatrix}$$

with respect to the standard basis in $\mathbb{R}^3$. Show that in cylindrical coordinates a normal form is given by

$$\begin{aligned} \dot{r} &= a_1 rz + a_2 r^3 + a_3 rz^2 + \mathcal{O}(4), \\ \dot{z} &= b_1 r^2 + b_2 z^2 + b_3 r^2 z + b_4 z^3 + \mathcal{O}(4), \\ \dot{\theta} &= \omega + c_1 z + \mathcal{O}(2), \end{aligned}$$

where a_1, a_2, a_3, b_1, b_2, b_3, b_4, and c_1 are constants. (*Hint:* lump the two coordinates associated with the block

$$\begin{pmatrix} 0 & -\omega \\ \omega & 0 \end{pmatrix}$$

into a single complex coordinate.)

2.10 Consider a four-dimensional $\mathbf{C}^r$ (r as large as necessary) vector field having a fixed point where the matrix associated with the linearization is given by

$$\begin{pmatrix} 0 & -\omega_1 & 0 & 0 \\ \omega_1 & 0 & 0 & 0 \\ 0 & 0 & 0 & -\omega_2 \\ 0 & 0 & \omega_2 & 0 \end{pmatrix}.$$

Compute the normal form through third order. (*Hint:* use two complex variables.) You should find that certain "resonance" problems arise; namely, the normal form will depend on $m\omega_1 + n\omega_2 \neq 0$, $|m| + |n| \leq 4$. Give the normal form for the cases

a) $m\omega_1 + n\omega_2 = 0, \qquad |m| + |n| = 1..$

b) $m\omega_1 + n\omega_2 = 0, \qquad |m| + |n| = 2.$

c) $m\omega_1 + n\omega_2 = 0, \qquad |m| + |n| = 3.$

d) $m\omega_1 + n\omega_2 = 0, \qquad |m| + |n| = 4.$

e) $m\omega_1 + n\omega_2 \neq 0, \qquad |m| + |n| \leq 4.$

2.11 Consider the normal form for a map of $\mathbb{R}^2$ in the neighborhood of a fixed point where the eigenvalues of the matrix associated with the linearization, denoted λ_1 and λ_2, are complex conjugates, i.e., $\lambda_1 = \overline{\lambda}_2$, and have modulus one, i.e., $|\lambda_1| = |\lambda_2| \equiv |\lambda| = 1$ (cf. Example 2.2.4). Compute the normal form for the cases.

a) $\lambda = 1$.

b) $\lambda^2 = 1$.

c) $\lambda^3 = 1$.

d) $\lambda^4 = 1$.

2.12 Compute the normal form of a map of $\mathbb{R}^2$ in the neighborhood of a fixed point where the matrix associated with the linearization has the following form

a) $\begin{pmatrix} 1 & 1 \\ 0 & 1 \end{pmatrix}.$

b) $\begin{pmatrix} 1 & 0 \\ 0 & 1 \end{pmatrix}.$

c) $\begin{pmatrix} -1 & 1 \\ 0 & -1 \end{pmatrix}.$

d) $\begin{pmatrix} -1 & 0 \\ 0 & -1 \end{pmatrix}.$

Compare your normal forms with those obtained in parts a) and b) of Exercise 2.11.

2.13 Consider a $\mathbf{C}^r$ $(r \geq 2)$ vector field

$$\dot{x} = f(x, \mu), \qquad x \in \mathbb{R}^2, \quad \mu \in \mathbb{R}^1,$$

defined on a sufficiently large open set in $\mathbb{R}^2 \times \mathbb{R}^1$. Suppose that $(x, \mu) = (0, 0)$ is a fixed point of this vector field and that $D_x f(0, 0)$ has a pair of purely imaginary eigenvalues.

a) Show that there exists a curve of fixed points of the vector field, denoted $x(\mu)$, $x(0) = 0$, for μ sufficiently small.

b) By using this curve of fixed points as a parameter-dependent coordinate transformation, show that one can choose coordinates so that the origin in phase space remains a fixed point for μ sufficiently small.

2.14 Consider the $\mathbf{C}^r$ $(r \geq 1)$ vector field

$$\begin{aligned} \dot{x} &= Ax + f(x, y), \\ \dot{y} &= By + g(x, y), \end{aligned} \qquad (x, y) \in \mathbb{R}^n \times \mathbb{R}^m, \tag{E2.1}$$

where A is an $n \times n$ matrix, B is an $m \times m$ matrix, $f(0,0) = 0$, $g(0,0) = 0$, $Df(0,0) = 0$, and $Dg(0,0) = 0$. Let

$$\phi_t(x, y) \equiv \left(\phi_t^x(x, y), \phi_t^y(x, y)\right) \tag{E2.2}$$

denote the flow generated by (E2.1), which we will assume exists for all $t \in \mathbb{R}$, $(x, y) \in \mathbb{R}^n \times \mathbb{R}^m$.

Consider the $\mathbf{C}^r$ $(r \geq 1)$ function

$$\begin{aligned} h: U \subset \mathbb{R}^n &\longrightarrow \mathbb{R}^m, \\ x &\longmapsto h(x), \end{aligned}$$

where U is a compact set in $\mathbb{R}^n$. Then

$$M \equiv \text{graph } h = \left\{(x, y) \in \mathbb{R}^n \times \mathbb{R}^m \mid y = h(x), x \in U\right\}$$

is an n-dimensional surface in $\mathbb{R}^n \times \mathbb{R}^m$.

Heuristically, M is said to be *invariant* under the dynamics generated by (E2.1) if points in M remain on M under evolution by the flow generated by (E2.1).

We want to explore and quantify this idea further.

a) Argue that this definition of invariance is equivalent to having the vector field (E2.1) tangent to M. Show that this implies

$$Dh(x)[Ax + f(x, h(x))] - Bh(x) - g(x, h(x)) = 0. \tag{E2.3}$$

b) Argue that the following condition also implies invariance

$$\phi_t^y(x, y) = h(\phi_t^x(x, y)), \qquad (x, y) \in M. \tag{E2.4}$$

Show that (E2.3) can be obtained by differentiating (E2.4) with respect to t.

c) We have avoided a slight technical difficulty, namely, that M has a boundary. Describe the boundary of M, denoted ∂M. What precautions must be taken in order to define invariant manifolds with boundary? In particular, describe the vector field on ∂M. Must a) and b) above be modified in any way? (Note: the reader can find discussions of invariant manifolds with boundary in Carr [1981], Fenichel [1971], [1974], [1977], [1979], Henry [1981], or Wiggins [1988].)

The remainder of this problem will be somewhat more specialized. Namely, we will consider the stable and center manifolds of the fixed point $(x, y) = (0, 0)$. For this we must add the following two assumptions concerning the matrices A and B.

Assumption 1. The n eigenvalues of A have zero real parts.

Assumption 2. The m eigenvalues of B have negative real parts.

d) For the linearized vector field

$$\begin{aligned} \dot{x} &= Ax, \\ \dot{y} &= By, \end{aligned} \qquad (x, y) \in \mathbb{R}^n \times \mathbb{R}^m,$$

$y = 0$ is the center manifold and $x = 0$ is the stable manifold of the fixed point $(x, y) = (0, 0)$. We seek local center and stable manifolds of the origin as graphs over the respective linear invariant manifolds as follows

$$W^c_{\text{loc}}(0) = \{(x, y) \in \mathbb{R}^n \times \mathbb{R}^m \mid y = h(x); h(0) = 0, Dh(0) = 0\},$$

$$W^s_{\text{loc}}(0) = \{(x, y) \in \mathbb{R}^n \times \mathbb{R}^m \mid x = v(y); v(0) = 0, Dv(0) = 0\}.$$

Describe the geometrical and dynamical meaning of the conditions

$$\begin{aligned} h(0) = 0, Dh(0) = 0, \\ v(0) = 0, Dv(0) = 0. \end{aligned} \tag{E2.5}$$

Show that $h(x)$ and $v(x)$ are solutions of the following quasilinear partial differential equations

$$Dh(x)[Ax + f(x, h(x))] - Bh(x) - g(x, h(x)) = 0, \tag{E2.6a}$$

$$Dv(y)[By + g(v(y), y)] - Av(y) - f(v(y), y) = 0. \tag{E2.6b}$$

e) Equations (E2.6a) and (E2.6b) can be solved by the method of characteristics. Describe the characteristic equations for (E2.6a) and (E2.6b). Does the initial data (E2.5) pose any problems in applying the method of characteristics?

f) Using (E2.6a) and (E2.6b), the existence of center and stable manifolds of fixed points is reduced to an existence problem for solutions of partial differential equations. Discuss this issue. In particular, since we are interested in *local* stable and unstable manifolds, "small" solutions of (E2.6a) and (E2.6b) are sufficient. This should indicate that some type of perturbative or iterative procedure could be used to solve (E2.6a) and (E2.6b). Discuss these possibilities. (*Hint:* you may want to consult the two classic papers of Moser [1966a], [1966b] on this subject.)

g) Suppose one finds a solution, $v(y)$, of (E2.6b). How can we then conclude (as we know must be the case) that solutions on the graph of $v(y)$ approach the origin at an exponential rate? Can we say anything about this exponential rate? Along these same lines, can we say anything about the growth or decay rate of solutions on the graph of $h(x)$?

h) Based on general properties of solutions of quasilinear partial differential equations, what can one conclude about the differentiability of $W^s_{\text{loc}}(0)$ and $W^c_{\text{loc}}(0)$? (*Hint:* you may again want to consult Moser [1966a], [1966b].)

i) Stable manifolds of fixed points are unique, but center manifolds may not be unique. Discuss this in the context of uniqueness (or nonuniqueness) or solutions of (E2.6a) and (E2.6b). Is this issue associated with properties of the matrices A and B?

j) Partial differential equations such as (E2.6) may exhibit shocks. Can this phenomenon arise in this context? If so, how would it be interpreted in terms of the dynamics of (E2.1)?

2.15 Consider the $\mathbf{C}^r$ $(r \geq 1)$ map

$$\begin{aligned} \bar{x} &= Ax + f(x,y), \\ \bar{y} &= By + g(x,y), \end{aligned} \qquad (x,y) \in \mathbb{R}^n \times \mathbb{R}^m, \qquad \text{(E2.7)}$$

with $f(0,0) = 0$, $Df(0,0) = 0$, $g(0,0) = 0$, and $Dg(0,0) = 0$ and where A is an $n \times n$ matrix whose eigenvalues all have modulus one and B is an $m \times m$ matrix whose eigenvalues all have modulus less than one. We are interested in the orbit structure near the fixed point $(x,y) = (0,0)$.

The linearized map is given by

$$\begin{aligned} \bar{x} &= Ax, \\ \bar{y} &= By, \end{aligned}$$

for which $y = 0$ is the center manifold and $x = 0$ is the stable manifold of the fixed point $(x,y) = (0,0)$. We seek local center and stable manifolds of the origin as graphs over the respective linear invariant manifolds as follows

$$W^c_{\text{loc}}(0) = \big\{(x,y) \in \mathbb{R}^n \times \mathbb{R}^m \mid y = h(x); h(0) = 0, Dh(0) = 0\big\},$$

$$W^s_{\text{loc}}(0) = \big\{(x,y) \in \mathbb{R}^n \times \mathbb{R}^m \mid x = v(y); v(0) = 0, Dv(0) = 0\big\}.$$

a) Describe the geometrical and dynamical meaning of the conditions

$$\begin{aligned} h(0) = 0, \ Dh(0) = 0, \\ v(0) = 0, \ Dv(0) = 0. \end{aligned} \tag{E2.8}$$

b) Show that the condition of invariance implies that $h(x)$ and $v(y)$ satisfy the following functional equations

$$Bh(x) + g(x, h(x)) = h(Ax + f(x, h(x))), \tag{E2.9a}$$

$$Av(y) + f(v(y), y) = v(By + g(v(y), y)). \tag{E2.9b}$$

c) (E2.9a) and (E2.9b) can be rewritten as

$$h(x) = B^{-1}[h(Ax + f(x, h(x))) - g(x, h(x))] \equiv T^c(h(x)), \tag{E2.10a}$$

$$v(y) = A^{-1}[v(By + g(v(y), y)) - f(v(y), y)] \equiv T^s(v(y)), \tag{E2.10b}$$

provided B^{-1} and A^{-1} exist (what can you say about this?). T^c and T^s are nonlinear operators defined on an appropriately chosen function space. Thus, from (E2.10a) and (E2.10b) the existence of local center and stable manifolds of $(x, y) = (0, 0)$ is equivalent to the existence of a fixed point for T^c and T^s, respectively. T^c and T^s are referred to as *graph transforms.*

Set up the problem so that a fixed point of T^s can be found using the contraction mapping principle. In this context address the question of the uniqueness of $W^s_{\text{loc}}(0)$ as well as the decay rates of orbits on $W^s_{\text{loc}}(0)$. Can the same approach be taken with T^c to prove the existence of $W^c_{\text{loc}}(0)$? What about questions of smoothness of the invariant manifold? (*Hint:* see Shub [1987] for help.)

2.16 Consider the $\mathbf{C}^r$ $(r \geq 1)$ vector field

$$\dot{x} = f(x), \qquad x \in \mathbb{R}^n. \tag{E2.11}$$

We denote the flow generated by this vector field by $\phi_t(x)$, and we assume that it exists for all $t \in \mathbb{R}$. The "time one" map generated by this flow is denoted $\phi_1(x)$.

a) Suppose M is an invariant manifold of the vector field (see Exercise 2.14). Is M also invariant under $\phi_1(x)$?

b) Suppose M is an invariant manifold for the map $\phi_1(x)$ (see Exercise 2.15). Is M also invariant under the flow generated by the vector field (E2.11)?

2.17 Consider the following planar Hamiltonian system

$$\begin{aligned} \dot{x} &= \frac{\partial H}{\partial y}(x,y), \\ \dot{y} &= -\frac{\partial H}{\partial x}(x,y), \end{aligned} \qquad (x,y) \in \mathbb{R}^2, \tag{E2.12}$$

where $H(x,y)$ is a $\mathbf{C}^{r+1}$ function (with r as large as is needed). Suppose (E2.12) has a fixed point at $(x,y)=(0,0)$. Our goal is to transform (E2.12) to normal form and maintain the Hamiltonian structure at each step. We might think of two methods to accomplish this goal.

Method 1. Taylor expand (E2.12) and use the method for vector fields described in Section 2.2a. The Hamiltonian structure can be enforced at each order of the Taylor expansion.

Method 2. Method 1 does not take advantage of the fact that the vector field is derived from a single scalar function. In this method simplify the Hamiltonian function directly as described in, e.g., Arnold [1978] or van der Meer [1985].

Suppose the matrix associated with the linearization of (E2.12) about the fixed point assumes the form

a) $\begin{pmatrix} 0 & -\omega \\ \omega & 0 \end{pmatrix}, \qquad \omega > 0,$

b) $\begin{pmatrix} \lambda & 0 \\ 0 & -\lambda \end{pmatrix}, \qquad \lambda > 0,$

where ω and λ are real numbers.

Using both Methods 1 and 2, compute the normal forms for Case a) and Case b).

2.18 Consider the Hamiltonian vector field

$$\dot{x} = JDH(x), \qquad x \in \mathbb{R}^{2n}, \tag{E2.13}$$

where $H(x)$ is a $\mathbf{C}^{r+1}$ function (with r as large as is needed) and

$$J = \begin{pmatrix} 0 & -\mathrm{id} \\ \mathrm{id} & 0 \end{pmatrix},$$

where "id" denotes the $n \times n$ identity matrix. Suppose (E2.13) has a fixed point at $x = 0$ and that the matrix $JD^2H(0)$ has $2n-2$ eigenvalues having nonzero real parts and two pure imaginary eigenvalues. Then $x = 0$ has a two-dimensional center manifold, an $(n-1)$-dimensional stable manifold, and an $(n-1)$-dimensional unstable manifold. Is the reduced vector field on the center manifold also Hamiltonian? What about the vector field restricted to the stable manifold? The unstable manifold? What can you conclude about the dynamics on the center manifold near $x = 0$?

Compare this with the nonHamiltonian case. (*Hint:* to begin, you might want to consider the simplest case, $n = 2$.)

2.19 Consider the $\mathbf{C}^r$ $(r \geq 1)$ map

$$x \mapsto f(x), \qquad x \in \mathbb{R}^n. \tag{E2.14}$$

Suppose that the map has a fixed point at $x = x_0$, i.e.,

$$x_0 = f(x_0).$$

Next consider the vector field

$$\dot{x} = f(x) - x. \tag{E2.15}$$

Clearly (E2.15) has a fixed point, and $x = x_0$. What can you determine about the orbit structure near the fixed point of the map (E2.14) based on knowledge of the orbit structure near the fixed point $x = x_0$ of the vector field (E2.15)?

2.20 Consider the $\mathbf{C}^r$ map

$$f: \mathbb{R}^1 \to \mathbb{R}^1$$

and denote the Taylor expansion of f by

$$f(x) = a_0 + a_1 x + \cdots + a_{r-1} x^{r-1} + \mathcal{O}(|x|^r).$$

Suppose f is identically zero. Then show that $a_i = 0$, $i = 0, \ldots, r-1$. Does the same result hold for the $\mathbf{C}^r$ map

$$f: \mathbb{R}^n \to \mathbb{R}^n, \qquad n > 1?$$

3

Local Bifurcations

In this chapter we study local bifurcations of vector fields and maps. By the term "local" we mean bifurcations occurring in a neighborhood of a fixed point. The term "bifurcation of a fixed point" will be defined after we have considered several examples. We begin by studying bifurcations of fixed points of vector fields.

3.1 Bifurcation of Fixed Points of Vector Fields

Consider the parameterized vector field

$$\dot{y} = g(y, \lambda), \qquad y \in \mathbb{R}^n, \quad \lambda \in \mathbb{R}^p, \tag{3.1.1}$$

where g is a $\mathbf{C}^r$ function on some open set in $\mathbb{R}^n \times \mathbb{R}^p$. The degree of differentiability will be determined by our need to Taylor expand (3.1.1). Usually $\mathbf{C}^5$ will be sufficient.

Suppose (3.1.1) has a fixed point at $(y, \lambda) = (y_0, \lambda_0)$, i.e.,

$$g(y_0, \lambda_0) = 0. \tag{3.1.2}$$

Two questions immediately arise.

1. Is the fixed point stable or unstable?
2. How is the stability or instability affected as λ is varied?

To answer Question 1, the first step to take is to examine the linear vector field obtained by linearizing (3.1.1) about the fixed point $(y, \lambda) = (y_0, \lambda_0)$. This linear vector field is given by

$$\dot{\xi} = D_y g(y_0, \lambda_0)\xi, \qquad \xi \in \mathbb{R}^n. \tag{3.1.3}$$

If the fixed point is hyperbolic (i.e., none of the eigenvalues of $D_y g(y_0, \lambda_0)$ lie on the imaginary axis), we know that the stability of (y_0, λ_0) in (3.1.1) is determined by the linear equation (3.1.3) (cf. Section 1.1A). This also enables us to answer Question 2, because since hyperbolic fixed points are structurally stable (cf. Section 1.2C), varying λ slightly does not change the nature of the stability of the fixed point. This should be clear intuitively, but let us belabor the point slightly.

We know that

$$g(y_0, \lambda_0) = 0, \tag{3.1.4}$$

and that

$$D_y g(y_0, \lambda_0) \tag{3.1.5}$$

has no eigenvalues on the imaginary axis. Therefore, $D_y g(y_0, \lambda_0)$ is invertible. By the implicit function theorem, there thus exists a *unique* $\mathbf{C}^r$ function, $y(\lambda)$, such that

$$g(y(\lambda), \lambda) = 0 \tag{3.1.6}$$

for λ sufficiently close to λ_0 with

$$y(\lambda_0) = y_0. \tag{3.1.7}$$

Now, by continuity of the eigenvalues with respect to parameters, for λ sufficiently close to λ_0,

$$D_y g(y(\lambda), \lambda) \tag{3.1.8}$$

has no eigenvalues on the imaginary axis. Therefore, for λ sufficiently close to λ_0, the hyperbolic fixed point (y_0, λ_0) of (3.1.1) persists and its stability type remains unchanged. To summarize, in a neighborhood of λ_0 an isolated fixed point of (3.1.1) persists and always has the same stability type.

The real fun starts when the fixed point (y_0, λ_0) of (3.1.1) is not hyperbolic, i.e., when $D_y g(y_0, \lambda_0)$ has some eigenvalues on the imaginary axis. In this case, for λ very close to λ_0 (and for y close to y_0), radically new dynamical behavior can occur. For example, fixed points can be created or destroyed and time-dependent behavior such as periodic, quasiperiodic, or even chaotic dynamics can be created. In a certain sense (to be clarified later), the more eigenvalues on the imaginary axis, the more exotic the dynamics will be.

We will begin our study by considering the simplest way in which $D_y g(y_0, \lambda_0)$ can be nonhyperbolic. This is the case where $D_y g(y_0, \lambda_0)$ has a single zero eigenvalue with the remaining eigenvalues having nonzero real parts. The question we ask in this situation is what is the nature of this nonhyperbolic fixed point for λ close to λ_0? It is under these circumstances where the real power of the center manifold theory becomes apparent, since we know that this question can be answered by studying the vector field (3.1.1) restricted to the associated center manifold (cf. Section 2.1). In this case the vector field on the center manifold will be a p-parameter family of one-dimensional vector fields. This represents a vast simplification of (3.1.1).

3.1A A Zero Eigenvalue

Suppose that $D_y g(y_0, \lambda_0)$ has a single zero eigenvalue with the remaining eigenvalues having nonzero real parts; then the orbit structure near (y_0, λ_0)

is determined by the associated center manifold equation, which we write as

$$\dot{x} = f(x, \mu), \qquad x \in \mathbb{R}^1, \quad \mu \in \mathbb{R}^p, \tag{3.1.9}$$

where $\mu = \lambda - \lambda_0$. Furthermore, we know that (3.1.9) must satisfy

$$f(0,0) = 0, \tag{3.1.10}$$

$$\frac{\partial f}{\partial x}(0,0) = 0. \tag{3.1.11}$$

Equation (3.1.10) is simply the fixed point condition and (3.1.11) is the zero eigenvalue condition. We remark that (3.1.9) is $\mathbf{C}^r$ if (3.1.1) is $\mathbf{C}^r$. Let us begin by studying a few specific examples. In these examples we will assume

$$\mu \in \mathbb{R}^1.$$

If there are more parameters in the problem (i.e., $\mu \in \mathbb{R}^p$, $p > 1$), we will consider all, except one, as fixed. Later we will consider more carefully the role played by the number of parameters in the problem. We remark also that we have not yet precisely defined what we mean by the term "bifurcation." We will consider this after the following series of examples.

i) Examples

Example 3.1.1 Consider the vector field

$$\dot{x} = f(x, \mu) = \mu - x^2, \qquad x \in \mathbb{R}^1, \quad \mu \in \mathbb{R}^1. \tag{3.1.12}$$

It is easy to verify that

$$f(0,0) = 0 \tag{3.1.13}$$

and

$$\frac{\partial f}{\partial x}(0,0) = 0, \tag{3.1.14}$$

but in this example we can determine much more. The set of all fixed points of (3.1.12) is given by

$$\mu - x^2 = 0$$

or

$$\mu = x^2. \tag{3.1.15}$$

This represents a parabola in the $\mu - x$ plane as shown in Figure 3.1.1. In the figure the arrows along the vertical lines represent the flow generated by (3.1.12) along the x-direction. Thus, for $\mu < 0$, (3.1.12) has no fixed points, and the vector field is decreasing in x. For $\mu > 0$, (3.1.12) has two fixed points. A simple linear stability analysis shows that one of the fixed points is stable (represented by the solid branch of the parabola), and the other fixed point is unstable (represented by the broken branch of the parabola).

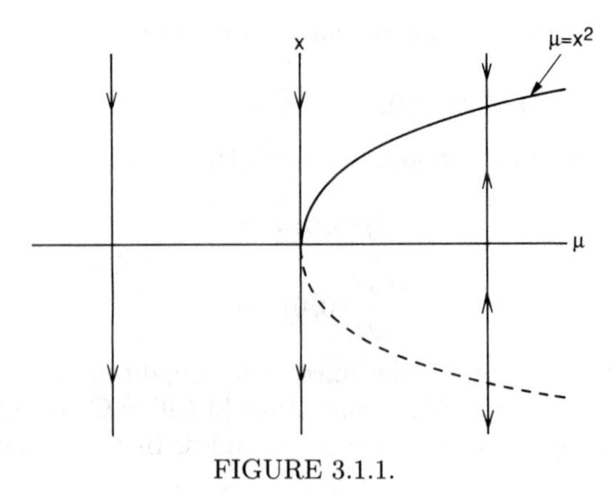

FIGURE 3.1.1.

However, we hope that it is obvious to the reader that, given a $\mathbf{C}^r$ $(r \geq 1)$ vector field on $\mathbb{R}^1$ having only two *hyperbolic* fixed points, one must be stable and the other unstable.

This is an example of *bifurcation.* We refer to $(x, \mu) = (0, 0)$ as a *bifurcation point* and the parameter value $\mu = 0$ as a *bifurcation value.*

Figure 3.1.1 is referred to as a *bifurcation diagram.* This particular type of bifurcation (i.e., where on one side of a parameter value there are no fixed points and on the other side there are two fixed points) is referred to as a *saddle-node bifurcation.* Later on we will worry about seeking precise conditions on the vector field on the center manifold that define the saddle-node bifurcation unambiguously.

EXAMPLE 3.1.2 Consider the vector field

$$\dot{x} = f(x, \mu) = \mu x - x^2, \qquad x \in \mathbb{R}^1, \quad \mu \in \mathbb{R}^1. \tag{3.1.16}$$

It is easy to verify that

$$f(0, 0) = 0 \tag{3.1.17}$$

and

$$\frac{\partial f(0, 0)}{\partial x} = 0. \tag{3.1.18}$$

Moreover, the fixed points of (3.1.16) are given by

$$x = 0 \tag{3.1.19}$$

and

$$x = \mu \tag{3.1.20}$$

and are plotted in Figure 3.1.2. Hence, for $\mu < 0$, there are two fixed points; $x = 0$ is stable and $x = \mu$ is unstable. These two fixed points coalesce at $\mu = 0$ and, for $\mu > 0$, $x = 0$ is unstable and $x = \mu$ is stable. Thus, an

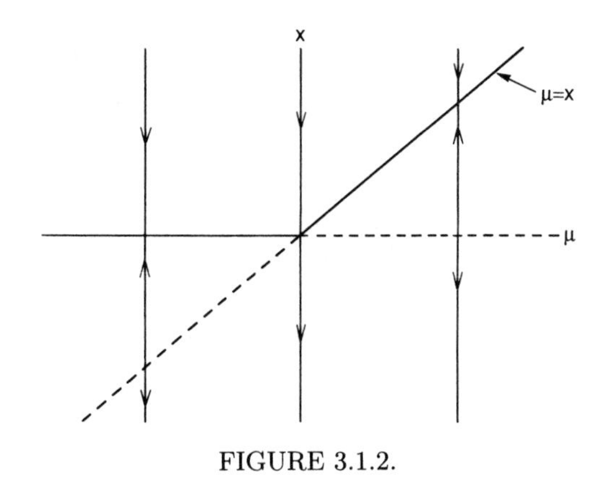

FIGURE 3.1.2.

exchange of stability has occurred at $\mu = 0$. This type of bifurcation is called a *transcritical bifurcation.*

EXAMPLE 3.1.3 Consider the vector field

$$\dot{x} = f(x,\mu) = \mu x - x^3, \qquad x \in \mathbb{R}^1, \quad \mu \in \mathbb{R}^1. \tag{3.1.21}$$

It is clear that we have

$$f(0,0) = 0, \tag{3.1.22}$$

$$\frac{\partial f}{\partial x}(0,0) = 0. \tag{3.1.23}$$

Moreover, the fixed points of (3.1.21) are given by

$$x = 0 \tag{3.1.24}$$

and

$$x^2 = \mu \tag{3.1.25}$$

and are plotted in Figure 3.1.3. Hence, for $\mu < 0$, there is one fixed point, $x = 0$, which is stable. For $\mu > 0$, $x = 0$ is still a fixed point, but two new fixed points have been created at $\mu = 0$ and are given by $x^2 = \mu$. In the process, $x = 0$ has become unstable for $\mu > 0$, with the other two fixed points stable. This type of bifurcation is called a *pitchfork bifurcation.*

EXAMPLE 3.1.4 Consider the vector field

$$\dot{x} = f(x,\mu) = \mu - x^3, \qquad x \in \mathbb{R}^1, \quad \mu \in \mathbb{R}^1. \tag{3.1.26}$$

It is trivial to verify that

$$f(0,0) = 0 \tag{3.1.27}$$

and

$$\frac{\partial f}{\partial x}(0,0) = 0. \tag{3.1.28}$$

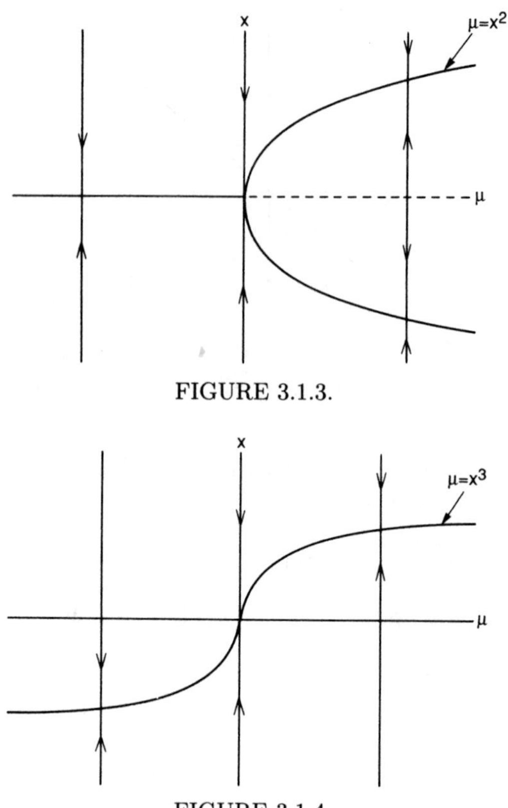

FIGURE 3.1.3.

FIGURE 3.1.4.

Moreover, all fixed points of (3.1.26) are given by

$$\mu = x^3 \tag{3.1.29}$$

and are shown in Figure 3.1.4. However in this example, despite (3.1.27) and (3.1.28), the dynamics of (3.1.26) are qualitatively the same for $\mu > 0$ and $\mu < 0$. Namely, (3.1.26) possesses a unique, stable fixed point.

ii) What Is A "Bifurcation of a Fixed Point"?

The term "bifurcation" is extremely general. We will begin to learn its uses in dynamical systems by understanding its use in describing the orbit structure near nonhyperbolic fixed points. Let us consider what we learned from the previous examples.

In all four examples we had

$$f(0,0) = 0$$

and

$$\frac{\partial f}{\partial x}(0,0) = 0,$$

and yet the orbit structure near $\mu = 0$ was different in all four cases. Hence, knowing that a fixed point has a zero eigenvalue for $\mu = 0$ is not sufficient to determine the orbit structure for λ near zero. Let us consider each example individually.

1. (*Example 3.1.1*). In this example a *unique* curve (or branch) of fixed points passed through the orgin. Moreover, the curve lay entirely on one side of $\mu = 0$ in the $\mu - x$ plane.

2. (*Example 3.1.2*). In this example two curves of fixed points intersected at the origin in the $\mu - x$ plane. Both curves existed on either side of $\mu = 0$. However, the stability of the fixed point along a given curve changed on passing through $\mu = 0$.

3. (*Example 3.1.3*). In this example two curves of fixed points intersected at the origin in the $\mu - x$ plane. Only one curve ($x = 0$) existed on both sides of $\mu = 0$; however, its stability changed on passing through $\mu = 0$. The other curve of fixed points lay entirely to one side of $\mu = 0$ and had a stability type that was the opposite of $x = 0$ for $\mu > 0$.

4. (*Example 3.1.4*). This example had a unique curve of fixed points passing through the origin in the $\mu - x$ plane and existing on both sides of $\mu = 0$. Moreover, all fixed points along the curve had the same stability type. Hence, despite the fact that the fixed point $(x, \mu) = (0, 0)$ was nonhyperbolic, the orbit structure was qualitatively the same for all μ.

We want to apply the term "bifurcation" to Examples 3.1.1, 3.1.2, and 3.1.3 but not to Example 3.1.4 to describe the change in orbit structure as μ passes through zero. We are therefore led to the following definition.

DEFINITION 3.1.1 A fixed point $(x, \mu) = (0, 0)$ of a one-parameter family of one-dimensional vector fields is said to undergo a *bifurcation* at $\mu = 0$ if the flow for μ near zero and x near zero is *not* qualitatively the same as the flow near $x = 0$ at $\mu = 0$.

Several remarks are now in order concerning this definition.

Remark 1. The phrase "qualitatively the same" is a bit vague. It can be made precise by substituting the term "$\mathbf{C}^0$-equivalent" (cf. Section 2.2D), and this is perfectly adequate for the study of the bifurcation of fixed points of *one-dimensional* vector fields. However, we will see that as we explore higher dimensional phase spaces and global bifurcations, how to make mathematically precise the statement "two dynamical systems have qualitatively the same dynamics" becomes more and more ambiguous.

Remark 2. Practically speaking, a fixed point (x_0, μ_0) of a one-dimensional vector field is a bifurcation point if either more than one curve of fixed points passes through (x_0, μ_0) in the $\mu - x$ plane or if only one curve of fixed points passes (x_0, μ_0) in the $\mu - x$ plane; then it (locally) lies entirely on one side of the line $\mu = \mu_0$ in the $\mu - x$ plane.

Remark 3. It should be clear from Example 3.1.4 that the condition that a fixed point is nonhyperbolic is a necessary but not sufficient condition for bifurcation to occur in one-parameter families of vector fields.

We next turn to deriving general conditions on one-parameter families of one-dimensional vector fields which exhibit bifurcations exactly as in Examples 3.1.1, 3.1.2, and 3.1.3.

iii) The Saddle-Node Bifurcation

We now want to derive conditions under which a general one-parameter family of one-dimensional vector fields will undergo a saddle-node bifurcation exactly as in Example 3.1.1. These conditions will involve derivatives of the vector field evaluated at the bifurcation point and are obtained by a consideration of the geometry of the curve of fixed points in the $\mu - x$ plane in a neighborhood of the bifurcation point.

Let us recall Example 3.1.1. In this example a *unique* curve of fixed points, parameterized by x, passed through $(\mu, x) = (0, 0)$. We denote the curve of fixed points by $\mu(x)$. The curve of fixed points satisfied two properties.

1. It was tangent to the line $\mu = 0$ at $x = 0$, i.e.,

$$\frac{d\mu}{dx}(0) = 0. \tag{3.1.30}$$

2. It lay entirely to one side of $\mu = 0$. Locally, this will be satisfied if we have

$$\frac{d^2\mu}{dx^2}(0) \neq 0. \tag{3.1.31}$$

Now let us consider a general, one-parameter family of one-dimensional vector fields.

$$\dot{x} = f(x, \mu), \qquad x \in \mathbb{R}^1, \quad \mu \in \mathbb{R}^1. \tag{3.1.32}$$

Suppose (3.1.32) has a fixed point at $(x, \mu) = (0, 0)$, i.e.,

$$f(0, 0) = 0. \tag{3.1.33}$$

Furthermore, suppose that the fixed point is not hyperbolic, i.e.,

$$\frac{\partial f}{\partial x}(0, 0) = 0. \tag{3.1.34}$$

Now, if we have

$$\frac{\partial f}{\partial \mu}(0,0) \neq 0, \tag{3.1.35}$$

then, by the implicit function theorem, there exists a unique function

$$\mu = \mu(x), \qquad \mu(0) = 0 \tag{3.1.36}$$

defined for x sufficiently small such that $f(x, \mu(x)) = 0$. (Note: the reader should check that (3.1.35) holds in Example 3.1.1.) Now we want to derive conditions in terms of derivatives of f evaluated at $(\mu, x) = (0,0)$ so that we have

$$\frac{d\mu}{dx}(0) = 0, \tag{3.1.37}$$

$$\frac{d^2\mu}{dx^2}(0) \neq 0. \tag{3.1.38}$$

Equations (3.1.37) and (3.1.38), along with (3.1.33), (3.1.34), and (3.1.35), imply that $(\mu, x) = (0,0)$ is a bifurcation point at which a saddle-node bifurcation occurs.

We can derive expressions for (3.1.37) and (3.1.38) in terms of derivatives of f at the bifurcation point by implicitly differentiating f along the curve of fixed points.

Using (3.1.35), we have

$$f(x, \mu(x)) = 0. \tag{3.1.39}$$

Differentiating (3.1.39) with respect to x gives

$$\frac{df}{dx}(x, \mu(x)) = 0 = \frac{\partial f}{\partial x}(x, \mu(x)) + \frac{\partial f}{\partial \mu}(x, \mu(x))\frac{d\mu}{dx}(x). \tag{3.1.40}$$

Evaluating (3.1.40) at $(\mu, x) = (0,0)$, we obtain

$$\frac{d\mu}{dx}(0) = \frac{-\frac{\partial f}{\partial x}(0,0)}{\frac{\partial f}{\partial \mu}(0,0)}; \tag{3.1.41}$$

thus we see that (3.1.34) and (3.1.35) imply that

$$\frac{d\mu}{dx}(0) = 0, \tag{3.1.42}$$

i.e., the curve of fixed points is tangent to the line $\mu = 0$ at $x = 0$.

Next, let us differentiate (3.1.40) once more with respect to x to obtain

$$\begin{aligned} \frac{d^2 f}{dx^2}(x, \mu(x)) = 0 = \frac{\partial^2 f}{\partial x^2}(x, \mu(x)) &+ 2\frac{\partial^2 f}{\partial x \partial \mu}(x, \mu(x))\frac{d\mu}{dx}(x) \\ &+ \frac{\partial^2 f}{\partial \mu^2}(x, \mu(x))\left(\frac{d\mu}{dx}(x)\right)^2 \\ &+ \frac{\partial f}{\partial \mu}(\mu, \mu(x))\frac{d^2\mu}{dx^2}(x). \end{aligned} \tag{3.1.43}$$

Evaluating (3.1.43) at $(\mu, x) = (0,0)$ and using (3.1.41) gives

$$\frac{\partial^2 f}{\partial x^2}(0,0) + \frac{\partial f}{\partial \mu}(0,0)\frac{d^2\mu}{dx^2}(0) = 0$$

or

$$\frac{d^2\mu}{dx^2}(0) = \frac{-\dfrac{\partial^2 f}{\partial x^2}(0,0)}{\dfrac{\partial f}{\partial \mu}(0,0)}. \tag{3.1.44}$$

Hence, (3.1.44) is nonzero provided we have

$$\frac{\partial^2 f}{\partial x^2}(0,0) \neq 0. \tag{3.1.45}$$

Let us summarize. In order for (3.1.32) to undergo a saddle-node bifurcation we must have

$$\left.\begin{array}{l} f(0,0) = 0 \\ \dfrac{\partial f}{\partial x}(0,0) = 0 \end{array}\right\} \quad \text{nonhyperbolic fixed point} \tag{3.1.46}$$

and

$$\frac{\partial f}{\partial \mu}(0,0) \neq 0, \tag{3.1.47}$$

$$\frac{\partial^2 f}{\partial x^2}(0,0) \neq 0. \tag{3.1.48}$$

Equation (3.1.47) implies that a unique curve of fixed points passes through $(\mu, x) = (0,0)$, and (3.1.48) implies that the curve lies locally on one side of $\mu = 0$. It should be clear that the sign of (3.1.44) determines on which side of $\mu = 0$ the curve lies. In Figure 3.1.5 we show both cases without indicating stability and leave it as an exercise for the reader to verify the stability types of the different branches of fixed points emanating from the bifurcation point (see Exercise 3.2).

Let us end our discussion of the saddle-node bifurcation with the following remark. Consider a general one-parameter family of one-dimensional vector fields having a nonhyperbolic fixed point at $(x, \mu) = (0,0)$. The Taylor expansion of this vector field is given as follows

$$f(x,\mu) = a_0\mu + a_1 x^2 + a_2\mu x + a_3\mu^2 + \mathcal{O}(3). \tag{3.1.49}$$

Our computations show that the dynamics of (3.1.49) near $(\mu, x) = (0,0)$ are qualitatively the same as one of the following vector fields

$$\dot{x} = \mu \pm x^2. \tag{3.1.50}$$

Hence, (3.1.50) can be viewed as the *normal form* for saddle-node bifurcations.

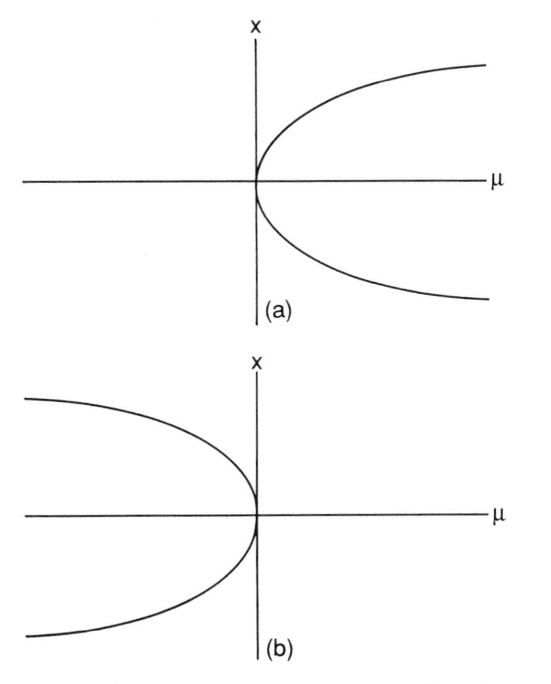

FIGURE 3.1.5. a) $\left(-\frac{\partial^2 f}{\partial x^2}(0,0)/\frac{\partial f}{\partial \mu}(0,0)\right) > 0$; b) $\left(-\frac{\partial^2 f}{\partial x^2}(0,0)/\frac{\partial f}{\partial \mu}(0,0)\right) < 0$.

This brings up another important point. In applying the method of normal forms there is always the question of truncation of the normal form; namely, how are the dynamics of the normal form including only the $\mathcal{O}(k)$ terms modified when the higher order terms are included? We see that, in the study of the saddle-node bifurcation, all terms of $\mathcal{O}(3)$ and higher could be neglected and the dynamics would not be qualitatively changed. The implicit function theorem was the tool that enabled us to verify this fact.

iv) The Transcritical Bifurcation

We want to follow the same strategy as in our discussion and derivation of general conditions for the saddle-node bifurcation given in the previous section, namely, to use the implicit function theorem to characterize the geometry of the curves of fixed points passing through the bifurcation point in terms of derivatives of the vector field evaluated at the bifurcation point.

For the example of transcritical bifurcation discussed in Example 3.1.2, the orbit structure near the bifurcation point was characterized as follows.

1. Two curves of fixed points passed through $(x, \mu) = (0, 0)$, one given by $x = \mu$, the other by $x = 0$.

2. Both curves of fixed points existed on both sides of $\mu = 0$.

3. The stability along each curve of fixed points changed on passing through $\mu = 0$.

Using these three points as a guide, let us consider a general one-parameter family of one-dimensional vector fields

$$\dot{x} = f(x, \mu), \qquad x \in \mathbb{R}^1, \quad \mu \in \mathbb{R}^1. \tag{3.1.51}$$

We assume that at $(x, \mu) = (0, 0)$, (3.1.51) has a nonhyperbolic fixed point, i.e.,

$$f(0, 0) = 0 \tag{3.1.52}$$

and

$$\frac{\partial f}{\partial x}(0, 0) = 0. \tag{3.1.53}$$

Now, in Example 3.1.2 we had two curves of fixed points passing through $(\mu, x) = (0, 0)$. In order for this to occur it is necessary to have

$$\frac{\partial f}{\partial \mu}(0, 0) = 0, \tag{3.1.54}$$

or else, by the implicit function theorem, only one curve of fixed points could pass through the origin.

Equation (3.1.54) presents a problem if we wish to proceed as in the case of the saddle-node bifurcation; in that situation we used the condition $\frac{\partial f}{\partial \mu}(0, 0) \neq 0$ in order to conclude that a unique curve of fixed points, $\mu(x)$, passed through the bifurcation point. We then evaluated the vector field on the curve of fixed points and used implicit differentiation to derive local characteristics of the geometry of the curve of fixed points based on properties of the derivatives of the vector field evaluated at the bifurcation point. However, if we use Example 3.1.2 as a guide, we can extricate ourselves from this difficulty.

In Example 3.1.2, $x = 0$ was a curve of fixed points passing through the bifurcation point. We will *require* that to be the case for (3.1.51), so that (3.1.51) has the form

$$\dot{x} = f(x, \mu) = xF(x, \mu), \qquad x \in \mathbb{R}^1, \quad \mu \in \mathbb{R}^1, \tag{3.1.55}$$

where, by definition, we have

$$F(x, \mu) \equiv \left\{ \begin{array}{ll} \frac{f(x,\mu)}{x}, & x \neq 0 \\ \frac{\partial f}{\partial x}(0, \mu), & x = 0 \end{array} \right\}. \tag{3.1.56}$$

Since $x = 0$ is a curve of fixed points for (3.1.55), in order to obtain an additional curve of fixed points passing through $(\mu, x) = (0, 0)$ we need to seek conditions on F whereby F has a curve of zeros passing through

$(\mu, x) = (0,0)$ (that is not given by $x = 0$). These conditions will be in terms of derivatives of F which, using (3.1.56), can be expressed as derivatives of f.

Using (3.1.56), it is easy to verify the following

$$F(0,0) = 0, \tag{3.1.57}$$

$$\frac{\partial F}{\partial x}(0,0) = \frac{\partial^2 f}{\partial x^2}(0,0), \tag{3.1.58}$$

$$\frac{\partial^2 F}{\partial x^2}(0,0) = \frac{\partial^3 f}{\partial x^3}(0,0), \tag{3.1.59}$$

and (most importantly)

$$\frac{\partial F}{\partial \mu}(0,0) = \frac{\partial^2 f}{\partial x \partial \mu}(0,0). \tag{3.1.60}$$

Now let us assume that (3.1.60) is *not* zero; then by the implicit function theorem there exists a function, $\mu(x)$, defined for x sufficiently small, such that

$$F(x, \mu(x)) = 0. \tag{3.1.61}$$

Clearly, $\mu(x)$ is a curve of fixed points of (3.1.55). In order for $\mu(x)$ to not coincide with $x = 0$ and to exist on both sides of $\mu = 0$, we must require that

$$0 < \left| \frac{d\mu}{dx}(0) \right| < \infty.$$

Implicitly differentiating (3.1.61) exactly as in the case of the saddle-node bifurcation we obtain

$$\frac{d\mu}{dx}(0) = \frac{-\frac{\partial F}{\partial x}(0,0)}{\frac{\partial F}{\partial \mu}(0,0)}. \tag{3.1.62}$$

Using (3.1.57), (3.1.58), (3.1.59), and (3.1.60), (3.1.62) becomes

$$\frac{d\mu}{dx}(0) = \frac{-\frac{\partial^2 f}{\partial x^2}(0,0)}{\frac{\partial^2 f}{\partial x \partial \mu}(0,0)}. \tag{3.1.63}$$

We now summarize our results. In order for a vector field

$$\dot{x} = f(x, \mu), \qquad x \in \mathbb{R}^1, \quad \mu \in \mathbb{R}^1, \tag{3.1.64}$$

to undergo a transcritical bifurcation, we must have

$$\left.\begin{aligned} f(0,0) &= 0 \\ \frac{\partial f}{\partial x}(0,0) &= 0 \end{aligned}\right\} \quad \text{nonhyperbolic fixed point} \tag{3.1.65}$$

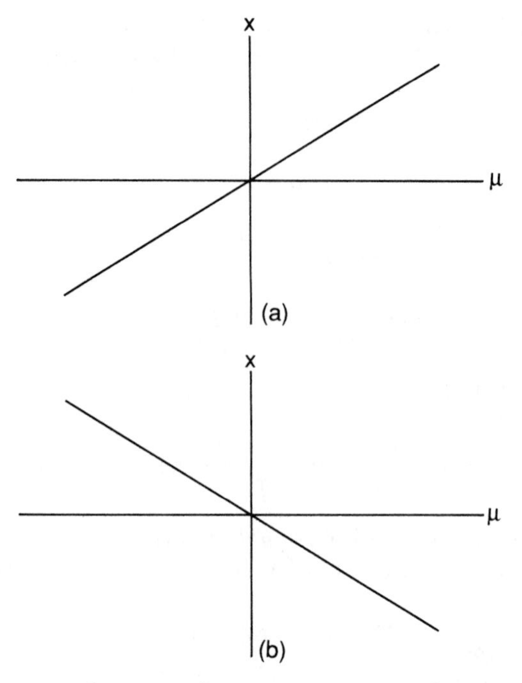

FIGURE 3.1.6. a) $\left(-\frac{\partial^2 f}{\partial x^2}(0,0)/\frac{\partial^2 f}{\partial x \partial \mu}(0,0)\right) > 0$; b) $\left(-\frac{\partial^2 f}{\partial x^2}(0,0)/\frac{\partial^2 f}{\partial x \partial \mu}(0,0)\right) < 0$.

and

$$\frac{\partial f}{\partial \mu}(0,0) = 0, \tag{3.1.66}$$

$$\frac{\partial^2 f}{\partial x \partial \mu}(0,0) \neq 0, \tag{3.1.67}$$

$$\frac{\partial^2 f}{\partial x^2}(0,0) \neq 0. \tag{3.1.68}$$

We note that the slope of the curve of fixed points not equal to $x = 0$ is given by (3.1.63). These two cases are shown in Figure 3.1.6; however, we do not indicate stabilities of the different branches of fixed points. We leave it as an exercise to the reader to verify the stability types of the different curves of fixed points emanating from the bifurcation point (see Exercise 3.3).

Thus, (3.1.65), (3.1.66), (3.1.67), and (3.1.68) show that the orbit structure near $(x, \mu) = (0,0)$ is qualitatively the same as the orbit structure near $(x, \mu) = (0,0)$ of

$$\dot{x} = \mu x \mp x^2. \tag{3.1.69}$$

Equation (3.1.69) can be viewed as a normal form for the transcritical bifurcation.

v) THE PITCHFORK BIFURCATION

The discussion and derivation of conditions under which a general one-parameter family of one-dimensional vector fields will undergo a bifurcation of the type shown in Example 3.1.3 follows very closely our discussion of the transcritical bifurcation.

The geometry of the curves of fixed points associated with the bifurcation in Example 3.1.3 had the following characteristics.

1. Two curves of fixed points passed through $(\mu, x) = (0,0)$, one given by $x = 0$, the other by $\mu = x^2$.

2. The curve $x = 0$ existed on both sides of $\mu = 0$; the curve $\mu = x^2$ existed on one side of $\mu = 0$.

3. The fixed points on the curve $x = 0$ had different stability types on opposite sides of $\mu = 0$. The fixed points on $\mu = x^2$ all had the same stability type.

Now we want to consider conditions on a general one-parameter family of one-dimensional vector fields having two curves of fixed points passing through the bifurcation point in the $\mu - x$ plane that have the properties given above.

We denote the vector field by

$$\dot{x} = f(x, \mu), \qquad x \in \mathbb{R}^1, \quad \mu \in \mathbb{R}^1, \tag{3.1.70}$$

and we suppose

$$f(0,0) = 0, \tag{3.1.71}$$

$$\frac{\partial f}{\partial x}(0,0) = 0. \tag{3.1.72}$$

As in the case of the transcritical bifurcation, in order to have more than one curve of fixed points passing through $(\mu, x) = (0,0)$ we must have

$$\frac{\partial f}{\partial \mu}(0,0) = 0. \tag{3.1.73}$$

Proceeding further along these lines, we *require* $x = 0$ to be a curve of fixed points for (3.1.70) by assuming the vector field (3.1.70) has the form

$$\dot{x} = xF(x, \mu), \qquad x \in \mathbb{R}^1, \quad \mu \in \mathbb{R}^1, \tag{3.1.74}$$

where

$$F(x, \mu) \equiv \left\{ \begin{array}{ll} \frac{f(x,\mu)}{x}, & x \neq 0 \\ \frac{\partial f}{\partial x}(0, \mu), & x = 0 \end{array} \right\}. \tag{3.1.75}$$

In order to have a second curve of fixed points passing through $(\mu, x) = (0,0)$ we must have

$$F(0,0) = 0 \tag{3.1.76}$$

with

$$\frac{\partial F}{\partial \mu}(0,0) \neq 0. \tag{3.1.77}$$

Equation (3.1.77) insures that only *one* additional curve of fixed points passes through $(\mu, x) = (0,0)$. Also, using (3.1.77), the implicit function theorem implies that for x sufficiently small there exists a unique function $\mu(x)$ such that

$$F(x, \mu(x)) = 0. \tag{3.1.78}$$

In order for the curve of fixed points, $\mu(x)$, to satisfy the above-mentioned characteristics, it is sufficient to have

$$\frac{d\mu}{dx}(0) = 0 \tag{3.1.79}$$

and

$$\frac{d^2\mu}{dx^2}(0) \neq 0. \tag{3.1.80}$$

The conditions for (3.1.79) and (3.1.80) to hold in terms of the derivatives of F evaluated at the bifurcation point can be obtained via implicit differentiation of (3.1.78) along the curve of fixed points exactly as in the case of the saddle-node bifurcation. They are given by

$$\frac{d\mu}{dx}(0) = \frac{-\frac{\partial F}{\partial x}(0,0)}{\frac{\partial F}{\partial \mu}(0,0)} = 0 \tag{3.1.81}$$

and

$$\frac{d^2\mu}{dx^2}(0) = \frac{-\frac{\partial^2 F}{\partial x^2}(0,0)}{\frac{\partial F}{\partial \mu}(0,0)} \neq 0. \tag{3.1.82}$$

Using (3.1.75), (3.1.81) and (3.1.82) can be expressed in terms of derivatives of f as follows

$$\frac{d\mu}{dx}(0) = \frac{-\frac{\partial^2 f}{\partial x^2}(0,0)}{\frac{\partial^2 f}{\partial x \partial \mu}(0,0)} = 0 \tag{3.1.83}$$

and

$$\frac{d^2\mu}{dx^2}(0) = \frac{-\frac{\partial^3 f}{\partial x^3}(0,0)}{\frac{\partial^2 f}{\partial x \partial \mu}(0,0)} \neq 0. \tag{3.1.84}$$

We summarize as follows. In order for the vector field

$$\dot{x} = f(x, \mu), \qquad x \in \mathbb{R}^1, \quad \mu \in \mathbb{R}^1, \tag{3.1.85}$$

to undergo a pitchfork bifurcation at $(x, \mu) = (0,0)$, it is sufficient to have

$$\left.\begin{array}{l} f(0,0) = 0 \\ \frac{\partial f}{\partial x}(0,0) = 0 \end{array}\right\} \quad \text{nonhyperbolic fixed point} \tag{3.1.86}$$

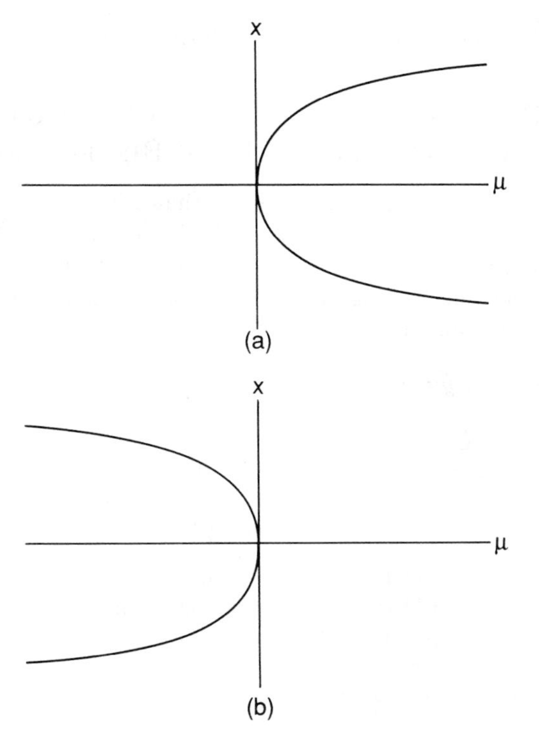

FIGURE 3.1.7. a) $\left(-\frac{\partial^3 f}{\partial x^3}(0,0)/\frac{\partial^2 f}{\partial x \partial \mu}(0,0)\right) > 0$; b) $\left(-\frac{\partial^3 f}{\partial x^3}(0,0)/\frac{\partial^2 f}{\partial x \partial \mu}(0,0)\right) < 0$.

with

$$\frac{\partial f}{\partial \mu}(0,0) = 0, \tag{3.1.87}$$

$$\frac{\partial^2 f}{\partial x^2}(0,0) = 0, \tag{3.1.88}$$

$$\frac{\partial^2 f}{\partial x \partial \mu}(0,0) \neq 0, \tag{3.1.89}$$

$$\frac{\partial^3 f}{\partial x^3}(0,0) \neq 0. \tag{3.1.90}$$

There are two possibilities for the disposition of the two branches of fixed points depending on the sign of (3.1.84). These two possibilities are shown in Figure 3.1.7 without indicating stabilities. We leave it as an exercise for the reader to verify the stability types for the different branches of fixed points emanating from the bifurcation point (see Exercise 3.4).

We conclude by noting that (3.1.86), (3.1.87), (3.1.88), (3.1.89), and (3.1.90) imply that the orbit structure near $(x, \mu) = (0,0)$ is qualitatively the same as the orbit structure near $(x, \mu) = (0,0)$ in the vector field

$$\dot{x} = \mu x \mp x^3. \tag{3.1.91}$$

Thus, (3.1.91) can be viewed as a normal form for the pitchfork bifurcation.

3.1B A Pure Imaginary Pair of Eigenvalues: The Poincare-Andronov-Hopf Bifurcation

We now turn to the next most simple way that a fixed point can be nonhyperbolic; namely, that the matrix associated with the vector field linearized about the fixed point has a pair of purely imaginary eigenvalues, with the remaining eigenvalues having nonzero real parts. Let us be more precise.

Recall (3.1.1), which we restate here;

$$\dot{y} = g(y, \lambda), \qquad y \in \mathbb{R}^n, \quad \lambda \in \mathbb{R}^p, \tag{3.1.92}$$

where g is $\mathbf{C}^r$ $(r \geq 5)$ on some sufficiently large open set containing the fixed point of interest. The fixed point is denoted by $(y, \lambda) = (y_0, \lambda_0)$, i.e.,

$$0 = g(y_0, \lambda_0). \tag{3.1.93}$$

We are interested in how the orbit structure near y_0 changes as λ is varied. In this situation the first thing to examine is the linearization of the vector field about the fixed point, which is given by

$$\dot{\xi} = D_y g(y_0, \lambda_0)\xi, \qquad \xi \in \mathbb{R}^n. \tag{3.1.94}$$

Suppose that $D_y g(y_0, \lambda_0)$ has two purely imaginary eigenvalues with the remaining $n-2$ eigenvalues having nonzero real parts. We know (cf. Sections 1.1A, 1.1B, 1.1C, and the remarks at the beginning of Section 3.1) that since the fixed point is not hyperbolic, the orbit structure of the linearized vector field near $(y, \lambda) = (y_0, \lambda_0)$ may reveal little (and, possibly, even incorrect) information concerning the nature of the orbit structure of the nonlinear vector field (3.1.92) near $(y, \lambda) = (y_0, \lambda_0)$.

Fortunately, we have a systematic procedure for analyzing this problem. By the center manifold theorem, we know that the orbit structure near $(y, \lambda) = (y_0, \lambda_0)$ is determined by the vector field (3.1.92) restricted to the center manifold. This restriction gives us a p-parameter family of vector fields on a two-dimensional center manifold. For now we will assume that we are dealing with a single, scalar parameter, i.e., $p = 1$. If there is more than one parameter in the problem, we will consider all but one of them as fixed.

On the center manifold the vector field (3.1.92) has the following form

$$\begin{pmatrix} \dot{x} \\ \dot{y} \end{pmatrix} = \begin{pmatrix} \text{Re } \lambda(\mu) & -\text{Im } \lambda(\mu) \\ \text{Im } \lambda(\mu) & \text{Re } \lambda(\mu) \end{pmatrix} \begin{pmatrix} x \\ y \end{pmatrix} + \begin{pmatrix} f^1(x, y, \mu) \\ f^2(x, y, \mu) \end{pmatrix},$$
$$(x, y, \mu) \in \mathbb{R}^1 \times \mathbb{R}^1 \times \mathbb{R}^1, \tag{3.1.95}$$

where f^1 and f^2 are nonlinear in x and y and $\lambda(\mu)$, $\overline{\lambda(\mu)}$ are the eigenvalues of the vector field linearized about the fixed point at the origin.

Equation (3.1.95) was first discussed in Section 2.1D, Example 2.2.3. The reader should recall that in performing the center manifold reduction to obtain (3.1.95), several preliminary steps were first implemented. Namely, first we transformed the fixed point to the origin and, then, if necessary, performed a linear transformation of the coordinates so that the vector field (3.1.92) was in the form of (3.1.95). We further remark that the eigenvalue, denoted $\lambda(\mu)$, should not be confused with the general vector of parameters in (3.1.92), denoted $\lambda \in \mathbb{R}^p$, which we subsequently restricted to a scalar and labeled μ. We will henceforth denote

$$\lambda(\mu) = \alpha(\mu) + i\omega(\mu), \tag{3.1.96}$$

and note that by our assumptions we have

$$\begin{aligned} \alpha(0) &= 0, \\ \omega(0) &\neq 0. \end{aligned} \tag{3.1.97}$$

The next step is to transform (3.1.95) into normal form. This was done in Example 2.2.3 of Section 2.1B. The normal form was found to be

$$\begin{aligned} \dot{x} &= \alpha(\mu)x - \omega(\mu)y + (a(\mu)x - b(\mu)y)(x^2 + y^2) + \mathcal{O}(|x|^5, |y|^5), \\ \dot{y} &= \omega(\mu)x + \alpha(\mu)y + (b(\mu)x + a(\mu)y)(x^2 + y^2) + \mathcal{O}(|x|^5, |y|^5). \end{aligned} \tag{3.1.98}$$

We will find it more convenient to work with (3.1.98) in polar coordinates. In polar coordinates (3.1.98) is given by

$$\begin{aligned} \dot{r} &= \alpha(\mu)r + a(\mu)r^3 + \mathcal{O}(r^5), \\ \dot{\theta} &= \omega(\mu) + b(\mu)r^2 + \mathcal{O}(r^4). \end{aligned} \tag{3.1.99}$$

Because we are interested in the dynamics near $\mu = 0$, it is natural to Taylor expand the coefficients in (3.1.99) about $\mu = 0$. Equation (3.1.99) thus becomes

$$\begin{aligned} \dot{r} &= \alpha'(0)\mu r + a(0)r^3 + \mathcal{O}(\mu^2 r, \mu r^3, r^5), \\ \dot{\theta} &= \omega(0) + \omega'(0)\mu + b(0)r^2 + \mathcal{O}(\mu^2, \mu r^2, r^4), \end{aligned} \tag{3.1.100}$$

where " $'$ " denotes differentiation with respect to μ and we have used the fact that $\alpha(0) = 0$.

Our goal is to understand the dynamics of (3.1.100) for r small and μ small. This will be accomplished in two steps.

Step 1. Neglect the higher order terms of (3.1.100) and study the resulting "truncated" normal form.

Step 2. Show that the dynamics exhibited by the truncated normal form are qualitatively unchanged when one considers the influence of the previously neglected higher order terms.

Step 1. Neglecting the higher order terms in (3.1.100) gives

$$\begin{aligned} \dot{r} &= d\mu r + ar^3, \\ \dot{\theta} &= \omega + c\mu + br^2, \end{aligned} \tag{3.1.101}$$

where, for ease of notation, we define

$$\begin{aligned} \alpha'(0) &\equiv d, \\ a(0) &\equiv a, \\ \omega(0) &\equiv \omega, \\ \omega'(0) &\equiv c, \\ b(0) &\equiv b. \end{aligned} \tag{3.1.102}$$

In analyzing the dynamics of vector fields we have always started with the simplest situation; namely, we have found the fixed points and studied the nature of their stability. In regard to (3.1.101), however, we proceed slightly differently because of the nature of the coordinate system. To be precise, values of $r > 0$ and μ for which $\dot{r} = 0$, but $\dot{\theta} \neq 0$, correspond to periodic orbits of (3.1.101). We highlight this in the following lemma.

Lemma 3.1.1 *For* $-\infty < \frac{\mu d}{a} < 0$ *and* μ *sufficiently small*

$$(r(t), \theta(t)) = \left(\sqrt{\frac{-\mu d}{a}}, \left[\omega + \left(c - \frac{bd}{a}\right)\mu\right] t + \theta_0\right) \tag{3.1.103}$$

is a periodic orbit for (3.1.101).

Proof: In order to interpret (3.1.103) as a periodic orbit, we need only to insure that $\dot{\theta}$ is not zero. Since ω is a constant independent of μ, this immediately follows by taking μ sufficiently small. □

We address the question of stability in the following lemma.

Lemma 3.1.2 *The periodic orbit is*

i) *asymptotically stable for* $a < 0$;

ii) *unstable for* $a > 0$.

Proof: The way to prove this lemma is to construct a one-dimensional Poincaré map along the lines of Section 1.2, Case 1. We have done so for this problem in Example 1.2.1, from which the results of this lemma follow. □

We note that since we must have $r > 0$, (3.1.103) is the only periodic orbit possible for (3.1.101). Hence, for $\mu \neq 0$, (3.1.101) possesses a unique periodic orbit having amplitude $\mathcal{O}(\sqrt{\mu})$. Concerning the details of stability of the periodic orbit and whether it exists for $\mu > 0$ or $\mu < 0$, from (3.1.103) it is easy to see that there are four possibilities:

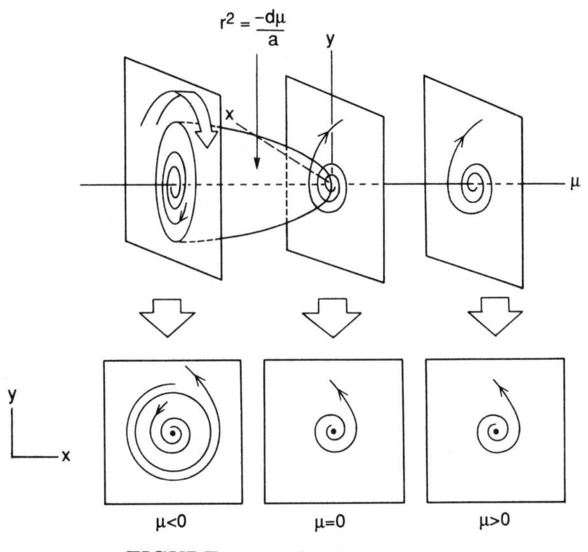

FIGURE 3.1.8. $d > 0$, $a > 0$.

1. $d > 0$, $a > 0$;
2. $d > 0$, $a < 0$;
3. $d < 0$, $a > 0$;
4. $d < 0$, $a < 0$.

We will examine each case individually; however, we note that in all cases the origin is a fixed point which is

$$\text{stable at } \mu = 0 \quad \text{for } a < 0,$$

$$\text{unstable at } \mu = 0 \quad \text{for } a > 0.$$

Case 1: $d > 0$, $a > 0$. In this case the origin is an unstable fixed point for $\mu > 0$ and an asymptotically stable fixed point for $\mu < 0$, with an unstable periodic orbit for $\mu < 0$ (note: the reader should realize that if the origin is stable for $\mu < 0$, then the periodic orbit should be unstable); see Figure 3.1.8.

Case 2: $d > 0$, $a < 0$. In this case the origin is an asymptotically stable fixed point for $\mu < 0$ and an unstable fixed point for $\mu > 0$, with an asymptotically stable periodic orbit for $\mu > 0$; see Figure 3.1.9.

Case 3: $d < 0$, $a > 0$. In this case the origin is an unstable fixed point for $\mu < 0$ and an asymptotically stable fixed point for $\mu > 0$, with an unstable periodic orbit for $\mu > 0$; see Figure 3.1.10.

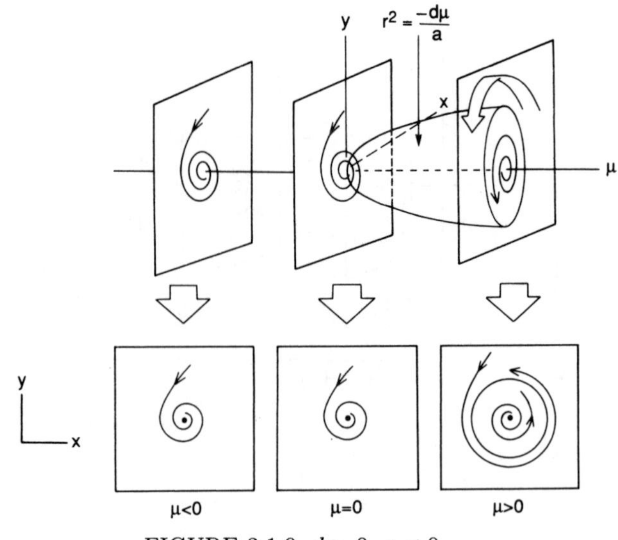

FIGURE 3.1.9. $d > 0$, $a < 0$.

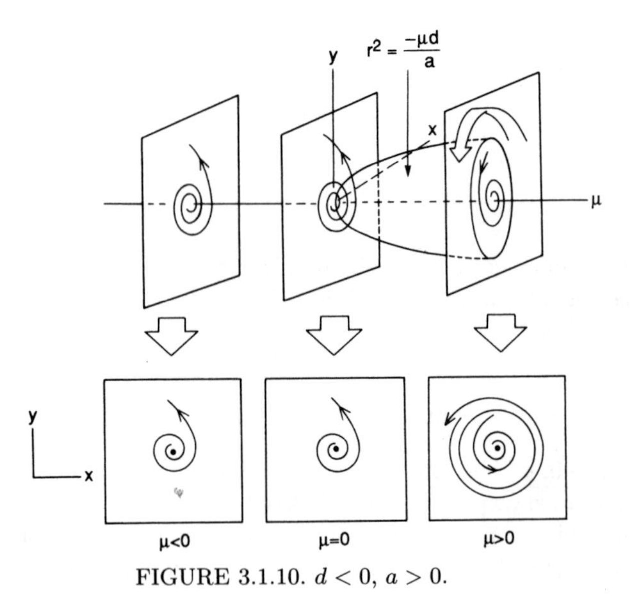

FIGURE 3.1.10. $d < 0$, $a > 0$.

Case 4: $d < 0$, $a < 0$. In this case the origin is an asymptotically stable fixed point for $\mu < 0$ and an unstable fixed point for $\mu > 0$, with an asymptotically stable periodic orbit for $\mu < 0$; see Figure 3.1.11.

From these four cases we can make the following general remarks.

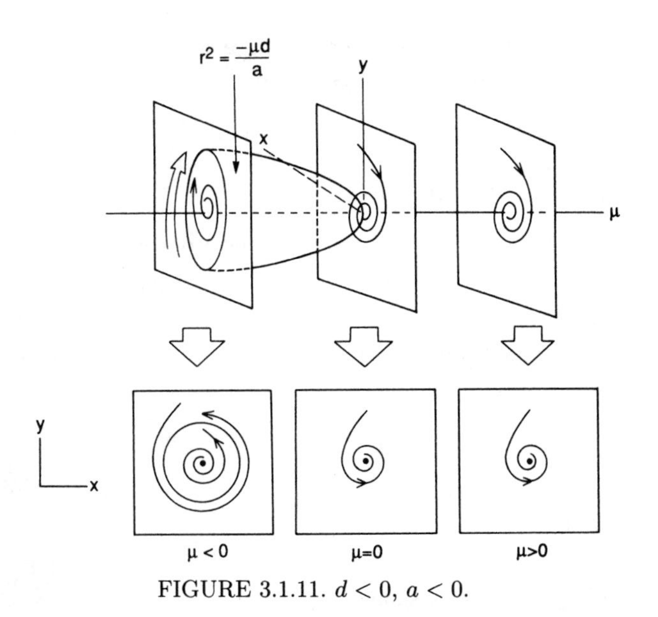

FIGURE 3.1.11. $d < 0$, $a < 0$.

Remark 1. For $a < 0$ it is possible for the periodic orbit to exist for either $\mu > 0$ (Case 2) or $\mu < 0$ (Case 4); however, in each case the periodic orbit is asymptotically stable. Similarly, for $a > 0$ it is possible for the periodic orbit to exist for either $\mu > 0$ (Case 3) or $\mu < 0$ (Case 1); however, in each case the periodic orbit is unstable. Thus, the number a tells us whether the bifurcating periodic orbit is stable ($a < 0$) or unstable ($a > 0$). The case $a < 0$ is referred to as a *supercritical* bifurcation, and the case $a > 0$ is referred to as a *subcritical* bifurcation.

Remark 2. Recall that

$$d = \frac{d}{d\mu}(\mathrm{Re}\lambda(\mu))\bigg|_{\mu=0}.$$

Hence, for $d > 0$, the eigenvalues cross from the left half-plane to the right half-plane as μ increases and, for $d < 0$, the eigenvalues cross from the right half-plane to the left half-plane as μ increases. For $d > 0$, it follows that the origin is asymptotically stable for $\mu < 0$ and unstable for $\mu > 0$. Similarly, for $d < 0$, the origin is unstable for $\mu < 0$ and asymptotically stable for $\mu > 0$.

Step 2. At this point we have a fairly complete analysis of the orbit structure of the truncated normal form near $(r, \mu) = (0, 0)$. We now must consider Step 2 in our analysis of the normal form (3.1.100); namely, are the dynamics that we have found in the truncated normal form changed when the effects of the neglected higher order term are considered? Fortunately, the answer to this question is no and is the content of the following theorem.

Theorem 3.1.3 (Poincaré-Andronov-Hopf Bifurcation) *Consider the full normal form (3.1.100). Then, for μ sufficiently small, Case 1, Case 2, Case 3, and Case 4 described above hold.*

Proof: We will outline two proofs and leave most of the details to the reader.

Proof 1: Utilize the Poincaré-Bendixson Theorem. We begin by considering the truncated normal form (3.1.101) and the case $a < 0$, $d > 0$. In this case the periodic orbit is stable and exists for $\mu > 0$, and the r coordinate is given by

$$r = \sqrt{\frac{-d\mu}{a}}.$$

We next choose $\mu > 0$ sufficiently small and consider the annulus in the plane, A, given by

$$A = \{(r, \theta) | \; r_1 \leq r \leq r_2\},$$

where r_1 and r_2 are chosen such that

$$0 < r_1 < \sqrt{\frac{-d\mu}{a}} < r_2.$$

By (3.1.101), it is easy to verify that on the boundary of A, the vector field given by the truncated normal form (3.1.101) is pointing *strictly* into the interior of A. Hence, A is a positive invariant region (cf. Definition 1.1.4); see Figure 3.1.12. It is also easy to verify that A contains no fixed points so, by the Poincaré-Bendixson theorem, A contains a stable periodic orbit. Of course we already knew this; our goal is to show that this situation still holds when the full normal form (3.1.100) is considered.

Now consider the full normal form (3.1.100). By taking μ and r sufficiently small, the $\mathcal{O}(\mu^2 r, \mu r^3, r^5)$ terms can be made much smaller than the rest of the normal form (i.e., the truncated normal form (3.1.101). Therefore, by taking r_1 and r_2 sufficiently small, A is still a positive invariant region containing no fixed points. Hence, by the Poincaré-Bendixson theorem, A contains a stable periodic orbit. The remaining three cases can be treated similarly; however, in the cases where $a > 0$, the time-reversed flow (i.e., letting $t \to -t$) must be considered.

Proof 2: Utilize the Method of Averaging. Consider the following rescaling of (3.1.100)

$$\begin{aligned} r &\to \varepsilon r, \\ \theta &\to \frac{\theta}{\varepsilon}, \\ t &\to \frac{t}{\varepsilon}, \\ \mu &\to \varepsilon^2 \mu. \end{aligned} \qquad (3.1.104)$$

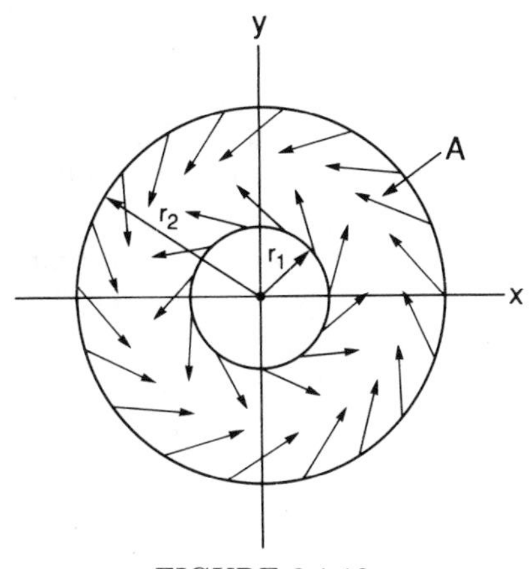

FIGURE 3.1.12.

Equation (3.1.100) then becomes

$$\begin{aligned} \dot{r} &= \varepsilon(d\mu r + ar^3) + \mathcal{O}(\varepsilon^2), \\ \dot{\theta} &= \omega + \mathcal{O}(\varepsilon^2). \end{aligned} \tag{3.1.105}$$

Equation (3.1.105) is in the standard form for applying the method of averaging (cf. Theorem 1.2.11 and Exercise 1.2.14). The proof of Theorem 3.1.3 then follows directly. We leave the details to the reader. □

To apply this theorem to specific systems, we need to know d (which is easy) and a. In principle, a is relatively straightforward to calculate. We simply carefully keep track of the coefficients in the normal form transformation in terms of our original vector field. However, in practice, the algebraic manipulations are horrendous. The explicit calculation can be found in Hassard, Kazarinoff, and Wan [1980], Marsden and McCracken[1976], and Guckenheimer and Holmes [1983]; here we will just state the result.

At bifurcation (i.e., $\mu = 0$), (3.1.95) becomes

$$\begin{pmatrix} \dot{x} \\ \dot{y} \end{pmatrix} = \begin{pmatrix} 0 & -\omega \\ \omega & 0 \end{pmatrix} \begin{pmatrix} x \\ y \end{pmatrix} + \begin{pmatrix} f^1(x,y,0) \\ f^2(x,y,0) \end{pmatrix}, \tag{3.1.106}$$

and the coefficient $a(0) \equiv a$ is given by

$$\begin{aligned} a = {} & \frac{1}{16}\left[f^1_{xxx} + f^1_{xyy} + f^2_{xxy} + f^2_{yyy}\right] \\ & + \frac{1}{16\omega}\left[f^1_{xy}\left(f^1_{xx} + f^1_{yy}\right) - f^2_{xy}\left(f^2_{xx} + f^2_{yy}\right)\right. \\ & \qquad \left. - f^1_{xx}f^2_{xx} + f^1_{yy}f^2_{yy}\right], \end{aligned} \tag{3.1.107}$$

where all partial derivatives are evaluated at the bifurcation point, i.e., $(x, y, \mu) = (0, 0, 0)$.

We end this section with some historical remarks. Usually Theorem 3.1.3 goes by the name of the "Hopf bifurcation theorem." However, as has been pointed out repeatedly by V. Arnold [1983], this is inaccurate, since examples of this type of bifurcation can be found in the work of Poincaré [1892]. The first specific study and formulation of a theorem was due to Andronov [1929]. However, this is not to say that E. Hopf did not make an important contribution; while the work of Poincaré and Andronov was concerned with two-dimensional vector fields, the theorem due to E. Hopf [1942] is valid in n dimensions (note: this was before the discovery of the center manifold theorem). For these reasons we refer to Theorem 3.1.3 as the Poincaré-Andronov-Hopf bifurcation theorem.

3.1c Stability of Bifurcations Under Perturbations

Let us recall the central motivational question raised at the beginning of this chapter; namely, what is the nature of the orbit structure *near* a nonhyperbolic fixed point of a vector field? The key word to focus on in this question is "near." We have seen that a nonhyperbolic fixed point can be either asymptotically stable or unstable. However, most importantly, we have seen that "nearby vector fields" can have very different orbit structures. The phrase "nearby vector fields" was made concrete by considering parameterized families of vector fields; at a certain parameter value the fixed point was not hyperbolic, and a qualitatively different orbit structure existed for nearby parameter values (i.e., new solutions were created as the parameter was varied). There is an important, general lesson to be learned from this, which we state as follows.

Pure Mathematical Lesson

From the point of view of stability of nonhyperbolic fixed points of vector fields, one should not only study the orbit structure near the fixed point but also the local orbit structure of nearby vector fields.

Applied Mathematical Lesson

From the point of view of "robustness" of mathematical models, suppose one has a vector field possessing a nonhyperbolic fixed point. The vector field should then possess enough (independent) parameters so that, as the parameters are varied, all possible *local* dynamical behavior is realized in this particular parameterized family of vector fields.

Before making these somewhat vague ideas more precise, let us consider how they are manifested in the saddle-node, transcritical, pitchfork, and Poincaré-Andronov-Hopf bifurcations of vector fields that we have already studied.

EXAMPLE 3.1.5: THE SADDLE-NODE BIFURCATION Consider the one-parameter family of one-dimensional vector fields

$$\dot{x} = f(x, \mu), \qquad y \in \mathbb{R}^1, \quad \mu \in \mathbb{R}^1, \tag{3.1.108}$$

with

$$f(0,0) = 0, \tag{3.1.109}$$

$$\frac{\partial f}{\partial x}(0,0) = 0. \tag{3.1.110}$$

We saw in Section 3.1A that the conditions

$$\frac{\partial f}{\partial \mu}(0,0) \neq 0, \tag{3.1.111}$$

$$\frac{\partial^2 f}{\partial x^2}(0,0) \neq 0, \tag{3.1.112}$$

were sufficient conditions in order for the vector field (3.1.108) to undergo a saddle-node bifurcation at $\mu = 0$. The question we ask is the following.

> If a one-parameter family of one-dimensional vector fields satisfying (3.1.109), (3.1.110), (3.1.111), and (3.1.112) is "perturbed," will the resulting family of one-dimensional vector fields have qualitatively the same dynamics?

We will have essentially answered this question once we have explained what we mean by the term "perturbed."

We do this by first eliminating the parameters entirely. Consider a one-dimensional vector field

$$\dot{x} = f(x) = a_0 x^2 + \mathcal{O}(x^3), \qquad x \in \mathbb{R}^1, \tag{3.1.113}$$

where, in the Taylor expansion of $f(x)$, we have omitted the constant and $\mathcal{O}(x)$ terms, since we want (3.1.113) to have a nonhyperbolic fixed point at $x = 0$. Because $x = 0$ is a nonhyperbolic fixed point, the orbit structure near $x = 0$ of vector fields near (3.1.113) may be very different. We consider vector fields close to (3.1.113) by embedding (3.1.113) in a one-parameter family of vector fields as follows

$$\dot{x} = f(x, \mu) = \mu + a_0 x^2 + \mathcal{O}(x^3). \tag{3.1.114}$$

The addition of the term "μ" in (3.1.114) can be viewed as a perturbation of (3.1.113) via adding lower-order terms in the Taylor expansion of the vector field about the nonhyperbolic fixed point (note: "lower-order terms" means terms of order lower than the first nonvanishing term in the Taylor expansion). Clearly, (3.1.114) satisfies (3.1.109), (3.1.110), (3.1.111), and

(3.1.112); hence, $(x, \mu) = (0,0)$ is a saddle-node bifurcation point. What about further perturbations of (3.1.114)? If we add terms of $\mathcal{O}(x^3)$ and larger, we see that this has no effect on the nature of the bifurcation, since the saddle-node bifurcation is completely determined by (3.1.109), (3.1.110), (3.1.111), and (3.1.112), i.e., by terms of $\mathcal{O}(x^2)$ and lower. We could perturb (3.1.114) further by adding lower-order terms. For example,

$$\dot{x} = f(x, \mu, \varepsilon) = \mu + \varepsilon x + a_0 x^2 + \mathcal{O}(x^3). \tag{3.1.115}$$

In this case we have a two-parameter family of one-dimensional vector fields having a nonhyperbolic fixed point at $(x, \mu, \varepsilon) = (0,0,0)$. However, the nature of the saddle-node bifurcation (i.e., the geometry of the curve(s) of fixed points passing through the bifurcation point) is completely determined by (3.1.109), (3.1.110), (3.1.111), and (3.1.112). Hence, the addition of the term "εx" in (3.1.115) does not introduce any new dynamical phenomena into (3.1.114) (provided $\mu \neq 0$).

EXAMPLE 3.1.6: THE TRANSCRITICAL BIFURCATION Consider the one-parameter family of one-dimensional vector fields

$$\dot{x} = f(x, \mu), \qquad x \in \mathbb{R}^1, \quad \mu \in \mathbb{R}^1, \tag{3.1.116}$$

with

$$f(0,0) = 0, \tag{3.1.117}$$

$$\frac{\partial f}{\partial x}(0,0) = 0. \tag{3.1.118}$$

We saw in Section 3.1A that if (3.1.116) also satisfies

$$\frac{\partial f}{\partial \mu}(0,0) = 0, \tag{3.1.119}$$

$$\frac{\partial^2 f}{\partial \mu \partial x}(0,0) \neq 0, \tag{3.1.120}$$

$$\frac{\partial^2 f}{\partial x^2}(0,0) \neq 0, \tag{3.1.121}$$

then a transcritical bifurcation occurs at $(x, \mu) = (0,0)$. The conditions (3.1.119), (3.1.120), and (3.1.121) imply that, in the study of the orbit structure near the bifurcation point, terms of $\mathcal{O}(x^3)$ and larger in the Taylor expansion of the vector field about the bifurcation point do not qualitatively affect the nature of the bifurcation (i.e., they do not affect the geometry of the curves of fixed points passing through the bifurcation point). From this we concluded that a normal form for the transcritical bifurcation was given by

$$\dot{x} = \mu x \pm x^2. \tag{3.1.122}$$

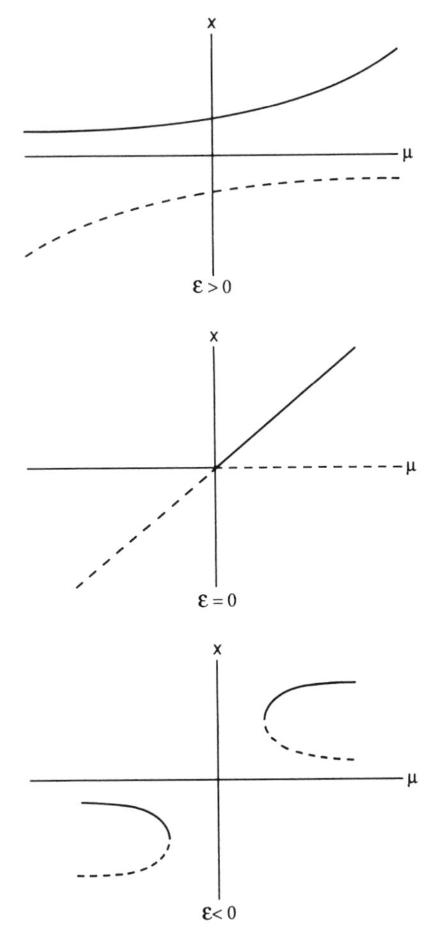

FIGURE 3.1.13.

Now let us consider a perturbation of the transcritical bifurcation by perturbing this normal form. Following our discussion of the perturbation of the saddle-node bifurcation and upon examining the defining conditions for the transcritical bifurcation given in (3.1.119), (3.1.120), and (3.1.121), we see that the only way to perturb (3.1.39) that may lead to qualitatively new dynamics is as follows

$$\dot{x} = \varepsilon + \mu x \mp x^2. \tag{3.1.123}$$

In Figure 3.1.13 we show what becomes of the transcritical bifurcation for $\varepsilon < 0$, $\varepsilon = 0$, and $\varepsilon > 0$. From this we see that the two curves of fixed points which pass through the origin for $\varepsilon = 0$ break apart into either a pair of curves of fixed points on which no bifurcation happens on passing through $\mu = 0$ or a pair of saddle-node bifurcations.

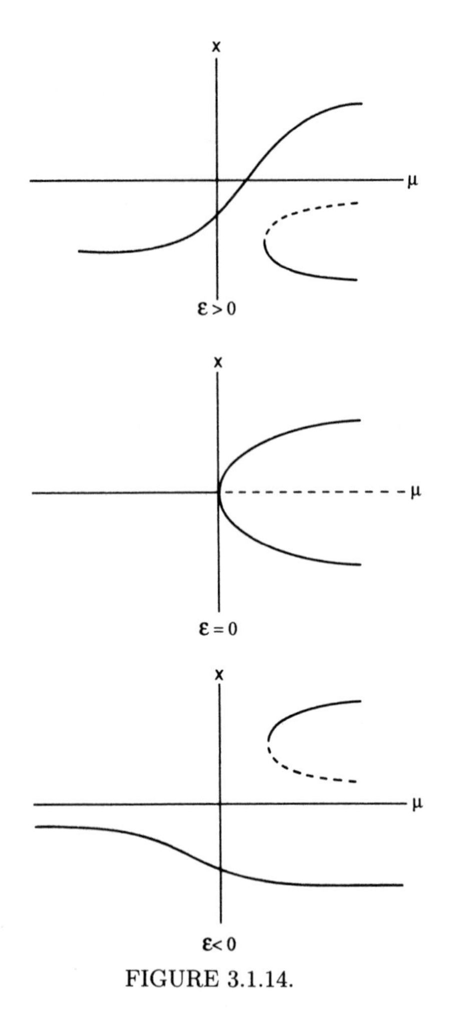

FIGURE 3.1.14.

EXAMPLE 3.1.7: THE PITCHFORK BIFURCATION From (3.1.91), the normal form for the pitchfork bifurcation was found to be

$$\dot{x} = \mu x \mp x^3, \qquad x \in \mathbb{R}^1, \quad \mu \in \mathbb{R}^1. \tag{3.1.124}$$

Using arguments exactly like those used in Examples 3.1.5 and 3.1.6, we can see that the only perturbations able to affect the orbit structure near $\mu = 0$ of (3.1.124) are

$$\dot{x} = \varepsilon + \mu x \mp x^3, \qquad \varepsilon \in \mathbb{R}^1. \tag{3.1.125}$$

In Figure 3.1.14 we show bifurcation diagrams for $\varepsilon < 0$, $\varepsilon = 0$, and $\varepsilon > 0$. As in the case of transcritical bifurcation, we see that, upon perturbation, the two curves of fixed points which pass through $(x, \mu) = (0, 0)$ for $\varepsilon = 0$

break up into either curves of fixed points exhibiting no bifurcation as μ varies through 0 or saddle-node bifurcations for $\varepsilon \neq 0$.

EXAMPLE 3.1.8 Recall the one-parameter family of one-dimensional vector fields discussed in Example 3.1.4

$$\dot{x} = \mu - x^3, \qquad x \in \mathbb{R}^1, \quad \mu \in \mathbb{R}^1. \tag{3.1.126}$$

The vector fields in this example have a nonhyperbolic fixed point at $(x, \mu) = (0, 0)$, but the orbit structure is qualitatively the same for all μ, i.e., no bifurcation occurs at $(x, \mu) = (0, 0)$.

Now consider the following perturbation of (3.1.126)

$$\dot{x} = \mu + \varepsilon x - x^3, \qquad \varepsilon \in \mathbb{R}^1. \tag{3.1.127}$$

From Figure 3.1.14 (with the roles of ε and μ reversed), it should be evident that (3.1.127) does exhibit saddle-node bifurcations for $\varepsilon \neq 0$.

EXAMPLE 3.1.9: THE POINCARÉ-ANDRONOV-HOPF BIFURCATION From Theorem 3.1.3, the normal form for the Poincaré-Andronov-Hopf bifurcation was given by

$$\dot{r} = \mu d r + a r^3, \qquad (r, \theta) \in \mathbb{R}^+ \times S^1, \quad \mu \in \mathbb{R}^1, \tag{3.1.128}$$

$$\dot{\theta} = \omega + c\mu + b r^2. \tag{3.1.129}$$

We want to consider how the bifurcation near $(r, \mu) = (0, 0)$ studied in Section 3.1B changes as (3.1.128) is perturbed. Three points should be considered.

1. Theorem 3.1.3 tells us that, for $a \neq 0$, $d \neq 0$, higher order terms (i.e., $\mathcal{O}(r^4)$) will not affect the dynamics of (3.1.128) near $(r, \mu) = (0, 0)$.

2. Since ω is a constant, for (r, μ) small, in order to determine the nature of solutions bifurcating from the origin we need only worry about the $\dot{r}$ component of the vector field (3.1.128).

3. Due to the structure of the linear part of the vector field, $r = 0$ is a fixed point for μ sufficiently small and *no terms of even order in* r are present in the $\dot{r}$ component of the normal form.

Using these three points, we see that, for $a \neq 0$, $d \neq 0$, no perturbations allowed by the structure of the vector field (cf. the third point above) will qualitatively alter the nature of the bifurcation near $(r, \mu) = (0, 0)$.

From Examples 3.1.5–3.1.9 we might conclude that, in one-parameter families of vector fields, the most "typical" bifurcations are saddle-node and Poincaré-Andronov-Hopf. This is indeed the case, as we will show in Section

3.1D. Moreover, these examples show that some nonhyperbolic fixed points are more degenerate than others in the sense that more parameters are needed in order to capture all possible nearby behavior. In Section 3.1D we explore the idea of the codimension of a bifurcation, which will enable us to quantify these ideas. A complete theory is given by Golubitsky and Schaeffer [1985] and Golubitsky, Stewart, and Schaeffer [1988].

3.1D The Idea of the Codimension of a Bifurcation

We have seen that some types of bifurcations (e.g., transcritical, pitchfork) are more degenerate than others (e.g., saddle-node). In this section we will attempt to make this more precise by introducing the idea of the codimension of a bifurcation. We will do this by starting with a heuristic discussion of the "big picture" of bifurcation theory. This will serve to show just how little is actually understood about bifurcation theory at this stage of the mathematical development of nonlinear dynamical systems theory.

Before beginning our discussion, however, we provide an outline of this section to help orient the reader.

i) The "big picture" for bifurcation theory

ii) The approach to local bifurcation theory

- iia) Ideas and results from singularity theory
- iib) The codimension of a local bifurcation
- iic) Construction of versal deformations

iii) Practical remarks on the location of parameters

Appendix 1. Versal deformations of families of matrices.

i) The "Big Picture" for Bifurcation Theory

The first step is to eliminate the consideration of all parameters from the problem; instead, we consider the infinite-dimensional space of all dynamical systems, either vector fields or maps. Within this space we consider the subset of all *structurally stable* dynamical systems, which we denote as S. By the definition of structural stability (cf. Section 1.2C), perturbations of structurally stable dynamical systems do not yield qualitatively new dynamical phenomena. Thus, from the point of view of bifurcation theory, it is not dynamical systems in S that are of interest but rather dynamical systems in the complement of S, denoted S^c, since perturbations of dynamical systems in S^c can result in systems exhibiting radically different dynamical behavior. Thus, in order to understand the types of bifurcations that may occur in a class of dynamical systems, it is necessary to understand the structure of S^c.

Presumably, in order for a dynamical system to be in S^c, the system must satisfy a certain number of extra conditions or constraints. When viewed geometrically in the infinite-dimensional function space setting, this can be interpreted as implying that S^c is a lower-dimensional "surface" contained in the space of dynamical systems. Here we use the word "surface" in a heuristic sense. More specifically, it would be nice if we could show that S^c is a codimension one submanifold. In practice, however, S^c may have singular regions and therefore be more appropriately described as an *algebraic variety* (see Arnold [1983]). In any case, for our heuristic discussion, it does no harm for the reader to visualize S^c as a surface. Before proceeding with this picture, let us first make a slight digression and define the notion of the "codimension of a submanifold."

The Codimension of a Submanifold

Let M be an m-dimensional manifold and let N be an n-dimensional submanifold contained in M; then the codimension of N is defined to be $m-n$. Equivalently, in a coordinate setting, the codimension of N is the number of independent equations needed to define N. Thus, the codimension of a submanifold is a measure of the avoidability of the submanifold as one moves about the ambient space; in particular, the codimension of a submanifold N is equal to the minimum dimension of a submanifold $P \subset M$ that intersects N such that the intersection is transversal. We have defined codimension in a finite-dimensional setting, which permits some intuition to be gained; now we move to the infinite-dimensional setting. Let M be an infinite-dimensional manifold and let N be a submanifold contained in M. (Note: for the definition of an infinite-dimensional manifold see Hirsch [1976]. Roughly speaking, an infinite-dimensional manifold is a set which is locally diffeomorphic to an infinite-dimensional Banach space. Because infinite-dimensional manifolds are discussed in this section only, and then mainly in a heuristic fashion, we refer the reader to the literature for the proper definitions.) We say that N is of codimension k if every point of N is contained in some open set in M which is diffeomorphic to $U \times \mathbb{R}^k$, where U is an open set in N. This implies that k is the smallest dimension of a submanifold $P \subset M$ that intersects N such that the intersection is transversal. Thus, the definition of codimension in the infinite-dimensional case has the same geometrical connotations as in the finite-dimensional case. Now we return to our main discussion.

Suppose S^c is a codimension one submanifold or, more generally, an algebraic variety. We might think of S^c as a surface dividing the infinite-dimensional space of dynamical systems as depicted in Figure 3.1.15. Bifurcations (i.e., topologically distinct orbit structures) occur as one passes through S^c. Thus, in the infinite-dimensional space of dynamical systems, one might define a *bifurcation point* as being any dynamical system which is structurally unstable.

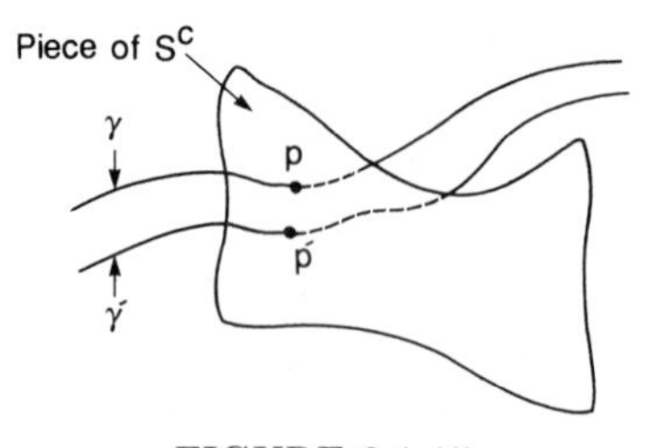

FIGURE 3.1.15.

In this setting one might initially conclude that bifurcations seldom occur and are unimportant, since any point p on S^c may be perturbed to S by (most) arbitrarily small perturbations. One might also conclude from a practical point of view that dynamical systems contained in S^c might not be very good models for physical systems, since any model is only an approximation to reality and therefore should be structurally stable. However, suppose we have a curve γ of dynamical systems transverse to S^c, i.e., a one-parameter family of dynamical systems. Then any sufficiently small perturbation of this curve γ still results in a curve γ' transverse to S^c. Thus, although any particular point on S^c may be removed from S^c by (most) arbitrarily small perturbations, a curve transverse to S^c remains transverse to S^c under perturbation. Bifurcation may therefore be unavoidable in a parameterized family of dynamical systems. This is an important point.

Now, even if we are able to show that S^c is a codimension one submanifold or algebraic variety, S^c itself may be divided up into objects of higher codimension corresponding to more degenerate types of bifurcations. A particular type of codimension k bifurcation in S^c would then be persistent in a k-parameter family of dynamical systems transverse to the codimension k submanifold.

This is essentially the program for bifurcation theory originally outlined by Poincaré. In order to utilize it in practice one would proceed as follows.

1. Given a specific dynamical system, determine whether or not it is structurally stable.

2. If it is not structurally stable, compute the codimension of the bifurcation.

3. Embed the system in a parameterized family of systems transverse to the bifurcation surface with the number of parameters equal to the codimension of the bifurcation. These parameterized systems are called *unfoldings* or *deformations* and, if they contain all possible qualitative dynamics that can occur near the bifurcation, they are called *universal unfoldings* or *versal deformations*; see Arnold [1983] and our discussion to follow.

4. Study the dynamics of the parameterized systems.

In this way one obtains structurally stable *families* of systems. Moreover, this provides a method for gaining a complete understanding of the qualitative dynamics of the space of dynamical systems with as little work as possible; namely, one uses the degenerate bifurcation points as "organizing centers" around which one studies the dynamics. Because elsewhere the dynamical systems are structurally stable, there is no need to worry about the details of their dynamics; qualitatively, they will be topologically conjugate to the structurally stable dynamical systems in a neighborhood of the bifurcation point.

This program is far from complete, and many of the problems associated with its completion are exactly those encountered in our discussion of structural stability in Section 1.2C. First, we must specify what we mean by the "infinite-dimensional space of dynamical systems." (Usually this does not present any major difficulties.) Next, we must equip the space with a topology in order to define what we mean by a perturbation of a dynamical system. We have already seen (cf. Example 1.2.5) that there can be problems with this if the phase space is unbounded; nevertheless, these difficulties can usually be brought under control. The real difficulty is the following. Given a dynamical system, what does one need to know about it in order to determine whether it is in S or S^c? For vector fields on compact, boundaryless two-dimensional manifolds Peixoto's theorem gives us an answer to this question (see Theorem 1.2.8 in Section 1.2C), but, in higher dimensions, we do not have nice analogs of this theorem. Moreover, the detailed structure of S^c is certainly beyond our reach at this time. Although the situation appears hopeless, some progress has been made along two fronts:

1. Local bifurcations;

2. Global bifurcations of specific orbits.

Since the subject of this chapter is local bifurcations we will discuss only this aspect. In Chapter 4 we will see examples of global bifurcations; for more information see Wiggins [1988].

Local bifurcation theory is concerned with the bifurcation of fixed points of vector fields and maps or with situations in which the problem can be cast into this form, such as in the study of bifurcations of periodic motions. For vector fields one can construct a local Poincaré map (see Section 1.2A) near the periodic orbit, thus reducing the problem to one of studying the bifurcation of a fixed point of a map, and for maps with a k periodic orbit one can consider the kth iterate of the map, thus reducing the problem to one of studying the bifurcation of a fixed point of the kth iterate of the map (see Section 1.1A). Utilizing a procedure such as the center manifold theorem or the Lyapunov-Schmidt reduction (see Chow and Hale [1982]), one can usually reduce the problem to that of studying an equation of the form

$$f(x, \lambda) = 0, \tag{3.1.130}$$

where $x \in \mathbb{R}^n$, $\lambda \in \mathbb{R}^p$ are the system parameters and $f : \mathbb{R}^n \times \mathbb{R}^p \to \mathbb{R}^n$ is assumed to be sufficiently smooth. The goal is to study the nature of the solutions of (3.1.130) as λ varies. In particular, it would be interesting to know for what parameter values solutions disappear or are created. These particular parameters are called *bifurcation values*, and there exists an extensive mathematical machinery called *singularity theory* (see Golubitsky and Guillemin [1973]) that deals with such questions. Singularity theory is concerned with the local properties of smooth functions near a zero of the function. It provides a classification of the various cases based on codimension in a spirit similar to that described in the beginning of the section. The reason this is possible is that the codimension k submanifolds in the space of all smooth functions having zeroes can be described algebraically by imposing conditions on derivatives of the functions. This gives us a way of classifying the various possible bifurcations and of computing the proper unfoldings or deformations. From this one might be led to believe that local bifurcation theory is a well-understood subject; however, this is not the case. The problem arises in the study of degenerate local bifurcations, specifically, in codimension k ($k \geq 2$) bifurcations of vector fields. Fundamental work of Takens [1974], Langford [1979], and Guckenheimer [1981] has shown that, arbitrarily near these degenerate bifurcation points, complicated global dynamical phenomena such as invariant tori and Smale horseshoes may arise. These phenomena cannot be described or detected via singularity theory techniques. Nevertheless, when one reads or hears the phrase "codimension k bifurcation" in the context of bifurcations of fixed points of dynamical systems, it is the singularity theory recipe that is used to compute the codimension. For this reason we want to describe the singularity theory approach.

ii) The Approach to Local Bifurcation Theory

iia) *Ideas and Results from Singularity Theory*

We now want to give a brief account of the techniques from singularity theory that are used to determine the codimension of a *local* bifurcation and the correspondingly appropriate unfolding or versal deformation. Our discussion follows closely Arnold [1983].

We begin by specifying the infinite-dimensional space of dynamical systems of interest. This will be the set of $\mathbf{C}^\infty$ maps of $\mathbb{R}^n$ into $\mathbb{R}^m$, denoted $\mathbf{C}^\infty(\mathbb{R}^n, \mathbb{R}^m)$. At this stage we can think of the elements of $\mathbf{C}^\infty(\mathbb{R}^n, \mathbb{R}^m)$ as either vector fields or maps; we will draw a distinction only when required by context. Several technical issues involving $\mathbf{C}^\infty(\mathbb{R}^n, \mathbb{R}^m)$ must now be addressed.

1. Throughout most of this book we have specified the minimum amount of differentiability needed for our dynamical systems. For example,

$\mathbf{C}^1$ was sufficient for local existence and uniqueness of solutions to ordinary differential equations and $\mathbf{C}^4$ was sufficient for deriving the Poincaré-Andronov-Hopf normal form. Now we have suddenly jumped to considering $\mathbf{C}^\infty$ functions. The reason for this is that we will need to utilize the Thom transversality theorem, which is proven in a $\mathbf{C}^\infty$ context (cf. Golubitsky and Guillemin [1973]).

2. Since we are interested only in local behavior, our maps need not be defined on all of $\mathbb{R}^n$, but rather they need only be defined on open sets containing the region of interest (i.e., the fixed point). Along these same lines, we can be more general and consider maps of $\mathbf{C}^\infty$ manifolds. However, since we are concerned with local questions this will not be an issue; the reader should consult Arnold [1983].

3. As mentioned in our discussion of structural stability in Section 1.2C, there can be technical difficulties in deciding when two dynamical systems are "close" when the phase space is unbounded (as is $\mathbb{R}^n$). Since we are interested only in the behavior near fixed points, we will be able to avoid this unpleasant issue.

4. As we have mentioned several times, we will be interested in the orbit structure of dynamical systems in a *sufficiently small neighborhood of a fixed point.* This phrase is a bit ambiguous, so at this point, we want to spend a little effort to try to make it clearer.

We begin by studying an example which will illustrate some of the salient points. Consider the vector field

$$\dot{x} = \mu - x^2 + \varepsilon x^3, \qquad x \in \mathbb{R}^1, \quad \mu \in \mathbb{R}^1, \tag{3.1.131}$$

where we view the term "εx^3" in (3.1.131) as a perturbation term. It should be evident (see Section 3.1A, iii)) that (3.1.131) undergoes a saddle-node bifurcation at $(x, \mu) = (0, 0)$ so that, in the $x - \mu$ plane, *in a sufficiently small neighborhood of the origin*, the curve of fixed points of (3.1.131) appear as in Figure 3.1.16. However, (3.1.131) is so simple that we can actually compute the global curve of fixed points, which is given by

$$\mu = x^2 - \varepsilon x^3 \tag{3.1.132}$$

and is shown in Figure 3.1.16. Thus, we see that, besides the saddle-node bifurcation point at $(x, \mu) = (0, 0)$, (3.1.131) has an additional saddle-node bifurcation point at $(x, \mu) = (2/3\varepsilon, 4/27\varepsilon^2)$. Clearly these two saddle-node bifurcation points are far apart for ε small; however, this example shows that the size of a "sufficiently small neighborhood of a point" can vary from situation to situation. In this example, the "sufficiently small neighborhood" of $(x, \mu) = (0, 0)$ shrinks to a point as $\varepsilon \to \infty$. The idea of a "germ" of a differentiable function has been invented in order to handle

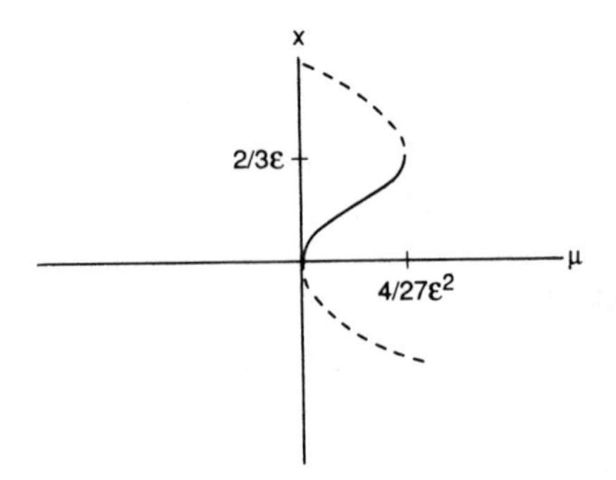

FIGURE 3.1.16.

this ambiguity, and we refer the interested reader to Arnold [1983] for an introduction to this formalism. However, in this book we will not utilize the idea of germs but rather the less mathematically precise and more verbose approach of reminding the reader that we are always working in a *sufficiently small neighborhood of a fixed point.*

Now that we have specified the infinite-dimensional space of dynamical systems, next we need some way of understanding the geometry of the "degenerate" dynamical systems. This is afforded by the notions of jets and jet spaces. We will first give five definitions and then explanations.

DEFINITION 3.1.2 Consider $f \in \mathbf{C}^\infty(\mathbb{R}^n, \mathbb{R}^m)$. The *k-jet of f at x* is given by the following $(k+2)$-tuple

$$(x, f(x), Df(x), \cdots, D^k f(x)).$$

Thus we see that the k-jet of a map at a point is simply the Taylor coefficients through order k plus the point at which they are evaluated. We denote the k-jet of f at x by

$$J_x^k(f).$$

Although it may seem a bit silly to dress up something as commonplace as a Taylor expansion in a new formalism, the reader will see that there is a definite payoff. We next consider the set of all k-jets of $\mathbf{C}^\infty$ maps of $\mathbb{R}^n$ into $\mathbb{R}^m$ at all points of $\mathbb{R}^n$.

DEFINITION 3.1.3 The set of all k-jets of $\mathbf{C}^\infty$ maps of $\mathbb{R}^n$ into $\mathbb{R}^m$ at all points of $\mathbb{R}^n$ is called the *space of k-jets* and is denoted by

$$J^k(\mathbb{R}^n, \mathbb{R}^m) = \{\text{The space of } k\text{-jets of } \mathbf{C}^\infty \text{ maps of } \mathbb{R}^n \text{ into } \mathbb{R}^m\}.$$

The spaces $J^k(\mathbb{R}^n, \mathbb{R}^m)$ have a nice linear vector space structure. In fact, we can identify $J^k(\mathbb{R}^n, \mathbb{R}^m)$ with $\mathbb{R}^p$ for an appropriate choice of p. We illustrate this in the following examples.

EXAMPLE 3.1.10 $J^0(\mathbb{R}^n, \mathbb{R}^m) = \mathbb{R}^n \times \mathbb{R}^m$.

EXAMPLE 3.1.11 $J^1(\mathbb{R}^1, \mathbb{R}^1)$ is three dimensional, because points in $J^1(\mathbb{R}^1, \mathbb{R}^1)$ can be assigned the coordinates

$$\left(x, f(x), \frac{\partial f}{\partial x}(x)\right).$$

EXAMPLE 3.1.12 $J^1(\mathbb{R}^2, \mathbb{R}^2)$ is eight dimensional, because points in $J^1(\mathbb{R}^2, \mathbb{R}^2)$ can be assigned the coordinates

$$(x, f(x), Df(x)),$$

where $Df(x)$ is a 2×2 matrix.

In a certain sense, the spaces $J^k(\mathbb{R}^n, \mathbb{R}^m)$ can be thought of as finite-dimensional approximations of $\mathbf{C}^\infty(\mathbb{R}^n, \mathbb{R}^m)$.

We now introduce a map that will play an important role when we discuss the notion of versal deformations.

DEFINITION 3.1.4 For any map $f \in \mathbf{C}^\infty(\mathbb{R}^n, \mathbb{R}^m)$ we define a map

$$\hat{f} : \mathbb{R}^n \to J^k(\mathbb{R}^n, \mathbb{R}^m),$$

$$x \mapsto J_x^k(f) \equiv \hat{f}(x),$$

which we call the *k-jet extension of f*.

Thus, the k-jet extension of f merely associates to each point in the phase space of the dynamical system (f) the k-jet of f at the point. We also remark that the k-jet extension of f can be viewed as a map of the phase space of the dynamical system (f) into $J^k(\mathbb{R}^n, \mathbb{R}^m)$. This remark will be important later on. We next introduce a new notion of transversality.

Recall from Section 1.2C the notion of two manifolds being transversal. We now introduce a similar idea: the notion of a map being transverse to a manifold.

DEFINITION 3.1.5 Consider a map $f \in \mathbf{C}^\infty(\mathbb{R}^n, \mathbb{R}^m)$ and a submanifold $M \subset \mathbb{R}^m$. The map is said to be *transversal to M at a point $x \in \mathbb{R}^n$* if either $f(x) \notin M$ or the tangent space to M at $f(x)$ and the image of the tangent space to $\mathbb{R}^n$ at x under $Df(x)$ are transversal, i.e.,

$$Df(x) \cdot T_x\mathbb{R}^n + T_{f(x)}M = T_{f(x)}\mathbb{R}^m.$$

DEFINITION 3.1.6 The map is said to be *transversal to M* if it is transversal to M at any point $x \in \mathbb{R}^n$.

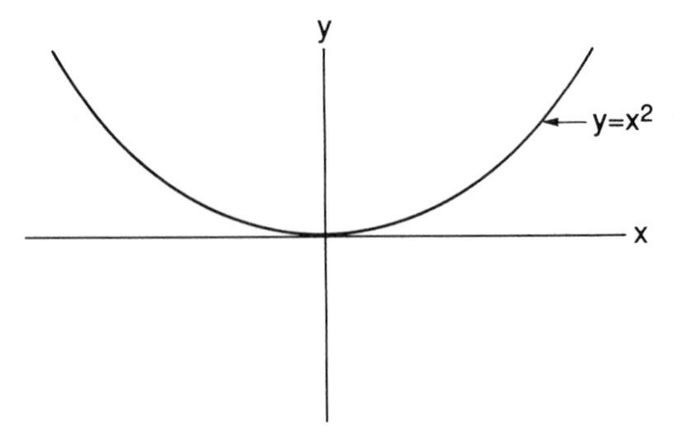

FIGURE 3.1.17.

Let us consider an example.

EXAMPLE 3.1.13 Consider the map

$$f : \mathbb{R}^1 \rightarrow \mathbb{R}^2,$$
$$x \mapsto (x, x^2).$$

When viewed geometrically, the image of $\mathbb{R}^1$ under f is only the parabola $y = x^2$; see Figure 3.1.17. We ask the following questions.

1. Thinking of the x-axis as a one-dimensional submanifold of $\mathbb{R}^2$, is f transverse to the x-axis?

2. Similarly for the y-axis, is f transverse to the y-axis?

The only point at which the image of $\mathbb{R}^1$ under f intersects either the x-axis or y-axis is at the origin; therefore, this is the only point at which we need to check transversality. Let us denote the x-axis by X and the y-axis by Y. Then, using the standard coordinates on $\mathbb{R}^2$, we can take

$$T_{(0,0)}X = (\mathbb{R}^1, 0), \tag{3.1.133}$$

$$T_{(0,0)}Y = (0, \mathbb{R}^1). \tag{3.1.134}$$

We have that $Df(0) = (1, 0)$; thus, on recalling Definition 3.1.5 and examining (3.1.133) and (3.1.134), it follows that f is transverse to Y but not transverse to X.

We now state the Thom transversality theorem, which will guide the construction of candidates for versal deformations of degenerate dynamical systems.

Theorem 3.1.4 (Thom) *Let* $\mathbf{C}$ *be a submanifold of* $J^k(\mathbb{R}^n, \mathbb{R}^m)$. *The set of maps* $f : \mathbb{R}^n \rightarrow \mathbb{R}^m$ *whose k-jet extensions are transversal to* $\mathbf{C}$ *is an everywhere dense countable intersection of open sets in* $\mathbf{C}^\infty(\mathbb{R}^n, \mathbb{R}^m)$.

Proof: See Arnold [1983]. □

This theorem allows us to work in finite dimensions, where geometrical intuition is more apparent, and to draw conclusions concerning the geometry in the infinite-dimensional space of dynamical systems, $\mathbf{C}^\infty(\mathbb{R}^n, \mathbb{R}^m)$. At this point we have developed enough machinery from singularity theory to discuss the idea of the codimension of a local bifurcation. We now turn to this question.

iib) *The Codimension of a Local Bifurcation*

As mentioned in our discussion of the "big picture" of bifurcation theory, the problem we are immediately faced with is determining whether a dynamical system is structurally stable. If we limit ourselves to the study of the behavior of fixed points, then the problem is much easier, because we know how to characterize structurally unstable fixed points—they are simply the nonhyperbolic fixed points. The techniques from singularity theory we developed in 3.1D will enable us to characterize the degree of "structural instability" of a fixed point in a way that is similar in spirit to our discussion of the "big picture" of bifurcation theory given in that section. However, the reader should realize that these ideas are concerned with fixed point behavior only and as such we must often do additional work to determine whether nearby *dynamical* phenomena are properly taken into account. We will see several examples of this as we go along.

Thus far we have been using the term "dynamical system" to refer to vector fields and maps interchangeably. At this stage we need to draw a slight distinction. We will be studying the fixed points of dynamical systems. For vector fields

$$\dot{x} = f(x), \qquad x \in \mathbb{R}^n, \tag{3.1.135}$$

this means studying the equation

$$f(x) = 0, \tag{3.1.136}$$

and for maps

$$x \mapsto g(x), \qquad x \in \mathbb{R}^n, \tag{3.1.137}$$

this means studying the equation

$$g(x) - x = 0. \tag{3.1.138}$$

It should be clear that from an analytic point of view, (3.1.136) and (3.1.138) are essentially the same. However, for the remainder of this section we will deal solely with vector fields and leave the trivial modifications for maps as exercises.

For vector fields the space of dynamical systems will be $\mathbf{C}^\infty(\mathbb{R}^n, \mathbb{R}^n)$ and the associated jet spaces $J^k(\mathbb{R}^n, \mathbb{R}^n)$. In particular, we will work solely in the finite-dimensional space $J^k(\mathbb{R}^n, \mathbb{R}^n)$. Within $J^k(\mathbb{R}^n, \mathbb{R}^n)$ we will be interested in the subset consisting of k-jets of vector fields having fixed points. We denote this subset by F (for fixed points) and note that F has codimension n. Within F we will be interested in the k-jets of vector fields having nonhyperbolic fixed points. We denote this subset by B (for bifurcation) and note the B has codimension $n+1$. We remark that the codimension of F and B is independent of k; however, the dimension of F and B is *not* independent of k. Let us now consider some specific examples.

EXAMPLE 3.1.14: ONE-DIMENSIONAL VECTOR FIELDS As our space of vector fields we take $\mathbf{C}^\infty(\mathbb{R}^1, \mathbb{R}^1)$. In this case $J^k(\mathbb{R}^1, \mathbb{R}^1)$ is $(k+2)$-dimensional with coordinates of points in $J^k(\mathbb{R}^n, \mathbb{R}^n)$ given by the $(k+2)$-tuples

$$\left(x, f(x), \frac{\partial f}{\partial x}(x), \frac{\partial^2 f}{\partial x^2}(x), \cdots, \frac{\partial^k f}{\partial x^k}(x)\right), \quad x \in \mathbb{R}^1, \ f \in \mathbf{C}^\infty(\mathbb{R}^1, \mathbb{R}^1),$$

F is $(k+1)$-dimensional (codimension 1) with coordinates of points in F given by

$$\left(x, 0, \frac{\partial f}{\partial x}(x), \frac{\partial^2 f}{\partial x^2}(x), \cdots, \frac{\partial^k f}{\partial x^k}(x)\right), \qquad x \in \mathbb{R}^1, \ f \in \mathbf{C}^\infty(\mathbb{R}^1, \mathbb{R}^1),$$

and B is $(k+2)$-dimensional (codimension 2) with coordinates of points in B given by

$$\left(x, 0, 0, \frac{\partial^2 f}{\partial x^2}(x), \cdots, \frac{\partial^k f}{\partial x^k}(x)\right), \qquad x \in \mathbb{R}^1, \ f \in \mathbf{C}^\infty(\mathbb{R}^1, \mathbb{R}^1).$$

EXAMPLE 3.1.15: n-DIMENSIONAL VECTOR FIELDS As our space of vector fields we take $\mathbf{C}^\infty(\mathbb{R}^n, \mathbb{R}^n)$. In this case coordinates of $J^k(\mathbb{R}^n, \mathbb{R}^n)$ are given by

$$\left(x, f(x), Df(x), \cdots, D^k f(x)\right), \qquad x \in \mathbb{R}^n, \quad f \in \mathbf{C}^\infty(\mathbb{R}^n, \mathbb{R}^n).$$

F is codimension n, with points in F having coordinates

$$\left(x, 0, Df(x), \cdots, D^k f(x)\right), \qquad x \in \mathbb{R}^n, \quad f \in \mathbf{C}^\infty(\mathbb{R}^n, \mathbb{R}^n),$$

and B is codimension $n+1$, with points in B having coordinates

$$\left(x, 0, \widetilde{Df(x)}, \cdots, D^k f(x)\right), \qquad x \in \mathbb{R}^n, \quad f \in \mathbf{C}^\infty(\mathbb{R}^n, \mathbb{R}^n),$$

where $\widetilde{Df(x)}$ represents a nonhyperbolic matrix which lies on a surface of codimension 1 in the n^2-dimensional space of $n \times n$ matrices. This is explained in great detail in Appendix 1.

We are now at the point where we can define the *codimension of a fixed point.* There are two possibilities for the choice of this number, and we closely follow the discussion of Arnold [1972].

DEFINITION 3.1.7: THE CODIMENSION OF A FIXED POINT Consider $J^k(\mathbb{R}^n, \mathbb{R}^n)$ and the subset of $J^k(\mathbb{R}^n, \mathbb{R}^n)$ consisting of k-jets of elements of $\mathbf{C}^\infty(\mathbb{R}^n, \mathbb{R}^n)$ that have fixed points. We denote this subset by F and note that F has codimension n in $J^k(\mathbb{R}^n, \mathbb{R}^n)$. Consider the k-jet of an element of $\mathbf{C}^\infty(\mathbb{R}^n, \mathbb{R}^n)$ that has a nonhyperbolic fixed point. Then this k-jet lies in a subset of F defined by conditions on the derivatives. Suppose this subset of F has codimension b in $J^k(\mathbb{R}^n, \mathbb{R}^n)$. Then we define the codimension of the fixed point to be $b - n$.

Before giving examples we want to make a few general remarks concerning this definition.

Remark 1. Evidently k must be taken sufficiently large so that the degree of degeneracy of the nonhyperbolic fixed point can be specified.

Remark 2. This definition says that the codimension of a fixed point is equal to the codimension of the subset of F specified by the degeneracy of the fixed point minus n. Thus, hyperbolic fixed points have codimension zero by this definition. This seems reasonable, since the notion of the codimension of a fixed point should somehow specify the degree of "nongenericity" of the fixed point.

Remark 3. There is another way of describing the reasons why we chose this way of defining the codimension of a fixed point. The dynamical system induces a map of the phase space, $\mathbb{R}^n$, into $J^k(\mathbb{R}^n, \mathbb{R}^n)$. This map is only the k-jet extension map (cf. Definition 3.1.4). At hyperbolic fixed points this map is transverse to the subset F of $J^k(\mathbb{R}^n, \mathbb{R}^n)$. Thus, hyperbolic fixed points cannot be destroyed by small perturbations. Hence, if the notion of codimension is to quantify the amount of "degeneracy" of nonhyperbolic fixed points, then the generic elements of F should be regarded as codimension zero. Practically speaking, this also implies that *generically* the fixed points move if they are perturbed.

Now we consider several examples where we will compute the codimension.

EXAMPLE 3.1.16 Consider the vector field

$$\dot{x} = f(x) = ax^2 + \mathcal{O}(x^3), \qquad x \in \mathbb{R}^1. \tag{3.1.139}$$

We are interested in studying (3.1.139) near the nonhyperbolic fixed point $x = 0$. At this point we see that the k-jet of (3.1.139) is a typical element of the set B described in Example 3.1.14. Hence, using Definition 3.1.7, we conclude that $x = 0$ is codimension 1.

EXAMPLE 3.1.17 Consider the vector field

$$\dot{x} = f(x) = ax^3 + \mathcal{O}(x^4), \qquad x \in \mathbb{R}^1. \tag{3.1.140}$$

We are interested in (3.1.140) near the nonhyperbolic fixed point $x = 0$. As in Example 3.1.16, the k-jet of (3.1.140) is contained in the set B described in Example 3.1.14. However, (3.1.140) is more degenerate than the typical element of B in that (3.1.140) also has

$$\frac{\partial^2 f}{\partial x^2}(0) = 0.$$

Hence, the k-jet of (3.1.140) lies in a lower dimensional subset of B, denoted B', which has codimension 3 in $J^k(\mathbb{R}^1, \mathbb{R}^1)$, with points in B' having coordinates

$$\left(x, 0, 0, 0, \frac{\partial^3 f}{\partial x^2}(x), \cdots, \frac{\partial^k f}{\partial x^k}(x)\right), \qquad x \in \mathbb{R}^n, \quad f \in \mathbf{C}^\infty(\mathbb{R}^n, \mathbb{R}^n).$$

Using Definition 3.1.7 we can thus conclude that $x = 0$ is codimension 2.

EXAMPLE 3.1.18 In Appendix 1 we show the following.

1. In the four-dimensional space of 2×2 matrices the matrices

$$\begin{pmatrix} 0 & -\omega \\ \omega & 0 \end{pmatrix} \tag{3.1.141}$$

and

$$\begin{pmatrix} 0 & 1 \\ 0 & 0 \end{pmatrix} \tag{3.1.142}$$

lie on surfaces of codimension 1 and codimension 2, respectively.

2. In the nine-dimensional space of 3×3 matrices the matrix

$$\begin{pmatrix} 0 & -\omega & 0 \\ \omega & 0 & 0 \\ 0 & 0 & 0 \end{pmatrix} \tag{3.1.143}$$

lies on a surface of codimension 2.

Hence, using Example 3.1.15 and Definition 3.1.7, we conclude that the codimension of a fixed point of a vector field whose 1-jet in (real) Jordan canonical form is given by (3.1.141), (3.1.142), or (3.1.143) is 1, 2, and 2, respectively.

We end this section with two remarks.

Remark 1. The codimensions computed in these examples are for generic vector fields. In particular, we have not considered the possibility of symmetries which would put extra constraints on eigenvalues and of derivatives which would result in a modification of the codimension; see Golubitsky and Schaeffer [1985] and Golubitsky, Stewart, and Schaeffer [1988].

Remark 2. We have referred to the sets F and B as "subsets" of $J^k(\mathbb{R}^n, \mathbb{R}^n)$. In applying the Thom transversality theorem the question of whether or not they are actually submanifolds arises. The same question is also of interest for the higher codimension subsets of B corresponding to more degenerate fixed points. In general, these subsets may have singular points and, thus, they will not have the structure of a submanifold. However, these singularities can be removed by slight perturbations and, hence, for our purposes, they can be treated as submanifolds. This technical point is treated in great detail in Gibson [1979].

iic) *Construction of Versal Deformations*

We now want to develop the necessary definitions in order to discuss *versal deformations* of vector fields. We follow Arnold [1972], [1983] very closely. We remark the the phrase "unfolding" is often used to refer to a similar procedure; see, e.g., Guckenheimer and Holmes [1983] (where the term "universal unfolding" is taken to be virtually synonymous with the phrase "versal deformation") or Golubitsky and Schaeffer [1985] (where various subtleties in the definitions are thoroughly explored).

Consider the following $\mathbf{C}^\infty$, parameter-dependent vector fields

$$\dot{x} = f(x, \lambda), \qquad x \in \mathbb{R}^n, \quad \lambda \in \mathbb{R}^\ell, \tag{3.1.144}$$

$$\dot{y} = g(y, \mu), \qquad y \in \mathbb{R}^n, \quad \mu \in \mathbb{R}^m. \tag{3.1.145}$$

We will be concerned with local behavior of these vector fields near fixed points. Therefore, we assume that (3.1.144) and (3.1.145) have fixed points at (x_0, λ_0) and (y_0, μ_0), respectively, and we will be concerned with the dynamics in a *sufficiently small neighborhood* of these points. We now give a parametric version of the notion of $\mathbf{C}^0$-equivalence given in Definition 2.2.1.

DEFINITION 3.1.8 Equations (3.1.144) and (3.1.145) are said to be $\mathbf{C}^0$-equivalent (or topologically equivalent) if there exists a continuous map

$$h : U \to V,$$

with U a neighborhood of (x_0, λ_0) and V a neighborhood of y_0 such that, for λ sufficiently close to λ_0,

$$h(\cdot, \lambda),$$

with $h(x_0, \lambda_0) = y_0$, is a homeomorphism that takes orbits of the flow generated by (3.1.144) onto orbits of the flow generated by (3.1.145), preserving orientation but not necessarily parameterization by time. If h does preserve parameterization by time, then (3.1.144) and (3.1.145) are said to be $\mathbf{C}^0$ conjugate (or topologically conjugate).

In the construction of versal deformations we will be concerned with having the minimum number of parameters (for reasons to be discussed later), not too many or too few. The following definition is the first step toward formalizing this notion.

Consider the following $\mathbf{C}^\infty$ vector field

$$\dot{x} = u(x, \mu), \qquad x \in \mathbb{R}^n, \quad \mu \in \mathbb{R}^m, \tag{3.1.146}$$

with

$$u(x_0, \mu_0) = 0.$$

Then we have the following definition.

DEFINITION 3.1.9 Let $\lambda = \phi(\mu)$ be a continuous map defined for μ sufficiently close to μ_0 with $\lambda_0 = \phi(\mu_0)$. We say that (3.1.146) is induced from (3.1.144) if

$$u(x, \mu) = f(x, \phi(\mu)).$$

We now can give our main definition.

DEFINITION 3.1.10 Equation (3.1.144) is called a $\mathbf{C}^0$-*equivalent versal deformation* (or just versal deformation) of

$$\dot{x} = f(x, \lambda_0) \tag{3.1.147}$$

at the point x_0 if every other parameterized family of $\mathbf{C}^\infty$ vector fields that reduces to (3.1.147) for a particular choice of parameters is equivalent to a family of vector fields induced from (3.1.144).

At this point we again remind the reader that we are working in a *sufficiently small neighborhood* of (x_0, λ_0). We are now at the stage where we can construct versal deformations of dynamical systems. We will deal with vector fields and merely state the simple modifications needed for dealing with maps.

Once we have a vector field having a nonhyperbolic fixed point there are four steps necessary in order to construct a versal deformation.

Step 1. Put the vector field in normal form to reduce the number of cases that need to be considered.

Step 2. Truncate the normal form and embed the resulting k-jet of the normal form in a parameterized family having the number of parameters equal to the codimension of the bifurcation such that the parameterized family of k-jets is transverse to the appropriate subset of degenerate k-jets.

Step 3. Appeal to the Thom transversality theorem (Theorem 3.1.4) to argue that transversality holds in the full infinite-dimensional space of dynamical systems, $\mathbf{C}^\infty(\mathbb{R}^n, \mathbb{R}^n)$.

Step 4. Prove that the parameterized family constructed in this manner is actually a versal deformation.

It should be apparent that Step 4 is by far the most difficult. Steps 1 through 3 merely give us a procedure for constructing a parameterized family that we hope will be a versal deformation. The reason why it may not be is related to *dynamics.* The whole procedure is static only in nature; it takes into account only the nature of the fixed point. For one-dimensional vector fields we will see that the method will yield versal deformations, since the only possible orbits distinguished by $\mathbf{C}^0$ equivalence are fixed points. However, for higher dimensional vector fields we will see that the method may not yield versal deformations and, indeed, no such versal deformation may exist. Let us now consider some examples. In all of the following examples the origin will be the degenerate fixed point.

EXAMPLE 3.1.19: ONE-DIMENSIONAL VECTOR FIELDS

a) Consider the vector field

$$\dot{x} = ax^2 + \mathcal{O}(x^3). \tag{3.1.148}$$

We follow the steps described above for constructing a versal deformation of (3.1.148).

Step 1. Equation (3.1.148) is already in a sufficient normal form.

Step 2. We truncate (3.1.148) to obtain

$$\dot{x} = ax^2. \tag{3.1.149}$$

From Example 3.1.16, we know that $x = 0$ is a degenerate fixed point of (3.1.149) having *codimension* 1. Now recall the definitions of the subsets F and B of $J^2(\mathbb{R}^1, \mathbb{R}^1)$ described in Section 3.1D. For reference, we denote below these submanifolds of $J^2(\mathbb{R}^1, \mathbb{R}^1)$ and to their right a typical coordinate in the submanifold

$$J^2(\mathbb{R}^1, \mathbb{R}^1) - \left(x, f(x), \frac{\partial f}{\partial x}(x), \frac{\partial^2 f}{\partial x^2}(x)\right), \tag{3.1.150}$$

$$F - \left(x, 0, \frac{\partial f}{\partial x}(x), \frac{\partial^2 f}{\partial x^2}(x)\right), \tag{3.1.151}$$

$$B - \left(x, 0, 0, \frac{\partial^2 f}{\partial x^2}(x)\right). \tag{3.1.152}$$

Now the 2-jet of (3.1.148) at $x = 0$ is a typical point in B, and we are seeking to embed (3.1.149) in a one-parameter family transverse to B.

Upon examining (3.1.150), (3.1.151), and (3.1.152), it appears that there are two possibilities, namely,

$$\dot{x} = \mu x + a x^2 \tag{3.1.153}$$

or

$$\dot{x} = \mu + a x^2, \qquad \mu \in \mathbb{R}^1. \tag{3.1.154}$$

However, recall the remarks immediately following the definition of codimension (Definition 3.1.7). We noted that typical points of F are considered to be codimension zero since hyperbolic fixed points are structurally stable (and generic). Practically speaking, this means that as the parameter is varied the fixed point may move (and may even disappear). Therefore, we choose a one-parameter family that is transverse to B and F. Hence, (3.1.154) is our candidate for a versal deformation of (3.1.149). Finally, let us remark that the two-parameter vector field

$$\dot{x} = \mu_1 + \mu_2 x + a x^2 \tag{3.1.155}$$

is also a transverse family; however, our goal is to find the minimum number of parameters.

Step 3. It follows from the Thom transversality theorem (Theorem 3.1.4) that (3.1.154) is generic. The reader should perform the simple calculation necessary to verify this fact.

Step 4. In Section 3.1A, iii) we showed that higher order terms do not affect the dynamics of (3.1.154) near $(x, \mu) = (0, 0)$. Hence, the deformation is versal.

We can conclude from this that the saddle-node bifurcation is generic in one-parameter families of vector fields.

b) Consider the vector field

$$\dot{x} = a x^3 + \mathcal{O}(x^4). \tag{3.1.156}$$

Step 1. Equation (3.1.156) is already in a sufficient normal form.

Step 2. We truncate (3.1.156) to obtain

$$\dot{x} = a x^3. \tag{3.1.157}$$

From Example 3.1.17, $x = 0$ is a codimension 2 fixed point. For reference, we denote below the subsets B', B, and F of $J^3(\mathbb{R}^1, \mathbb{R}^1)$ discussed in Example 3.1.17 with typical coordinates immediately to the right.

$$J^3(\mathbb{R}^1, \mathbb{R}^1) - \left(x, f(x), \frac{\partial f}{\partial x}(x), \frac{\partial^2 f}{\partial x^2}(x), \frac{\partial^3 f}{\partial x^3}(x)\right), \tag{3.1.158}$$

$$F - \left(x, 0, \frac{\partial f}{\partial x}(x), \frac{\partial^2 f}{\partial x^2}(x), \frac{\partial^3 f}{\partial x^3}(x)\right), \tag{3.1.159}$$

$$B - \left(x, 0, 0, \frac{\partial^2 f}{\partial x^2}(x), \frac{\partial^3 f}{\partial x^3}(x)\right), \tag{3.1.160}$$

$$B' - \left(x, 0, 0, 0, \frac{\partial^3 f}{\partial x^3}(x)\right). \tag{3.1.161}$$

We want to embed (3.1.157) in a two-parameter family transverse to B'. The family must also be transverse to B and F. On examining (3.1.158), (3.1.1259), (3.1.1260), and (3.1.161), it is easy to see that such a family is given by

$$\dot{x} = \mu_1 + \mu_2 x + a x^3. \tag{3.1.162}$$

Step 3. It is an immediate consequence of the Thom transversality theorem (Theorem 3.1.4) that (3.1.162) is a generic family. The reader should perform the calculations necessary to verify this statement.

Step 4. In Section 3.1A,v), we proved that higher order terms do not qualitatively effect the local dynamics of (3.1.162). Hence, we have found a versal deformation of (3.1.156).

EXAMPLE 3.1.20: TWO-DIMENSIONAL VECTOR FIELDS

a) *The Poincaré-Andronov-Hopf Bifurcation*

Consider the vector field

$$\begin{aligned} \dot{x} &= -\omega y + \mathcal{O}(2), \\ \dot{y} &= \omega x + \mathcal{O}(2). \end{aligned} \tag{3.1.163}$$

Step 1. From Example 2.2.3, the normal form for this vector field is

$$\begin{aligned} \dot{x} &= -\omega y + (ax - by)(x^2 + y^2) + \mathcal{O}(5), \\ \dot{y} &= \omega x + (bx + ay)(x^2 + y^2) + \mathcal{O}(5). \end{aligned} \tag{3.1.164}$$

We note the important fact that due to the structure of the linear part of the vector field the normal form (3.1.164) contains no even-order terms.

Step 2. We take as the truncated normal form

$$\begin{aligned} \dot{x} &= -\omega y + (ax - by)(x^2 + y^2), \\ \dot{y} &= \omega x + (bx + ay)(x^2 + y^2). \end{aligned} \tag{3.1.165}$$

From Example 3.1.18 we recall that $(x, y) = (0, 0)$ is a codimension 1 fixed point. From Example 3.1.15 we recall the subsets B and F of $J^2(\mathbb{R}^2, \mathbb{R}^2)$ and denote typical coordinates on these subsets to the right.

$$J^2(\mathbb{R}^2, \mathbb{R}^2) - \left(x, f(x), Df(x), D^2 f(x)\right), \tag{3.1.166}$$

$$F - \left(x, 0, Df(x), D^2 f(x)\right), \tag{3.1.167}$$

$$B - \left(x, 0, \widetilde{Df}(x), D^2 f(x)\right), \tag{3.1.168}$$

where $\widetilde{Df}(x)$ represents nonhyperbolic matrices. Now we want to embed (3.1.165) in a one-parameter family transverse to B, but due to the structure of the linear part of the vector field, we want the origin to remain a fixed point. From Appendix 1 the matrix

$$\begin{pmatrix} 0 & -\omega \\ \omega & 0 \end{pmatrix} \tag{3.1.169}$$

lies on a surface of codimension one in the four-dimensional space of 2×2 matrices. Moreover, a versal deformation for (3.1.169) is given by

$$\begin{pmatrix} \mu & -\omega \\ \omega & \mu \end{pmatrix}. \tag{3.1.170}$$

Using (3.1.170), we take as our transverse family

$$\begin{aligned} \dot{x} &= \mu x - \omega y + (ax - by)(x^2 + y^2), \\ \dot{y} &= \omega x + \mu y + (bx + ay)(x^2 + y^2). \end{aligned} \tag{3.1.171}$$

Step 3. It follows from the Thom transversality theorem (Theorem 3.1.4) that (3.1.171) is generic (when we take into account that the origin must remain a fixed point in the one-parameter family). The reader should perform the necessary calculations to verify this statement.

Step 4. Theorem 3.1.3 implies that the higher order terms in the normal form do not qualitatively change the dynamics of (3.1.171). Hence, we have constructed a versal deformation of (3.1.163).

Thus, we can conclude that, like saddle-node bifurcations, Poincaré-Andronov-Hopf bifurcations are also generic in one-parameter families of vector fields.

Due to the importance of the Poincaré-Andronov-Hopf bifurcation, before ending our discussion let us reexamine it from a slightly different point of view.

In polar coordinates, the normal form is given by (cf. (3.1.100))

$$\begin{aligned} \dot{r} &= ar^3 + \mathcal{O}(r^5), \\ \dot{\theta} &= \omega + \mathcal{O}(r^2). \end{aligned} \tag{3.1.172}$$

Our goal is to construct a versal deformation of (3.1.172). It is in this context that we see yet another example of the power and conceptual clarity that results upon transforming the system to normal form.

In our study of the dynamics of (3.1.172) we have seen (see Section 3.1B, Lemma 3.1.1) that, for r sufficiently small, we need only to study

$$\dot{r} = ar^3 + \mathcal{O}(r^5) = 0, \tag{3.1.173}$$

since $\theta(t)$ merely increases monotonically in t.

Equation (3.1.173) looks very much like the degenerate one-dimensional vector field studied in Example 3.1.19 whose versal deformation yielded the pitchfork bifurcation. However, there is an important difference; namely, due to the structure of the linear part of the vector field in (3.1.163), the $\dot{r}$ component of the vector field must have no even-order terms in r and must be zero at $r = 0$. Hence, this degenerate fixed point is codimension 1 rather than codimension 2 as in the pitchfork bifurcation, and a natural candidate for a versal deformation is

$$\begin{aligned} \dot{r} &= \mu r + a r^3, \\ \dot{\theta} &= \omega. \end{aligned} \tag{3.1.174}$$

It is the content of Theorem 3.1.3 that (3.1.174) is indeed a versal deformation.

b) *A Double-Zero Eigenvalue*

Consider the vector field

$$\begin{aligned} \dot{x} &= y + \mathcal{O}(2), \\ \dot{y} &= \mathcal{O}(2). \end{aligned} \tag{3.1.175}$$

Step 1. From Example 2.2.2, the normal form for (3.1.175) is given by

$$\begin{aligned} \dot{x} &= y + \mathcal{O}(3), \\ \dot{y} &= axy + by^2 + \mathcal{O}(3). \end{aligned} \tag{3.1.176}$$

Step 2. We take as the truncated normal form

$$\begin{aligned} \dot{x} &= y, \\ \dot{y} &= axy + by^2. \end{aligned} \tag{3.1.177}$$

From Example 3.1.18, recall that $(x, y) = (0, 0)$ is a codimension 2 fixed point. From Example 3.1.15 we denote the subsets B and F of $J^2(\mathbb{R}^2, \mathbb{R}^2)$ with typical coordinates of points in these sets to the right.

$$J^1(\mathbb{R}^2, \mathbb{R}^2) - (x, f(x), Df(x)), \tag{3.1.178}$$

$$F - (x, 0, Df(x)), \tag{3.1.179}$$

$$B - \left(x, 0, \widetilde{Df}(x)\right), \tag{3.1.180}$$

where $\widetilde{Df}(x)$ represent nonhyperbolic matrices. In Appendix 1 we show that these matrices form a three-dimensional surface in the four-dimensional space of 2×2 matrices. We also show in this appendix that matrices with Jordan canonical form given by

$$\begin{pmatrix} 0 & 1 \\ 0 & 0 \end{pmatrix} \tag{3.1.181}$$

form a two-dimensional surface, B', with $B' \subset B \subset F \subset J^1(\mathbb{R}^1, \mathbb{R}^1)$. We now seek to embed (3.1.177) in a two-parameter family transverse to B'. In Appendix 1 we show that a versal deformation of (3.1.181) is given by

$$\begin{pmatrix} 0 & 1 \\ \mu_1 & \mu_2 \end{pmatrix}. \tag{3.1.182}$$

Thus, one might take as a transverse family

$$\begin{aligned} \dot{x} &= y, \\ \dot{y} &= \mu_1 x + \mu_2 y + ax^2 + bxy. \end{aligned} \tag{3.1.183}$$

However, recall the remarks following the definition of codimension (Definition 3.1.7). Generically, we expect the fixed points to move as the parameters are varied. This does not happen in (3.1.183); the origin always remains a fixed point. This situation is easy to remedy.

Notice from the form of (3.1.183) that any fixed point must have $y = 0$. If we make the coordinate transformation

$$\begin{aligned} x &\to x - x_0, \\ y &\to y, \end{aligned} \tag{3.1.184}$$

and take as a versal deformation of the linear part

$$\begin{pmatrix} 0 & 1 \\ \mu_1 & \mu_2 \end{pmatrix} \begin{pmatrix} x - x_0 \\ y \end{pmatrix}, \tag{3.1.185}$$

then a simple rescaling and reparameterization allows us to transform (3.1.183) into

$$\begin{aligned} \dot{x} &= y, \\ \dot{y} &= \mu_1 + \mu_2 y + axy + by^2 \end{aligned} \tag{3.1.186}$$

(cf. Exercise 3.20). We remark that in some cases it may be necessary for the origin to remain a fixed point as the parameters are varied; for example, in the case where the normal form is invariant under the transformation $(x, y) \to (-x, -y)$; see Exercise 3.32.

Step 3. It follows from the Thom transversality theorem that (3.1.186) is a generic family. The reader should perform the necessary calculation to verify this statement.

Step 4. Bogdanov [1975] proved that (3.1.186) is a versal deformation. We will consider this question in great detail in Section 3.1E.

iii) Practical Remarks on the Location of Parameters

For a specific dynamical system arising in applications the number of parameters and their locations in the equation are usually fixed. The theory developed in this section tells us that too many parameters (i.e., more than the codimension of the fixed point) are permissible provided they result in a transverse family. Having more parameters than the codimension of the fixed point will simply require more work in enumerating all the cases. However, one must verify that the parameters are in the correct location so as to form a transverse family. We have seen that there is a lot of freedom in where the parameters may be; in one dimension it is fairly obvious, in higher dimensions it is not as obvious, but the reader should keep in mind that transversality is the typical situation.

Appendix 1: Versal Deformations of Families of Matrices

In this appendix we develop the theory of versal deformations of matrices which we use in computing versal deformations of fixed points of dynamical systems. Our discussion follows Arnold [1983]. Let M be the space of $n \times n$ matrices with complex entries. The relation of similarity of matrices partitions the entire space into manifolds consisting of matrices having the same eigenvalues and dimensions of Jordan blocks; this partitioning is continuous, since the eigenvalues vary continuously.

Suppose we have a matrix having some identical eigenvalues, and we want to reduce it to Jordan canonical form. This process is not stable, because the slightest perturbation might destroy the Jordan canonical form completely. Thus, if the matrix is only approximately known (or the reduction is attempted by computer), then the procedure may yield nonsense.

We give an example. Consider the matrix

$$A(\lambda) = \begin{pmatrix} 0 & \lambda \\ 0 & 0 \end{pmatrix}.$$

The Jordan canonical form is given by

$$\begin{pmatrix} 0 & 1 \\ 0 & 0 \end{pmatrix}, \qquad \lambda \neq 0,$$

with conjugating matrix

$$C(\lambda) = \begin{pmatrix} 1 & 0 \\ 0 & 1/\lambda \end{pmatrix}, \qquad \lambda \neq 0.$$

Now, at $\lambda = 0$, the Jordan canonical form of $A(\lambda)$ is given by

$$\begin{pmatrix} 0 & 0 \\ 0 & 0 \end{pmatrix},$$

with conjugating matrix

$$C(0) = \begin{pmatrix} 1 & 0 \\ 0 & 1 \end{pmatrix}.$$

Therefore, $C(\lambda)$ is discontinuous at $\lambda = 0$.

However, even though multiple eigenvalues is an unstable situation for individual matrices, it is stable for parametrized families of matrices, i.e., perturbing the family does not remove the multiple eigenvalue matrix from the family. Thus, while we can reduce every member of the family to Jordan canonical form (as in the example above), in general, the transformation will depend discontinuously on the parameter. The problem we address is the following:

What is the simplest form to which a family of matrices can be reduced depending smoothly (say, analytically) on the parameters by a change of parameters depending smoothly (analytically) on the parameters?

In the following we will construct such families and determine the minimum number of parameters, but first we must begin with some definitions and develop some necessary machinery.

We will consider $n \times n$ matrices whose entries are complex numbers. Let A_0 be such a matrix.

A *deformation* of A_0 is an analytic mapping

$$\begin{aligned} A : \Lambda &\to \mathbb{C}^{n^2}, \\ \lambda &\to A(\lambda), \end{aligned}$$

where $\Lambda \in \mathbb{C}^{\ell}$ is some parameter space and

$$A(\lambda_0) = A_0.$$

A deformation is also called a family, the variables λ_i, $i = 1, \cdots, \ell$, are called the parameters, and Λ is called the *base* of the family.

Two deformations $A(\lambda)$, $B(\lambda)$ of A_0 are called *equivalent* if there exists a deformation of the identity matrix $C(\lambda)$ ($C(\lambda_0) = \text{id}$) with the same base such that

$$B(\lambda) = C(\lambda)A(\lambda)C^{-1}(\lambda).$$

Let $\Sigma \subset \mathbb{C}^m$, $\Lambda \subset \mathbb{C}^{\ell}$ be open sets. Consider the analytic mapping

$$\begin{aligned} \phi : \Sigma &\to \Lambda, \\ \mu &\to \phi(\mu), \end{aligned}$$

with $\phi(\mu_0) = \lambda_0$.

The *family induced from A by the mapping ϕ* is called $(\phi^* A)(\mu)$ and is defined by

$$(\phi^* A)(\mu) \equiv A(\phi(\mu)), \qquad \mu \in \mathbb{C}^m.$$

This will be useful for reparametrizing families of matrices in order to reduce the number of parameters.

A deformation $A(\lambda)$ of a matrix A_0 is said to be *versal* if any deformation $B(\mu)$ of A_0 is equivalent to a deformation induced from A, i.e.,

$$B(\mu) = C(\mu)A(\phi(\mu))C^{-1}(\mu)$$

for some change of parameters

$$\phi : \Sigma \to \Lambda,$$

with $C(\mu_0) = \text{id}$ and $\phi(\mu_0) = \lambda_0$.

A versal deformation is said to be *universal* if the inducing mapping (i.e., change of parameters map) is determined uniquely by the deformation B.

A versal deformation is said to be *miniversal* if the dimension of the parameter space is the smallest possible for a versal deformation.

At this stage it is useful to consider an example. Consider the matrix

$$A_0 = \begin{pmatrix} 0 & 1 \\ 0 & 0 \end{pmatrix}.$$

It should be clear that a versal deformation of A_0 is given by

$$B(\mu) \equiv \begin{pmatrix} 0 & 1 \\ 0 & 0 \end{pmatrix} + \begin{pmatrix} \mu_1 & \mu_2 \\ \mu_3 & \mu_4 \end{pmatrix},$$

where $\mu \equiv (\mu_1, \mu_2, \mu_3, \mu_4) \in \mathbb{C}^4$. However, $B(\mu)$ is not miniversal; a miniversal deformation is given by

$$A(\lambda) = \begin{pmatrix} 0 & 1 \\ 0 & 0 \end{pmatrix} + \begin{pmatrix} 0 & 0 \\ \lambda_1 & \lambda_2 \end{pmatrix},$$

where $\lambda = (\lambda_1, \lambda_2) \in \mathbb{C}^2$. This can be seen by showing that $B(\mu)$ is equivalent to a deformation induced from $A(\lambda)$. If we let

$$C(\mu) = \begin{pmatrix} 1 + \mu_2 & 0 \\ -\mu_1 & 1 \end{pmatrix}, \qquad C^{-1}(\mu) = \frac{1}{1 + \mu_2} \begin{pmatrix} 1 & 0 \\ \mu_1 & 1 + \mu_2 \end{pmatrix},$$

then it follows that

$$\begin{aligned} A(\lambda) = A(\phi(\mu)) &= C^{-1}(\mu)B(\mu)C(\mu) \\ &= \begin{pmatrix} 0 & 1 \\ 0 & 0 \end{pmatrix} + \begin{pmatrix} 0 & 0 \\ \mu_3(1 + \mu_2) - \mu_1\mu_4 & \mu_1 + \mu_4 \end{pmatrix}, \end{aligned}$$

where we take

$$\phi(\mu) = (\phi_1(\mu), \phi_2(\mu)) = (\mu_3(1 + \mu_2) - \mu_1\mu_4, \mu_1 + \mu_4) \equiv (\lambda_1, \lambda_2) \equiv \lambda$$

as the inducing mapping.

Now that we have the necessary definitions out of the way we can proceed toward our goal, which is to construct normal forms (miniversal deformations) of matrices having multiple eigenvalues. It is important to know the number of parameters necessary and to know the conditions that the normal form must satisfy for versality. To reach that point we must develop some machinery in order that the result does not appear to be "pulled out of the air."

We denote the set of all $n \times n$ matrices with complex entries by M. M is isomorphic to $\mathbb{C}^{n^2}$; however, we will simply write $M = \mathbb{C}^{n^2}$.

Now let us consider the Lie group $G = GL(n, \mathbb{C})$ of all nonsingular $n \times n$ matrices with complex entries. $GL(n, \mathbb{C})$ is a submanifold of $\mathbb{C}^{n^2}$.

The group G acts on M according to the formula

$$Ad_g m = gmg^{-1}, \qquad (m \in M, \quad g \in G) \tag{A.1.1}$$

(Ad stands for adjoint).

Consider the orbit of an arbitrary fixed matrix $A_0 \in M$ under the action of G; this is the set of points $m \in M$ such that $m = gA_0g^{-1}$ for all $g \in G$. The orbit of A_0 under G forms a smooth submanifold of M, which we denote by N. Thus, from (A.1.1), the orbit, N, of A_0 consists of all matrices similar to A_0.

We next restate the notion of *transversality of a map.* (cf. Definition 1.2.6).

Let $N \subset M$ be a smooth submanifold of a manifold M. Consider a smooth mapping of another manifold Λ into M and let λ be a point in Λ such that $A(\lambda) \in N$. Then the mapping A is called *transversal to N at λ* if the tangent space to M at $A(\lambda)$ is the sum

$$TM_{A(\lambda)} = TN_{A(\lambda)} + DA(\lambda) \cdot T\Lambda_\lambda, \tag{A.1.2}$$

where $DA(\lambda)$ denotes the derivative of A at λ; see Figure A.1.1.

With these two notions we can state and prove the proposition that provides the key for constructing miniversal deformations of Jordan matrices.

Proposition A.1.1 *If the mapping A is transversal to the orbit of A_0 at $\lambda = \lambda_0$, then $A(\lambda)$ is a versal deformation. If the dimension of the parameter space is equal to the codimension of the orbit of A_0, then the deformation is miniversal.*

Proof: Unfortunately we cannot proceed to the proof directly but must go in a rather roundabout way through several steps and definition. First note, however, that a geometrical picture of Proposition A.1.1 is given in Figure A.1.2.

In Figure A.1.2 N is codimension 2; hence, we choose the dimension of λ to be 2. Since $A(\lambda)$ is transverse to N at $\lambda = \lambda_0$, we thus represent it as a two-dimensional surface passing through A_0. We want to show that $A(\lambda)$ satisfying this geometrical picture is actually a miniversal deformation of A_0. To do this we will need to develop a local coordinate structure near A_0 which describes points along the orbit of A_0 and points off of the orbit of A_0. We begin with a definition.

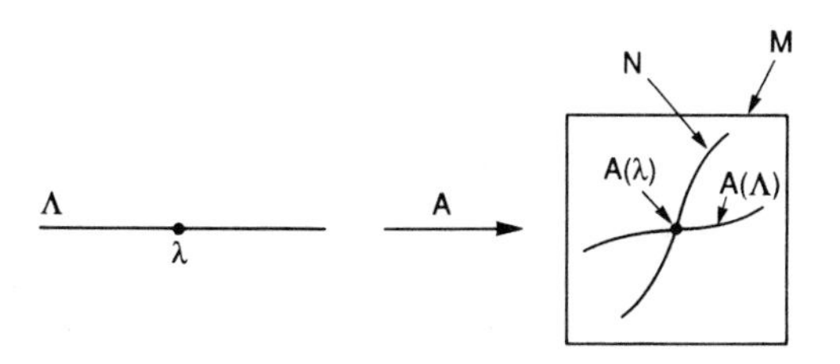

FIGURE A.1.1

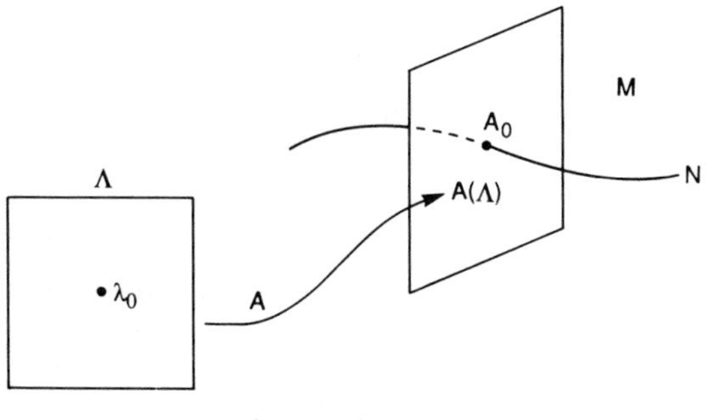

FIGURE A.1.2

The *centralizer of a matrix* u is the set of all matrices commuting with u denoted

$$Z_u = \{v : [u,v] = 0\}, \qquad [u,v] \equiv uv - vu. \tag{A.1.3}$$

It is easy to show that the centralizer of any matrix of order n is a linear subspace of $M = \mathbb{C}^{n^2}$. We leave this as an exercise for the reader.

Now we want to develop the geometrical structure of M near $A(\lambda) \equiv A_0$. Let $\mathcal{Z}$ be the centralizer of the matrix A_0. Consider the set of nonsingular matrices that contain the identity matrix (which we denote by "id"). Clearly this set has dimension n^2. Within this set consider a smooth submanifold, P, intersecting the subspace id$+\mathcal{Z}$ transversely at id and having dimension equal to the codimension of the centralizer; see Figure A.1.3.

With Figure A.1.3 in mind, consider the mapping

$$\Phi : P \times \Lambda \to \mathbb{C}^{n^2},$$

$$\Phi : (p, \lambda) \to pA(\lambda)p^{-1} \equiv \Phi(p, \lambda) \tag{A.1.4}$$

(we will worry about dimensions shortly).

The following lemma will provide us with local coordinates near A_0.

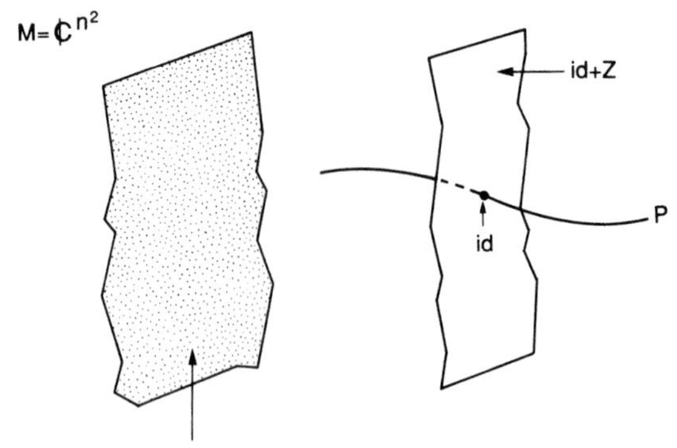

FIGURE A.1.3

Lemma A.1.2 *In a neighborhood of* (id, λ_0) Φ *is a local diffeomorphism.*

Proof: Before proving lemma this we need to state several facts.

1) Consider the mapping

$$\begin{aligned}\psi : G &\to \mathbb{C}^{n^2},\\ b &\to bA_0b^{-1} \equiv \psi(b).\end{aligned}$$

The derivative of ψ at the identity is a linear mapping of $T_{\mathrm{id}}G$ onto $T_{A_0}\mathbb{C}^{n^2}$. Without loss of generality, we can take $T_{\mathrm{id}}G = \mathbb{C}^{n^2}$ and $T_{A_0}\mathbb{C}^{n^2} = \mathbb{C}^{n^2}$. Denoting the derivative of ψ at id by $D\psi(\mathrm{id})$, we want to show that, for $u \in \mathbb{C}^{n^2}$, $D\psi(\mathrm{id})u$ is given by the operation of commutation of u with A_0, i.e., $D\psi(\mathrm{id})u = [u, A_0]$.

This is easily calculated as follows

$$\begin{aligned}D\psi(\mathrm{id})u &= \lim_{\varepsilon\to 0} \frac{(\mathrm{id} + \varepsilon u)A_0(\mathrm{id} + \varepsilon u)^{-1} - A_0}{\varepsilon}\\ &= \lim_{\varepsilon\to 0} \frac{(A_0 + \varepsilon u A_0)(\mathrm{id} - \varepsilon u) + \mathcal{O}(\varepsilon^2) - A_0}{\varepsilon}\\ &= \lim_{\varepsilon\to 0} \frac{A_0 - \varepsilon A_0 u + \varepsilon u A_0 + \mathcal{O}(\varepsilon^2) - A_0}{\varepsilon}\\ &= uA_0 - A_0u \equiv [u, A_0];\end{aligned}$$

therefore,

$$\begin{aligned}D\psi(\mathrm{id}) : \mathbb{C}^{n^2} &\to \mathbb{C}^{n^2},\\ u &\to [u, A_0].\end{aligned} \tag{A.1.5}$$

We make the following observation. Since $\dim G = \dim M = n^2$, the dimension of the centralizer is equal to the codimension of the orbit of A_0.

This is because, roughly speaking, from (A.1.1) and (A.1.4) we can think of the centralizer of A_0 to be the matrices that do not change A_0. Thus, we have

$$\dim \mathcal{Z} = \dim \Lambda, \tag{A.1.6}$$

$$\dim P = \dim N, \tag{A.1.7}$$

and

$$\dim \Lambda + \dim N = n^2.$$

Now, returning to Φ,

$$\Phi : P \times \Lambda \to \mathbb{C}^{n^2}.$$

From (A.1.6) and (A.1.7) we see that the dimensions are consistent (i.e., $\dim(P \times \Lambda) = \dim \mathbb{C}^{n^2}$) for Φ to be a diffeomorphism.

We can now finally prove Lemma A.1.2. We compute the derivative of Φ at (id, λ_0), denoted $D\Phi(\mathrm{id}, \lambda_0)$, and examine how it acts on a typical element of $T_{(\mathrm{id},\lambda_0)}(P \times \Lambda)$. Let $(u, \lambda) \in T_{(\mathrm{id},\lambda_0)}(P \times \Lambda)$; then we have

$$D\Phi(\mathrm{id}, \lambda_0)(u, \lambda) = (D_p\Phi(\mathrm{id}, \lambda_0), D_\lambda\Phi(\mathrm{id}, \lambda_0))(u, \lambda). \tag{A.1.8}$$

Using (A.1.4) and (A.1.5) it is easy to see that (A.1.8) is given by

$$D\Phi(\mathrm{id}, \lambda_0)(u, \lambda) = ([u, A_0], DA(\lambda_0)\lambda). \tag{A.1.9}$$

By construction of the submanifold P, $D_p\Phi(\mathrm{id}, \lambda_0)$ maps $T_{\mathrm{id}}P$ isomorphically to a space tangent to the orbit N at A_0 (check dimensions and the fact that $[u, A_0] \neq 0$). Also by the hypothesis of Proposition A.1.1, $DA(\lambda_0)$ maps $T_{\lambda_0}\Lambda$ isomorphically to a space transverse to N at $A(\lambda_0) = A_0$. Consequently, $D\Phi(\mathrm{id}, \lambda_0)$ is an isomorphism between linear spaces of dimension n^2; thus, by the inverse function theorem, Φ is a local diffeomorphism. This completes our proof of Lemma A.1.2. □

This lemma tells us that we have a local product structure (in terms of coordinates) near A_0 in M (note: this is for a sufficiently small neighborhood of (id, λ_0), since the inverse function theorem is only a local result); see Figure A.1.4.

Now we can finish the proof of Proposition A.1.1.

Let $B(\mu)$ for some fixed $\mu \in \Sigma \subset \mathbb{C}^m$ be an arbitrary deformation of A_0 (i.e., for some $\mu_0 \in \Sigma \subset \mathbb{C}^m$, $B(\mu_0) = A_0$). In the local coordinates near A_0, we know that any matrix sufficiently close to A_0 can be represented as

$$\Phi(p, \lambda) = pA(\lambda)p^{-1}, \qquad p \in P, \quad \lambda \in \Lambda \subset \mathbb{C}^\ell.$$

Hence, for $\mu - \mu_0$ sufficiently small, $B(\mu)$ has the representation

$$B(\mu) = \Phi(p, \lambda) = pA(\lambda)p^{-1}$$

for some $p \in P$, $\lambda \in \Lambda \subset \mathbb{C}^\ell$.

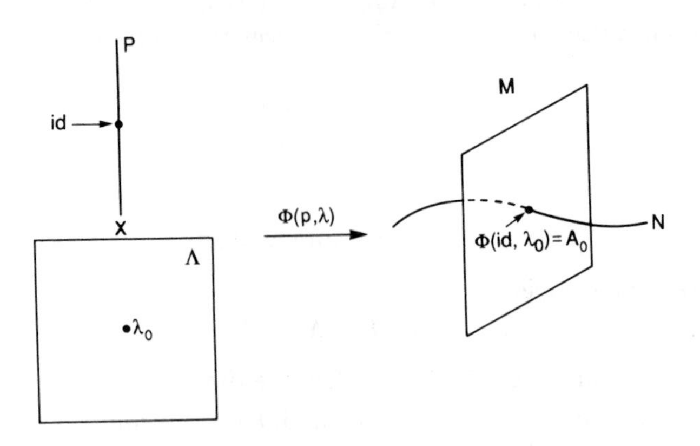

FIGURE A.1.4

Now let π_1 and π_2 be the projections onto P and Λ of $P \times \Lambda$, respectively. Then it follows by definition of Φ that

$$\lambda = \pi_2 \Phi^{-1}(B(\mu)),$$
$$p = \pi_1 \Phi^{-1}(B(\mu)).$$

Thus, letting

$$\phi(\mu) = \pi_2 \Phi^{-1}(B(\mu)),$$
$$C(\mu) = \pi_1 \Phi^{-1}(B(\mu)),$$

it follows that

$$B(\mu) = C(\mu) A(\phi(\mu)) C^{-1}(\mu).$$

This proves Proposition A.1.1. □

We remark that it should be clear from the argument that the deformation is miniversal, i.e., we have used the smallest number of parameters.

Thus, the proposition tells us that, in order to construct a miniversal deformation of A_0, we may take the family of matrices

$$A_0 + B,$$

where B is in the orthogonal complement of the orbit of A_0. An obvious question therefore is how do you compute B?

Lemma A.1.3 *A vector B in the tangent space of $\mathbb{C}^{n^2}$ at the point A_0 is perpendicular to the orbit of A_0 if and only if*

$$[B^*, A_0] = 0$$

where B^* denotes the adjoint of B.

Proof: Vectors tangent to the orbit are matrices representable in the form

$$[x, A_0], \qquad x \in M.$$

Orthogonality of B to the orbit of A_0 means that, for any $x \in M$,

$$\langle [x, A_0], B \rangle = 0, \tag{A.1.10}$$

where $\langle \ , \ \rangle$ is the inner product on the space of matrices, which we take as

$$\langle A, B \rangle = \mathrm{tr}(AB^*). \tag{A.1.11}$$

Using (A.1.10), (A.1.11) becomes

$$\begin{aligned} 0 &= \mathrm{tr}\left([x, A_0]B^*\right) \\ &= \mathrm{tr}\left(xA_0B^* - A_0xB^*\right). \end{aligned}$$

Using the fact that

$$\begin{aligned} \mathrm{tr}(AB) &= \mathrm{tr}(BA), \\ \mathrm{tr}(A+B) &= \mathrm{tr}A + \mathrm{tr}B, \end{aligned}$$

we obtain

$$\begin{aligned} \mathrm{tr}(xA_0B^* - A_0xB^*) &= \mathrm{tr}(xA_0B^*) - \mathrm{tr}(A_0xB^*) \\ &= \mathrm{tr}(A_0B^*x) - \mathrm{tr}(xB^*A_0) \\ &= \mathrm{tr}(A_0B^*x) - \mathrm{tr}(B^*A_0x) \\ &= \mathrm{tr}((A_0B^* - B^*A_0)x) \\ &= \mathrm{tr}([A_0, B^*]x) \\ &= \langle [A_0, B^*], x^* \rangle = 0. \end{aligned}$$

Since x was arbitrary, this implies

$$[A_0, B^*] = 0.$$

□

This lemma actually allows us to "read off" the form of B if A_0 is in Jordan canonical form.

Suppose A_0 has been transformed to Jordan canonical form and has distinct eigenvalues

$$\alpha_i, \qquad i = 1, \cdots, s,$$

and to each eigenvalue there corresponds a finite number of Jordan blocks of order n_i

$$n_1(\alpha_i) \geq n_2(\alpha_i) \geq \cdots .$$

For the moment, to simplify our arguments, we will assume that our matrix has only one distinct eigenvalue, α, and let us say three Jordan blocks $n_1(\alpha) \geq n_2(\alpha) \geq n_3(\alpha)$. The matrices which commute with A_0 then

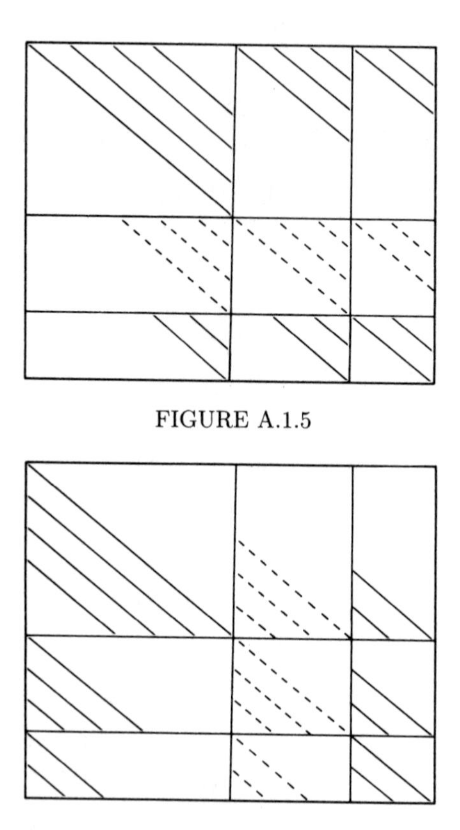

FIGURE A.1.5

FIGURE A.1.6

have the structure shown in Figure A.1.5, where *each oblique segment in each separate Jordan block* denotes a sequence of equal entries. Thus, a matrix B^* in the orthogonal complement of A_0 has the structure shown in Figure A.1.6. The general proofs of these statements can be found in Gantmacher [1977], [1989] for the case of an arbitrary number of eigenvalues and Jordan blocks; however, we will not need such generality, since we will be concerned with 2×2, 3×3, and 4×4 matrices only and, in these cases, it is relatively easy to verify the structures shown in Figures A.1.5 and A.1.6 by direct calculation. We will do this shortly.

Therefore, a matrix of the structure shown in Figure A.1.6 is orthogonal to A_0. Now in general we only desire transversality, not orthogonality, and it may simplify matters (i.e., reduce the number of matrix elements) to choose a basis not orthogonal but transverse to A_0 so that B^* appears as simply as possible. This can be accomplished by taking a matrix of the form shown in Figure A.1.6 and B^* and replacing every slanted line by one independent parameter and the rest of the entries on the slanted line by

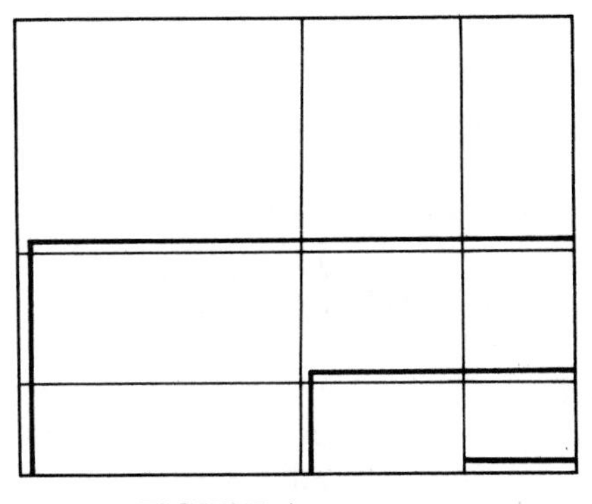

FIGURE A.1.7

zeros. The nonzero entry (independent parameter) can be placed at any position along the slanted line. Thus, matrices transverse to A_0 would have the structure shown in Figure A.1.7, where the horizontal and vertical lines in the figure represent the positions where the required number of independent parameters are placed. From the above form of the matrices commuting with A_0, and only *one distinct eigenvalue*, the number of parameters needed for a miniversal deformation is given by the formula

$$n_1(\alpha) + 3n_2(\alpha) + 5n_3(\alpha) + \cdots . \tag{A1.1.12}$$

We now will state the general case for an arbitrary number of eigenvalues α_i, $i = 1, \cdots, s$, and then work out some explicit examples.

Theorem A.1.4 *The smallest number of parameters of a versal deformation of the matrix A_0 is equal to*

$$d = \sum_{i=1}^{s} [n_1(\alpha_i) + 3n_2(\alpha_i) + 5n_3(\alpha_i) + \cdots] . \tag{A.1.13}$$

Proof: See Arnold [1983] and Gantmacher [1977], [1989]. □

We now sum up everything in our main theorem.

Theorem A.1.5 *Every matrix A_0 has a miniversal deformation; the number of its parameters is equal to the codimension of the orbit of A_0 or, equivalently, to the dimension of the centralizer of A_0.*

If A_0 is in Jordan normal form, then for a miniversal deformation we may take a d-parameter normal form (with d given in Theorem A.1.4) $A_0 + B$, where the blocks of B have the previously described form.

In other words, any complex matrix close to a given matrix can be reduced to the above d-parameter normal form $A_0 + B$ (where A_0 is the Jordan *canonical form of the given matrix), so that the reducing mapping and the parameters of the normal form depend analytically on the elements of the original matrix.*

Now we will compute some examples.

EXAMPLE A.1.1. Consider the matrix

$$A_0 = \begin{pmatrix} \alpha & 1 \\ 0 & \alpha \end{pmatrix}, \qquad \alpha^2. \tag{A.1.14}$$

This matrix is denoted α^2, where α refers to the eigenvalue and two refers to the size of the Jordan block. In the context of bifurcations of fixed points of vector fields, we take $\alpha = 0$ and, for fixed points of maps, we take $\alpha = 1$. From (A.1.12) it follows that a versal deformation of (A.1.14) has at least two parameters; hence, matrices having Jordan canonical form A_0 (i.e., from (A.1.1), the orbit of A_0) form a codimension 2 submanifold of $\mathbb{C}^4$.

We now want to compute a versal deformation for A_0. First we compute a matrix which commutes with A_0.

$$\begin{aligned} &\begin{pmatrix} a & b \\ c & d \end{pmatrix}\begin{pmatrix} \alpha & 1 \\ 0 & \alpha \end{pmatrix} - \begin{pmatrix} \alpha & 1 \\ 0 & \alpha \end{pmatrix}\begin{pmatrix} a & b \\ c & d \end{pmatrix} \\ &= \begin{pmatrix} a\alpha & a + b\alpha \\ c\alpha & c + d\alpha \end{pmatrix} - \begin{pmatrix} a\alpha + c & b\alpha + d \\ \alpha c & \alpha d \end{pmatrix} \\ &= \begin{pmatrix} -c & a - d \\ 0 & c \end{pmatrix} = \begin{pmatrix} 0 & 0 \\ 0 & 0 \end{pmatrix}; \end{aligned}$$

thus,

$$c = 0, \qquad a = d, \qquad \text{and} \quad b = \text{arbitrary}.$$

Therefore, we obtain

$$B^* = \begin{pmatrix} a & b \\ 0 & a \end{pmatrix}.$$

From Lemma A.1.3, we have that a matrix *orthogonal* to A_0 is given by

$$B = \begin{pmatrix} \bar{a} & 0 \\ \bar{b} & \bar{a} \end{pmatrix},$$

where $\bar{a}, \bar{b}$ are arbitrary complex numbers.

A family of matrices that is transverse to A_0 and that simplifies our expression for a versal deformation would be

$$\tilde{B} = \begin{pmatrix} 0 & 0 \\ \bar{b} & \bar{a} \end{pmatrix}.$$

We can check whether or not $\tilde{B}$ is transverse to A_0 by showing that

$$\langle \tilde{B}, B \rangle \equiv \operatorname{tr}(\tilde{B}B^*) \neq 0.$$

In our case we have

$$\tilde{B}B^* = \begin{pmatrix} 0 & 0 \\ \bar{b} & \bar{a} \end{pmatrix} \begin{pmatrix} a & 0 \\ b & a \end{pmatrix} = \begin{pmatrix} 0 & 0 \\ \bar{b}a + \bar{a}b & |a|^2 \end{pmatrix}$$

so that

$$\langle \tilde{B}, B \rangle \equiv \text{tr}(\tilde{B}B^*) = |a|^2 \neq 0.$$

Relabeling by letting $\bar{b} = \lambda_1$ and $\bar{a} = \lambda_2$, we obtain the following versal deformation

$$\begin{pmatrix} \alpha & 1 \\ 0 & \alpha \end{pmatrix} + \begin{pmatrix} 0 & 0 \\ \lambda_1 & \lambda_2 \end{pmatrix}.$$

EXAMPLE A.1.2 Consider the matrix

$$A_0 = \begin{pmatrix} \alpha & 0 \\ 0 & \alpha \end{pmatrix}, \qquad \alpha\alpha. \tag{A.1.15}$$

From (A.1.12), it follows that a versal deformation of (A.1.15) has four parameters; hence, matrices having Jordan normal form A_0 form a codimension 4 submanifold of $\mathbb{C}^4$.

We next compute a family orthogonal to the orbit of A_0 as follows

$$\begin{pmatrix} a & b \\ c & d \end{pmatrix} \begin{pmatrix} \alpha & 0 \\ 0 & \alpha \end{pmatrix} - \begin{pmatrix} \alpha & 0 \\ 0 & \alpha \end{pmatrix} \begin{pmatrix} a & b \\ c & d \end{pmatrix} = \begin{pmatrix} 0 & 0 \\ 0 & 0 \end{pmatrix}.$$

or

$$\begin{pmatrix} a\alpha & b\alpha \\ c\alpha & d\alpha \end{pmatrix} - \begin{pmatrix} \alpha a & \alpha b \\ \alpha c & \alpha d \end{pmatrix} = \begin{pmatrix} 0 & 0 \\ 0 & 0 \end{pmatrix}.$$

Thus, a, b, c, d can be anything; this situation is therefore codimension 4 with a versal deformation given by

$$\begin{pmatrix} \alpha & 0 \\ 0 & \alpha \end{pmatrix} + \begin{pmatrix} \lambda_1 & \lambda_2 \\ \lambda_3 & \lambda_4 \end{pmatrix},$$

where we have relabeled the parameters as in Example A.1.1.

EXAMPLE A.1.3 Consider the matrix

$$A_0 = \begin{pmatrix} \alpha & 1 & 0 \\ 0 & \alpha & 0 \\ 0 & 0 & \alpha \end{pmatrix}, \qquad \alpha^2\alpha. \tag{A.1.16}$$

From (A.1.12), it follows that a versal deformation of (A.1.16) has at least five parameters; hence, matrices having Jordan canonical form A_0 form a codimension 5 submanifold of $\mathbb{C}^9$.

We compute a family of matrices orthogonal to the orbit of A_0 as follows

$$\begin{pmatrix} a & b & c \\ d & e & f \\ g & h & i \end{pmatrix}\begin{pmatrix} \alpha & 1 & 0 \\ 0 & \alpha & 0 \\ 0 & 0 & \alpha \end{pmatrix} \quad \begin{pmatrix} \alpha & 1 & 0 \\ 0 & \alpha & 0 \\ 0 & 0 & \alpha \end{pmatrix}\begin{pmatrix} a & b & c \\ d & e & f \\ g & h & i \end{pmatrix} = \begin{pmatrix} 0 & 0 & 0 \\ 0 & 0 & 0 \\ 0 & 0 & 0 \end{pmatrix}$$

$$\begin{pmatrix} a\alpha & a+b\alpha & c\alpha \\ d\alpha & d+e\alpha & f\alpha \\ g\alpha & g+h\alpha & i\alpha \end{pmatrix} - \begin{pmatrix} a\alpha+d & \alpha b+e & \alpha c+f \\ \alpha d & \alpha e & \alpha f \\ \alpha g & \alpha h & \alpha i \end{pmatrix} = \begin{pmatrix} -d & a-e & -f \\ 0 & d & 0 \\ 0 & g & 0 \end{pmatrix} = \begin{pmatrix} 0 & 0 & 0 \\ 0 & 0 & 0 \\ 0 & 0 & 0 \end{pmatrix};$$

we thus obtain $d=0$, $g=0$, $f=0$, $a=e$, and $b,c,h,i=$ arbitrary.

Therefore, we obtain

$$B = \begin{pmatrix} \overline{a} & 0 & \overline{g} \\ \overline{b} & \overline{a} & \overline{h} \\ \overline{c} & 0 & \overline{i} \end{pmatrix}$$

or a simpler form, transverse to the orbit of A_0, given by

$$\tilde{B} = \begin{pmatrix} 0 & 0 & 0 \\ \lambda_1 & \lambda_2 & \lambda_3 \\ \lambda_4 & 0 & \lambda_5 \end{pmatrix},$$

where we have relabeled the parameters as in Example A.1.1. Hence, a versal deformation of A_0 is given by

$$\begin{pmatrix} \alpha & 1 & 0 \\ 0 & \alpha & 0 \\ 0 & 0 & \alpha \end{pmatrix} + \begin{pmatrix} 0 & 0 & 0 \\ \lambda_1 & \lambda_2 & \lambda_3 \\ \lambda_4 & 0 & \lambda_5 \end{pmatrix}.$$

We leave it to the reader to verify the transversality of $\tilde{B}$ to the orbit of A_0.

To summarize, the first few low codimension matrices have versal deformations given by

$\boxed{\alpha^2}$
$$\begin{pmatrix} \alpha & 1 \\ 0 & \alpha \end{pmatrix} + \begin{pmatrix} 0 & 0 \\ \lambda_1 & \lambda_2 \end{pmatrix},$$

$\boxed{\alpha\alpha}$
$$\begin{pmatrix} \alpha & 0 \\ 0 & \alpha \end{pmatrix} + \begin{pmatrix} \lambda_1 & \lambda_2 \\ \lambda_3 & \lambda_4 \end{pmatrix},$$

$\boxed{\alpha^2\alpha}$
$$\begin{pmatrix} \alpha & 1 & 0 \\ 0 & \alpha & 0 \\ 0 & 0 & \alpha \end{pmatrix} + \begin{pmatrix} 0 & 0 & 0 \\ \lambda_1 & \lambda_2 & \lambda_3 \\ \lambda_4 & 0 & \lambda_5 \end{pmatrix}.$$

These are the simplest forms to which these parametrized families of matrices containing multiple eigenvalues can be reduced by a *transformation depending analytically on the parameters.*

Before leaving this appendix there is a very important point we should address, namely, that all of our work in this appendix has dealt with complex numbers. The reason for this is simple; it is much easier to deal with the diagonalization problem when dealing with matrices of complex numbers. However, throughout this book we are mainly interested in real-valued vector fields. Thus it is fortunate that the results for versal deformations of matrices of complex numbers go over immediately to the situation of versal deformations of matrices of real numbers. The main idea is the following (Arnold [1983]).

> The decomplexification of a versal deformation with the minimum number of parameters of a complex matrix, $\tilde{A}_0$, can be chosen to be a versal deformation with the minimum number of parameters of the real matrix A_0, where A_0 is the decomplexification of $\tilde{A}_0$.

This statement should be almost obvious after reviewing some definitions and terminology. We will discuss only what is necessary for our purposes and refer the reader to Arnold [1973] or Hirsch and Smale [1974] for more information.

It should be clear that the decomplexification of $\mathbb{C}^n$; is $\mathbb{R}^{2n}$. Moreover, if $e_1, \cdots, e_n$ is a basis of $\mathbb{C}^n$, then $e_1, \cdots, e_n$, $ie_1, \ldots, ie_n$ is a basis for the decomplexification of $\mathbb{C}^n$, $\mathbb{R}^{2n}$. Now, let $A = A_r + iA_i$ be a matrix representation of some complex linear operator mapping $\mathbb{C}^n$ into itself. Then the decomplexification of this matrix is given by the $2n \times 2n$ matrix

$$\begin{pmatrix} A_r & -A_i \\ A_i & A_r \end{pmatrix}. \tag{A.1.17}$$

Consider the following versal deformations of complex matrices

$$\begin{pmatrix} \alpha & 1 \\ 0 & \alpha \end{pmatrix} + \begin{pmatrix} 0 & 0 \\ \lambda_1 & \lambda_2 \end{pmatrix}, \tag{A.1.18a}$$

$$\begin{pmatrix} \alpha & 0 \\ 0 & \alpha \end{pmatrix} + \begin{pmatrix} \lambda_1 & \lambda_2 \\ \lambda_3 & \lambda_4 \end{pmatrix}. \tag{A.1.18b}$$

Letting

$$\alpha = \rho + i\tau,$$
$$\lambda_i = \mu_i + i\gamma_i,$$

where ρ, τ, μ_i, and γ_i are real, (A.1.17) implies that the decomplexification of these matrices is given by

$$\begin{pmatrix} \rho & 1 & 0 & -\tau \\ 0 & \rho & -\tau & 0 \\ 0 & \tau & \rho & 1 \\ \tau & 0 & 0 & \rho \end{pmatrix} + \begin{pmatrix} 0 & 0 & 0 & 0 \\ \mu_1 & \mu_2 & -\gamma_1 & -\gamma_2 \\ 0 & 0 & 0 & 0 \\ \gamma_1 & \gamma_2 & \mu_1 & \mu_2 \end{pmatrix}, \tag{A.1.19a}$$

$$\begin{pmatrix} \rho & 0 & 0 & -\tau \\ 0 & \rho & -\tau & 0 \\ 0 & \tau & \rho & 0 \\ \tau & 0 & 0 & \rho \end{pmatrix} + \begin{pmatrix} \mu_1 & \mu_2 & -\gamma_1 & -\gamma_2 \\ \mu_3 & \mu_4 & -\gamma_3 & -\gamma_4 \\ \gamma_1 & \gamma_2 & \mu_1 & \mu_2 \\ \gamma_3 & \gamma_4 & \mu_3 & \mu_4 \end{pmatrix}. \tag{A.1.19b}$$

As an example, in the study of bifurcations of fixed points of vector fields, we will be interested in the case where the eigenvalues are zero; hence, from (A.1.19a) and (A.1.198b) we conclude that versal deformations of

$$\begin{pmatrix} 0 & 1 \\ 0 & 0 \end{pmatrix} \tag{A.1.20a}$$

and

$$\begin{pmatrix} 0 & 0 \\ 0 & 0 \end{pmatrix} \tag{A.1.20b}$$

are given by

$$\begin{pmatrix} 0 & 1 \\ 0 & 0 \end{pmatrix} + \begin{pmatrix} 0 & 0 \\ \mu_1 & \mu_2 \end{pmatrix} \tag{A.1.21a}$$

and

$$\begin{pmatrix} 0 & 0 \\ 0 & 0 \end{pmatrix} + \begin{pmatrix} \mu_1 & \mu_2 \\ \mu_3 & \mu_4 \end{pmatrix}, \tag{A.1.21b}$$

respectively.

Finally, let us consider a trivial case which will be important in our study of the Poincaré-Andronov-Hopf bifurcation. Consider the complex 1×1 matrix

$$(\alpha), \qquad \alpha \in \mathbb{C}^1; \tag{A.1.22}$$

then the decomplexification of (A.1.22) is given by

$$\begin{pmatrix} \rho & -\tau \\ \tau & \rho \end{pmatrix}, \tag{A.1.23}$$

where $\alpha = \rho + i\tau$.

We can thus see that a versal deformation of the real matrix

$$\begin{pmatrix} 0 & -\tau \\ \tau & 0 \end{pmatrix} \tag{A.1.24}$$

is given by

$$\begin{pmatrix} 0 & -\tau \\ \tau & 0 \end{pmatrix} + \begin{pmatrix} \rho & 0 \\ 0 & \rho \end{pmatrix}. \tag{A.1.25}$$

We will treat some important remaining cases in the exercises following Chapter 3.

Finally, we remark that in the study of bifurcations in Hamiltonian systems, it is necessary to consider versal deformations of symplectic matrices. This situation has been studied by Galin [1982] and Kocak [1984].

3.1E The Double-Zero Eigenvalue

Suppose we have a vector field on $\mathbb{R}^n$ having a fixed point at which the matrix associated with the linearization of the vector field about the fixed point has two zero eigenvalues bounded away from the imaginary axis. In this case we know that the study of the dynamics near this nonhyperbolic fixed point can be reduced to the study of the dynamics of the vector field restricted to the associated two-dimensional center manifold (cf. Chapter 2).

We assume that the reduction to the two-dimensional center manifold has been made, and the Jordan canonical form of the linear part of the vector field is given by

$$\begin{pmatrix} 0 & 1 \\ 0 & 0 \end{pmatrix}. \tag{3.1.187}$$

Our goal is to study the dynamics near a nonhyperbolic fixed point having linear part given by (3.1.187). The procedure is fairly systematic and will be accomplished in the following steps.

1. Compute a normal form and truncate.
2. Rescale the normal form so as to reduce the number of cases to be studied.
3. Embed the truncated normal form in an appropriate two-parameter family (see Example 3.1.20b)
4. Study the local dynamics of the two-parameter family of vector fields.
   - 4a. Find the fixed points and study the nature of their stability.
   - 4b. Study the bifurcations associated with the fixed points.
   - 4c. Based on a consideration of the local dynamics, infer if global bifurcations must be present.
5. Analyze the global bifurcations.
6. Study the effect of the neglected higher order terms in the normal form on the dynamics of the truncated normal form.

We remark that Step 4c is a new phenomenon. However, we will see that it is not uncommon for global effects to be associated with *local* codimension k ($k \geq 2$) bifurcations. Moreover, we will see that it is often possible to "guess" their existence from a thorough local analysis. We will discuss this in more detail later on. Now we begin our analysis with Step 1.

Step 1: The Normal Form. In Example 2.2.2 we saw that a normal form associated with a fixed point of a vector field having linear part (3.1.187) is given by

$$\begin{aligned} \dot{x} &= y + \mathcal{O}(|x|^3, |y|^3), \\ \dot{y} &= ax^2 + bxy + \mathcal{O}(|x|^3, |y|^3), \qquad (x, y) \in \mathbb{R}^2. \end{aligned} \tag{3.1.188}$$

At this stage we will neglect the $\mathcal{O}(3)$ terms in (3.1.188) and study the resulting truncated normal form

$$\begin{aligned} \dot{x} &= y, \\ \dot{y} &= ax^2 + bxy. \end{aligned} \tag{3.1.189}$$

Step 2: Rescaling. Letting

$$\begin{aligned} x &\to \alpha x, \\ y &\to \beta y, \\ t &\to \gamma t, \qquad \gamma > 0, \end{aligned}$$

(3.1.189) becomes

$$\begin{aligned} \dot{x} &= \left(\frac{\gamma\beta}{\alpha}\right) y, \\ \dot{y} &= \left(\frac{\gamma a \alpha^2}{\beta}\right) x^2 + (\gamma b \alpha) xy. \end{aligned} \tag{3.1.190}$$

Now we want to choose γ, β, and α so that the coefficients of (3.1.190) are as simple as possible. Ideally, they would all be unity; we will see that this is not possible but that we can come close.

We will require

$$\frac{\gamma\beta}{\alpha} = 1 \tag{3.1.191}$$

or

$$\gamma = \frac{\alpha}{\beta}.$$

Equation (3.1.191) fixes γ. We require α and β to have the same signs so that stability will not be affected under the rescaling (since γ scales time).

Next, we require that

$$\frac{\gamma a \alpha^2}{\beta} = 1. \tag{3.1.192}$$

Using (3.1.191), (3.1.192) becomes

$$\frac{a\alpha^3}{\beta^2} = a\alpha\left(\frac{\alpha^2}{\beta^2}\right) = 1. \tag{3.1.193}$$

Equation (3.1.193) fixes α/β.

We finally require

$$\gamma b \alpha = 1 \tag{3.1.194}$$

but, using (3.1.191), (3.1.194) becomes

$$\frac{b\alpha^2}{\beta} = b\beta\left(\frac{\alpha^2}{\beta^2}\right) = 1. \tag{3.1.195}$$

We can see that a and b can have either sign and that α and β must have the same sign. From (3.1.193) we can further see that a and α must have the same sign. Therefore, if (3.1.195) is to hold, we conclude that b and a have the same sign. This is too restrictive—the best we can do and still retain full generality is to require

$$b\beta\left(\frac{\alpha^2}{\beta^2}\right) = \pm 1, \tag{3.1.196}$$

so that, in the rescaled variables, the normal form is

$$\begin{aligned} \dot{x} &= y, \\ \dot{y} &= x^2 \pm xy. \end{aligned} \tag{3.1.197}$$

Step 3: Construct a Candidate for a Versal Deformation. From Example 3.1.20b, a likely candidate for a versal deformation is

$$\begin{aligned} \dot{x} &= y, \\ \dot{y} &= \mu_1 + \mu_2 y + x^2 + bxy, \qquad b = \pm 1. \end{aligned} \tag{3.1.198}$$

Step 4: Study the Local Dynamics of (3.1.198). We take the case $b = +1$.

Step 4a: Fixed Points and Their Stability. It is easy to see that the fixed points of (3.1.198) are given by

$$(x, y) = (\pm\sqrt{-\mu_1}, 0). \tag{3.1.199}$$

In particular, there are no fixed points for $\mu_1 > 0$.

Next we check the stability of these fixed points.

The Jacobian of the vector field evaluated at the fixed point is given by

$$\left.\begin{pmatrix} 0 & 1 \\ 2x & \mu_2 + x \end{pmatrix}\right|_{(\pm\sqrt{-\mu_1}, 0)} = \begin{pmatrix} 0 & 1 \\ \pm 2\sqrt{-\mu_1} & \mu_2 \pm \sqrt{-\mu_1} \end{pmatrix}. \tag{3.1.200}$$

The eigenvalues are given by

$$\lambda_{1,2} = \frac{\mu_2 \pm \sqrt{-\mu_1}}{2} \pm \frac{1}{2}\sqrt{(\mu_2 \pm \sqrt{-\mu_1})^2 \pm 8\sqrt{-\mu_1}}. \tag{3.1.201}$$

If we denote the two branches of fixed points by $(x^+, 0) \equiv (+\sqrt{-\mu_1}, 0)$ and $(x^-, 0) = (-\sqrt{-\mu_1}, 0)$, we see from (3.1.201) that $(x^+, 0)$ is a *saddle* for $\mu_1 < 0$ and all μ_2, while for $\mu_1 = 0$ the eigenvalues on $(x^+, 0)$ are given by

$$\lambda_{1,2} = \mu_2, 0.$$

The fixed point $(x^-, 0)$ is a *source* for $\{\mu_2 > \sqrt{-\mu_1}, \mu_1 < 0\}$ and a sink for $\{\mu_2 < \sqrt{-\mu_1}, \mu_1 < 0\}$; for $\mu_1 = 0$, the eigenvalues on $(x^-, 0)$ are given by

$$\lambda_{1,2} = \mu_2, 0$$

and, for $\mu_2 = \sqrt{-\mu_1},\ \ \mu_1 < 0$, the eigenvalues on $(x^-, 0)$ are given by

$$\lambda_{1,2} = \pm i\sqrt{2\sqrt{-\mu_1}}.$$

Thus, we might expect that $\mu_1 = 0$ is a bifurcation curve on which $(x^{\pm}, 0)$ are born in a saddle-node bifurcation and $\mu_2 = \sqrt{-\mu_1},\ \mu_1 < 0$, is a bifurcation curve on which $(x^-, 0)$ undergoes a Poincaré–Andronov–Hopf bifurcation. We now turn to verifying this and studying the orbit structure associated with these bifurcations.

Step 4b: The Bifurctations of the Fixed Points. We begin by examining the orbit structure near $\mu_1 = 0$, μ_2 arbitrary. We will use the center manifold theorem (Theorem 2.1.1).

First we put the system into the "normal form" for the center manifold theorem. We treat μ_2 as a fixed constant in the problem and think of μ_1 as a parameter, and we examine bifurcations from $\mu_1 = 0$.

To transform (3.1.198) into the form in which the center manifold can be applied, we use the following linear transformation

$$\begin{pmatrix} x \\ y \end{pmatrix} = \begin{pmatrix} 1 & 1 \\ 0 & \mu_2 \end{pmatrix} \begin{pmatrix} u \\ v \end{pmatrix}, \qquad \begin{pmatrix} u \\ v \end{pmatrix} = \frac{1}{\mu_2} \begin{pmatrix} \mu_2 & -1 \\ 0 & 1 \end{pmatrix} \begin{pmatrix} x \\ y \end{pmatrix}, \tag{3.1.202}$$

which transforms (3.1.198) into

$$\begin{pmatrix} \dot{u} \\ \dot{v} \end{pmatrix} = \begin{pmatrix} 0 & 0 \\ 0 & \mu_2 \end{pmatrix} \begin{pmatrix} u \\ v \end{pmatrix} + \frac{1}{\mu_2} \begin{pmatrix} -\mu_1 \\ \mu_1 \end{pmatrix} + \frac{1}{\mu_2} \begin{pmatrix} -(u^2 + (2+\mu_2)uv + (1+\mu_2)v^2) \\ u^2 + (2+\mu_2)uv + (1+\mu_2)v^2 \end{pmatrix}$$

or

$$\begin{aligned} \dot{u} &= -\frac{\mu_1}{\mu_2} - \frac{1}{\mu_2}\left[u^2 + (2+\mu_2)uv + (1+\mu_2)v^2\right], \\ \dot{v} &= \mu_2 v + \frac{\mu_1}{\mu_2} + \frac{1}{\mu_2}\left[u^2 + (2+\mu_2)uv + (1+\mu_2)v^2\right]. \end{aligned} \tag{3.1.203}$$

Without actually computing the center manifold, we can argue as follows.

The center manifold will be given as a graph over u and μ_1, $v(u, \mu_1)$, and be at least $\mathcal{O}(2)$. Thus, from this we can immediately see that the reduced system is given by

$$\dot{u} = -\frac{1}{\mu_2}(\mu_1 + u^2) + \mathcal{O}(3), \tag{3.1.204}$$

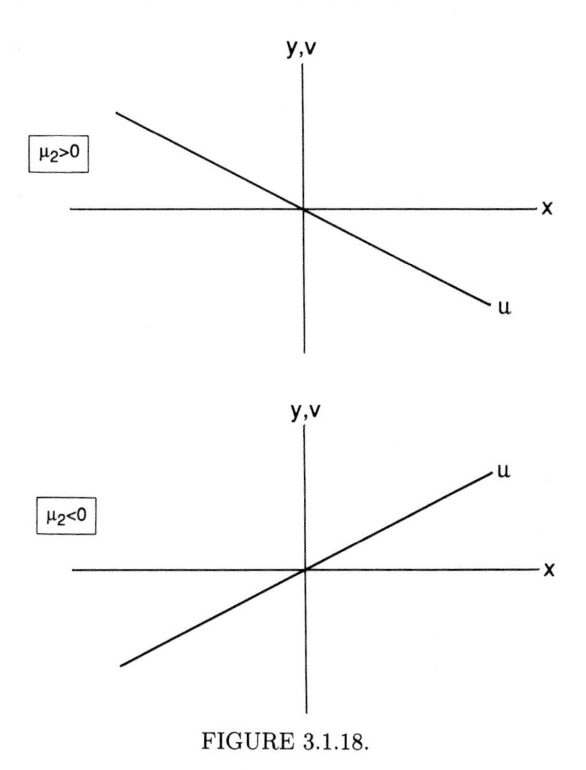

FIGURE 3.1.18.

and thus it undergoes a saddle-node bifurcation at $\mu_1 = 0$.

We can immediately conclude that the stable (unstable) manifold of the node connects to the unstable (stable) manifold of the saddle, since this always occurs for one-dimensional flows. This points out another advantage of the center manifold analysis, since such results are nontrival in dimensions ≥ 2.

We next want to examine in more detail the nature of the flow on the center manifold and what it implies for the full two-dimensional flow. Recall that the eigenvalues of the linearized two-dimensional vector field on the bifurcation curve are given by

$$\lambda_{1,2} = \mu_2, 0.$$

In transforming (3.1.198) to (3.1.203), we see from (3.1.202) that the coordinate axes have be transformed as in Figure 3.1.18.

Now the flow on the center manifold is given by

$$\dot{u} = -\frac{1}{\mu_2}(\mu_1 + u^2) + \mathcal{O}(3), \tag{3.1.205}$$

and in (u, μ_1) coordinates appear as in Figure 3.1.19.

Using the information in Figures 3.1.18 and 3.1.19 and recalling that the eigenvalues of the vector field linearized about the fixed point at $\mu_1 = 0$

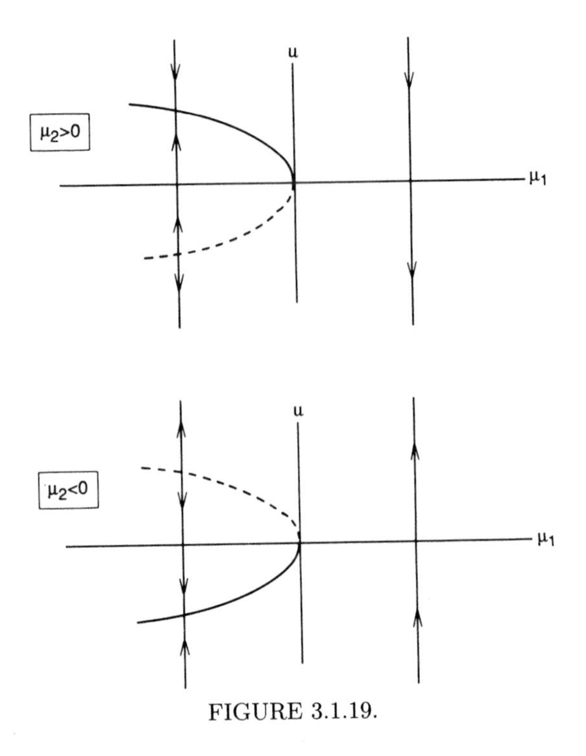

FIGURE 3.1.19.

are $\lambda_{1,2} = \mu_2, 0$, we can easily obtain phase portraits near the origin that show the bifurcation in the two-dimensional phase space on crossing the μ_2-axis; see Figure 3.1.20.

In the cases μ_1 slightly negative, notice the reversals of position for the stable and unstable manifolds of the saddle for $\mu_2 > 0$ and $\mu_2 < 0$.

We next examine the change of stability of the fixed points $(x^-, 0)$ on $\mu_2 = \sqrt{-\mu_1}$, $\mu_1 < 0$. From (3.1.201), the eigenvalues associated with the linearization about this curve of fixed points are

$$\lambda_{1,2} = \pm i\sqrt{2\sqrt{-\mu_1}}.$$

If we view μ_2 as a parameter, then using (3.1.201) we obtain

$$\frac{d}{d\mu_2} \operatorname{Re} \lambda_{1,2}\bigg|_{\mu_2=\sqrt{-\mu_1}} = \frac{1}{2} \neq 0.$$

Thus, it appears that a Poincaré-Andronov-Hopf bifurcation occurs on $\mu_2 = \sqrt{-\mu_1}$.

Next we check the stability of the bifurcating periodic orbits. Recall from Theorem 3.1.3 that this involves putting the equation in a certain "normal form" and then computing a coefficient, a, which is given by derivatives of functions occuring in this normal form.

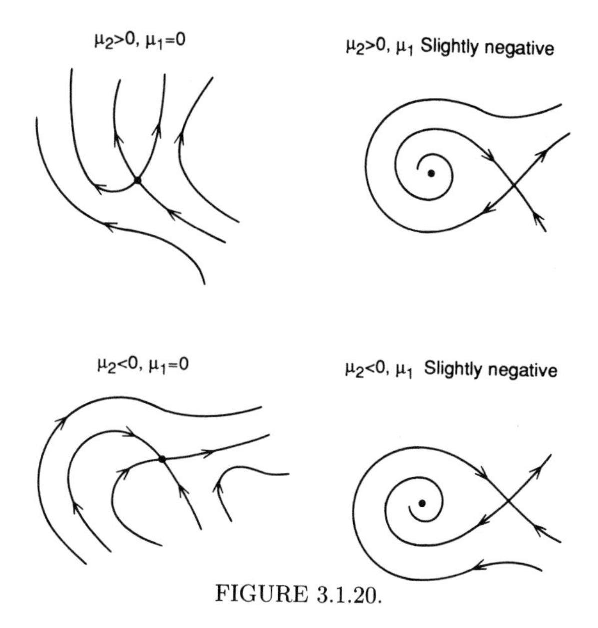

FIGURE 3.1.20.

First we transform the fixed point to the origin via

$$\bar{x} = x - x^-,$$
$$\bar{y} = y,$$

so that (3.1.198) becomes

$$\begin{pmatrix} \dot{\bar{x}} \\ \dot{\bar{y}} \end{pmatrix} = \begin{pmatrix} 0 & 1 \\ -2\sqrt{-\mu_1} & 0 \end{pmatrix} \begin{pmatrix} \bar{x} \\ \bar{y} \end{pmatrix} + \begin{pmatrix} 0 \\ \bar{x}\bar{y} + \bar{x}^2 \end{pmatrix}. \tag{3.1.206}$$

Then we put the linear part of (3.1.206) in normal form via the linear transformation

$$\begin{pmatrix} \bar{x} \\ \bar{y} \end{pmatrix} = \begin{pmatrix} 0 & 1 \\ \sqrt{-2\sqrt{-\mu_1}} & 0 \end{pmatrix} \begin{pmatrix} u \\ v \end{pmatrix}, \tag{3.1.207}$$

under which (3.1.206) becomes

$$\begin{pmatrix} \dot{u} \\ \dot{v} \end{pmatrix} = \begin{pmatrix} 0 & -\sqrt{-2\sqrt{-\mu_1}} \\ \sqrt{-2\sqrt{-\mu_1}} & 0 \end{pmatrix} \begin{pmatrix} u \\ v \end{pmatrix} + \begin{pmatrix} uv + \frac{1}{\sqrt{-2\sqrt{-\mu_1}}} v^2 \\ 0 \end{pmatrix}. \tag{3.1.208}$$

Notice that (3.1.208) is exactly in the form of (3.1.106), in which the coefficient a was given as follows

$$a = \frac{1}{16}\left[f_{uuu} + f_{uvv} + g_{uuv} + g_{vvv}\right] + \frac{1}{16\sqrt{-2\sqrt{-\mu_1}}}\left[f_{uv}(f_{uu} + f_{vv}) - g_{uv}(g_{uu} + g_{vv}) - f_{uu}g_{uu} + f_{vv}g_{vv}\right],$$

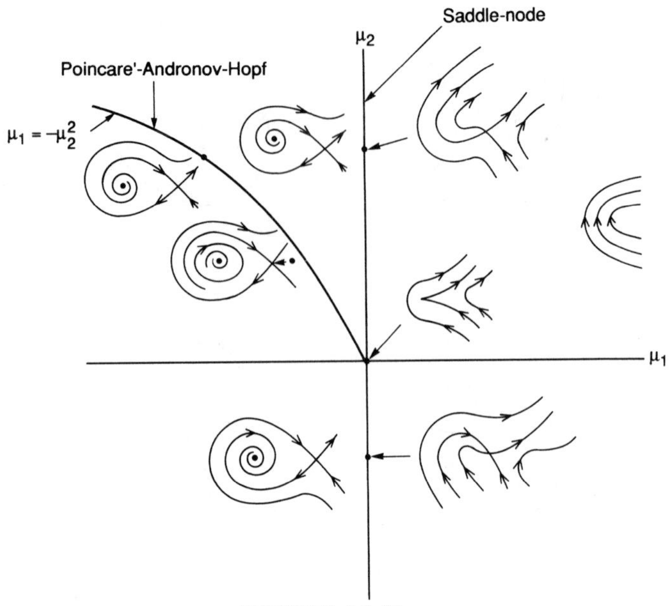

FIGURE 3.1.21.

where all partial derivatives are evaluated at the origin. In our case,

$$f = uv + \frac{1}{\sqrt{-2\sqrt{-\mu_1}}} v^2,$$

$$g = 0;$$

thus, an easy calculation gives

$$a = \frac{1}{16\sqrt{-\mu_1}} > 0,$$

indicating a *subcritical* Poincaré-Andronov-Hopf bifurcation to an unstable periodic orbit below $\mu_2^2 = -\mu_1$.

This completes the local analysis, and we summarize the results in the bifurcation diagram in Figure 3.1.21.

Step 4c: Global Dynamics. At this stage we have analyzed all possible local bifurcations; however, a careful study of Figure 3.1.21 reveals that there must be additional bifurcations. This conclusion is based on the following facts.

1. Note the stable and unstable manifolds of the saddle point. For the case $\mu_2 > \sqrt{-\mu_1}$, $\mu_1 < 0$, the stable and unstable manifolds have the opposite "orientation" compared with the case $\mu_2 < 0$, $\mu_1 < 0$. It appears as if the manifolds have "passed through each other" as μ_2 decreases.

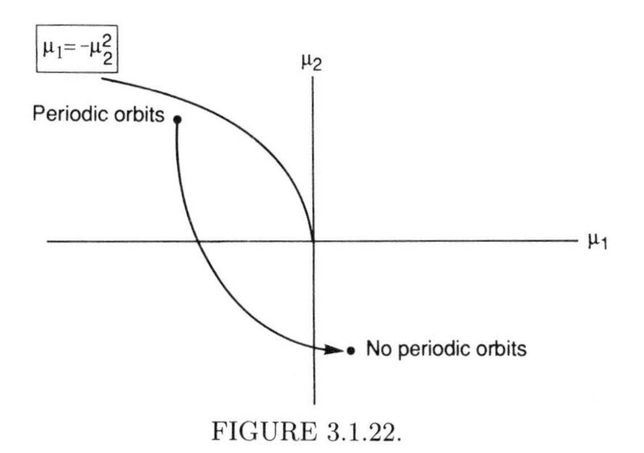

FIGURE 3.1.22.

2. Using index theory, it is easy to verify that (3.1.198) ($b = +1$) has no periodic orbits for $\mu_1 > 0$ (since there are no fixed points in this region). Hence, in traversing the $\mu_1 - \mu_2$ plane in an arc around the origin starting on the curve $\mu_2 = -\mu_1^2$ and ending in $\mu_1 > 0$ (see Figure 3.1.22), we must somehow cross a bifurcation curve(s) which results in the annihilation of all periodic orbits. This cannot be a local bifurcation, because these have all been taken into account.

Step 5: Analysis of Global Bifurcations. We postpone an analysis of the global bifurcation in this case until Chapter 4 and merely state the result for now.

In this case a likely candidate for the global bifurcation which will complete the bifurcation diagram is a *saddle-connection* or *homoclinic bifurcation.* In Chapter 4 we show that this occurs on the curve

$$\mu_1 = -\frac{49}{25}\mu_2^2 + \mathcal{O}(\mu_2^{5/2}),$$

which we shown in Figure 3.1.23. From this figure one can see that the saddle-connection or homoclinic bifurcation is described by the periodic orbit created in the subcritical Poincaré-Andronov-Hopf bifurcation growing in amplitude as μ_2 is decreased until it collides with the saddle point, creating a homoclinic orbit. As μ_2 is further decreased, the homoclinic orbit breaks. This explains the reversal in orientation of the stable and unstable manifolds of the saddle point described above, and it also explains how the periodic orbits created in the subcritical Poincaré-Andronov-Hopf bifurcation are destroyed.

Step 6: Effects of Higher Order Terms in the Normal Form. Bogdanov [1975] proved that the dynamics of (3.1.198) are not qualitatively changed by the higher order terms in the normal form. Hence, (3.1.198) is a versal deformation. We will discuss the issues involved with proving this more thoroughly in Chapter 4.

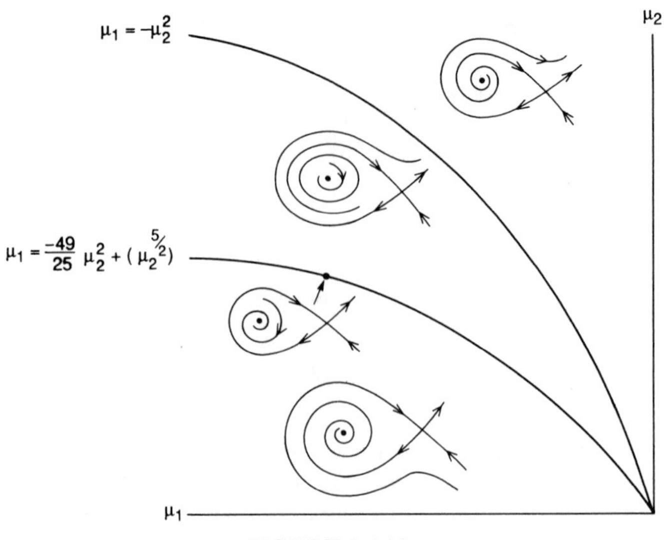

FIGURE 3.1.23.

This completes our analysis of the case $b = +1$; the case $b = -1$ is very similar, so we leave it as an exercise. Before leaving the double-zero eigenvalue, we want to make some final remarks.

Remark 1. The reader should note the generality of this analysis. The normal form is completely determined by the structure of the linear part of the vector field.

Remark 2. Global dynamics arose from a local bifurcation analysis. For two-dimensional vector fields these dynamics cannot be very complicated, but for three-dimensional vector fields chaotic dynamics may occur.

Remark 3. When one speaks of the "double-zero eigenvalue" for vector fields, one usually means a vector field whose linear part (in Jordan canonical form) is given by

$$\begin{pmatrix} 0 & 1 \\ 0 & 0 \end{pmatrix}.$$

However, the linear part

$$\begin{pmatrix} 0 & 0 \\ 0 & 0 \end{pmatrix}$$

is also a double-zero eigenvalue. This case is *codimension 4* and is consequently more difficult to analyze.

3.1F A Zero and a Pure Imaginary Pair of Eigenvalues

Suppose that the linear part of the vector field (after a possible center manifold reduction) has the following form

$$\begin{pmatrix} 0 & -\omega & 0 \\ \omega & 0 & 0 \\ 0 & 0 & 0 \end{pmatrix}\begin{pmatrix} x \\ y \\ z \end{pmatrix}. \tag{3.1.209}$$

Using this to compute the normal form and subsequently transforming to cylindrical coordinates gives the following normal form (cf. Exercise 2.9)

$$\begin{aligned}
\dot{r} &= a_1 rz + a_2 r^3 + a_3 rz^2 + \mathcal{O}(|r|^4, |z|^4),\\
\dot{z} &= b_1 r^2 + b_2 z^2 + b_3 r^2 z + \mathcal{O}(|r|^4, |z|^4),\\
\dot{\theta} &= \omega + c_1 z + \mathcal{O}(|r|^2, |z|^2).
\end{aligned} \tag{3.1.210}$$

This is the vector field we will study. Notice that the θ-dependence in the r and z components of the vector field can be removed to order k for k arbitrarily large (note: exactly the same thing occurred when analyzing the normal form for the Poincaré-Andronov-Hopf bifurcation). This is important. because it is a major tool in facilitating the analysis of this system. Specifically, recall that our analysis is only local (i.e., r, z sufficiently small), so that we have, for r, z sufficiently small, $\dot{\theta} \neq 0$. Thus, we will truncate our equation at some order and, ignoring the $\dot{\theta}$ part of our vector field, perform a phase plane analysis on the r, z part of the vector field. For r, z sufficiently small, in some sense (to be made precise later) the $r - z$ phase plane can be thought of as a Poincaré map for the full three-dimensional system. Also, we must consider the effects of higher order terms on our analysis, since it is not necessarily true that in the actual vector field the (r, z) components are independent of θ; we only push the θ-dependence up to higher order with the method of normal forms.

Our analysis will follow the same steps as our analysis of the double-zero eigenvalue in Section 3.1E.

Step 1: Compute and Truncate the Normal Form. The normal form is given by (3.1.210). For now we will neglect terms of $\mathcal{O}(3)$ and higher and, as described above, the $\dot{\theta}$ component of (3.1.210). Thus, the vector field we will study is

$$\begin{aligned}
\dot{r} &= a_1 rz,\\
\dot{z} &= b_1 r^2 + b_2 z^2.
\end{aligned} \tag{3.1.211}$$

Step 2: Rescaling to Reduce the Number of Cases. Rescaling by letting $\bar{r} = \alpha r$ and $\bar{z} = \beta z$, we obtain

$$\dot{\bar{r}} = \alpha \left[a_1 \frac{\bar{r}\bar{z}}{\alpha\beta} \right],$$

$$\dot{\bar{z}} = \beta \left[\frac{b_1}{\alpha^2} \bar{r}^2 + \frac{b_2}{\beta^2} \bar{z}^2 \right].$$

Now, letting $\beta = -b_2$, $\alpha = -\sqrt{|b_1 b_2|}$, and dropping the bars on $\bar{r}, \bar{z}$, we obtain

$$\begin{aligned}
\dot{r} &= -\frac{a_1}{b_2} r z, \\
\dot{z} &= \frac{-b_1 b_2}{|b_1 b_2|} r^2 - z^2,
\end{aligned}$$

or

$$\begin{aligned}
\dot{r} &= a r z, \\
\dot{z} &= b r^2 - z^2,
\end{aligned} \tag{3.1.212}$$

where $a = \frac{-a_1}{b_2}$ is arbitrary (except that it is nonzero and bounded) and $b = \pm 1$.

Next we want to determine the topologically distinct phase portraits of (3.1.212) which occur for the various choices of a and b. We will find that there are six different types, which (following Guckenheimer and Holmes [1983]) we label I, IIa, IIb, III, IVa, IVb, because the versal deformation of IIa and IIb as well as the versal deformation of IVa and IVb are essentially the same.

The key idea in determining these classifications involves finding certain invariant lines (separatrices) for the flow given by $z = kr$ (note that $r \geq 0$). Substituting this into our equation gives

$$\frac{dz}{dr} = k = \frac{br^2 - k^2 r^2}{akr^2} = \frac{b - k^2}{ak}$$

or

$$k = \pm \sqrt{\frac{b}{a+1}}; \tag{3.1.213}$$

hence, the condition for such invariant lines to exist is $\frac{b}{a+1} > 0$.

Note that $r = 0$ *is always invariant, and the equation is invariant under the transformation* $z \to -z$, $t \to -t$.

Therefore, for $b = 1$ there are two distinct cases

$$a \leq -1, \qquad a > -1,$$

and, for $b = -1$, there are two distinct cases

$$a < -1, \qquad a \geq -1.$$

The direction of the flow on these invariant lines can be calculated by taking the dot product of the vector field with a radial vector field evaluated on the invariant line.

$$\begin{aligned}
s &\equiv (arz, br^2 - z^2) \cdot (r, z)\big|_{z=kr} \\
&= r^3 k (a + b - k^2).
\end{aligned} \tag{3.1.214}$$

Substituting (3.1.213) into (3.1.214) (and taking the '+' sign in (3.1.213), which will give the direction of flow along $z = kr$ in the first quadrant) gives

$$s = \frac{ar^2 z}{1+a}(a+b+1). \tag{3.1.215}$$

If this quantity s is > 0 (take $z, k > 0$), then the flow is directed radially outward for $z > 0$. If $s < 0$, then the flow is directed inward for $z > 0$. The opposite case occurs for $z, k < 0$. We summarize this information below.

$b = +1,\ a \le -1$. There are no invariant lines except $r = 0$.

$b = +1,\ a > -1$. From (3.1.213) we see that, in this case, we do have an invariant line and, from (3.1.215), that the direction of flow along this line is governed by the sign of

$$\frac{a}{1+a}.$$

Hence, we have two cases

$$\frac{a}{1+a} > 0 \qquad \text{for} \quad a > 0,$$

$$\frac{a}{1+a} < 0 \qquad \text{for} \quad -1 < a < 0.$$

We will not consider the degenerate $a = 0$ case, since this will necessitate the consideration of higher order terms in the normal form.

$b = -1,\ a \ge -1$. There are no invariant lines except $r = 0$.

$b = -1,\ a < -1$. From (3.1.213) and (3.1.215) we see that, in this case, we do have an invariant line with $s < 0$.

We summarize the information obtained concerning the orbit structure of (3.1.212) in Figure 3.1.24. There are thus six topologically distinct cases. However, we are still not quite finished with their phase portraits. Notice that (3.1.212) has the first integral

$$I(r, z) = \frac{a}{2} r^{2/a} \left[\frac{br^2}{1+a} - z^2 \right]. \tag{3.1.216}$$

The reader can check this by showing

$$\frac{\partial I}{\partial r}\dot{r} + \frac{\partial I}{\partial z}\dot{z} = 0,$$

where $\dot{r}$ and $\dot{z}$ are obtained from (3.1.212).

Now this first integral can give us information concerning whether or not there are closed orbits in our phase portraits and, of course, the level curves give us all the trajectories. We will thus examine the level curves of $I(r, z)$ for each of our six cases. Also, we will only analyze the $r \ge 0,\ z \ge 0$

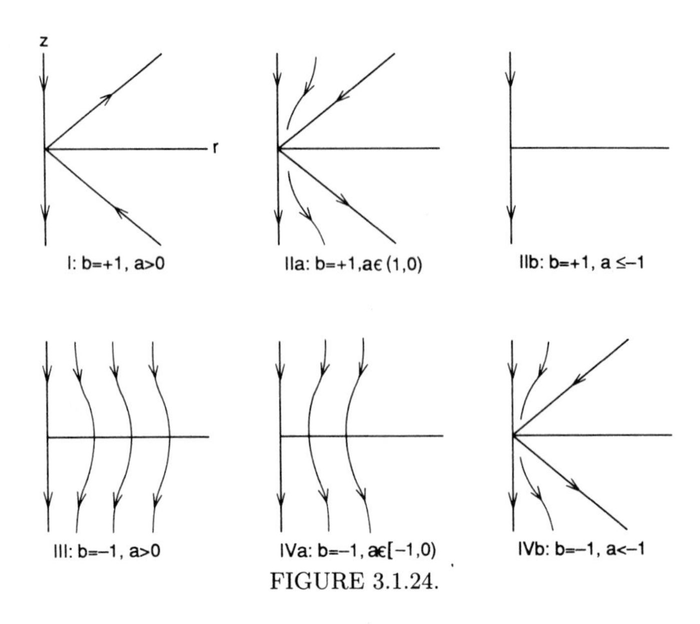

FIGURE 3.1.24.

quadrant of the (r, z) plane, since knowledge of the flow in this quadrant is sufficient due to the symmetry $z \to -z$, $t \to -t$.

Case I. We begin with Case I for which we have $b = +1$ and $a > 0$, which implies that $k = \sqrt{\frac{1}{1+a}} < 1$.

Recall that, in this case, the vector field is given by

$$\begin{aligned} \dot{r} &= arz, \\ \dot{z} &= r^2 - z^2, \end{aligned}$$

from which we see that, in $r \geq 0$, $z \geq 0$, we have

$$\dot{r} > 0 \Rightarrow r \qquad \text{is increasing on orbits,}$$

$$\dot{z} = 0 \qquad \text{on the line} \quad r = z.$$

For $z > 0, z$ below the line $r = z$, we have $\dot{z} > 0$, which implies that z is increasing on orbits with initial z values below the line $r = z$. The opposite conclusion holds for orbits starting above $r = z$. Also, since the line $z = kr$ is invariant and thus cannot be crossed by trajectories, and since $z = kr$ lies below $z = r$, we conclude that trajectories below the line $z = kr$ must have z and r components increasing montonically. These observations allow us to sketch the phase portrait shown in Figure 3.1.25.

Case IIa. We have $b = +1$, $a \in (-1, 0)$, which implies $k = \sqrt{\frac{1}{1+a}} > 1$.

1. In this case the line $z = r$ lies below the invariant line $z = kr$; therefore, the only place where $\dot{z}$ can vanish (besides the origin) is below $z = kr$.

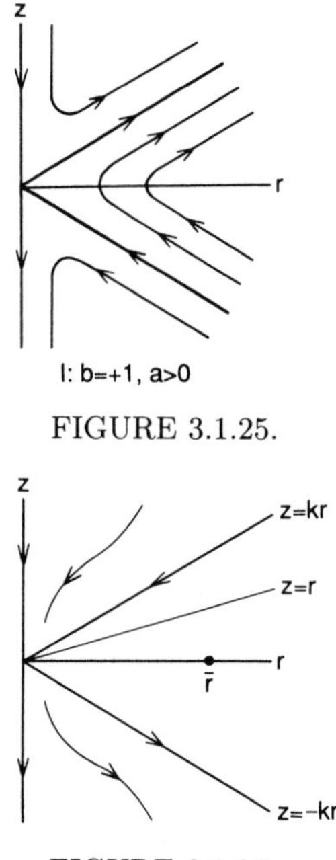

FIGURE 3.1.25.

FIGURE 3.1.26.

2. Also, due to the fact that $a \in (-1, 0)$, in the quadrant $r > 0$, $z > 0$, we always have $\dot{r} < 0$; hence, r is always decreasing on orbits.

Now we will consider our first integral

$$I(r, z) = \frac{a}{2} r^{2/a} \left[\frac{r^2}{1+a} - z^2 \right].$$

The level curves of this function are trajectories of (3.1.212). The following lemma will prove useful.

Lemma 3.1.5 *A level curve of $I(r, z)$ may intersect the line $z = r$ only once in $r > 0$, $z > 0$.*

Proof: At this stage of our analysis of Case IIa the orbit structure shown in Figure 3.1.26 has been verified. Note that at the point $(\bar{r}, 0)$ shown in Figure 3.1.26 we have

$$\dot{r} = 0, \qquad \dot{z} = \bar{r}^2 \qquad \text{at} \quad (\bar{r}, 0).$$

By the comments 1 and 2 above, since $\dot{r} < 0$ everywhere in $z > 0$, $r > 0$, and $\dot{z} > 0$ at $(\bar{r}, 0)$, the trajectory starting at $(\bar{r}, 0)$ must eventually cross the line $z = r$.

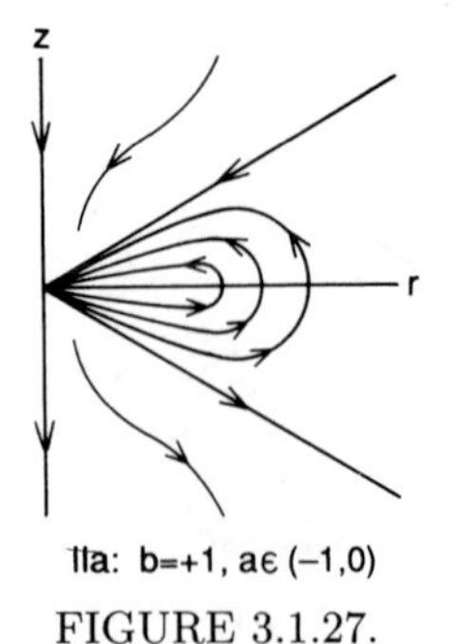

FIGURE 3.1.27.

Now the trajectory starting at $(\bar{r}, 0)$ lies on the level curve given by

$$I(\bar{r}, 0) = \frac{a\bar{r}^{2+\frac{2}{a}}}{2(1+a)} \equiv \bar{c}. \tag{3.1.217}$$

It intersects the line $z = r$, and we can compute the r coordinate of the intersection as follows on $z = r$

$$\begin{aligned} I(r, r) = \bar{c} &= \frac{a}{2} r^{2/a} \left[\frac{r^2}{1+a} - \frac{r^2(1+a)}{1+a} \right] \\ &= \frac{a}{2} r^{2/a} \left[-\frac{ar^2}{1+a} \right] = -\frac{a^2}{2(1+a)} r^{2+(2/a)}. \end{aligned} \tag{3.1.218}$$

We can compute the r coordinate of the intersection point in terms of the starting point $\bar{r}$ by equating (3.1.217) and (3.1.218)

$$-\frac{a^2}{2(1+a)} r^{2+(2/a)} = \frac{a\bar{r}^{2+(2/a)}}{2(1+a)},$$

and solving for

$$r = \left(-\frac{1}{a} \right)^{a/(2a+2)} \bar{r}.$$

Thus, given $(\bar{r}, 0)$ as an initial condition for a trajectory, we conclude that this trajectory can intersect the line $z = r$ only at the unique value of r given above (and hence also at a unique value of z). This proves the lemma. $\square$

This lemma therefore tells us that once a trajectory crosses the $z = r$ line it is forever trapped between the lines $z = r$, $z = kr$ and, since $\dot{r} < 0$, it must approach $(0, 0)$ asymptotically. Putting together these results and using the symmetry $z \to -z$, $t \to -t$, we obtain the phase portrait for Case IIa shown in Figure 3.1.27.

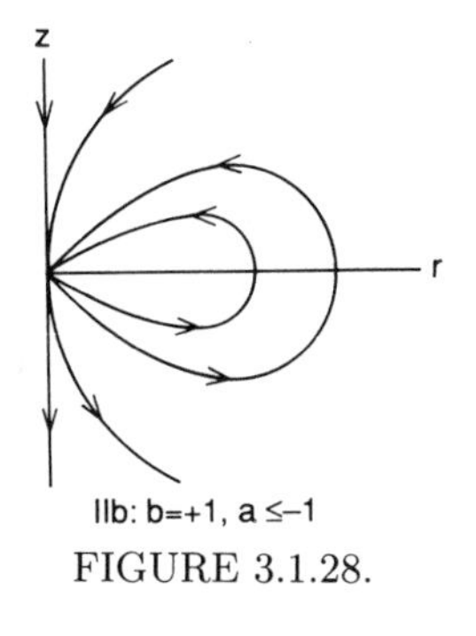

FIGURE 3.1.28.

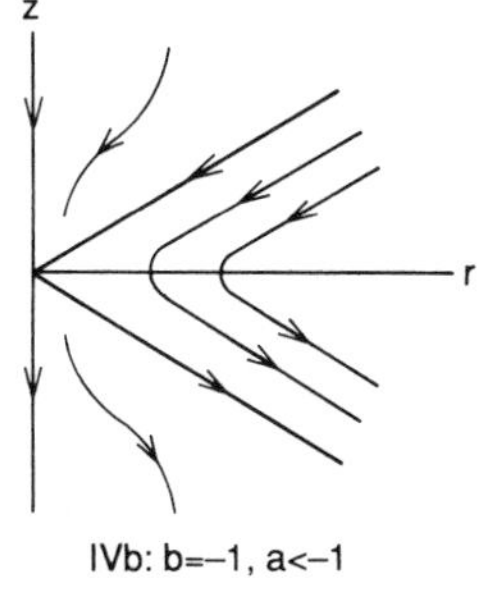

FIGURE 3.1.29.

Case IIb. There are no invariant lines in this case; however, the arguments given in Case IIa can be slightly modified to yield the phase portrait shown in Figure 3.1.28.

We now proceed to the $b = -1$ cases. In these cases z is always decreasing.

Cases III and IVa. These cases are easy since there are no invariant lines; hence, there is no additional orbit structure beyond that shown in Figure 3.1.24 (note: the reader should verify the different "dimples" exhibited by phase curves in these figures upon crossing the r-axis).

Case IVb. We have $b = -1$ and $a < -1$ with $k = \sqrt{\frac{-1}{1+a}}$. Since $\dot{z}$ is decreasing and $\dot{r}$ is decreasing we have the phase portrait shown in Figure 3.1.29. This completes the classification of the possible local phase portraits. We now show them together in Figure 3.1.30 for comparative purposes.

Step 3: Construct a Candidate for a Versal Deformation. From Section 3.1D, Appendix 1, a candidate for a versal deformation is given by

$$\begin{aligned} \dot{r} &= \mu_1 r + arz, \\ \dot{z} &= \mu_2 + br^2 - z^2, \qquad b = \pm 1. \end{aligned} \tag{3.1.219}$$

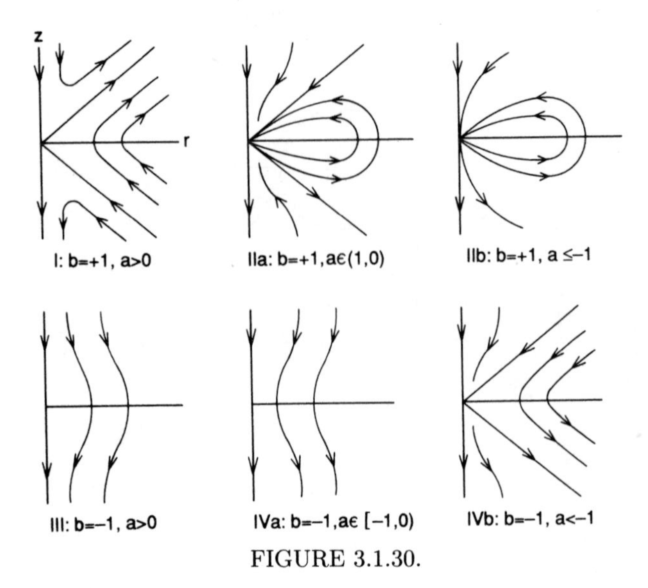

FIGURE 3.1.30.

Step 4: Study the Local Dynamics of (3.1.220).

Step 4a. Fixed Points and Their Stability. It is easy to see that there are three branches of fixed points of (3.1.219) given by

$$
\begin{aligned}
(r, z) &= (0, \pm\sqrt{\mu_2}), \\
(r, z) &= \left(\sqrt{\frac{1}{b}(\frac{\mu_1^2}{a^2} - \mu_2)}, \frac{-\mu_1}{a} \right)
\end{aligned}
\tag{3.1.220}
$$

(note: $r \geq 0$).

We next examine the stability of these fixed points. The matrix associated with the linearized vector field is given by

$$
J = \begin{pmatrix} \mu_1 + az & ar \\ 2br & -2z \end{pmatrix}. \tag{3.1.221}
$$

Before analyzing the stability of each branch of fixed points, we want to work out a general result that will save much time.

The eigenvalues of (3.1.221) are given by

$$
\lambda_{1,2} = \frac{\operatorname{tr} J}{2} \pm \frac{1}{2}\sqrt{(\operatorname{tr} J)^2 - 4 \det J}. \tag{3.1.222}
$$

Using (3.1.222), we notice the following facts

$$
\begin{aligned}
&\text{if } \operatorname{tr} J > 0, \ \det J > 0, \qquad \text{then } \lambda_1 > 0, \lambda_2 > 0 \Rightarrow \text{ source}, \\
&\text{if } \operatorname{tr} J > 0, \ \det J < 0, \qquad \text{then } \lambda_1 > 0, \lambda_2 < 0 \Rightarrow \text{ saddle},
\end{aligned}
$$

$$\begin{array}{lll}
\text{if } \operatorname{tr} J < 0,\ \det J > 0, & \text{then } \lambda_1 < 0, \lambda_2 < 0 \Rightarrow \text{ sink,} \\
\text{if } \operatorname{tr} J < 0,\ \det J < 0, & \text{then } \lambda_1 > 0, \lambda_2 < 0 \Rightarrow \text{ saddle,} \\
\text{if } \operatorname{tr} J = 0,\ \det J < 0, & \text{then } \lambda_{1,2} = \pm\sqrt{|\det J|} \Rightarrow \text{ saddle,} \\
\text{if } \operatorname{tr} J = 0,\ \det J > 0, & \text{then } \lambda_{1,2} = \pm i\sqrt{|\det J|} \Rightarrow \text{ center,} \\
\text{if } \operatorname{tr} J > 0,\ \det J = 0, & \text{then } \lambda_1 = \operatorname{tr} J, \lambda_2 = 0, \\
\text{if } \operatorname{tr} J < 0,\ \det J = 0, & \text{then } \lambda_1 = \operatorname{tr} J, \lambda_2 = 0.
\end{array} \tag{3.1.223}$$

We now analyze each branch of fixed points individually.

$(0, \sqrt{\mu_2})$. On this branch of fixed points we have

$$\begin{aligned}
\operatorname{tr} J &= (\mu_1 + a\sqrt{\mu_2}) - 2\sqrt{\mu_2}, \\
\det J &= -2\sqrt{\mu_2}(\mu_1 + a\sqrt{\mu_2}),
\end{aligned}$$

from which we conclude

$$\begin{aligned}
\operatorname{tr} J > 0 &\Rightarrow \quad \mu_1 + a\sqrt{\mu_2} > 2\sqrt{\mu_2}, \\
\operatorname{tr} J < 0 &\Rightarrow \quad \mu_1 + a\sqrt{\mu_2} < 2\sqrt{\mu_2}, \\
\det J > 0 &\Rightarrow \quad \mu_1 + a\sqrt{\mu_2} < 0, \\
\det J < 0 &\Rightarrow \quad \mu_1 + a\sqrt{\mu_2} > 0.
\end{aligned}$$

Appealing to (3.1.222) and using (3.1.223), we make the following conclusions concerning stability of the branch of fixed points $(0, \sqrt{\mu_2})$

$$\begin{array}{ll}
\operatorname{tr} J > 0,\ \det J > 0, & \text{cannot occur,} \\
\text{if } \operatorname{tr} J > 0,\ \det J < 0, & \text{then } \mu_1 + a\sqrt{\mu_2} > 2\sqrt{\mu_2} \Rightarrow \text{ saddle,} \\
\text{if } \operatorname{tr} J < 0,\ \det J > 0, & \text{then } \mu_1 + a\sqrt{\mu_2} < 0 \Rightarrow \text{ sink,} \\
\text{if } \operatorname{tr} J < 0,\ \det J < 0, & \text{then } 0 < \mu_1 + a\sqrt{\mu_2} < 2\sqrt{\mu_2}, \\
\text{if } \operatorname{tr} J = 0,\ \det J < 0, & \text{then } \begin{array}{l} \mu_1 + a\sqrt{\mu_2} = 2\sqrt{\mu_2} \\ \mu_1 + a\sqrt{\mu_2} > 0 \end{array} \Rightarrow \text{ saddle,} \\
\operatorname{tr} J = 0,\ \det J > 0, & \text{cannot occur,} \\
\text{if } \operatorname{tr} J > 0,\ \det J = 0, & \mu_2 = 0, \mu_1 > 0 \Rightarrow \text{ bifurcation,} \\
\text{if } \operatorname{tr} J < 0,\ \det J = 0, & \text{then } \mu_1 + a\sqrt{\mu_2} = 0 \text{ or } \mu_2 = 0, \\
 & \mu_1 < 0 \Rightarrow \text{ bifurcation.}
\end{array}$$

Thus, $(0, \sqrt{\mu_2})$ is a

$$\begin{array}{ll}
\text{sink} & \text{for } \mu_1 + a\sqrt{\mu_2} < 0, \\
\text{saddle} & \text{for } \mu_1 + a\sqrt{\mu_2} > 0.
\end{array}$$

Later we will examine the nature of the bifurcation occurring on $\mu_1 + a\sqrt{\mu_2} = 0$ and $\mu_2 = 0$.

Next we examine the branch $(0, -\sqrt{\mu_2})$.

$(0, -\sqrt{\mu_2})$. On this branch we have

$$\begin{aligned}
\operatorname{tr} J &= \mu_1 - a\sqrt{\mu_2} + 2\sqrt{\mu_2}, \\
\det J &= 2\sqrt{\mu_2}(\mu_1 - a\sqrt{\mu_2}),
\end{aligned}$$

from which we conclude

$$\begin{aligned}
\operatorname{tr} J > 0 &\Rightarrow \ \mu_1 - a\sqrt{\mu_2} > -2\sqrt{\mu_2},\\
\operatorname{tr} J < 0 &\Rightarrow \ \mu_1 - a\sqrt{\mu_2} < -2\sqrt{\mu_2},\\
\det J > 0 &\Rightarrow \ \mu_1 - a\sqrt{\mu_2} > 0,\\
\det J < 0 &\Rightarrow \ \mu_1 - a\sqrt{\mu_2} < 0.
\end{aligned}$$

Appealing to (3.1.222) and using (3.1.223), we make the following conclusions concerning the stability of the branch of fixed points $(0, -\sqrt{\mu_2})$

$$\begin{array}{ll}
\text{if } \operatorname{tr} J > 0,\ \det J > 0, & \text{then } \mu_1 - a\sqrt{\mu_2} > 0 \Rightarrow \quad \text{source},\\
\text{if } \operatorname{tr} J > 0,\ \det J < 0, & \text{then } 0 > \mu_1 - a\sqrt{\mu_2} > -2\sqrt{\mu_2} \Rightarrow \quad \text{saddle},\\
\operatorname{tr} J < 0,\ \det J > 0, & \text{cannot occur},\\
\text{if } \operatorname{tr} J < 0,\ \det J < 0, & \text{then } \mu_1 - a\sqrt{\mu_2} < -2\sqrt{\mu_2} \Rightarrow \quad \text{saddle},\\
\operatorname{tr} J = 0,\ \det J > 0 & \text{cannot occur},\\
\text{if } \operatorname{tr} J = 0,\ \det J < 0, & \text{then } \mu_1 - a\sqrt{\mu_2} = -2\sqrt{\mu_2} \Rightarrow \quad \text{saddle},\\
\text{if } \operatorname{tr} J > 0,\ \det J = 0, & \text{then } \mu_1 = a\sqrt{\mu_2} \Rightarrow \quad \text{bifurcation},\\
\operatorname{tr} J < 0,\ \det J = 0 & \text{cannot occur}.
\end{array}$$

We thus conclude that $(0, -\sqrt{\mu_2} > 0)$ has the following stability characteristics.

$$\begin{array}{ll}
\text{source} & \text{for } \mu_1 - a\sqrt{\mu_2} > 0,\\
\text{saddle} & \text{for } \mu_1 - a\sqrt{\mu_2} < 0.
\end{array}$$

Later we will examine the bifurcation occuring on $\mu_1 - a\sqrt{\mu_2} = 0$.

Now we turn to an examination of the remaining branch of fixed points (note that our previous analysis did not depend on b)

$$(r, z) = \left(\sqrt{\frac{1}{b}\left(\frac{\mu_1^2}{a^2} - \mu_2\right)}, \frac{-\mu_1}{a}\right).$$

We examine the cases $b = +1$ and $b = -1$ separately.

$b = +1, \left(\sqrt{\frac{\mu_1^2}{a^2} - \mu_2}, \frac{-\mu_1}{a}\right)$. This branch exists only for $\frac{\mu_1^2}{a^2} > \mu_2$, and on this branch we have

$$\operatorname{tr} J = \frac{2\mu_1}{a}, \tag{3.1.224a}$$

$$\det J = -2a\left(\frac{\mu_1^2}{a^2} - \mu_2\right). \tag{3.1.224b}$$

We now must consider Cases I and IIa and b individually.

Case I: $a > 0$. From (3.1.224b), for $a > 0$ we have $\det J \le 0$. Using (3.1.223), for $\det J \le 0$, the fixed point is always a saddle.

Cases IIa,b: $a < 0$. From (3.1.224b), for $a < 0$ we have $\det J \geq 0$. Hence, using (3.1.223), we conclude the following

$$\mu_1 > 0, \qquad \frac{\mu_1^2}{a^2} - \mu_2 > 0 \Rightarrow \text{ source},$$
$$\mu_1 < 0, \qquad \frac{\mu_1^2}{a^2} - \mu_2 > 0 \Rightarrow \text{ sink}.$$

We might guess that on $\mu_1 = 0$, $\mu_2 < 0$, a Poincaré-Andronov-Hopf bifurcation occurs.

Next we examine the case $b = -1$.

b= -1, $\left(\sqrt{\mu_2 - \frac{\mu_1^2}{a^2}}, \frac{-\mu_1}{a}\right)$. For this case we have

$$\operatorname{tr} J = \frac{2\mu_1}{a}, \tag{3.1.225a}$$
$$\det J = 2a\left(\mu_2 - \frac{\mu_1^2}{a^2}\right). \tag{3.1.225b}$$

(Note that $\mu_2 - \frac{\mu_1^2}{a^2} \geq 0$.)

We will examine Cases III and IVa and b individually.

Case III: $a > 0$. Using (3.1.225b), it follows that $\det J \geq 0$ which, when used with (3.1.225a) and (3.1.223), allows us to conclude that

$$\mu_1 > 0 \Rightarrow \text{ source},$$
$$\mu_1 < 0 \Rightarrow \text{ sink}.$$

Hence, we might guess that a Poincaré-Andronov-Hopf bifurcation is possible on $\mu_1 = 0$, $\mu_2 > 0$.

Cases IVa,b: $a < 0$. Using (3.1.225b), we see that $\det J \leq 0$, which, when used with (3.1.225a) and (3.1.223), allows us to conclude that

$$\mu_1 < 0 \Rightarrow \text{ saddle},$$
$$\mu_1 > 0 \Rightarrow \text{ saddle}.$$

Hence, no Poincaré-Andronov-Hopf bifurcations occur.

This completes the stability analysis of the fixed points. We next examine the nature of the various possible bifurcations.

Step 4b: The Bifurcations of the Fixed Points. First we examine the two branches

$$(0, \pm\sqrt{\mu_2}).$$

These branches exist only for $\mu_2 \geq 0$, coalescing at $\mu_2 = 0$. We thus expect them to bifurcate from $(0,0)$ in a *saddle-node* bifurcation. Since these

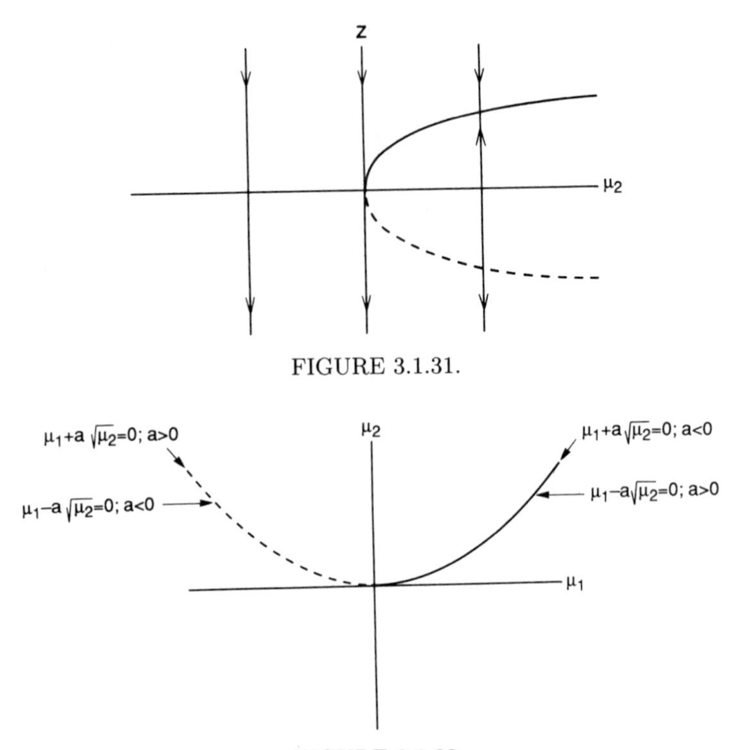

FIGURE 3.1.31.

FIGURE 3.1.32.

branches start on $r = 0$ and remain on $r = 0$, the center manifold analysis is particulary simple—we simply set $r = 0$ in our original equations and obtain

$$\dot{z} = \mu_2 - z^2.$$

(Note that the equation is independent of b with bifurcation diagram shown in Figure 3.1.31.)

Next we examine the bifurcation of the branches $(0, \pm\sqrt{\mu_2})$, $\mu_2 > 0$, which occurs on $\mu_1 \pm a\sqrt{\mu_2} = 0$.

We will do a center manifold analysis. First we transform the fixed point to the origin.

Letting $\xi = z \mp \sqrt{\mu_2}$, (3.1.219) becomes

$$\begin{aligned} \dot{r} &= \mu_1 r + ar(\xi \pm \sqrt{\mu_2}), \\ \dot{\xi} &= \mu_2 + br^2 - (\xi \pm \sqrt{\mu_2})^2, \qquad \mu_2 > 0. \end{aligned} \tag{3.1.226}$$

We are interested in the flow in a neighborhood of the curve $\mu_1 \pm a\sqrt{\mu_2} = 0$; we illustrate this curve in the $(\mu_1 - \mu_2)$-parameter plane in Figure 3.1.32.

Therefore, we will set $\mu_1 =$ constant and $\sqrt{\mu_2} = \mp\frac{\mu_1}{a} - \varepsilon$. This corresponds to crossing the parabola vertically for fixed μ_1. We will have to pay close attention to the direction in which we cross the curve by varying ε; we will come back to this later. Substituting $\sqrt{\mu_2} = \mp\frac{\mu_1}{a} - \varepsilon$ into (3.1.226) gives

$$\dot{r} = \mu_1 r + ar\left(\xi \pm \left(\mp\frac{\mu_1}{a} - \varepsilon\right)\right),$$
$$\dot{\xi} = \left(\mp\frac{\mu_1}{a} - \varepsilon\right)^2 + br^2 - \left(\xi \pm \left(\mp\frac{\mu_1}{a} - \varepsilon\right)\right)^2,$$

or

$$\dot{r} = ar\xi \mp ar\varepsilon,$$
$$\dot{\xi} = 2\left(\frac{\mu_1}{a} \pm \varepsilon\right)\xi + br^2 - \xi^2;$$

in matrix form (including the parameters as a dependent variable in anticipation of applying center manifold theory), it gives

$$\begin{pmatrix} \dot{r} \\ \dot{\xi} \end{pmatrix} = \begin{pmatrix} 0 & 0 \\ 0 & \frac{2\mu_1}{a} \end{pmatrix} + \begin{pmatrix} ar\xi \mp ar\varepsilon \\ \pm 2\varepsilon\xi + br^2 - \xi^2 \end{pmatrix},$$
$$\dot{\varepsilon} = 0. \tag{3.1.227}$$

Fortunately, (3.1.227) is already in the standard form for application of the center manifold theorem. From Theorem 2.1.1, the center manifold can be represented as follows

$$W^c = \{(r, \varepsilon, \xi) \,|\, \xi = h(r, \varepsilon), h(0,0) = 0; Dh(0,0) = 0\}$$

for r and ε sufficiently small where h satisfies

$$Dh(x)[Bx + f(x, h(x))] - Ch(x) - g(x, h(x)) = 0, \tag{3.1.228}$$

with

$$x \equiv (r, \varepsilon), \qquad B = \begin{pmatrix} 0 & 0 \\ 0 & 0 \end{pmatrix}, \qquad C = \frac{2\mu_1}{a},$$
$$f = \begin{pmatrix} ar\xi \mp ar\varepsilon \\ 0 \end{pmatrix}, \qquad g = \pm 2\varepsilon\xi + br^2 - \xi^2.$$

Substituting $h(r, \varepsilon) = \alpha r^2 + \beta r\varepsilon + \gamma\varepsilon^2 + \mathcal{O}(3)$ into (3.1.228) gives

$$(2\alpha r + \beta\varepsilon + \mathcal{O}(3), \beta r + 2\gamma\varepsilon + \mathcal{O}(3)) \begin{pmatrix} arh \mp ar\varepsilon \\ 0 \end{pmatrix}$$
$$-\frac{2\mu_1}{a}(\alpha r^2 + \beta r\varepsilon + \gamma\varepsilon^2 + \mathcal{O}(3)) - (\pm 2\varepsilon h + br^2 - h^2) = 0. \tag{3.1.229}$$

Balancing coefficients on powers of r and ε in (3.1.229) gives

$$r^2 : -\frac{2\mu_1}{a}\alpha - b = 0 \Rightarrow \alpha = -\frac{ab}{2\mu_1},$$
$$\varepsilon r : \frac{2\mu_1}{a}\beta = 0 \Rightarrow \beta = 0,$$
$$\varepsilon^2 : \frac{2\mu_1}{a}\gamma = 0 \Rightarrow \gamma = 0;$$

hence, the center manifold is the graph of

$$h(r, \varepsilon) = -\frac{ab}{2\mu_1} r^2 + \mathcal{O}(3)$$

and, therefore, the vector field (3.1.227) restricted to the center manifold is given by

$$\dot{r} = ar\left(-\frac{ab}{2\mu_1} r^2 + \mathcal{O}(3)\right) \mp ar\varepsilon$$

or

$$\dot{r} = r\left(-\frac{a^2 b}{2\mu_1} r^2 \mp a\varepsilon\right) + \cdots . \qquad (3.1.230)$$

This equation indicates that pitchfork bifurcations occur at $\varepsilon = 0$, but note that, for us, the only bifurcating solution that has meaning is the $r > 0$ solution.

We now use (3.1.230) to derive bifurcation diagrams for each branch of fixed points.

$(0, +\sqrt{\mu_2})$. The bifurcation curve is given by

$$\sqrt{\mu_2} = -\frac{\mu_1}{a},$$

and ε increasing from negative to positive corresponds to decreasing μ_2 across the bifurcation curve.

The vector field restricted to the center manifold is

$$\dot{r} = r\left(-\frac{a^2 b}{2\mu_1} r^2 - a\varepsilon\right) + \cdots ,$$

from which we easily obtain the bifurcation diagrams shown in Figure 3.1.33.

$(0, -\sqrt{\mu_2})$. The bifurcation curve is given by

$$\sqrt{\mu_2} = -\frac{\mu_1}{a}.$$

The vector field restricted to the center manifold is

$$\dot{r} = r\left(-\frac{a^2 b}{2\mu_1} r^2 + a\varepsilon\right) + \cdots ,$$

from which we easily obtain the bifurcation diagrams shown in Figure 3.1.34.

Now we want to translate these center manifold pictures into phase portraits for the two-dimensional flows. Recall that the eigenvalues for the direction normal to the center manifold is given by $\frac{+2\mu_1}{a}$. We draw separate diagrams for each branch in Figures 3.1.35 and 3.1.36.

At this point we will summarize our results thus far.

A *saddle-node* bifurcation occurs at $\mu_2 = 0$, giving us two branches of fixed points $(0, \pm\sqrt{\mu_2})$.

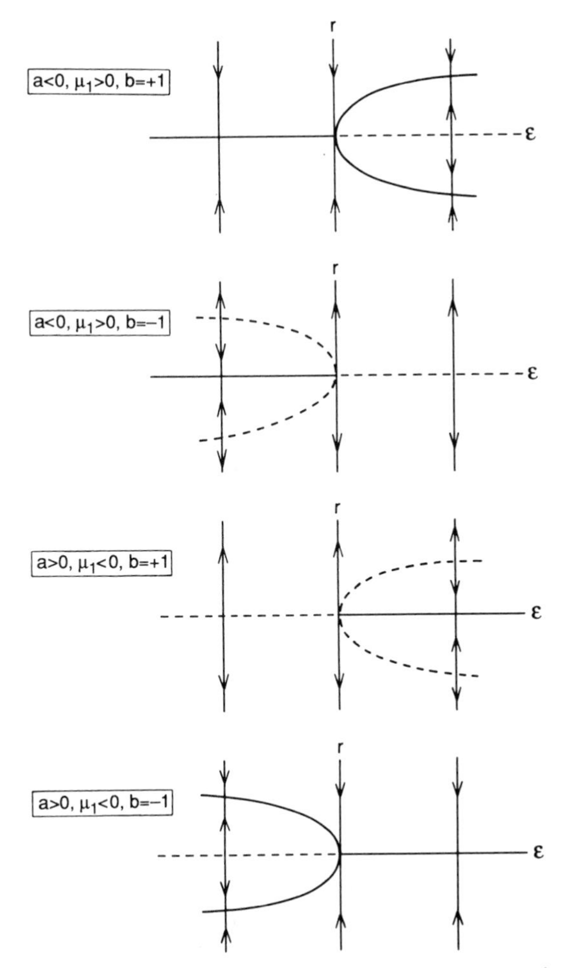

FIGURE 3.1.33. Bifurcations on the center manifold for $(0, +\sqrt{\mu_2})$.

1. $(0, +\sqrt{\mu_2})$ undergoes a pitchfork bifurcation on $\sqrt{\mu_2} + \frac{\mu_1}{a} = 0$ with a new fixed point being born above the curve for $b = -1$ and below the curve for $b = +1$.

2. $(0, -\sqrt{\mu_2})$ undergoes a pitchfork bifurcation on $\sqrt{\mu_2} - \frac{\mu_1}{a} = 0$ with a new fixed point being born above the curve for $b = -1$ and below the curve for $b = +1$.

Detailed stability diagrams can be obtained from the previously given diagrams.

To complete the local analysis we must examine the nature of the possible Poincaré-Andronov-Hopf bifurcations in Cases IIa,b and III.

We examine each case individually.

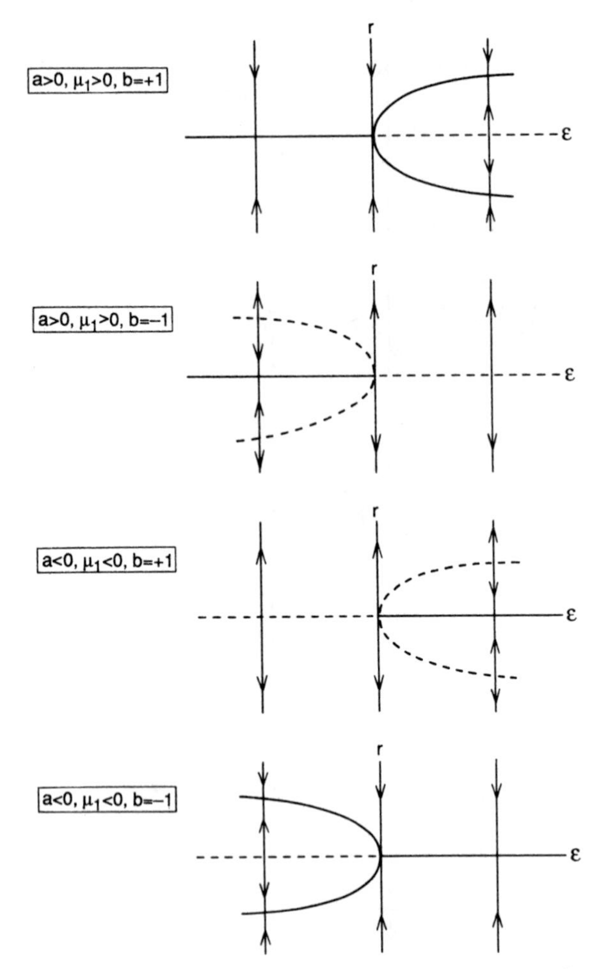

FIGURE 3.1.34. Bifurcations on the center manifold for $(0, -\sqrt{\mu_2})$.

Cases IIa,b. The branch of fixed points is given by

$$(r, z) = \left(+\sqrt{\frac{\mu_1^2}{a^2} - \mu_2}, \frac{-\mu_1}{a} \right), \tag{3.1.231}$$

where $a < 0$ and $\dfrac{\mu_1^2}{a^2} \geq \mu_2$.

Our candidate for the Poincaré-Andronov-Hopf bifurcation curve is given by

$$\mu_1 = 0, \qquad \mu_2 < 0.$$

On this curve the eigenvalues of the vector field linearized about a fixed

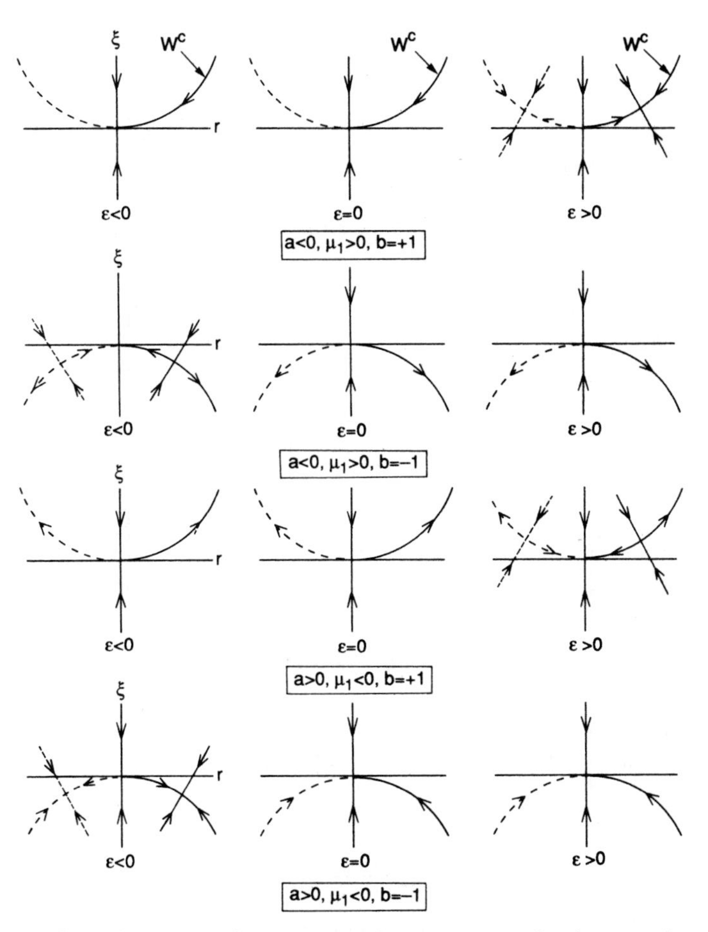

FIGURE 3.1.35. Bifurcations in the $r - \xi$ plane for $(0, +\sqrt{\mu_2})$.

point on the branch (3.1.231) are

$$\lambda_{1,2} = \pm i\sqrt{|2a\mu_2|}.$$

The reader should recall from Theorem 3.1.3 that there are two quantities which need to be determined.

1. The eigenvalues cross the imaginary axis transversely.

2. The coefficient a in the Poincaré-Andronov-Hopf normal form is nonzero. (Note: this should not be confused with the coefficient a in the normal form that we are presently studying, which we have assumed to be nonzero.)

We begin by verifying statement 1.

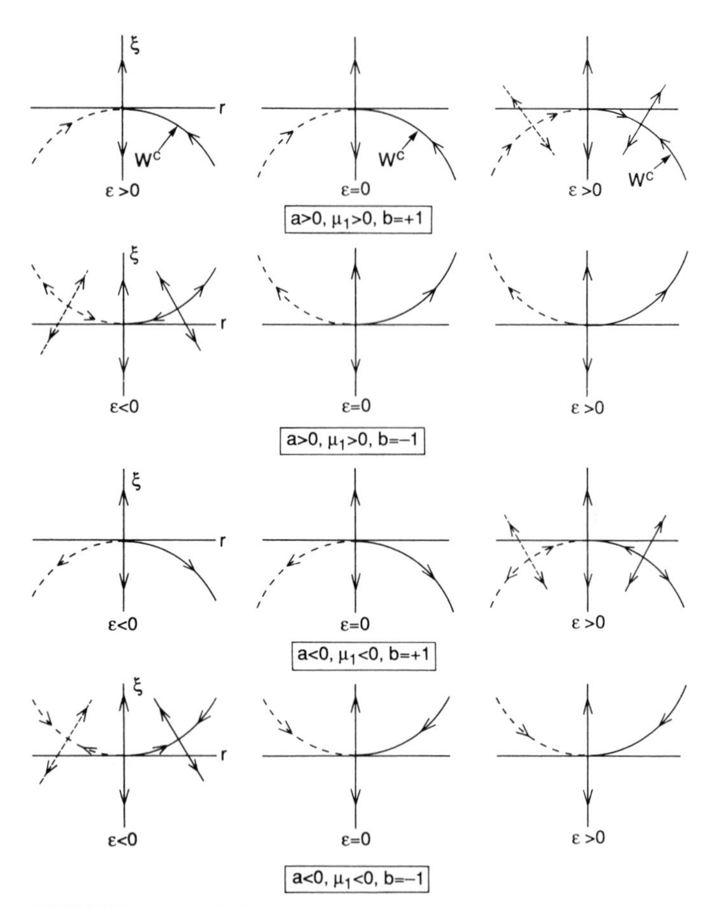

FIGURE 3.1.36. Bifurcations in the $r - \xi$ plane for $(0, -\sqrt{\mu_2})$.

The general expression for the eigenvalues is

$$\lambda_{1,2} = \frac{\operatorname{tr} J}{2} \pm \frac{1}{2}\sqrt{(\operatorname{tr} J)^2 - 4 \det J},$$

where in our case

$$\operatorname{tr} J = \frac{2\mu_1}{a},$$

$$\det J = -2a\left(\frac{\mu_1^2}{a^2} - \mu_2\right).$$

We will view μ_2 as fixed and μ_1 as a parameter; since we are interested only in the behavior of the eigenvalues near $\mu_1 = 0$, $\mu_2 < 0$, we can take the real part of $\lambda_{1,2}$ to be $\frac{\operatorname{tr} J}{2}$ and thus obtain

$$\frac{d}{d\mu_1}\operatorname{Re} \lambda = \frac{2}{a} \neq 0.$$

Next we check statement 2.

We set $b = +1$ in (3.1.219) and obtain

$$\begin{aligned}\dot{r} &= \mu_1 r + arz,\\ \dot{z} &= \mu_2 + r^2 - z^2.\end{aligned}$$

We next put this system into the normal form so that the coefficient a in the Poincaré-Andronov-Hopf normal form can be computed. First we translate the fixed point to the origin by letting

$$\rho = r - \sqrt{\frac{\mu_1^2}{a} - \mu_2}, \qquad \xi = z + \frac{\mu_1}{a},$$

and hence obtain

$$\begin{aligned}\dot{\rho} &= \mu_1\left(\rho + \sqrt{\frac{\mu_1^2}{a^2} - \mu_2}\right) + a\left(\rho + \sqrt{\frac{\mu_1^2}{a^2} - \mu_2}\right)\left(\xi - \frac{\mu_1}{a}\right),\\ \dot{\xi} &= \mu_2 + \left(\rho + \sqrt{\frac{\mu_1^2}{a^2} - \mu_2}\right)^2 - \left(\xi - \frac{\mu_1}{a}\right)^2,\end{aligned}$$

or

$$\begin{aligned}\dot{\rho} &= a\xi\sqrt{\frac{\mu_1^2}{a^2} - \mu_2} + a\rho\xi,\\ \dot{\xi} &= 2\rho\sqrt{\frac{\mu_1^2}{a^2} - \mu_2} + \frac{2\mu_1}{a}\xi + \rho^2 - \xi^2.\end{aligned}$$

We next evaluate this equation on the bifurcation curve $\mu_1 = 0$, $\mu_2 < 0$ and get

$$\begin{aligned}\dot{\rho} &= a\sqrt{|\mu_2|}\xi + a\rho\xi,\\ \dot{\xi} &= 2\sqrt{|\mu_2|}\rho + \rho^2 - \xi^2.\end{aligned}$$

The matrix associated with the linear part of this equation is given by

$$\begin{pmatrix} 0 & a\sqrt{|\mu_2|} \\ 2\sqrt{|\mu_2|} & 0 \end{pmatrix}.$$

Introducing the linear tranformation

$$\begin{pmatrix}\rho\\ \xi\end{pmatrix} = \begin{pmatrix} 0 & -\sqrt{\frac{|a|}{2}} \\ 1 & 0 \end{pmatrix}\begin{pmatrix}u\\ v\end{pmatrix};$$

$$\begin{pmatrix}u\\ v\end{pmatrix} = \frac{1}{\sqrt{\frac{|a|}{2}}}\begin{pmatrix} 0 & \sqrt{\frac{|a|}{2}} \\ -1 & 0 \end{pmatrix}\begin{pmatrix}\rho\\ \xi\end{pmatrix},$$

the equation becomes

$$\begin{pmatrix} \dot{u} \\ \dot{v} \end{pmatrix} = \begin{pmatrix} 0 & -\sqrt{|2a\mu_2|} \\ \sqrt{|2a\mu_2|} & 0 \end{pmatrix} \begin{pmatrix} u \\ v \end{pmatrix} + \begin{pmatrix} \frac{|a|}{2}v^2 - u^2 \\ -|a|uv \end{pmatrix}.$$

This is the standard form given in (3.1.106) from which the coefficient a in the Poincaré-Andronov-Hopf normal form can be computed.

From (3.1.107), this coefficient is given by

$$\frac{1}{16}\left[f_{uuu} + f_{uvv} + g_{uuv} + g_{vvv}\right] + \frac{1}{16\sqrt{|2a\mu_2|}}\left[f_{uv}(f_{uu} + f_{vv})\right.$$
$$\left. - g_{uv}(g_{uu} + g_{vv}) - f_{uu}g_{uu} + f_{vv}g_{vv}\right],$$

and all partial derivatives are evaluated at $(0,0)$.

In our case

$$f \equiv \frac{|a|}{2}v^2 - u^2,$$
$$g \equiv -|a|uv.$$

Now we work out the partial derivatives (note all third derivatives vanish)

$$\begin{aligned} f_{uu} &= -2, & g_{uu} &= 0, \\ f_{uv} &= 0, & g_{uv} &= -|a|, \\ f_{vv} &= |a|, & g_{vv} &= 0. \end{aligned}$$

Thus, the coefficient a is identically zero. This tells us that we must retain (at least) cubic terms in our normal form in order to get any stability information concerning the Poincaré-Andronov-Hopf bifurcation. (Note: we might have guessed that this would be the case. Why?) We will complete this analysis in Chapter 4.

Now we must examine Poincaré-Andronov-Hopf bifurcations in the remaining case.

Case III, $a > 0$. It is straightforward to verify that Poincaré-Andronov-Hopf bifurcations occur on $\mu_1 = 0$, $\mu_2 > 0$ for the branch of fixed points given by

$$\left(\sqrt{\mu_2 - \frac{\mu_1^2}{a^2}}, \frac{-\mu_1}{a}\right)$$

and that, unfortunately, in this case also, the coefficient in the Poincaré-Andronov-Hopf normal form is identically zero.

We now want to summarize these results in the following bifurcation diagrams.

Case I: $b = +1$, $a > 0$. In Figure 3.1.37 we show phase portraits for different regions in the $\mu_1 - \mu_2$ plane. Note that by index theory there can be no periodic orbits in Case I (which might arise via some global bifurcation).

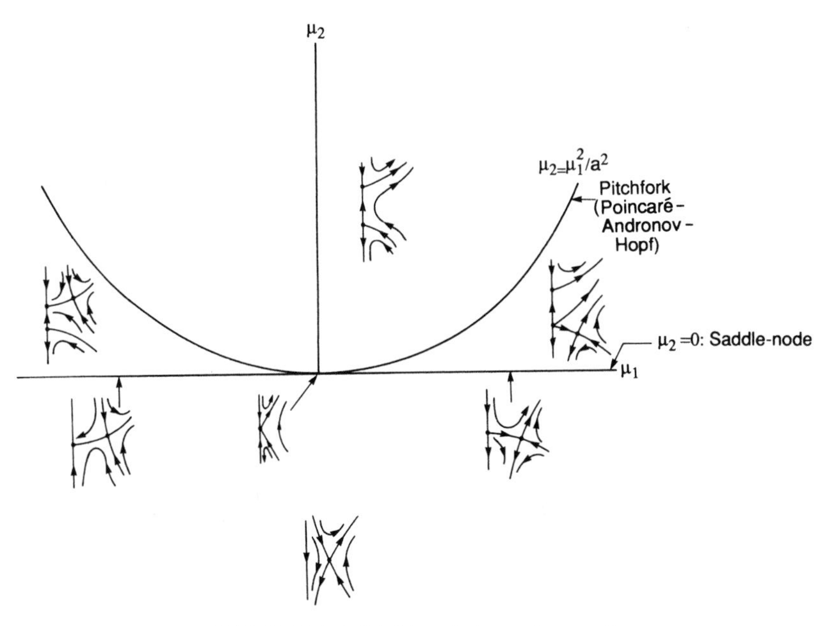

FIGURE 3.1.37. Case I: $b = +1$, $a > 0$.

This is because we must have $r > 0$, and the only fixed point in $r > 0$ is a saddle point. Thus, Figure 3.1.37 represents the complete story for Case I. It remains only to interpret the $r-z$ phase plane results in terms of the full three-dimensional vector field and consider the effects of the higher order terms of the normal form. We will do this later in this section.

Case IIa,b: $b = +1$, $a > 0$. We show phase portraits for this case in different regions of the $\mu_1-\mu_2$ plane in Figure 3.1.38. Note that the eigenvalues of the matrix associated with the vector field linearized about $\left(\sqrt{\frac{\mu_1^2}{a^2}-\mu_2}, \frac{-\mu_1}{a}\right)$ are given by

$$\lambda_{1,2} = \frac{\mu_1}{a} \pm \sqrt{\frac{\mu_1^2}{a^2} + \frac{2}{a}(\mu_1^2 - a^2\mu_2)},$$

and that these eigenvalues have nonzero imaginary part for

$$\mu_2 < \mu_1^2\left(2 + \frac{1}{a}\right)/2a^2, \qquad \mu_2 < 0.$$

This gives us a better idea of the local orbit structure near these fixed points, and we illustrate this curve with a dotted line in Figure 3.1.38. We caution, however, that it is *not* a bifurcation curve.

Note that on $\mu_1 = 0$ the truncated normal form has the first integral

$$F(r,z) = \frac{a}{2}r^{2/a}\left[\mu_2 + \frac{r^2}{1+a} - z^2\right]. \tag{3.1.232}$$

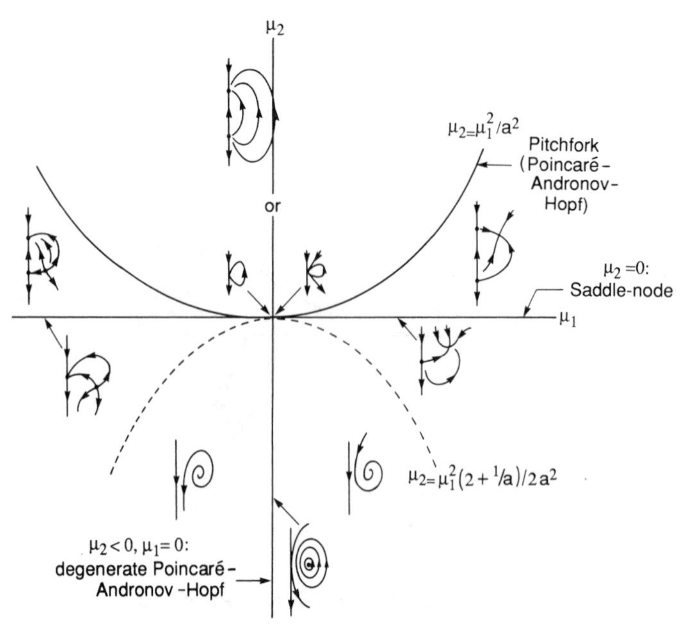

FIGURE 3.1.38. Case IIa,b: $b = +1$, $a > 0$.

This should give some insight into the "degenerate" Poincaré-Andronov-Hopf bifurcation, since (3.1.232) implies that on $\mu_1 = 0$ the truncated normal form has a one-parameter family of periodic orbits. We expect that this degenerate situation will dramatically change when the effects of the higher order terms in the normal form are taken into account. In particular, we would expect that a finite number of these periodic orbits survive. Exactly how many is a delicate issue that we will examine in Chapter 4.

Case III: $b = -1$, $a > 0$. We show phase portraits in different regions of the $\mu_1 - \mu_2$ plane in Figure 3.1.39. This case suffers from many of the same difficulties of Cases IIa,b. In particular, the truncated normal form has the first integral

$$G(r,z) = \frac{a}{2} r^{2/a} \left(\mu_2 - \frac{1}{1+a} r^2 - z^2 \right) \tag{3.1.233}$$

on $\mu_1 = 0$. An examination of the first integral shows that, for $\mu_2 > 0$, the truncated normal form has a one-parameter family of periodic orbits which limit on a heteroclinic cycle as shown in Figure 3.1.39. In Chapter 4 we will consider the effects of the higher order terms in the normal form on this degenerate phase portrait.

Case IVa,b: $b = -1$, $a < 0$. We show phase portraits in the different regions in the $\mu_1 - \mu_2$ plane in Figure 3.1.40. Using index theory, it is easy to argue that these cases have no periodic orbits. Hence, Figure 3.1.40 represents the complete story for the $r - z$ phase plane analysis of the truncated normal form.

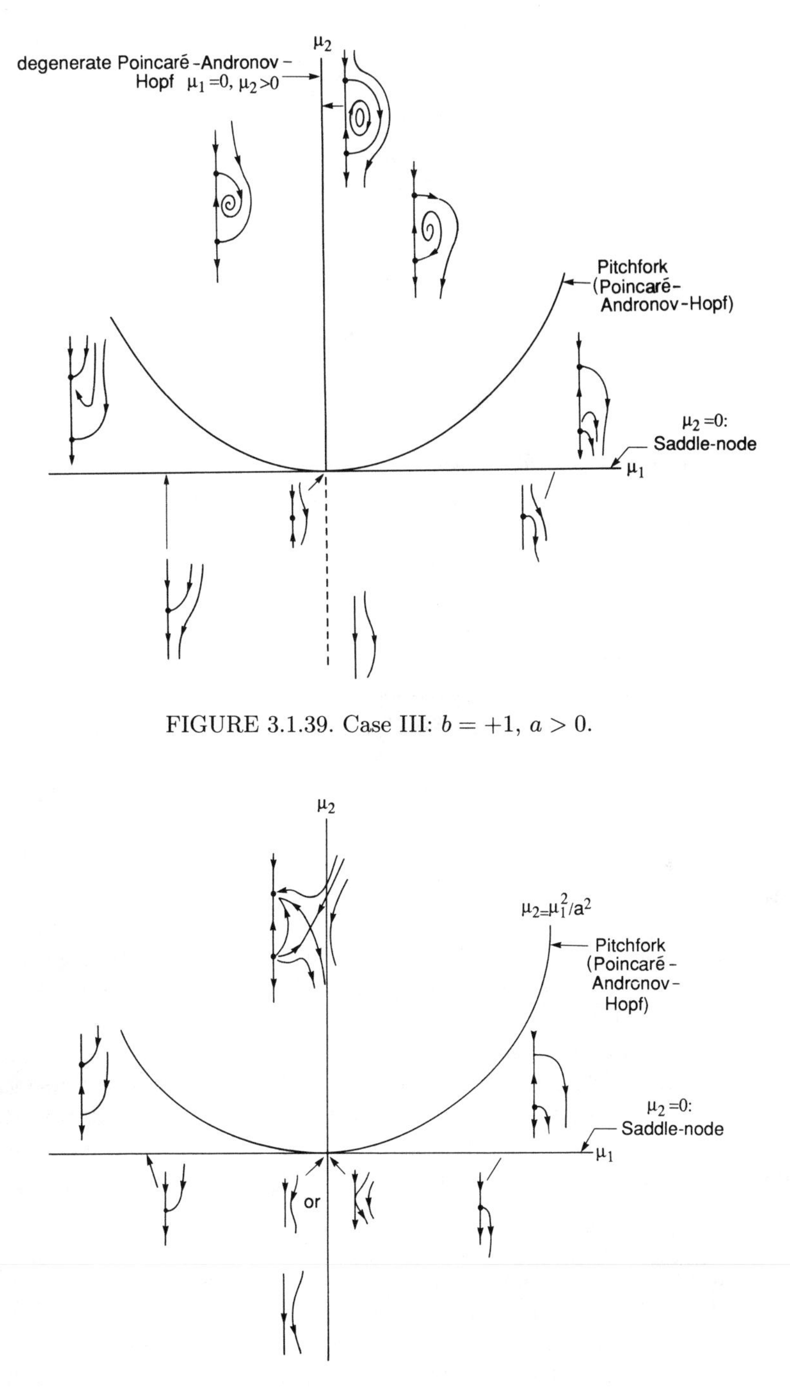

FIGURE 3.1.39. Case III: $b=+1$, $a>0$.

FIGURE 3.1.40. Case IVa,b: $b=-1$, $a<0$.

Relation of the Dynamics in the $r - z$ Phase Plane to the Full Three-Dimensional Vector Field

We now want to discuss how the dynamics of

$$\begin{aligned} \dot{r} &= \mu_1 r + arz, \\ \dot{z} &= \mu_2 + br^2 - z^2, \end{aligned} \tag{3.1.234}$$

relate to

$$\begin{aligned} \dot{r} &= \mu_1 r + arz, \\ \dot{z} &= \mu_2 + br^2 - z^2, \\ \dot{\theta} &= \omega + \cdots . \end{aligned} \tag{3.1.235}$$

We are interested in three types of invariant sets studied in (3.1.233). They are fixed points, periodic orbits, and heteroclinic cycles. We consider each case separately.

Fixed Points

There are two cases, fixed points with $r = 0$ and fixed points with $r > 0$. It is easy to see (returning to the definition of r and θ in terms of the original Cartesian coordinates) that fixed points of (3.1.234) with $r = 0$ correspond to fixed points of (3.1.235). *Hyperbolic* fixed points of (3.1.234) with $r > 0$ correspond to periodic orbits of (3.1.235). This follows immediately by applying the method of averaging. See Figure 3.1.41 for a geometrical description.

Periodic Orbits

We have not developed the theoretical tools to treat this situation rigorously; we refer the reader to Section 4.9. However, for now we will give a heuristic description of what is happening. Notice that the r and z components of (3.1.235) are independent of θ. This implies that the periodic orbit in the $r - z$ plane is manifested as an invariant two-torus in the $r - z - \theta$ phase space; see Figure 3.1.42. This is a very delicate situation regarding the higher order terms in the normal form, since they could dramatically affect the flow on the torus, in particular, whether or not we get quasiperiodic motion or phase locking (periodic motion).

Heteroclinic Cycles

Following the discussion for periodic orbits given above, the part of the heteroclinic cycle in Case III ($\mu_2 > 0$) on the z axis is manifested as an invariant line in $r - z - \theta$ space, and the part of the heteroclinic cycle having $r > 0$ is manifested as an invariant sphere in $r - z - \theta$ space; see Figure 3.1.43. This is a very degenerate situation and could be dramatically affected by the higher order terms of the normal form. Indeed, we will see

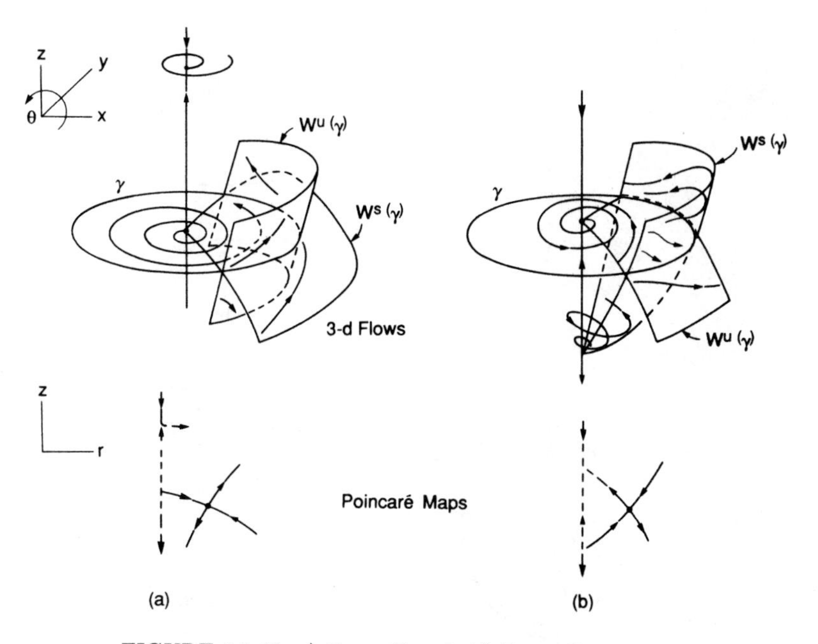

FIGURE 3.1.41. a) From Case I, b) From Case IVa,b.

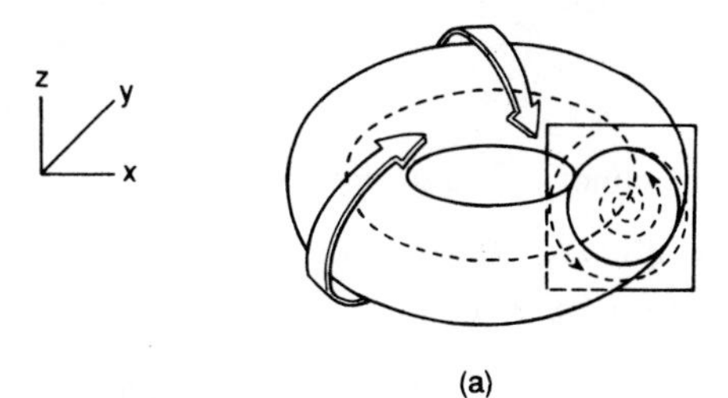

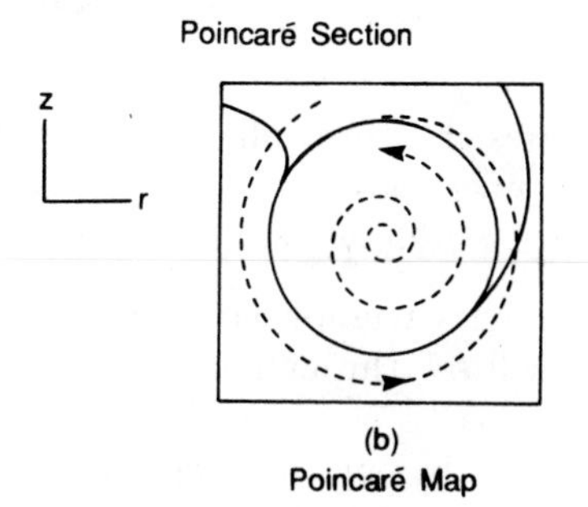

FIGURE 3.1.42. a) Three-dimensional flow, b) Poincaré section.

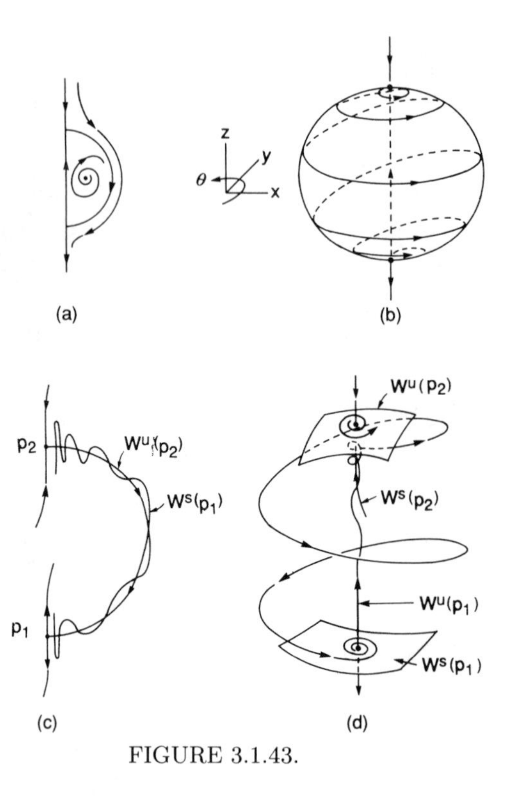

FIGURE 3.1.43.

in Chapter 4 that chaotic dynamics may result.

Step 5: Analysis of Global Bifurcations. As we have mentioned, this will be completed in Section 4.9 after we have developed the necessary theoretical tools.

Step 6: Effects of the Higher Order Terms in the Normal Form. In Cases I and IVa,b the method of averaging essentially enables us to conclude that the higher order terms do not qualitatively change the dynamics. Thus we have found a versal deformation. The details of proving this, however, are left to the exercises.

The remaining cases are more difficult and, ultimately, we will argue that versal deformations may not exist in some circumstances.

Before leaving this section we want to make some final remarks.

Remark 1. This analysis reemphasizes the power of the method of normal forms. As we will see throughout the remainder of this book, vector fields having phase spaces of dimension three or more can exhibit very complicated dynamics. In our case the method of normal forms utilized the structure of the vector field to naturally "separate" the variables. This enabled us to "get our foot in the door" by using powerful phase plane techniques.

Remark 2. From the double-zero eigenvalue and now this case, a lesson to be learned is that Poincaré-Andronov-Hopf bifurcations always cause us trouble in the sense of how they relate to global bifurcations and/or how they are affected by the consideration of the higher order terms of the normal form.

3.2 Bifurcations of Fixed Points of Maps

The theory for bifurcations of fixed points of maps is very similar to the theory for vector fields. Therefore, we will not include as much detail but merely highlight the differences when they occur.

Consider a p-parameter family of maps of $\mathbb{R}^n$ into $\mathbb{R}^n$

$$y \mapsto g(y, \lambda), \qquad y \in \mathbb{R}^n, \quad \lambda \in \mathbb{R}^p \tag{3.2.1}$$

where g is $\mathbf{C}^r$ (with r to be specified later, usually $r \geq 5$ is sufficient) on some sufficiently large open set in $\mathbb{R}^n \times \mathbb{R}^p$. Suppose (3.2.1) has a fixed point at $(y, \lambda) = (y_0, \lambda_0)$, i.e.,

$$g(y_0, \lambda_0) = y_0. \tag{3.2.2}$$

Then, just as in the case for vector fields, two questions naturally arise.

1. Is the fixed point stable or unstable?
2. How is the stability or instability affected as λ is varied?

As in the case for vector fields, an examination of the associated linearized map is the first place to start in order to answer these questions. The associated linearized map is given by

$$\xi \mapsto D_y g(y_0, \lambda_0)\xi, \qquad \xi \in \mathbb{R}^n, \tag{3.2.3}$$

and, from Sections 1.1A and 1.1C, we know that if the fixed point is hyperbolic (i.e., none of the eigenvalues of $D_y g(y_0, \lambda_0)$ have unit modulus), then stability (resp. instability) in the linear approximation implies stability (resp. instability) of the fixed point of the nonlinear map. Moreover, using an implicit function theorem argument exactly like that given at the beginning of Section 3.1, it can be shown that, in a sufficiently small neighborhood of (y_0, λ_0), for each λ there is a unique fixed point having the same stability type as (y_0, λ_0). Thus, hyperbolic fixed points are locally dynamically dull!

The fun begins when we consider Questions 1 and 2 above in the situation when the fixed point is *not hyperbolic*. Just as in the case for vector fields, the linear approximation cannot be used to determine stability, and varying λ can result in the creation of new orbits (i.e., bifurcation). The simplest ways in which a fixed point of a map can be nonhyperbolic are the following.

1. $D_y g(y_0, \lambda_0)$ has a single eigenvalue equal to 1 with the remaining $n-1$ eigenvalues having moduli not equal to 1.

2. $D_y g(y_0, \lambda_0)$ has a single eigenvalue equal to -1 with the remaining $n-1$ eigenvalues having moduli not equal to 1.

3. $D_y g(y_0, \lambda_0)$ has two complex conjugate eigenvalues having modulus 1 (which are *not* one of the first four roots of unity) with the remaining $n-2$ eigenvalues having moduli not equal to 1.

Using the center manifold theory, the analysis of the above situations can be reduced to the analysis of a p-parameter family of one-, one-, and two-dimensional maps, respectively. We begin with the first case.

3.2A An Eigenvalue of 1

In this case, the study of the orbit structure near the fixed point can be reduced to the study of a parametrized family of maps on the one-dimensional center manifold. We suppose that the map on the center manifold is given by

$$x \mapsto f(x, \mu), \qquad x \in \mathbb{R}^1, \quad \mu \in \mathbb{R}^1, \tag{3.2.4}$$

where, for now, we will consider only one parameter (if there is more than one parameter in the problem, we will consider all but one as fixed constants). In making the reduction to the center manifold, the fixed point $(y_0, \lambda_0) \in \mathbb{R}^n \times \mathbb{R}^p$ has been transformed to the origin in $\mathbb{R}^1 \times \mathbb{R}^1$ (cf. Section 2.1A) so that we have

$$f(0,0) = 0, \tag{3.2.5}$$

$$\frac{\partial f}{\partial x}(0,0) = 1. \tag{3.2.6}$$

i) The Saddle-Node Bifurcation

Consider the map

$$x \mapsto f(x, \mu) = x + \mu \mp x^2, \qquad x \in \mathbb{R}^1, \quad \mu \in \mathbb{R}^1. \tag{3.2.7}$$

It is easy to verify that $(x, \mu) = (0,0)$ is a nonhyperbolic fixed point of (3.2.7) with eigenvalue 1, i.e.,

$$f(0,0) = 0, \tag{3.2.8}$$

$$\frac{\partial f}{\partial x}(0,0) = 1. \tag{3.2.9}$$

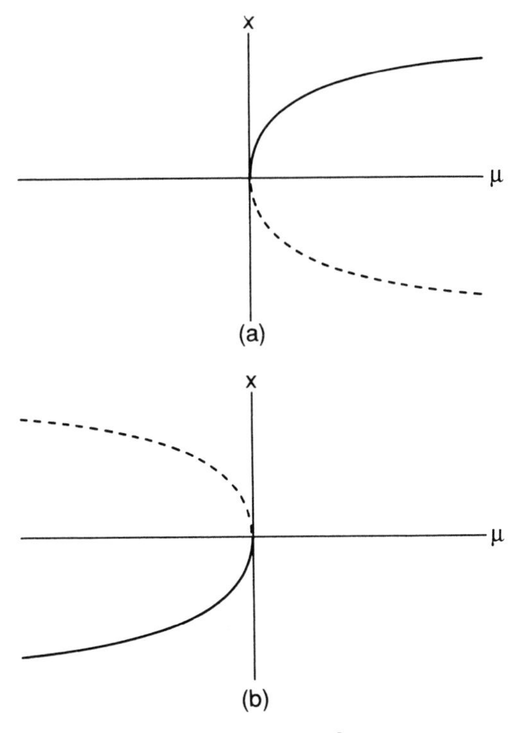

FIGURE 3.2.1. a) $f(x, \mu) = x + \mu - x^2$; b) $f(x, \mu) = x + \mu + x^2$.

We are interested in the nature of the fixed points for (3.2.7) near $(x, \mu) = (0, 0)$. Since (3.2.7) is so simple, we can solve for the fixed points directly as follows

$$f(x, \mu) - x = \mu \mp x^2 = 0. \tag{3.2.10}$$

We show the two curves of fixed points in Figure 3.2.1 and leave it as an exercise for the reader to verify the stability types of the different branches of fixed points shown in this figure. We refer to the bifurcation occuring at $(x, \mu) = (0, 0)$ as a *saddle-node* bifurcation.

In analogy with the situation for vector fields (see Section 3.1A) we want to find general conditions (in terms of derivatives evaluated at the bifurcation point) under which a map will undergo a saddle-node bifurcation, i.e.,

> the map possesses a unique curve of fixed points in the $x - \mu$ plane passing through the bifurcation point which locally lies on one side of $\mu = 0$.

We proceed using the implicit function theorem exactly as in the case for vector fields.

Consider a general one-parameter family of one-dimensional maps

$$x \mapsto f(x,\mu), \qquad x \in \mathbb{R}^1, \quad \mu \in \mathbb{R}^1, \tag{3.2.11}$$

with

$$f(0,0) = 0, \tag{3.2.12}$$

$$\frac{\partial f}{\partial x}(0,0) = 1. \tag{3.2.13}$$

The fixed points of (3.2.11) are given by

$$f(x,\mu) - x \equiv h(x,\mu) = 0. \tag{3.2.14}$$

We seek conditions under which (3.2.14) defines a curve in the $x-\mu$ plane with the properties described above. By the implicit function theorem,

$$\frac{\partial h}{\partial \mu}(0,0) = \frac{\partial f}{\partial \mu}(0,0) \neq 0 \tag{3.2.15}$$

implies that a single curve of fixed points passes through $(x,\mu) = (0,0)$; moreover, for x sufficiently small, this curve of fixed points can be represented as a graph over the x variables, i.e., there exists a unique $\mathbf{C}^r$ function, $\mu(x)$, x sufficiently small, such that

$$h(x,\mu(x)) \equiv f(x,\mu(x)) - x = 0. \tag{3.2.16}$$

Now we simply require that

$$\frac{d\mu}{dx}(0) = 0, \tag{3.2.17}$$

$$\frac{d^2\mu}{dx^2}(0) \neq 0. \tag{3.2.18}$$

As was the case for vector fields (Section 3.1A), we obtain (3.2.17) and (3.2.18) in terms of derivatives of the map at the bifurcation point by implicitly differentiating (3.2.16). Following (3.1.40) and (3.1.43), we obtain

$$\frac{d\mu}{dx}(0) = \frac{-\frac{\partial h}{\partial x}(0,0)}{\frac{\partial h}{\partial \mu}(0,0)} = -\frac{\left(\frac{\partial f}{\partial x}(0,0) - 1\right)}{\frac{\partial f}{\partial \mu}(0,0)} = 0, \tag{3.2.19}$$

$$\frac{d^2\mu}{dx^2}(0) = \frac{-\frac{\partial^2 h}{\partial x^2}(0,0)}{\frac{\partial h}{\partial \mu}(0,0)} = \frac{-\frac{\partial^2 f}{\partial x^2}(0,0)}{\frac{\partial f}{\partial \mu}(0,0)}. \tag{3.2.20}$$

To summarize, a general one-parameter family of $\mathbf{C}^r$ $(r \geq 2)$ one-dimensional maps

$$x \mapsto f(x,\mu), \qquad x \in \mathbb{R}^1, \quad \mu \in \mathbb{R}^1$$

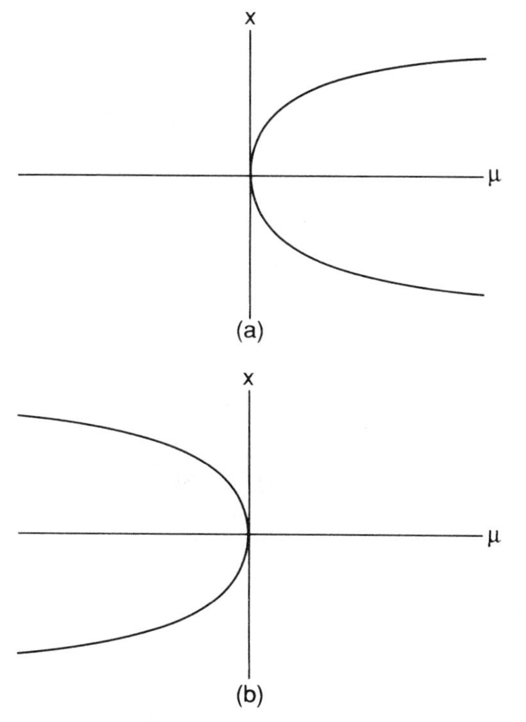

FIGURE 3.2.2. a) $\left(-\frac{\partial^2 f}{\partial x^2}(0,0)/\frac{\partial f}{\partial \mu}(0,0)\right) > 0$; b) $\left(-\frac{\partial^2 f}{\partial x^2}(0,0)/\frac{\partial f}{\partial \mu}(0,0)\right) < 0$.

undergoes a *saddle-node* bifurcation at $(x, \mu) = (0, 0)$ if

$$\left.\begin{aligned} f(0,0) &= 0 \\ \frac{\partial f}{\partial \mu}(0,0) &= 1 \end{aligned}\right\} \qquad \text{nonhyperbolic fixed point} \tag{3.2.21}$$

with

$$\frac{\partial f}{\partial \mu}(0,0) \neq 0, \tag{3.2.22}$$

$$\frac{\partial^2 f}{\partial x^2}(0,0) \neq 0. \tag{3.2.23}$$

Moreover, the sign of (3.2.20) tells us on which side of $\mu = 0$ the curve of fixed points is located; we show the two cases in Figure 3.2.2 and leave it as an exercise for the reader to compute the possible stability types of the branches of fixed points shown in the figure (see Exercise 3.5). Thus, (3.2.7) can be viewed as a normal form for the saddle-node bifurcation of maps. Notice that, with the exception of the condition $\frac{\partial f}{\partial x}(0,0) = 1$, the conditions for a one-parameter family of one-dimensional maps to undergo a saddle-node bifurcation in terms of derivatives of the map at the bifurcation point are exactly the same as those for vector fields (cf. (3.1.46), (3.1.47) and (3.1.48)). The reader should consider the implications of this.

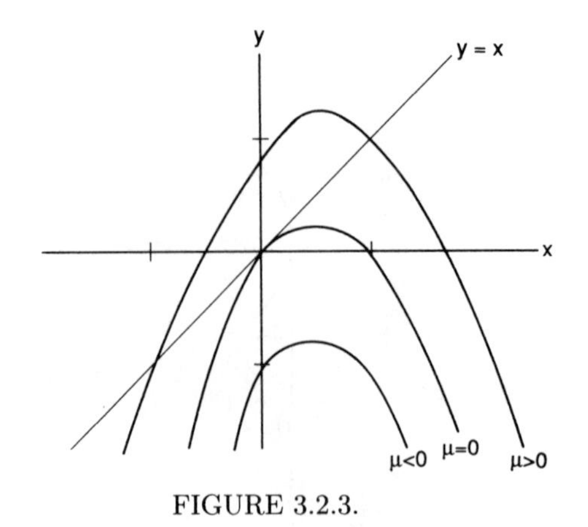

FIGURE 3.2.3.

Before finishing our discussion of the saddle-node bifurcation we want to describe a way of geometrically visualizing the bifurcation which will be useful later on. In the $x-y$ plane, the graph of $f(x,\mu)$ (thinking of μ as fixed) is given by

$$\text{graph } f(x,\mu) = \{(x,y) \in \mathbb{R}^2 \,|\, y = f(x,\mu)\}.$$

The graph of the function $g(x) = x$ is given by

$$\text{graph } g(x) = \{(x,y) \in \mathbb{R}^2 \,|\, y = x\}.$$

The intersection of these two graphs is given by

$$\{(x,y) \in \mathbb{R}^2 \,|\, y = x = f(x,\mu)\},$$

i.e., this is simply the set of fixed points for the map $x \mapsto f(x,\mu)$. The latter is simply a more mathematically complete way of saying that we draw the curve $y = f(x,\mu)$ (μ fixed) and the line $y = x$ in the $x-y$ plane and look for their intersections. We illustrate this for the map

$$x \mapsto x + \mu - x^2$$

in Figure 3.2.3 for different values of μ that graphically demonstrate the saddle-node bifurcation.

ii) THE TRANSCRITICAL BIFURCATION

Consider the maps

$$x \mapsto f(x,\mu) = x + \mu x \mp x^2, \qquad x \in \mathbb{R}^1, \quad \mu \in \mathbb{R}^1. \tag{3.2.24}$$

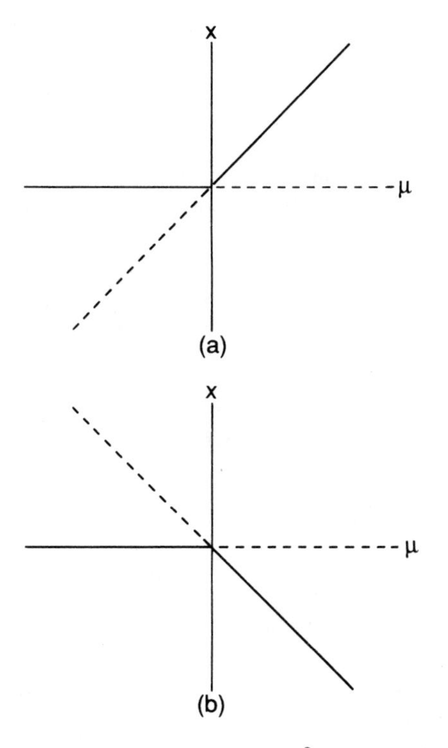

FIGURE 3.2.4. a) $f(x,\mu) = x + \mu x - x^2$; b) $f(x,\mu) = x + \mu x + x^2$.

It is easy to verify that $(x,\mu) = (0,0)$ is a nonhyperbolic fixed point of (3.2.24) with eigenvalue 1, i.e.,

$$f(0,0) = 0, \tag{3.2.25}$$

$$\frac{\partial f}{\partial x}(0,0) = 1. \tag{3.2.26}$$

The simplicity of (3.2.24) allows us to calculate all the fixed points relatively easily. They are given by

$$f(x,\mu) - x = \mu x \mp x^2 = 0. \tag{3.2.27}$$

Hence, there are two curves of fixed points passing through the bifurcation point,

$$x = 0 \tag{3.2.28}$$

and

$$\mu = \pm x. \tag{3.2.29}$$

We illustrate the two cases in Figure 3.2.4 and leave it as an exercise for the reader to compute the stability types of the different curves of fixed points shown in this figure. We refer to this type of bifurcation as a *transcritical* bifurcation.

We now want to find conditions for a general one-parameter family of $\mathbf{C}^r$ ($r \geq 2$) one-dimensional maps to undergo a transcritical bifurcation, i.e.,

> in the $x - \mu$ plane the map has two curves of fixed points passing through the origin and existing on both sides of $\mu = 0$.

Consider a $\mathbf{C}^r$ ($r \geq 2$) map

$$x \mapsto f(x, \mu), \qquad x \in \mathbb{R}^1, \quad \mu \in \mathbb{R}^1 \tag{3.2.30}$$

with

$$\left.\begin{array}{l} f(0,0) = 0 \\ \frac{\partial f}{\partial x}(0,0) = 1 \end{array}\right\} \qquad \text{nonhyperbolic fixed point.} \tag{3.2.31}$$

The fixed points of (3.2.30) are given by

$$f(x, \mu) - x \equiv h(x, \mu) = 0. \tag{3.2.32}$$

Henceforth the argument is very similar to that for the transcritical bifurcation of one-parameter families of one-dimensional vector fields; see Section 3.1A. We want two curves of fixed points to pass through the bifurcation point $(x, \mu) = (0, 0)$, so we require that

$$\frac{\partial h}{\partial \mu}(0,0) = \frac{\partial f}{\partial \mu}(0,0) = 0. \tag{3.2.33}$$

Next, we want one of these curves of fixed points to be given by

$$x = 0; \tag{3.2.34}$$

we thus take (3.2.32) of the form

$$h(x, \mu) = xH(x, \mu) = x(F(x, \mu) - 1), \tag{3.2.35}$$

where

$$F(x, \mu) = \left\{ \begin{array}{ll} \frac{f(x,\mu)}{x}, & x \neq 0 \\ \frac{\partial f}{\partial x}(0, \mu), & x = 0 \end{array} \right\} \tag{3.2.36}$$

and, hence,

$$H(x, \mu) = \left\{ \begin{array}{ll} \frac{h(x,\mu)}{x}, & x \neq 0 \\ \frac{\partial h}{\partial x}(0, \mu), & x = 0 \end{array} \right\}. \tag{3.2.37}$$

Now we require $H(x, \mu)$ to have a unique curve of zeros passing through $(x, \mu) = (0, 0)$ and existing on both sides of $\mu = 0$. For this it is sufficient to have

$$\frac{\partial H}{\partial \mu}(0,0) = \frac{\partial F}{\partial \mu}(0,0) \neq 0 \tag{3.2.38}$$

and, using (3.2.36), (3.2.38) is the same as

$$\frac{\partial^2 f}{\partial x \partial \mu}(0,0) \neq 0. \tag{3.2.39}$$

By the implicit function theorem, (3.2.39) implies that there exists a unique $\mathbf{C}^r$ function $\mu(x)$ (x sufficiently small) such that

$$H(x, \mu(x)) = F(x, \mu(x)) - 1 = 0. \tag{3.2.40}$$

Hence, we require

$$\frac{d\mu}{dx}(0) \neq 0. \tag{3.2.41}$$

Implicitly differentiating (3.2.40) gives

$$\frac{d\mu}{dx}(0) = \frac{-\frac{\partial H}{\partial x}(0,0)}{\frac{\partial H}{\partial \mu}(0,0)} = \frac{-\frac{\partial F}{\partial x}(0,0)}{\frac{\partial F}{\partial \mu}(0,0)}. \tag{3.2.42}$$

Using (3.2.36), (3.2.42) becomes

$$\frac{d\mu}{dx}(0) = \frac{-\frac{\partial^2 f}{\partial x^2}(0,0)}{\frac{\partial^2 f}{\partial x \partial \mu}(0,0)}. \tag{3.2.43}$$

We now summarize the results. A one-parameter family of $\mathbf{C}^r$ ($r \geq 2$) one-dimensional maps

$$x \mapsto f(x, \mu), \qquad x \in \mathbb{R}^1, \quad \mu \in \mathbb{R}^1 \tag{3.2.44}$$

having a nonhyperbolic fixed point, i.e.,

$$f(0,0) = 0, \tag{3.2.45}$$

$$\frac{\partial f}{\partial x}(0,0) = 1, \tag{3.2.46}$$

undergoes a transcritical bifurcation at $(x, \mu) = (0,0)$ if

$$\frac{\partial f}{\partial \mu}(0,0) = 0, \tag{3.2.47}$$

$$\frac{\partial^2 f}{\partial x \partial \mu}(0,0) \neq 0, \tag{3.2.48}$$

and

$$\frac{\partial^2 f}{\partial x^2}(0,0) \neq 0. \tag{3.2.49}$$

We remark that the sign of (3.2.43) gives us the slope of the curve of fixed points that is not $x = 0$. In Figure 3.2.5 we show the two cases and leave it

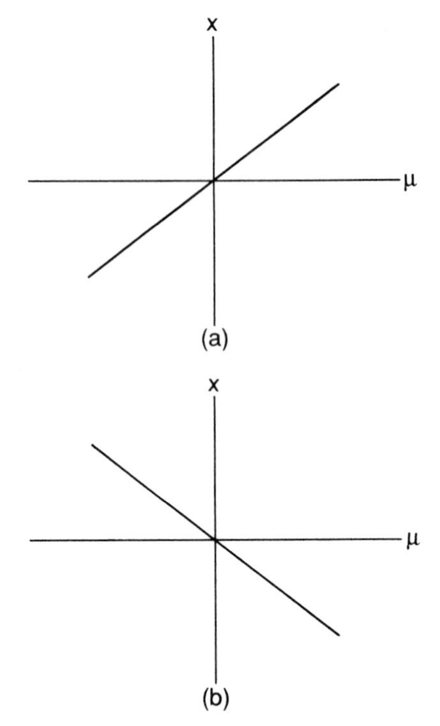

FIGURE 3.2.5. a) $\left(-\frac{\partial^2 f}{\partial x^2}(0,0)/\frac{\partial^2 f}{\partial x \partial \mu}(0,0)\right) > 0$; b) $\left(-\frac{\partial^2 f}{\partial x^2}(0,0)/\frac{\partial^2 f}{\partial x \partial \mu}(0,0)\right) < 0$.

as an exercise for the reader to compute the possible stability types for the different curves of fixed points shown in the figure; see Exercise 3.6. Thus (3.2.24) can be viewed as a normal form for the transcritical bifurcation.

We end our discussion of the transcritical bifurcation by graphically showing the transcritical bifurcation in Figure 3.2.6 for the map

$$x \mapsto x + \mu x - x^2;$$

cf. the discussion at the end of Section 3.2A, i).

iii) The Pitchfork Bifurcation

Consider the maps

$$x \mapsto f(x,\mu) = x + \mu x \mp x^3, \qquad x \in \mathbb{R}^1, \quad \mu \in \mathbb{R}^1. \tag{3.2.50}$$

It is easy to verify that $(x,\mu) = (0,0)$ is a nonhyperbolic fixed point of (3.2.50) with eigenvalue 1, i.e.,

$$f(0,0) = 0, \tag{3.2.51}$$

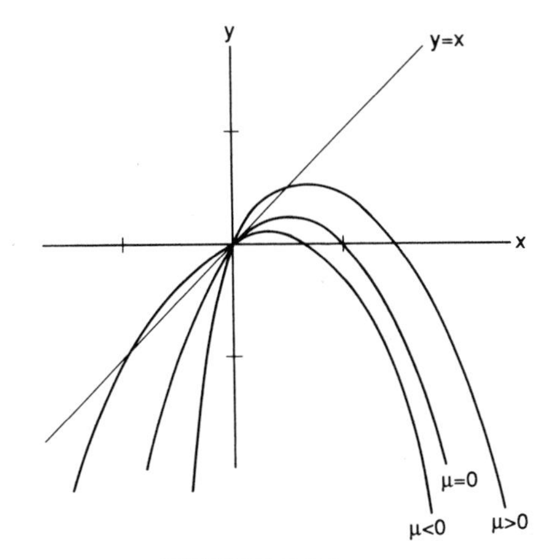

FIGURE 3.2.6.

$$\frac{\partial f}{\partial x}(0,0) = 1. \tag{3.2.52}$$

The fixed points of (3.2.50) are given by

$$f(x,\mu) - x = \mu x \mp x^3 = 0. \tag{3.2.53}$$

Thus, there are two curves of fixed points passing through the bifurcation point,

$$x = 0 \tag{3.2.54}$$

and

$$\mu = \pm x^2 \tag{3.2.55}$$

We illustrate the two cases in Figure 3.2.7 and leave it as an exercise for the reader to verify the stability types of the different branches of fixed points shown in this figure. We refer to this type of bifurcation as a *pitchfork* bifurcation for maps.

We now seek general conditions for a one-parameter family of $\mathbf{C}^r$ $(r \geq 3)$ one-dimensional maps to undergo a pitchfork bifurcation, i.e.,

> in the $x - \mu$ plane the map has two curves of fixed points passing through the bifurcation point; one curve exists on both sides of $\mu = 0$ and the other lies locally to one side of $\mu = 0$.

Consider a $\mathbf{C}^r$ $(r \geq 3)$ map

$$x \mapsto f(x,\mu), \qquad x \in \mathbb{R}^1, \quad \mu \in \mathbb{R}^1 \tag{3.2.56}$$

with

$$\left.\begin{aligned} f(0,0) &= 0 \\ \tfrac{\partial f}{\partial x}(0,0) &= 1 \end{aligned}\right\} \quad \text{nonhyperbolic fixed point.} \tag{3.2.57}$$

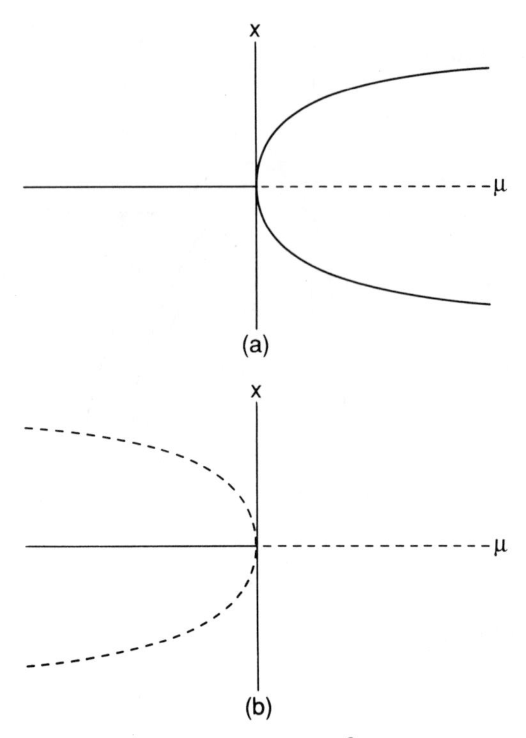

FIGURE 3.2.7. a) $f(x,\mu) = x + \mu x - x^3$, b) $f(x,\mu) = x + \mu x + x^3$.

The fixed points of (3.2.56) are given by

$$f(x,\mu) - x \equiv h(x,\mu) = 0. \tag{3.2.58}$$

Henceforth, the discussion is very similar to the discussion of the pitchfork bifurcation for vector fields (see Section 3.1A). In order to have more than one curve of fixed points passing through $(x,\mu) = (0,0)$, we must have

$$\frac{\partial h}{\partial \mu}(0,0) = \frac{\partial f}{\partial \mu}(0,0) = 0. \tag{3.2.59}$$

Since we want one curve of fixed points to be $x = 0$, we take (3.2.58) of the form

$$h(x,\mu) = xH(x,\mu) = x(F(x,\mu) - 1), \tag{3.2.60}$$

where

$$H(x,\mu) = \left\{ \begin{array}{ll} \frac{h(x,\mu)}{x}, & x \neq 0 \\ \frac{\partial h}{\partial x}(0,\mu), & x = 0 \end{array} \right\} \tag{3.2.61}$$

and, hence,

$$F(x,\mu) = \left\{ \begin{array}{ll} \frac{f(x,\mu)}{x}, & x \neq 0 \\ \frac{\partial f}{\partial x}(0,\mu), & x = 0 \end{array} \right\}. \tag{3.2.62}$$

Since we want only one additional curve of fixed points to pass through $(x, \mu) = (0, 0)$, we require

$$\frac{\partial H}{\partial \mu}(0,0) = \frac{\partial F}{\partial \mu}(0,0) \neq 0. \tag{3.2.63}$$

Using (3.2.62), (3.2.63) becomes

$$\frac{\partial^2 f}{\partial x \partial \mu}(0,0) \neq 0. \tag{3.2.64}$$

The implicit function theorem and (3.2.64) imply that there is a unique $\mathbf{C}^r$ function, $\mu(x)$ (x sufficiently small), such that

$$H(x, \mu(x)) \equiv F(x, \mu(x)) - 1 = 0. \tag{3.2.65}$$

We require

$$\frac{d\mu}{dx}(0) = 0 \tag{3.2.66}$$

and

$$\frac{d^2\mu}{dx^2}(0) \neq 0. \tag{3.2.67}$$

Implicitly differentiating (3.2.65) gives

$$\frac{d\mu}{dx}(0) = \frac{-\frac{\partial H}{\partial x}(0,0)}{\frac{\partial H}{\partial \mu}(0,0)} = \frac{-\frac{\partial F}{\partial x}(0,0)}{\frac{\partial F}{\partial \mu}(0,0)}, \tag{3.2.68}$$

$$\frac{d^2\mu}{dx^2}(0) = \frac{-\frac{\partial^2 H}{\partial x^2}(0,0)}{\frac{\partial H}{\partial \mu}(0,0)} = \frac{-\frac{\partial^2 F}{\partial x^2}(0,0)}{\frac{\partial F}{\partial \mu}(0,0)}. \tag{3.2.69}$$

Using (3.2.62), (3.2.68) and (3.2.69) become

$$\frac{d\mu}{dx}(0) = \frac{-\frac{\partial^2 f}{\partial x^2}(0,0)}{\frac{\partial^2 f}{\partial x \partial \mu}(0,0)}, \tag{3.2.70}$$

$$\frac{d^2\mu}{dx^2}(0) = \frac{-\frac{\partial^3 f}{\partial x^3}(0,0)}{\frac{\partial^2 f}{\partial x \partial \mu}(0,0)}. \tag{3.2.71}$$

To summarize, a one-parameter family of $\mathbf{C}^r$ $(r \geq 3)$ one-dimensional maps

$$x \mapsto f(x, \mu), \qquad x \in \mathbb{R}^1, \quad \mu \in \mathbb{R}^1 \tag{3.2.72}$$

having a nonhyperbolic fixed point, i.e.,

$$f(0,0) = 0, \tag{3.2.73}$$

$$\frac{\partial f}{\partial x}(0,0) = 1, \tag{3.2.74}$$

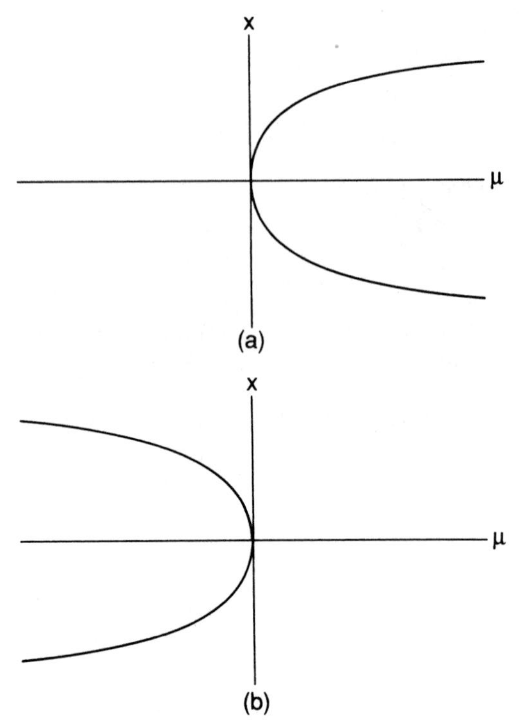

FIGURE 3.2.8. a) $\left(-\frac{\partial^3 f}{\partial x^3}(0,0)/\frac{\partial^2 f}{\partial x \partial \mu}(0,0)\right) > 0$; b) $\left(-\frac{\partial^3 f}{\partial x^3}(0,0)/\frac{\partial^2 f}{\partial x \partial \mu}(0,0)\right) < 0$.

undergoes a pitchfork bifurcation at $(x, \mu) = (0, 0)$ if

$$\frac{\partial f}{\partial \mu}(0,0) = 0, \tag{3.2.75}$$

$$\frac{\partial^2 f}{\partial x^2}(0,0) = 0, \tag{3.2.76}$$

$$\frac{\partial^2 f}{\partial x \partial \mu}(0,0) \neq 0, \tag{3.2.77}$$

$$\frac{\partial^3 f}{\partial x^3}(0,0) \neq 0. \tag{3.2.78}$$

Moreover, the sign of (3.2.71) tells us on which side of $\mu = 0$ that one of the curves of fixed points lies. We illustrate both cases in Figure 3.2.8 and leave it as an exercise (see Exercise 3.7) for the reader to compute the possible stability types of the different branches shown in Figure 3.2.8. Thus, we can view (3.2.50) as a normal form for the pitchfork bifurcation.

We end our discussion of the pitchfork bifurcation by graphically showing the bifurcation for

$$x \mapsto x + \mu x - x^3$$

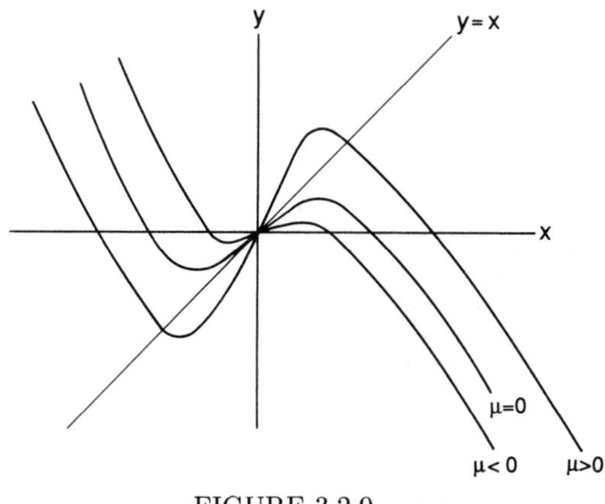

FIGURE 3.2.9.

in Figure 3.2.9 in the manner discussed at the end of Section 3.2A,i).

3.2B An Eigenvalue of -1

Suppose that our one-parameter family of $\mathbf{C}^r$ ($r \geq 3$) one-dimensional maps has a nonhyperbolic fixed point, and the eigenvalue associated with the linearization of the map about the fixed point is -1 rather than 1. Up to this point the bifurcations of one-parameter families of one-dimensional maps have been very much the same as the analogous cases for vector fields. However, the case of an eigenvalue equal to -1 is fundamentally different and does not have an analog with *one-dimensional* vector field dynamics. We begin by studying a specific example.

i) Example

Consider the following one-parameter family of one-dimensional maps

$$x \mapsto f(x,\mu) = -x - \mu x + x^3, \qquad x \in \mathbb{R}^1, \quad \mu \in \mathbb{R}^1. \tag{3.2.79}$$

It is easy to verify that (3.2.79) has a nonhyperbolic fixed point at $(x,\mu) = (0,0)$ with eigenvalue -1, i.e.,

$$f(0,0) = 0, \tag{3.2.80}$$

$$\frac{\partial f}{\partial x}(0,0) = -1. \tag{3.2.81}$$

The fixed points of (3.2.79) can be calculated directly and are given by

$$f(x,\mu) - x = x(x^2 - (2+\mu)) = 0. \tag{3.2.82}$$

Thus, (3.2.79) has two curves of fixed points,

$$x = 0 \tag{3.2.83}$$

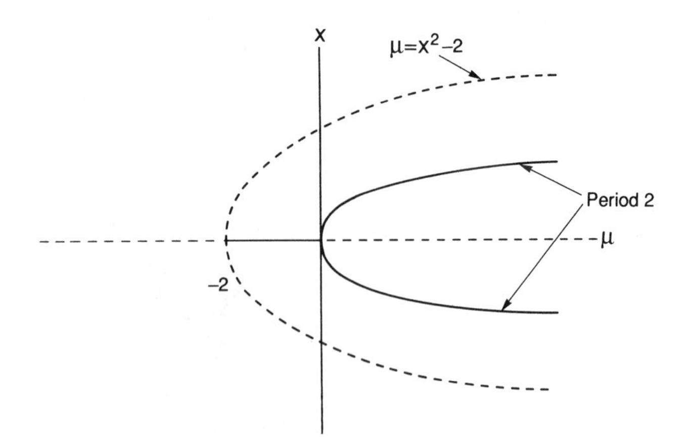

FIGURE 3.2.10.

and

$$x^2 = 2 + \mu, \tag{3.2.84}$$

but only (3.2.83) passes through the bifurcation point $(x, \mu) = (0, 0)$. In Figure 3.2.10 we illustrate the two curves of fixed points and leave it as an exercise for the reader to verify the stability types for the different curves of fixed points shown in the figure. In particular we have

$$x = 0 \text{ is} \qquad \begin{cases} \text{unstable for } \mu \leq -2, \\ \text{stable for } -2 < \mu < 0, \\ \text{unstable for } \mu > 0, \end{cases} \tag{3.2.85}$$

and

$$x^2 = 2 + \mu \text{ is} \qquad \begin{cases} \text{unstable for } \mu \geq -2, \\ \text{does not exist for } \mu < -2. \end{cases} \tag{3.2.86}$$

From (3.2.85) and (3.2.86) we can immediately see there is a problem, namely, that for $\mu > 0$, the map has exactly three fixed points and all are unstable. (Note: this situation could not occur for one-dimensional vector fields.) A way out of this difficulty would be provided if stable periodic orbits bifurcated from $(x, \mu) = (0, 0)$. We will see that this is indeed the case.

Consider the *second iterate* of (3.2.79), i.e.,

$$x \mapsto f^2(x, \mu) = x + \mu(2 + \mu)x - 2x^3 + \mathcal{O}(4). \tag{3.2.87}$$

It is easy to verify that (3.2.87) has a nonhyperbolic fixed point at $(x, \mu) = (0, 0)$ having an eigenvalue of 1, i.e.,

$$f^2(0, 0) = 0, \tag{3.2.88}$$

$$\frac{\partial f^2}{\partial x}(0, 0) = 1. \tag{3.2.89}$$

Moreover,

$$\frac{\partial f^2}{\partial \mu}(0,0) = 0, \tag{3.2.90}$$

$$\frac{\partial^2 f^2}{\partial x \partial \mu}(0,0) = 2, \tag{3.2.91}$$

$$\frac{\partial^2 f^2}{\partial x^2}(0,0) = 0, \tag{3.2.92}$$

$$\frac{\partial^3 f^2}{\partial x^3}(0,0) = -12. \tag{3.2.93}$$

Hence, from (3.2.75), (3.2.76), (3.2.77), and (3.2.78), (3.2.90), (3.2.91), (3.2.92), and (3.2.93) imply that the second iterate of (3.2.79) undergoes a pitchfork bifurcation at $(x, \mu) = (0,0)$. Since the new fixed points of $f^2(x, \mu)$ are not fixed points of $f(x, \mu)$, they must be period two points of $f(x, \mu)$. Hence, $f(x, \mu)$ is said to have undergone a *period-doubling bifurcation* at $(x, \mu) = (0,0)$.

ii) The Period-Doubling Bifurcation

Consider a one-parameter family of $\mathbf{C}^r$ $(r \geq 3)$ one-dimensional maps

$$x \mapsto f(x, \mu), \qquad x \in \mathbb{R}^1, \quad \mu \in \mathbb{R}^1. \tag{3.2.94}$$

We seek conditions for (3.2.94) to undergo a period-doubling bifurcation. The previous example will be our guide. It should be clear from the example that conditions sufficient for (3.2.94) to undergo a period-doubling bifurcation are for the map to have a nonhyperbolic fixed point with eigenvalue -1 and for the second iterate of the map to undergo a pitchfork bifurcation at the same nonhyperbolic fixed point. To summarize, using (3.2.73), (3.2.74), (3.2.75), (3.2.76), (3.2.77), and (3.2.78), it is sufficient for (3.2.94) to satisfy

$$f(0,0) = 0, \tag{3.2.95}$$

$$\frac{\partial f}{\partial x}(0,0) = -1, \tag{3.2.96}$$

$$\frac{\partial f^2}{\partial \mu}(0,0) = 0, \tag{3.2.97}$$

$$\frac{\partial^2 f^2}{\partial x^2}(0,0) = 0, \tag{3.2.98}$$

$$\frac{\partial^2 f^2}{\partial x \partial \mu}(0,0) \neq 0, \tag{3.2.99}$$

$$\frac{\partial^3 f^2}{\partial x^3}(0,0) \neq 0. \tag{3.2.100}$$

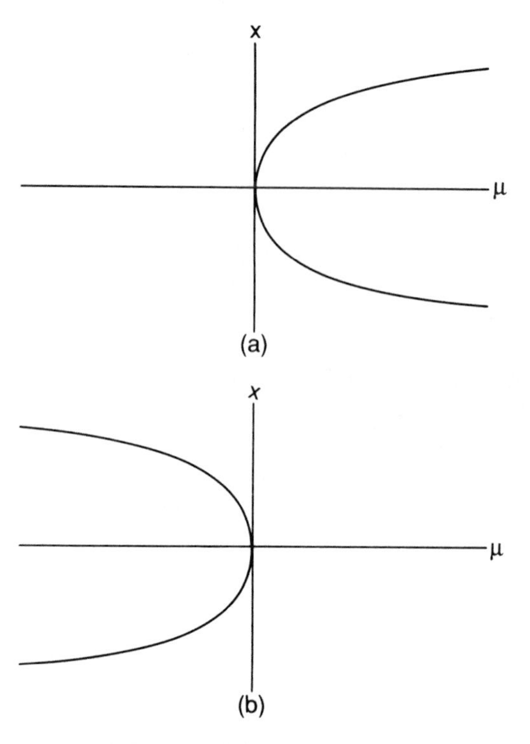

FIGURE 3.2.11. a) $(-\frac{\partial^3 f^2}{\partial x^3}(0,0)/\frac{\partial^2 f^2}{\partial x \partial \mu}(0,0)) > 0$; b) $(-\frac{\partial^3 f^2}{\partial x^3}(0,0)/\frac{\partial^2 f^2}{\partial x \partial \mu}(0,0)) < 0$.

Moreover, the sign of $\left(-\frac{\partial^3 f^2(0,0)}{\partial x^3} \Big/ \frac{\partial^2 f^2(0,0)}{\partial x \partial \mu}\right)$ tells us on which side of $\mu = 0$ the period two points lie. We show both cases in Figure 3.2.11 and leave it as an exercise for the reader to compute the possible stability types for the different curves of fixed points shown in the figure; see Exercise 3.8.

Finally, we demonstrate graphically the period-doubling bifurcation for

$$x \mapsto -x - \mu x + x^3 \equiv f(x, \mu)$$

and the associated pitchfork bifurcation for $f^2(x, \mu)$ in the graphical manner discribed at the end of Section 3.2A, i) in Figure 3.2.12.

3.2c A Pair of Eigenvalues of Modulus 1: The Naimark-Sacker Bifurcation

This section describes the map analog of the Poincaré-Andronov-Hopf bifurcation for vector fields but with some very different twists. Although this bifurcation often goes by the name of "Hopf bifurcation for maps," this is misleading because the bifurcation theorem was first proved independently by Naimark [1959] and Sacker [1965]. Consequently, we will use the term

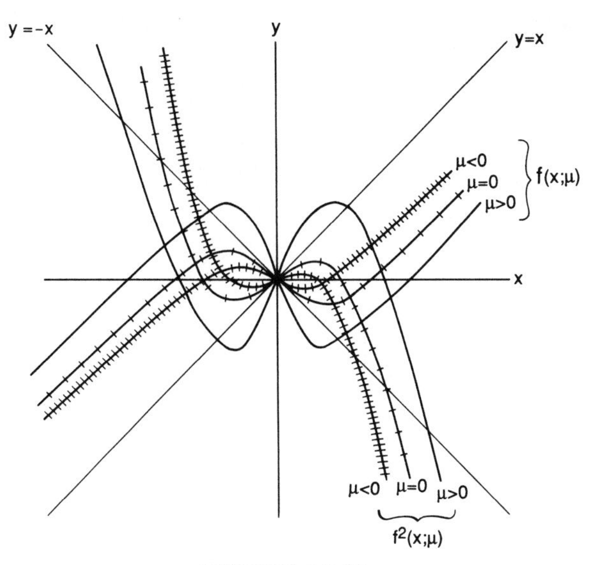

FIGURE 3.2.12.

"Naimark-Sacker bifurcation."

We know that in this situation the study of the dynamics of (3.2.1) near the fixed point $(y_0, \lambda_0) \in \mathbb{R}^n \times \mathbb{R}^p$ can be reduced to the study of (3.2.1) restricted to a p-parameter family of two-dimensional center manifolds. We assume that the reduced map has been calculated and is given by

$$x \mapsto f(x, \mu), \qquad x \in \mathbb{R}^2, \quad \mu \in \mathbb{R}^1, \tag{3.2.101}$$

where we take $p = 1$. If there is more than one parameter, we consider all but one as fixed and denote it as μ. In restricting the map to the center manifold, some preliminary transformations have been made so that the fixed point of (3.2.101) is given by $(x, \mu) = (0, 0)$, i.e., we have

$$f(0,0) = 0, \tag{3.2.102}$$

with the matrix

$$D_x f(0,0) \tag{3.2.103}$$

having two complex conjugate eigenvalues, denoted $\lambda(0), \bar{\lambda}(0)$, with

$$|\lambda(0)| = 1. \tag{3.2.104}$$

We will also require that

$$\lambda^n(0) \neq 1, \qquad n = 1, 2, 3, 4. \tag{3.2.105}$$

(Note: if $\lambda(0)$ satisfies (3.2.105), then so does $\bar{\lambda}(0)$, and vice versa.)

We showed in Example 2.2.4 that under these conditions a normal form for (3.2.101) is given by

$$z \mapsto \lambda(\mu)z + c(\mu)z^2\bar{z} + \mathcal{O}(4), \qquad z \in \mathbb{C}, \quad \mu \in \mathbb{R}^1. \tag{3.2.106}$$

We transform (3.2.106) into polar coordinates by letting

$$z = re^{2\pi i\theta},$$

and obtain

$$\begin{aligned} r &\mapsto |\lambda(\mu)| \left(r + \left(\text{Re} \left(\frac{c(\mu)}{\lambda(\mu)} \right) \right) r^3 + \mathcal{O}(r^4) \right), \\ \theta &\mapsto \theta + \phi(\mu) + \frac{1}{2\pi} \left(\text{Im} \left(\frac{c(\mu)}{\lambda(\mu)} \right) \right) r^2 + \mathcal{O}(r^3), \end{aligned} \tag{3.2.107}$$

where

$$\phi(\mu) \equiv \frac{1}{2\pi} \tan^{-1} \frac{\omega(\mu)}{\alpha(\mu)} \tag{3.2.108}$$

and

$$c(\mu) = \alpha(\mu) + i\omega(\mu). \tag{3.2.109}$$

We then Taylor expand the coefficients of (3.2.107) about $\mu = 0$ and obtain

$$\begin{aligned} r &\mapsto \left(1 + \frac{d}{d\mu}|\lambda(\mu)|\bigg|_{\mu=0} \mu \right) r + \left(\text{Re} \left(\frac{c(0)}{\lambda(0)} \right) \right) r^3 + \mathcal{O}(\mu^2 r, \mu r^3, r^4), \\ \theta &\mapsto \theta + \phi(0) + \frac{d}{d\mu}(\phi(\mu))\bigg|_{\mu=0} \mu + \frac{1}{2\pi} \left(\text{Im} \frac{c(0)}{\lambda(0)} \right) r^2 \\ &\quad + \mathcal{O}(\mu^2, \mu r^2, r^3), \end{aligned} \tag{3.2.110}$$

where we have used the condition that $|\lambda(0)| = 1$. Note that, since $\lambda^n(0) \neq 1$, where $n = 1, 2, 3, 4$, from (3.2.108) we see that $\phi(0) \neq 0$. We simplify the notation associated with (3.2.110) by setting

$$\begin{aligned} d &\equiv \frac{d}{d\mu}|\lambda(\mu)|\bigg|_{\mu=0}, \\ a &\equiv \text{Re} \left(\frac{c(0)}{\lambda(0)} \right), \\ \phi_0 &\equiv \phi(0), \\ \phi_1 &\equiv \frac{d}{d\mu}(\phi(\mu))\bigg|_{\mu=0}, \\ b &\equiv \frac{1}{2\pi} \text{Im} \, \frac{c(0)}{\lambda(0)}; \end{aligned}$$

hence, (3.2.110) becomes

$$\begin{aligned} r &\mapsto r + (d\mu + ar^2)r + \mathcal{O}(\mu^2 r, \mu r^3, r^4), \\ \theta &\mapsto \theta + \phi_0 + \phi_1\mu + br^2 + \mathcal{O}(\mu^2, \mu r^2, r^3). \end{aligned} \tag{3.2.111}$$

We are interested in the dynamics of (3.2.111) for r small, μ small. Our strategy for understanding this will be the same as in our study of the Poincaré-Andronov-Hopf bifurcation for vector fields (cf. Section 3.1B); namely, we will study the dynamics of (3.2.111) *with the higher order terms neglected* (i.e., the truncated normal form) and then try to understand how the dynamics of the truncated normal form are affected by the higher order terms.

The truncated normal form is given by

$$\begin{aligned} r &\mapsto r + (d\mu + ar^2)r, \\ \theta &\mapsto \theta + \phi_0 + \phi_1\mu + br^2. \end{aligned} \tag{3.2.112}$$

Note that $r = 0$ is a fixed point of (3.2.112) that is

asymptotically stable for $d\mu < 0$,
unstable for $d\mu > 0$,
unstable for $\mu = 0, a > 0$,

and

asymptotically stable for $\mu = 0, a < 0$.

We recall our study of the truncated normal form for the Poincaré-Andronov-Hopf bifurcation for vector fields (see Section 3.1B). In that case, fixed points of the $\dot{r}$ component of the truncated normal form, with $r > 0$, corresponded to periodic orbits. Something geometrically (but *not* dynamically) similar happens for maps also.

Lemma 3.2.1 $\left\{ (r,\theta) \in \mathbb{R}^+ \times S^1 \,\middle|\, r = \sqrt{\frac{-\mu d}{a}} \right\}$ *is a circle which is invariant under the dynamics generated by (3.2.112).*

Proof: That this set of points is a circle is obvious. The fact that it is invariant under the dynamics generated by (3.2.112) follows from the fact that the r coordinate of points starting on the circle do not change under iteration by (3.2.112). □

It should be clear that the invariant circle can exist for either $\mu > 0$ or $\mu < 0$ depending on the signs of d and a and that there will be only one invariant circle at a distance $\mathcal{O}(\sqrt{\mu})$ from the origin. Stability of the invariant circle is determined by the sign of a. This is a new concept of stability not previously discussed in this book, namely, the stability of an invariant set. Its meaning, hopefully, is intuitively clear. The invariant circle is stable if initial conditions "sufficiently near" the circle stay near the circle

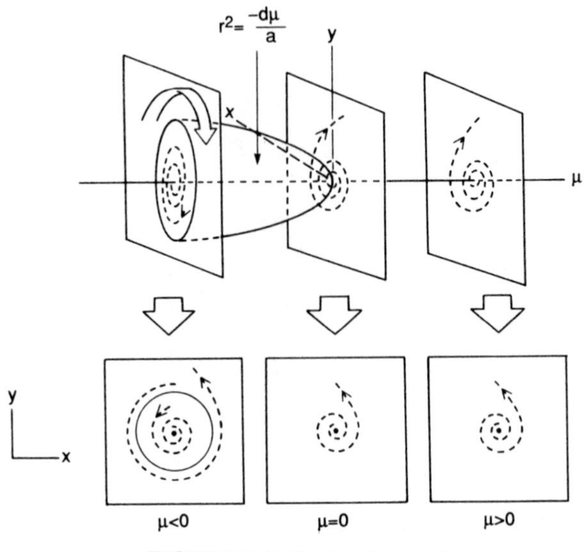

FIGURE 3.2.13. $d > 0$, $a > 0$.

under all forward iterations by (3.2.112). It is asymptotically stable if the points actually approach the circle. We summarize this in the following lemma.

Lemma 3.2.2 *The invariant circle is asymptotically stable for $a < 0$ and unstable for $a > 0$.*

Proof: Since the r component of (3.2.112) is independent of θ, this problem reduces to the study of the stability of a fixed point of a one-dimensional map (i.e., the θ dynamics are irrelevant). We leave the details as an exercise for the reader. □

We now describe the four possible cases for the bifurcation of an invariant circle from a fixed point.

Case 1: $d > 0$, $a > 0$. In this case, the origin is an unstable fixed point for $\mu > 0$ and an asymptotically stable fixed point for $\mu < 0$ with an unstable invariant circle for $\mu < 0$; see Figure 3.2.13.

Case 2: $d > 0$, $a < 0$. In this case, the origin is an unstable fixed point for $\mu > 0$ and an asymptotically stable fixed point for $\mu < 0$ with an asymptotically stable invariant circle for $\mu > 0$; see Figure 3.2.14.

Case 3: $d < 0$, $a > 0$. In this case, the origin is an asymptotically stable fixed point for $\mu > 0$ and an unstable fixed point for $\mu < 0$ with an unstable invariant circle for $\mu > 0$; see Figure 3.2.15.

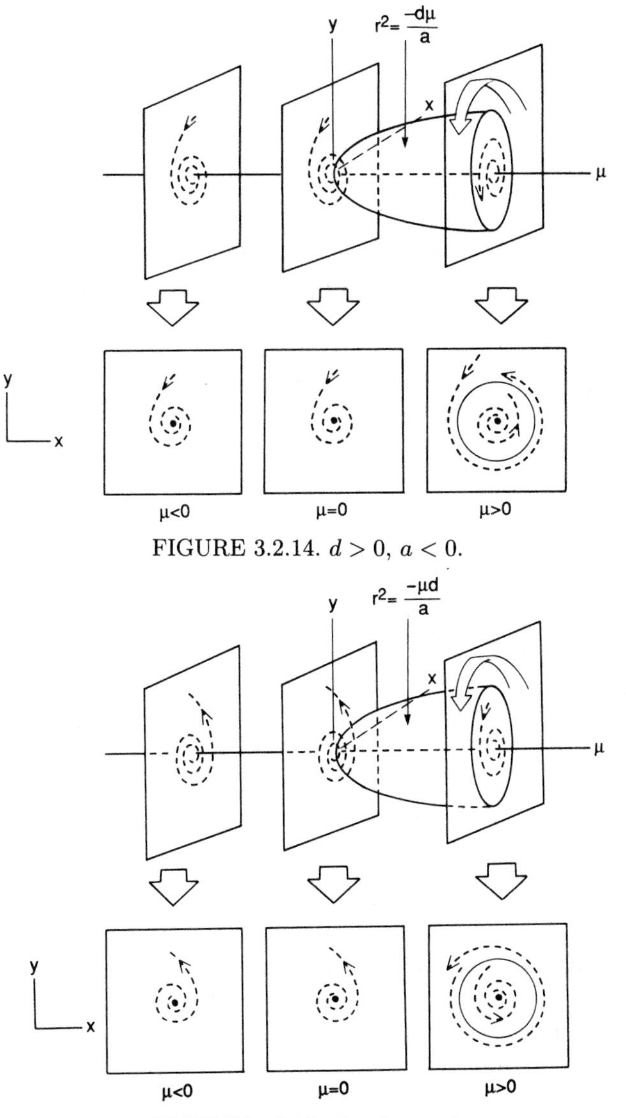

FIGURE 3.2.14. $d > 0$, $a < 0$.

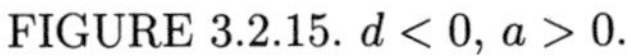

FIGURE 3.2.15. $d < 0$, $a > 0$.

Case 4: $d < 0$, $a < 0$. In this case, the origin is an asymptotically stable fixed point for $\mu > 0$ and an unstable fixed point for $\mu < 0$ with an asymptotically stable invariant circle for $\mu < 0$; see Figure 3.2.16.

We make the following general remarks.

Remark 1. For $a > 0$, the invariant circle can exist for either $\mu < 0$ (Case 1) or $\mu > 0$ (Case 3) and, in each case, the invariant circle is unstable.

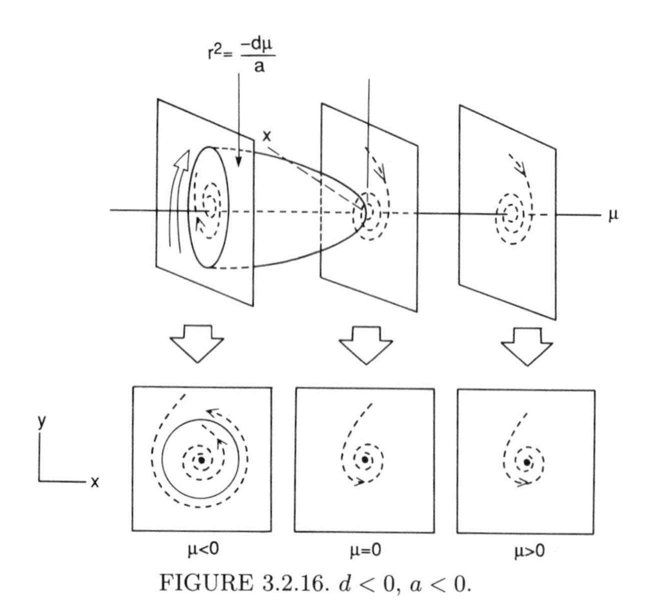

FIGURE 3.2.16. $d < 0$, $a < 0$.

Similarly, for $a < 0$, the invariant circle can exist for either $\mu < 0$ (Case 4) or $\mu > 0$ (Case 2) and, in each case, the invariant circle is asymptotically stable. Hence, the quantity a determines the stability of the invariant circle, but it does not tell us on which side of $\mu = 0$ the invariant circle exists.

Remark 2. Recall that

$$d = \frac{d}{d\mu} \left|\lambda(\mu)\right|_{\mu=0}.$$

Hence, for $d > 0$, the eigenvalues cross from inside to outside the unit circle as μ increases through zero and, for $d < 0$, the eigenvalues cross from outside to inside the unit circle as μ increases through zero. Thus, for $d > 0$, it follows that the origin is asymptotically stable for $\mu < 0$ and unstable for $\mu > 0$. Similarly, for $d < 0$, the origin is unstable for $\mu < 0$ and asymptotically stable for $\mu > 0$.

At this point the reader is probably struck by the similarities between the analysis of the truncated normal form (3.2.112) and the analysis for the normal form associated with the Poincaré-Andronov-Hopf bifurcation (see Section 3.1B). However, we want to stress that the situation for maps is fundamentally different indeed from this and from all other bifurcations we have studied thus far. In all other bifurcations (either in vector fields or maps) the invariant sets that are created consist of *single orbits* while, in this case, the bifurcation consists of an invariant surface (i.e., a circle) which contains many different orbits. We can study the dynamics on the invariant circle by studying the dynamics of (3.2.112) restricted to the invariant circle (i.e., by considering only initial conditions that start on the invariant circle). Points on the invariant circle have initial r coordinates given by

$$r = \sqrt{\frac{-\mu d}{a}},$$

so that the associated *circle map* is given by

$$\theta \mapsto \theta + \phi_0 + \left(\phi_1 - \frac{d}{a}\right)\mu. \tag{3.2.113}$$

The dynamics of (3.2.112) are easy to understand and depend entirely on the quantity $\phi_0 + (\phi_1 - \frac{d}{a})\mu$. If $\phi_0 + (\phi_1 - \frac{d}{a})\mu$ is rational, then all orbits on the invariant circle are periodic. If $\phi_0 + (\phi_1 - \frac{d}{a})\mu$ is irrational, then all orbits on the invariant circle densely fill the circle. We proved these statements in Example 1.2.3. Thus, as μ is varied, the orbit structure on the invariant circle continually alternates between periodic and quasiperiodic.

Although all of this analysis is for the truncated normal form (3.2.112), our real interest is in the full normal form (3.2.111). For this we have the following theorem.

Theorem 3.2.3 (Naimark-Sacker bifurcation) *Consider the full normal form (3.2.111). Then, for μ sufficiently small, Cases 1, 2, 3, and 4 described above hold.*

Proof: The proof requires more technical machinery than the proof of the Poincaré-Andronov-Hopf bifurcation for vector fields. This should come as no surprise, since the Poincaré-Bendixson theorem and the method of averaging (as developed in Section 1.2D,i) do not immediately apply to maps. For this reason, we will not state this proof here; excellent expositions of the proof may be found in, e.g., Iooss [1979]. □

We must take care to interpret this theorem correctly. Roughly speaking, it tells us that the "tail" of the normal form does not qualitatively affect the bifurcation of the *invariant circle* exhibited by the truncated normal form. However, it tells us nothing about the dynamics *on the invariant circle.* Indeed, we would expect the higher order terms of the normal form to have a very important effect on the circle map associated with the invariant circle of the full normal form (3.2.111). This is because the circle map associated with the invariant circle of the truncated normal form is structurally unstable; see Exercise 3.47.

3.2D The Codimension of Local Bifurcations of Maps

The degeneracy of fixed points of maps can be specified by the concept of codimension as described in Section 3.1D. The codimension of a case is computed using exactly the same procedure (cf. Definition 3.1.7). Indeed, for a $\mathbf{C}^\infty$ vector field

$$\dot{x} = f(x), \qquad x \in \mathbb{R}^n, \tag{3.2.114}$$

we are interested in the local structure of

$$f(x) = 0, \tag{3.2.115}$$

and, for a $\mathbf{C}^\infty$ map,

$$x \mapsto g(x), \qquad x \in \mathbb{R}^n, \tag{3.2.116}$$

we are interested in the local structure of

$$g(x) - x = 0. \tag{3.2.117}$$

Thus, it should be clear that, mathematically, (3.2.114) and (3.2.116) are the same. We will, therefore, state the results for maps only and leave the verification to the reader.

One-Dimensional Maps

The maps

$$x \mapsto x + ax^2 + \mathcal{O}(x^3), \tag{3.2.118}$$

$$x \mapsto -x + ax^3 + \mathcal{O}(x^4), \tag{3.2.119}$$

have nonhyperbolic fixed points at $x = 0$. Using the techniques of Section 3.1D it is easy to see that these are *codimension one* fixed points with versal deformations given by

$$x \mapsto x + \mu \mp x^2, \tag{3.2.120}$$

$$x \mapsto -x + \mu x \mp x^3. \tag{3.2.121}$$

Hence, the generic bifurcations of fixed points of one-parameter families of one-dimensional maps are saddle-nodes and period-doublings.

Similarly, the map

$$x \mapsto x + ax^3 + \mathcal{O}(x^4) \tag{3.2.122}$$

has a nonhyperbolic fixed point at $x = 0$. It is easy to show that $x = 0$ is a *codimension two* fixed point, and a versal deformation is given by

$$x \mapsto x + \mu_1 + \mu_2 x \mp x^3. \tag{3.2.123}$$

Two-Dimensional Maps

Suppose we have a two-dimensional map having a fixed point at the origin with the two eigenvalues of the associated linear map being complex conjugates with modulus one. We denote the two eigenvalues by λ and $\bar{\lambda}$. One can assign a codimension to this nonhyperbolic fixed point using the methods of Section 3.1D; the number obtained will depend on λ and $\bar{\lambda}$ (we will do this shortly). We can then construct a candidate for a versal deformation using the same methods. However, these will not give versal deformations. One of the obstructions to this involves the dynamics on invariant circles. As we have seen earlier (see Section 3.2C), all of the higher order terms in the normal form may affect the dynamics on the invariant circle. Nevertheless, we will give the codimension and the associated parametrized families using the techniques of Section 3.1D for the different cases.

The case

$$\lambda^n \neq 1, \qquad n = 1, 2, 3, 4, \tag{3.2.124}$$

is *codimension one.* The associated one-parameter family of normal forms for this bifurcation is

$$z \mapsto (1 + \mu)z + cz^2\bar{z}, \qquad z \in \mathbb{C} \tag{3.2.125}$$

(see Example 2.2.4). If we ignore the dynamics on the invariant circle, then (3.2.125) captures all local dynamics.

The cases ruled out by (3.2.124) are referred to as the *strong resonances.* The cases $n = 3$ *and* $n = 4$ are *codimension one* with associated one-parameter families of normal forms given by

$$z \mapsto (1 + \mu)z + c_1\bar{z}^2 + c_2z^2\bar{z}, \qquad n = 3, \qquad z \in \mathbb{C}, \quad \mu \in \mathbb{R}^1, \tag{3.2.126}$$

$$z \mapsto (1 + \mu)z + c_1\bar{z}^3 + c_2z^2\bar{z}, \qquad n = 4, \qquad z \in \mathbb{C}, \quad \mu \in \mathbb{R}^1. \tag{3.2.127}$$

The cases $n = 1$ and $n = 2$ correspond to double 1 and double -1 eigenvalues, respectively. If the matrices associated with the linear parts (in Jordan canonical form) are given by

$$\begin{pmatrix} 1 & 1 \\ 0 & 1 \end{pmatrix}, \qquad n = 1, \tag{3.2.128}$$

and

$$\begin{pmatrix} -1 & 1 \\ 0 & -1 \end{pmatrix}, \qquad n = 2, \tag{3.2.129}$$

then these cases are *codimension two* with associated two-parameter families of normal forms given by

$$\begin{aligned} x &\mapsto x + y, \\ y &\mapsto \mu_1 + \mu_2 x + y + ax^2 + bxy, \end{aligned} \qquad n = 1, \tag{3.2.130}$$

$$\begin{aligned} x &\mapsto x + y, \\ y &\mapsto \mu_1 x + \mu_2 y + y + ax^3 + bx^2 y \end{aligned} \qquad n = 2. \tag{3.2.131}$$

Arnold [1983] studied (3.2.126), (3.2.127), (3.2.130), and (3.2.131) in detail by using the important local technique of interpolating a discrete map by a flow. We will work out some of the details in the exercises. The strongly resonant cases arise in applications in, for example, the situation of applying a periodic external force to a system which would freely oscillate with a given frequency. Strong resonance would occur when the forcing frequency and natural frequency were commensurate in the ratios 1/1, 1/2, 1/3, and 1/4. The surprising fact is that, in the cases $n = 1$ and $n = 2$, chaotic motions of the Smale–horseshoe type (see Gambaudo [1985]) may arise.

3.3 On the Interpretation and Application of Bifurcation Diagrams: A Word of Caution

At this point, we have seen enough examples so that it should be clear that the term bifurcation refers to the phenomenon of a system exhibiting qualitatively new dynamical behavior as parameters are varied. However, the phrase "as parameters are varied" deserves careful consideration. Let us consider a highly idealized example.

The situation we imagine is wind blowing through Venetian blinds hanging in an open window. The "parameter" in this system will be the wind speed. From experience, most people have observed that nothing much happens when the wind speed is low enough, but, for high enough wind speeds, the blinds begin to oscillate or "flutter." Thus, at some critical parameter value, a Poincaré-Andronov-Hopf bifurcation occurs. However, we must be careful here. In all of our analyses thus far the parameters have been *constant*. Therefore, in order to apply the Poincaré-Andronov-Hopf bifurcation theorem to this problem, the wind speed must be constant. At low constant speeds, the blinds lie flat; at constant speeds above a certain critical value, the blinds oscillate. The point is that we cannot think of the parameter as varying in time, e.g., wind speed increasing over time, even though this is what happens in practice. Dynamical systems having parameters that change in time (no matter how slowly!) and that pass through bifurcation values often exhibit behavior that is very different from the analogous situation where the parameters are constant. Let us consider a more mathematical example due to Schecter [1985].

Consider the vector field

$$\dot{x} = f(x, \mu), \qquad x \in \mathbb{R}^1, \quad \mu \in \mathbb{R}^1. \tag{3.3.1}$$

We suppose that

$$f(0, \mu) = 0 \tag{3.3.2}$$

so that $x = 0$ is always a fixed point, and that

$$f(x, \mu) = 0 \tag{3.3.3}$$

intersects $x = 0$ at $\mu = b$ and appears as in Figure 3.3.1. We further assume that

$$\frac{\partial f}{\partial x}(0, \mu) \quad \text{is} \qquad \begin{cases} < 0 & \text{for } x < b, \\ > 0 & \text{for } x > b, \end{cases} \tag{3.3.4}$$

so that the stability of the fixed points is as shown in Figure 3.3.1. Thus, (3.3.1) undergoes a transcritical bifurcation at $\mu = b$. Now, if we think of (3.3.1) as modelling a physical system, we would expect to observe the system in a stable equilibrium state. For $\mu < b$, this would be $x = 0$, and for μ slightly larger than b, x small enough, this would be the upper branch of fixed points bifurcating from the transcritical bifurcation point.

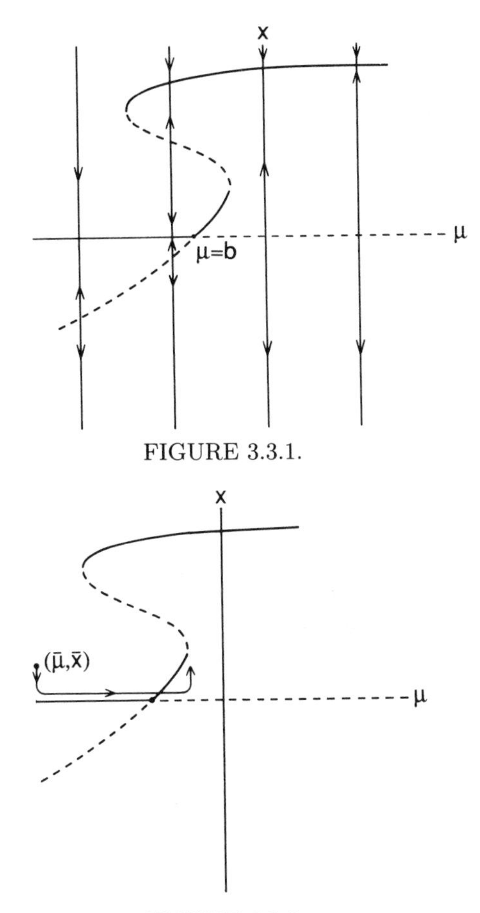

FIGURE 3.3.1.

FIGURE 3.3.2.

Let us consider the situation in which the parameter is allowed to drift slowly in time as follows

$$\dot{x} = f(x, \mu), \tag{3.3.5}$$

$$\dot{\mu} = \varepsilon, \tag{3.3.6}$$

where ε is viewed as small and positive (so that trajectories always move toward the right). (Note: Schecter [1985] considers a much more general situation where $\dot{\mu}$ may depend on x and μ.) Now let us consider the fate of an initial condition $(\bar{\mu}, \bar{x})$ with $\bar{\mu} < b$ and $\bar{x} > 0$ sufficiently small; see Figure 3.3.2. This point is attracted strongly toward $x = 0$ (but it can never cross $x = 0$. Why?) and drifts slowly toward the right. Schecter proves that, on passing through $\mu = b$, rather than being repelled from $x = 0$ as would be the case for $\varepsilon = 0$, the trajectory follows closely $x = 0$ (which is an unstable invariant manifold for $\mu > b$) for awhile before ultimately being repelled away; see Figure 3.3.2. Thus, what one would observe for $\varepsilon = 0$ differs from that for $\varepsilon \neq 0$; for $\varepsilon \neq 0$, certain trajectories tend to remain in the neighborhood (for awhile) of what are unstable fixed points for $\varepsilon = 0$.

A detailed analysis of problems of this type is beyond the scope of this book (such problems fit very nicely into the context of singular perturbation theory). The point we wish to make is that, within a given system, the behavior of that system on either side of a bifurcation value may be very different in cases when the parameter varies slowly in time (no matter how slowly) through the bifurcation value, as opposed to cases when the parameter is constant. The reader will find detailed analyses of such problems in Mitropol'skii [1965], Lebovitz and Schaar [1975], [1977], Haberman [1979], Neishtadt [1987], [1988], Baer et al. [1989], Erneux and Mandel [1986], and Mandel and Erneux [1987].

Exercises

3.1 Consider a $\mathbf{C}^r$ $(r \geq 1)$ autonomous vector field on $\mathbb{R}^1$ having precisely two *hyperbolic* fixed points. Can you infer the nature of the stability of the two fixed points? How does the situation change if one of the fixed points is not hyperbolic? Can both fixed points be nonhyperbolic? Construct explicit examples illustrating each situation.

3.2 Consider the saddle-node bifurcation for vector fields and Figure 3.1.5. For the case $\left(-\frac{\partial^2 f}{\partial x^2}(0,0)/\frac{\partial f}{\partial \mu}(0,0)\right) > 0$, give conditions under which the upper part of the curve of fixed points is stable and the lower part is unstable. Alternatively, give conditions under which the upper part of the curve of fixed points is unstable and the lower part is stable.

Repeat the exercise for the case $\left(-\frac{\partial^2 f}{\partial x^2}(0,0)/\frac{\partial f}{\partial \mu}(0,0)\right) < 0$.

3.3 Consider the transcritical bifurcation for vector fields and Figure 3.1.6. For the case $\left(-\frac{\partial^2 f}{\partial x^2}(0,0)/\frac{\partial^2 f}{\partial x \partial \mu}(0,0)\right) > 0$, give conditions for $x = 0$ to be stable for $\mu > 0$ and unstable for $\mu < 0$. Alternatively, give conditions for $x = 0$ to be unstable for $\mu > 0$ and stable for $\mu < 0$.

Repeat the exercise for the case $\left(-\frac{\partial^2 f}{\partial x^2}(0,0)/\frac{\partial^2 f}{\partial x \partial \mu}(0,0)\right) < 0$.

3.4 Consider the pitchfork bifurcation for vector fields and Figure 3.1.7. For the case $\left(-\frac{\partial^3 f}{\partial x^3}(0,0)/\frac{\partial^2 f}{\partial x \partial \mu}(0,0)\right) > 0$, give conditions for $x = 0$ to be stable for $\mu > 0$ and unstable for $\mu < 0$. Alternatively, give conditions for $x = 0$ to be unstable for $\mu > 0$ and stable for $\mu < 0$.

Repeat the exercise for the case $\left(-\frac{\partial^3 f}{\partial x^3}(0,0)/\frac{\partial^2 f}{\partial x \partial \mu}(0,0)\right) < 0$.

3.5 Consider the saddle-node bifurcation for maps and Figure 3.2.2. For the case $\left(-\frac{\partial^2 f}{\partial x^2}(0,0)/\frac{\partial f}{\partial \mu}(0,0)\right) > 0$, give conditions under which the upper part of the curve of fixed points is stable and the lower part is unstable. Alternatively, give conditions under which the upper part of the curve of fixed points is unstable and the lower part is stable.

Repeat the exercise for the case $\left(-\frac{\partial^2 f}{\partial x^2}(0,0)/\frac{\partial f}{\partial \mu}(0,0)\right) < 0$.

3.6 Consider the transcritical bifurcation for maps and Figure 3.2.5. For the case $\left(-\frac{\partial^2 f}{\partial x^2}(0,0)/\frac{\partial^2 f}{\partial x \partial \mu}(0,0)\right) > 0$, give conditions for $x = 0$ to be stable for $\mu > 0$ and unstable for $\mu < 0$. Alternatively, give conditions for $x = 0$ to be unstable for $\mu > 0$ and stable for $\mu < 0$.

Repeat the exercise for the case $\left(-\frac{\partial^2 f}{\partial x^2}(0,0)/\frac{\partial^2 f}{\partial x \partial \mu}(0,0)\right) < 0$.

3.7 Consider the pitchfork bifurcation for maps and Figure 3.2.8. For the case $\left(-\frac{\partial^3 f}{\partial x^3}(0,0)/\frac{\partial^2 f}{\partial x \partial \mu}(0,0)\right) > 0$, give conditions for $x = 0$ to be stable for $\mu > 0$ and unstable for $\mu < 0$. Alternatively, give conditions for $x = 0$ to be unstable for $\mu > 0$ and stable for $\mu < 0$.

Repeat the exercise for the case $\left(-\frac{\partial^3 f}{\partial x^3}(0,0)/\frac{\partial^2 f}{\partial x \partial \mu}(0,0)\right) < 0$.

3.8 Consider the period-doubling bifurcation for maps and Figure 3.2.11. For the case $\left(-\frac{\partial^3 f^2}{\partial x^3}(0,0)/\frac{\partial^2 f^2}{\partial x \partial \mu}(0,0)\right) > 0$, give conditions for the period one points, $x = 0$, to be stable for $\mu > 0$ and unstable for $\mu < 0$. Alternatively, give conditions for $x = 0$ to be unstable for $\mu > 0$ and stable for $\mu < 0$.

Repeat the exercise for the case $\left(-\frac{\partial^3 f^2}{\partial x^3}(0,0)/\frac{\partial^2 f^2}{\partial x \partial \mu}(0,0)\right) < 0$.

3.9 In Exercise 2.2 we computed center manifolds near the origin for the following one-parameter families of vector fields. Describe the bifurcations of the origin. In, for example, a) and a′) the parameter ε multiplies a linear and nonlinear term, respectively. In terms of bifurcations, is there a qualitative difference in the two cases? What kinds of general statements can you make?

a) $$\begin{aligned} \dot{\theta} &= -\theta + \varepsilon v + v^2, \\ \dot{v} &= -\sin\theta, \end{aligned} \qquad (\theta, v) \in S^1 \times \mathbb{R}^1.$$

a′) $$\begin{aligned} \dot{\theta} &= -\theta + v^2 + \varepsilon v^2, \\ \dot{v} &= -\sin\theta. \end{aligned}$$

b) $$\begin{aligned} \dot{x} &= \frac{1}{2}x + y + x^2 y, \\ \dot{y} &= x + 2y + \varepsilon y + y^2, \end{aligned} \qquad (x, y) \in \mathbb{R}^2.$$

b′) $$\begin{aligned} \dot{x} &= \frac{1}{2}x + y + x^2 y, \\ \dot{y} &= x + 2y + y^2 + \varepsilon y^2. \end{aligned}$$

c) $$\begin{aligned} \dot{x} &= x - 2y + \varepsilon x, \\ \dot{y} &= 3x - y - x^2, \end{aligned} \qquad (x, y) \in \mathbb{R}^2.$$

c′) $\begin{aligned}\dot{x} &= x - 2y + \varepsilon x^2,\\ \dot{y} &= 3x - y - x^2.\end{aligned}$

d) $\begin{aligned}\dot{x} &= 2x + 2y + \varepsilon y,\\ \dot{y} &= x + y + x^4,\end{aligned} \qquad (x, y) \in \mathbb{R}^2.$

d′) $\begin{aligned}\dot{x} &= 2x + 2y,\\ \dot{y} &= x + y + x^4 + \varepsilon y^2.\end{aligned}$

e) $\begin{aligned}\dot{x} &= -y - \varepsilon x - y^3,\\ \dot{y} &= 2x,\end{aligned} \qquad (x, y) \in \mathbb{R}^2.$

e′) $\begin{aligned}\dot{x} &= -y - y^3,\\ \dot{y} &= 2x + \varepsilon x^2.\end{aligned}$

f) $\begin{aligned}\dot{x} &= -2x + 3y + \varepsilon x + y^3,\\ \dot{y} &= 2x - 3y + x^3,\end{aligned} \qquad (x, y) \in \mathbb{R}^2.$

f′) $\begin{aligned}\dot{x} &= -2x + 3y + y^3 + \varepsilon x^2,\\ \dot{y} &= 2x - 3y + x^3.\end{aligned}$

g) $\begin{aligned}\dot{x} &= -x - y + \varepsilon x - xy,\\ \dot{y} &= 2x + y + 2xy,\end{aligned} \qquad (x, y) \in \mathbb{R}^2.$

g′) $\begin{aligned}\dot{x} &= -x - y - xy + \varepsilon x^2,\\ \dot{y} &= 2x + y + 2xy.\end{aligned}$

h) $\begin{aligned}\dot{x} &= -x + y,\\ \dot{y} &= -e^x + e^{-x} + 2x + \varepsilon y,\end{aligned} \qquad (x, y) \in \mathbb{R}^2.$

h′) $\begin{aligned}\dot{x} &= -x + y + \varepsilon x^2,\\ \dot{y} &= -e^x + e^{-x} + 2x.\end{aligned}$

i) $\begin{aligned}\dot{x} &= -2x + y + z + \varepsilon x + y^2 z,\\ \dot{y} &= x - 2y + z + \varepsilon x + xz^2,\\ \dot{z} &= x + y - 2z + \varepsilon x + x^2 y,\end{aligned} \qquad (x, y, z) \in \mathbb{R}^3.$

i′) $\begin{aligned}\dot{x} &= -2x + y + z + \varepsilon x^2 + y^2 z,\\ \dot{y} &= x - 2y + z + \varepsilon xy + xz^2,\\ \dot{z} &= x + y - 2z + x^2 y.\end{aligned}$

j) $\begin{aligned}\dot{x} &= -x - y + z^2,\\ \dot{y} &= 2x + y + \varepsilon y - z^2,\\ \dot{z} &= x + 2y - z,\end{aligned} \qquad (x, y, z) \in \mathbb{R}^3.$

j′) $\begin{aligned}\dot{x} &= -x - y + \varepsilon x^2 + z^2,\\ \dot{y} &= 2x + y - z^2 + \varepsilon y^2,\\ \dot{z} &= x + 2y - z.\end{aligned}$

k) $\begin{aligned}\dot{x} &= -x - y - z + \varepsilon x - yz,\\ \dot{y} &= -x - y - z - xz,\\ \dot{z} &= -x - y - z - yz,\end{aligned} \qquad (x, y, z) \in \mathbb{R}^3.$

k′) $$\begin{aligned}\dot{x} &= -x - y - z - yz + \varepsilon x^2,\\ \dot{y} &= -x - y - z - xz,\\ \dot{z} &= -x - y - z - xy.\end{aligned}$$

l) $$\begin{aligned}\dot{x} &= y + x^2 + \varepsilon y,\\ \dot{y} &= -y - x^2,\end{aligned} \qquad (x, y) \in \mathbb{R}^2.$$

l′) $$\begin{aligned}\dot{x} &= y + x^2 + \varepsilon y^2,\\ \dot{y} &= -y - x^2.\end{aligned}$$

m) $$\begin{aligned}\dot{x} &= x^2 + \varepsilon y,\\ \dot{y} &= -y - x^2,\end{aligned} \qquad (x, y) \in \mathbb{R}^2.$$

m′) $$\begin{aligned}\dot{x} &= x^2 + \varepsilon y^2,\\ \dot{y} &- -y - x^2.\end{aligned}$$

3.10 In Exercise 2.4 we computed center manifolds near the origin for the following one-parameter families of maps. Describe the bifurcations of the origin. In, for example, a) and a′) the parameter ε multiplies a linear and nonlinear term, respectively. In terms of bifurcations, is there a qualitative difference in the two cases? What kind of general statements can you make?

a) $$\begin{aligned}x &\mapsto -\frac{1}{2}x - y - xy^2,\\ y &\mapsto -\frac{1}{2}x + \varepsilon y + x^2,\end{aligned} \qquad (x, y) \in \mathbb{R}^2.$$

a′) $$\begin{aligned}x &\mapsto -\frac{1}{2}x - y - xy^2,\\ y &\mapsto -\frac{1}{2}y + \varepsilon y^2 + x^2.\end{aligned}$$

b) $$\begin{aligned}x &\mapsto x + 2y + x^3,\\ y &\mapsto 2x + y + \varepsilon y,\end{aligned} \qquad (x, y) \in \mathbb{R}^2.$$

b′) $$\begin{aligned}x &\mapsto x + 2y + x^3,\\ y &\mapsto 2x + y + \varepsilon y^2.\end{aligned}$$

c) $$\begin{aligned}x &\mapsto -x + y - xy^2,\\ y &\mapsto y + \varepsilon y + x^2 y,\end{aligned} \qquad (x, y) \in \mathbb{R}^2.$$

c′) $$\begin{aligned}x &\mapsto -x + y - xy^2,\\ y &\mapsto y + \varepsilon y^2 + x^2 y.\end{aligned}$$

d) $$\begin{aligned}x &\mapsto 2x + y,\\ y &\mapsto 2x + 3y + \varepsilon x + x^4,\end{aligned} \qquad (x, y) \in \mathbb{R}^2.$$

d′) $$\begin{aligned}x &\mapsto 2x + y + \varepsilon x^2,\\ y &\mapsto 2x + 3y + x^4.\end{aligned}$$

e) $$\begin{aligned}x &\mapsto x + \varepsilon y,\\ y &\mapsto x + 2y + y^2,\end{aligned} \qquad (x, y) \in \mathbb{R}^2.$$

e′) $\begin{aligned} x &\mapsto x + \varepsilon y^2, \\ y &\mapsto x + 2y + y^2. \end{aligned}$

f) $\begin{aligned} x &\mapsto 2x + 3y, \\ y &\mapsto x + \varepsilon y + x^2 + xy^2, \end{aligned} \qquad (x, y) \in \mathbb{R}^2.$

f′) $\begin{aligned} x &\mapsto 2x + 3y, \\ y &\mapsto x + x^2 + \varepsilon y^2 + xy^2. \end{aligned}$

g) $\begin{aligned} x &\mapsto x - z^3, \\ y &\mapsto 2x - y + \varepsilon y, \\ z &\mapsto x + \frac{1}{2}z + x^3, \end{aligned} \qquad (x, y, z) \in \mathbb{R}^3.$

g′) $\begin{aligned} x &\mapsto x - z^3, \\ y &\mapsto 2x - y + \varepsilon y^2, \\ z &\mapsto x + \frac{1}{2}z + x^3. \end{aligned}$

h) $\begin{aligned} x &\mapsto x + \varepsilon z^4, \\ y &\mapsto -x - 2y - x^3, \\ z &\mapsto y - \frac{1}{2}z + y^2, \end{aligned} \qquad (x, y, z) \in \mathbb{R}^3.$

h′) $\begin{aligned} x &\mapsto x + \varepsilon x + z^4, \\ y &\mapsto -x - 2y - x^3, \\ z &\mapsto y - \frac{1}{2}z + y^2. \end{aligned}$

i) $\begin{aligned} x &\mapsto y + \varepsilon x + x^2, \\ y &\mapsto y + xy, \end{aligned} \qquad (x, y) \in \mathbb{R}^2.$

i′) $\begin{aligned} x &\mapsto y + x^2, \\ y &\mapsto y + xy + \varepsilon x^2. \end{aligned}$

j) $\begin{aligned} x &\mapsto \varepsilon x + x^2, \\ y &\mapsto y + xy, \end{aligned} \qquad (x, y) \in \mathbb{R}^2.$

j′) $\begin{aligned} x &\mapsto x^2 + \varepsilon y, \\ y &\mapsto y + xy. \end{aligned}$

3.11 *Center Manifolds at a Saddle-node Bifurcation Point for Vector Fields*

In developing the center manifold theory for parametrized families of vector fields, we dealt with equations of the following form

$$\begin{aligned} \dot{x} &= Ax + f(x, y, \varepsilon), \\ \dot{y} &= By + g(x, y, \varepsilon), \end{aligned} \qquad (x, y, \varepsilon) \in \mathbb{R}^c \times \mathbb{R}^s \times \mathbb{R}^p, \tag{E3.1}$$

where A is a $c \times c$ matrix whose eigenvalues all have zero real parts, B is an $s \times s$ matrix whose eigenvalues all have negative real parts, and

$$\begin{aligned} f(0,0,0) &= 0, \qquad & Df(0,0,0) &= 0, \\ g(0,0,0) &= 0, \qquad & Dg(0,0,0) &= 0. \end{aligned} \tag{E3.2}$$

The conditions $Df(0,0,0) = 0$, $Dg(0,0,0) = 0$ imply that $(x,y) = (0,0)$ is a fixed point of (E3.1) for *all* ε in an open set containing zero. Clearly, this *cannot* be the case at a saddle-node bifurcation point, and we want to consider this issue in this exercise. Although this could have been done in Chapter 2, in that chapter we were introducing only center manifold theory and were not really concerned with bifurcations. In this case the form of the equations given by (E3.1) and (E3.2) was the "cleanest and quickest" way to introduce the notion of parametrized families of center manifolds.

We will start at a very basic level. Consider the $\mathbf{C}^r$ (r as large as necessary) vector field

$$\dot{z} = F(z,\varepsilon), \qquad (z,\varepsilon) \in \mathbb{R}^{c+s} \times \mathbb{R}^p. \tag{E3.3}$$

Suppose that $(z,\varepsilon) = (0,0)$ is a fixed point of (E3.3) at which the matrix

$$D_z F(0,0) \tag{E3.4}$$

has c eigenvalues with zero real parts and s eigenvalues with negative real parts. Our goal is to apply the center manifold theory in order to examine the dynamics of (E3.3) near $(z,\varepsilon) = (0,0)$.

We rewrite Equation (E3.3) as follows

$$\dot{z} = D_z F(0,0)z + D_\varepsilon F(0,0)\varepsilon + G(z,\varepsilon), \tag{E3.5}$$

where

$$G(z,\varepsilon) = \left[F(z,\varepsilon) - D_z F(0,0)z - D_\varepsilon F(0,0)\varepsilon\right] = \mathcal{O}(2) \tag{E3.6}$$

in z and ε. Note that the term "$D_\varepsilon F(0,0)\varepsilon$" in (E3.5) is the new wrinkle—it was zero under our previous assumptions. For notational purposes we let

$$\begin{aligned} D_z F(0,0) &\equiv M \qquad -(c+s)\times(c+s) \text{ matrix},\\ D_\varepsilon F(0,0) &\equiv \Lambda \qquad -(c+s)\times p \text{ matrix}, \end{aligned}$$

so that (E3.5) becomes

$$\dot{z} = Mz + \Lambda\varepsilon + G(z,\varepsilon). \tag{E3.7}$$

Now let T be the $(s+c)\times(s+c)$ matrix that puts M into the following block diagonal form

$$T^{-1}MT = \begin{pmatrix} A & 0 \\ 0 & B \end{pmatrix}, \tag{E3.8}$$

where A is a $(c\times c)$ matrix with all eigenvalues having zero real parts and B is an $(s\times s)$ matrix with all eigenvalues having negative real parts. If we let

$$z = Tw, \qquad (x,y) \in \mathbb{R}^c \times \mathbb{R}^s, \tag{E3.9}$$

where $w = (x, y)$, and apply this linear transformation to (E3.7), we obtain

$$\begin{pmatrix} \dot{x} \\ \dot{y} \end{pmatrix} = \begin{pmatrix} A & 0 \\ 0 & B \end{pmatrix} \begin{pmatrix} x \\ y \end{pmatrix} + \overline{\Lambda}\varepsilon + \begin{pmatrix} f(x,y,\varepsilon) \\ g(x,y,\varepsilon) \end{pmatrix}, \tag{E3.10}$$

where

$$\overline{\Lambda} \equiv T^{-1}\Lambda,$$

$$\begin{pmatrix} f(x,y,\varepsilon) \\ g(x,y,\varepsilon) \end{pmatrix} \equiv T^{-1}G(T(x,y),\varepsilon).$$

Note that $f(0,0,0) = 0$, $g(0,0,0) = 0$, $Df(0,0,0) = 0$, and $Dg(0,0,0) = 0$. Next, let

$$\overline{\Lambda} = \begin{pmatrix} \overline{\Lambda}_c \\ \overline{\Lambda}_s \end{pmatrix},$$

where Λ_c corresponds to the first c rows of Λ, and Λ_s corresponds to the last s rows of Λ. Then (E3.10) can be rewritten as

$$\begin{pmatrix} \dot{x} \\ \dot{\varepsilon} \\ \dot{y} \end{pmatrix} = \begin{pmatrix} A & \overline{\Lambda}_c & 0 \\ 0 & 0 & 0 \\ 0 & \overline{\Lambda}_s & B \end{pmatrix} \begin{pmatrix} x \\ \varepsilon \\ y \end{pmatrix} + \begin{pmatrix} f(x,y,\varepsilon) \\ 0 \\ g(x,y,\varepsilon) \end{pmatrix}. \tag{E.3.11}$$

The reader should recognize that (E3.11) is "almost" in the standard normal form for application of the center manifold theory. The final step would be to introduce a linear transformation that block diagonalizes the linear part of (E3.11) into a $(c+p) \times (c+p)$ matrix with eigenvalues all having zero real parts (and p identically zero) and an $(s \times s)$ matrix with all eigenvalues having negative real parts (see Exercise 3.13).

a) Carry out this final step and discuss applying the center manifold theorem to the resulting system. In particular, do the relevant theorems from Chapter 2 go through?

Before we work out some specific problems, let us first answer an example.

Consider the vector field

$$\begin{aligned} \dot{x} &= \varepsilon + x^2 + y^2, \\ \dot{y} &= -y + x^2, \end{aligned} \qquad (x, y, \varepsilon) \in \mathbb{R}^3. \tag{E3.12}$$

It should be clear that $(x, y, \varepsilon) = (0, 0, 0)$ is a fixed point of (E3.12). We want to study the orbit structure near this fixed point for ε small. Rewriting (E3.12) in the form of (E3.11) gives

$$\begin{pmatrix} \dot{x} \\ \dot{\varepsilon} \\ \dot{y} \end{pmatrix} = \begin{pmatrix} 0 & 1 & 0 \\ 0 & 0 & 0 \\ 0 & 0 & -1 \end{pmatrix} \begin{pmatrix} x \\ \varepsilon \\ y \end{pmatrix} + \begin{pmatrix} x^2 + y^2 \\ 0 \\ x^2 \end{pmatrix}. \tag{E3.13}$$

We seek a center manifold of the form

$$h(x,\varepsilon) = ax^2 + bx\varepsilon + c\varepsilon^2 + \mathcal{O}(3).$$

Utilizing the usual procedure for calculating the center manifold, we obtain

$$h(x,\varepsilon) = x^2 - 2x\varepsilon + 2\varepsilon^2 + \mathcal{O}(3).$$

The vector field restricted to the center manifold is then given by

$$\begin{aligned} \dot{x} &= \varepsilon + x^2 + \mathcal{O}(4), \\ \dot{\varepsilon} &= 0. \end{aligned}$$

Hence, a saddle-node bifurcation occurs at $\varepsilon = 0$.

Now consider the following vector fields

b) $\begin{aligned} \dot{x} &= \varepsilon + x^4 + y^2, \\ \dot{y} &= -y + x^3, \end{aligned}$ $\qquad (x, y, \varepsilon) \in \mathbb{R}^3.$

c) $\begin{aligned} \dot{x} &= \varepsilon + x^2 - y^3, \\ \dot{y} &= \varepsilon - y + x^2. \end{aligned}$

d) $\begin{aligned} \dot{x} &= \varepsilon + \varepsilon x + x^2, \\ \dot{y} &= -y + x^2. \end{aligned}$

e) $\begin{aligned} \dot{x} &= \varepsilon + \varepsilon x + x^2, \\ \dot{y} &= \varepsilon - y + x^2. \end{aligned}$

f) $\begin{aligned} \dot{x} &= \varepsilon + \frac{1}{2}x + y + x^3, \\ \dot{y} &= x + 2y - xy. \end{aligned}$

g) $\begin{aligned} \dot{x} &= 2\varepsilon + 2x + 2y, \\ \dot{y} &= \varepsilon + x + y + y^2. \end{aligned}$

h) $\begin{aligned} \dot{x} &= \varepsilon - 2x + 2y - x^4, \\ \dot{y} &= 2x - 2y. \end{aligned}$

i) $\begin{aligned} \dot{x} &= \varepsilon - 2x + y + z + yz, \\ \dot{y} &= x - 2y + z + zx, \\ \dot{z} &= x + y - 2z + xy, \end{aligned}$ $\qquad (x, y, z, \varepsilon) \in \mathbb{R}^4.$

For each vector field, construct the center manifold and discuss the dynamics near the origin for ε small. What types of bifurcations occur?

3.12 *Center Manifolds at a Saddle-node Bifurcation Point for Maps*

Following the discussion in Exercise 3.11, develop the center manifold theory for maps so that it applies at a saddle-node bifurcation point.

Apply the resulting theory to the following maps. In each case, compute the center manifold and describe the dynamics near the origin for ε small. Discuss the bifurcations that occur (if any) at $\varepsilon = 0$.

a) $\begin{aligned} x &\mapsto \varepsilon + x + x^2 - y^2, \\ y &\mapsto x^2 + y^2, \end{aligned} \qquad (x, y, \varepsilon) \in \mathbb{R}^3.$

b) $\begin{aligned} x &\mapsto \varepsilon + \varepsilon x + x^2 - y^2, \\ y &\mapsto x^3 + y^2. \end{aligned}$

c) $\begin{aligned} x &\mapsto \varepsilon + x + x^2 - y^2, \\ y &\mapsto \varepsilon + x^2 + y^2. \end{aligned}$

d) $\begin{aligned} x &\mapsto \varepsilon + \frac{1}{2}x - y - x^2, \\ y &\mapsto \frac{1}{2}x + y^2. \end{aligned}$

e) $\begin{aligned} x &\mapsto \varepsilon - x + y + x^3, \\ y &\mapsto y + \varepsilon x - x^2. \end{aligned}$

f) $\begin{aligned} x &\mapsto \varepsilon + x + \varepsilon x + y^3, \\ y &\mapsto x + 2y - x^2. \end{aligned}$

g) $\begin{aligned} x &\mapsto \varepsilon - x + xy + y^2, \\ y &\mapsto 2x - xy - y^2. \end{aligned}$

h) $\begin{aligned} x &\mapsto \varepsilon + 2x + y + x^2 y, \\ y &\mapsto 12x + 3y - xy^2. \end{aligned}$

3.13 Consider the block diagonal "normal form" of (E3.1) to which we first transformed the vector field in order to apply the center manifold theory. Discuss why (or why not) this preliminary transformation was necessary. Is this preliminary transformation necessary for equations of the form of (E3.11) in order to apply the center manifold theory? Work out several examples to support your views and illustrate the relevant points. (*Hint:* consider the coordinatization of the center manifold and how the invariance condition is manifested in those coordinates.)

3.14 This exercise comes from Marsden and McCracken [1976]. Consider the following vector fields

a) $\begin{aligned} \dot{r} &= -r(r-\mu)^2, \\ \dot{\theta} &= 1, \end{aligned} \qquad (r, \theta) \in \mathbb{R}^+ \times S^1.$

b) $\begin{aligned} \dot{r} &= r(\mu - r^2)(2\mu - r^2)^2, \\ \dot{\theta} &= 1. \end{aligned}$

c) $\begin{aligned} \dot{r} &= r(r+\mu)(r-\mu), \\ \dot{\theta} &= 1. \end{aligned}$

d) $\begin{aligned} \dot{r} &= \mu r(r+\mu)^2, \\ \dot{\theta} &= 1. \end{aligned}$

e) $\begin{aligned} \dot{r} &= -\mu^2 r(r+\mu)^2(r-\mu)^2, \\ \dot{\theta} &= 1. \end{aligned}$

Match each of these vector fields to the appropriate phase portrait in Figure E3.1 and explain which hypotheses (if any) of the Poincaré–Andronov–Hopf bifurcation theorem are violated.

3.15 Consider the Poincaré–Andronov–Hopf bifurcation theorem (Theorem 3.1.3).

a) Work out all of the details of Proof 1, which utilizes the Poincaré–Bendixson theorem.

b) Work out all the details of Proof 2, which utilizes the method of averaging.

3.16 For the Poincaré–Andronov–Hopf bifurcation, compute the expression for the coefficient a given in (3.1.107).

3.17 For the Naimark–Sacker bifurcation, compute the expression for the coefficient a analogous to (3.1.107) for vector field. (*Hint:* the answer can be found in Guckenheimer and Holmes [1983].)

3.18 Consider the following one-parameter family of two-dimensional $\mathbf{C}^r$ (r as large as necessary) vector fields

$$\dot{x} = f(x;\mu), \qquad (x,\mu) \in \mathbb{R}^2 \times \mathbb{R}^1,$$

where $f(0;0) = 0$ and $D_x f(0,0)$ has a zero eigenvalue and a negative eigenvalue. Suppose the vector field has the following symmetry

$$f(x,\mu) = -f(-x,\mu).$$

What can you then conclude concerning the symmetry of the vector field restricted to the center manifold for x and μ small? Can the vector field undergo a saddle-node bifurcation at $(x,\mu) = (0,0)$? Can the vector field undergo a saddle-node bifurcation at other points $(x,\mu) \in \mathbb{R}^2 \times \mathbb{R}^1$?

3.19 a) Consider the following three-parameter family of one-dimensional vector fields

$$\dot{x} = x^3 + \mu_3 x^2 + \mu_2 x + \mu_1, \qquad x \in \mathbb{R}^1. \tag{E3.14}$$

Show that by a parameter-dependent shift of x, (E3.14) can be written as a two-parameter family

$$\dot{\overline{x}} = \overline{x}^3 + \overline{\mu}_2 \overline{x} + \overline{\mu}_1.$$

What are $\overline{x}$, $\overline{\mu}_2$, and $\overline{\mu}_1$ in terms of x, μ_1, μ_2, and μ_3?

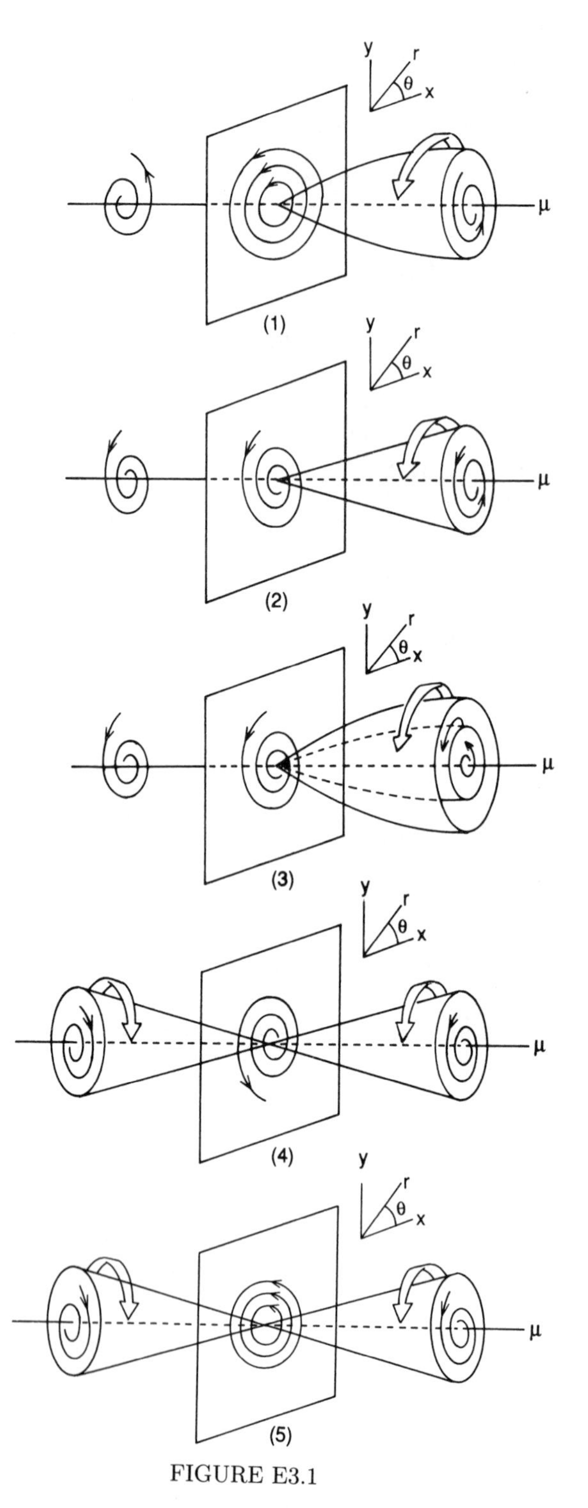

FIGURE E3.1

b) Consider the following two-parameter family of one-dimensional vector fields

$$\dot{x} = x^2 + \mu_2 x + \mu_1, \qquad x \in \mathbb{R}^1. \tag{E3.15}$$

Show that by a parameter-dependent shift of x, (E3.15) can be written as a one-parameter family

$$\dot{\overline{x}} = \overline{x}^2 + \overline{\mu}_1.$$

What are $\overline{x}$ and $\overline{\mu}_1$ in terms of x, μ_1, and μ_2?

Discuss the results of a) and b) in terms of the codimension of a bifurcation and the number of parameters.

In particular, consider part b) and address these issues in relation to a comparison of the saddle-node bifurcation

$$\dot{x} = x^2 + \mu$$

and the transcritical bifurcation

$$\dot{x} = x^2 + \mu x.$$

3.20 Consider the following two-parameter family of planar vector fields

$$\begin{aligned} \dot{x} &= y, \\ \dot{y} &= \mu_1 x + \mu_2 y + a x^2 + b x y. \end{aligned} \tag{E3.16}$$

Under the shift of coordinates

$$\begin{aligned} x &\to x + x_0, \\ y &\to y, \end{aligned}$$

(E3.16) becomes

$$\begin{aligned} \dot{x} &= y, \\ \dot{y} &= -\mu_1 x_0 + (\mu_1 + a y) x_0 + (\mu_2 - a x_0) y + b y^2. \end{aligned} \tag{E3.17}$$

Show that by a parameter-dependent shift of y (but not x), (E3.17) can be transformed to the form

$$\begin{aligned} \dot{x} &= \overline{y}, \\ \dot{\overline{y}} &= \overline{\mu}_1 + \overline{\mu}_2 \overline{y} + x \overline{y} + \overline{b} y^2. \end{aligned} \tag{E3.18}$$

What are $\overline{y}$, $\overline{\mu}_1$, $\overline{\mu}_2$, and $\overline{b}$ in terms of y, x_0, μ_1, μ_2, a, and b?

3.21 Let M denote the set of all $n \times n$ matrices with complex entries. Show that M can be identified as $\mathbb{C}^{n^2}$ (note: see Dubrovin, Fomenko, and Novikov [1984] for an excellent discussion of matrix groups as surfaces).

3.22 Show that $GL(n, \mathbb{C})$ is a submanifold of $\mathbb{C}^{n^2}$.

3.23 Show that the orbit of a matrix $A_0 \in M$ under the action of $GL(n, \mathbb{C})$ defined by

$$gA_0g^{-1}, \qquad g \in GL(n, \mathbb{C}),$$

is a submanifold of M.

3.24 Show that the centralizer of any matrix of order n (with complex entries) is a linear subspace of $\mathbb{C}^{n^2}$.

3.25 Prove that the dimension of the centralizer is equal to the codimension of the orbit of A_0.

3.26 Explain why that, in the local coordinates near A_0 constructed in Lemma A1.2, any matrix sufficiently close to A_0 can be represented in the form

$$pA(\lambda)p^{-1}, \qquad p \in P, \quad \lambda \in \Lambda \subset \mathbb{C}^{\ell}.$$

Can you give a more intuitive explanation of this based on elementary notions from linear algebra?

3.27 Explain why the deformation constructed in Proposition A1.1 is miniversal.

3.28 Prove the following statement:

> The decomplexification of a versal deformation with the minimum number of parameters of a complex matrix, $\tilde{A}_0$, can be chosen to be a versal deformation with the minimum number of parameters of the real matrix A_0, where A_0 is the decomplexification of $\tilde{A}_0$.

3.29 Prove that the decomplexification of $\mathbb{C}^n$ is $\mathbb{R}^{2n}$ and that if $e_1, \cdots, e_n$ is a basis of $\mathbb{C}^n$, then $e_1, \cdots e_n, ie_1, \cdots, ie_n$ is a basis for the decomplexification of $\mathbb{C}^n$, $\mathbb{R}^{2n}$.

3.30 Suppose $A = A_r + iA_i$ is the matrix representation of some linear mapping of $\mathbb{C}^n$ into $\mathbb{C}^n$. Then show that

$$\begin{pmatrix} A_r & -A_i \\ A_i & A_r \end{pmatrix}$$

is the decomplexification of this matrix.

3.31 Compute miniversal deformations of the following *real* matrices.

a) $\begin{pmatrix} 0 & -\omega & 0 \\ \omega & 0 & 0 \\ 0 & 0 & 0 \end{pmatrix}$

b) $\begin{pmatrix} 0 & -\omega_1 & 0 & 0 \\ \omega_1 & 0 & 0 & 0 \\ 0 & 0 & 0 & -\omega_2 \\ 0 & 0 & \omega_2 & 0 \end{pmatrix}$

c) $\begin{pmatrix} 1 & 0 \\ 0 & 1 \end{pmatrix}$

d) $\begin{pmatrix} 1 & 0 \\ 0 & -1 \end{pmatrix}$

e) $\begin{pmatrix} 0 & 1 & 0 \\ 0 & 0 & 0 \\ 0 & 0 & 0 \end{pmatrix}$

f) $\begin{pmatrix} 0 & 1 & 0 \\ 0 & 0 & 1 \\ 0 & 0 & 0 \end{pmatrix}$

g) $\begin{pmatrix} 0 & 0 & 0 \\ 0 & 0 & 1 \\ 0 & 0 & 0 \end{pmatrix}$

h) $\begin{pmatrix} 0 & -\omega & 0 \\ \omega & 0 & 0 \\ 0 & 0 & 1 \end{pmatrix}$

i) $\begin{pmatrix} 0 & 0 \\ 1 & 0 \end{pmatrix}$

j) $\begin{pmatrix} 0 & -\omega \\ \omega & 1 \end{pmatrix}$.

3.32 *The Double-Zero Eigenvalue with Symmetry.* Consider a $\mathbf{C}^r$ (r as large as necessary) vector field on $\mathbb{R}^2$ having a fixed point at which the matrix associated with the linearization has the following Jordan canonical form

$$\begin{pmatrix} 0 & 1 \\ 0 & 0 \end{pmatrix}.$$

Let (x, y) denote coordinates on $\mathbb{R}^2$, and suppose further that the vector field is invariant under the coordinate transformation

$$(x, y) \mapsto (-x, -y).$$

This exercise is concerned with the bifurcations near such a degenerate fixed point.

a) Show that a normal form for this vector field near this nonhyperbolic fixed point is given by

$$\begin{aligned} \dot{x} &= y + \mathcal{O}(5), \\ \dot{y} &= ax^3 + bx^2 y + \mathcal{O}(5). \end{aligned}$$

b) Following the procedure outlined in Section 3.1D, show that a candidate for a versal deformation is given by

$$\begin{aligned}\dot{x} &= y + \mathcal{O}(5),\\ \dot{y} &= \mu_1 x + \mu_2 y + ax^3 + bx^2 y + \mathcal{O}(5).\end{aligned}$$

In the following we will be concerned with the dynamics of the truncated normal form

$$\begin{aligned}\dot{x} &= y,\\ \dot{y} &= \mu_1 x + \mu_2 y + ax^3 + bx^2 y.\end{aligned}$$

c) Show that by rescaling, the number of cases to be considered can be reduced to the following

$$\begin{aligned}\dot{x} &= y,\\ \dot{y} &= \mu_1 x + \mu_2 y + cx^3 - x^2 y,\end{aligned} \tag{E3.19}$$

where $c = \pm 1$.

d) For $\mu_1 = \mu_2 = 0$, show that the flow near the origin appears as in Figure E3.2 for $c = +1$ and as in Figure E3.3 for $c = -1$.

e) Show that (E3.19) has the following fixed points

$$\begin{aligned}&\underline{c = +1:} \qquad (0,0), \quad (\pm\sqrt{-\mu_1}, 0),\\ &\underline{c = -1:} \qquad (0,0), \quad (\pm\sqrt{\mu_1}, 0).\end{aligned}$$

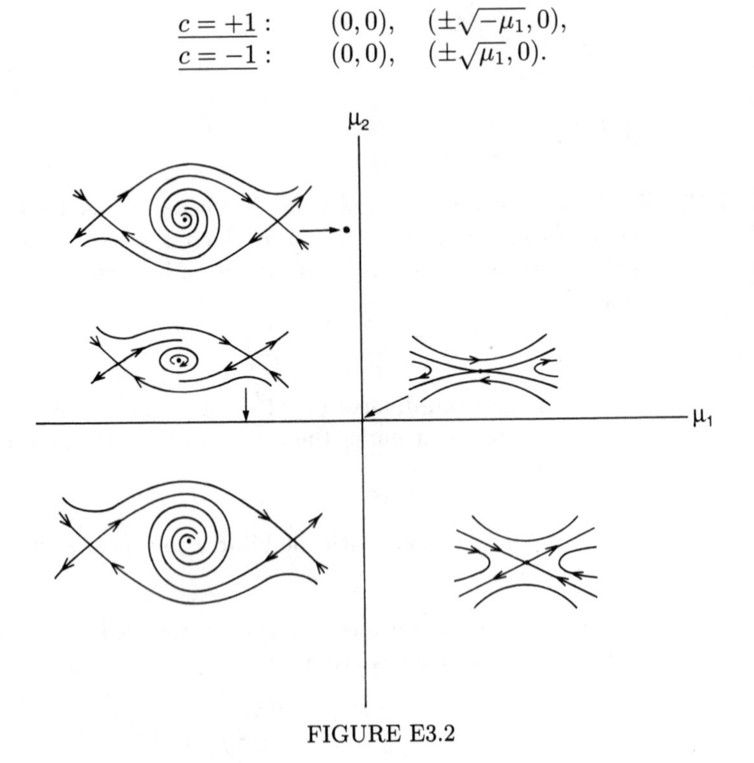

FIGURE E3.2

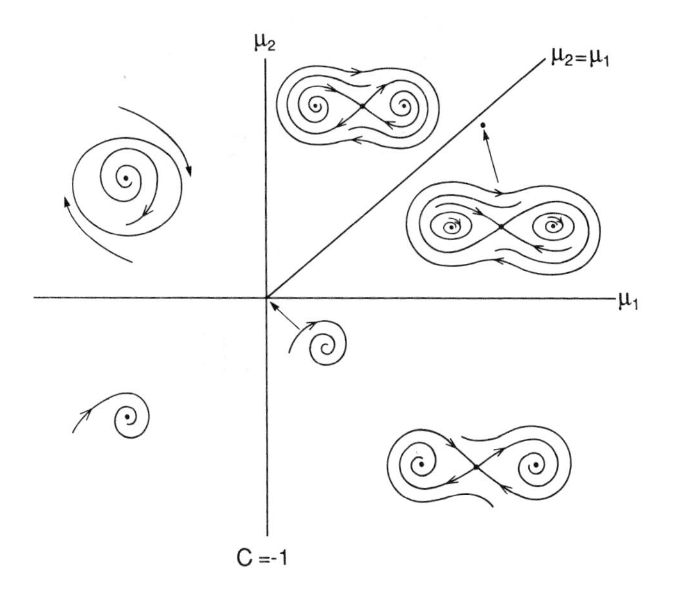

FIGURE E3.3

f) Compute the linearized stability for the fixed points for both $c = +1$ and $c = -1$ and show that the following bifurcations occur.

$\underline{c = +1}$: pitchfork on $\mu_1 = 0$,
supercritical Poincaré–Andronov–Hopf on $\mu_1 < 0,\ \mu_2 = 0$.

$\underline{c = -1}$: pitchfork on $\mu_1 = 0$,
subcritical Poincaré–Andronov–Hopf on $\mu_1 = \mu_2,\ \mu_1 > 0$.

g) Show that (E3.19) has no periodic orbits for

$\underline{c = +1}$: $\mu_1 > 0$;
$\mu_1 < 0,\ \mu_2 < 0$;
$\mu_2 > -\mu_1/5,\ \mu_1 < 0$.

$\underline{c = -1}$: $\mu_2 < 0$.

(*Hint:* use Bendixson's criterion and index theory.)

h) Use the results obtained in d) $\rightarrow$ e) and *completely justify* the local bifurcation diagrams shown in Figure E3.2 for $c = +1$ and in Figure E3.3 for $c = -1$.

i) Based on an examination of Figures E3.2 and E3.3, can you infer the necessity of the existence of global bifurcations? What scenarios are most likely?

We will return to this exercise in Chapter 4 to study possible global bifurcations in more detail.

3.33 Consider a three-dimensional autonomous $\mathbf{C}^r$ (r as large as necessary) vector field having a fixed point where the linear part, in Cartesian coordinates, takes the form

$$\begin{pmatrix} 0 & -\omega & 0 \\ \omega & 0 & 0 \\ 0 & 0 & 0 \end{pmatrix} \begin{pmatrix} x \\ y \\ z \end{pmatrix}.$$

The versal deformation of this nonhyperbolic fixed point was studied in some detail in Section 3.1F. Suppose now we assume that the vector field is unchanged under the coordinate transformation

$$(x, y, z) \mapsto (x, y, -z).$$

a) Show that the normal form in cylindrical coordinates is given by

$$\begin{aligned} \dot{r} &= r(a_1 r^2 + a_2 z^2) + \cdots, \\ \dot{z} &= z(b_1 r^2 + b_2 z^2) + \cdots, \\ \dot{\theta} &= \omega + \cdots. \end{aligned}$$

b) Show that a candidate for a versal deformation is given by

$$\begin{aligned} \dot{r} &= r(\mu_1 + a_1 r^2 + a_2 z^2), \\ \dot{z} &= z(\mu_2 + b_1 r^2 + b_2 z^2), \\ \dot{\theta} &= \omega + \cdots. \end{aligned}$$

c) Following the steps in the analysis of the nonsymmetric case in Section 3.1F, analyze this versal deformation completely, addressing all issues discussed in Section 3.1F.

For an excellent review and bibliography of this nonhyperbolic fixed point with various symmetries see Langford [1985].

3.34 Consider a four-dimensional autonomous $\mathbf{C}^r$ (r as large as necessary) vector field having a nonhyperbolic fixed point at which the linear part has the form

$$\begin{pmatrix} 0 & -\omega_1 & 0 & 0 \\ \omega_1 & 0 & 0 & 0 \\ 0 & 0 & 0 & -\omega_2 \\ 0 & 0 & \omega_2 & 0 \end{pmatrix} \begin{pmatrix} w \\ x \\ y \\ z \end{pmatrix}.$$

a) Suppose $m\omega_1 + n\omega_2 \neq 0$, $|m| + |n| \leq 4$. Then show that in polar coordinates a normal form is given by

$$\begin{aligned}
\dot{r}_1 &= a_1 r_1^3 + a_2 r_1 r_2^2 + \cdots, \\
\dot{r}_2 &= b_1 r_1^2 r_2 + b_2 r_1^3 + \cdots, \\
\dot{\theta}_1 &= \omega_1 + \cdots, \\
\dot{\theta}_2 &= \omega_2 + \cdots.
\end{aligned}$$

(See Exercise 2.10.)

b) Show that a candidate for a versal deformation is given by

$$\begin{aligned}
\dot{r}_1 &= \mu_1 r_1 + a_1 r_1^3 + a_2 r_1 r_2^2, \\
\dot{r}_2 &= \mu_2 r_2 + b_1 r_1^2 r_2 + b_2 r_1^3, \\
\dot{\theta}_1 &= \omega_1 + \cdots, \\
\dot{\theta}_2 &= \omega_2 + \cdots.
\end{aligned}$$

c) Following the steps outlined in Sections 3.1E and 3.1F, analyze this versal deformation completely and address all issues raised in this section. In particular, under what conditions may "three-tori" arise?

d) For each of the resonant cases

$$m\omega_1 + n\omega_2 = 0, \qquad |m| + |n| \leq 4,$$

discuss the codimension of the bifurcation and candidates for versal deformations.

3.35 Consider a two-dimensional autonomous $\mathbf{C}^r$ (r as large as necessary) *Hamiltonian* vector field (i.e., $\dot{x} = \frac{\partial H}{\partial y}(x, y)$, $\dot{y} = -\frac{\partial H}{\partial x}(x, y)$ for some $\mathbf{C}^{r+1}$ scalar function $H(x, y)$) having a nonhyperbolic fixed point at which the linear part takes the form

a) $\begin{pmatrix} 0 & -\omega \\ \omega & 0 \end{pmatrix}$

b) $\begin{pmatrix} 0 & 1 \\ 0 & 0 \end{pmatrix}$

c) $\begin{pmatrix} 0 & 0 \\ 0 & 0 \end{pmatrix}$.

Discuss the codimension of the bifurcation and derive candidates for versal deformations. Analyze each versal deformation completely. Discuss global phenomena that arise and how the Hamiltonian structure can be used for the analysis of global phenomena. What types of symmetry may occur, and how do they affect the situations? (For help, see Golubitsky and Stewart [1987], Galin [1982] and Kocak [1984].)

3.36 Consider the vector field

$$\begin{aligned}\dot{x} &= \varepsilon + x^2 + y^2, \\ \dot{y} &= x^2 - y^2,\end{aligned} \qquad (x, y, \varepsilon) \in \mathbb{R}^3.$$

For this vector field the tangent space approximation is sufficient for approximating the center manifold of the origin. Verify this statement and discuss conditions under which the tangent space approximation might work in general. Consider your ideas in the context of the following examples.

a) $$\begin{aligned}\dot{x} &= \varepsilon x + x^2 + y^2, \\ \dot{y} &= x^2 - y^2.\end{aligned}$$

b) $$\begin{aligned}\dot{x} &= \varepsilon + x^2 + xy, \\ \dot{y} &= x^2 - y^2.\end{aligned}$$

c) $$\begin{aligned}\dot{x} &= \varepsilon + y^2, \\ \dot{y} &= x^2 - y^2.\end{aligned}$$

d) $$\begin{aligned}\dot{x} &= \varepsilon + xy + y^2, \\ \dot{y} &= x^2 - y^2.\end{aligned}$$

3.37 Consider the following $\mathbf{C}^r$ ($r \geq 1$) two-dimensional, time-periodic vector field

$$\begin{aligned}\dot{x} &= y, \\ \dot{y} &= f(x, t),\end{aligned} \qquad (x, y) \in \mathbb{R}^2,$$

where $f(x, t)$ has period T in t.

a) Show that the vector field has a (time-dependent) first integral and that the first integral is actually a Hamiltonian for the system.

b) Suppose that the vector field is (constantly) linearly damped as follows

$$\begin{aligned}\dot{x} &= y, \\ \dot{y} &= -\delta y + f(x, t),\end{aligned} \qquad \delta > 0.$$

Show that the associated Poincaré map *cannot* undergo Naimark–Sacker bifurcations.

3.38 Consider the following ordinary differential equation

$$\begin{aligned}\dot{x} &= \frac{\omega}{\sqrt{3}}(y - z) + [\varepsilon - \mu(x^2 - yz)]x, \\ \dot{y} &= \frac{\omega}{\sqrt{3}}(z - x) + [\varepsilon - \mu(y^2 - xz)]y, \qquad (x, y, z) \in \mathbb{R}^3 \qquad \text{(E.3.20)} \\ \dot{z} &= \frac{\omega}{\sqrt{3}}(x - y) + [\varepsilon - \mu(z^2 - xy)]z,\end{aligned}$$

where $\varepsilon > 0$, $\mu > 0$, and ω are parameters. This system is useful for modeling and simulating synchronous machine systems in the study of power system dynamics; see Kaplan and Yardeni [1989] and Kaplan and Kottick [1983], [1985], [1987].

It should be obvious that

$$(x, y, z) = (0, 0, 0)$$

is a fixed point of (E3.20) for all parameter values. We are interested in studying the bifurcations associated with this fixed point.

a) Show that for $\varepsilon = 0$ the eigenvalues of the matrix associated with the linearized vector field are given by

$$\varepsilon, \ \varepsilon \pm i\omega.$$

b) Study the bifurcations associated with this fixed point for $\varepsilon = 0$, $\omega \neq 0$, and $\varepsilon = 0$, $\omega = 0$.

3.39 Consider the following class of feedback control systems studied by Holmes [1985].

$$\begin{aligned} \ddot{x} + \delta\dot{x} + g(x) &= -z, \\ \dot{z} + \alpha z &= \alpha\gamma(x - r), \end{aligned} \tag{E3.21}$$

where x and $\dot{x}$ represent the displacement and velocity, respectively, of an oscillatory system with nonlinear stiffness $g(x)$ and linear damping $\delta\dot{x}$ subject to negative feedback control z. The controller has first-order dynamics with time constant $\frac{1}{\alpha}$ and gain γ. A constant or time-varying bias r can be applied. This system provides the simplest possible model for a nonlinear elastic system whose position is controlled by a servomechanism with negligible inertia; see Holmes and Moon [1983] for details.

For this exercise we will assume

$$g(x) = x(x^2 - 1)$$

and

$$r = 0.$$

Rewriting (E3.21) as a system gives

$$\begin{aligned} \dot{x} &= y, \\ \dot{y} &= x - x^3 - \delta y - z, \qquad (x, y, z) \in \mathbb{R}^3 \\ \dot{z} &= \alpha\gamma x - \alpha z, \end{aligned} \tag{E3.22}$$

with scalar parameters δ, α, $\gamma > 0$. This exercise is concerned with studying local bifurcations of (E3.22).

a) Show that (E3.22) has fixed points at

$$(x, y, z) = (0, 0, 0) \equiv \underline{0}$$

and

$$(x, y, z) = (\pm\sqrt{1-\gamma}, 0, \pm\gamma\sqrt{1-\gamma}) \equiv \underline{p}_{\pm}, \qquad (\gamma < 1).$$

b) Linearize about these three fixed points and show that (E3.22) has the following *bifurcation surfaces* in (α, δ, γ) space.

$\gamma = 1$	one eigenvalue is zero for $\underline{0}$
$\gamma = \frac{\delta}{\alpha}(\alpha^2 + \alpha\delta - 1)$, $\gamma > 1$	a pair of eigenvalues is pure imaginary for $\underline{0}$
$\gamma = \frac{\delta}{\alpha+3\delta}(\alpha^2 + \alpha\delta + 2)$, $0 < \gamma < 1$	a pair of eigenvalues is pure imaginary for $\underline{p}_{\pm}$.

c) Show that these three surfaces meet on the curve

$$\gamma = 1, \quad \delta = \frac{1}{\alpha},$$

where there is a double-zero eigenvalue with the third eigenvalue being $-(1 + \alpha^2)/\alpha$.

d) Fix $\alpha > 0$ and study the bifurcations from the double-zero eigenvalue in the (δ, γ) plane. (*Hint:* use normal form and center manifold theory. Exercise 3.32 will also be useful.)

e) Describe all attractors as a function of δ and γ. Discuss the implications for the control problem.

We remark that, although this exercise is concerned with local nonlinear analysis, global techniques for studying problems of the form (E3.21) have been developed in Wiggins and Holmes [1987a], [1987b] and Wiggins [1988].

3.40 Consider a map of $\mathbb{R}^2$ having a fixed point at the origin where the eigenvalues associated with the linearized map are complex conjugate and of unit modulus. We denote the two eigenvalues by λ and $\overline{\lambda}$. The goal of this exercise is to study the dynamics near the origin in the cases

$$\lambda^q = 1, \qquad q = 1, 2, 3, 4.$$

We will begin by developing a very powerful local technique; namely, interpolating a map by a flow.

a) Prove the following lemma (Arnold [1983]).

Consider a $\mathbf{C}^r$ *(r as large as necessary) mapping* $f\colon \mathbb{R}^2 \to \mathbb{R}^2$ *having a fixed point at the origin with the eigenvalues of the linearization at the origin given by* $e^{\pm 2\pi i p/q}$ *(and with a Jordan block of order* 2 *if* $q = 1$ *or* 2*). In a sufficiently small neighborhood of the origin the iterate* f^q *can be represented as follows*

$$f^q(z) = \varphi_1(z) + \mathcal{O}(|z|^N), \qquad z \in \mathbb{R}^2,$$

where $\varphi_1(z)$ *is the time one map obtained from the flow generated by a vector field,* $v(z)$*. Moreover, the vector field is invariant under rotations about the origin through the angle* $2\pi/q$*.*

(*Hint:* first put f^q in normal form

$$f^q = \Lambda z + F_2^r(z) + F_3^r(z) + \cdots + F_{N-1}^r(z) + \mathcal{O}(|z|^N).$$

Next consider the vector field

$$\dot{z} = (\Lambda - \mathrm{id})z + F_2^r(z) + F_3^r(z) + \cdots + F_{N-1}^r(z).)$$

Approximate the time one map of this vector field via Picard iteration (justify the use of this method), and show that it gives the desired result. This will show why it is first necessary to put the map in normal form. Moser [1968] gives an alternate proof of this lemma as well as a nice discussion of interpolation of maps by flows.

Next we must deal with versal deformations of these maps.

b) Prove the following lemma (Arnold [1983]).

Consider a deformation f_λ, $\lambda \in \mathbb{R}^p$, *of a mapping* $f_0 = f$ *satisfying the hypotheses of the lemma in Part* a)*. In a sufficiently small neighborhood of the origin, the iterate* f_λ^q *can be represented as*

$$f_\lambda^q(z) = \varphi_{1,\lambda}(z) + \mathcal{O}(|z|^N), \qquad z \in \mathbb{R}^2,$$

where $\varphi_{1,\lambda}(z)$ *is the time one map obtained from the flow generated by a vector field,* v_λ*, that is invariant under rotations about the origin through the angle* $2\pi/q$*. Moreover,* $\varphi_{1,0}(z) = \varphi_1(z)$ *and* $v_0(z) = v(z)$*.*

(*Hint:* this lemma is a consequence of the fact that the normalizing transformations, up to order N, depend differentiably on the parameters.)

c) A vector field on the plane may have

1. Fixed points (hyperbolic and nonhyperbolic).
2. Periodic orbits.
3. Homoclinic orbits.
4. Heteroclinic orbits.

Suppose the time one map of a vector field having all of these orbits approximates the map f^q in the sense of the lemmas in Part a) and b). How would each of these orbits be affected by the higher order terms (i.e., the $\mathcal{O}(|z|^N)$ terms)?

Now we return to the main problem.

d) Show that the normal forms in the cases $\lambda^q = 1$, $q = 1, 2, 3, 4$, are given by

$$q=1: \quad \begin{aligned} x &\mapsto x + y + \cdots, \\ y &\mapsto y + ax^2 + bxy + \cdots, \end{aligned} \qquad (x,y) \in \mathbb{R}^2.$$

$$q=2: \quad \begin{aligned} x &\mapsto x + y + \cdots, \\ y &\mapsto y + ax^3 + bx^2y + \cdots, \end{aligned} \qquad (x,y) \in \mathbb{R}^2.$$

$$q=3: \quad z \mapsto z + c_1\bar{z}^2 + c_2 z^2\bar{z} + \cdots, \qquad z \in \mathbb{C}.$$

$$q=4: \quad z \mapsto z + c_1\bar{z}^3 + c_2 z^2\bar{z} + \cdots, \qquad z \in \mathbb{C}.$$

e) Compute the codimension in each case. Argue that candidates for versal deformations (using the ideas in Section 3.1D) are given by

$$q=1: \quad \begin{aligned} x &\mapsto x + y, \\ y &\mapsto \mu_1 + \mu_2 y + y + ax^2 + bxy, \end{aligned} \qquad (x,y) \in \mathbb{R}^2.$$

$$q=2: \quad \begin{aligned} x &\mapsto x + y, \\ y &\mapsto \mu_1 x + (1+\mu_2)y + ax^3 + bx^2y, \end{aligned} \qquad (x,y) \in \mathbb{R}^2.$$

$$q=3: \quad z \mapsto (1+\mu)z + c_1\bar{z}^2 + c_2 z^2\bar{z}, \qquad z \in \mathbb{C}.$$

$$q=4: \quad z \mapsto (1+\mu)z + c_1\bar{z}^3 + c_2 z^2\bar{z}, \qquad z \in \mathbb{C}.$$

f) Show that the vector fields that interpolate f^q through the order given in Part e) are

$$q=1: \quad \begin{aligned} \dot{x} &= y, \\ \dot{y} &= \mu_1 + \mu_2 y + ax^2 + bxy, \end{aligned} \qquad (x,y) \in \mathbb{R}^2.$$

$$q=2: \quad \begin{aligned} \dot{x} &= y, \\ \dot{y} &= \mu_1 x + \mu_2 y + ax^3 + bx^2y, \end{aligned} \qquad (x,y) \in \mathbb{R}^2.$$

$$q=3: \quad \dot{z} = \mu z + c_1\bar{z}^2 + c_2 z^2\bar{z}, \qquad z \in \mathbb{C}.$$

$$q=4: \quad \dot{z} = \mu z + c_1\bar{z}^3 + c_2 z^2\bar{z}, \qquad z \in \mathbb{C}.$$

g) Describe the complete dynamics of each of the vector fields in Part f).

h) Using the results from Parts g) and c), describe the dynamics of f^q near the origin for $q = 1, 2, 3$, and 4.

We remark that the results of this problem were first obtained by Arnold [1977], [1983]. A very interesting application to a vector field undergoing a Poincaré–Andronov–Hopf bifurcation that is subjected to an external time-periodic perturbation can be found in Gambaudo [1985].

Exercises 3.41 and 3.42 deal with bifurcations and the subharmonic Melnikov theory. The setting is as follows. Suppose (1.2.158) depends on a single scalar parameter μ

$$\begin{aligned} \dot{x} &= \frac{\partial H}{\partial y}(x, y) + \varepsilon g_1(x, y, t; \mu, \varepsilon), \\ \dot{y} &= -\frac{\partial H}{\partial x}(x, y) + \varepsilon g_2(x, y, t; \mu, \varepsilon), \qquad \mu \in \mathbb{R}^1. \end{aligned}$$

Then the mth iterate of the Poincaré map that we derived from (1.2.158) in Section 1.2D, ii) will also depend on μ through the subharmonic Melnikov vector as follows.

$$\begin{aligned} P_\varepsilon^m(I, \theta; \mu) &= (I, \theta + mT\Omega(I)) \\ &\quad + \varepsilon\big(M_1^{m/n}(I, \theta, \varphi_0; \mu), M_2^{m/n}(I, \theta, \varphi_0; \mu)\big) + \mathcal{O}(\varepsilon^2). \end{aligned}$$

3.41 Prove the following *saddle-node* bifurcation theorem.

For the parametrized Poincaré map above, suppose there exists a point $(\overline{I}, \overline{\theta}, \overline{\mu})$ such that $nT(\overline{I}) = mT$ and one of the following conditions hold:

FP1)

i) $\left.\dfrac{\partial \Omega}{\partial I}\right|_{\overline{I}} \neq 0,$

ii) $M_1^{m/n}(\overline{I}, \overline{\theta}, \varphi_0, \overline{\mu}) = 0$

iii) $\dfrac{\partial M_1^{m/n}}{\partial \theta}(\overline{I}, \overline{\theta}, \varphi_0, \overline{\mu}) = 0,$

iv) $\dfrac{\partial M_1^{m/n}}{\partial \mu}(\overline{I}, \overline{\theta}, \varphi_0, \overline{\mu}) \neq 0,$

v) $\dfrac{\partial^2 M_1^{m}}{\partial \theta^2}(\overline{I}, \overline{\theta}, I_0, \overline{\mu}) \neq 0,$

or FP2)

i) $\left.\dfrac{\partial \Omega}{\partial I}\right|_{\overline{I}} = 0,$

ii) $M^{m/n}(\overline{I}, \overline{\theta}, \varphi_0, \overline{\mu}) = 0,$

iii) $$\left(\frac{\partial M_1^{m/n}}{\partial I}\frac{\partial M_2^{m/n}}{\partial \theta} - \frac{\partial M_1^{m/n}}{\partial \theta}\frac{\partial M_2^{m/n}}{\partial I}\right)\Bigg|_{(\overline{I},\overline{\theta},\overline{\mu})}$$
$$\equiv \left.\frac{\partial(M_1^{m/n}, M_2^{m/n})}{\partial(I,\theta)}\right|_{(\overline{I},\overline{\theta},\overline{\mu})} = 0,$$

with one of the following holding

i) $$\left.\frac{\partial(M_1^{m/n}, M_2^{m/n})}{\partial(I,\mu)}\right|_{(\overline{I},\overline{\theta},\overline{\mu})} \neq 0, \quad \frac{d}{d\theta}\left(\frac{\partial(M_1^{m/n}, M_2^{m/n})}{\partial(I,\theta)}\right)\Bigg|_{(\overline{I},\overline{\theta},\overline{\mu})} \neq 0$$

ii) $$\left.\frac{\partial(M_1^{m/n}, M_2^{m/n})}{\partial(\theta,\mu)}\right|_{(\overline{I},\overline{\theta},\overline{\mu})} \neq 0, \quad \frac{d}{dI}\left(\frac{\partial(M_1^{m/n}, M_2^{m/n})}{\partial(I,\theta)}\right)\Bigg|_{(\overline{I},\overline{\theta},\overline{\mu})} \neq 0$$

Then $(\overline{I}, \overline{\theta}, \overline{\mu}) + \mathcal{O}(\varepsilon)$ *is a saddle-node bifurcation point for* P_ε^m.

Hint: for Case FP1, consider ε fixed. Then the equations

$$M_1^{m/n}(I,\theta,\varphi_0,\mu) + \mathcal{O}(\varepsilon) = 0,$$

$$mT\Omega(I) - 2\pi n + \varepsilon M_2^{m/n}(I,\theta,\varphi_0,\mu) + \mathcal{O}(\varepsilon^2) = 0,$$

define a curve of fixed points in (I,θ,μ) space. Seek conditions under which the curve is locally parabolic at $(\overline{I}, \overline{\theta}, \overline{\mu}) + \mathcal{O}(\varepsilon)$ so that it represents a saddle-node bifurcation. Case FP2 is proven similarly.

We remark that Case FP1 is equivalent to Theorem 4.6.3 from Guckenheimer and Holmes [1983].

The next part of the exercise is trivial, merely an exercise in notation.

For FP1, this saddle-node bifurcation theorem can be rewritten in our alternate notation as follows.

Suppose we have $\overline{\alpha} \in [-1, 0)$ such that

$$mT = nT^{\overline{\alpha}}$$

and a point $(\overline{t}_0, \overline{\mu})$ such that

a) $\overline{M}_1^{m/n}(\overline{t}_0, \varphi_0, \overline{\mu}) = 0,$

b) $\dfrac{\partial \overline{M}_1^{m/n}}{\partial t_0}(\bar{t}_0, \varphi_0, \overline{\mu}) = 0,$

c) $\dfrac{\partial \overline{M}_1^{m/n}}{\partial \mu}(\bar{t}_0, \varphi_0, \overline{\mu}) \neq 0,$

d) $\dfrac{\partial^2 \overline{M}_1^{m/n}}{\partial t_0^2}(\bar{t}_0, \varphi_0, \overline{\mu}) \neq 0.$

Then $(\overline{\alpha}, \bar{t}_0, \overline{\mu})$ is a bifurcation point at which saddle nodes occur for P_ε^m.

Describe in detail the translation between the notation.

3.42 Prove the following *Naimark–Sacker* bifurcation theorem.

> *Let $(I(\mu), \theta(\mu), \mu)$ be a smooth curve of fixed points for the parametrized Poincaré map, P_ε^m, where $\mu \in J$ and J is some open interval in $\mathbb{R}$. Suppose there exists some $\overline{\mu} \in J$ such that*

FP1)

i) $$\left.\frac{\partial \Omega}{\partial I}\frac{\partial M_1^{m/n}}{\partial \theta}\right|_{(I(\overline{\mu}),\theta(\overline{\mu}),\overline{\mu})} < 0,$$

ii) $$\left.\left(\frac{\partial M_1^{m/n}}{\partial I} + \frac{\partial M_2^{m/n}}{\partial \theta} - mT\frac{\partial \Omega}{\partial I}\frac{\partial M_1^{m/n}}{\partial \theta}\right)\right|_{(I(\overline{\mu}),\theta(\overline{\mu}),\overline{\mu})} = 0,$$

iii) $$\left.\frac{d}{d\mu}\left[\frac{\partial M_1^{m/n}}{\partial I} + \frac{\partial M_2^{m/n}}{\partial \theta} - mT\frac{\partial \Omega}{\partial I}\frac{\partial M_1^{m/n}}{\partial \theta}\right]\right|_{I(\overline{\mu}),\theta(\overline{\mu}),\overline{\mu})} \neq 0,$$

or FP2)

i) $$\left.\frac{\partial \Omega}{\partial I}\right|_{I(\overline{\mu})} = 0,$$

ii) $$\left.\left(\frac{\partial M_1^{m/n}}{\partial I}\frac{\partial M_2^{m/n}}{\partial \theta} - \frac{\partial M_1^{m/n}}{\partial \theta}\frac{\partial M_2^{m/n}}{\partial I}\right)\right|_{(I(\overline{\mu}),\theta(\overline{\mu}),\overline{\mu})} > 0,$$

iii) $$\left.\left(\frac{\partial M_1^{m/n}}{\partial I} + \frac{\partial M_2^{m/n}}{\partial \theta}\right)\right|_{(I(\overline{\mu}),\theta(\overline{\mu}),\overline{\mu})} = 0,$$

iv) $$\left.\frac{d}{d\mu}\left(\frac{\partial M_1^{m/n}}{\partial I} + \frac{\partial M_2^{m/n}}{\partial \theta}\right)\right|_{I(\overline{\mu}),\theta(\overline{\mu}),\overline{\mu})} \neq 0.$$

Then $\overline{\mu} + \mathcal{O}(\varepsilon)$ is a bifurcation value at which invariant circles for P_ε^m are born.

Hint: show that the three conditions for Case FP1 and the four conditions for FP2 imply that the hypotheses of the Naimark–Sacker bifurcation theorem (Theorem 3.2.3) are satisfied by P_ε^m.

We make the following remark concerning these two theorems. The Case FP1 is defined by $\frac{\partial \Omega}{\partial I} \neq 0$. For saddle-node bifurcations of P_ε^m, if $\frac{\partial \Omega}{\partial I} \neq 0$, one needs information concerning $M_1^{m/n}(I, \theta, \varphi_0; \mu)$ only. However, for the Naimark–Sacker bifurcation, even if $\frac{\partial \Omega}{\partial I} \neq 0$, one needs information concerning both $M_1^{m/n}(I, \theta, \varphi_0; \mu)$ and $M_2^{m/n}(I, \theta, \varphi_0; \mu)$.

3.43 Return to our discussion of the periodically forced, damped Duffing oscillator and discuss the types of bifurcations that occur on the resonance bands as γ and δ is varied. What role does ω play? Can this system undergo Naimark–Sacker bifurcations?

3.44 Consider the following $\mathbf{C}^r$ (r as large as necessary) vector field

$$\dot{x} = \varepsilon f(x, t; \mu) + \varepsilon^2 g(x, t; \mu), \quad x \in \mathbb{R}^n, \quad f, g \; T\text{-periodic in } t \tag{E3.23}$$

and the associated averaged equation

$$\dot{y} = \varepsilon \overline{f}(y, \mu), \qquad \overline{f} = \frac{1}{T}\int_0^T f(y, t; \mu)dt \tag{E3.24}$$

where $\mu \in \mathbb{R}^1$ is a parameter.

Prove the following theorem from Guckenheimer and Holmes [1983].

Theorem: *If at $\mu = \mu_0$* (E3.24) *undergoes a saddle-node or Poincaré–Andronov–Hopf bifurcation, then, for μ near μ_0 and ε sufficiently small, the Poincaré map of* (E3.23) *undergoes a saddle-node or Naimark–Sacker bifurcation.*

(*Hint:* use the implicit function theorem.)

What can you say about the dynamics on the invariant circle that are created in the Naimark–Sacker bifurcation?

3.45 Recall the averaged equations for the forced van der Pol oscillator derived in Exercise 1.2.17

$$\begin{aligned} \dot{u} &= u - \sigma v - u(u^2 + v^2), \\ \dot{v} &= \sigma u + v - v(u^2 + v^2) - \gamma. \end{aligned} \tag{E3.25}$$

Consider the bifurcation diagram in Figure E3.4.

The object of this exercise is to derive the bifurcation diagram.

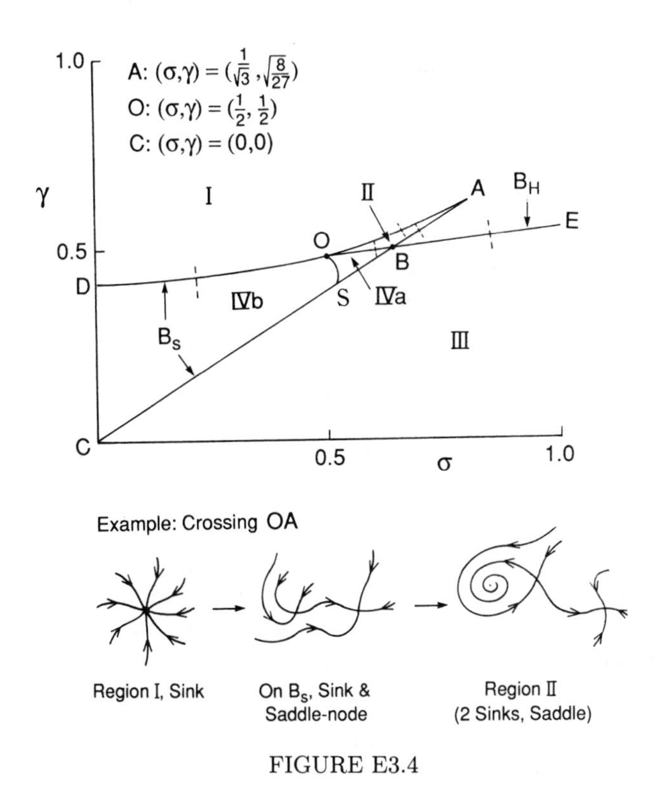

FIGURE E3.4

a) Show that (E3.25) has a single fixed point in regions I and III (a sink in I, a source in III). Show that in region II there are two sinks and a saddle, and in region IVa ∪ IVb there is a sink, a saddle, and a source.

b) Show that (E3.25) undergoes a saddle-node bifurcation on

$$\frac{\gamma^4}{4} - \frac{\gamma^2}{27}(1 + 9\sigma^2) + \frac{\sigma^2}{27}(1 + \sigma^2)^2 = 0.$$

This is the curve DAC marked B_S in Figure E3.4.

c) Show that (E3.25) undergoes a Poincaré–Andronov–Hopf bifurcation on

$$8\gamma^2 = 4\sigma^2 + 1, \qquad |\sigma| > \frac{1}{2}.$$

This is the curve OE marked B_H in Figure E3.4.

d) In Figure E3.4, consider the broken lines − − − crossing the curves OA, OD, AB, BE, and OB. Draw phase portraits representing the flow on and to each side of the indicated curve; see the example in Figure E3.4.

e) OS is a curve on which homoclinic orbits occur (sometimes called saddle connections). Give an intuitive argument as to why such a curve should exist. (Is it obvious that it should be a smooth curve?)

f) Discuss the nature of (E3.25) near the points A, O, and C.

g) (E3.25) is an autonomous equation whose flow gives an approximation to the Poincaré map of the original forced van der Pol equation in a sense made precise by the averaging theorem. Using the previously obtained results, interpret Parts i) → vi) in terms of the dynamics of the original forced van der Pol equation. In particular, list the structurally stable motions and bifurcations along with the structurally unstable bifurcations.

If you need help you may consult Holmes and Rand [1978], where these results were first worked out.

3.46 Return to Exercise 1.2.18 (see also Exercises 1.2.36 and 1.2.37) and discuss the bifurcations associated with the passage through the $1:1$ and $1:3$ resonances in the periodically forced, damped Duffing oscillator. You can use either the method of averaging or the subharmonic Melnikov theory. See, e.g., Holmes and Holmes [1981] or Morozov and Silnikov [1984] for help.

3.47 The purpose of this exercise is to give some idea of how the neglected higher order terms in the normal form can affect the dynamics on the invariant circle arising in a Naimark–Sacker bifurcation.

Consider the two-parameter family of maps

$$x \mapsto x + \mu + \varepsilon \cos 2\pi x \equiv f(x, \mu, \varepsilon), \qquad x \in \mathbb{R}^1, \quad \varepsilon \geq 0, \quad \text{(E3.26)}$$

where we identify points in $\mathbb{R}_1$ that differ by an integer so that (E3.26) can be regarded as a map defined on the circle $S^1 = \mathbb{R}^1/\mathbb{Z}$.

a) Discuss the orbit structure of (E3.26) for $\varepsilon = 0$. In particular, what is the Lebesgue measure of the set of parameter values for which (E3.26) has periodic orbits?

b) Consider the following regions in the $\mu - \varepsilon$ plane (see Figure E3.5).

$$\begin{aligned}
\mu &= 1 \pm \varepsilon, \\
\mu &= \frac{1}{2} \pm \varepsilon^2 \frac{\pi}{2} + \mathcal{O}(\varepsilon^3), \\
\mu &= \frac{1}{3} + \varepsilon^2 \frac{\sqrt{3}}{6}\pi \pm \varepsilon^3 \frac{\sqrt{7}\pi}{6} + \mathcal{O}(\varepsilon^4).
\end{aligned}$$

Show that, for parameter values in the interior of these regions, (E3.26) has a period 1, period 2, and period 3 point, respectively.

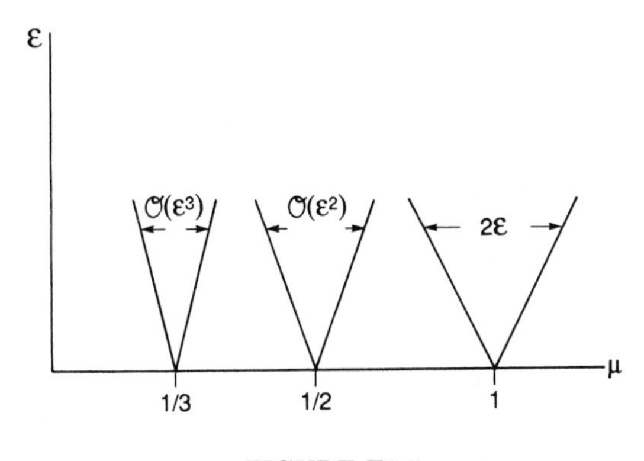

FIGURE E3.5

Hint: we outline the procedure for the period 2 points.

1. If x is a period 2 point of (E3.26), then

$$f^2(x, \mu, \varepsilon) - x - 1 = G(x, \mu, \varepsilon) = 0.$$

2. If $\frac{\partial G}{\partial \mu} \neq 0$, then we have a function $\mu = \mu(x, \varepsilon)$ such that

$$G(x, \mu(x, \varepsilon), \varepsilon) = 0. \tag{E3.27}$$

3. Expand the function $\mu(x, \varepsilon)$ as follows

$$\mu(x, \varepsilon) = \mu(x, 0) + \varepsilon \frac{\partial \mu}{\partial \varepsilon}(x, 0) + \frac{\varepsilon^2}{2} \frac{\partial^2 \mu}{\partial \varepsilon^2}(x, 0) + \mathcal{O}(\varepsilon^3).$$

4. Implicitly differentiating (E3.27), show that

$$\begin{aligned} \mu(x, 0) &= \frac{1}{2}, \\ \frac{\partial \mu}{\partial \varepsilon}(x, 0) &= 0, \\ \frac{\partial^2 \mu}{\partial \varepsilon^2}(x, 0) &= 2\pi \sin 4\pi x. \end{aligned}$$

5. Taking the infimum and supremum in Step 4, we obtain

$$\mu = \mu(x, \varepsilon) = \frac{1}{2} \pm \varepsilon^2 \frac{\pi}{2} + \mathcal{O}(\varepsilon^3).$$

Justify all steps completely.

These regions in the $\mu - \varepsilon$ plane are called *Arnold tongues.* It can actually be shown that, given any rational number p/q, there exists an Arnold tongue given by $\mu = \frac{p}{q} + \cdots$. These results follow from the general theory of circle maps; see Devaney [1986] for an excellent introduction.

Now let us return to the setting of the Naimark–Sacker bifurcation. For $\varepsilon = 0$, (E3.26) has the form of the truncated Naimark–Sacker normal form restricted to the invariant circle. The term $\varepsilon \cos 2\pi x$ could be viewed as illustrating the possible effects of higher order terms in the normal form. For (E3.26), at $\varepsilon = 0$ the map has periodic orbits for all rational μ (i.e., a set of Lebesgue measure zero). For ε small and fixed, our results show that the measure of the set of μ values for which (E3.26) has a periodic orbit is positive. Thus, based on this example, we might expect the higher order terms in the Naimark–Sacker normal form to have a dramatic influence on the dynamics restricted to the bifurcated invariant circle. See Iooss [1979] for more details.

3.48 Consider the following partial differential equation known as the *complex Ginzbury–Landau* (CGL) *equation*

$$iA_t + \hat{\alpha} A_{xx} = \hat{\beta} A - \hat{\gamma} |A|^2 A, \tag{E3.28}$$

where $(x, t) \in \mathbb{R}^1 \times \mathbb{R}^1$, $A(x, t)$ is complex and $\hat{\alpha} = \alpha_R + i\alpha_I$, $\hat{\beta} = \beta_R + i\beta_I$, and $\gamma = \gamma_R + i\gamma_I$ are complex numbers.

If we set $\alpha_I = 0$, $\gamma_I = 0$, and $\hat{\beta} = 0$, (E3.28) reduces to

$$iA_t + \alpha_R A_{xx} = -\gamma_R |A|^2 A, \tag{E3.29}$$

which is a famous *completely integrable* partial differential equation known as the *nonlinear Schrödinger* (NLS) *equation.* We refer the reader to Newell [1985] for background material and a discussion of the physical circumstances in which (E3.28) and (E3.29) arise. We will comment on this in more detail at the end of this exercise.

a) Show that (E3.28) is invariant under translations in space and time, i.e., under the transformation

$$(x, t) \mapsto (x + x_0, t + t_0).$$

Show also that (E3.28) is invariant under multiplication by a complex number of unit modulus, i.e., under the transformation

$$A \mapsto Ae^{i\psi_0}.$$

Our goal in this exercise will be to study solutions of (E3.28) that have the form

$$A(x, t) = a(x)e^{i\omega t}. \tag{E3.30}$$

b) Substitute (E3.30) into (E3.28) and show that $a(x)$ satisfies the following *complex Duffing equation*

$$a'' - (\alpha + i\beta)a + (\gamma + i\delta)|a|^2 a = 0, \tag{E3.31}$$

where

$$\begin{aligned}
\alpha &= [\alpha_R(\omega + \beta_R + \alpha_I\beta_I]/\Delta, \\
\beta &= [\alpha_R\beta_I - \alpha_I(\omega + \beta_R)]/\Delta, \\
\gamma &= [\alpha_R\gamma_R + \alpha_I\gamma_I]/\Delta, \\
\delta &= [\alpha_R\gamma_I - \alpha_I\gamma_R]/\Delta,
\end{aligned}$$

and

$$\Delta = \alpha_R^2 + \alpha_I^2.$$

c) Letting $a = b + ic$, show that (E3.31) can be written as

$$\begin{aligned}
b' &= d, \\
d' &= \alpha b - \beta c - (\gamma b - \delta c)(b^2 + c^2), \\
c' &= e, \\
e' &= \beta b + \alpha c - (\delta b + \gamma c)(b^2 + c^2).
\end{aligned} \tag{E3.32}$$

d) Show that, for $\beta = \delta = 0$, (E3.32) is a completely integrable Hamiltonian system with integrals

$$\begin{aligned}
H &= \frac{d^2 + e^2}{2} - \frac{\alpha}{2}(a^2 + b^2) + \frac{\gamma}{4}(a^2 + b^2)^2, \\
m &= be - cd.
\end{aligned}$$

e) Using the transformation

$$a = \rho e^{i\varphi},$$

show that (E3.31) can be written in the form

$$\begin{aligned}
\rho'' - \rho(\varphi')^2 &= \alpha\rho - \gamma\rho^3, \\
(\rho^2\varphi')' &= (\beta - \delta\rho^2)\rho^2.
\end{aligned} \tag{E3.33}$$

f) Let

$$\begin{aligned}
r &= \rho^2, \\
v &= \rho'/\rho,
\end{aligned}$$

and

$$m = \rho^2\varphi',$$

and show that (E3.33) can be written in the form

$$\begin{aligned}
r' &= 2rv, \\
v' &= \frac{m^2}{r^2} - v^2 + \alpha - \gamma r, \\
m' &= (\beta - \delta r)r.
\end{aligned} \tag{E3.34}$$

g) For $\beta = \delta = 0$, show that (E3.34) has the form of a one-parameter family (with m playing the role of the parameter) of two-dimensional Hamiltonian systems with Hamiltonian function

$$H(r, v; m) = rv^2 + \frac{m^2}{r} - \alpha r + \frac{\gamma}{2}r^2. \tag{E3.35}$$

h) Using (E3.35), give a complete description of the orbit structure of (E3.34) for $\beta = \delta = 0$.

i) Consider the symmetries of the CGL equation described in a). Discuss how these symmetries are manifested in (E3.31), (E3.32), (E3.33), and (E3.34).

j) For $\beta\gamma = \alpha\delta$, the point

$$(r, v, m) = \left(\frac{\alpha}{\gamma} = \frac{\beta}{\delta}, 0, 0\right)$$

is a fixed point of (E3.34) where the eigenvalues of the matrix associated with the linearization are given by

$$0, \pm i\sqrt{2\alpha}.$$

Study the bifurcations associated with this fixed point (take $\alpha > 0$).

k) For β and δ small, study the bifurcations of large amplitude (i.e., away from fixed points) periodic orbits by applying the method of reduction discussed in Exercise 1.2.32 and the subharmonic Melnikov theory.

l) Discuss the implications of the results obtained concerning the dynamics of the ordinary differential equations for the spatial and temporal structure of solutions to the CGL equation.

m) In our original discussion of the CGL and NLS equations, we did not mention initial or boundary conditions. Discuss this issue in the context of the solutions we found.

The CGL equation is a fundamental equation that arises in a variety of physical situations. See Newell [1985], where it is derived in the context of nonlinear waves, and Landman [1987], where it is used to understand the transition to turbulence in Poiseuille flow. Most of this exercise is based on results in Holmes [1986b]; see also Holmes and Wood [1985] and Newton and Sirovich [1986a,b].

3.49 Consider the following $\mathbf{C}^r$ (r as large as necessary) one-parameter family of vector fields

$$\dot{x} = f(x, \mu), \qquad (x, \mu) \in \mathbb{R}^n \times \mathbb{R}^1.$$

Suppose this vector field has a fixed point at $(x, \mu) = (0, 0)$.

a) Suppose $n = 1$; can $(x, \mu) = (0, 0)$ undergo a period-doubling bifurcation?

b) Suppose $n = 2$; can $(x, \mu) = (0, 0)$ undergo a period-doubling bifurcation?

c) Suppose $n = 3$; can $(x, \mu) = (0, 0)$ undergo a period-doubling bifurcation?

(*Hint:* consider a linear vector field

$$\dot{x} = Ax, \qquad x \in \mathbb{R}^n,$$

where the flow is given by

$$x = e^{At}x_0,$$

and $\det e^{At} > 0$ for finite t. Use these facts.)

3.50 In our development of the transcritical and pitchfork bifurcations we assumed that $x = 0$ was a trivial solution. Was this necessary? In particular, would the conditions for transcritical and pitchfork bifurcations change if this were not the case?

4

Some Aspects of Global Bifurcations and Chaos

In this chapter we will develop some techniques for describing what is meant by the term "chaos" as applied to *deterministic* dynamical systems. We will study the mechanisms that give rise to chaotic dynamics as well as develop analytical techniques for predicting (in terms of the system parameters) when these mechanics occur in specific dynamical systems.

4.1 The Smale Horseshoe

We will begin our study of "chaotic dynamics" by describing and analyzing a two-dimensional map possessing an invariant set having a delightfully complicated structure. The discussion is virtually the same as the discussion in Wiggins [1988]. Our map is a simplified version of a map first studied by Smale [1963], [1980] and, due to the shape of the image of the domain of the map, is called a *Smale horseshoe.*

We will see that the Smale horseshoe is the prototypical map possessing a chaotic invariant set (note: the phrase "chaotic invariant set" will be precisely defined later on in the discussion). Therefore, we feel that a thorough understanding of the Smale horseshoe is absolutely essential for understanding what is meant by the term "chaos" as it is applied to the dynamics of specific physical systems. For this reason we will first endeavor to define as simple a two-dimensional map as possible that still contains the necessary ingredients for possessing a complicated and chaotic dynamical structure so that the reader may get a feel for what is going on in the map with a minimum of distractions. As a result, our construction may not appeal to those interested in applications, since it may appear rather artificial. However, following our discussion of the simplified Smale horseshoe map, we will give sufficient conditions for the existence of Smale horseshoe–like dynamics in two-dimensional maps that are of a very general nature. We will begin by defining the map and then proceed to a geometrical construction of the invariant set of the map. We will utilize the nature of the geometrical construction in such a way as to motivate a description of the dynamics of the map on its invariant set by symbolic dynamics, following which we will make precise the idea of chaotic dynamics.

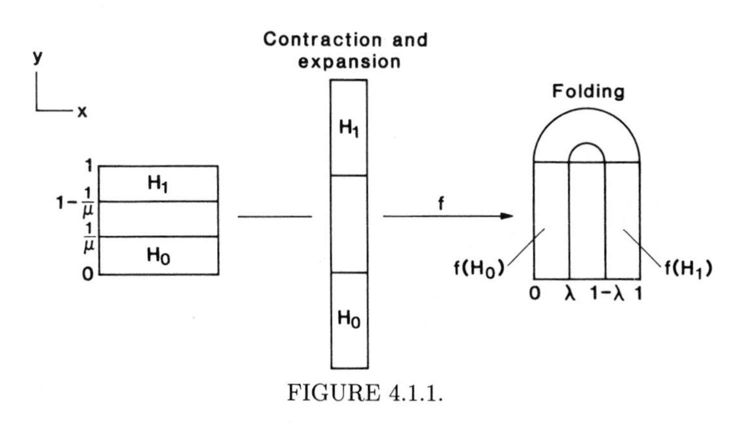

FIGURE 4.1.1.

4.1A Definition of the Smale Horseshoe Map

We will give a combination geometrical-analytical definition of the map. Consider a map, f, from the square having sides of unit length into $\mathbb{R}^2$

$$f : D \to \mathbb{R}^2, \qquad D = \{(x,y) \in \mathbb{R}^2 \,|\, 0 \le x \le 1, 0 \le y \le 1\}, \tag{4.1.1}$$

which contracts the x-direction, expands the y-direction, and folds D around, laying it back on itself as shown in Figure 4.1.1.

We will assume that f acts affinely on the "horizontal" rectangles

$$H_0 = \{(x,y) \in \mathbb{R}^2 \,|\, 0 \le x \le 1, 0 \le y \le 1/\mu\} \tag{4.1.2a}$$

and

$$H_1 = \{(x,y) \in \mathbb{R}^2 \,|\, 0 \le x \le 1, 1 - 1/\mu \le y \le 1\}, \tag{4.1.2b}$$

taking them to the "vertical" rectangles

$$f(H_0) \equiv V_0 = \{(x,y) \in \mathbb{R}^2 \,|\, 0 \le x \le \lambda, 0 \le y \le 1\} \tag{4.1.3a}$$

and

$$f(H_1) \equiv V_1 = \{(x,y) \in \mathbb{R}^2 \,|\, 1 - \lambda \le x \le 1, 0 \le y \le 1\}, \tag{4.1.3b}$$

with the form of f on H_0 and H_1 given by

$$\begin{aligned} H_0 &: \begin{pmatrix} x \\ y \end{pmatrix} \mapsto \begin{pmatrix} \lambda & 0 \\ 0 & \mu \end{pmatrix} \begin{pmatrix} x \\ y \end{pmatrix}, \\ H_1 &: \begin{pmatrix} x \\ y \end{pmatrix} \mapsto \begin{pmatrix} -\lambda & 0 \\ 0 & -\mu \end{pmatrix} \begin{pmatrix} x \\ y \end{pmatrix} + \begin{pmatrix} 1 \\ \mu \end{pmatrix}, \end{aligned} \tag{4.1.4}$$

and with $0 < \lambda < 1/2$, $\mu > 2$ (note: the fact that, on H_1, the matrix elements are negative means that, in addition to being contracted in the x-direction by a factor λ and expanded in the y-direction by a factor μ, H_1 is also rotated 180°). Additionally, it follows that f^{-1} acts on D as shown in

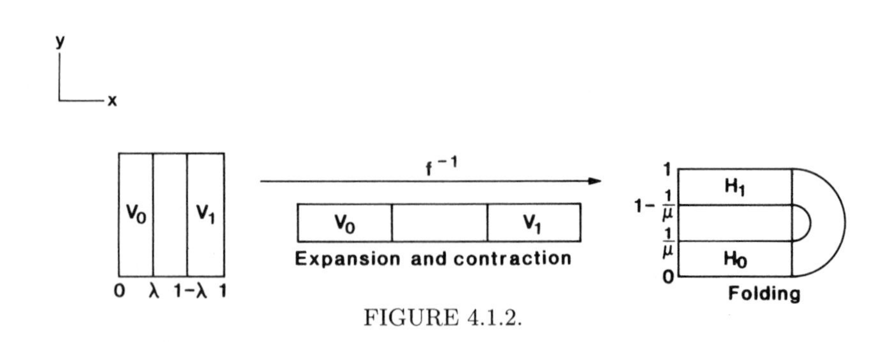

FIGURE 4.1.2.

Figure 4.1.2, taking the "vertical" rectangles V_0 and V_1 to the "horizontal" rectangles H_0 and H_1, respectively (note: by "vertical rectangle" we will mean a rectangle in D whose sides parallel to the y axis each have length one, and by "horizontal rectangle" we will mean a rectangle in D whose sides parallel to the x axis each have length one). This serves to define f; however, before proceeding to study the dynamics of f on D, there is a consequence of the definition of f which we want to single out, since it will be very important later.

Lemma 4.1.1 a) *Suppose V is a vertical rectangle; then $f(V) \cap D$ consists of precisely two vertical rectangles, one in V_0 and one in V_1, with their widths each being equal to a factor of λ times the width of V.* b) *Suppose H is a horizontal rectangle; then $f^{-1}(H) \cap D$ consists of precisely two horizontal rectangles, one in H_0 and one in H_1, with their widths being a factor of $1/\mu$ times the width of H.*

Proof: We will prove Case a). Note that from the definition of f, the horizontal and vertical boundaries of H_0 and H_1 are mapped to the horizontal and vertical boundaries of V_0 and V_1, respectively. Let V be a vertical rectangle; then V intersects the horizontal boundaries of H_0 and H_1, and hence, $f(V) \cap D$ consists of two vertical rectangles, one in H_0 and one in H_1. The contraction of the width follows from the form of f on H_0 and H_1, which indicates that the x-direction is contracted uniformly by a factor λ on H_0 and H_1. Case b) is proved similarly. See Figure 4.1.3. □

We make the following remarks concerning this lemma.

Remark 1. The qualitative features of Lemma 4.1.1 are independent of the particular analytical form for f given in (4.1.4); rather, they are more geometrical in nature. This will be important in generalizing the results of this section to arbitrary maps.

Remark 2. Lemma 4.1.1 is concerned only with the behavior of f and f^{-1}. However, we will see in the construction of the invariant set that the behavior described in Lemma 4.1.1 allows us to understand the behavior of f^n for all n.

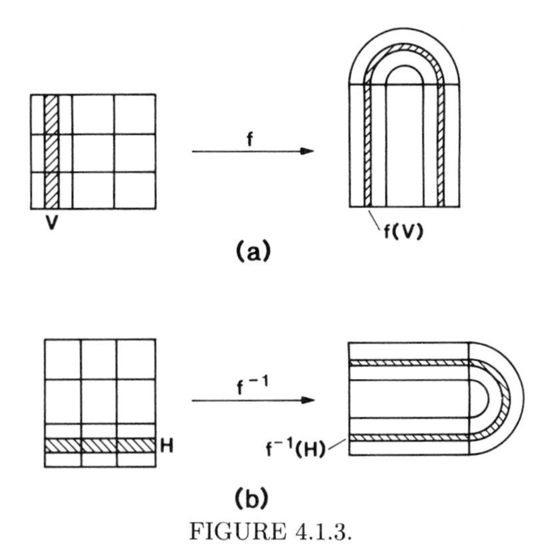

FIGURE 4.1.3.

We now turn to the construction of the invariant set for f.

4.1B Construction of the Invariant Set

We now will geometrically construct the set of points, Λ, which remain in D under all possible iterations by f; thus Λ is defined as

$$\cdots \cap f^{-n}(D) \cap \cdots \cap f^{-1}(D) \cap D \cap f(D) \cap \cdots \cap f^{n}(D) \cap \cdots$$

or

$$\bigcap_{n=-\infty}^{\infty} f^n(D).$$

We will construct this set inductively, and it will be convenient to construct separately the "halves" of Λ corresponding to the positive iterates and the negative iterates and then take their intersections to obtain Λ. Before proceeding with the construction, we need some notation in order to keep track of the iterates of f at each step of the inductive process. Let $S = \{0, 1\}$ be an index set, and let s_i denote one of the two elements of S, i.e., $s_i \in S$, $i = 0, \pm 1, \pm 2, \cdots$ (note: the reason for this notation will become apparent later on).

We will construct $\bigcap_{n=0}^{\infty} f^n(D)$ by constructing $\bigcap_{n=0}^{n=k} f^n(D)$ and then determining the nature of the limit as $k \to \infty$.

$\underline{D \cap f(D)}$. By the definition of f, $D \cap f(D)$ consists of the two vertical rectangles V_0 and V_1, which we denote as follows

$$D \cap f(D) = \bigcup_{s_{-1} \in S} V_{s_{-1}} = \{p \in D \,|\, p \in V_{s_{-1}},\ s_{-1} \in S\}, \qquad (4.1.5)$$

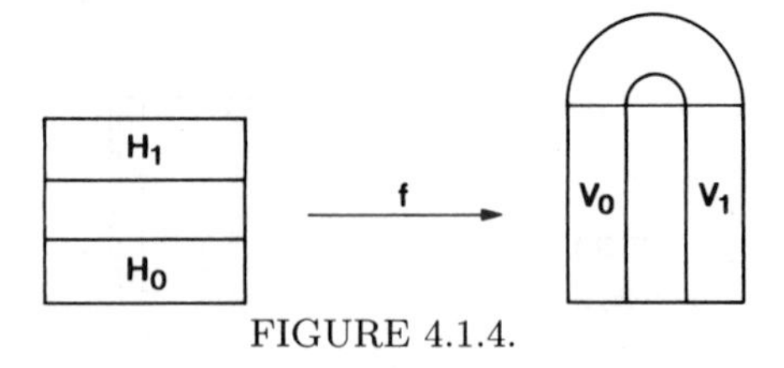

FIGURE 4.1.4.

where $V_{s_{-1}}$ is a vertical rectangle of width λ; see Figure 4.1.4.

$\underline{D \cap f(D) \cap f^2(D)}$. It is easy to see that this set is obtained by acting on $D \cap f(D)$ with f and taking the intersection with D, since $D \cap f(D \cap f(D)) = D \cap f(D) \cap f^2(D)$. Thus, by Lemma 4.1.1, since $D \cap f(D)$ consists of the vertical rectangles V_0 and V_1 with each intersecting H_0 and H_1 and their respective horizontal boundaries in two components, then $D \cap f(D) \cap f^2(D)$ corresponds to four vertical rectangles, two each in V_0 and V_1, with each of width λ^2. Let us write this out more explicitly. Using (4.1.5) we have

$$D \cap f(D) \cap f^2(D) = D \cap f(D \cap f(D)) = D \cap f\Bigl(\bigcup_{s_{-2}\in S} V_{s_{-2}}\Bigr), \tag{4.1.6}$$

where, in substituting (4.1.5) into (4.1.6), we have changed the subscript s_{-1} on $V_{s_{-1}}$ to $V_{s_{-2}}$. As we will see, this is a notational convenience which will be a counting aid. It should be clear that this causes no problems, since s_{-i} is merely a dummy variable. Using a few set-theoretic manipulations, (4.1.6) becomes

$$D \cap f\Bigl(\bigcup_{s_{-2}\in S} V_{s_{-2}}\Bigr) = \bigcup_{s_{-2}\in S} D \cap f(V_{s_{-2}}). \tag{4.1.7}$$

Now, from Lemma 4.1.1, $f(V_{s_{-2}})$ cannot intersect all of D but only $V_0 \cup V_1$, so (4.1.7) becomes

$$\bigcup_{s_{-2}\in S} D \cap f(V_{s_{-2}}) = \bigcup_{\substack{s_{-i}\in S\\ i=1,2}} V_{s_{-1}} \cap f(V_{s_{-2}}). \tag{4.1.8}$$

Putting this all together, we have shown that

$$\begin{aligned} D &\cap f(D) \cap f^2(D) \\ &= \bigcup_{\substack{s_{-i}\in S\\ i=1,2}} (f(V_{s_{-2}}) \cap V_{s_{-1}}) \equiv \bigcup_{\substack{s_{-i}\in S\\ i=1,2}} V_{s_{-1}s_{-2}} \\ &= \bigl\{p \in D \,|\, p \in V_{s_{-1}}, f^{-1}(p) \in V_{s_{-2}}, s_{-i} \in S, i = 1,2\bigr\}. \end{aligned} \tag{4.1.9}$$

Pictorially, this set is described in Figure 4.1.5.

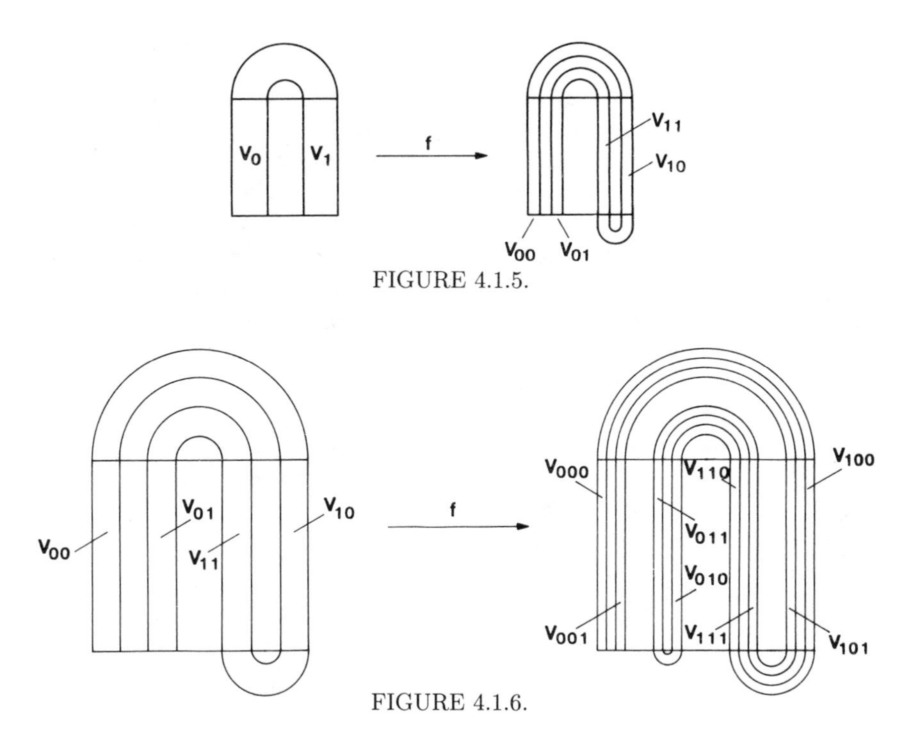

FIGURE 4.1.5.

FIGURE 4.1.6.

$D \cap f(D) \cap f^2(D) \cap f^3(D)$. Using the same reasoning as in the previous steps, this set consists of eight vertical rectangles, each having width λ^3, which we denote as follows

$$\begin{aligned} D &\cap f(D) \cap f^2(D) \cap f^3(D) \\ &= \bigcup_{\substack{s_{-1}\in S\\ i=1,2,3}} (f(V_{s_{-2}s_{-3}}) \cap V_{s_{-1}}) \equiv \bigcup_{\substack{s_{-i}\in S\\ i=1,2,3}} V_{s_{-1}s_{-2}s_{-3}} \\ &= \{p \in D \,|\, p \in V_{s-1}, f^{-1}(p) \in V_{s_{-2}}, \\ &\qquad f^{-2}(p) \in V_{s_{-3}}, s_{-i} \in S, i = 1,2,3\}, \end{aligned} \tag{4.1.10}$$

and is represented pictorially in Figure 4.1.6.

If we continually repeat this procedure, we almost immediately encounter extreme difficulty in trying to represent this process pictorially, as in Figures 4.1.4 through 4.1.6. However, using Lemma 4.1.1 and our labeling scheme developed above, it is not hard to see that at the kth step we obtain

$$\begin{aligned} D &\cap f(D) \cap \cdots \cap f^k(D) \\ &= \bigcup_{\substack{s_{-i}\in S\\ i=1,2,\ldots,k}} (f(V_{s_{-2}\cdots s_{-k}}) \cap V_{s_{-1}}) \equiv \bigcup_{\substack{s_{-i}\in S\\ i=1,2,\ldots,k}} V_{s_{-1}\cdots s_{-k}} \end{aligned}$$

$$= \{p \in D \,|\, f^{-i+1}(p) \in V_{s_{-i}}, s_{-i} \in S, i = 1, \cdots, k\} \qquad (4.1.11)$$

and that this set consists of 2^k vertical rectangles, each of width λ^k.

Before proceeding to discuss the limit as $k \to \infty$, we want to make the following important observation concerning the nature of this construction process. Note that at the kth stage, we obtain 2^k vertical rectangles, and that each vertical rectangle can be labeled by a sequence of 0's and 1's of length k. The important point to realize is that there are 2^k possible distinct sequences of 0's and 1's having length k and that each of these is realized in our construction process; thus, the labeling of each vertical rectangle is unique at each step. This fact follows from the geometric definition of f and the fact that V_0 and V_1 are disjoint.

Letting $k \to \infty$, since a decreasing intersection of compact sets is non-empty, it is clear that we obtain an infinite number of vertical rectangles and that the width of each of these rectangles is zero, since $\lim_{k\to\infty} \lambda^k = 0$ for $0 < \lambda < 1/2$. Thus, we have shown that

$$\begin{aligned}\bigcap_{n=0}^{\infty} f^n(D) &= \bigcup_{\substack{s_{-i}\in S\\ i=1,2,\ldots}} \left(f(V_{s_{-2}\cdots s_{-k}\cdots}) \cap V_{s_{-1}}\right)\\ &\equiv \bigcup_{\substack{s_{-i}\in S\\ i=1,2,\ldots}} V_{s_{-1}\cdots s_{-k}\cdots}\\ &= \{p \in D \,|\, f^{-i+1}(p) \in V_{s_{-i}}, s_{-i} \in S, i = 1, 2, \cdots\}\end{aligned} \qquad (4.1.12)$$

consists of an infinite number of vertical lines and that each line can be labeled by a unique infinite sequence of 0's and 1's (note: we will give a more detailed set-theoretic description of $\cap_{n=0}^{\infty} f^n(D)$ later on).

Next we will construct $\bigcap_{-\infty}^{n=0} f^n(D)$ inductively.

$\underline{D \cap f^{-1}(D)}$. From the definition of f, this set consists of the two horizontal rectangles H_0 and H_1 and is denoted as follows

$$D \cap f^{-1}(D) = \bigcup_{s_0\in S} H_{s_0} = \{p \in D \,|\, p \in H_{s_0}, s_0 \in S\}. \qquad (4.1.13)$$

See Figure 4.1.7.

$\underline{D \cap f^{-1}(D) \cap f^{-2}(D)}$. We obtain this set from the previously constructed set, $D \cap f^{-1}(D)$, by acting on $D \cap f^{-1}(D)$ with f^{-1} and taking the intersection with D, since $D \cap f^{-1}(D \cap f^{-1}(D)) = D \cap f^{-1}(D) \cap f^{-2}(D)$. Also, by Lemma 4.1.1, since H_0 intersects both vertical boundaries of V_0 and V_1, as does H_1, $D \cap f^{-1}(D) \cap f^{-2}(D)$ consists of four horizontal rectangles,

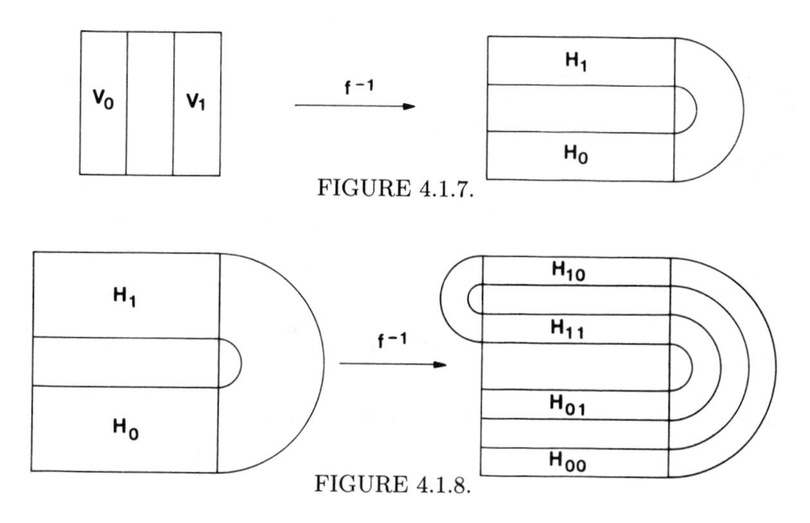

FIGURE 4.1.7.

FIGURE 4.1.8.

each of width $1/\mu^2$. Let us write this out more explicitly. Using (4.1.13), we have

$$\begin{aligned} D \cap f^{-1}(D \cap f^{-1}(D)) &= D \cap f^{-1}\Big(\bigcup_{s_1 \in S} H_{s_1}\Big) \\ &= \bigcup_{s_1 \in S} D \cap f^{-1}(H_{s_1}), \end{aligned} \tag{4.1.14}$$

where in substituting (4.1.13) into (4.1.14) we have changed the subscript s_0 on H_{s_0} to s_1. This has no real effect, since s_i is simply a dummy variable. The reason for doing so is that it will provide a useful counting aid.

From Lemma 4.1.1, it follows that $f^{-1}(H_{s_1})$ cannot intersect all of D, only $H_0 \cup H_1$, so that (4.1.14) becomes

$$\bigcup_{s_1 \in S} D \cap f^{-1}(H_{s_1}) = \bigcup_{\substack{s_i \in S \\ i=0,1}} H_{s_0} \cap f^{-1}(H_{s_1}). \tag{4.1.15}$$

Putting everything together, we have shown that

$$\begin{aligned} &D \cap f^{-1}(D) \cap f^{-2}(D) \\ &\quad = \bigcup_{\substack{s_i \in S \\ i=0,1}} (f^{-1}(H_{s_1}) \cap H_{s_0}) \equiv \bigcup_{\substack{s_i \in S \\ i=0,1}} H_{s_0 s_1} \\ &\quad = \big\{p \in D \,|\, p \in H_{s_0}, f(p) \in H_{s_1}, s_i \in S, i = 0, 1\big\}. \end{aligned} \tag{4.1.16}$$

See Figure 4.1.8.

$\underline{D \cap f^{-1}(D) \cap f^{-2}(D) \cap f^{-3}(D)}$. Using the same arguments as those given in the previous steps, it is not hard to see that this set consists of eight

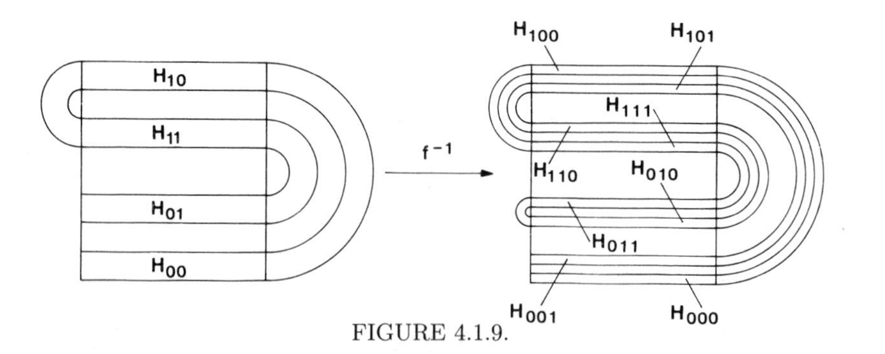

FIGURE 4.1.9.

horizontal rectangles each having width $1/\mu^3$ and that it can be denoted as

$$\begin{aligned} D \cap f^{-1}(D) \cap f^{-2}(D) \cap f^{-3}(D) &= \bigcup_{\substack{s_i \in S \\ i=0,1,2}} \left(f^{-1}(H_{s_1 s_2}) \cap H_{s_0}\right) \equiv \bigcup_{\substack{s_i \in S \\ i=0,1,2}} H_{s_0 s_1 s_2} \\ &= \{p \in D \,|\, p \in H_{s_0}, f(p) \in H_{s_1}, \\ &\qquad f^2(p) \in H_{s_2}, s_i \in S, i = 0,1,2\}. \end{aligned} \tag{4.1.17}$$

See Figure 4.1.9.

Continuing this procedure, at the kth step we obtain $D \cap f^{-1}(D) \cap \cdots \cap f^{-k}(D)$, which consists of 2^k horizontal rectangles each having width $1/\mu^k$. This set is denoted by

$$\begin{aligned} D \cap f^{-1}(D) \cap \cdots \cap f^{-k}(D) &= \bigcup_{\substack{s_i \in S \\ i=0,\cdots,k-1}} \left(f^{-1}(H_{s_1 \cdots s_{k-1}}) \cap H_{s_0}\right) \equiv \bigcup_{\substack{s_i \in S \\ i=0,\cdots,k-1}} H_{s_0 \cdots s_{k-1}} \\ &= \{p \in D \,|\, f^i(p) \in H_{s_i}, s_i \in S, i = 0, \cdots, k-1\}. \end{aligned} \tag{4.1.18}$$

As in the case of vertical rectangles, we note the important fact that at the kth step of the inductive process, each one of the 2^k vertical rectangles can be labeled uniquely with a sequence of 0's and 1's of length k. Now, as we take the limit as $k \to \infty$, we arrive at $\bigcap_{-\infty}^{n=0} f^n(D)$, which is an infinite set of horizontal lines, since a decreasing intersection of compact sets is nonempty and the width of each component of the intersection is given by $\lim_{k\to\infty}(1/\mu^k) = 0$, $\mu > 2$. Each line is labeled by a unique infinite sequence of 0's and 1's as follows

$$\begin{aligned} \bigcap_{-\infty}^{n=0} f^n(D) &= \bigcup_{\substack{s_i \in S \\ i=0,1,\cdots}} \left(f(H_{s_1 \cdots s_k \cdots}) \cap H_{s_0}\right) \equiv \bigcup_{\substack{s_i \in S \\ i=0,1,\cdots}} H_{s_0 \cdots s_k \cdots} \\ &= \{p \in D \,|\, f^i(p) \in H_{s_i}, s_i \in S, i = 0, 1, \cdots\}. \end{aligned} \tag{4.1.19}$$

Thus, we have

$$\Lambda = \bigcap_{n=-\infty}^{\infty} f^n(D) = \left[\bigcap_{n=-\infty}^{0} f^n(D)\right] \cap \left[\bigcap_{n=0}^{\infty} f^n(D)\right], \tag{4.1.20}$$

which consists of an infinite set of points, since each vertical line in $\bigcap_{n=0}^{\infty} f^n(D)$ intersects each horizontal line in $\bigcap_{-\infty}^{n=0} f^n(D)$ in a unique point. Furthermore, each point $p \in \Lambda$ can be labeled *uniquely* by a bi-infinite sequence of 0's and 1's which is obtained by concatenating the sequences associated with the respective vertical and horizontal lines that serve to define p. Stated more precisely, let $s_{-1} \cdots s_{-k} \cdots$ be a particular infinite sequence of 0's and 1's; then $V_{s_{-1} \cdots s_{-k} \cdots}$ corresponds to a unique vertical line. Let $s_0 \cdots s_k \cdots$ likewise be a particular infinite sequence of 0's and 1's; then $H_{s_0 \cdots s_k \cdots}$ corresponds to a unique horizontal line. Now a horizontal line and vertical line intersect in a unique point p; thus, we have a well-defined map from points $p \in \Lambda$ to bi-infinite sequences of 0's and 1's which we call ϕ.

$$p \stackrel{\phi}{\longmapsto} \cdots s_{-k} \cdots s_{-1} s_0 \cdots s_k \cdots .$$

Notice that because

$$\begin{aligned} V_{s_{-1} \cdots s_{-k} \cdots} &= \{p \in D \mid f^{-i+1}(p) \in V_{s_{-i}}, i = 1, \cdots\} \\ &= \{p \in D \mid f^{-i}(p) \in H_{s_{-i}}, i = 1, \cdots\} \\ &\qquad \text{since } f(H_{s_i}) = V_{s_i} \end{aligned} \tag{4.1.21}$$

and

$$H_{s_0 \cdots s_k \cdots} = \{p \in D \mid f^i(p) \in H_{s_i}, i = 0, \cdots\}, \tag{4.1.22}$$

we have

$$\begin{aligned} p &= V_{s_{-1} \cdots s_{-k} \cdots} \cap H_{s_0 \cdots s_k \cdots} \\ &= \{p \in D \mid f^i(p) \in H_{s_i}, i = 0, \pm 1, \pm 2, \cdots\}. \end{aligned} \tag{4.1.23}$$

Therefore, we see that the unique sequence of 0's and 1's we have associated with p contains information concerning the behavior of p under iteration by f. In particular, the $s_{k\text{th}}$ element in the sequence associated with p indicates that $f^k(p) \in H_{s_k}$. Now, note that for the bi-infinite sequence of 0's and 1's associated with p, the decimal point separates the past iterates from the future iterates; thus, the sequence of 0's and 1's associated with $f^k(p)$ is obtained from the sequence associated with p merely by shifting the decimal point in the sequence associated with p k places to the right if k is positive or k places to the left if k is negative, until s_k is the symbol immediately to the right of the decimal point. We can define a map of bi-infinite sequences of 0's and 1's, called the shift map, σ, which takes a sequence and shifts the decimal point one place to the right. Therefore, if we consider a point $p \in \Lambda$ and its associated bi-infinite sequence of 0's and 1's, $\phi(p)$, we can take any iterate of p, $f^k(p)$, and we can immediately obtain its associated bi-infinite sequence of 0's and 1's given by $\sigma^k(\phi(p))$. Hence, there is a direct relationship between iterating any point $p \in \Lambda$ under f and iterating the sequence of 0's and 1's associated with p under the shift map σ.

Now, at this point, it is not clear where we are going with this analogy between points in Λ and bi-infinite sequences of 0's and 1's since, although the sequence associated with a given point $p \in \Lambda$ contains information on the entire future and past as to whether or not it is in H_0 or H_1 for any given iterate, it is not hard to imagine different points, both contained in the same horizontal rectangle after any given iteration, whose orbits are completely different. The fact that this cannot happen for our map and that the dynamics of f on Λ are completely modeled by the dynamics of the shift map acting on sequences of 0's and 1's is an amazing fact which, to justify, we must digress into symbolic dynamics.

4.1c Symbolic Dynamics

Let $S = \{0, 1\}$ be the set of nonnegative integers consisting of 0 and 1. Let Σ be the collection of all bi-infinite sequences of elements of S, i.e., $s \in \Sigma$ implies

$$s = \left\{\cdots s_{-n} \cdots s_{-1}.s_0 \cdots s_n \cdots\right\}, \quad s_i \in S \quad \forall i.$$

We will refer to Σ as the space of bi-infinite sequences of two symbols. We wish to introduce some structure on Σ in the form of a metric, $d(\cdot,\cdot)$, which we do as follows. Consider

$$\begin{aligned} s &= \{\cdots s_{-n} \cdots s_{-1}.s_0 \cdots s_n \cdots\}, \\ \overline{s} &= \{\cdots \overline{s}_{-n} \cdots \overline{s}_{-1}.\overline{s}_0 \cdots \overline{s}_n \cdots\} \in \Sigma; \end{aligned}$$

we define the distance between s and $\overline{s}$, denoted $d(s, \overline{s})$, as follows

$$d(s, \overline{s}) = \sum_{i=-\infty}^{\infty} \frac{\delta_i}{2^{|i|}} \qquad \text{where } \delta_i = \begin{cases} 0 & \text{if } s_i = \overline{s}_i, \\ 1 & \text{if } s_i \neq \overline{s}_i. \end{cases} \tag{4.1.24}$$

Thus, two sequences are "close" if they agree on a long central block. (Note: the reader should check that $d(\cdot,\cdot)$ does indeed satisfy the properties of a metric. See Devaney [1986] for a proof.)

We consider a map of Σ into itself, which we shall call the shift map, σ, defined as follows: For $s = \{\cdots s_{-n} \cdots s_{-1}.s_0 s_1 \cdots s_n \cdots\} \in \Sigma$, we define

$$\sigma(s) = \{\cdots s_{-n} \cdots s_{-1} s_0.s_1 \cdots s_n \cdots\}$$

or $\sigma(s)_i = s_{i+1}$. Also, σ is continuous; we give a proof of this later in Section 4.2. Next, we want to consider the dynamics of σ on Σ (note: for our purposes the phrase "dynamics of σ on Σ" refers to the orbits of points in Σ under iteration by σ). It should be clear that σ has precisely two fixed points, namely, the sequence whose elements are all zeros and the sequence whose elements are all ones (notation: bi-infinite sequences which periodically repeat after some fixed length will be denoted by the finite length sequence with an overbar, e.g., $\{\cdots 101010.101010 \cdots\}$ is denoted by $\{\overline{10.10}\}$).

In particular, it is easy to see that the orbits of sequences which periodically repeat are periodic under iteration by σ. For example, consider the sequence $\{\overline{10.10}\}$. We have

$$\sigma\{\overline{10.10}\} = \{\overline{01.01}\}$$

and

$$\sigma\{\overline{01.10}\} = \{\overline{10.10}\};$$

thus,

$$\sigma^2\{\overline{10.10}\} = \{\overline{10.10}\}.$$

Therefore, the orbit of $\{\overline{10.10}\}$ is an orbit of period two for σ. So, from this particular example, it is easy to see that for any fixed k, the orbits of σ having period k correspond to the orbits of sequences made up of periodically repeating blocks of 0's and 1's with the blocks having length k. Thus, since for any fixed k the number of sequences having a periodically repeating block of length k is finite, we see that σ has a countable infinity of periodic orbits having all possible periods. We list the first few below.

$$\begin{array}{ll}
\text{Period 1}: & \{\overline{0.0}\}, \{\overline{1.1}\} \\
\text{Period 2}: & \{\overline{01.01}\} \xrightarrow{\sigma} \{\overline{10.10}\} \xrightarrow{\sigma} \{\overline{01.01}\} \\
\text{Period 3}: & \{\overline{001.001}\} \xrightarrow{\sigma} \{\overline{010.010}\} \xrightarrow{\sigma} \{\overline{100.100}\} \xrightarrow{\sigma} \{\overline{001.001}\} \\
: & \{\overline{110.110}\} \xrightarrow{\sigma} \{\overline{101.101}\} \xrightarrow{\sigma} \{\overline{011.011}\} \xrightarrow{\sigma} \{\overline{110.110}\} \\
\vdots & \\
\text{etc.} &
\end{array}$$

Also, σ has an uncountable number of nonperiodic orbits. To show this, we need only construct a nonperiodic sequence and show that there are an uncountable number of such sequences. A proof of this fact goes as follows: we can easily associate an infinite sequence of 0's and 1's with a given bi-infinite sequence by the following rule

$$\cdots s_{-n} \cdots s_{-1}.s_0 \cdots s_n \cdots \quad \longrightarrow \quad .s_0 s_1 s_{-1} s_2 s_{-2} \cdots .$$

Now, we will take it as a known fact that the irrational numbers in the closed unit interval $[0,1]$ constitute an uncountable set, and that every number in this interval can be expressed in base 2 as a binary expansion of 0's and 1's with the irrational numbers corresponding to nonrepeating sequences. Thus, we have a one-to-one correspondence between an uncountable set of points and nonrepeating sequences of 1's and 0's. As a result, the orbits of these sequences are the nonperiodic orbits of σ, and there are an uncountable number of such orbits.

Another interesting fact concerning the dynamics of σ on Σ is that there exists an element, say $s \in \Sigma$, whose orbit is dense in Σ, i.e., for any given $s' \in \Sigma$ and $\varepsilon > 0$, there exists some integer n such that $d(\sigma^n(s), s') < \varepsilon$. This is easiest to see by constructing s directly. We do this by first constructing all possible sequences of 0's and 1's having length $1, 2, 3, \ldots$. This process is well defined in a set-theoretic sense, since there are only a finite number of possibilities at each step (more specifically, there are 2^k distinct sequences of 0's and 1's of length k). The first few of these sequences would be as follows

$$\begin{array}{ll}
\text{length } 1: & \{0\}, \{1\} \\
\text{length } 2: & \{00\}, \{01\}, \{10\}, \{11\} \\
\text{length } 3: & \{000\}, \{001\}, \{010\}, \{011\}, \{100\}, \{101\}, \{110\}, \{111\} \\
\vdots & \vdots \\
 & \text{etc.}
\end{array}$$

We can now introduce an ordering on the collection of sequences of 0's and 1's in order to keep track of the different sequences in the following way. Consider two finite sequences of 0's and 1's

$$s = \{s_1 \cdots s_k\}, \qquad \bar{s} = \{\bar{s}_1 \cdots \bar{s}_{k'}\}.$$

We can then say

$$s < \bar{s} \qquad \text{if} \quad k < k'.$$

If $k = k'$, then

$$s < \bar{s} \qquad \text{if} \quad s_i < \bar{s}_i,$$

where i is the *first* integer such that $s_i \neq \bar{s}_i$. For example, using this ordering we have

$$\begin{aligned}
\{0\} &< \{1\}, \\
\{0\} &< \{00\}, \\
\{00\} &< \{01\}, \qquad \text{etc.}
\end{aligned}$$

This ordering gives us a systematic way of distinguishing different sequences that have the same length. Thus, we will denote the sequences of 0's and 1's having length k as follows

$$s_1^k < \cdots < s_{2^k}^k,$$

where the superscript refers to the length of the sequence and the subscript refers to a particular sequence of length k which is uniquely specified by the above ordering scheme. This will give us a systematic way of writing down our candidate for a dense orbit.

Now consider the following sequence

$$s = \{\cdots s_8^3 s_6^3 s_4^3 s_2^3 s_4^2 s_2^2 s_2^1 . s_1^1 s_1^2 s_3^2 s_1^3 s_3^3 s_5^3 s_7^3 \cdots\}.$$

Thus, s contains all possible sequences of 0's and 1's of any fixed length. To show that the orbit of s is dense in Σ, we argue as follows: let s' be an arbitrary point in Σ and let $\varepsilon > 0$ be given. An ε-neighborhood of s' consists of all points $s'' \in \Sigma$ such that $d(s', s'') < \varepsilon$, where d is the metric given in (4.1.24). Therefore, by definition of the metric on Σ, there must be some integer $N = N(\varepsilon)$ such that $s_i' = s_i''$, $|i| \leq N$ (note: a proof of this statement can be found in Devaney [1986] or in Section 4.2). By construction, the finite sequence $\{s'_{-N} \cdots s'_{-1}.s'_0 \cdots s'_N\}$ is contained somewhere in s; therefore, there must be some integer $\tilde{N}$ such that $d\big(\sigma^{\tilde{N}}(s), s'\big) < \epsilon$, and we can then conclude that the orbit of s is dense in Σ.

We summarize these facts concerning the dynamics of σ on Σ in the following theorem.

Theorem 4.1.2 *The shift map σ acting on the space of bi-infinite sequences of 0's and 1's, Σ, has*

i) *a countable infinity of periodic orbits of arbitrarily high period;*

ii) *an uncountable infinity of nonperiodic orbits;*

iii) *a dense orbit.*

4.1D The Dynamics on the Invariant Set

At this point we want to relate the dynamics of σ on Σ, about which we have a great deal of information, to the dynamics of the Smale horseshoe f on its invariant set Λ, about which we know little except for its complicated geometric structure. Recall that we have shown the existence of a well-defined map ϕ which associates to each point, $p \in \Lambda$, a bi-infinite sequence of 0's and 1's, $\phi(p)$. Furthermore, we noted that the sequence associated with any iterate of p, say $f^k(p)$, can be found merely by shifting the decimal point in the sequence associated with p k places to the right if k is positive or k places to the left if k is negative. In particular, the relation $\sigma \circ \phi(p) = \phi \circ f(p)$ holds for every $p \in \Lambda$. Now, if ϕ were invertible and continuous (continuity is necessary since f is continuous), the following relationship would hold

$$\phi^{-1} \circ \sigma \circ \phi(p) = f(p) \qquad \forall\, p \in \Lambda. \tag{4.1.25}$$

Thus, if the orbit $p \in \Lambda$ under f is denoted by

$$\{\cdots f^{-n}(p), \cdots, f^{-1}(p), p, f(p), \cdots, f^n(p), \cdots\}, \tag{4.1.26}$$

then, since $\phi^{-1} \circ \sigma \circ \phi(p) = f(p)$, we see that

$$\begin{aligned} f^n(p) &= (\phi^{-1} \circ \sigma \circ \phi) \circ (\phi^{-1} \circ \sigma \circ \phi) \circ \cdots \circ (\phi^{-1} \circ \sigma \circ \phi(p)) \\ &= \phi^{-1} \circ \sigma^n \circ \phi(p), \quad n \geq 0. \end{aligned} \tag{4.1.27}$$

Also, from (4.1.25) we have

$$f^{-1}(p) = \phi^{-1} \circ \sigma^{-1} \circ \phi(p) \qquad \forall p \in \Lambda,$$

from which we see that

$$\begin{aligned} f^{-n}(p) &= (\phi^{-1} \circ \sigma^{-1} \circ \phi) \circ (\phi^{-1} \circ \sigma^{-1} \circ \phi)) \circ \cdots \circ (\phi^{-1} \circ \sigma^{-1} \circ \phi(p)) \\ &= \phi^{-1} \circ \sigma^{-n} \circ \phi(p), \quad n \geq 0. \end{aligned} \tag{4.1.28}$$

Therefore, using (4.1.26), (4.1.27), and (4.1.28), we see that the orbit of $p \in \Lambda$ under f would correspond directly to the orbit of $\phi(p)$ under σ in Σ. In particular, the entire orbit structure of σ on Σ would be identical to the structure of f on Λ. Hence, in order to verify that this situation holds, we need to show that ϕ is a homeomorphism of Λ and Σ.

Theorem 4.1.3 *The map $\phi : \Lambda \to \Sigma$ is a homeomorphism.*

Proof: We need only show that ϕ is one-to-one and continuous, since continuity of the inverse will follow from the fact that one-to-one, onto, and continuous maps from compact sets into Hausdorff spaces are homeomorphisms (see Dugundji [1966]). We prove each condition separately.

ϕ is one-to-one: This means that given $p, p' \in \Lambda$, if $p \neq p'$, then $\phi(p) \neq \phi(p')$.

We give a proof by contradiction. Suppose $p \neq p'$ and

$$\phi(p) = \phi(p') = \{\cdots s_{-n} \cdots s_{-1}.s_0 \cdots s_n \cdots\}.$$

Then, by construction of Λ, p and p' lie in the intersection of the vertical line $V_{s_{-1} \cdots s_{-n} \cdots}$ and the horizontal line $H_{s_0 \cdots s_n \cdots}$. However, the intersection of a horizontal line and a vertical line consists of a unique point; therefore $p = p'$, contradicting our original assumption. This contradiction is due to the fact that we have assumed $\phi(p) = \phi(p')$; thus, for $p \neq p'$, $\phi(p) \neq \phi(p')$.

ϕ is onto: This means that given any bi-infinite sequence of 0's and 1's in Σ, say $\{\cdots s_{-n} \cdots s_{-1}.s_0 \cdots s_n \cdots\}$, there is a point $p \in \Lambda$ such that $\phi(p) = \{\cdots s_{-n} \cdots s_{-1}.s_0 \cdots s_n \cdots\}$.

The proof goes as follows: Recall the construction of $\bigcap_{n=0}^{\infty} f^n(D)$ and $\bigcap_{-\infty}^{n=0} f^n(D)$; given any infinite sequence of 0's and 1's, $\{\cdots s_{-n} \cdots s_{-1}.\}$, there is a *unique* vertical line in $\bigcap_{n=0}^{\infty} f^n(D)$ corresponding to this sequence. Similarly, given any infinite sequence of 0's and 1's, $\{.s_0 \cdots s_n \cdots\}$, there is a unique horizontal line in $\bigcap_{-\infty}^{n=0} f^n(D)$ corresponding to this sequence. Therefore, we see that for a given horizontal and

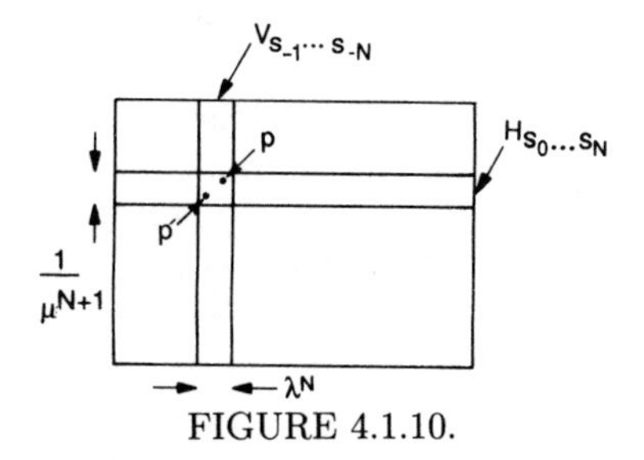

FIGURE 4.1.10.

vertical line, we can associate a unique bi-infinite sequence of 0's and 1's, $\{\cdots s_{-n}\cdots s_{-1}.s_0\cdots s_n\cdots\}$ and, since a horizontal and vertical line intersect in a unique point, p, to every bi-infinite sequence of 0's and 1's, there corresponds a unique point in Λ.

ϕ is continuous: This means that, given any point $p\in\Lambda$ and $\varepsilon>0$, we can find a $\delta=\delta(\varepsilon,p)$ such that

$$|p-p'|<\delta \qquad \text{implies} \quad d\big(\phi(p),\phi(p')\big)<\varepsilon,$$

where $|\cdot|$ is the usual distance measurement in $\mathbb{R}^2$ and $d(\cdot,\cdot)$ is the metric on Σ introduced earlier.

Let $\varepsilon>0$ be given; then, if we are to have $d(\phi(p),\phi(p'))<\varepsilon$, there must be some integer $N=N(\epsilon)$ such that if

$$\phi(p)=\{\cdots s_{-n}\cdots s_{-1}.s_0\cdots s_n\cdots\},$$
$$\phi(p')=\{\cdots s'_{-n}\cdots s'_{-1}.s'_0\cdots s'_n\cdots\},$$

then $s_i=s'_i$, $i=0,\pm1,\ldots,\pm N$. Thus, by construction of Λ, p and p' lie in the rectangle defined by $H_{s_0\cdots s_N}\cap V_{s_{-1}\cdots s_{-N}}$; see Figure 4.1.10. Recall that the width and height of this rectangle are λ^N and $1/\mu^{N+1}$, respectively. Thus we have $|p-p'|\le(\lambda^N+1/\mu^{N+1})$. Therefore, if we take $\delta=\lambda^N+1/\mu^{N+1}$, continuity is proved. $\square$

We make the following remarks.

Remark 1. Recall from Section 1.2B that the dynamical systems f acting on Λ and σ acting on Σ are said to be *topologically conjugate* if $\phi\circ f(p)=\sigma\circ\phi(p)$. (Note: the equation $\phi\circ f(p)=\sigma\circ\phi(p)$ is also expressed by saying that the following diagram "commutes.")

$$\begin{array}{ccc}\Lambda & \xrightarrow{f} & \Lambda\\ \phi\downarrow & & \downarrow\phi\\ \Sigma & \xrightarrow{\sigma} & \Sigma\end{array}$$

Remark 2. The fact that Λ and Σ are homeomorphic allows us to make several conclusions concerning the set-theoretic nature of Λ. We have already shown that Σ is uncountable, and we state without proof that Σ is a closed, perfect (meaning every point is a limit point), totally disconnected set and that these properties carry over to Λ via the homeomorphism ϕ. A set having these properties is called a *Cantor set.* We will give more detailed information concerning symbolic dynamics and Cantor sets in Section 4.2.

Now we can state a theorem regarding the dynamics of f on Λ that is almost precisely the same as Theorem 4.1.2, which describes the dynamics of σ on Σ.

Theorem 4.1.4 *The Smale horseshoe, f, has*

i) *a countable infinity of periodic orbits of arbitrarily high period. These periodic orbits are all of saddle type;*

ii) *an uncountable infinity of non-periodic orbits;*

iii) *a dense orbit.*

Proof: This is an immediate consequence of the topological conjugacy of f on Λ with σ on Σ, except for the stability result. The stability result follows from the form of f on H_0 and H_1 given in (4.1.4). □

4.1E Chaos

Now we can make precise the statement that the dynamics of f on Λ is chaotic.

Let $p \in \Lambda$ with corresponding symbol sequence

$$\phi(p) = \left\{\cdots s_{-n} \cdots s_{-1}.s_0 \cdots s_n \cdots\right\}.$$

We want to consider points close to p and how they behave under iteration by f as compared with p. Let $\varepsilon > 0$ be given; then we consider an ε-neighborhood of p determined by the usual topology of the plane. Hence there also exists an integer $N = N(\varepsilon)$ such that the corresponding neighborhood of $\phi(p)$ includes the set of sequences $s' = \left\{\cdots s'_{-n} \cdots s'_{-1}.s'_0 \cdots s'_n \cdots\right\} \in \Sigma$ such that $s_i = s'_i$, $|i| \le N$. Now suppose the $N+1$ entry in the sequence corresponding to $\phi(p)$ is 0, and the $N+1$ entry in the sequence corresponding to some s' is 1. Thus, after N iterations, no matter how small ε, the point p is in H_0; the point, say p', corresponding to s' under ϕ^{-1} is in H_1, and they are at least a distance $1 - 2\lambda$ apart. Therefore, for any point $p \in \Lambda$, no matter how small a neighborhood of p we consider, there is at least one point in this neighborhood such that, after a finite number of iterations, p and this point have separated by some fixed distance. A system displaying such behavior is said to exhibit *sensitive dependence on initial*

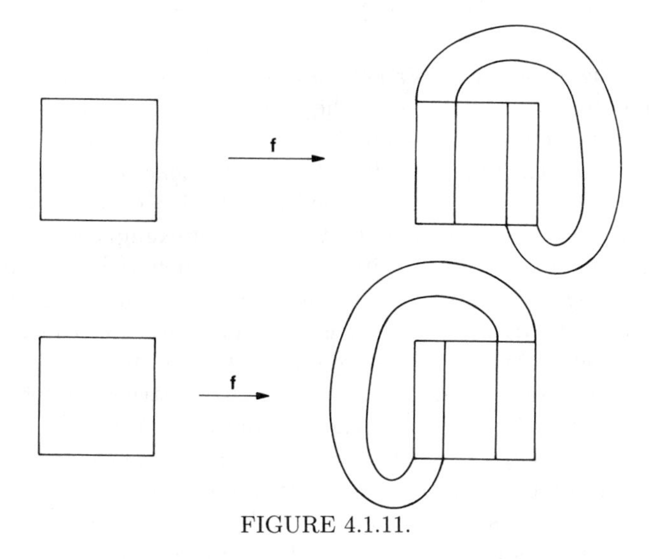

FIGURE 4.1.11.

conditions. A dynamical system displaying sensitive dependence on initial conditions on a closed invariant set (which consists of more than one orbit) will be called *chaotic.* In Section 4.11 we will explore more fully the nature of chaotic dynamical systems.

Now we want to end our discussion of this simplified version of the Smale horseshoe with some final observations.

1. If you consider carefully the main ingredients of f which led to Theorem 4.1.4, you will see that there are two key elements.

   (a) The square is contracted, expanded, and folded in such a way that we can find disjoint regions that are mapped over themselves.

   (b) There exists "strong" stretching and contraction in complementary directions.

2. From observation 1), the fact that the image of the square appears in the shape of a horseshoe is not important. Other possible scenarios are shown in Figure 4.1.11.

Notice that, in our study of the invariant set of f, we do not consider the question of the geometry of the points which escape from the square. We remark that this could be an interesting research topic, since this more global question may enable one to determine conditions under which the horseshoe becomes an attractor.

4.2 Symbolic Dynamics

In the previous section we saw an example of a two-dimensional map which possessed an invariant Cantor set. The map, restricted to its invariant set, was shown to have a countable infinity of periodic orbits of all periods, an uncountable infinity of nonperiodic orbits, and a dense orbit. Now, in general, the determination of such detailed information concerning the orbit structure of a map is not possible. However, in our example we were able to show that the map restricted to its invariant set behaved the same as the shift map acting on the space of bi-infinite sequences of 0's and 1's (more precisely, these two dynamical systems were shown to be topologically conjugate; thus their orbit structures are identical). The shift map was no less complicated than our original map but, due to its structure, many of the features concerning its dynamics (e.g., the nature and number of its periodic orbits) were more or less obvious. The technique of characterizing the orbit structure of a dynamical system via infinite sequences of "symbols" (in our case 0's and 1's) is known as *symbolic dynamics.* The technique is not new and appears to have originally been applied by Hadamard [1898] in the study of geodesics on surfaces of negative curvature and Birkhoff [1927], [1935] in his studies of dynamical systems. The first exposition of symbolic dynamics as an independent subject was given by Morse and Hedlund [1938]. Applications of this idea to differential equations can be found in Levinson's work on the forced van der Pol equation (Levinson [1949]), from which came Smale's inspiration for his construction of the horseshoe map (Smale [1963], [1980]), and also in the work of Alekseev [1968], [1969], who gives a systematic account of the technique and applies it to problems arising from celestial mechanics. These references by no means represent a complete account of the history of symbolic dynamics or of its applications, and we refer the reader to the bibliographies of the above listed references or to Moser [1973] for a more complete list of references on the subject and its applications. In recent times (say from about 1965 to the present) there has been a flood of applications of the technique.

Symbolic dynamics will play a key role in explaining the dynamical phenomena we encounter in this chapter. For this reason, we now want to describe some aspects of symbolic dynamics viewed as an independent subject. Our discussion follows Wiggins [1988].

We let $S = \{1, 2, 3, \cdots, N\}$, $N \geq 2$ be our collection of symbols. We will build our sequences from elements of S. Note that for the purpose of constructing sequences, the elements of S could be anything, e.g., letters of the alphabet, Chinese characters, etc. We will use positive integers since they are familiar, easy to write down, and we have as many of them as we desire.

4.2A The Structure of the Space of Symbol Sequences

We now want to construct the space of all symbol sequences, which we will refer to as Σ^N, from elements of S and derive some properties of Σ^N. It will be convenient to construct Σ^N as a Cartesian product of infinitely many copies of S. This construction will allow us to make some conclusions concerning the properties of Σ^N based only on our knowledge of S and the structure which we give to S.

We now want to give some structure to S; specifically, we want to make S into a metric space. Since S is a finite set of points consisting of the first N positive integers, it is very natural to define the distance between two elements of S to be the absolute value of the difference of the two elements. We denote this as follows

$$d(a,b) \equiv |a-b| \qquad \forall\, a,b \in S. \tag{4.2.1}$$

Thus S is a discrete space (i.e., the open sets in S defined by the metric consist of the individual points which make up S so that all subsets of S are open) and, hence, it is totally disconnected. We summarize the properties of S in the following proposition.

Proposition 4.2.1 *The set S equipped with the metric (4.2.1) is a compact, totally disconnected metric space.*

We remark that compact metric spaces are automatically complete metric spaces (see Dugundji [1966]).

Now we will construct Σ^N as a bi-infinite Cartesian product of copies of S

$$\Sigma^N \equiv \cdots \times S \times S \times S \times S \times \cdots \equiv \prod_{i=-\infty}^{\infty} S^i \quad \text{where} \quad S^i = S \quad \forall i. \tag{4.2.2}$$

Thus, a point in Σ^N is represented as a "bi-infinity-tuple" of elements of S

$$s \in \Sigma^N \Rightarrow s = \left\{\cdots, s_{-n}, \cdots, s_{-1}, s_0, s_1, \cdots, s_n, \cdots\right\} \qquad \text{where} \;\; s_i \in S \quad \forall i,$$

or, more succintly, we will write s as

$$s = \left\{\cdots s_{-n} \cdots s_{-1}.s_0 s_1 \cdots s_n \cdots\right\} \qquad \text{where} \quad s_i \in S \quad \forall i.$$

A word should be said about the "decimal point" that appears in each symbol sequence and has the effect of separating the symbol sequence into two parts, with both parts being infinite (hence the reason for the phrase "bi-infinite sequence"). At present it does not play a major role in our discussion and could easily be left out with all of our results describing the structure of Σ^N going through just the same. In some sense, it serves as a starting point for constructing the sequences by giving us a natural way of subscripting each element of a sequence. This notation will prove convenient shortly when we define a metric on Σ^N. However, the real significance of the decimal point will become apparent when we define and discuss the shift map acting on Σ^N and its orbit structure.

In order to discuss limit processes in Σ^N, it will be convenient to define a metric on Σ^N. Since S is a metric space, it is also possible to define a metric on Σ^N. There are many possible choices for a metric on Σ^N; however, we will utilize the following. For

$$\begin{aligned} s &= \{\cdots s_{-n} \cdots s_{-1}.s_0 s_1 \cdots s_n \cdots\}, \\ \overline{s} &= \{\cdots \overline{s}_{-n} \cdots \overline{s}_{-1}.\overline{s}_0 \overline{s}_1 \cdots \overline{s}_n \cdots\} \in \Sigma^N, \end{aligned}$$

the distance between s and $\overline{s}$ is defined as

$$d(s, \overline{s}) = \sum_{i=-\infty}^{\infty} \frac{1}{2^{|i|}} \frac{|s_i - \overline{s}_i|}{1 + |s_i - \overline{s}_i|}.$$

(Note: the reader should check that $d(\cdot, \cdot)$ satisfies the four properties which, by definition, a metric must possess.) Intuitively, this choice of metric implies that two symbol sequences are "close" if they agree on a long central block. The following lemma makes this precise.

Lemma 4.2.2 *For* $s, \overline{s} \in \Sigma^N$,

i) *Suppose* $d(s, \overline{s}) < 1/(2^{M+1})$; *then* $s_i = \overline{s}_i$ *for all* $|i| \leq M$.

ii) *Suppose* $s_i = \overline{s}_i$ *for* $|i| \leq M$; *then* $d(s, \overline{s}) \leq 1/(2^{M-1})$.

Proof: The proof of i) is by contradiction. Suppose the hypothesis of i) holds and there exists some j with $|j| \leq M$ such that $s_j \neq \overline{s}_j$. Then there exists a term in the sum defining $d(s, \overline{s})$ of the form

$$\frac{1}{2^{|j|}} \frac{|s_j - \overline{s}_j|}{1 + |s_j - \overline{s}_j|}.$$

However,

$$\frac{|s_j - \overline{s}_j|}{1 + |s_j - \overline{s}_j|} \geq \frac{1}{2},$$

and each term in the sum defining $d(s, \overline{s})$ is positive so that we have

$$d(s, \overline{s}) \geq \frac{1}{2^{|j|}} \frac{|s_j - \overline{s}_j|}{1 + |s_j - \overline{s}_j|} \geq \frac{1}{2^{|j|+1}} \geq \frac{1}{2^{M+1}},$$

but this contradicts the hypothesis of i).

We now prove ii). If $s_i = \bar{s}_i$ for $|i| \leq M$, we have

$$d(s, \bar{s}) = \sum_{-\infty}^{i=-(M+1)} \frac{1}{2^{|i|}} \frac{|s_i - \bar{s}_i|}{1 + |s_i - \bar{s}_i|} + \sum_{i=M+1}^{\infty} \frac{1}{2^{|i|}} \frac{|s_i - \bar{s}_i|}{1 + |s_i - \bar{s}_i|};$$

however, $\big(|s_i - \bar{s}_i|/(1 + |s_i - \bar{s}_i|)\big) \leq 1$, so we obtain

$$d(s, \bar{s}) \leq 2 \sum_{i=M+1}^{\infty} \frac{1}{2^i} = \frac{1}{2^{M-1}}. \qquad \square$$

Armed with our metric, we can define neighborhoods of points in Σ^N and describe limit processes. Suppose we are given a point

$$\bar{s} = \{\cdots \bar{s}_{-n} \cdots \bar{s}_{-1}.\bar{s}_0 \bar{s}_1 \cdots \bar{s}_n \cdots\} \in \Sigma^N, \qquad \bar{s}_i \in S \quad \forall i,$$

and a positive real number $\varepsilon > 0$, and we wish to describe the "ε-neighborhood of $\bar{s}$", i.e., the set of $s \in \Sigma^N$ such that $d(s, \bar{s}) < \varepsilon$. Then, by Lemma 4.2.2, given $\varepsilon > 0$, we can find a positive integer $M = M(\varepsilon)$ such that $d(s, \bar{s}) < \varepsilon$ implies $s_i = \bar{s}_i \ \forall |i| \leq M$. Thus, our notation for an ε-neighborhood of an arbitrary $\bar{s} \in \Sigma^N$ will be as follows

$$\mathcal{N}^{M(\varepsilon)}(\bar{s}) = \{s \in \Sigma^N \mid s_i = \bar{s}_i \,\forall\, |i| \leq M, \ s_i, \bar{s}_i \in S \,\forall\, i\}.$$

Before stating our theorem concerning the structure of Σ^N we need the following definition.

DEFINITION 4.2.1 A set is called *perfect* if it is closed and every point in the set is a limit point of the set.

The following theorem of Cantor gives us information concerning the cardinality of perfect sets.

Theorem 4.2.3 *Every perfect set in a complete space has at least the cardinality of the continuum.*

Proof: See Hausdorff [1957]. $\square$

We are now ready to state our main theorem concerning the structure of Σ^N.

Proposition 4.2.4 *The space Σ^N equipped with the metric (4.2.1) is*

i) *compact,*

ii) *totally disconnected, and*

iii) *perfect.*

Proof: i) Since S is compact, Σ^N is compact by Tychonov's theorem (Dugundji [1966]). ii) By Proposition 4.2.1, S is totally disconnected, and therefore Σ^N is totally disconnected, since the product of totally disconnected spaces is likewise totally disconnected (Dugundji [1966]). iii) Σ^N is closed, since it is a compact metric space. Let $\overline{s} \in \Sigma^N$ be an arbitrary point in Σ^N; then, to show that $\overline{s}$ is a limit point of Σ^N, we need only show that every neighborhood of $\overline{s}$ contains a point $s \neq \overline{s}$ with $s \in \Sigma^N$. Let $\mathcal{N}^{M(\varepsilon)}(\overline{s})$ be a neighborhood of $\overline{s}$ and let $\hat{s} = \overline{s}_{M(\varepsilon)+1} + 1$ if $\overline{s}_{M(\varepsilon)+1} \neq N$, and $\hat{s} = \overline{s}_{M(\varepsilon)+1} - 1$ if $\overline{s}_{M(\varepsilon)+1} = N$. Then the sequence

$$\{\cdots \overline{s}_{-M(\varepsilon)-2}\hat{s}\overline{s}_{-M(\varepsilon)} \cdots \overline{s}_{-1}.\overline{s}_0\overline{s}_1 \cdots \overline{s}_{M(\varepsilon)}\hat{s}\overline{s}_{M(\varepsilon)+2} \cdots\}$$

is contained in $\mathcal{N}^{M(\varepsilon)}(\overline{s})$ and is not equal to $\overline{s}$; thus Σ^N is perfect. □

We remark that the three properties of Σ^N stated in Proposition 4.2.4 are often taken as the defining properties of a *Cantor set* of which the classical Cantor "middle-thirds" set is a prime example.

Next we want to make a remark which will be of interest later when we use Σ^N as a "model space" for the dynamics of maps defined on more "normal" domains than Σ^N (i.e., by "normal" domain we mean the type of domain which might arise as the phase space of a specific physical system). Recall that a map, $h: X \to Y$, of two topological spaces X and Y is called a *homeomorphism* if h is continuous, one-to-one, and onto and h^{-1} is also continuous. Now there are certain properties of topological spaces which are invariant under homeomorphisms. Such properties are called *topological invariants;* compactness, connectedness, and perfectness are three examples of topological invariants (see Dugundji [1966] for a proof). We summarize this in the following proposition.

Proposition 4.2.5 *Let Y be a topological space and suppose that Σ^N and Y are homeomorphic, i.e., there exists a homeomorphism h taking Σ^N to Y. Then Y is compact, totally disconnected, and perfect.*

4.2B The Shift Map

Now that we have established the structure of Σ^N, we want to define a map of Σ^N into itself, denoted by σ, as follows.

For $s = \{\cdots s_{-n} \cdots s_{-1}.s_0 s_1 \cdots s_n \cdots\} \in \Sigma^N$, we define

$$\sigma(s) \equiv \{\cdots s_{-n} \cdots s_{-1} s_0.s_1 \cdots s_n \cdots\}$$

or $[\sigma(s)]_i \equiv s_{i+1}$.

The map, σ, is referred to as the *shift map,* and when the domain of σ is taken to be all of Σ^N, it is often referred to as a *full shift on N symbols.* We have the following proposition concerning some properties of σ.

Proposition 4.2.6 i) $\sigma(\Sigma^N) = \Sigma^N$. ii) σ *is continuous.*

Proof: The proof of i) is obvious. To prove ii) we must show that, given $\varepsilon > 0$, there exists a $\delta(\varepsilon)$ such that $d(s, \bar{s}) < \delta$ implies $d(\sigma(s), \sigma(\bar{s})) < \varepsilon$ for $s, \bar{s} \in \Sigma^N$. Suppose $\varepsilon > 0$ is given; then choose M such that $1/(2^{M-2}) < \varepsilon$. If we then let $\delta = 1/2^{M+1}$, we see by Lemma 4.2.2 that $d(s, \bar{s}) < \delta$ implies $s_i = \bar{s}_i$ for $|i| \leq M$; hence, $[\sigma(s)]_i = [\sigma(\bar{s})]_i$, $|i| \leq M - 1$. Then, also by Lemma 4.2.2, we have $d(\sigma(s), \sigma(\bar{s})) < 1/2^{M-2} < \varepsilon$. $\square$

We now want to consider the orbit structure of σ acting on Σ^N. We have the following proposition.

Proposition 4.2.7 *The shift map σ has*

i) *a countable infinity of periodic orbits consisting of orbits of all periods;*

ii) *an uncountable infinity of nonperiodic orbits; and*

iii) *a dense orbit.*

Proof: i) This is proven in exactly the same way as the analogous result obtained in our discussion of the symbolic dynamics for the Smale horseshoe map in Section 4.1C. In particular, the orbits of the periodic symbol sequences are periodic, and there is a countable infinity of such sequences. ii) By Theorem 4.2.3 Σ^N is uncountable; thus, removing the countable infinity of periodic symbol sequences leaves an uncountable number of nonperiodic symbol sequences. Since the orbits of the nonperiodic sequences never repeat, this proves ii). iii) This is proven in exactly the same way as the analogous result obtained in our discussion of the Smale horseshoe map in Section 4.1; namely, we form a symbol sequence by stringing together all possible symbol sequences of any finite length. The orbit of this sequence is dense in Σ^N since, by construction, some iterate of this symbol sequence will be arbitrarily close to any given symbol sequence in Σ^N. $\square$

4.3 The Conley–Moser Conditions, or "How to Prove That a Dynamical System is Chaotic"

In this section we will give sufficient conditions in order for a two-dimensional invertible map to have an invariant Cantor set on which the dynamics are topologically conjugate to a full shift on N symbols ($N \geq 2$). These conditions were first given by Conley and Moser (see Moser [1973]), and we give slight improvements on their estimates. Alekseev [1968], [1969] developed similar criteria. Generalizations to n-dimensions can be found in Wiggins [1988].

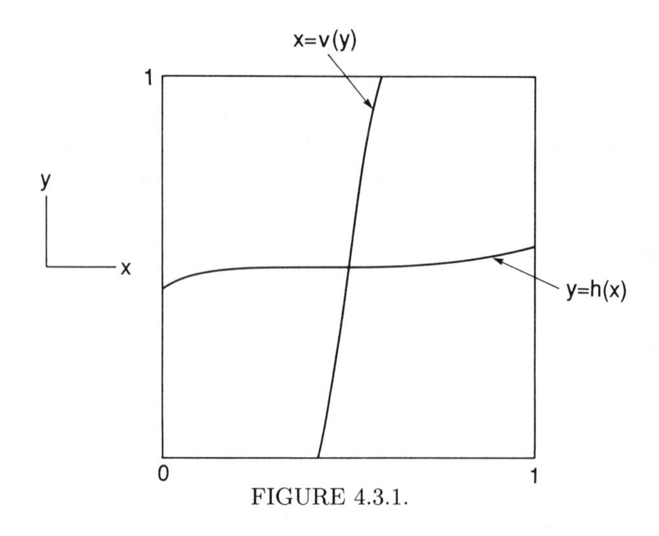

FIGURE 4.3.1.

4.3A The Main Theorem

We begin with several definitions.

Definition 4.3.1 A *μ_v-vertical curve* is the graph of a function $x = v(y)$ for which

$$0 \leq v(y) \leq 1, \qquad |v(y_1) - v(y_2)| \leq \mu_v |y_1 - y_2| \quad \text{for} \quad 0 \leq y_1,\ y_2 \leq 1.$$

Similarly, a *μ_h-horizontal curve* is the graph of a function $y = h(x)$ for which

$$0 \leq h(x) \leq 1, \qquad |h(x_1) - h(x_2)| \leq \mu_h |x_1 - x_2| \quad \text{for } 0 \leq x_1,\ x_2 \leq 1;$$

see Figure 4.3.1.

We make the following remarks concerning Definition 4.3.1.

Remark 1. Functions $x = v(y)$ and $y = h(x)$ satisfying Definition 4.3.1 are called *Lipschitz functions* with Lipschitz constants μ_v and μ_h, respectively.

Remark 2. The constant μ_h can be interpreted as a bound on the slope of the curve defined by the graph of $y = h(x)$. A similar interpretation holds for μ_v and the graph of $x = v(y)$.

Remark 3. For $\mu_v = 0$, the graph of $x = v(y)$ is a vertical line and, for $\mu_h = 0$, the graph of $y = h(x)$ is a horizontal line.

Remark 4. At this point we have put no restrictions on the relationship or magnitudes of μ_v and μ_h.

Next we want to "fatten up" these μ_v-vertical curves and μ_h-horizontal curves into μ_v-vertical strips and μ_h-horizontal strips, respectively.

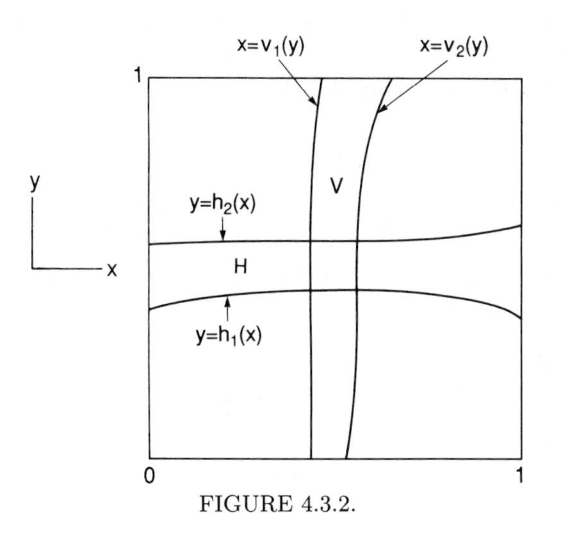

FIGURE 4.3.2.

DEFINITION 4.3.2 Given two nonintersecting μ_v-vertical curves $v_1(y) < v_2(y)$, $y \in [0,1]$, we define a *μ_v-vertical strip* as

$$V = \{(x,y) \in \mathbb{R}^2 \mid x \in [v_1(y), v_2(y)];\ y \in [0,1]\}.$$

Similarly, given two nonintersecting μ_h-horizontal curves $h_1(x) < h_2(x)$, $x \in [0,1]$, we define a *μ_h-horizontal strip* as

$$H = \{(x,y) \in \mathbb{R}^2 \mid y \in [h_1(x), h_2(x)];\ x \in [0,1]\};$$

see Figure 4.3.2. The width of horizontal and vertical strips is defined as

$$d(H) = \max_{x \in [0,1]} |h_2(x) - h_1(x)|, \tag{4.3.1a}$$

$$d(V) = \max_{y \in [0,1]} |v_2(y) - v_1(y)|. \tag{4.3.1b}$$

The following two lemmas will play an important role in the inductive process of constructing the invariant set for the map f.

Lemma 4.3.1 i) *If $V^1 \supset V^2 \supset \cdots \supset V^k \supset \cdots$ is a nested sequence of μ_v-vertical strips with $d(V^k) \underset{k\to\infty}{\longrightarrow} 0$, then $\bigcap_{k=1}^{\infty} V^k \equiv V^\infty$ is a μ_v-vertical curve.*

ii) *If $H^1 \supset H^2 \supset \cdots \supset H^k \supset \cdots$ is a nested sequence of μ_h-horizontal strips with $d(H^k) \underset{k\to\infty}{\longrightarrow} 0$, then $\bigcap_{k=1}^{\infty} H^k \equiv H^\infty$ is a μ_h-horizontal curve.*

Proof: We will prove i) only, since the proof of ii) requires only trivial modifications.

Let $C_{\mu_v}[0,1]$ denote the set of Lipschitz functions with Lipschitz constant μ_v defined on the interval $[0,1]$. Then with the metric defined by the maximum norm, $C_{\mu_v}[0,1]$ is a complete metric space (see Arnold [1973] for a proof). Let $x = v_1^k(y)$ and $x = v_2^k(y)$ form the vertical boundaries of the μ_v-vertical strip V^k. Now consider the sequence

$$\{v_1^1(y), v_2^1(y), v_1^2(y), v_2^2(y), \cdots, v_1^k(y), v_2^k(y), \cdots\}. \tag{4.3.2}$$

By definition of the V^k, (4.3.2) is a sequence of elements of $C_{\mu_v}[0,1]$, and since $d(V^k) \underset{k\to\infty}{\longrightarrow} 0$, it is a Cauchy sequence. Therefore, since $C_{\mu_v}[0,1]$ is a complete metric space, the Cauchy sequence converges to a unique μ_v-vertical curve. This proves i). □

Lemma 4.3.2 *Suppose* $0 \leq \mu_v\mu_h < 1$. *Then a* μ_v*-vertical curve and a* μ_h*-horizontal curve intersect in a unique point.*

Proof: Let the μ_h-horizontal curve be given by the graph of

$$y = h(x),$$

and let the μ_v-vertical curve be given by the graph of

$$x = v(y).$$

The condition for intersection is that there exists a point (x, y) in the unit square satisfying each relation, i.e., we have

$$y = h(x), \tag{4.3.3}$$

where x in (4.3.3) satisfies

$$x = v(y),$$

in other words, the equation

$$y = h(v(y)) \tag{4.3.4}$$

has a solution. We want to show that this solution is *unique*. We will use the contraction mapping theorem (see Arnold [1973]).

Let us give some background. Consider a map

$$g: M \longrightarrow M,$$

where M is a complete metric space. Then g is said to be a *contraction map* if

$$|g(m_1) - g(m_2)| \leq k|m_1 - m_2|, \qquad m_1, m_2 \in M,$$

for some constant $0 \leq k < 1$, where $|\cdot|$ denotes the metric on M. The contraction mapping theorem says that g has a unique fixed point, i.e., there exists *one* point $\overline{m} \in M$ such that

$$g(\overline{m}) = \overline{m}.$$

We now apply this to our situation.

Let I denote the closed unit interval, i.e.,

$$I = \{y \in \mathbb{R}^1 \,|\, 0 \leq y \leq 1\}.$$

Clearly I is a complete metric space. Also, it should be evident that

$$h \circ v : I \longrightarrow I. \tag{4.3.5}$$

Hence, if we show that $h \circ v$ is a contraction map, then by the contraction mapping theorem (4.3.5) has a *unique* solution and we are done. This is just a simple computation. For $y_1, y_2 \in I$ we have

$$\begin{aligned} |h(v(y_1)) - h(v(y_2))| &\leq \mu_h |v(y_1) - v(y_2)| \\ &\leq \mu_h \mu_v |y_1 - y_2|. \end{aligned}$$

Since we have assumed $0 \leq \mu_v \mu_h < 1$, $h \circ v$ is a contraction map. □

We consider a map

$$f: D \longrightarrow \mathbb{R}^2,$$

where D is the unit square in $\mathbb{R}^2$, i.e.,

$$D = \big\{(x, y) \in \mathbb{R}^2 \,|\, 0 \leq x \leq 1,\ 0 \leq y \leq 1\big\}.$$

Let

$$S = \{1, 2, \cdots, N\}, \qquad (N \geq 2),$$

be an index set, and let

$$H_i, \qquad i = 1, \cdots, N$$

be a set of *disjoint* μ_h-horizontal strips. Finally, let

$$V_i, \qquad i = 1, \cdots, N,$$

be a set of *disjoint* μ_v-vertical strips. Suppose that f satisfies the following two conditions.

Assumption 1. $0 \leq \mu_v \mu_h < 1$ and f maps H_i *homeomorphically* onto V_i, $(f(H_i) = V_i)$ for $i = 1, \cdots, N$. Moreover, the horizontal boundaries of H_i map to the horizontal boundaries of V_i and the vertical boundaries of H_i map to the vertical boundaries of V_i.

Assumption 2. Suppose H is a μ_h-horizontal strip contained in $\bigcup_{i \in S} H_i$. Then

$$f^{-1}(H) \cap H_i \equiv \tilde{H}_i$$

is a μ_h-horizontal strip for every $i \in S$. Moreover,

$$d(\tilde{H}_i) \leq \nu_h d(H) \qquad \text{for some } \ 0 < \nu_h < 1.$$

Similarly, suppose V is a μ_v-vertical strip contained in $\bigcup_{i\in S} V_i$. Then

$$f(V) \cap V_i \equiv \tilde{V}_i$$

is a μ_v-vertical strip for every $i \in S$. Moreover,

$$d(\tilde{V}_i) \leq \nu_v d(V) \qquad \text{for some } \ 0 < \nu_v < 1.$$

Now we can state our main theorem.

Theorem 4.3.3 *Suppose f satisfies Assumptions 1 and 2. Then f has an invariant Cantor set, Λ, on which it is topologically conjugate to a full shift on N symbols, i.e., the following diagram commutes*

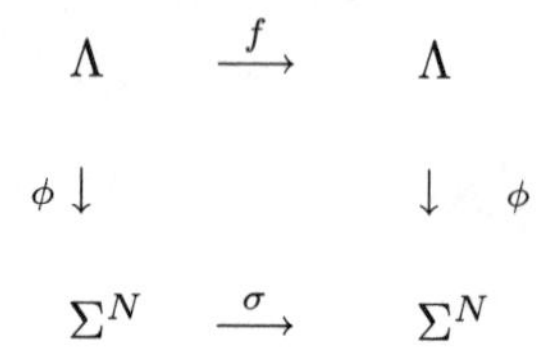

$$\begin{array}{ccc} \Lambda & \xrightarrow{f} & \Lambda \\ \phi \downarrow & & \downarrow \phi \\ \Sigma^N & \xrightarrow{\sigma} & \Sigma^N \end{array}$$

where ϕ is a homeomorphism mapping Λ onto Σ^N.

The proof has four steps.

Step 1. Construct Λ.

Step 2. Define the map $\phi: \Lambda \longrightarrow \Sigma^N$.

Step 3. Show that ϕ is a homeomorphism.

Step 4. Show that $\phi \circ f = \sigma \circ \phi$.

Proof: Step 1: Construction of the Invariant Set. The construction of the invariant set of the map is very similar to the construction of the invariant set for the Smale horseshoe in Section 4.1. We first construct a set of points that remains in $\bigcup_{i\in S} V_i$ under all backward iterates. This will turn out to be an uncountable infinity of μ_v-vertical curves. Next we construct a set of points that remains in $\bigcup_{i\in S} H_i$ under all forward iterates. This will turn out to be an uncountable infinity of μ_h-horizontal curves. Then the intersection of these two sets is clearly an invariant set contained in $(\bigcup_{i\in S} H_i) \cap (\bigcup_{i\in S} V_i) \subset D$.

The reader may wonder why our terminology here is different than that used in the discussion of the construction of *the* invariant set for the Smale horseshoe. In that case *the* invariant set, Λ, was given by

$$\Lambda = \bigcap_{n=-\infty}^{\infty} f^n(D).$$

However, for the Smale horseshoe we knew how the map acted on all of D. Namely, the part of D not contained in $H_0 \cup H_1$ was "thrown out" of D under the action of f. We have not assumed such behavior in the situation presently under consideration. We know only how the map f acts on $\bigcup_{i\in S} H_i$ and how f^{-1} acts on $\bigcup_{i\in S} V_i$. We will comment more on this following the proof of the theorem.

We begin by inductively constructing the set of points in $\bigcup_{i\in S} V_i$ that remain in $\bigcup_{i\in S} V_i$ under all backwards iterations by f. We denote this set by $\Lambda_{-\infty}$ and Λ_{-n}, $n = 1, 2, \cdots$ denoting the set of points in $\bigcup_{i\in S} V_i$ that remain in $\bigcup_{i\in S} V_i$ under $n-1$ backwards iterations by f.

$\underline{\Lambda_{-1}}$. Λ_{-1} is obvious.

$$\Lambda_{-1} = \bigcup_{s_{-1}\in S} V_{s_{-1}}. \tag{4.3.6}$$

$\underline{\Lambda_{-2}}$. It should be clear that

$$\Lambda_{-2} = f(\Lambda_{-1}) \cap \left(\bigcup_{s_{-1}\in S} V_{s_{-1}} \right) \tag{4.3.7}$$

is the set of points in $\bigcup_{s_{-1}\in S} V_{s_{-1}}$ that are mapped into Λ_{-1} under f^{-1}. Then, using (4.3.6), (4.3.7) becomes

$$\begin{aligned} \Lambda_{-2} &= \left(\bigcup_{s_{-2}\in S} f(V_{s_{-2}}) \right) \cap \left(\bigcup_{s_{-1}\in S} V_{s_{-1}} \right) \\ &= \bigcup_{\substack{s_{-i}\in S \\ i=1,2}} f(V_{s_{-2}}) \cap V_{s_{-1}} \equiv \bigcup_{\substack{s_{-i}\in S \\ i=1,2}} V_{s_{-1}s_{-2}}. \end{aligned} \tag{4.3.8}$$

We note the following.

i) $V_{s_{-1}s_{-2}} = \{p \in D \,|\, p \in V_{s_{-1}}, f^{-1}(p) \in V_{s_{-2}}\}$ with $V_{s_{-1}s_{-2}} \subset V_{s_{-1}}$.

ii) It follows from Assumptions 1 and 2 that $V_{s_{-1}s_{-2}}$, $s_{-i} \in S$, $i = 1, 2$, is N^2 μ_v-vertical strips with N of them in each of the V_i, $i \in S$. Note that there are N^2 sequences of length two that are made up of elements of S and that the $V_{s_{-1}s_{-2}}$ can be put in one-to-one correspondence with these sequences.

iii) It follows from Assumption 2 that

$$d(V_{s_{-1}s_{-2}}) \le \nu_v d(V_{s_{-1}}) \le \nu_v. \tag{4.3.9}$$

$\underline{\Lambda_{-3}}$. We construct Λ_{-3} from Λ_{-2} as follows

$$\Lambda_{-3} = f(\Lambda_{-2}) \cap \left(\bigcup_{s_{-1}\in S} V_{s_{-1}} \right). \tag{4.3.10}$$

Hence, (4.3.10) is the set of points in $\bigcup_{s_{-1}\in S} V_{s_{-1}}$ that are mapped into Λ_{-2} under f^{-1}. Using (4.3.8), (4.3.10) becomes

$$\begin{aligned}\Lambda_{-3} &= f\left(\bigcup_{\substack{s_{-i}\in S\\ i=2,3}} f(V_{s_{-3}})\cap V_{s_{-2}}\right)\cap\left(\bigcup_{s_{-1}\in S} V_{s_{-1}}\right)\\ &= \bigcup_{\substack{s_{-i}\in S\\ i=1,2,3}} f^2(V_{s_{-3}})\cap f(V_{s_{-2}})\cap V_{s_{-1}}\\ &\equiv \bigcup_{\substack{s_{-i}\in S\\ i=1,2,3}} V_{s_{-1}s_{-2}s_{-3}},\end{aligned} \tag{4.3.11}$$

where we have the following.

i) $V_{s_{-1}s_{-2}s_{-3}} = \{p\in D\,|\,p\in V_{s_{-1}}, f^{-1}(p)\in V_{s_{-2}}, f^{-2}(p)\in V_{s_{-3}}\}$ with $V_{s_{-1}s_{-2}s_{-3}}\subset V_{s_{-1}s_{-2}}\subset V_{s_{-1}}$.

ii) It follows from Assumptions 1 and 2 that $V_{s_{-1}s_{-2}s_{-3}}$, $s_{-i}\in S$, $i=1,2,3$, is N^3 μ_v-vertical strips with N^2 of them in each of the V_i, $i\in S$. Note that there are N^3 sequences of length three made up of elements of S and that the $V_{s_{-1}s_{-2}s_{-3}}$ can be put in one-to-one correspondence with these sequences.

iii) It follows from Assumption 2 that

$$d(V_{s_{-1}s_{-2}s_{-3}})\le \nu_v d(V_{s_{-1}s_{-2}})\le \nu_v^2 d(V_{s_{-1}})\le \nu_v^2. \tag{4.3.12}$$

This procedure can be carried on indefinitely. At the $(k+1)^{\text{th}})$ step we have

$$\begin{aligned}\Lambda_{-k-1} &= f(\Lambda_{-k})\cap\left(\bigcup_{s_{-1}\in S} V_{s_{-1}}\right)\\ &= f\left(\bigcup_{\substack{s_{-i}\in S\\ i=2,\ldots,-k-1}} f^{k-1}(V_{s_{-k-1}})\cap\cdots\cap f(V_{s_{-3}})\cap V_{s_{-2}}\right)\\ &\quad\cap\left(\bigcup_{s_{-1}\in S} V_{s_{-1}}\right)\\ &= \bigcup_{\substack{s_{-i}\in S\\ i=1,\ldots,-k-1}} f^{k}(V_{s_{-k-1}})\cap\cdots\cap f^2(V_{s_{-3}})\cap f(V_{s_{-2}})\cap V_{s_{-1}}\\ &\equiv \bigcup_{\substack{s_{-i}\in S\\ i=1,\ldots,-k-1}} V_{s_{-1}\cdots s_{-k-1}},\end{aligned} \tag{4.3.13}$$

where we have the following.

i) $V_{s_{-1}\ldots s_{-k-1}} = \{p \in D \,|\, f^{-i+1}(p) \in V_{s_{-i}}, i = 1, 2, \ldots, k+1\}$ with $V_{s_{-1}\cdots s_{-k-1}} \subset V_{s_{-1}\cdots s_{-k}} \subset \cdots \subset V_{s_{-1}s_{-2}} \subset V_{s_{-1}}$.

ii) It follows from Assumptions 1 and 2 that $V_{s_{-1}\cdots s_{-k-1}}$, $s_{-i} \in S$, $i = 1, \cdots, k+1$, is N^{k+1} μ_v-vertical strips with N^k of them in each of the V_i, $i \in S$. Note that there are N^{k+1} sequences of length $k+1$ constructed from elements of S and that these sequences can be put in one-to-one correspondence with the $V_{s_{-1}\cdots s_{-k-1}}$.

iii) From Assumption 2 it follows that

$$\begin{aligned} d(V_{s_{-1}\cdots s_{-k-1}}) &\leq \nu_v d(V_{s_{-1}\cdots s_{-k}}) \leq \nu_v^2 d(V_{s_{-1}\cdots s_{-k+1}}) \\ &\leq \nu_v^3 d(V_{s_{-1}\cdots s_{-k+2}}) \leq \cdots \leq \nu_v^k d(V_{s_{-1}}) \\ &\leq \nu_v^k. \end{aligned} \tag{4.3.14}$$

It follows from Assumptions 1 and 2 that in passing to the limit as $k \to \infty$ we obtain

$$\begin{aligned} \Lambda_{-\infty} &\equiv \bigcup_{\substack{s_{-i}\in S \\ i=1,2,\ldots}} \cdots \cap f^k(V_{s_{-k-1}}) \cap \cdots \cap f(V_{s_{-2}}) \cap V_{s_{-1}} \\ &\equiv \bigcup_{\substack{s_{-i}\in S \\ i=1,2,\cdots}} V_{s_{-1}\cdots s_{-k}\cdots}, \end{aligned} \tag{4.3.15}$$

which, from Lemma 4.3.1, consists of an infinite number of μ_v-vertical curves. This follows from the fact that given *any* infinite sequence made up of elements of S, say

$$s_{-1}s_{-2}\cdots s_{-k}\cdots,$$

we have (by the construction process) an element of $\Lambda_{-\infty}$ which we denote by

$$V_{s_{-1}s_{-2}\cdots s_{-k}\cdots}.$$

Now, by construction, $V_{s_{-1}\cdots s_{-k}\cdots}$ is the intersection of the following nested sequence of sets

$$V_{s_{-1}} \supset V_{s_{-1}s_{-2}} \supset \cdots \supset V_{s_{-1}s_{-2}\cdots s_{-k}} \supset \cdots,$$

where from (4.3.14) it follows that

$$d(V_{s_{-1}\cdots s_{-k}}) \underset{k\to\infty}{\longrightarrow} 0.$$

Thus, by Lemma 4.3.1,

$$V_{s_{-1}s_{-2}\cdots s_{-k}\cdots} = \bigcap_{k=1}^{\infty} V_{s_{-1}\cdots s_{-k}}$$

consists of a μ_v-vertical curve. It also follows by construction that

$$V_{s_{-1}\cdots s_{-k}\cdots} = \{p \in D \mid f^{-i+1}(p) \in V_{s_{-i}}, i = 1, 2, \cdots\}. \tag{4.3.16}$$

We next construct Λ_∞, the set of points in $\bigcup_{i\in S} H_i$ that remain in $\bigcup_{i\in S} H_i$ under all forward iterations by f. We denote the set of points that remain in $\bigcup_{i\in S} H_i$ under n iterations by f by Λ_n. Since the construction of Λ_∞ is very similar to the construction of $\Lambda_{-\infty}$ we will leave out many details that we explicitly noted in the construction of $\Lambda_{-\infty}$.

We have

$$\Lambda_0 = \bigcup_{s_0 \in S} H_{s_0}. \tag{4.3.17}$$

$\underline{\Lambda_1}$.

$$\Lambda_1 = f^{-1}(\Lambda_0) \cap \left(\bigcup_{s_0\in S} H_{s_0} \right) \tag{4.3.18}$$

is the set of points in $\bigcup_{s_0\in S} H_{s_0}$ that map into Λ_0 under f. Therefore, using (4.3.17), (4.3.18) becomes

$$\begin{aligned} \Lambda_1 &= f^{-1}\left(\bigcup_{s_1\in S} H_{s_1} \right) \cap \left(\bigcup_{s_0\in S} H_{s_0} \right) \\ &= \bigcup_{\substack{s_i\in S \\ i=0,1}} f^{-1}(H_{s_1}) \cap H_{s_0} \equiv \bigcup_{\substack{s_i\in S \\ i=0,1}} H_{s_0 s_1}, \end{aligned} \tag{4.3.19}$$

where we have the following.

i) $H_{s_0 s_1} = \{p \in D \mid p \in H_{s_0}, f(p) \in H_{s_1}\}$.

ii) It follows from Assumptions 1 and 2 that Λ_1 consists of N^2 μ_h-horizontal strips with N of them in each of the H_i, $i \in S$. Moreover, there are N^2 sequences of length two made up of elements of S and these can be put in one-to-one correspondence with the $H_{s_0 s_1}$.

iii) From Assumption 2 we have

$$d(H_{s_0 s_1}) \le \nu_h d(H_{s_0}) \le \nu_h. \tag{4.3.20}$$

Continuing the construction in this manner and repeatedly appealing to Assumptions 1 and 2 allows us to conclude that

$$\begin{aligned} \Lambda_k &= f^{-1}(\Lambda_{k-1}) \cap \left(\bigcup_{s_0\in S} H_{s_0} \right) \\ &= \bigcup_{\substack{s_i\in S \\ i=1,\cdots,k}} f^{-k}(H_{s_k}) \cap \cdots \cap f^{-1}(H_{s_1}) \cap H_{s_0} \\ &\equiv \bigcup_{\substack{s_i\in S \\ i=1,\cdots,k}} H_{s_0\cdots s_k} \end{aligned} \tag{4.3.21}$$

consists of N^{k+1} μ_h-horizontal strips with N^k of them in each of the H_i, $i \in S$. Moreover,

$$d(H_{s_0 \cdots s_k}) \le \nu_h^{k+1}.$$

It should also be clear that

$$H_{s_0 \cdots s_k} = \{p \in D \,|\, f^i(p) \in H_{s_i}, i = 0, 1, \cdots, k\}$$

and that there are N^{k+1} sequences of length $k+1$ made up of elements of S which can be put in a one-to-one correspondence with the $H_{s_0 \cdots s_k}$.

Thus, in passing to the limit as $k \to \infty$, we obtain

$$\begin{aligned}\Lambda_\infty &= \bigcup_{\substack{s_i \in S\\ i=0,1,\cdots}} \cdots \cap f^{-k}(H_{s_k}) \cap \cdots \cap f^{-1}(H_{s_1}) \cap H_{s_0} \\ &= \bigcup_{\substack{s_i \in S\\ i=0,1,\cdots}} H_{s_0 s_1 \cdots s_k \cdots} \end{aligned} \tag{4.3.22}$$

and

$$H_{s_0 s_1 \cdots s_k \cdots} = \{p \in D \,|\, f^i(p) \in H_{s_i}, i = 0, 1, \cdots\}. \tag{4.3.23}$$

By Lemma 4.3.1, Λ_∞ consists of an infinite number of μ_h-horizontal curves. This follows from the fact that given any infinite sequence made up of elements of S, say

$$s_0 s_1 \cdots s_k \cdots ,$$

the construction process implies that there is an element of Λ_∞ which we denote by

$$H_{s_0 s_1 \cdots s_k \cdots}.$$

Now, by construction, $H_{s_0 s_1 \cdots s_k \cdots}$ is the intersection of the following nested sequence of sets

$$H_{s_0} \supset H_{s_0 s_1} \supset \cdots \supset H_{s_0 s_1 \cdots s_k} \supset \cdots ,$$

and from (4.3.22) it follows that

$$d(H_{s_0 s_1 \cdots s_k}) \underset{k \to \infty}{\longrightarrow} 0.$$

Hence, by Lemma 4.3.1, $H_{s_0 s_1 \cdots s_k \cdots}$ is a μ_h-horizontal curve.

It follows that an invariant set, i.e., a set of points that remains in D under *all* iterations by f is given by

$$\Lambda = \{\Lambda_{-\infty} \cap \Lambda_\infty\} \supset \left\{ \left(\bigcup_{i \in S} H_i \right) \cap \left(\bigcup_{i \in S} V_i \right) \right\} \subset D.$$

Moreover, by Lemma 4.3.2, since $0 \le \mu_v \mu_h < 1$, Λ is a set of discrete points. It should be clear that Λ is uncountable, and shortly we will show that it is a Cantor set.

Step 2: The Definition of $\phi: \Lambda \to \Sigma^N$. Choose any point $p \in \Lambda$; then by construction there exist two (and only two) infinite sequences

$$\begin{array}{l} s_0 s_1 \cdots s_k \cdots , \\ s_{-1} s_{-2} \cdots s_{-k} \cdots , \end{array} \qquad s_i \in S, \quad |i| = 0, \pm 1, \pm 2, \cdots ,$$

such that

$$p = V_{s_{-1}s_{-2}\cdots s_{-k}\cdots} \cap H_{s_0 s_1 \cdots s_k \cdots}. \tag{4.3.24}$$

We thus associate with every point $p \in \Lambda$ a bi-infinite sequence made up of elements of S, i.e., an element of Σ^N, as follows

$$\begin{aligned} \phi: \Lambda &\longrightarrow \Sigma^N, \\ p &\longmapsto (\cdots s_{-k} \cdots s_{-1}.s_0 s_1 \cdots s_k \cdots) \end{aligned} \tag{4.3.25}$$

where the bi-infinite sequence associated with p, $\phi(p)$, is constructed by concatenating the infinite sequence associated with the μ_v-vertical curve and the infinite sequence associated with the μ_h-horizontal curve whose intersection gives p, as indicated in (4.3.24). Since a μ_h-horizontal curve and a μ_v-vertical curve can only intersect in one point (for $0 \le \mu_v \mu_h < 1$ by Lemma 4.3.2), the map ϕ is well defined.

Now recall from (4.3.16) that

$$V_{s_{-1}s_{-2}\cdots s_{-k}\cdots} = \{p \in D \,|\, f^{-i+1}(p) \in V_{s_{-i}}, i = 1, 2, \cdots\}, \tag{4.3.26}$$

and by assumption we have

$$f(H_{s_i}) = V_{s_i};$$

thus, (4.3.26) is the same as

$$V_{s_{-1}s_{-2}\cdots s_{-k}\cdots} = \{p \in D \,|\, f^{-i}(p) \in H_{s_{-i}}, i = 1, 2, \cdots\}. \tag{4.3.27}$$

Also, by (4.3.23) we have

$$H_{s_0 s_1 \cdots s_k \cdots} = \{p \in D \,|\, f^{i}(p) \in H_{s_i}, i = 0, 1, 2, \cdots\}. \tag{4.3.28}$$

Therefore, from (4.3.25), (4.3.27), and (4.3.28) we see that the bi-infinite sequence associated with any point $p \in \Lambda$ contains information concerning the behavior of the orbit of p. In particular, from $\phi(p)$ we can determine which H_i, $i \in S$ contains $f^k(p)$, i.e., $f^k(p) \in H_{s_k}$.

Alternatively, we could have arrived at the definition of ϕ in a slightly different manner. By construction, the orbit of any $p \in \Lambda$ must remain in $\bigcup_{i \in S} H_i$. Hence, we can associate with any $p \in \Lambda$ a bi-infinite sequence made up of elements of S, i.e., an element of Σ^N, as follows

$$p \longrightarrow \phi(p) = \{\cdots s_{-k} \cdots s_{-1}.s_0 s_1 \cdots s_k \cdots\}$$

by the rule that the kth element of the sequence $\phi(p)$ is chosen to be the subscript of the H_i, $i \in S$, which contains $f^k(p)$, i.e., $f^k(p) \in H_{s_k}$. This gives a well-defined map of Λ into Σ^N since the H_i are disjoint.

Step 3: ϕ is a Homeomorphism. We must show that ϕ is one-to-one, onto, and continuous. Continuity of ϕ^{-1} will follow from the fact that one-to-one, onto, and continuous maps from compact sets into Hausdorff spaces are homeomorphisms (see Dugundji [1966]). The proof is virtually the same as for the analogous situation in Theorem 4.1.3; however, continuity presents a slight twist so, for the sake of completeness, we will give all of the details.

ϕ is One-to-One. This means that given $p, p' \in \Lambda$, if $p \neq p'$, then $\phi(p) \neq \phi(p')$.

We give a proof by contradiction. Suppose $p \neq p$ and

$$\phi(p) = \phi(p') = \{\cdots s_{-n} \cdots s_{-1}.s_0 \cdots s_n \cdots\},$$

then, by construction of Λ, p and p' lie in the intersection of a μ_v-vertical curve $V_{s_{-1}\cdots s_{-n}\cdots}$ and a μ_h-horizontal curve $H_{s_0\cdots s_n\cdots}$. However, by Lemma 4.3.2, the intersection of a μ_h-horizontal curve and a μ_v-vertical curve consists of a unique point; therefore, $p = p'$, contradicting our original assumption. This contradiction is due to the fact that we have assumed $\phi(p) = \phi(p')$; thus, for $p \neq p'$, $\phi(p) \neq \phi(p')$.

ϕ is Onto. This means that, given any bi-infinite sequence in Σ^N, say $\{\cdots s_{-n} \cdots s_{-1}.s_0 \cdots s_n \cdots\}$, there is a point $p \in \Lambda$ such that $\phi(p) = \{\cdots s_{-n} \cdots s_{-1}.s_0 \cdots s_n \cdots\}$.

The proof goes as follows. Choose $\{\cdots s_{-k} \cdots s_{-1}.s_0 s_1 \cdots s_k \cdots\} \in \Sigma^N$. Then, by construction of $\Lambda = \Lambda_{-\infty} \cap \Lambda_{\infty}$, we can find a μ_h-horizontal curve in Λ_∞ denoted $H_{s_0 s_1 \cdots s_k \cdots}$ and a μ_v-vertical curve in $\Lambda_{-\infty}$ denoted $V_{s_{-1}s_{-2}\cdots s_{-k}\cdots}$. Now, by Lemma 4.3.2, $H_{s_0 s_1 \cdots s_k \cdots}$ and $V_{s_{-1}\cdots s_{-k}\cdots}$ intersect in a unique point $p \in \Lambda$ and, by definition of ϕ, the sequence associated with p, $\phi(p)$, is given by $\{\cdots s_{-k} \cdots s_{-1}.s_0 s_1 \cdots s_k \cdots\}$.

ϕ is Continuous. This means that given any point $p \in \Lambda$ and $\varepsilon > 0$, we can find a $\delta = \delta(\varepsilon, p)$ such that

$$|p - p'| < \delta \qquad \text{implies} \quad d(\phi(p), \phi(p')) < \varepsilon,$$

where $|\cdot|$ is the usual distance measurement in $\mathbb{R}^2$ and $d(\cdot,\cdot)$ is the metric on Σ^N introduced in Section 4.2A.

Let $\varepsilon > 0$ be given; then, by Lemma 4.2.2, if we are to have $|\phi(p) - \phi(p')| < \varepsilon$ there must be some integer $N = N(\varepsilon)$ such that if

$$\begin{aligned} \phi(p) &= \{\cdots s_{-n} \cdots s_{-1}.s_0 \cdots s_n \cdots\}, \\ \phi(p') &= \{\cdots s'_{-n} \cdots s'_{-1}.s'_0 \cdots s'_n \cdots\}, \end{aligned}$$

then $s_i = s'_i$, $i = 0, \pm 1, \cdots, \pm N$. Thus, by construction of Λ, p and p' lie in the set defined by $H_{s_0\cdots s_N} \cap V_{s_{-1}\cdots s_{-N}}$. We denote the μ_v-vertical curves defining the boundary of $V_{s_{-1}\cdots s_{-N}}$ by the graphs of $x = v_1(y)$

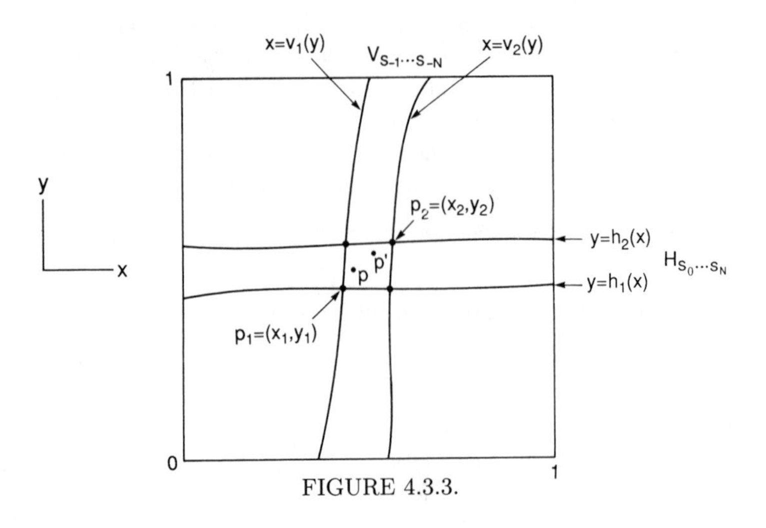

FIGURE 4.3.3.

and $x = v_2(y)$. Similarly, we denote the μ_h-horizontal curves defining the boundary of $H_{s_0 \cdots s_N}$ by the graphs of $y = h_1(x)$ and $y = h_2(x)$; see Figure 4.3.3. Note from (4.3.22) and (4.3.14) that we have

$$d(H_{s_0 \cdots s_N}) \leq \nu_h^N, \tag{4.3.29a}$$

$$d(V_{s_{-1} \cdots s_{-N}}) \leq \nu_v^{N-1}. \tag{4.3.29b}$$

Hence, from Definition 4.3.2 we have

$$\max_{y \in [0,1]} |v_1(y) - v_2(y)| \equiv \|v_1 - v_2\| \leq \nu_v^{N-1}, \tag{4.3.30a}$$

$$\max_{x \in [0,1]} |h_1(x) - h_2(x)| \equiv \|h_1 - h_2\| \leq \nu_h^N. \tag{4.3.30b}$$

The following lemma will prove useful in proving continuity of ϕ.

Lemma 4.3.4

$$|x_1 - x_2| \leq \frac{1}{1 - \mu_v \mu_h} \big[\|v_1 - v_2\| + \mu_v \|h_1 - h_2\|\big], \tag{4.3.31a}$$

$$|y_1 - y_2| \leq \frac{1}{1 - \mu_v \mu_h} \big[\|h_1 - h_2\| + \mu_h \|v_1 - v_2\|\big]. \tag{4.3.31b}$$

Proof: This follows from the following simple calculations.

$$\begin{aligned} |x_1 - x_2| &= |v_1(y_1) - v_2(y_2)| \\ &\leq |v_1(y_1) - v_1(y_2)| + |v_1(y_2) - v_2(y_2)| \\ &\leq \mu_v |y_1 - y_2| + \|v_1 - v_2\| \end{aligned} \tag{4.3.32a}$$

and

$$\begin{aligned}|y_1 - y_2| &= |h_1(x_1) - h_2(x_2)| \\ &\le |h_1(x_1) - h_1(x_2)| + |h_1(x_2) - h_2(x_2)| \\ &\le \mu_h |x_1 - x_2| + \|h_1 - h_2\|. \end{aligned} \tag{4.3.32b}$$

Substituting (4.3.32b) into (4.3.32a) gives (4.3.31a), and substituting (4.3.32a) into (4.3.32b) gives (4.3.31b). Note that these algebraic manipulations require $1 - \mu_v \mu_h > 0$. This proves Lemma 4.3.4. □

Now we can complete the proof that ϕ is continuous. Let p_1 denote the intersection of the graph of $h_1(x)$ with $v_1(y)$ and p_2 denote the intersection of the graph of $h_2(x)$ with $v_2(y)$. Now it follows that

$$|p - p'| \le |p_1 - p_2|. \tag{4.3.33}$$

We denote the coordinates of p_1 and p_2 by (x_1, y_1) and (x_2, y_2), respectively. Using (4.3.33), we obtain

$$|p - p'| \le |x_1 - x_2| + |y_1 - y_2|, \tag{4.3.34}$$

and using Lemma 4.3.4, we obtain

$$\begin{aligned}|x_1 - x_2| + |y_1 - y_2| \le \frac{1}{1 - \mu_v \mu_h} \big[&(1 + \mu_h)\|v_1 - v_2\| \\ &+ (1 + \mu_v)\|h_1 - h_2\| \big]. \end{aligned} \tag{4.3.35}$$

Using (4.3.34), (4.3.35), and (4.3.30), we obtain

$$|p - p'| \le \frac{1}{1 - \mu_v \mu_h} \left[(1 + \mu_h)\nu_v^{N-1} + (1 + \mu_v)\nu_h^N \right].$$

Hence, if we take

$$\delta = \frac{1}{1 - \mu_v \mu_h} \left[(1 + \mu_h)\nu_v^{N-1} + (1 + \mu_v)\nu_h^N \right],$$

continuity is proved.

Step 4: $\phi \circ f = \sigma \circ \phi$. Choose any $p \in \Lambda$ and let

$$\phi(p) = \{\cdots s_{-k} \cdots s_{-1}.s_0 s_1 \cdots s_k \cdots\};$$

then

$$\sigma \circ \phi(p) = \{\cdots s_{-k} \cdots s_{-1} s_0 . s_1 \cdots s_k \cdots\}. \tag{4.3.36}$$

Now, by definition of ϕ, it follows that

$$\phi \circ f(p) = \{\cdots s_{-k} \cdots s_{-1} s_0 . s_1 \cdots s_k \cdots\}; \tag{4.3.37}$$

hence, from (4.3.36) and (4.3.37), we see that

$$\phi \circ f(p) = \sigma \circ \phi(p)$$

and p is arbitrary. This completes our proof of Theorem 4.3.3. □

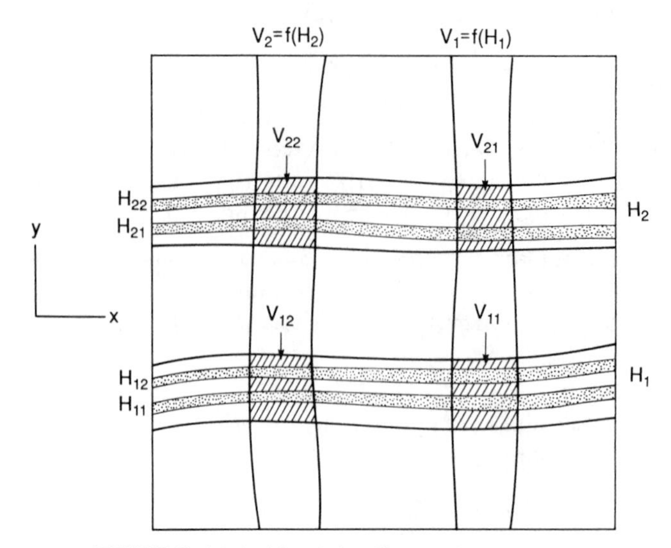

FIGURE 4.3.4. $N = 2$ for illustrative purposes.

4.3B Sector Bundles

In Sections 4.4 and 4.8 we will see that certain orbits called homoclinic orbits give rise to the geometrical conditions that allow Assumption 1 to hold in a two-dimensional map. However, a direct verification of Assumption 2 is not easy. When one thinks of stretching and contraction rates of maps, it is natural to think of the properties of the derivative of the map at different points. We now want to derive a condition that is equivalent to Assumption 2 that is based solely on properties of the derivative of f (hence we must assume that f is at least $\mathbf{C}^1$). We begin by establishing some notation.

We define

$$f(H_i) \cap H_j \equiv V_{ji} \tag{4.3.38}$$

and

$$H_i \cap f^{-1}(H_j) \equiv H_{ij} = f^{-1}(V_{ji}) \tag{4.3.39}$$

for $i, j \in S$ where $S = \{1, \dots, N\}$ $(N \geq 2)$ is an index set; see Figure 4.3.4. We further define

$$\mathcal{H} = \bigcup_{i,j \in S} H_{ij}$$

and

$$\mathcal{V} = \bigcup_{i,j \in S} V_{ji}.$$

It should be clear that

$$f(\mathcal{H}) = \mathcal{V}.$$

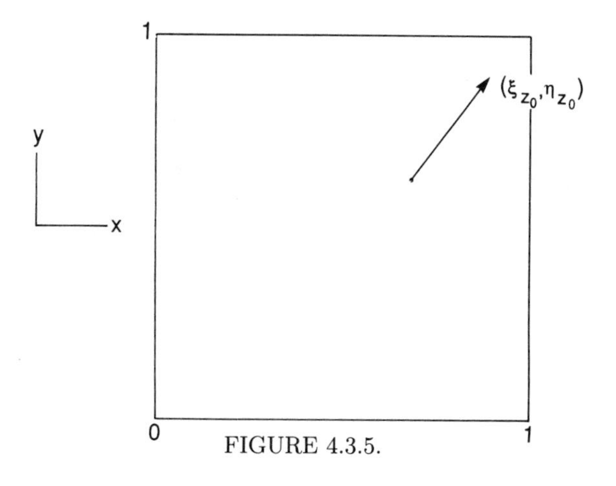

FIGURE 4.3.5.

We now want to strengthen our requirements on f by assuming that f *maps* $\mathcal{H}$ $\mathbf{C}^1$ *diffeomorphically onto* $\mathcal{V}$.

For any point $z_0 = (x_0, y_0) \in \mathcal{H} \bigcup \mathcal{V}$, we denote a vector emanating from this point by $(\xi_{z_0}, \eta_{z_0}) \in \mathbb{R}^2$; see Figure 4.3.5. We define the *stable sector at* z_0 as follows

$$\mathcal{S}^s_{z_0} = \left\{ (\xi_{z_0}, \eta_{z_0}) \in \mathbb{R}^2 \mid |\eta_{z_0}| \leq \mu_h |\xi_{z_0}| \right\}. \tag{4.3.40}$$

Geometrically, $\mathcal{S}^s_{z_0}$ defines a cone of vectors emanating from z_0, where μ_h is the maximum of the absolute value of the slope of any vector in the cone and slope is measured with respect to the x-axis; see Figure 4.3.6. Similarly, the *unstable sector at* z_0 is defined as

$$\mathcal{S}^u_{z_0} = \left\{ (\xi_{z_0}, \eta_{z_0}) \in \mathbb{R}^2 \mid |\xi_{z_0}| \leq \mu_v |\eta_{z_0}| \right\}. \tag{4.3.41}$$

Geometrically, $\mathcal{S}^u_{z_0}$ defines a cone of vectors emanating from z_0, where μ_v is the maximum of the absolute value of the slope of any vector in the cone and slope is measured with respect to the y-axis; see Figure 4.3.6. We will put restrictions on μ_v and μ_h shortly.

We take the union of the stable and unstable sectors over points in $\mathcal{H}$ and $\mathcal{V}$ to form *sector bundles* as follows

$$\mathcal{S}^s_{\mathcal{H}} = \bigcup_{z_0 \in \mathcal{H}} \mathcal{S}^s_{z_0},$$

$$\mathcal{S}^s_{\mathcal{V}} = \bigcup_{z_0 \in \mathcal{V}} \mathcal{S}^s_{z_0},$$

$$\mathcal{S}^u_{\mathcal{H}} = \bigcup_{z_0 \in \mathcal{H}} \mathcal{S}^u_{z_0},$$

$$\mathcal{S}^u_{\mathcal{V}} = \bigcup_{z_0 \in \mathcal{H}} \mathcal{S}^u_{z_0}.$$

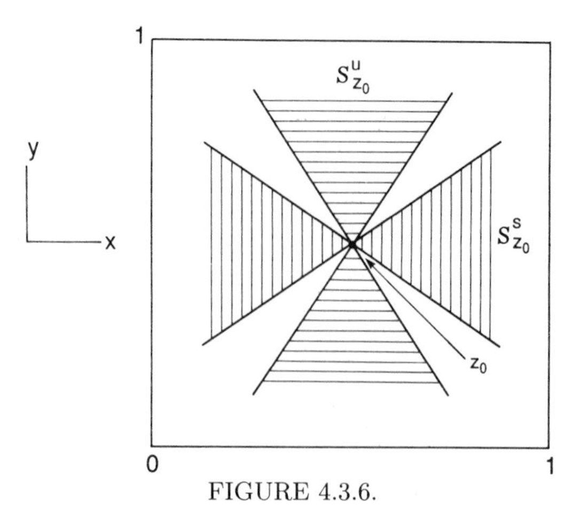

FIGURE 4.3.6.

We refer to $\mathcal{S}^s_{\mathcal{H}}$ as the *stable sector bundle over* $\mathcal{H}$, $\mathcal{S}^s_{\mathcal{V}}$ as the *stable sector bundle over* $\mathcal{V}$, $\mathcal{S}^u_{\mathcal{H}}$ as the *unstable sector bundle over* $\mathcal{H}$, and $\mathcal{S}^u_{\mathcal{V}}$ as the *unstable sector bundle over* $\mathcal{V}$.

Now we can state our alternative to Assumption 2.

Assumption 3. $Df(\mathcal{S}^u_{\mathcal{H}}) \subset \mathcal{S}^u_{\mathcal{V}}$ and $Df^{-1}(\mathcal{S}^s_{\mathcal{V}}) \subset \mathcal{S}^s_{\mathcal{H}}$.

Moreover, if $(\xi_{z_0}, \eta_{z_0}) \in \mathcal{S}^u_{z_0}$ and $Df(z_0)(\xi_{z_0}, \eta_{z_0}) \equiv (\xi_{f(z_0)}, \eta_{f(z_0)}) \in \mathcal{S}^u_{f(z_0)}$, then we have

$$|\eta_{f(z_0)}| \geq \left(\frac{1}{\mu}\right) |\eta_{z_0}|.$$

Similarly, if $(\xi_{z_0}, \eta_{z_0}) \in \mathcal{S}^s_{z_0}$ and $Df^{-1}(z_0)(\xi_{z_0}, \eta_{z_0}) \equiv (\xi_{f^{-1}(z_0)}, \eta_{f^{-1}(z_0)}) \in \mathcal{S}^s_{f^{-1}(z_0)}$, then

$$|\xi_{f^{-1}(z_0)}| \geq \left(\frac{1}{\mu}\right) |\xi_{z_0}|,$$

where $0 < \mu < 1 - \mu_v \mu_h$; see Figure 4.3.7. We remark that the notation $Df(\mathcal{S}^u_{\mathcal{H}}) \subset \mathcal{S}^u_{\mathcal{V}}$ is somewhat abbreviated. More completely, it means that for every $z_0 \in \mathcal{H}$, $(\xi_{z_0}, \eta_{z_0}) \in \mathcal{S}^u_{z_0}$, we have $Df(z_0)(\xi_{z_0}, \eta_{z_0}) \equiv (\xi_{f(z_0)}, \eta_{f(z_0)}) \in \mathcal{S}^u_{f(z_0)}$; similarly for $Df^{-1}(\mathcal{S}^s_{\mathcal{V}}) \subset \mathcal{S}^s_{\mathcal{H}}$. We now state the main theorem.

Theorem 4.3.5 *If Assumptions 1 and 3 hold with* $0 < \mu < 1 - \mu_v \mu_h$, *then Assumption 2 holds with* $\nu_h = \nu_v = \mu/(1 - \mu_v \mu_h)$.

Proof: We will prove only the part concerning horizontal strips; the part concerning vertical strips is proven similarly. The proof consists of several steps.

Step 1. Let $\overline{H}$ be a μ_h-horizontal curve contained in $\bigcup_{j \in S} H_j$. Then show $f^{-1}(\overline{H}) \cap H_i \equiv \overline{H}_i$ is a μ_h-horizontal curve contained in H_i for all $i \in S$.

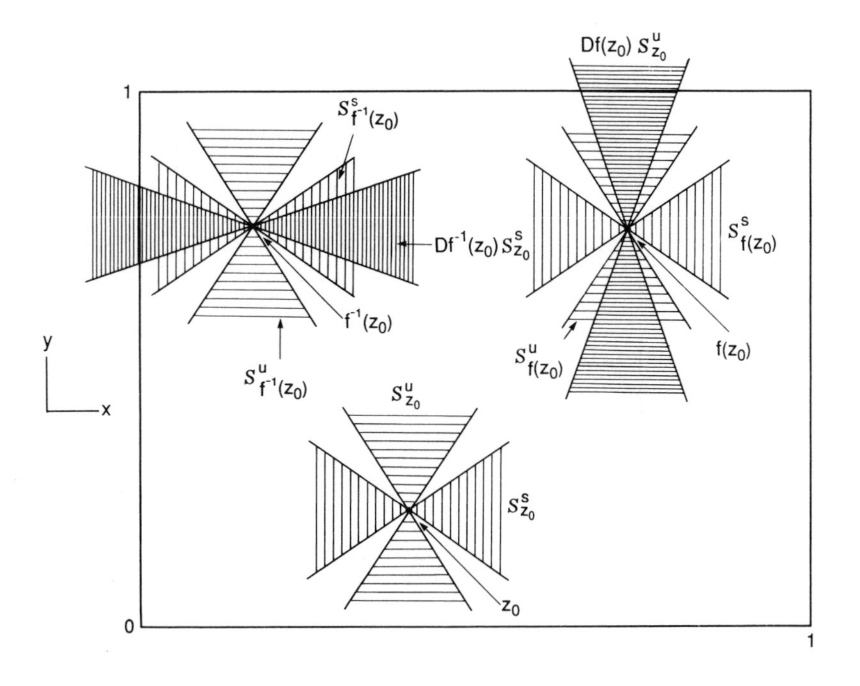

FIGURE 4.3.7.

Step 2. Let H be a μ_h-horizontal strip contained in $\bigcup_{j\in S} H_j$. Then use Step 1 to show that $f^{-1}(H) \cap H_i \equiv \tilde{H}_i$ is a μ_h-horizontal strip for each $i \in S$.

Step 3. Show that $d(\tilde{H}_i) \leq (\mu/(1-\mu_v\mu_h))\, d(H)$. We begin with Step 1.

Step 1. Let $\overline{H} \subset \bigcup_{j\in S} H_j$ be a μ_h-horizontal curve. Then $\overline{H}$ intersects both vertical boundaries of each V_i $\forall\, i \in S$. Hence, $f^{-1}(\overline{H}) \cap H_i$ is a curve for each $i \in S$ by Assumption 1.

Next we argue that $f^{-1}(\overline{H}) \cap H_i$ is a μ_h-horizontal curve $\forall\, i \in S$.

This follows from Assumption 3, since Df^{-1} maps $\mathcal{S}^s_{\mathcal{V}}$ into $\mathcal{S}^s_{\mathcal{H}}$. Let (x_1, y_1), (x_2, y_2) be any two points on $f^{-1}(\overline{H}) \cap H_i$, i fixed; then by the mean value theorem

$$|y_1 - y_2| \leq \mu_h |x_1 - x_2|.$$

Thus, $f^{-1}(\overline{H}) \cap H_i$ is the graph of a μ_h-horizontal curve $y = h(x)$.

Step 2. Let $H \subset \bigcup_{j\in S} H_j$ be a μ_h-horizontal strip. Then applying Step 1 to the horizontal boundaries of H shows that $f^{-1}(H) \cap H_i$ is a μ_h-horizontal strip for every $i \in S$.

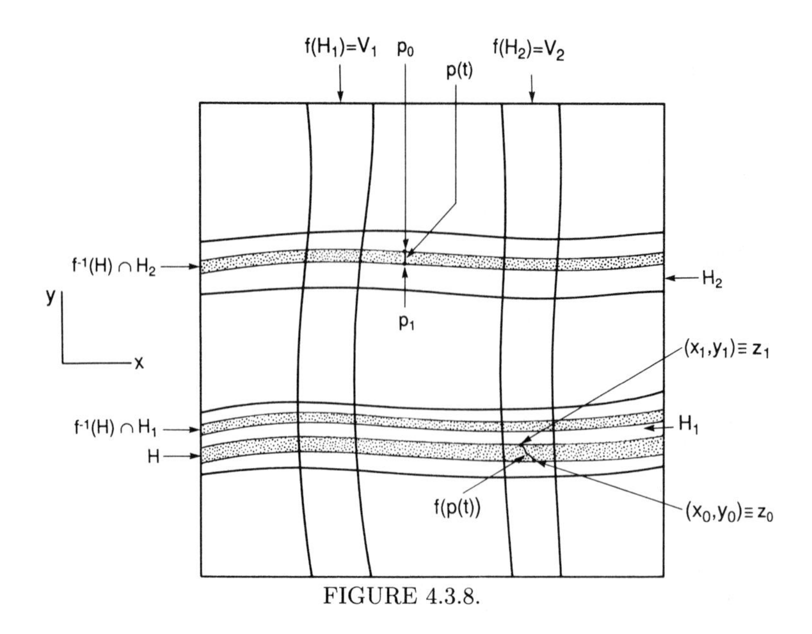

FIGURE 4.3.8.

Step 3. Fix i and choose points p_0 and p_1 on the horizontal boundary of $\tilde{H}_i$ *having the same x-components* such that

$$d(\tilde{H}_i) = |p_0 - p_1|. \tag{4.3.42}$$

Consider the vertical line connecting p_0 and p_1 defined as follows

$$p(t) = tp_1 + (1-t)p_0, \qquad 0 \le t \le 1;$$

see Figure 4.3.8. Then it should be obvious that

$$\dot{p}(t) = p_1 - p_0 \in \mathcal{S}^u_{\mathcal{H}} \qquad \forall\, 0 \le t \le 1.$$

Next we consider the image of $p(t)$ under f, which we denote by

$$f(p(t)) \equiv z(t) = (x(t), y(t)), \qquad 0 \le t \le 1.$$

It should be clear the $z(t)$ is a curve connecting the two horizontal boundaries of H as shown in Figure 4.3.8. We denote the endpoints of the curve by

$$f(p(0)) \equiv z_0 = (x_0, y_0)$$

and

$$f(p(1)) \equiv z_1 = (x_1, y_1).$$

Moreover, since H is μ_h-horizontal, z_0 lies on a μ_h-horizontal curve that we denote by $y = h_0(x)$ and z_1 lies on a μ_h-horizontal curve that we denote by $y = h_1(x)$. Tangent vectors to $z(t)$ are given by

$$\dot{z}(t) = Df(p(t))\dot{p}(t). \tag{4.3.43}$$

Using (4.3.43) and the fact that from Assumption 3 $Df(\mathcal{S}^u_{\mathcal{H}}) \subset \mathcal{S}^u_{\mathcal{V}}$, we conclude that $z(t)$ is a μ_v-vertical curve. Therefore, applying Lemma 4.3.4 to $z(t)$, $y = h_0(x)$, and $y = h_1(x)$ we obtain

$$|y_0 - y_1| \leq \frac{1}{1 - \mu_v \mu_h} \|h_0 - h_1\| = \frac{1}{1 - \mu_v \mu_h} d(H). \tag{4.3.44}$$

Also by Assumption 3 we have

$$|\dot{y}(t)| \geq \frac{1}{\mu} |\dot{p}(t)| = \frac{1}{\mu} |p_1 - p_0|. \tag{4.3.45}$$

Integrating (4.3.45) gives

$$|p_1 - p_0| \leq \mu \int_0^1 |\dot{y}(t)| dt \leq \mu |y_1 - y_0|.. \tag{4.3.46}$$

Finally, (4.3.44) and (4.3.46) along with (4.3.42) give

$$d(\tilde{H}_i) \leq \frac{\mu}{1 - \mu_v \mu_h} d(H). \qquad \square$$

4.3c Hyperbolic Invariant Sets

We now want to show that the invariant Cantor set Λ has a very special structure. In particular, it is an example of a *hyperbolic invariant set.* We have the following definition.

Definition 4.3.3 Let $f: \mathbb{R}^2 \to \mathbb{R}^2$ be a $\mathbf{C}^r$ $(r \geq 1)$ diffeomorphism and let $\Lambda \subset \mathbb{R}^2$ be a compact set that is invariant under f. Then we say that Λ is a *hyperbolic invariant set* if

1. At each point $z_0 \in \Lambda$ there exists a pair of lines, $E^s_{z_0}$ and $E^u_{z_0}$, that are invariant under $Df(z_0)$ in the sense that

$$Df(z_0) E^s_{z_0} = E^s_{f(z_0)}$$

and

$$Df(z_0) E^u_{z_0} = E^u_{f(z_0)}.$$

2. There exists a constant $0 < \lambda < 1$ such that if

$$\zeta_{z_0} = (\xi_{z_0}, \eta_{z_0}) \in E^s_{z_0}, \qquad \text{then } |Df(z_0)\zeta_{z_0}| < \lambda |\zeta_{z_0}|$$

and if

$$\zeta_{z_0} = (\xi_{z_0}, \eta_{z_0}) \in E^u_{z_0}, \qquad \text{then } |Df^{-1}(z_0)\zeta_{z_0}| < \lambda |\zeta_{z_0}|,$$

where $|\zeta_{z_0}| = \sqrt{(\xi_{z_0})^2 + (\eta_{z_0})^2}$.

3. $E^s_{z_0}$ and $E^u_{z_0}$ vary continuously with $z_0 \in \Lambda$.

We make the following remarks concerning this definition.

Remark 1. It should be clear that hyperbolic fixed points and hyperbolic periodic orbits are examples of hyperbolic invariant sets.

Remark 2. $E^s \equiv \bigcup_{z_0 \in \Lambda} E^s_{z_0}$ and $E^u \equiv \bigcup_{z_0 \in \Lambda} E^u_{z_0}$ are called the invariant stable and unstable line bundles over Λ, respectively.

We now state the theorem that Λ is a hyperbolic invariant set.

Theorem 4.3.6 (Moser [1973]) *Consider the* $\mathbf{C}^r$ $(r \geq 1)$ *diffeomorphism* f *and its invariant set* Λ *described in Theorem 4.3.5. Let* $\Delta = \sup_\Lambda(\det Df)$*. Then if*

$$\Delta, \Delta^{-1} \leq \mu^{-2},$$

where $0 < \mu < 1 - \mu_v\mu_h$*,* Λ *is a hyperbolic invariant set.*

Proof: We begin by constructing the unstable invariant line bundle over Λ, $E^u \equiv \bigcup_{z_0 \in \Lambda} E^u_{z_0}$. First, we want to recall some important points from the hypotheses of Theorem 4.3.5.

i)
$$\mathcal{S}^u_{z_0} = \left\{(\xi_{z_0}, \eta_{z_0}) \in \mathbb{R}^2 \mid |\xi_{z_0}| \leq \mu_v |\eta_{z_0}|\right\}. \tag{4.3.47}$$

ii)
$$Df(\mathcal{S}^u_{\mathcal{H}}) \subset \mathcal{S}^u_{\mathcal{V}}. \tag{4.3.48}$$

iii) For $(\xi_{z_0}, \eta_{z_0}) \in \mathcal{S}^u_{z_0}$, $Df(z_0)(\xi_{z_0}, \eta_{z_0}) \equiv (\xi_{f(z_0)}, \eta_{f(z_0)}) \in \mathcal{S}^u_{f(z_0)}$, we have
$$|\eta_{f(z_0)}| \geq \frac{1}{\mu}|\eta_{z_0}|, \tag{4.3.49}$$
where $0 < \mu < 1 - \mu_v\mu_h$.

The construction of E^u will be by the contraction mapping principle. We define

$$\mathcal{L}^u_\Lambda = \{\text{continuous line bundles over } \Lambda \text{ contained in } \mathcal{S}^u_\Lambda\}.$$

"Points" in $\mathcal{L}^u_\Lambda$ will be denoted by

$$\begin{aligned}
\mathcal{L}^u_\Lambda(\alpha(z_0)) &\equiv \bigcup_{z_0 \in \Lambda} L^u_{\alpha(z_0)}, \\
\mathcal{L}^u_\Lambda(\beta(z_0)) &\equiv \bigcup_{z_0 \in \Lambda} L^u_{\beta(z_0)},
\end{aligned} \tag{4.3.50}$$

where

$$\begin{aligned}
L^u_{\alpha(z_0)} &= \left\{(\xi_{z_0}, \eta_{z_0}) \in \mathbb{R}^2 \mid \xi_{z_0} = \alpha(z_0)\eta_{z_0}\right\}, \\
L^u_{\beta(z_0)} &= \left\{(\xi_{z_0}, \eta_{z_0}) \in \mathbb{R}^2 \mid \xi_{z_0} = \beta(z_0)\eta_{z_0}\right\},
\end{aligned} \tag{4.3.51}$$

with $\alpha(z_0)$, $\beta(z_0)$ continuous functions on Λ and

$$\sup_{z_0 \in \Lambda} |\alpha(z_0)| \leq \mu_v,$$

$$\sup_{z_0 \in \Lambda} |\beta(z_0)| \leq \mu_v.$$

As notation for a line in a line bundle, say $\mathcal{L}_\Lambda^u(\alpha(z_0))$, at a point $z_0 \in \Lambda$ we have

$$\left(\mathcal{L}_\Lambda^u(\alpha(z_0))\right)_{z_0} \equiv L_{\alpha(z_0)}^u.$$

$\mathcal{L}_\Lambda^u$ is a complete metric space with metric defined by

$$\|\mathcal{L}_\Lambda^u(\alpha(z_0)) - \mathcal{L}_\Lambda^u(\beta(z_0))\| \equiv \sup_{z_0 \in \Lambda} |\alpha(z_0) - \beta(z_0)|. \tag{4.3.52}$$

From (4.3.52), the geometrical meaning of the continuity of a line bundle should be clear.

We define a map on $\mathcal{L}_\Lambda^u$ as follows. For any $\mathcal{L}_\Lambda^u(\alpha(z_0)) \in \mathcal{L}_\Lambda^u$ we have

$$\left(F(\mathcal{L}_\Lambda^u(\alpha(z_0)))\right)_{z_0} \equiv Df(f^{-1}(z_0)) L_{\alpha(f^{-1}(z_0))}^u. \tag{4.3.53}$$

From the fact that $Df(\mathcal{S}_\mathcal{H}^u) \subset \mathcal{S}_\mathcal{V}^u$, it follows that

$$F(\mathcal{L}_\Lambda^u) \subset \mathcal{L}_\Lambda^u. \tag{4.3.54}$$

We now show that F is a contraction map. Choose $\mathcal{L}_\Lambda^u(\alpha(z_0)), \mathcal{L}_\Lambda^u(\beta(z_0)) \in \mathcal{L}_\Lambda^u$; then we must show that

$$\|F(\mathcal{L}_\Lambda^u(\alpha(z_0))) - F(\mathcal{L}_\Lambda^u(\beta(z_0)))\| \leq k\|\mathcal{L}_\Lambda^u(\alpha(z_0)) - \mathcal{L}_\Lambda^u(\beta(z_0))\|, \tag{4.3.55}$$

where $0 < k < 1$.

From (4.3.54),

$$F(\mathcal{L}_\Lambda^u(\alpha(z_0))) = \mathcal{L}_\Lambda^u(\alpha^*(z_0)) \in \mathcal{L}_\Lambda^u,$$
$$F(\mathcal{L}_\Lambda^u(\beta(z_0))) = \mathcal{L}_\Lambda^u(\beta^*(z_0)) \in \mathcal{L}_\Lambda^u,$$

and we must compute $\alpha^*(z_0)$ and $\beta^*(z_0)$ in order to verify (4.3.55).

Let us denote, for simplicity of notation,

$$Df \equiv \begin{pmatrix} a & b \\ c & d \end{pmatrix}, \tag{4.3.56}$$

where, of course, a, b, c, and d are the appropriate partial derivatives of $f \equiv (f_1, f_2)$ and are therefore functions of z_0. However, to carry this along in the formulae would result in very cumbersome expressions. Hence, we remind the reader to think of the partial derivatives a, b, c, and d as being evaluated at the same point as $Df(\cdot)$ even though we will not explicitly display this dependence.

Using (4.3.56), we have

$$\begin{aligned} a\xi_{z_0} + b\eta_{z_0} &= \xi_{f(z_0)}, \\ c\xi_{z_0} + d\eta_{z_0} &= \eta_{f(z_0)}. \end{aligned} \tag{4.3.57}$$

Consider an arbitrary line

$$L^u_{\alpha(z_0)} = \{(\xi_{z_0}, \eta_{z_0}) \in \mathbb{R}^2 \mid \xi_{z_0} = \alpha(z_0)\eta_{z_0}\} \in \mathcal{L}^u_\Lambda(\alpha(z_0)); \tag{4.3.58}$$

then

$$Df(f^{-1}(z_0))L^u_{\alpha(f^{-1}(z_0))} \equiv L^u_{\alpha^*(z_0)} \tag{4.3.59}$$

and, using (4.3.57) and (4.3.58), we have

$$\xi_{z_0} = \left(\frac{a\alpha(f^{-1}(z_0)) + b}{c\alpha(f^{-1}(z_0)) + d}\right)\eta_{z_0} \equiv \alpha^*(z_0)\eta_{z_0}. \tag{4.3.60}$$

Thus,

$$\alpha^*(z_0) \equiv \frac{a\alpha(f^{-1}(z_0)) + b}{c\alpha(f^{-1}(z_0)) + d} \tag{4.3.61}$$

is also continuous over Λ. We have thus shown that

$$F(\mathcal{L}^u_\Lambda(\alpha(z_0))) = \mathcal{L}^u_\Lambda(\alpha^*(z_0)),$$

where

$$\alpha^*(z_0) = \frac{a\alpha(f^{-1}(z_0)) + b}{c\alpha(f^{-1}(z_0)) + d}.$$

Similarly,

$$F(\mathcal{L}^u_\Lambda(\beta(z_0))) = \mathcal{L}^u_\Lambda(\beta^*(z_0)),$$

where

$$\beta^*(z_0) = \frac{a\beta(f^{-1}(z_0)) + b}{c\beta(f^{-1}(z_0)) + d}. \tag{4.3.62}$$

Now, using (4.3.52), we have

$$\|F(\mathcal{L}^u_\Lambda(\alpha(z_0))) - F(\mathcal{L}^u_\Lambda(\beta(z_0)))\| = \sup_{z_0 \in \Lambda} |\alpha^*(z_0) - \beta^*(z_0)|. \tag{4.3.63}$$

Using (4.3.61) and (4.3.62), we obtain

$$|\alpha^*(z_0) - \beta^*(z_0)| \leq \Delta \frac{|\alpha(f^{-1}(z_0)) - \beta(f^{-1}(z_0))|}{|c\alpha(f^{-1}(z_0)) + d|\,|c\beta(f^{-1}(z_0)) + d|}. \tag{4.3.64}$$

From (4.3.49) and (4.3.57) we have

$$\frac{|\eta_{z_0}|}{|\eta_{f^{-1}(z_0)}|} = \left|c\frac{\xi_{f^{-1}(z_0)}}{\eta_{f^{-1}(z_0)}} + d\right| \geq \frac{1}{\mu} \tag{4.3.65}$$

and, therefore, since $\xi_{f^{-1}(z_0)} = \alpha(f^{-1}(z_0))\eta_{f^{-1}(z_0)}$, we have

$$|c\alpha(f^{-1}(z_0)) + d|, \quad |c\beta(f^{-1}(z_0)) + d| \geq \frac{1}{\mu} \tag{4.3.66}$$

where $0 < \mu < 1 - \mu_v\mu_h$.

Combining (4.3.64) and (4.3.66) gives

$$|\alpha^*(z_0) - \beta^*(z_0)| \leq \mu^2\Delta|\alpha(f^{-1}(z_0)) - \beta(f^{-1}(z_0))|. \tag{4.3.67}$$

Note that since Λ is invariant under f, we have

$$\sup_{z_0\in\Lambda} |\alpha(f^{-1}(z_0)) - \beta(f^{-1}(z_0))| = \sup_{z_0\in\Lambda} |\alpha(z_0) - \beta(z_0)|. \tag{4.3.68}$$

Taking the supremum of (4.3.67) over $z_0 \in \Lambda$ and using (4.3.68), (4.3.52), and (4.3.63) gives

$$\begin{aligned} &\|F\big(\mathcal{L}^u_\Lambda(\alpha(z_0))\big) - F\big(\mathcal{L}^u_\Lambda(\beta(z_0))\big)\| \\ &\qquad \leq \mu^2\Delta\|\mathcal{L}^u_\Lambda(\alpha(z_0)) - \mathcal{L}^u_\Lambda(\beta(z_0))\|; \end{aligned} \tag{4.3.69}$$

thus F is a contraction map provided

$$\mu^2\Delta < 1. \tag{4.3.70}$$

Therefore, by the contraction mapping principle, F has a unique, continuous fixed point. We denote this fixed point by

$$E^u = \bigcup_{z_0\in\Lambda} E^u_{z_0}. \tag{4.3.71}$$

Thus we have

$$F(E^u) = E^u$$

or, from (4.3.59),

$$(F(E^u))_{z_0} = Df(f^{-1}(z_0))E^u_{f^{-1}(z_0)} = E^u_{z_0},$$

which is what we wanted to construct. The construction of the stable line bundle over Λ is virtually identical and we leave it as an exercise for the reader.

This shows that Λ satisfies Part 1 of Definition 4.3.3. Part 3 follows from continuity of the fixed point of F (by the contraction mapping principle). Continuity is measured with respect to the metric (4.3.52) so the geometrical meaning should be clear. Part 2 of Definition 4.3.3, expansion and contraction rates, is a trivial consequence of Assumption 3 that we leave as an exercise for the reader (see Exercise 4.1). □

Theorem 4.3.6 gives us information on the linearization of f at each point of Λ, but we can also obtain information on f itself. Recall from the proof of Theorem 4.3.3 that

$$\Lambda = \Lambda_{-\infty} \cap \Lambda_{\infty},$$

with

$$\Lambda_{-\infty} = \bigcup_{\substack{s_i \in S \\ i=1,2,\cdots}} V_{s_{-1}\cdots s_{-k}\cdots},$$

$$\Lambda_{\infty} = \bigcup_{\substack{s_i \in S \\ i=0,1,\cdots}} H_{s_0\cdots s_k\cdots},$$

where, for each infinite sequence of elements of S, $V_{s_{-1}\cdots s_{-k}\cdots}$ and $H_{s_0\cdots s_k\cdots}$ are μ_v-vertical and μ_h-horizontal curves, respectively, with $0 \le \mu_v\mu_h < 1$. Thus, for any $z_0 \in \Lambda$, there exists a unique μ_v-vertical curve in $\Lambda_{-\infty}$, $V_{s_{-1}\cdots s_{-k}\cdots}$, and a unique μ_h-horizontal curve in Λ_∞, $H_{s_0\cdots s_k\cdots}$, such that

$$z_0 = V_{s_{-1}\cdots s_{-k}\cdots} \cap H_{s_0\cdots s_k\cdots}.$$

Moreover, we can prove the following theorem.

Theorem 4.3.7 (Moser [1973]) *Consider the* $\mathbf{C}^r$ $(r \ge 1)$ *diffeomorphism* f *and its invariant set* Λ *described in Theorem 4.3.5. Let* $\Delta = \sup_\Lambda(\det Df)$. *Then if*

$$0 < \mu \le \min\left(\sqrt{|\Delta|}, \frac{1}{\sqrt{|\Delta|}}\right),$$

the curves in $\Lambda_{-\infty}$ *and* Λ_∞ *are* $\mathbf{C}^1$ *curves whose tangents at points in* Λ *coincide with* E^u *and* E^s, *respectively.*

Proof: The proof follows the same ideas as Theorem 4.3.7; in Exercise 4.2 we outline the steps one must complete to establish the theorem. □

In some sense we can think of these curves as defining the stable and unstable manifolds of points in Λ. We must first give some definitions.

For any point $p \in \Lambda$, $\varepsilon > 0$, the stable and unstable sets of p of size ε are defined as follows

$$W^s_\varepsilon(p) = \{p' \in \Lambda \,|\, |f^n(p) - f^n(p')| \le \varepsilon \text{ for } n \ge 0\},$$
$$W^u_\varepsilon(p) = \{p' \in \Lambda \,|\, |f^{-n}(p) - f^{-n}(p')| \le \varepsilon \text{ for } n \ge 0\}.$$

From Section 1.1C we have seen that if p is a hyperbolic fixed point the following hold.

1. For ε sufficiently small, $W^s_\varepsilon(p)$ is a $\mathbf{C}^r$ manifold tangent to E^s_p at p and having the same dimension as E^s_p. $W^s_\varepsilon(p)$ is called the local stable manifold of p.

2. The stable manifold of p is defined as follows

$$W^s(p) = \bigcup_{n=0}^{\infty} f^{-n}(W^s_\varepsilon(p)).$$

Similar statements hold for $W^u_\varepsilon(p)$.

The invariant manifold theorem for hyperbolic invariant sets (see Hirsch, Pugh, and Shub [1977]) tells us that a similar structure holds for each point in Λ.

Theorem 4.3.8 *Let Λ be a hyperbolic invariant set of a $\mathbf{C}^r$ $(r \geq 1)$ diffeomorphism f. Then, for $\varepsilon > 0$ sufficiently small and for each point $p \in \Lambda$, the following hold.*

i) *$W^s_\varepsilon(p)$ and $W^u_\varepsilon(p)$ are $\mathbf{C}^r$ manifolds tangent to E^s_p and E^u_p, respectively, at p and having the same dimension as E^s_p and E^u_p, respectively.*

ii) *There are constants $C > 0$, $0 < \lambda < 1$, such that if $p' \in W^s_\varepsilon(p)$, then $|f^n(p) - f^n(p')| \leq C\lambda^n|p - p'|$ for $n \geq 0$ and if $p' \in W^u_\varepsilon(p)$, then $|f^{-n}(p) - f^{-n}(p')| \leq C\lambda^n|p - p'|$ for $n \geq 0$.*

iii)

$$f(W^s_\varepsilon(p)) \subset W^s_\varepsilon(f(p)),$$
$$f^{-1}(W^u_\varepsilon(p)) \subset W^u_\varepsilon(f^{-1}(p)).$$

iv) *$W^s_\varepsilon(p)$ and $W^u_\varepsilon(p)$ vary continuously with p.*

Proof: See Hirsch, Pugh, and Shub [1977]. □

We leave it as an exercise for the reader to verify that the curves in $\Lambda_{-\infty}$ and Λ_∞ play the role of $W^u_\varepsilon(p)$ and $W^s_\varepsilon(p)$, respectively, near points in the hyperbolic set constructed in Theorem 4.3.6. With Theorem 4.3.8 in hand, one can then define the global stable and unstable manifolds of any point $p \in \Lambda$ as follows

$$W^s(p) = \bigcup_{n=0}^{\infty} f^{-n}(W^s_\varepsilon(f^n(p))),$$
$$W^u(p) = \bigcup_{n=0}^{\infty} f^{n}(W^u_\varepsilon(f^{-n}(p))).$$

We refer the reader to the exercises for more detailed studies of the issues raised here in the context of the hyperbolic invariant set constructed in Theorem 4.3.6 and we end this section with some final remarks.

Remark 1. The notion of hyperbolic structure and stable and unstable manifolds of hyperbolic invariant sets goes through in arbitrary dimensions; see Hirsch, Pugh, and Shub [1977] and Shub [1987].

Remark 2. Hyperbolic invariant sets (like hyperbolic fixed points and periodic orbits) are structurally stable; see Hirsch, Pugh, and Shub [1977].

Remark 3. The reason for discussing the concept of hyperbolic invariant sets is that they have played a central role in the development of modern dynamical systems theory. For example, the ideas of Markov partitions, pseudoorbits, shadowing, etc., all utilize crucially the notion of a hyperbolic invariant set. Indeed, the existence of a hyperbolic invariant set is often assumed a priori. This has caused the applied scientist great difficulty since, in order to utilize many of the techniques or theorems of dynamical systems theory, he or she must first show that the system under study possesses a hyperbolic invariant set. The techniques developed here, specifically the preservation of sector bundles, allow one to explicitly construct hyperbolic invariant sets. These techniques generalize to higher dimensions; see Newhouse and Palis [1973] and Wiggins [1988]. For more information on the consequences and utilization of hyperbolic invariant sets see Smale [1967], Nitecki [1971], Bowen [1970], [1978], Conley [1978], Shub [1987], and Franks [1982].

4.4 Dynamics Near Homoclinic Points of Two-Dimensional Maps

In Section 4.3 we gave sufficient conditions for a two-dimensional map to possess an invariant Cantor set on which it is topologically conjugate to a full shift on N symbols. In this section we want to show that the existence of certain orbits of a two-dimensional map, specifically, transverse homoclinic orbits to a hyperbolic fixed point, imply that in a sufficiently small neighborhood of a point on the homoclinic orbit, the conditions given in Section 4.3 hold. There are two very similar theorems which deal with this situation; Moser's theorem (see Moser [1973]) and the Smale–Birkhoff homoclinic theorem (see Smale [1963]). We will prove Moser's theorem and describe how the Smale–Birkhoff theorem differs.

The situation that we are considering is as follows: let

$$f: \mathbb{R}^2 \longrightarrow \mathbb{R}^2$$

be a $\mathbf{C}^r$ $(r \geq 1)$ diffeomorphism satisfying the following hypotheses.

Hypothesis 1. f has a hyperbolic periodic point, p.

Hypothesis 2. $W^s(p)$ and $W^u(p)$ intersect transversely.

Without loss of generality we can assume that the hyperbolic periodic point, p, is a fixed point, for if p has period k, then $f^k(p) = p$ and the following arguments can be applied to f^k. A point that is in $W^s(p) \cap W^u(p)$ is said to be *homoclinic to p*. If $W^s(p)$ intersects $W^u(p)$ transversely in a point, then the point is called a *transverse homoclinic point*. Our goal is to show that in a neighborhood of a transverse homoclinic point there exists an invariant Cantor set on which the dynamics are topologically conjugate to a full shift on N symbols. Before formulating this as a theorem, a few preliminary steps need to be taken.

Step 1: Local Coordinates for f. Without loss of generality we can assume that the hyperbolic fixed point, p, is located at the origin (cf. Section 1.1C). Let U be a neighborhood of the origin. Then, in U, f can be written in the form

$$\begin{aligned} \xi &\longmapsto \lambda\xi + g_1(\xi,\eta), \\ \eta &\longmapsto \mu\eta + g_2(\xi,\eta), \end{aligned} \qquad (\xi,\eta) \in U \subset \mathbb{R}^2, \tag{4.4.1}$$

where $0 < |\lambda| < 1$, $|\mu| > 1$, and g_1, g_2 are $\mathcal{O}(2)$ in ξ and η. Hence, $\eta = 0$ and $\xi = 0$ are the stable and unstable manifolds, respectively, of the *linearized* map. However, in the proof of the theorem we will find it more convenient to use the local stable and unstable manifolds of the origin as coordinates. This can be done by utilizing a simple (nonlinear) change of coordinates.

We know from Theorem 1.1.3 that the local stable and unstable manifolds of the hyperbolic fixed point can be represented as the graphs of $\mathbf{C}^r$ functions, i.e.,

$$\begin{aligned} W^s_{\text{loc}}(0) &= \text{graph } h^s(\xi), \\ W^u_{\text{loc}}(0) &= \text{graph } h^u(\eta), \end{aligned} \tag{4.4.2}$$

where $h^s(0) = h^u(0) = Dh^s(0) = Dh^u(0) = 0$. If we define the coordinate transformation

$$(x,y) = (\xi - h^u(\eta), \eta - h^s(\xi)), \tag{4.4.3}$$

then (4.4.1) takes the form

$$\begin{aligned} x &\longmapsto \lambda x + f_1(x,y), \\ y &\longmapsto \mu y + f_2(x,y), \end{aligned} \tag{4.4.4}$$

with

$$\begin{aligned} f_1(0,y) &= 0, \\ f_2(x,0) &= 0. \end{aligned} \tag{4.4.5}$$

Equation (4.4.5) implies that $y = 0$ and $x = 0$ are the stable and unstable manifolds, respectively, of the origin. We emphasize that the transformation (4.4.3) is only locally valid; therefore (4.4.4) has meaning only in a "sufficiently small" neighborhood of the origin. Despite the bad notation, we refer to this neighborhood, as above, by U.

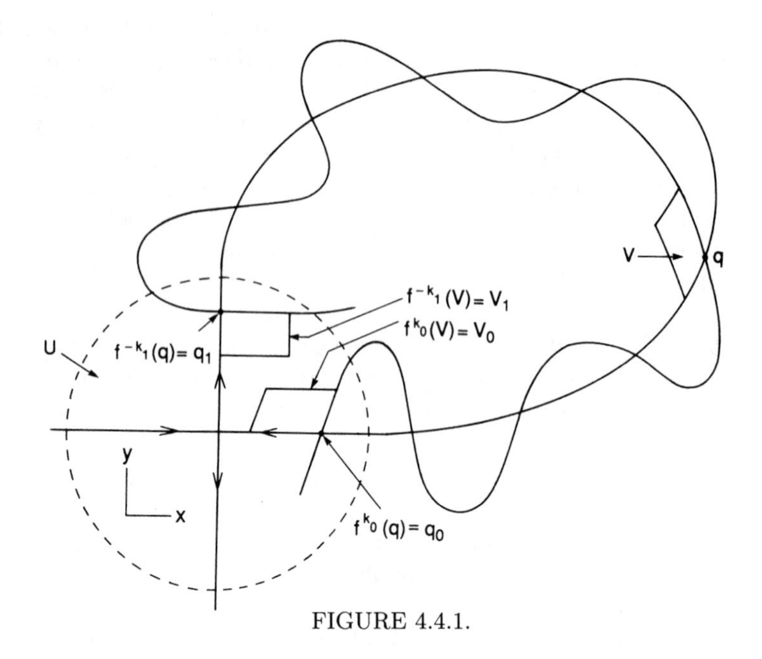

FIGURE 4.4.1.

Step 2: Global Consequences of a Homoclinic Orbit. By assumption, $W^s(0)$ and $W^u(0)$ intersect at, say, q. Then, since $q \in W^s(0) \cap W^u(0)$,

$$\begin{aligned} \lim_{n\to\infty} f^n(q) &= 0, \\ \lim_{n\to-\infty} f^n(q) &= 0. \end{aligned} \tag{4.4.6}$$

Therefore, we can find positive integers k_0 and k_1 such that

$$\begin{aligned} f^{k_0}(q) &\equiv q_0 \in U, \\ f^{-k_1}(q) &= q_1 \in U. \end{aligned} \tag{4.4.7}$$

In the coordinates in U we denote

$$\begin{aligned} q_0 &= (x_0, 0), \\ q_1 &= (0, y_1). \end{aligned}$$

It follows from (4.4.7) that

$$f^k(q_1) = q_0,$$

where $k = k_0 + k_1$; see Figure 4.4.1.

Next we choose a region V as shown in Figure 4.4.1 with one side along $W^s(0)$ emanating from q, one side along $W^u(0)$ emanating from q, and the remaining two sides parallel to the tangent vectors of $W^s(0)$ and $W^u(0)$

at q. Now V can be chosen to lie on the appropriate sides of $W^s(0)$ and $W^u(0)$ and taken sufficiently small so that

$$f^{-k_1}(V) \equiv V_1 \subset U$$

and

$$f^{k_0}(V) \equiv V_0 \subset U \tag{4.4.8}$$

appear as in Figure 4.4.1. From (4.4.8) it follows that we have

$$f^k(V_1) = V_0. \tag{4.4.9}$$

Let us make an important comment concerning Figure 4.4.1. The important aspect is that we can choose V and (large) positive integers k_0 and k_1 such that $f^{k_0}(V)$ and $f^{-k_1}(V)$ are *both in the first quadrant.* This can always be done and in Exercise 4.7 we will show how this fact can be proved. (Note: Certainly k depends on the size of U as U shrinks to a point $k \to \infty$.) We remark that in Figure 4.4.1 we depict $W^s(0)$ and $W^u(0)$ as winding amongst each other. We will discuss the geometrical aspects of this more fully later on; for the proof of this theorem a detailed knowledge of the geometry of the "homoclinic tangle" is not so important. However, the one aspect of the intersection of $W^s(0)$ and $W^u(0)$ *at* q that will be of importance is the assumption that the intersection is *transversal at* q. Since f is a diffeomorphism, this implies that $W^s(0)$ and $W^u(0)$ also intersect transversely at $f^{k_0}(q) = q_0$ and $f^{-k_1}(q) = q_1$ (or, more generally, at $f^k(q)$ for any integer k).

The next step involves gaining an understanding of the dynamics near the hyperbolic fixed point.

Step 3: Dynamics Near the Origin. We first state a well-known lemma that describes some geometric aspects of the dynamics of a curve as it passes near the hyperbolic fixed point under iteration by f.

Let $\bar{q} \in W^s(0) - \{0\}$, and let C be a curve intersecting $W^s(0)$ transversely at $\bar{q}$. Let C^N denote the connected component of $f^N(C) \cap U$ to which $f^N(\bar{q})$ belongs; see Figure 4.4.2. Then we have the following lemma.

Lemma 4.4.1 (the lambda lemma) *Given $\varepsilon > 0$ and U sufficiently small there exists a positive integer N_0 such that for $N \geq N_0$ C^N is $\mathbf{C}^1$ ε-close to $W^u(0) \cap U$.*

Proof: See Palis and de Melo [1982] or Newhouse [1980]. □

We make several remarks regarding the lambda lemma.

Remark 1. The lambda lemma is valid in n-dimensions (Palis and de Melo [1982], Newhouse [1980]) and even ∞-dimensions (Hale and Lin [1986], Lerman and Silnikov [1989]). However, the statement of the lemma requires some technical modifications.

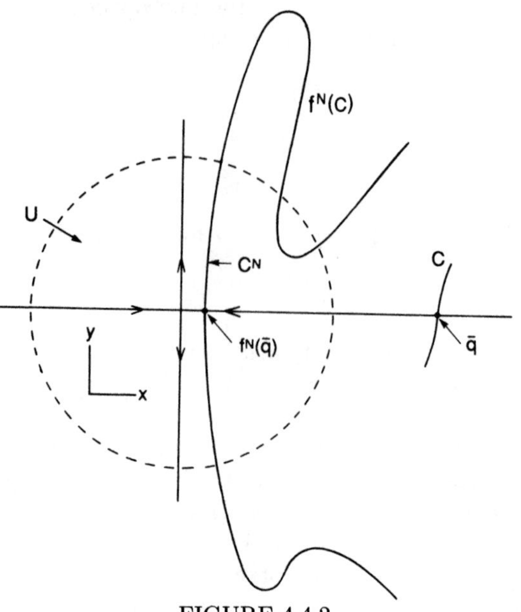

FIGURE 4.4.2.

Remark 2. The phrase $\mathbf{C}^1$ ε-close implies that tangent vectors on C^N are ε-close to tangent vectors on $W^s(0) \cap U$. By our choice of coordinates in U, all vectors tangent to $W^s(0) \cap U$ are parallel to $(0, 1)$.

Remark 3. The estimates involved in proving the lambda lemma give us information on the stretching of tangent vectors. In particular, let $z_0 \in f^{-N}(C^N)$ with (ξ_{z_0}, η_{z_0}) a vector tangent to $f^{-N}(C^N)$ at z_0. It follows that $Df^N(z_0)(\xi_{z_0}, \eta_{z_0}) \equiv (\xi_{f^N(z_0)}, \eta_{f^N(z_0)})$ is a vector tangent to C^N at $f^N(z_0)$. Then

$$|\xi_{f^N(z_0)}|$$

can be made arbitrarily *small* by taking N large enough, and

$$|\eta_{f^N(z_0)}|$$

can be made arbitrarily large by taking N large enough; see Figure 4.4.3.

We now define the *transversal map,* f^T, of V_0 into V_1 as follows. Let $D(f^T)$ denote the domain of f^T and choose $p \in V_0$. We then say that $p \in D(f^T)$ if there exists an integer $n > 0$ such that

$$f^n(p) \in V_1$$

and

$$f(p), f^2(p), \cdots, f^{n-1}(p) \in U. \tag{4.4.10}$$

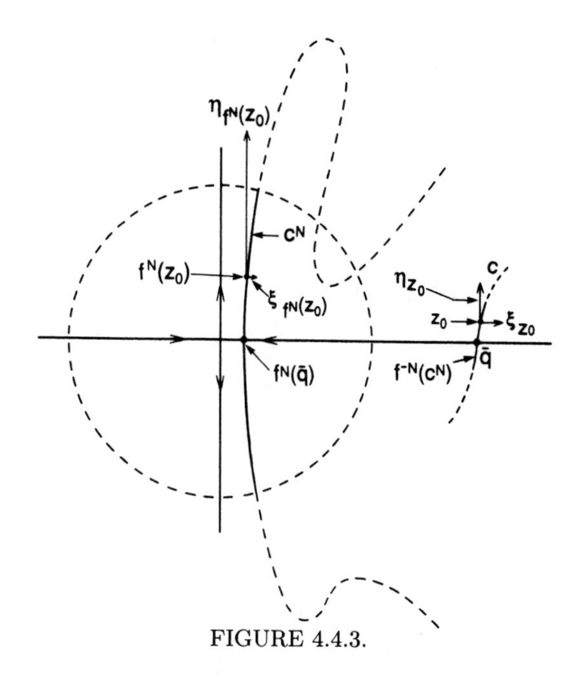

FIGURE 4.4.3.

Next we define

$$f^T(p) = f^n(p) \in V_1, \tag{4.4.11}$$

where n is the smallest integer such that (4.4.11) holds.

Step 4: The Dynamics Outside of U. Recall from (4.4.9) that we have

$$f^k(V_1) = V_0.$$

Hence, since $V_1 \subset U$, f^k can be represented in the $x-y$ coordinates as follows

$$f^k(x, \overline{y}) = \begin{pmatrix} x_0 \\ 0 \end{pmatrix} + \begin{pmatrix} a & b \\ c & d \end{pmatrix} \begin{pmatrix} x \\ \overline{y} \end{pmatrix} + \begin{pmatrix} \phi_1(x, \overline{y}) \\ \phi_2(x, \overline{y}) \end{pmatrix}, \quad (x, \overline{y}) \in V_1, \tag{4.4.12}$$

where $\overline{y} = y - y_1$, $\phi_1(x, \overline{y})$ and $\phi_2(x, \overline{y})$ are $\mathcal{O}(2)$ in x and $\overline{y}$, a, b, c, d are constants, and, from (4.4.7), $q_0 \equiv (x_0, 0)$ and $q_1 \equiv (0, y_1)$.

Step 5: The Transversal Map of V_0 into V_0. Using Steps 3 and 4 we have that

$$f^k \circ f^T : D(f^T) \subset V_0 \to V_0 \tag{4.4.13}$$

is a transversal map of $D(f^T) \subset V_0$ into V_0.

Now we can finally state Moser's theorem.

Theorem 4.4.2 (Moser [1973]) *The map $f^k \circ f^T$ has an invariant Cantor set on which it is topologically conjugate to a full shift on N symbols.*

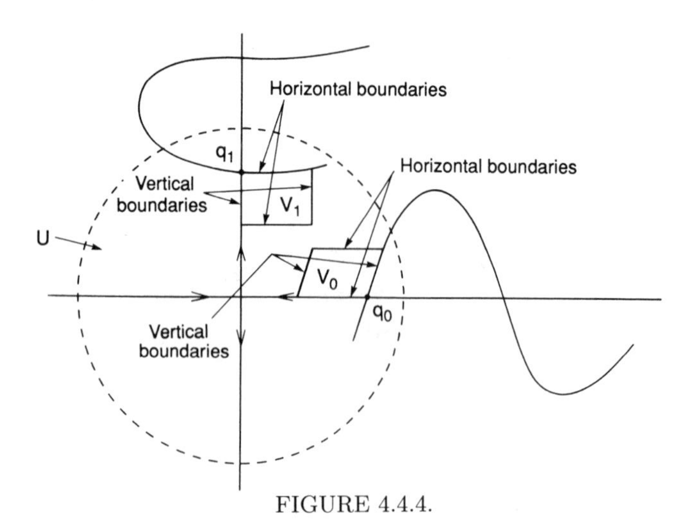

FIGURE 4.4.4.

Proof: The strategy is to find μ_h-horizontal strips in V_0 that are mapped homeomorphically onto μ_v-vertical strips in V_0 with proper behavior of the boundaries such that Assumptions 1 and 3 of Section 4.3 hold. Theorem 4.4.2 will then follow from Theorem 4.3.5. We remark that we will take as the horizontal boundaries of V_0 the two segments of the boundary of V_0 that are "parallel" to $W^s(0)$ and as the vertical boundary of V_0 the remaining two segments of the boundary of V_0. Similarly, the horizontal boundary of V_1 is taken to be the two segments of the boundary of V_1 parallel to $W^s(0)$, and the horizontal boundary of V_1 is taken to be the remaining two segments of the boundary of V_1; see Figure 4.4.4. □

We begin by choosing a set of μ_h-horizontal strips in V_0 such that Assumption 1 of Section 4.3A holds. First we make the observation that it follows from the lambda lemma that there exists a positive integer N_0 such that for $N \geq N_0$ both vertical boundaries of the component of $f^N(V_0) \cap U$ containing $f^N(q_0)$ intersect both horizontal boundaries of V_1 as shown in Figure 4.4.5. Let $\tilde{V}_N \equiv f^N(V_0) \cap V_1$ denote this set. Then, for N_0 sufficiently large, it follows by applying the lambda lemma to f^{-1} that $f^{-N}(\tilde{V}_N) \equiv H_N$ is a μ_h-horizontal strip stretching across V_0 with the vertical boundaries of H_N contained in the vertical boundaries of V_0 as shown in Figure 4.4.5.

(Note: since the tangent vectors at each point on the horizontal boundaries of H_N can be made arbitrarily close to the tangent vector of $W^s(0) \cap U$, it follows that the horizontal boundaries of H_N are graphs over x.) It should be clear from the definition of f^T that $H_N \subset D(f^T)$.

Now we choose a sequence of integers, $N_0 + 1, N_0 + 2, \cdots$ with

$$\tilde{V}_i \equiv f^{N_0+i}(V_0) \cap V_1 \tag{4.4.14}$$

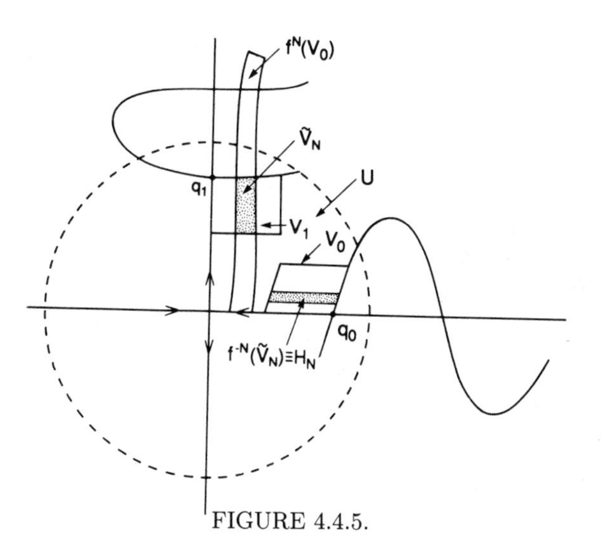

FIGURE 4.4.5.

as described above. Then

$$f^{-N_0-i}(\tilde{V}_i) \equiv H_i, \qquad i = 1, 2, \cdots, \tag{4.4.15}$$

are a set of μ_h-horizontal strips contained in V_0. It follows by applying the lambda lemma to f^{-1} that, for N_0 sufficiently large, μ_h is arbitrarily close to zero. We choose a finite number of the H_i

$$\{H_1, \cdots, H_N\}.$$

Then $f^k \circ f^T(H_i) \equiv V_i$, $i = 1, \cdots, N$ appear as in Figure 4.4.6. A consideration of the manner in which the boundary of V_1 maps to the boundary of V_0 under f^k shows that horizontal (resp. vertical) boundaries of the H_i map to horizontal (resp. vertical) boundaries of the V_i under $f^k \circ f^T$. We now need to argue that the V_i are μ_v-vertical strips with $0 \leq \mu_v\mu_h < 1$. This goes as follows. By the lambda lemma, for N_0 sufficiently large, the vertical boundaries of the V_i are arbitrarily close to $W^u(0) \cap U$. Hence, by the form of f^k given in Step 4, the vertical boundaries of $f^k(\tilde{V}_i) \equiv V_i$ are arbitrarily close to the tangent vector to $W^u(0)$ at q_0. Therefore, the vertical boundaries of the V_i can be represented as graphs over the y variables. Moreover, by the remark above, μ_h can be taken as small as we like (by taking N_0 large); it follows that we can satisfy $0 \leq \mu_h\mu_v < 1$, where μ_v is taken to be *twice the absolute value* of the slope of the tangent vector of $W^u(0)$ at q_0. (Note: the reason for taking this choice for μ_v will be apparent when we define stable sectors and verify Assumption 3.) Hence, Assumption 1 holds.

Next we need to show that Assumption 3 holds. We will show this for the unstable sectors and leave it as an exercise for the reader to verify the

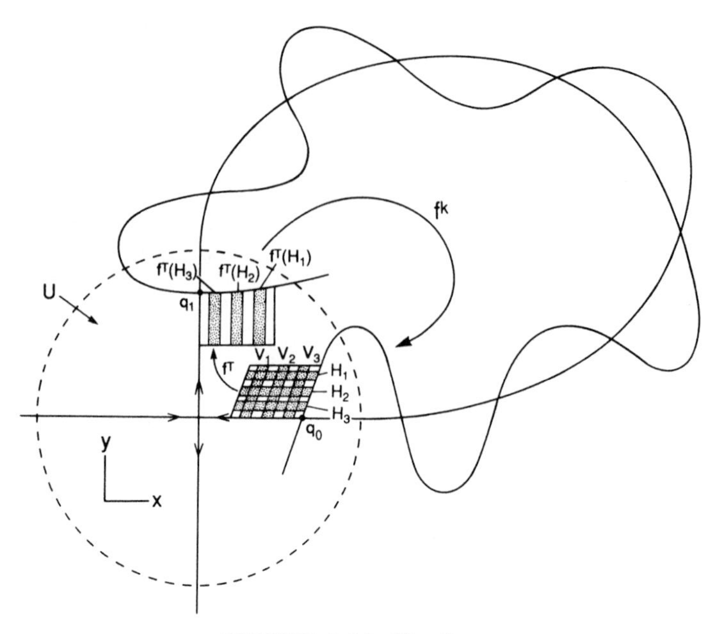

FIGURE 4.4.6. $N = 3$.

part of Assumption 3 dealing with the stable sectors. First we must define the unstable sector bundle.

Recall from Section 4.3B that

$$f^k \circ f^T(H_i) \cap H_j \equiv V_{ji},$$

$$H_i \cap (f^k \circ f^T)^{-1}(H_j) \equiv H_{ij} = (f^k \circ f^T)^{-1}(V_{ji}), \tag{4.4.16}$$

with

$$\mathcal{H} = \bigcup_{i,j \in S} H_{ij} \quad \text{and} \quad \mathcal{V} = \bigcup_{i,j \in S} v_{ji}, \tag{4.4.17}$$

where $S = \{1, \cdots, N\}$ $(N \geq 2)$ is the index set. We choose $z_0 \equiv (x_0, y) \in \mathcal{H} \cup \mathcal{V}$; then the unstable sector at z_0 is denoted by

$$\mathcal{S}_{z_0}^u = \left\{(\xi_{z_0}, \eta_{z_0}) \in \mathbb{R}^2 \mid |\xi_{z_0}| \leq \mu_v |\eta_{z_0}|\right\}, \tag{4.4.18}$$

and we have the sector bundles

$$\mathcal{S}_{\mathcal{H}}^u = \bigcup_{z_0 \in \mathcal{H}} \mathcal{S}_{z_0}^u \quad \text{and} \quad \mathcal{S}_{\mathcal{V}}^u = \bigcup_{z_0 \in \mathcal{V}} \mathcal{S}_{z_0}^u. \tag{4.4.19}$$

We make the important remark that by our choice of μ_v a vector parallel to the tangent vector of $W^u(0)$ at q_0 is contained in the interior of $\mathcal{S}_{z_0}^u$; see Figure 4.4.7.

We must show that

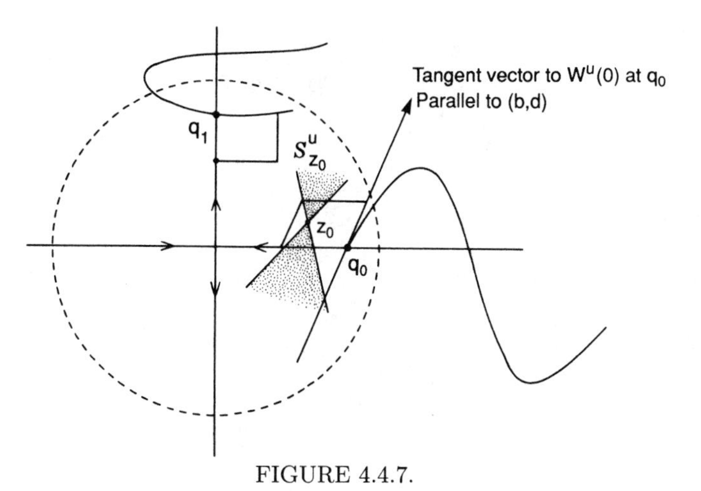

FIGURE 4.4.7.

$$1) \qquad D(f^k \circ f^T)(\mathcal{S}^u_{\mathcal{H}}) \subset \mathcal{S}^u_{\mathcal{V}} \tag{4.4.20}$$

and

$$2) \qquad |\eta_{f^k \circ f^T(z_0)}| \geq \left(\frac{1}{\mu}\right) |\eta_{z_0}|; \tag{4.4.21}$$

where $0 < \mu < 1 - \mu_v \mu_h$.

Before showing that 1) and 2) hold we make the following observations.

Observation 1.

$$D(f^k \circ f^T) = Df^k Df^T \tag{4.4.22}$$

and, from (4.4.12),

$$Df^k = \begin{pmatrix} a & b \\ c & d \end{pmatrix} + \begin{pmatrix} \phi_{1x} & \phi_{1\overline{y}} \\ \phi_{2x} & \phi_{2\overline{y}} \end{pmatrix}. \tag{4.4.23}$$

Now, since $\phi_1(x, \overline{y})$ and $\phi_2(x, \overline{y})$ are $\mathcal{O}(2)$ in x and $\overline{y}$, by choosing U sufficiently small ϕ_{1x}, $\phi_{1\overline{y}}$, ϕ_{2x}, and $\phi_{2\overline{y}}$ can be made arbitrarily small compared to a, b, c, and d.

Observation 2: Consequences of Transversality. From Step 2, $W^s(0)$ and $W^u(0)$ intersect transversely at q_0 and q_1. We have

$$f^k(q_1) = q_0,$$

$$Df^k(q_1) = \begin{pmatrix} a & b \\ c & d \end{pmatrix}, \tag{4.4.24}$$

and

$$Df^{-k}(q_0) = (Df^k(q_1))^{-1} = \frac{1}{ad - bc} \begin{pmatrix} d & -b \\ -c & a \end{pmatrix}. \tag{4.4.25}$$

In our choice of coordinates a vector tangent to $W^u(0)$ at q_1 is parallel to $(0,1)$, and a vector tangent to $W^s(0)$ at q_0 is parallel to $(1,0)$. Hence, if $W^s(0)$ and $W^u(0)$ intersect transversely at q_0 and q_1, we must have that

$$Df^k(q_1)\begin{pmatrix}0\\1\end{pmatrix} = \begin{pmatrix}a & b\\ c & d\end{pmatrix}\begin{pmatrix}0\\1\end{pmatrix} = \begin{pmatrix}b\\d\end{pmatrix}$$

is not parallel to

$$\begin{pmatrix}1\\0\end{pmatrix},$$

and

$$\begin{aligned} Df^{-k}(q_0)\begin{pmatrix}1\\0\end{pmatrix} &= \frac{1}{ad-bc}\begin{pmatrix}d & -b\\ -c & a\end{pmatrix}\begin{pmatrix}1\\0\end{pmatrix} \\ &= \frac{1}{ad-bc}\begin{pmatrix}d\\-c\end{pmatrix} \end{aligned}$$

is not parallel to

$$\begin{pmatrix}0\\1\end{pmatrix}.$$

It is easy to see that these conditions will be satisfied provided

$$d \neq 0.$$

Note that (b,d) is a vector parallel to the tangent vector to $W^u(0)$ at q_0. We will use this later on.

Now we return to showing that Assumption 3 holds for the unstable sector bundle. We must demonstrate the following.

1. $D(f^k \circ f^T)(\mathcal{S}^u_{\mathcal{H}}) \subset \mathcal{S}^u_{\mathcal{V}}$, and
2. $|\eta_{f^k \circ f^T(z_0)}| > \frac{1}{\mu}|\eta_{z_0}|$; $0 < \mu < 1 - \mu_v\mu_h$.

$\underline{D(f^k \circ f^T)(\mathcal{S}^u_{\mathcal{H}}) \subset \mathcal{S}^u_{\mathcal{V}}}$. We choose $z_0 \in \mathcal{H}$ and let $(\xi_{z_0}, \eta_{z_0}) \in \mathcal{S}^u_{z_0}$ and

$$Df^T(z_0)(\xi_{z_0}, \eta_{z_0}) = \big(\xi_{f^T(z_0)}, \eta_{f^T(z_0)}\big). \tag{4.4.26}$$

Then, using (4.4.22) and (4.4.23), we have

$$\begin{aligned} Df^k(f^T(z_0))Df^T(z_0)(\xi_{z_0}, \eta_{z_0}) &= \begin{pmatrix}(a+\phi_{1x})\xi_{f^T(z_0)} + (b+\phi_{1\overline{y}})\eta_{f^T(z_0)}\\ (c+\phi_{2x})\xi_{f^T(z_0)} + (d+\phi_{2\overline{y}})\eta_{f^T(z_0)}\end{pmatrix} \\ &\equiv \begin{pmatrix}\xi_{f^k\circ f^T(z_0)}\\ \eta_{f^k\circ f^T(z_0)}\end{pmatrix}, \end{aligned} \tag{4.4.27}$$

where all partial derivatives are evaluated at $z_0 = (x_0, y_0)$. We must show that

$$\begin{aligned} \frac{|\xi_{f^k\circ f^T(z_0)}|}{|\eta_{f^k\circ f^T(z_0)}|} &= \frac{|(a+\phi_{1x})(\xi_{f^T(z_0)})/(\eta_{f^T(z_0)}) + (b+\phi_{1\overline{y}})|}{|(c+\phi_{2x})(\xi_{f^T(z_0)})/(\eta_{f^T(z_0)}) + (d+\phi_{2\overline{y}})|} \\ &\in \mathcal{S}^u_{f^k\circ f^T(z_0)}. \end{aligned} \tag{4.4.28}$$

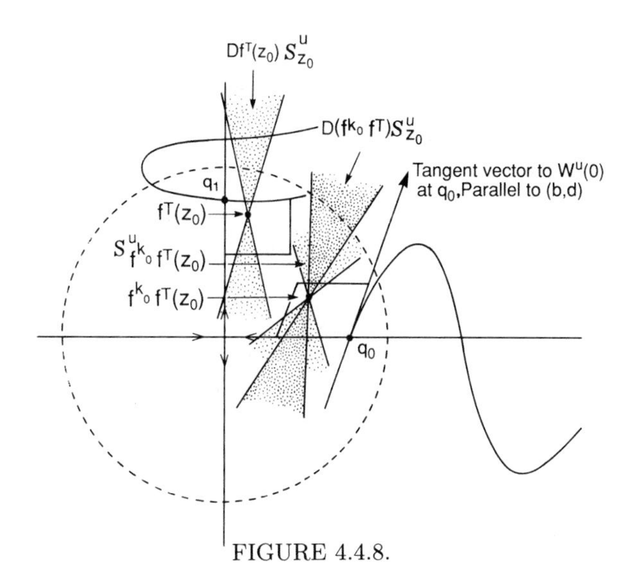

FIGURE 4.4.8.

From Remark 3 following the lambda lemma, for N_0 sufficiently large,

$$\frac{|\xi_{f^T(z_0)}|}{|\eta_{f^T(z_0)}|}$$

can be made arbitrarily small. Also, from Observation 1 above, for V sufficiently small,

$$\phi_{1x}, \phi_{1\overline{y}}, \phi_{2x}, \phi_{2\overline{y}}$$

can be made arbitrarily small. Hence, using these two results we see that

$$\frac{|\xi_{f^k \circ f^T(z_0)}|}{|\eta_{f^k \circ f^T(z_0)}|}$$

can be made arbitrarily close to

$$\frac{|b|}{|d|},$$

where, as shown in Observation 2 above, (b, d) is a vector parallel to the tangent vector of $W^u(0)$ at q_0; see Figure 4.4.8. Since this holds for any $z_0 \in \mathcal{H}$, we have $D(f^k \circ f^T)(\mathcal{S}^u_{\mathcal{H}}) \subset \mathcal{S}^u_{\mathcal{V}}$.

<u>$|\eta_{f^k \circ f^T(z_0)}| \geq \frac{1}{\mu}|\eta_{z_0}|; 0 < \mu < 1 - \mu_v \mu_h$.</u> Using (4.4.27) we have

$$\frac{|\eta_{f^k \circ f^T(z_0)}|}{|\eta_{z_0}|} = \frac{|(c + \phi_{2x})\xi_{f^T(z_0)} + (d + \phi_{2\overline{y}})\eta_{f^T(z_0)}|}{|\eta_{z_0}|}. \tag{4.4.29}$$

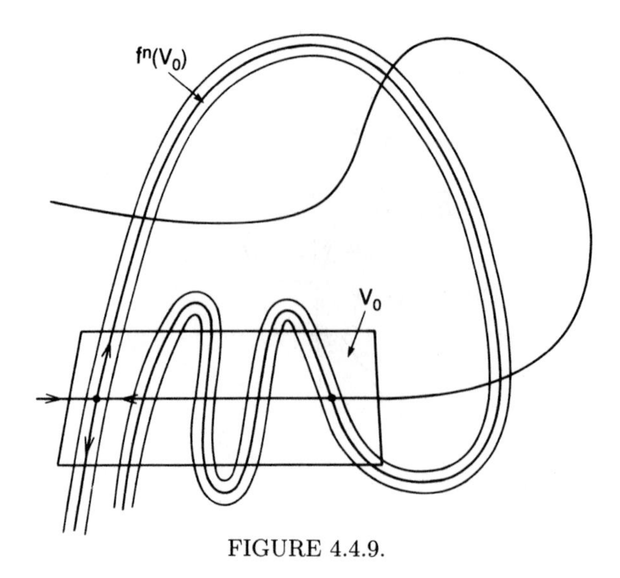

FIGURE 4.4.9.

Again, as a result of the lambda lemma, for N_0 sufficiently large, $|\eta_{f^T(z_0)}|$ can be made arbitrarily large, $|\xi_{f^T(z_0)}|$ can be made arbitrarily small, and by transversality of the intersection of $W^u(0)$ and $W^s(0)$ at q, $d \neq 0$ (with $\phi_{1\overline{y}}$ small compared to d). Thus, (4.4.29) can be made as large as we desire by choosing N_0 big enough.

The Smale–Birkhoff Homoclinic Theorem

The Smale–Birkhoff homoclinic theorem is very similar to Moser's theorem. We will state the theorem and describe briefly how it differs. The assumptions and set-up are the same as for Moser's theorem.

Theorem 4.4.3 (Smale [1963]) *There exists an integer $n \geq 1$ such that f^n has an invariant Cantor set on which it is topologically conjugate to a full shift on N symbols.*

Proof: We will only give the barest outline in order to show the difference between the Smale–Birkhoff homoclinic theorem and Moser's theorem and leave the details as an exercise for the reader (Exercise 4.8).

Choose a "rectangle," V_0, containing a homoclinic point *and* the hyperbolic fixed point as shown in Figure 4.4.9. Then, for n sufficiently large, $f^n(V_0)$ intersects V_0 a finite number of times as shown in Figure 4.4.9. Now, one can find μ_h-horizontal strips in V_0 that map over themselves in μ_v-vertical strips such that Assumptions 1 and 3 of Section 4.3 hold; see Figure 4.4.10. The details needed to prove these statements are very similar to those needed for the proof of Moser's theorem, and it will be an instructive exercise for the reader to give a rigorous proof. □

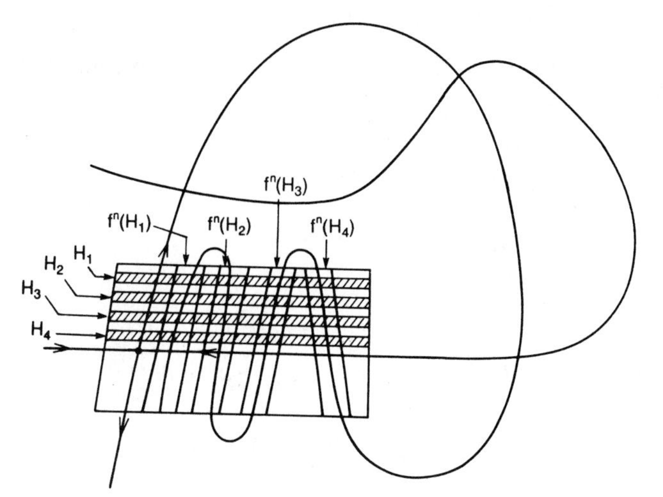

FIGURE 4.4.10. Horizontal strips $H_1, \cdots, H_4$ and their image under f^n.

From the outline of the proof of the Smale–Birkhoff homoclinic theorem, one can see how it differs from Moser's theorem. In both cases the invariant Cantor set is constructed near a homoclinic point sufficiently close to the hyperbolic fixed point. However, in the Smale–Birkhoff theorem, all points leave the Cantor set and return *at the same time* (i.e., after n iterates of f); in Moser's construction, points leave the Cantor set and may return *at different times* (recall the definition of f^T). What are the dynamical consequences of the two different constructions (see Exercise 4.9)?

4.5 Melnikov's Method for Homoclinic Orbits in Two-Dimensional, Time-Periodic Vector Fields

We have seen that transverse homoclinic orbits to hyperbolic periodic points of two-dimensional maps gives rise to chaotic dynamics in the sense of Theorems 4.4.2 and 4.4.3. We will now develop a perturbation method originally due to Melnikov [1963] for proving the existence of transverse homoclinic orbits to hyperbolic periodic orbits in a class of two-dimensional, time-periodic vector fields; then by considering a Poincaré map, Theorems 4.4.2 and 4.4.3 can be applied to conclude that the system possesses chaotic dynamics.

4.5A The General Theory

We study the same class of systems as in our development of the subharmonic Melnikov theory in Section 1.2D, ii).

$$\begin{aligned} \dot{x} &= \frac{\partial H}{\partial y}(x,y) + \varepsilon g_1(x,y,t,\epsilon), \\ \dot{y} &= -\frac{\partial H}{\partial x}(x,y) + \varepsilon g_2(x,y,t,\varepsilon), \end{aligned} \qquad (x,y) \in \mathbb{R}^2; \tag{4.5.1}$$

or, in vector form,

$$\dot{q} = JDH(q) + \varepsilon g(q,t,\varepsilon), \tag{4.5.2}$$

where $q = (x,y)$, $DH = (\frac{\partial H}{\partial x}, \frac{\partial H}{\partial y})$, $g = (g_1, g_2)$, and

$$J = \begin{pmatrix} 0 & 1 \\ -1 & 0 \end{pmatrix}.$$

We assume that (4.5.1) is sufficiently differentiable ($\mathbf{C}^r$, $r \geq 2$ will do) on the region of interest; see Section 1.2D, ii) where this is described in more detail. Most importantly, we also assume that g is periodic in t with period $T = 2\pi/\omega$.

We referred to (4.5.1) with $\varepsilon = 0$ as the unperturbed system

$$\begin{aligned} \dot{x} &= \frac{\partial H}{\partial y}(x,y), \\ \dot{y} &= -\frac{\partial H}{\partial x}(x,y), \end{aligned} \tag{4.5.3}$$

or, in vector form,

$$\dot{q} = JDH(q), \tag{4.5.4}$$

and we had the following assumptions on the structure of the phase space of the unperturbed system (see Figure 4.5.1).

Assumption 1. The unperturbed system possesses a hyperbolic fixed point, p_0, connected to itself by a homoclinic orbit $q_0(t) \equiv (x_0(t), y_0(t))$.

Assumption 2. Let $\Gamma_{p_0} = \{q \in \mathbb{R}^2 \mid q = q_0(t), t \in \mathbb{R}\} \cup \{p_0\} = W^s(p_0) \cap W^u(p_0) \cup \{p_0\}$. The interior of Γ_{p_0} is filled with a continuous family of periodic orbits $q^\alpha(t)$ with period T^α, $\alpha \in (-1,0)$. We assume that $\lim_{\alpha \to 0} q^\alpha(t) = q_0(t)$, and $\lim_{\alpha \to 0} T^\alpha = \infty$.

The subharmonic Melnikov theory enabled us to understand how the periodic orbits $q^\alpha(t)$ were affected by the perturbation; now we will develop a technique to see how the homoclinic orbit, Γ_{p_0}, is so affected. Geometrically, the homoclinic Melnikov method is a bit different from the subharmonic Melnikov method. However, there is an important relationship between the two as $\alpha \to 0$ (i.e., as the periodic orbits limit on the homoclinic orbit) that

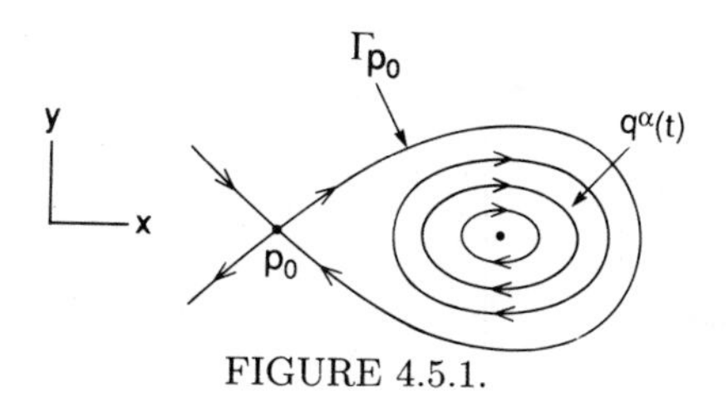

FIGURE 4.5.1.

we want to point out later on in this section. We remark that it is possible to develop the homoclinic Melnikov method for a more general class of two-dimensional, time-periodic systems than (4.5.1); in particular, we do not have to assume that the unperturbed system is Hamiltonian. We will deal with these generalizations in the exercises.

Our development of the homoclinic Melnikov method will consist of several steps, which we briefly describe.

Step 1. Develop a parametrization of the homoclinic "manifold" of the unperturbed system.

Step 2. Develop a measure of the "splitting" of the manifolds for the perturbed system using the unperturbed "homoclinic coordinates."

Step 3. Derive the Melnikov function and show how it is related to the distance between the manifolds.

Before beginning with Step 1 we want to rewrite (4.5.1) as an autonomous three-dimensional system (cf. Section 1.1G) as follows

$$\begin{aligned}
\dot{x} &= \frac{\partial H}{\partial y}(x, y) + \varepsilon g_1(x, y, \phi, \varepsilon), \\
\dot{y} &= -\frac{\partial H}{\partial x}(x, y) + \varepsilon g_2(x, y, \phi, \varepsilon), \qquad (x, y, \phi) \in \mathbb{R}^1 \times \mathbb{R}^2 \times S^1, \\
\dot{\phi} &= \omega,
\end{aligned} \tag{4.5.5}$$

or, in vector form,

$$\begin{aligned}
\dot{q} &= JDH(q) + \varepsilon g(q, \phi; \varepsilon), \\
\dot{\phi} &= \omega.
\end{aligned} \tag{4.5.6}$$

The unperturbed system is obtained from (4.5.6) by setting $\varepsilon = 0$, i.e.,

$$\begin{aligned}
\dot{q} &= JDH(q), \\
\dot{\phi} &= \omega.
\end{aligned} \tag{4.5.7}$$

We will see that this apparently trivial trick offers several geometrical advantages. In particular, the perturbed system is of a very different character than the unperturbed system, and this trick forces us to treat them on a more equal footing. Also, the relationship between the splitting of the manifolds "in time" and the splitting of the manifolds on a particular Poincaré

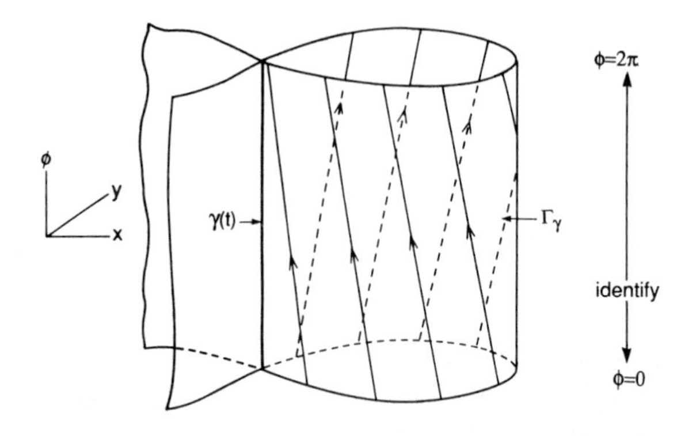

FIGURE 4.5.2. The homoclinic manifold, Γ_γ. The lines on Γ_γ represent a typical trajectory.

section and how this is manifested by the Melnikov function will be more apparent.

Step 1: Phase Space Geometry of the Unperturbed Vector Field: A Parametrization of the Homoclinic Manifold. When viewed in the three-dimensional phase space $\mathbb{R}^2 \times S^1$, the hyperbolic fixed point p_0 of the q component of the unperturbed system (4.5.7) becomes a periodic orbit

$$\gamma(t) = (p_0, \phi(t) = \omega t + \phi_0). \tag{4.5.8}$$

We denote the two-dimensional stable and unstable manifolds of $\gamma(t)$ by $W^s(\gamma(t))$ and $W^u(\gamma(t))$, respectively. Because of Assumption 1 above, $W^s(\gamma(t))$ and $W^u(\gamma(t))$ coincide along a two-dimensional *homoclinic manifold.* We denote this homoclinic manifold by Γ_γ; see Figure 4.5.2. We remark that the structure of this figure should not be surprising; it reflects the fact that the unperturbed phase space is independent of time (ϕ).

Our goal is to determine how Γ_γ "breaks up" under the influence of the perturbation. We now want to describe what we mean by this statement, which will serve to motivate the following discussion.

The homoclinic manifold Γ_γ is formed by the coincidence of two two-dimensional surfaces, a branch of $W^s(\gamma(t))$ and a branch of $W^u(\gamma(t))$. In three dimensions, one would not expect two two-dimensional surfaces to coincide in this manner but, rather, one would expect them to intersect in one-dimensional curves as shown in Figure 4.5.3. (Note: as mentioned in Section 1.2C, if two invariant manifolds of a *vector field* intersect, they must intersect along (at least) a one-dimensional trajectory of the vector field if we have uniqueness of solutions; we will explore this in more detail later.) Figure 4.5.3 illustrates what we mean by the term "break up" of Γ_γ. Now we want to analytically quantify Figure 4.5.3. In order to do this we will develop a measurement of the deviation of the perturbed stable and unstable manifolds of $\gamma(t)$ from Γ_γ. This will consist of measuring

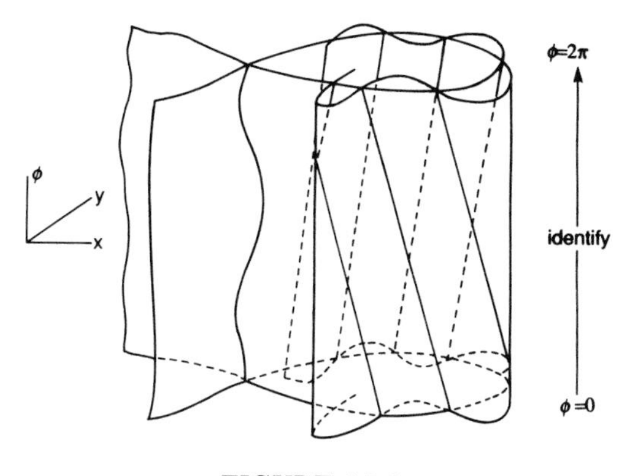

FIGURE 4.5.3.

the distance between the perturbed stable and unstable manifolds along the direction normal to Γ_γ. Evidently, this measurement will vary from point-to-point on Γ_γ so we first need to describe a parametrization of Γ_γ.

Parametrization of Γ_γ*: Homoclinic Coordinates.* Every point on Γ_γ can be represented by

$$(q_0(-t_0), \phi_0) \in \Gamma_\gamma \tag{4.5.9}$$

for $t_0 \in \mathbb{R}^1$, $\phi_0 \in (0, 2\pi]$. The interpretation of t_0 is the time of flight from the point $q_0(-t_0)$ to the point $q_0(0)$ along the unperturbed homoclinic trajectory $q_0(t)$. Since the time of flight from $q_0(-t_0)$ to $q_0(0)$ is unique, the map

$$(t_0, \phi_0) \longmapsto (q_0(-t_0), \phi_0) \tag{4.5.10}$$

is one-to-one so that for a given $(t_0, \phi_0) \in \mathbb{R}^1 \times S^1$, $(q_0(-t_0), \phi_0)$ corresponds to a unique point on Γ_γ (see Exercise 4.12). Hence, we have

$$\Gamma_\gamma = \{(q, \phi) \in \mathbb{R}^2 \times S^1 \mid q = q_0(-t_0), t_0 \in \mathbb{R}^1; \phi = \phi_0 \in (0, 2\pi]\}. \tag{4.5.11}$$

The geometrical meaning of the parameters t_0 and ϕ_0 should be clear from Figure 4.5.2.

At each point $p \equiv (q_0(-t_0), \phi_0) \in \Gamma_\gamma$ we construct a vector, π_p, normal to Γ_γ that is defined as follows

$$\pi_p = \left(\frac{\partial H}{\partial x}(x_0(-t_0), y_0(-t_0)), \frac{\partial H}{\partial y}(x_0(-t_0), y_0(-t_0)), 0\right) \tag{4.5.12}$$

or, in vector form,

$$\pi_p \equiv (DH(q_0(-t_0)), 0). \tag{4.5.13}$$

Thus, varying t_0 and ϕ_0 serves to move π_p to every point on Γ_γ; see Figure 4.5.4. We make the important remark that at each point $p \in \Gamma_\gamma$, $W^s(\gamma(t))$ and $W^u(\gamma(t))$ intersect π_p transversely at p. Finally, when considering the

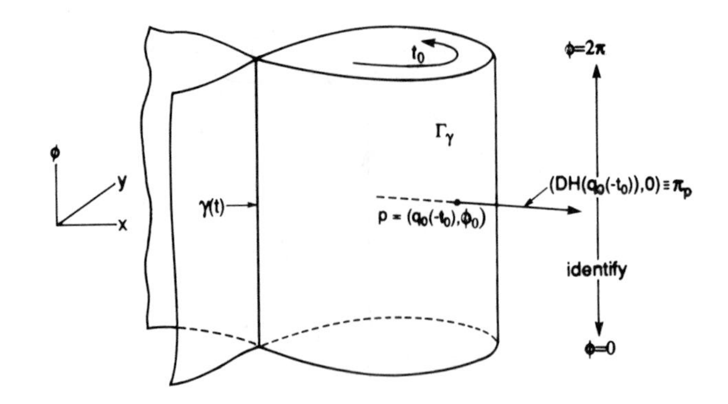

FIGURE 4.5.4. Homoclinic coordinates.

behavior of Γ_γ near p under perturbation, we will be interested only in the points on π_p that are $\mathcal{O}(\varepsilon)$ close to p. This will be further clarified in Step 2.

Step 2: Phase Space Geometry of the Perturbed Vector Field: "The Splitting of the Manifolds." We now turn our attention to describing how Γ_γ is affected by the perturbation. However, first we need some preliminary results concerning the persistence of $\gamma(t)$ along with its stable and unstable manifolds.

Proposition 4.5.1 *For ε sufficiently small, the periodic orbit $\gamma(t)$ of the unperturbed vector field (4.5.7) persists as a periodic orbit, $\gamma_\varepsilon(t) = \gamma(t) + \mathcal{O}(\varepsilon)$, of the perturbed vector field (4.5.6) having the same stability type as $\gamma(t)$ with $\gamma_\varepsilon(t)$ depending on ε in a $\mathbf{C}^r$ manner. Moreover, $W^s_{\text{loc}}(\gamma_\varepsilon(t))$ and $W^u_{\text{loc}}(\gamma_\varepsilon(t))$ are $\mathbf{C}^r$ ε-close to $W^s_{\text{loc}}(\gamma(t))$ and $W^u_{\text{loc}}(\gamma(t))$, respectively.*

Proof: Using the idea of a Poincaré map and appealing to the stable and unstable manifold theorem for maps, proof of this theorem is an easy exercise that we leave for the reader (see Exercise 4.13). □

The global stable and unstable manifolds of $\gamma_\varepsilon(t)$ can be obtained from the local stable and unstable manifolds of $\gamma_\varepsilon(t)$ by time evolution as follows. Let $\phi_t(\cdot)$ denote the flow generated by (4.5.6) (note: do not confuse the notation for the flow, $\phi_t(\cdot)$, with the angle ϕ). Then we define the global stable and unstable manifolds of $\gamma_\varepsilon(t)$ as follows

$$\begin{aligned} W^s(\gamma_\varepsilon(t)) &= \bigcup_{t\leq 0} \phi_t(W^s_{\text{loc}}(\gamma_\varepsilon(t)), \\ W^u(\gamma_\varepsilon(t)) &= \bigcup_{t\geq 0} \phi_t(W^u_{\text{loc}}(\gamma_\varepsilon(t)). \end{aligned} \tag{4.5.14}$$

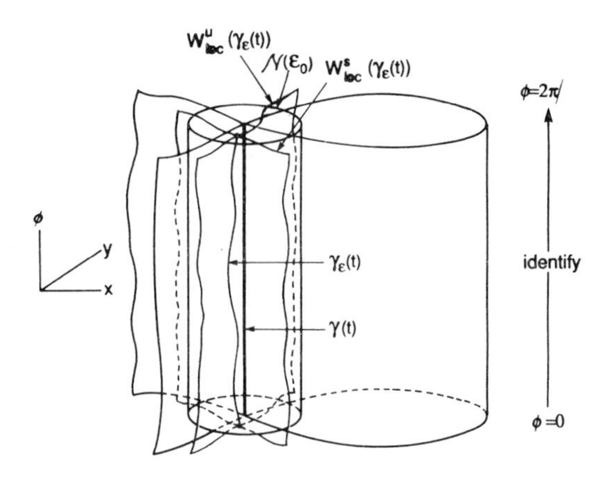

FIGURE 4.5.5.

If we restrict ourselves to compact sets in $\mathbb{R}^2 \times S^1$ containing $W^s(\gamma_\varepsilon(t))$ and $W^u(\gamma_\varepsilon(t))$, then $W^s(\gamma_\varepsilon(t))$ and $W^u(\gamma_\varepsilon(t))$ are $\mathbf{C}^r$ functions of ε on these compact sets. This follows from the fact that $\phi_t(\cdot)$ is a $\mathbf{C}^r$ diffeomorphism that is also $\mathbf{C}^r$ in ε (see Theorem 1.1.10). Our analysis of the splitting of the manifolds will be restricted to an $\mathcal{O}(\varepsilon)$ neighborhood of Γ_γ.

Let us describe Proposition 4.5.1 more geometrically. The content of the proposition is that, for some ε_0 small, we can find a neighborhood $\mathcal{N}(\varepsilon_0)$ in $\mathbb{R}^2 \times S^1$ containing $\gamma(t)$ with the distance from $\gamma(t)$ to the boundary of $\mathcal{N}(\varepsilon_0)$ being $\mathcal{O}(\varepsilon_0)$. Moreover, for $0 < \varepsilon < \varepsilon_0$, $\gamma_\varepsilon(t)$ is also contained in $\mathcal{N}(\varepsilon_0)$ with $W^s(\gamma(t)) \cap \mathcal{N}(\varepsilon_0) \equiv W^s_{\text{loc}}(\gamma(t))$ and $W^u(\gamma(t)) \cap \mathcal{N}(\varepsilon_0) \equiv W^u_{\text{loc}}(\gamma(t))$ $\mathbf{C}^r$ ε-close to $W^s(\gamma_\varepsilon(t)) \cap \mathcal{N}(\varepsilon_0) \equiv W^s_{\text{loc}}(\gamma_\varepsilon(t))$ and $W^u(\gamma_\varepsilon(t)) \cap \mathcal{N}(\varepsilon_0) \equiv W^u_{\text{loc}}(\gamma_\varepsilon(t))$, respectively. We can choose $\mathcal{N}(\varepsilon_0)$ to be a solid torus as follows

$$\mathcal{N}(\varepsilon_0) = \{(q, \phi) \in \mathbb{R}^2 \mid |q - p_0| \leq C\varepsilon_0, \phi \in (0, 2\pi]\}, \tag{4.5.15}$$

where C is some positive constant; see Figure 4.5.5.

In some of our geometrical arguments we will be comparing individual trajectories of the unperturbed vector field with trajectories of the perturbed vector field. For this it will often be easier to consider the projection of these trajectories into the q-plane or a plane parallel to the q-plane. Let us show how this is done. Consider the following cross-section of the phase space (don't think about Poincaré maps yet)

$$\Sigma^{\phi_0} = \{(q, \phi) \in \mathbb{R}^2 \mid \phi = \phi_0\}. \tag{4.5.16}$$

It should be clear that Σ^{ϕ_0} is parallel to the q-plane and coincides with the q-plane for $\phi_0 = 0$; see Figure 4.5.6. Note that

$$\gamma(t) \cap \Sigma^{\phi_0} = p_0 \tag{4.5.17}$$

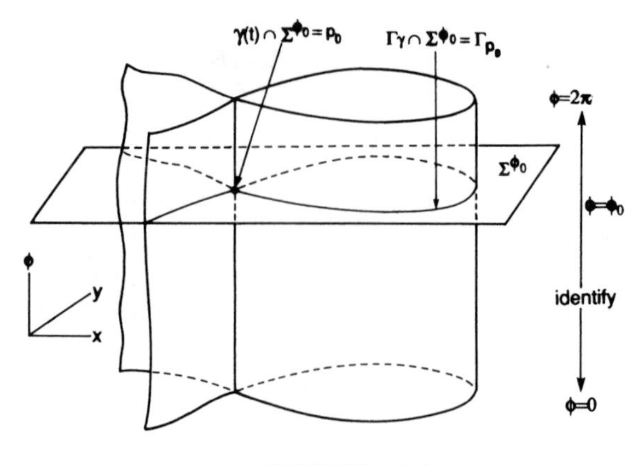

FIGURE 4.5.6.

and

$$\Gamma_\gamma \cap \Sigma^{\phi_0} = \left\{ q \in \mathbb{R}^2 \,|\, q = q_0(t), t \in \mathbb{R} \right\} = \Gamma_{p_0}. \tag{4.5.18}$$

In particular, (4.5.17) and (4.5.18) are independent of ϕ_0; this simply reflects the fact that the unperturbed vector field is autonomous. Now let $(q(t), \phi(t))$ and $(q_\varepsilon(t), \phi(t))$ be trajectories of the unperturbed and perturbed vector fields, respectively. Then the projections of these trajectories onto Σ^{ϕ_0} are given by

$$(q_0(t), \phi_0) \tag{4.5.19}$$

and

$$(q_\varepsilon(t), \phi_0). \tag{4.5.20}$$

We remark that $q_\varepsilon(t)$ actually depends on ϕ_0 (as opposed to $q(t)$), since the q-component of the perturbed vector field (4.5.6) depends on ϕ, i.e., the perturbed vector field is nonautonomous. Therefore, (4.5.20) could be a very complicated curve in Σ^{ϕ_0}, possibly intersecting itself many times. This tends to obscure much of the dynamical content of the trajectory. We will remedy this situation later on when we consider a Poincaré map constructed from the flow generated by the perturbed vector field; see Figures 4.5.7 and 4.5.8 for an illustration of the geometry behind the projections (4.5.19) and (4.5.20).

We are now at the point where we can define the splitting of $W^s(\gamma_\varepsilon(t))$ and $W^u(\gamma_\varepsilon(t))$. Choose any point $p \in \Gamma_\gamma$. Then $W^s(\gamma(t))$ and $W^u(\gamma(t))$ intersect π_p transversely at p. Hence, by the persistence of transversal intersections and the fact that $W^s(\gamma_\varepsilon(t))$ and $W^u(\gamma_\varepsilon(t))$ are $\mathbf{C}^r$ in ε, for ε sufficiently small $W^s(\gamma_\varepsilon(t))$ and $W^u(\gamma_\varepsilon(t))$ intersect π_p transversely in the points p_ε^s and p_ε^u, respectively. It is therefore natural to define *the distance between $W^s(\gamma_\varepsilon(t))$ and $W^u(\gamma_\varepsilon(t))$ at the point* p, denoted $d(p, \varepsilon)$, to be

$$d(p, \varepsilon) \equiv |p_\varepsilon^u - p_\varepsilon^s|; \tag{4.5.21}$$

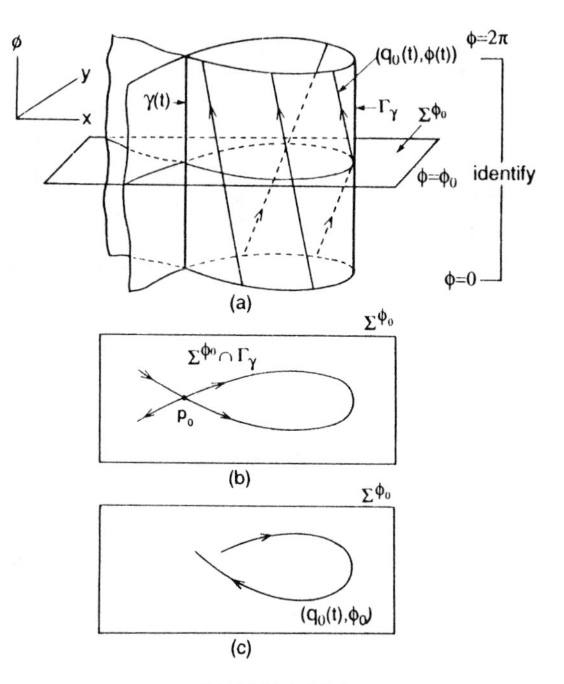

FIGURE 4.5.7.

see Figure 4.5.9. We will find it convenient in the next step to redefine (4.5.21) in an equivalent, but slightly less natural manner as follows

$$d(p,\varepsilon) = \frac{(p_\varepsilon^u - p_\varepsilon^s)\cdot(DH(q_0(-t_0)),0)}{\|DH(q_0(-t_0))\|} \tag{4.5.22}$$

where "$\cdot$" denotes the vector scalar product and

$$\|DH(q_0(-t_0))\| = \sqrt{\left(\frac{\partial H}{\partial x}(q_0(-t_0))\right)^2 + \left(\frac{\partial H}{\partial y}(q_0(-t_0))\right)^2}.$$

Because p_ε^u and p_ε^s are chosen to lie on the vector $(DH(q_0(-t_0)),0)$, it should be clear that the magnitude of (4.5.22) is equal to the magnitude of (4.5.21). However, (4.5.22) is a signed measure of the distance and reflects the relative orientations of $W^s(\gamma_\varepsilon(t))$ and $W^u(\gamma_\varepsilon(t))$ near p; see Figure 4.5.10. Note that since p_ε^u and p_ε^s lie on π_p, we can write

$$p_\varepsilon^u = (q_\varepsilon^u, \phi_0) \tag{4.5.23}$$

and

$$p_\varepsilon^s = (q_\varepsilon^s, \phi_0), \tag{4.5.24}$$

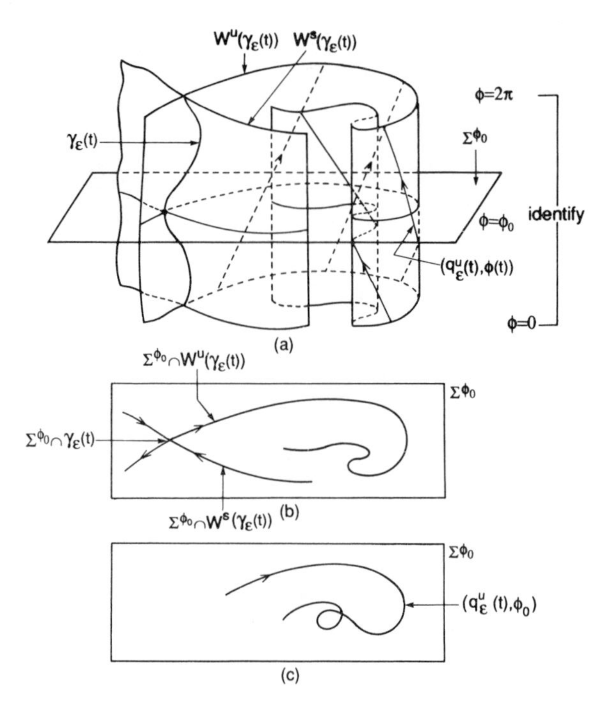

FIGURE 4.5.8.

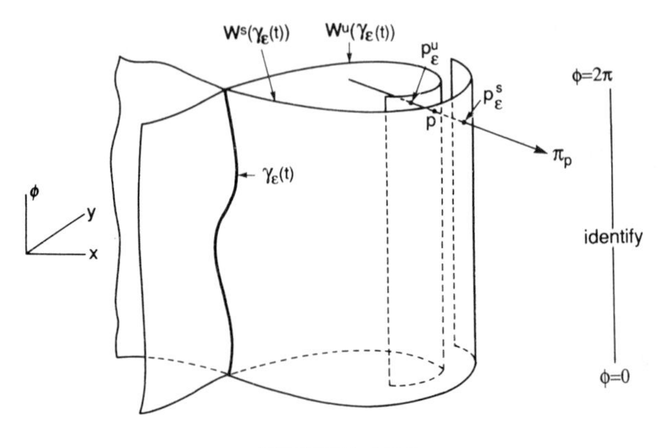

FIGURE 4.5.9.

i.e., p^u_ε and p^s_ε have the same ϕ_0 coordinate. Thus, (4.5.22) is the same as

$$d(t_0, \phi_0, \varepsilon) = \frac{DH(q_0(-t_0)) \cdot (q^u_\varepsilon - q^s_\varepsilon)}{\|DH(q_0(-t_0))\|}, \tag{4.5.25}$$

where we are now denoting $d(p, \varepsilon)$ by $d(t_0, \phi_0, \varepsilon)$, since every point $p \in \Gamma_\gamma$

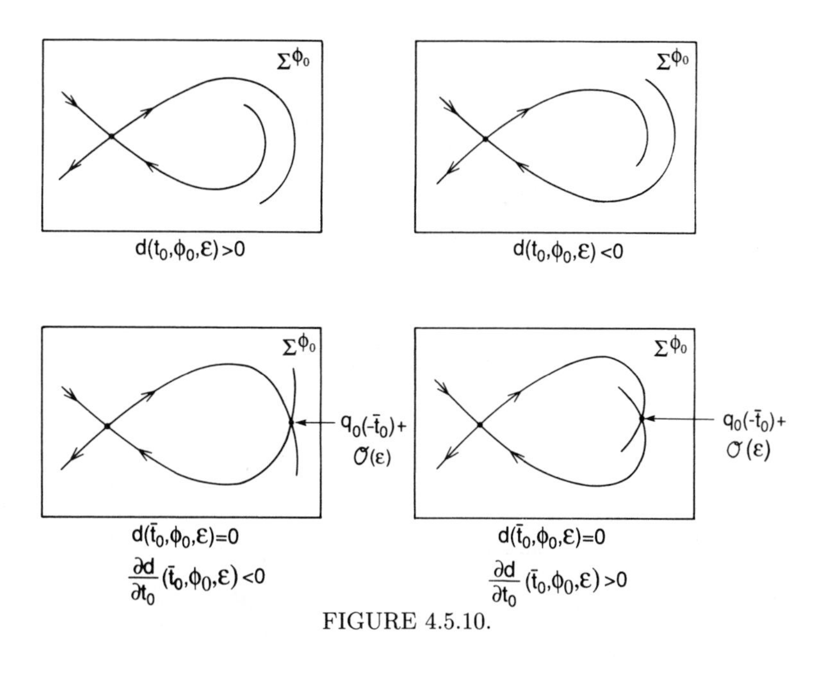

FIGURE 4.5.10.

can be uniquely represented by the parameters (t_0, ϕ_0), $t_0 \in \mathbb{R}$, $\phi_0 \in (0, 2\pi]$, according to the parametrization $p = (q_0(-t_0), \phi_0)$ described in Step 1.

Before deriving a computable approximation to (4.5.25) in Step 3, we want to address a technical issue involving the choice of p^u_ε and p^s_ε. Certainly by transversality and $\mathbf{C}^r$ dependence on ε, for ε sufficiently small, $W^s(\gamma_\varepsilon(t))$ and $W^u(\gamma_\varepsilon(t))$ intersect π_p. However, these manifolds may intersect π_p in more than one point (indeed, an infinite number of points is possible), as shown in Figure 4.5.11. The question then arises as to which points p^u_ε and p^s_ε are chosen so as to define (4.5.25). We first give a definition.

DEFINITION 4.5.1 Let $p^s_{\varepsilon,i} \in W^s(\gamma_\varepsilon(t)) \cap \pi_p$ and $p^u_{\varepsilon,i} \in W^u(\gamma_\varepsilon(t)) \cap \pi_p$, $i \in \mathcal{I}$, where $\mathcal{I}$ is some index set. Let $(q^s_{\varepsilon,i}(t), \phi(t)) \in W^s(\gamma_\varepsilon(t))$ and $(q^u_{\varepsilon,i}(t), \phi(t)) \in W^u(\gamma_\varepsilon(t))$ denote orbits of the perturbed vector field (4.5.6) satisfying $(q^s_{\varepsilon,i}(0), \phi(0)) = p^s_{\varepsilon,i}$ and $(q^u_{\varepsilon,i}(0), \phi(0)) = p^u_{\varepsilon,i}$, respectively. Then we have the following (see Figure 4.5.12).

1. For some $i = \bar{i} \in \mathcal{I}$ we say that $p^s_{\varepsilon,\bar{i}}$ is the point in $W^s(\gamma_\varepsilon(t)) \cap \pi_p$ that is closest to $\gamma_\varepsilon(t)$ in terms of positive time of flight along $W^s(\gamma_\varepsilon(t))$ if, for all $t > 0$, $(q^s_{\varepsilon,\bar{i}}(t), \phi_0) \cap \pi_p = \emptyset$.

2. For some $i = \bar{i} \in \mathcal{I}$ we say that $p^u_{\varepsilon,\bar{i}}$ is the point in $W^u(\gamma_\varepsilon(t)) \cap \pi_p$ that is closest to $\gamma_\varepsilon(t)$ in terms of negative time of flight along $W^u(\gamma_\varepsilon(t))$ if, for all $t < 0$, $(q^u_{\varepsilon,\bar{i}}(t), \phi_0) \cap \pi_p = \emptyset$.

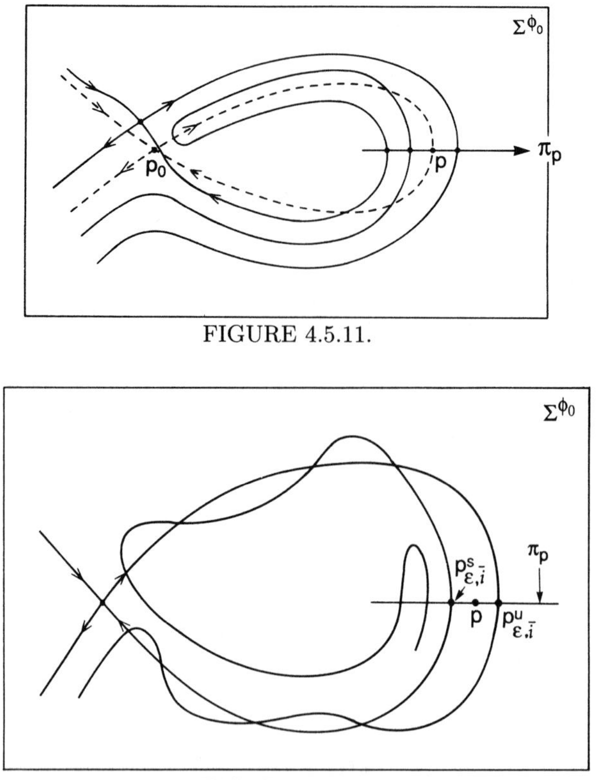

FIGURE 4.5.11.

FIGURE 4.5.12.

We make the following remarks regarding this definition.

Remark 1. For p fixed, we are interested only in the points in $W^s(\gamma_\varepsilon(t)) \cap \pi_p$ and $W^u(\gamma_\varepsilon(t)) \cap \pi_p$ that are $\mathcal{O}(\varepsilon)$ close to p. This is because our methods are perturbative.

Remark 2. For the unperturbed system, the orbit through p leaves π_p in positive and negative time and enters a neighborhood of the hyperbolic orbit without ever returning to π_p. The orbits through π_p described in Definition 4.5.1 are the perturbed orbits that most closely behave in this manner.

Remark 3. The points $p^s_{\varepsilon,\bar{i}}$ and $p^u_{\varepsilon,\bar{i}}$ are unique. This will follow from the proof of Lemma 4.5.2.

The points p^s_ε and p^u_ε used in defining (4.5.25) are chosen to be closest to $\gamma_\varepsilon(t)$ in the sense of positive time of flight along $W^s(\gamma_\varepsilon(t))$ and negative time of flight along $W^u(\gamma_\varepsilon(t))$, respectively, as described in Definition 4.5.1.

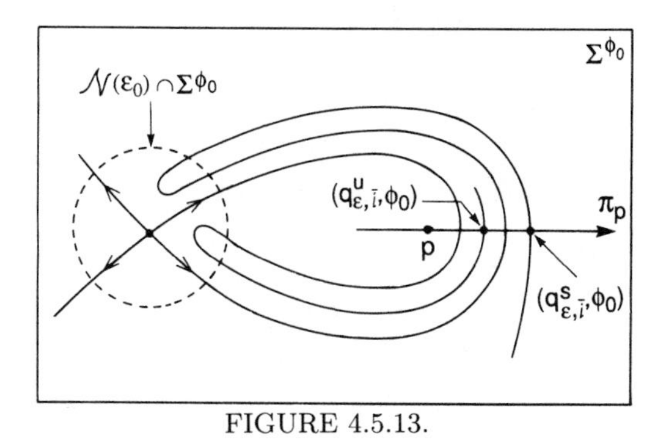

FIGURE 4.5.13.

Still, the question remains as to why this choice. The consequences of the following lemma will answer this question.

Lemma 4.5.2 *Let $p^s_{\varepsilon,\bar{i}}$ (resp. $p^u_{\varepsilon,\bar{i}}$) be a point on $W^s(\gamma_\varepsilon(t)) \cap \pi_p$ (resp. $W^u(\gamma_\varepsilon(t)) \cap \pi_p$) that is* not *closest to $\gamma_\varepsilon(t)$ in the sense of Definition 4.5.1, and let $\mathcal{N}(\varepsilon_0)$ denote the neighborhood of $\gamma(t)$ and $\gamma_\varepsilon(t)$ described following Proposition 4.5.1. Let $(q^s_{\varepsilon,\bar{i}}(t), \phi(t))$ (resp. $(q^u_{\varepsilon,\bar{i}}(t), \phi(t))$) be a trajectory in $W^s(\gamma_\varepsilon(t))$ (resp. $W^u(\gamma_\varepsilon(t))$) satisfying $(q^s_{\varepsilon,\bar{i}}(0), \phi(0)) = p^s_{\varepsilon,\bar{i}}$ (resp. $(q^u_{\varepsilon,\bar{i}}(0), \phi(0)) = p^u_{\varepsilon,\bar{i}}$). Then, for ε sufficiently small,* before *$(q^s_{\varepsilon,\bar{i}}(t), \phi_0)$, $t > 0$, (resp. $(q^u_{\varepsilon,\bar{i}}(t), \phi_0)$, $t < 0$) can intersect π_p (as it must by Definition 4.5.1), it must pass through $\mathcal{N}(\varepsilon_0)$* (see Figure 4.5.13).

Proof: We give the argument for trajectories in $W^s(\gamma_\varepsilon(t))$; the argument for trajectories in $W^u(\gamma_\varepsilon(t))$ will follow immediately by considering the time-reversed vector field.

First we consider the unperturbed vector field. Consider any point (q^s_0, ϕ_0) on $W^s(\gamma(t)) \cap \mathcal{N}(\varepsilon_0)$; see Figure 4.5.14. Let $(q^s_0(t), \phi(t)) \in W^s(\gamma(t))$ satisfy $(q^s_0(0), \phi(0)) = (q^s_0, \phi_0)$. Then there exists a *finite time*, $-\infty < T^s < 0$, such that $(q^s_0(T^s), \phi(T^s)) \in W^s(\gamma(t)) \cap \mathcal{N}(\varepsilon_0)$. In other words, T^s is the time that it takes a trajectory leaving $\mathcal{N}(\varepsilon_0)$ in $W^s(\gamma(t))$ to reenter $\mathcal{N}(\varepsilon_0)$; see Figure 4.5.14.

We now want to compare trajectories in $W^s(\gamma(t))$ with trajectories in $W^s(\gamma_\varepsilon(t))$. Choose points

$$(q^s_0, \phi_0) \in W^s_{\text{loc}}(\gamma(t)) \cap \mathcal{N}(\varepsilon_0)$$

and

$$(q^s_\varepsilon, \phi_0) \in W^s_{\text{loc}}(\gamma_\varepsilon(t)) \cap \mathcal{N}(\varepsilon_0),$$

and consider trajectories

$$(q^s_0(t), \phi(t)) \in W^s(\gamma(t))$$

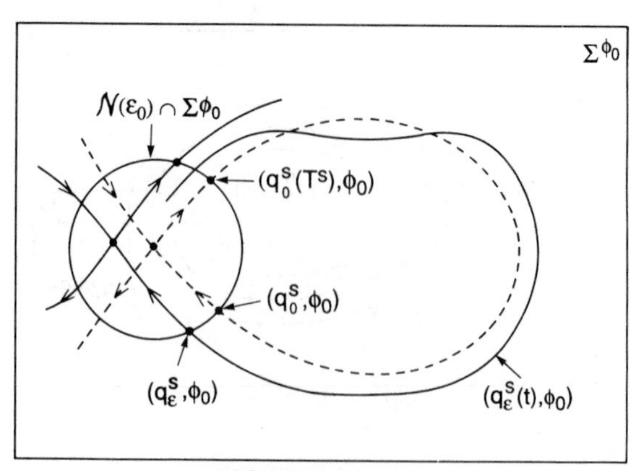

FIGURE 4.5.14.

and

$$(q_\varepsilon^s(t), \phi(t)) \in W^s(\gamma_\varepsilon(t))$$

satisfying

$$(q_0^s(0), \phi(0)) = (q_0^s, \phi_0)$$

and

$$(q_\varepsilon^s(0), \phi(0)) = (q_\varepsilon^s, \phi_0);$$

see Figure 4.5.14. Then

$$|(q_\varepsilon^s(t), \phi(t)) - (q_0^s(t), \phi(t))| = \mathcal{O}(\varepsilon_0) \tag{4.5.26}$$

for $0 \leq t \leq \infty$ and, by Gronwall's inequality (see Lemma 1.2.9),

$$|(q_\varepsilon^s(t), \phi(t)) - (q_0^s(t), \phi(t)| = \mathcal{O}(\varepsilon) \tag{4.5.27}$$

for $T^s \leq t \leq 0$. Therefore, a trajectory in $W^s(\gamma_\varepsilon(t))$ leaving $\mathcal{N}(\varepsilon_0)$ in *negative time* must follow $\mathcal{O}(\varepsilon)$ close to a trajectory in $W^s(\gamma(t))$ until it reenters $\mathcal{N}(\varepsilon_0)$ (since we only have a finite time estimate outside of $\mathcal{N}(\varepsilon_0)$).

Hence, this argument shows that $(q_0^s(t), \phi(t))$ and $(q_\varepsilon^s(t), \phi(t))$ remain ε-close for $T^s \leq t < \infty$, i.e., until $(q_\varepsilon^s(t), \phi(t))$ enters $\mathcal{N}(\varepsilon_0)$ under the negative time flow. However, this argument does not rule out the fact that $(q_\varepsilon^s(t), \phi(t))$ can develop "kinks" and therefore re-intersect π_p (while remaining ε-close to $(q_0^s(t), \phi(t))$), as shown in Figure 4.5.15. This *does not* happen since tangent vectors of $(q_\varepsilon^s(t), \phi(t))$ and $(q_0^s(t), \phi(t))$ are $\mathcal{O}(\varepsilon)$ close for $T^s \leq t < \infty$. This can be seen as follows; we have just shown that on $T^s \leq t < \infty$ we have

$$(q_\varepsilon^s(t), \phi(t)) = (q_0^s(t) + \mathcal{O}(\varepsilon), \phi(t)), \tag{4.5.28}$$

and a vector tangent to $(q_\varepsilon^s(t), \phi(t))$ is given by

$$\begin{aligned} \dot{q}_\varepsilon^s &= JDH(q_\varepsilon^s) + \varepsilon g(q_\varepsilon^s, \phi(t)), \\ \dot{\phi} &= \omega. \end{aligned} \tag{4.5.29}$$

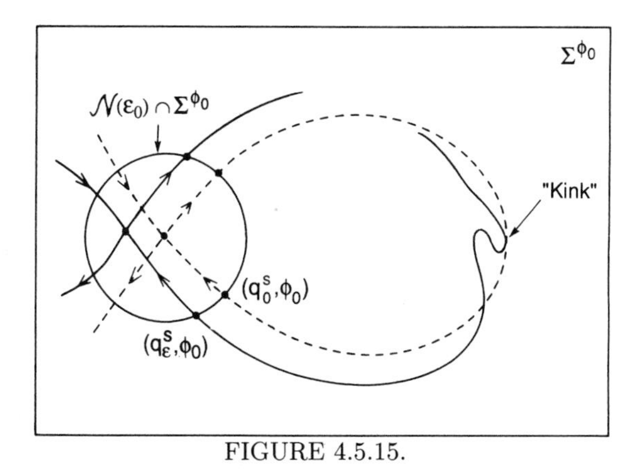

FIGURE 4.5.15.

Substituting (4.5.28) into the right-hand side of (4.5.29) and Taylor expanding about $\varepsilon = 0$ gives

$$\begin{aligned} \dot{q}_\varepsilon^s &= JDH(q_0^s) + \mathcal{O}(\varepsilon), \\ \dot{\phi} &= \omega. \end{aligned} \tag{4.5.30}$$

Now, a vector tangent to $(q_0^s(t), \phi(t))$ is given by

$$\begin{aligned} \dot{q}_0^s &= JDH(q_0^s), \\ \phi &= \omega. \end{aligned} \tag{4.5.31}$$

Clearly, (4.5.30) and (4.5.31) are $\mathcal{O}(\varepsilon)$ close on $T^s \leq t < \infty$ so that the situation shown in Figure 4.5.15 *cannot* occur on this time interval for ε sufficiently small.

Now let $p^s_{\varepsilon,\bar{i}} \in W^s(\gamma_\varepsilon(t)) \cap \pi_p$ be a point that is *not* closest to $\gamma_\varepsilon(t)$ in terms of positive time of flight as defined in Definition 4.5.1. Let $(q^s_{\varepsilon,\bar{i}}(t), \phi(t)) \in W^s(\gamma_\varepsilon(t))$ satisfy $(q^s_{\varepsilon,\bar{i}}(0), \phi(0)) = p^s_{\varepsilon,\bar{i}}$. Then, by Definition 4.5.1, for some $\bar{t} > 0$, $(q^s_{\varepsilon,\bar{i}}(\bar{t}), \phi_0) \in W^s(\gamma_\varepsilon(t)) \cap \pi_p$. Hence, by the argument given above, for ε sufficiently small, somewhere in $0 < t < \bar{t}$, $(q^s_{\varepsilon,\bar{i}}(t), \phi(t))$ must have entered $\mathcal{N}(\varepsilon_0)$. $\square$

We make the following remarks regarding Lemma 4.5.2.

Remark 1. The reader should note that $T^s = T^s(\varepsilon_0)$. That is why it was necessary to consider a fixed neighborhood containing $\gamma(t)$ and $\gamma_\varepsilon(t)$.

Remark 2. From the proof of Lemma 4.5.2 it follows that the points closest to $\gamma_\varepsilon(t)$ in the sense described in Definition 4.5.1 are unique. We leave the details to Exercise 4.15.

Remark 3. Let $p_\varepsilon^s = (q_\varepsilon^s, \phi_0) \in W^s(\gamma_\varepsilon(t)) \cap \pi_p$, and let $(q_\varepsilon^s(t), \phi(t)) \in W^s(\gamma_\varepsilon(t))$ satisfy $(q_\varepsilon^s(0), \phi(0)) = (q_\varepsilon^s, \phi_0)$. Then if p_ε^s is the point closest to $\gamma_\varepsilon(t)$ in the sense of Definition 4.5.1, it follows from the proof of Lemma 4.5.2 that

$$|q_\varepsilon^s(t) - q_0(t - t_0)| = \mathcal{O}(\varepsilon), \qquad t \in [0, \infty), \tag{4.5.32}$$

$$|\dot{q}_\varepsilon^s(t) - \dot{q}_0(t - t_0)| = \mathcal{O}(\varepsilon), \qquad t \in [0, \infty). \tag{4.5.33}$$

A similar statement can be made for points in $W^u(\gamma_\varepsilon(t)) \cap \pi_p$ and solutions in $W^u(\gamma_\epsilon(t))$. In deriving the Melnikov function in the next step we will need to approximate perturbed solutions in $W^s(\gamma_\varepsilon(t))$ and $W^u(\gamma_\varepsilon(t))$ by unperturbed solutions in $W^s(\gamma(t))$ and $W^u(\gamma(t))$ for semi-infinite time intervals with $\mathcal{O}(\varepsilon)$ accuracy. This is why the Melnikov function will only detect points on $W^s(\gamma_\varepsilon(t)) \cap \pi_p \cap W^u(\gamma_\varepsilon(t))$ that are closest to $\gamma_\varepsilon(t)$ in the sense of Definition 4.5.1.

Step 3: Derivation of the Melnikov Function. Taylor expanding (4.5.25) about $\varepsilon = 0$ gives

$$d(t_0, \phi_0, \varepsilon) = d(t_0, \phi_0, 0) + \varepsilon \frac{\partial d}{\partial \varepsilon}(t_0, \phi_0, 0) + \mathcal{O}(\varepsilon^2), \tag{4.5.34}$$

where

$$d(t_0, \phi_0, 0) = 0 \tag{4.5.35}$$

and

$$\frac{\partial d}{\partial \varepsilon}(t_0, \phi_0, 0) = \frac{DH(q_0(-t_0)) \cdot \left(\frac{\partial q_\varepsilon^u}{\partial \varepsilon}|_{\varepsilon=0} - \frac{\partial q_\varepsilon^s}{\partial \varepsilon}|_{\varepsilon=0}\right)}{\|DH(q_0(-t_0))\|}. \tag{4.5.36}$$

The *Melnikov function* is defined to be

$$M(t_0, \phi_0) \equiv DH(q_0(-t_0)) \cdot \left(\left.\frac{\partial q_\varepsilon^u}{\partial \varepsilon}\right|_{\varepsilon=0} - \left.\frac{\partial q_\varepsilon^s}{\partial \varepsilon}\right|_{\varepsilon=0} \right). \tag{4.5.37}$$

Now, since

$$DH(q_0(-t_0)) = \left(\frac{\partial H}{\partial x}(q_0(-t_0)), \frac{\partial H}{\partial y}(q_0(-t_0)) \right)$$

is not zero on $q_0(-t_0)$, for t_0 finite, we see that

$$M(t_0, \phi_0) = 0 \Rightarrow \frac{\partial d}{\partial \varepsilon}(t_0, \phi_0) = 0. \tag{4.5.38}$$

Therefore, up to a nonzero normalization factor ($\|DH(q_0(-t_0))\|$), the Melnikov function is the lowest order nonzero term in the Taylor expansion for the distance between $W^s(\gamma_\varepsilon(t))$ and $W^u(\gamma_\varepsilon(t))$ at the point p.

We now want to derive an expression for $M(t_0, \phi_0)$ that can be computed without needing to know the solution of the perturbed vector field. We do this by utilizing Melnikov's original trick. We define a time-dependent Melnikov function using the flow generated by both the unperturbed vector field and the perturbed vector field. We have to be careful here since we do not have any a priori knowledge of arbitrary orbits generated by the perturbed vector field. However, the persistence and differentiability of $\gamma(t)$, $W^s(\gamma(t))$, and $W^u(\gamma(t))$ as described in Proposition 4.5.1 are all that we need, since Definition 4.5.1 and Lemma 4.5.2 allow us to characterize the orbits of interest for determining the splitting of $W^s(\gamma_\varepsilon(t))$ and $W^u(\gamma_\varepsilon(t))$. We then derive an ordinary differential equation which the time-dependent Melnikov function must satisfy. The ordinary differential equation turns out to be first order and linear; hence it is trivially solvable. The solution evaluated at the appropriate time will yield the Melnikov function.

We begin by defining the time-dependent Melnikov function as follows

$$M(t;t_0,\phi_0) \equiv DH(q_0(t-t_0)) \cdot \left(\left.\frac{\partial q_\varepsilon^u(t)}{\partial \varepsilon}\right|_{\varepsilon=0} - \left.\frac{\partial q_\varepsilon^s(t)}{\partial \varepsilon}\right|_{\varepsilon=0} \right). \tag{4.5.39}$$

We want to take some care in describing precisely what we mean by (4.5.39). We denote orbits in $W^s(\gamma_\varepsilon(t))$ and $W^u(\gamma_\varepsilon(t))$ by $q_\varepsilon^s(t)$ and $q_\varepsilon^u(t)$, respectively. Then in (4.5.39) the expressions

$$\left.\frac{\partial q_\varepsilon^u(t)}{\partial \varepsilon}\right|_{\varepsilon=0} \tag{4.5.40}$$

and

$$\left.\frac{\partial q_\varepsilon^s(t)}{\partial \varepsilon}\right|_{\varepsilon=0} \tag{4.5.41}$$

are simply the derivatives with respect to ε (evaluated at $\varepsilon = 0$) of $q_\varepsilon^u(t)$ and $q_\varepsilon^s(t)$, respectively, where $q_\varepsilon^u(t)$ and $q_\varepsilon^s(t)$ satisfy

$$q_\varepsilon^u(0) = q_\varepsilon^u, \tag{4.5.42}$$

$$q_\varepsilon^s(0) = q_\varepsilon^s. \tag{4.5.43}$$

The expression $q_0(t-t_0)$ denotes the unperturbed homoclinic orbit. Thus, we see that (4.5.39) is a bit unusual; part of it, $(DH(q_0(t-t_0)))$, evolves in time under the dynamics of the unperturbed vector field with the remaining part

$$\left(\left.\frac{\partial q_\varepsilon^u(t)}{\partial \varepsilon}\right|_{\varepsilon=0} - \left.\frac{\partial q_\varepsilon^s(t)}{\partial \varepsilon}\right|_{\varepsilon=0} \right)$$

evolving in time under the dynamics of the perturbed vector field. It should be obvious that the relationship between the time-dependent Melnikov function and the Melnikov function is given by

$$M(0;t_0,\phi_0) = M(t_0,\phi_0). \tag{4.5.44}$$

Next we turn to deriving an ordinary differential equation that $M(t;t_0,\phi_0)$ must satisfy. The expressions we derive will be a bit cumbersome, so for the sake of a more compact notation we define

$$\left.\frac{\partial q_\varepsilon^u(t)}{\partial \varepsilon}\right|_{\varepsilon=0} \equiv q_1^u(t),$$

$$\left.\frac{\partial q_\varepsilon^s(t)}{\partial \varepsilon}\right|_{\varepsilon=0} \equiv q_1^s(t).$$

Then (4.5.39) can be rewritten as

$$M(t;t_0,\phi_0) = DH(q_0(t-t_0)) \cdot (q_1^u(t) - q_1^s(t)). \tag{4.5.45}$$

We want to introduce a further definition to compactify the notation as follows

$$M(t;t_0,\phi_0) \equiv \Delta^u(t) - \Delta^s(t), \tag{4.5.46}$$

where

$$\Delta^{u,s}(t) \equiv DH(q_0(t-t_0)) \cdot q_1^{u,s}(t). \tag{4.5.47}$$

Differentiating (4.5.47) with respect to t gives

$$\begin{aligned}\frac{d}{dt}(\Delta^{u,s}(t)) = {}& \left(\frac{d}{dt}(DH(q_0(t-t_0)))\right) \cdot q_1^{u,s}(t) \\ & + DH(q_0(t-t_0)) \cdot \frac{d}{dt} q_1^{u,s}(t).\end{aligned} \tag{4.5.48}$$

The term $\frac{d}{dt}(q_1^{u,s}(t))$ in (4.5.48) needs some explanation. Recall from above that we have defined

$$q_1^{u,s}(t) \equiv \left.\frac{\partial q_\varepsilon^{u,s}(t)}{\partial \varepsilon}\right|_{\varepsilon=0},$$

and $q_\varepsilon^{u,s}(t)$ solves

$$\frac{d}{dt}(q_\varepsilon^{u,s}(t)) = JDH(q_\varepsilon^{u,s}(t)) + \varepsilon g(q_\varepsilon^{u,s}(t), \phi(t), \varepsilon), \tag{4.5.49}$$

where $\phi(t) = \omega t + \phi_0$. Since $q_\varepsilon^{u,s}(t)$ is $\mathbf{C}^r$ in ε and t (see Theorem 1.1.10), we can differentiate (4.5.49) with respect to ε and interchange the order of the ε and t differentiations to obtain

$$\begin{aligned}\frac{d}{dt}\left(\left.\frac{\partial q_\varepsilon^{u,s}(t)}{\partial \varepsilon}\right|_{\varepsilon=0}\right) = {}& JD^2H(q_0(t-t_0)) \left.\frac{\partial q_\varepsilon^{u,s}(t)}{\partial \varepsilon}\right|_{\varepsilon=0(t)} \\ & + g(q_0(t-t_0), \phi(t), 0)\end{aligned} \tag{4.5.50}$$

or

$$\frac{d}{dt} q_1^{u,s}(t) = JD^2H(q_0(t-t_0)) q_1^{u,s}(t) + g(q_0(t-t_0), \phi(t), 0). \tag{4.5.51}$$

Equation (4.5.51) is referred to as the *first variational equation* (see Section 1.2D, ii) and Exercise 1.2.11). We remark that $q_1^u(t)$ solves (4.5.51) for $t \in (-\infty, 0]$, and $q_1^s(t)$ solves (4.5.51) for $t \in (0, \infty]$; see the remarks following Lemma 4.5.2. Substituting (4.5.51) into (4.5.48) gives

$$\begin{aligned}\frac{d}{dt}(\Delta^{u,s}(t)) = {}& \left(\frac{d}{dt}(DH(q_0(t-t_0)))\right) \cdot q_1^{u,s}(t) \\ & + DH(q_0(t-t_0)) \cdot JD^2H(q_0(t-t_0)) q_1^{u,s}(t) \\ & + DH(q_0(t-t_0)) \cdot g(q_0(t-t_0), \phi(t), 0).\end{aligned} \tag{4.5.52}$$

Now a wonderful thing happens.

Lemma 4.5.3

$$\frac{d}{dt}(DH(q_0(t-t_0)))\cdot q_1^{u,s}(t) + DH(q_0(t-t_0))\cdot\ JD^2H(q_0(t-t_0))q_1^{u,s}(t)=0.$$

Proof: First note that

$$\begin{aligned}\frac{d}{dt}\big(DH(q_0(t-t_0))\big) &= D^2H(q_0(t-t_0))\dot{q}_0(t-t_0)\\ &= \big(D^2H(q_0(t-t_0))\big)\big(JDH(q_0(t-t_0))\big).\end{aligned} \tag{4.5.53}$$

Let $q_1^{u,s}(t)=(x_1^{u,s}(t),y_1^{u,s}(t))$. Then we have

$$\begin{aligned}(D^2H)(JDH)\cdot q_1^{u,s} &= \begin{pmatrix}\frac{\partial^2 H}{\partial x^2} & \frac{\partial^2 H}{\partial x\partial y}\\ \frac{\partial^2 H}{\partial x\partial y} & \frac{\partial^2 H}{\partial y^2}\end{pmatrix}\begin{pmatrix}\frac{\partial H}{\partial y}\\ -\frac{\partial H}{\partial x}\end{pmatrix}\cdot\begin{pmatrix}x_1^{u,s}\\ y_1^{u,s}\end{pmatrix}\\ &= x_1^{u,s}\left[\frac{\partial^2 H}{\partial x^2}\frac{\partial H}{\partial y}-\frac{\partial^2 H}{\partial x\partial y}\frac{\partial H}{\partial x}\right]\\ &\quad + y_1^{u,s}\left[\frac{\partial^2 H}{\partial x\partial y}\frac{\partial H}{\partial y}-\frac{\partial^2 H}{\partial y^2}\frac{\partial H}{\partial x}\right]\end{aligned} \tag{4.5.54}$$

and

$$\begin{aligned}DH\cdot(JD^2H)q_1^{u,s} &= \begin{pmatrix}\frac{\partial H}{\partial x}\\ \frac{\partial H}{\partial y}\end{pmatrix}\cdot\begin{pmatrix}\frac{\partial^2 H}{\partial x\partial y} & \frac{\partial^2 H}{\partial y^2}\\ -\frac{\partial^2 H}{\partial x^2} & -\frac{\partial^2 H}{\partial x\partial y}\end{pmatrix}\begin{pmatrix}x_1^{u,s}\\ y_1^{u,s}\end{pmatrix}\\ &= x_1^{u,s}\left[\frac{\partial^2 H}{\partial x\partial y}\frac{\partial H}{\partial x}-\frac{\partial^2 H}{\partial x^2}\frac{\partial H}{\partial y}\right]\\ &\quad + y_1^{u,s}\left[\frac{\partial^2 H}{\partial y^2}\frac{\partial H}{\partial x}-\frac{\partial^2 H}{\partial x\partial y}\frac{\partial H}{\partial y}\right],\end{aligned} \tag{4.5.55}$$

where we have left out the argument $q_0(t-t_0)=(x_0(t-t_0),y_0(t-t_0))$ for the sake of a less cumbersome notation. Adding (4.5.54) and (4.5.55) gives the result. □

Therefore, using Lemma 4.5.3, (4.5.48) becomes

$$\frac{d}{dt}(\Delta^{u,s}(t)) = DH(q_0(t-t_0))\cdot g(q_0(t-t_0),\phi(t),0). \tag{4.5.56}$$

Integrating $\Delta^u(t)$ and $\Delta^s(t)$ individually from $-\tau$ to 0 and 0 to τ ($\tau > 0$), respectively, gives

$$\Delta^u(0) - \Delta^u(-\tau) = \int_{-\tau}^{0} DH(q_0(t-t_0)) \cdot g(q_0(t-t_0), \omega t + \phi_0, 0)dt \quad (4.5.57)$$

and

$$\Delta^s(\tau) - \Delta^s(0) = \int_{0}^{\tau} DH(q_0(t-t_0)) \cdot g(q_0(t-t_0), \omega t + \phi_0, 0)dt \quad (4.5.58)$$

where we have substituted $\phi(t) = \omega t + \phi_0$ into the integrand. Adding (4.5.57) and (4.5.58) and referring to (4.5.44) and (4.5.46) gives

$$\begin{aligned} M(t_0, \phi_0) &= M(0, t_0, \phi_0) = \Delta^u(0) - \Delta^s(0) \\ &= \int_{-\tau}^{\tau} DH(q_0(t-t_0)) \cdot g(q_0(t-t_0), \omega t + \phi_0, 0)dt \\ &\quad + \Delta^s(\tau) - \Delta^u(-\tau). \end{aligned} \quad (4.5.59)$$

We now want to consider the limit of (4.5.59) as $\tau \to \infty$.

Lemma 4.5.4

$$\lim_{\tau \to \infty} \Delta^s(\tau) = \lim_{\tau \to \infty} \Delta^u(-\tau) = 0.$$

Proof: Recall from (4.5.47) that

$$\Delta^{u,s}(t) = DH(q_0(t-t_0)) \cdot q_1^{u,s}(t).$$

Now, as $t \to \infty$ (resp. $-\infty$), $DH(q_0(t-t_0))$ goes to zero *exponentially* fast, since $q_0(t-t_0)$ approaches a *hyperbolic* fixed point. Also, as $t \to \infty$ (resp. $-\infty$), $q_1^s(t)$ (resp. $q_1^u(t)$) is bounded (see Exercises 1.2.12 and 4.17). Hence, $\Delta^s(\tau)$ (resp. $\Delta^u(-\tau)$) goes to zero as $\tau \to \infty$. □

Lemma 4.5.5 *The improper integral*

$$\int_{-\infty}^{\infty} DH(q_0(t-t_0)) \cdot g(q_0(t-t_0), \omega t + \phi_0, 0)dt$$

converges absolutely.

Proof: This result follows from the fact that $g(q_0(t-t_0), \omega t + \phi_0, 0)$ is bounded for all t and $DH(q_0(t-t_0))$ goes to zero exponentially fast as $t \to \pm\infty$. □

Hence, combining Lemma 4.5.4 and Lemma 4.5.5, (4.5.59) becomes

$$M(t_0, \phi_0) = \int_{-\infty}^{\infty} DH(q_0(t-t_0)) \cdot g(q_0(t-t_0), \omega t + \phi_0, 0)dt. \quad (4.5.60)$$

Before giving the main theorem we want to point out an interesting property of the Melnikov function. If we make the transformation

$$t \longrightarrow t + t_0,$$

then (4.5.60) becomes

$$M(t_0, \phi_0) = \int_{-\infty}^{\infty} DH(q_0(t)) \cdot g(q_0(t), \omega t + \omega t_0 + \phi_0, 0)dt. \qquad (4.5.61)$$

Recall that $g(q, \cdot, 0)$ is periodic, which implies that $M(t_0, \phi_0)$ is periodic in t_0 with period $2\pi/\omega$ *and* periodic in ϕ_0 with period 2π. The geometry of this will be explained shortly. However, it should be clear from (4.5.61) that varying t_0 and varying ϕ_0 have the same effect. Moreover, from (4.5.61) and the periodicity of $g(q, \cdot, 0)$, it follows that

$$\frac{\partial M}{\partial \phi_0}(t_0, \phi_0) = \omega \frac{\partial M}{\partial t_0}(t_0, \phi_0); \qquad (4.5.62)$$

hence, $\frac{\partial M}{\partial t_0} = 0$ if and only if $\frac{\partial M}{\partial \phi_0} = 0$. In Theorem 4.5.6 we will need to have $\frac{\partial M}{\partial t_0} \neq 0$ or $\frac{\partial M}{\partial \phi_0} \neq 0$. However, from (4.5.62), if one is nonzero, then so is the other; hence, we will state the theorem in terms of $\frac{\partial M}{\partial t_0} \neq 0$.

Theorem 4.5.6 *Suppose we have a point $(t_0, \phi_0) = (\bar{t}_0, \bar{\phi}_0)$ such that*

i) $M(\bar{t}_0, \bar{\phi}_0) = 0$ *and*

ii) $\left.\frac{\partial M}{\partial t_0}\right|_{(\bar{t}_0, \bar{\phi}_0)} \neq 0.$

Then $W^s(\gamma_\varepsilon(t))$ and $W^u(\gamma_\varepsilon(t))$ intersect transversely at $(q_0(-\bar{t}_0) + \mathcal{O}(\varepsilon), \bar{\phi}_0)$. Moreover, if $M(t_0, \phi_0) \neq 0$ for all $(t_0, \phi_0) \in \mathbb{R}^1 \times S^1$, then $W^s(\gamma_\varepsilon(t)) \cap W^u(\gamma_\varepsilon(t)) = \emptyset$.

Proof: Recall from (4.5.34), (4.5.36), and (4.5.37) that we have

$$d(t_0, \phi_0, \varepsilon) = \varepsilon \frac{M(t_0, \phi_0)}{\|DH(q_0(-t_0))\|} + \mathcal{O}(\varepsilon^2). \qquad (4.5.63)$$

Note that if we define

$$d(t_0, \phi_0, \varepsilon) = \varepsilon \tilde{d}(t_0, \phi_0, \varepsilon), \qquad (4.5.64)$$

where

$$\tilde{d}(t_0, \phi_0, \varepsilon) = \frac{M(t_0, \phi_0)}{\|DH(q_0(-t_0))\|} + \mathcal{O}(\varepsilon), \qquad (4.5.65)$$

then

$$\tilde{d}(t_0, \phi_0, \varepsilon) = 0 \Rightarrow d(t_0, \phi_0, \varepsilon) = 0. \qquad (4.5.66)$$

Therefore, we will work with $\tilde{d}(t_0, \phi_0, \varepsilon)$.

Now at $(t_0, \phi_0, \varepsilon) = (\bar{t}_0, \bar{\phi}_0, 0)$ we have

$$\tilde{d}(\bar{t}_0, \bar{\phi}_0, 0) = \frac{M(\bar{t}_0, \bar{\phi}_0)}{\|DH(q_0(-\bar{t}_0))\|} = 0, \tag{4.5.67}$$

with

$$\left.\frac{\partial \tilde{d}}{\partial t_0}\right|_{(\bar{t}_0, \bar{\phi}_0, 0)} = \frac{1}{\|DH(q_0(-\bar{t}_0))\|} \left.\frac{\partial M}{\partial t_0}\right|_{(\bar{t}_0, \bar{\phi}_0)} \neq 0. \tag{4.5.68}$$

By the implicit function theorem, there thus exists a function

$$t_0 = t_0(\phi_0, \varepsilon) \tag{4.5.69}$$

for $|\phi - \phi_0|$, ε sufficiently small, such that

$$\tilde{d}(t_0(\phi_0, \varepsilon), \phi_0, \varepsilon) = 0. \tag{4.5.70}$$

This shows that $W^s(\gamma_\varepsilon(t))$ and $W^u(\gamma_\varepsilon(t))$ intersect $\mathcal{O}(\varepsilon)$ close to $(q_0(-t_0), \phi_0)$. Next we need to worry about transversality.

Suppose that $W^s(\gamma_\varepsilon(t))$ and $W^u(\gamma_\varepsilon(t))$ intersect at some point p. Then recall from Section 1.2C that the intersection is said to be transversal if

$$T_pW^s(\gamma_\varepsilon(t)) + T_pW^u(\gamma_\varepsilon(t)) = \mathbb{R}^3. \tag{4.5.71}$$

Now, for ε sufficiently small, the points on $W^s(\gamma_\varepsilon(t))$ and $W^u(\gamma_\varepsilon(t))$ that are closest to $\gamma_\varepsilon(t)$ in the sense of Definition 4.5.1 can be parametrized by t_0 and ϕ_0. Hence,

$$\left(\frac{\partial q_\varepsilon^u}{\partial t_0}, \frac{\partial q_\varepsilon^u}{\partial \phi_0}\right) \tag{4.5.72}$$

and

$$\left(\frac{\partial q_\varepsilon^s}{\partial t_0}, \frac{\partial q_\varepsilon^s}{\partial \phi_0}\right) \tag{4.5.73}$$

are a basis for $T_pW^u(\gamma_\varepsilon(t))$ and $T_pW^s(\gamma_\varepsilon(t))$, respectively.

(Note: it is important for the reader to understand how (4.5.72) and (4.5.73) are computed. By definition, $p = (q_\varepsilon^s, \phi_0) = (q_\varepsilon^u, \phi_0)$, and q_ε^s and q_ε^u are the points satisfying $q_\varepsilon^s(0) = q_\varepsilon^s$, $q_\varepsilon^u(0) = q_\varepsilon^u$, where $q_\varepsilon^s(t)$ and $q_\varepsilon^u(t)$ are trajectories in $W^s(\gamma_\varepsilon(t))$ and $W^u(\gamma_\varepsilon(t))$, respectively. Since those trajectories depend parametrically on t_0 and ϕ_0, (4.5.72) and (4.5.73) are simply the derivatives of the respective trajectories with respect to t_0 and ϕ_0 evaluated at $t = 0$.)

$T_pW^s(\gamma_\varepsilon(t))$ and $T_pW^u(\gamma_\varepsilon(t))$ will not be tangent at p provided

$$\frac{\partial q_\varepsilon^u}{\partial t_0} - \frac{\partial q_\varepsilon^s}{\partial t_0} \neq 0 \tag{4.5.74}$$

or

$$\frac{\partial q_\varepsilon^u}{\partial \phi_0} - \frac{\partial q_\varepsilon^u}{\partial \phi_0} \neq 0. \tag{4.5.75}$$

Differentiating $d(t_0, \phi_0, \varepsilon)$ with respect to t_0 and ϕ_0 and evaluating at the intersection point given by $(\bar{t}_0 + \mathcal{O}(\varepsilon), \overline{\phi}_0)$ (where $M(\bar{t}_0, \overline{\phi}_0) = 0$) gives

$$\begin{aligned}\frac{\partial d}{\partial t_0}(\bar{t}_0, \overline{\phi}_0, \varepsilon) &= \frac{DH(q_0(-\bar{t}_0)) \cdot ((\partial q_\varepsilon^u)/(\partial t_0) - (\partial q_\varepsilon^s)/(\partial t_0))}{\|DH(q_0(-\bar{t}_0))\|} \\ &= \varepsilon \frac{(\partial M/\partial t_0)(\bar{t}_0, \overline{\phi}_0)}{\|DH(q_0(-t_0))\|} + \mathcal{O}(\varepsilon^2),\end{aligned} \tag{4.5.76}$$

$$\begin{aligned}\frac{\partial d}{\partial \phi_0}(\bar{t}_0, \overline{\phi}_0, \varepsilon) &= \frac{DH(q_0(-\bar{t}_0)) \cdot ((\partial q_\varepsilon^u)/(\partial \phi_0) - (\partial q_\varepsilon^s)/(\partial \phi_0))}{\|DH(q_0(-\bar{t}_0))\|} \\ &= \varepsilon \frac{(\partial M/\partial \phi_0)(\bar{t}_0, \overline{\phi}_0)}{\|DH(q_0(-\bar{t}_0))\|} + \mathcal{O}(\varepsilon^2).\end{aligned} \tag{4.5.77}$$

Hence, it should be clear from (4.5.76) and (4.5.77) that, for ε sufficiently small, a sufficient condition for transversality is

$$\frac{\partial M}{\partial \phi_0}(\bar{t}_0, \overline{\phi}_0) = \omega \frac{\partial M}{\partial t_0}(\bar{t}_0, \overline{\phi}_0) \neq 0. \tag{4.5.78}$$

Finally, we leave the fact that $M(t_0, \phi_0) \neq 0$ implies $W^s(\gamma_\varepsilon(t)) \cap W^u(\gamma_\varepsilon(t)) = \emptyset$ as an exercise for the reader (see Exercise 4.18). □

4.5B Poincaré Maps and the Geometry of the Melnikov Function

We now want to describe the geometry associated with the independent variables, t_0 and ϕ_0, of the Melnikov function.

Consider the following cross-section to the phase space $\mathbb{R}^2 \times S^1$

$$\Sigma^{\phi_0} = \left\{(q, \phi) \in \mathbb{R}^2 \times S^1 \mid \phi = \phi_0\right\}. \tag{4.5.79}$$

Since $\dot{\phi} = \omega > 0$, it follows that the vector field is transverse to Σ^{ϕ_0}. Then the Poincaré map of Σ^{ϕ_0} into itself defined by the flow generated by the perturbed vector field (4.5.6) is given by

$$\begin{aligned} P_\varepsilon : \Sigma^{\phi_0} &\longrightarrow \Sigma^{\phi_0}, \\ q_\varepsilon(0) &\longmapsto q_\varepsilon(2\pi/\omega), \end{aligned} \tag{4.5.80}$$

where $(q_\varepsilon(t), \phi(t) = \omega t + \phi_0)$ denotes the flow generated by the perturbed vector field (4.5.6). Now the periodic orbit $\gamma_\varepsilon(t)$ intersects Σ^{ϕ_0} in a point which we denote as

$$p_{\varepsilon,\phi_0} = \gamma_\varepsilon(t) \cap \Sigma^{\phi_0}. \tag{4.5.81}$$

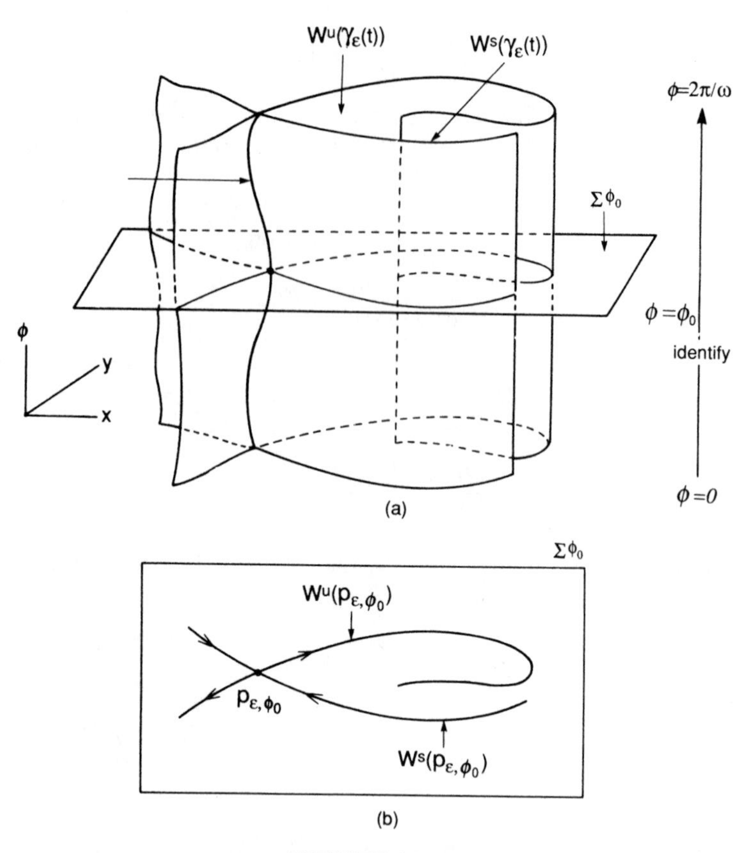

FIGURE 4.5.16.

It should be clear that p_{ε,ϕ_0} is a hyperbolic fixed point for the Poincaré map having a one-dimensional stable manifold, $W^s(p_{\varepsilon,\phi_0})$ and a one-dimensional unstable manifold, $W^u(p_{\varepsilon,\phi_0})$ given by

$$W^s(p_{\varepsilon,\phi_0}) \equiv W^s(\gamma_\varepsilon(t)) \cap \Sigma^{\phi_0}$$

and

$$W^u(p_{\varepsilon,\phi_0}) \equiv W^u(\gamma_\varepsilon(t)) \cap \Sigma^{\phi_0}, \tag{4.5.82}$$

respectively; see Figure 4.5.16.

Now let us return to the Melnikov function. From our parametrization of the homoclinic manifold, Γ_γ, it follows that *fixing* ϕ_0 and varying t_0 corresponds to restricting our distance measurement to a fixed cross-section Σ^{ϕ_0}. In this case, $M(t_0, \phi_0)$, ϕ_0 fixed, is a measurement of the distance between $W^s(p_{\varepsilon,\phi_0})$ and $W^u(p_{\varepsilon,\phi_0})$. In this case also zeros of the Melnikov function correspond to homoclinic points of a two-dimensional map, and Moser's theorem or the Smale–Birkhoff homoclinic theorem may be applied to conclude that the dynamics are chaotic (provided the homoclinic points are transverse).

Alternately, we can *fix* t_0 and vary ϕ_0 in the Melnikov function. This would correspond to fixing π_p at a specific point $(q_0(-t_0), \phi_0)$ on Γ_γ and varying the cross-section Σ^{ϕ_0}. The Melnikov function would be a measure of the distance between $W^s(\gamma_\varepsilon(t))$ and $W^u(\gamma_\varepsilon(t))$ at a fixed location in q but on different cross-sections Σ^{ϕ_0}. Since the vector field has no fixed points on $W^s(\gamma_\varepsilon(t)) \cup W^u(\gamma_\varepsilon(t))$, as the cross-section is varied all orbits in $W^s(\gamma_\varepsilon(t)) \cup W^u(\gamma_\varepsilon(t))$ must pass through π_p with $q_0(-t_0)$ fixed. Hence, no homoclinic orbits would be "missed" by the Melnikov function.

However, recall the form of the Melnikov function given in (4.5.61)

$$M(t_0, \phi_0) = \int_{-\infty}^{\infty} DH(q_0(t)) \cdot g(q_0(t), \omega t + \omega t_0 + \phi_0, 0) dt. \qquad (4.5.83)$$

It is clear from (4.5.83) that, analytically, the variation of t_0 with ϕ_0 fixed is equivalent to the variation of ϕ_0 with t_0 fixed. The underlying reason for this is that if $W^s(\gamma_\varepsilon(t))$ and $W^u(\gamma_\varepsilon(t))$ intersect, by uniqueness of solutions they cannot intersect at isolated points but, rather, must intersect along a trajectory (solution of (4.5.6)) that is asymptotic to $\gamma_\varepsilon(t)$ in both positive and negative time.

4.5c Some Properties of the Melnikov Function

Here we collect some basic properties and characteristics of the Melnikov function.

1. As mentioned earlier, $M(t_0, \phi_0)$ is a signed measure of the distance. We now want to explore this in more detail. Let us restrict ourselves to the Poincaré map defined on the cross-section Σ^{ϕ_0} and view ϕ_0 as fixed; then $d(t_0, \phi_0, \varepsilon)$ measures the distance between $W^s(p_{\varepsilon,\phi_0})$ and $W^u(p_{\varepsilon,\phi_0})$ (see Section 4.5b). We recall that the *distance between* $W^s(p_{\varepsilon,\phi_0})$ *and* $W^u(p_{\varepsilon,\phi_0})$ *at the point* $p = (q_0(-t_0), \phi_0)$ is given by

$$\begin{aligned} d(t_0, \phi_0, \varepsilon) &= \frac{DH(q_0(-t_0)) \cdot (q_\varepsilon^u - q_\varepsilon^s)}{\|DH(q_0(-t_0))\|} \\ &= \varepsilon \frac{M(t_0, \phi_0)}{\|DH(q_0(-t_0))\|} + \mathcal{O}(\varepsilon^2) \end{aligned} \qquad (4.5.84)$$

where q_ε^u and q_ε^s are defined in Definition 4.5.1. Hence, for ε sufficiently small

$$M(t_0, \phi_0) \begin{matrix} > 0 \\ < 0 \end{matrix} \Rightarrow d(t_0, \phi_0, \varepsilon) \begin{matrix} > 0 \\ < 0 \end{matrix}. \qquad (4.5.85)$$

Thus, using (4.5.84) and (4.5.85), Figure 4.5.10 holds if we replace $d(t_0, \phi_0, \varepsilon)$ with $M(t_0, \phi_0)$.

2. $M(t_0, \phi_0)$ is periodic in t_0 with period $2\pi/\omega$ and periodic in ϕ_0 with period 2π. This follows from the fact that the perturbation, $\varepsilon g(q, t, \varepsilon)$, is periodic in t with period $2\pi/\omega$ as well as from the form of $M(t_0, \phi_0)$ given in (4.5.61).

3. Recall from (4.5.84) that the distance between $W^s(\gamma_\varepsilon(t))$ and $W^u(\gamma_\varepsilon(t))$ is given by

$$d(t_0, \phi_0, \varepsilon) = \varepsilon \frac{M(t_0, \phi_0)}{\|DH(q_0(-t_0))\|} + \mathcal{O}(\varepsilon^2). \tag{4.5.86}$$

   We want to focus on the denominator, $\|DH(q_0(-t_0))\|$, in the $\mathcal{O}(\varepsilon)$ term of (4.5.86). Let us consider the situation with ϕ_0 fixed so that $d(t_0, \phi_0, \varepsilon)$ measures the distance between the stable and unstable manifolds of a hyperbolic fixed point of the Poincaré map (see Section 4.5B). Then, as $t_0 \to \pm\infty$, the measurement of distance is being made close to the hyperbolic fixed point (since $q_0(-t_0)$ approaches the unperturbed hyperbolic fixed point as $t_0 \to \pm\infty$). Also, as $t_0 \to \pm\infty$, $\|DH(q_0(-t_0))\| \to 0$, indicating that $d(t_0, \phi_0, \varepsilon) \to \infty$. Geometrically, this means that the distance between the manifolds is oscillating unboundedly near the hyperbolic fixed point; see Figure 4.6.18. The reader should compare this analytic result with the geometrical picture given by the lambda lemma (Lemma 4.4.1).

4. Suppose that the perturbation is autonomous, i.e., $\varepsilon g(q)$ does not depend explicitly on time. Then the Melnikov function is given by

$$M = \int_{-\infty}^{\infty} DH(q_0(t)) \cdot g(q_0(t)) dt. \tag{4.5.87}$$

   In this case M is just a number, i.e., it is not a function of t_0 and ϕ_0. This makes sense, since, for autonomous two-dimensional vector fields, either the stable and unstable manifolds of a hyperbolic fixed point coincide or they do not intersect at all. In Exercise 4.22 we will deal more fully with the geometry of the Melnikov function for autonomous problems.

5. Suppose the vector field is Hamiltonian, i.e., we have a $\mathbf{C}^{r+1}$ $(r \geq 2)$ function periodic in t with period $T = 2\pi/\omega$ given by

$$H_\varepsilon(x, y, t) = H(x, y) + \varepsilon H_1(x, y, t, \varepsilon) \tag{4.5.88}$$

   such that the perturbed vector field (4.5.1) is given by

$$\begin{aligned} \dot{x} &= \frac{\partial H}{\partial y}(x, y) + \varepsilon \frac{\partial H_1}{\partial y}(x, y, t, \varepsilon), \\ \dot{y} &= -\frac{\partial H}{\partial x}(x, y) - \varepsilon \frac{\partial H_1}{\partial x}(x, y, t, \varepsilon). \end{aligned} \tag{4.5.89}$$

In this case, using (4.5.61) and (4.5.89), it is easy to see that the Melnikov function is given by

$$M(t_0, \phi_0) = \int_{-\infty}^{\infty} \{H, H_1\}(q_0(t), \omega t + \omega t_0 + \phi_0, 0)dt, \tag{4.5.90}$$

where

$$\{H, H_1\} \equiv \frac{\partial H}{\partial x}\frac{\partial H_1}{\partial y} - \frac{\partial H}{\partial y}\frac{\partial H_1}{\partial x} \tag{4.5.91}$$

is the *Poisson bracket* of H with H_1.

4.5D Relationship with the Subharmonic Melnikov Function

Recall our development of the subharmonic Melnikov theory in Section 1.2D ii). This theory was concerned with the one-parameter family of periodic orbits, $q^\alpha(t)$, $\alpha \in [-1, 0)$, with period T^α, inside the homoclinic orbit. Using action-angle variables (see Section 1.2D ii)) the perturbed vector field (4.5.1) was transformed into

$$\begin{aligned} \dot{I} &= \varepsilon F(I, \theta, \phi, \varepsilon), \\ \dot{\theta} &= \Omega(I) + \varepsilon G(I, \theta, \phi, \varepsilon), \qquad (I, \theta, \phi) \in \mathbb{R}^+ \times S^1 \times S^1, \\ \dot{\phi} &= \omega, \end{aligned} \tag{4.5.92}$$

where

$$\begin{aligned} F &= \frac{\partial I}{\partial x} g_1 + \frac{\partial I}{\partial y} g_2, \\ G &= \frac{\partial \theta}{\partial x} g_1 + \frac{\partial \theta}{\partial y} g_2. \end{aligned} \tag{4.5.93}$$

Defining a cross-section to the phase space $\mathbb{R}^+ \times S^1 \times S^1$ as follows

$$\Sigma^{\phi_0} = \{(I, \theta, \phi) \in \mathbb{R}^+ \times S^1 \times S^1 \mid \phi = \phi_0\}, \tag{4.5.94}$$

we derived an approximation to the mth iterate of the Poincaré map generated by (4.5.92), which we give below

$$P_\varepsilon^m \colon \Sigma^{\phi_0} \longrightarrow \Sigma^{\phi_0} \tag{4.5.95}$$

$$(I, \theta) \mapsto (I, \theta + mT\Omega(I)) + \varepsilon(M_1^{m/n}(I, \theta, \phi_0), M_2^{m/n}(I, \theta, \phi_0)) + \mathcal{O}(\varepsilon^2).$$

The vector

$$M^{m/n}(I, \theta, \phi_0) \equiv \left(M_1^{m/n}(I, \theta, \phi_0), M_2^{m/n}(I, \theta, \phi_0)\right) \tag{4.5.96}$$

was defined to be the *subharmonic Melnikov vector.* In the (generic) situation of

$$\frac{\partial \Omega}{\partial I} \neq 0, \tag{4.5.97}$$

we saw (Theorem 1.2.13) that zeros in θ of the *first component* of the subharmonic Melnikov vector at values of I in which the resonance condition

$$mT = nT(I) \tag{4.5.98}$$

(m, n relatively prime integers) is satisfied corresponded to fixed points of the mth iterate of the Poincaré map. Moreover, using several transformations (see the remarks following Theorem 1.2.13), we showed that $M_1^{m/n}(I, \theta, \phi_0)$ could be computed without first transforming the vector field into action-angle coordinates. In particular, we showed

$$\begin{aligned} M_1^{m/n}&(t_0, \phi_0) \\ &= \frac{1}{\Omega(I)} \int_0^{mT} DH(q^\alpha(t)) \cdot g(q^\alpha(t), \omega t + \omega t_0 + \phi_0, 0) dt, \end{aligned} \tag{4.5.99}$$

where α replaced I (which is related to m/n via the resonance condition (4.5.98)) as the variable labeling the unperturbed periodic orbits and $\theta = \omega t_0$. The expression

$$\begin{aligned} \bar{M}_1^{m/n}&(t_0, \phi_0) \\ &= \int_0^{mT} DH(q^\alpha(t)) \cdot g(q^\alpha(t), \omega t + \omega t_0 + \phi_0, 0) dt \end{aligned} \tag{4.5.100}$$

was defined to be the *subharmonic Melnikov function.*

At this point the reader should begin to see the similarities between the homoclinic Melnikov theory and the subharmonic Melnikov theory. Recall from Section 4.5A that the distance between the stable and unstable manifolds of the hyperbolic fixed point of the Poincaré map defined on the cross-section Σ^{ϕ_0} is given by

$$d(t_0, \phi_0, \varepsilon) = \varepsilon \frac{M(t_0, \phi_0)}{\| DH(q_0(-t_0)) \|} + \mathcal{O}(\varepsilon^2), \tag{4.5.101}$$

with

$$M(t_0, \phi_0) = \int_{-\infty}^{\infty} DH(q_0(t)) \cdot (q_0(t), \omega t + \omega t_0 + \phi_0, 0) dt.$$

The reader should note the following on comparing (4.5.99) and (4.5.101).

1. $M(t_0, \phi_0)$ and $\bar{M}_1^{m/n}(t_0, \phi_0)$ are very similar. $M(t_0, \phi_0)$ is the function $DH(\cdot) \cdot g(\cdot, \omega t + \omega t_0 + \phi_0, 0)$ integrated around the unperturbed homoclinic orbit $q_0(t)$ and $\bar{M}_1^{m/n}(t_0, \phi_0)$ is the function $DH(\cdot) \cdot g(\cdot, \omega t + \omega t_0 + \phi_0, 0)$ integrated around the unperturbed periodic orbit, $q^\alpha(t)$, that satisfies the resonance relation $nT^\alpha = mT$.

2. The factor $\frac{1}{\Omega(I)} > 0$ in (4.5.99) is the same as $\frac{1}{\| DH(q_0(-t_0)) \|}$ in (4.5.101) written in action-angle variables.

Now let us consider the case $n = 1$. The resonance condition is given by

$$mT = T^\alpha; \tag{4.5.102}$$

hence, as $m \to \infty$, $\alpha \to 0$ (and vice-versa) in order for (4.5.102) to be satisfied. Since $q^\alpha(t)$ is periodic with period T^α with T^α satisfying (4.5.102), making the transformation $t \to t - \frac{mT}{2}$, (4.5.100) becomes

$$\bar{M}_1^{m/1}(t_0, \phi_0) = \int_{-mT/2}^{mT/2} DH(q^\alpha(t)) \cdot g(q^\alpha(t), \omega t + \omega t_0 + \phi_0, 0) dt. \tag{4.5.103}$$

Theorem 4.5.7

$$\lim_{m \to \infty} \bar{M}_1^{m/1}(t_0, \phi_0) = M(t_0, \phi_0).$$

Proof: Using (4.5.103) and (4.5.101), this proof is a simple problem in analysis that we leave as an exercise for the reader (see Exercise 4.19). □

4.5E Homoclinic and Subharmonic Bifurcations

Suppose the vector field (4.5.6) depends on a scalar parameter μ, i.e.,

$$\begin{aligned} \dot{q} &= JDH(q) + \varepsilon g(q, \phi, \mu, \varepsilon), \\ \dot{\phi} &= \omega, \end{aligned} \qquad (q, \phi, \mu) \in \mathbb{R}^2 \times S^1 \times \mathbb{R}^1. \tag{4.5.104}$$

If, in a specific problem, there is more than one parameter, then consider all but one as fixed. In this case the subharmonic and homoclinic Melnikov functions depend on the parameter μ. In particular, we write

$$\bar{M}_1^{m/n}(t_0, \phi_0, \mu) \tag{4.5.105}$$

and

$$M(t_0, \phi_0, \mu). \tag{4.5.106}$$

We will consider ϕ_0 as fixed, i.e., the Poincaré map associated with (4.5.104) is defined on the cross-section Σ^{ϕ_0}. We have the following bifurcation theorem for the homoclinic Melnikov function.

Theorem 4.5.8 *Suppose we have $(\bar{t}_0, \bar{\mu})$ such that*

i) $M(\bar{t}_0, \phi_0, \bar{\mu}) = 0$,

ii) $\dfrac{\partial M}{\partial t_0}(\bar{t}_0, \phi_0, \bar{\mu}) = 0$,

iii) $\dfrac{\partial M}{\partial \mu}(\bar{t}_0, \phi_0, \bar{\mu}) \neq 0$,

iv) $\dfrac{\partial^2 M}{\partial t_0^2}(\bar{t}_0, \phi_0, \bar{\mu}) \neq 0$.

Then the stable and unstable manifolds of the hyperbolic fixed point on the cross-section Σ^{ϕ_0} *are quadratically tangent at* $(q_0(-\bar{t}_0)) + \mathcal{O}(\varepsilon)$ *for* $\mu = \bar{\mu} + \mathcal{O}(\varepsilon)$.

Proof: Tangency of the manifolds implies

$$\begin{aligned} d(t_0, \phi_0, \mu, \varepsilon) &= 0, \\ \frac{\partial d}{dt_0}(t_0, \phi_0, \mu, \varepsilon) &= 0. \end{aligned} \tag{4.5.107}$$

Let $d(t_0, \phi_0, \mu, \varepsilon) = \varepsilon \tilde{d}(t_0, \phi_0, \mu, \varepsilon)$ with

$$\tilde{d}(t_0, \phi_0, \mu, \varepsilon) = \frac{M(t_0, \phi_0, \mu)}{\|DH(q_0(-t_0))\|} + \mathcal{O}(\varepsilon), \tag{4.5.108}$$

as in the proof of Theorem 4.5.6. Then a solution of

$$\begin{aligned} \tilde{d}(t_0, \phi_0, \mu, \varepsilon) &= 0, \\ \frac{\partial \tilde{d}}{\partial t_0}(t_0, \phi_0, \mu, \varepsilon) &= 0, \end{aligned} \tag{4.5.109}$$

is a solution of (4.5.107). Now by Hypotheses i) and ii) of the theorem, (4.5.109) has a solution at $(\bar{t}_0, \phi_0, \bar{\mu}, 0)$. Hypotheses iii) and iv) allow us to apply the implicit function theorem to show that the solution persits for ε sufficiently small; the details follow exactly as in the proof of Theorem 4.5.6 and are left as an exercise for the reader. □

We remark that the condition $\frac{\partial^2 M}{\partial t_0^2}(\bar{t}_0, \phi_0, \bar{\mu}) \neq 0$ implies that the tangency is quadratic.

In Exercise 3.41 we gave a theorem for subharmonic saddle-node bifurcations in terms of the subharmonic Melnikov vector. Here we reproduce the theorem for the case FP1 (i.e., $\frac{\partial \Omega}{\partial I} \neq 0$) in a slightly different notation (i.e., we substitute α for I).

Theorem 4.5.9 *Suppose we have* $\overline{\alpha} \in [-1, 0)$ *such that*

$$mT = nT^{\overline{\alpha}},$$

and a point $(\bar{t}_0, \bar{\mu})$ *such that*

i) $\bar{M}_1^{m/n}(\bar{t}_0, \phi_0, \bar{\mu}) = 0$,

ii) $\dfrac{\partial \bar{M}_1^{m/n}}{\partial t_0}(\bar{t}_0, \phi_0, \bar{\mu}) = 0$,

iii) $\dfrac{\partial \bar{M}_1^{m/n}}{\partial \mu}(\bar{t}_0, \phi_0, \bar{\mu}) \neq 0$,

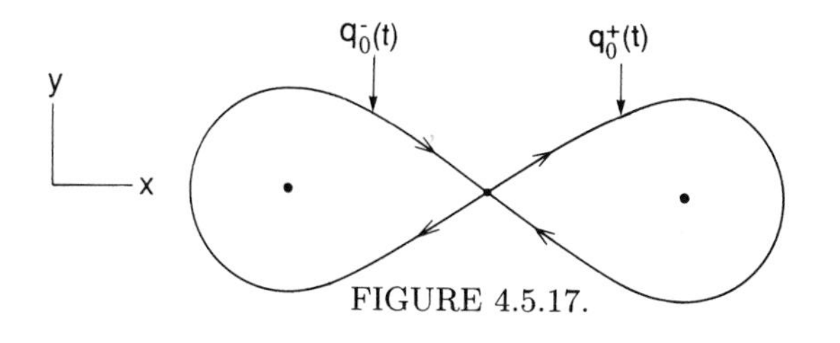

FIGURE 4.5.17.

iv) $\dfrac{\partial^2 \bar{M}_1^{m/n}}{\partial t_0^2}(\bar{t}_0, \phi_0, \overline{\mu}) \neq 0.$

Then $(\overline{\alpha}, \bar{t}_0, \overline{\mu}) + \mathcal{O}(\varepsilon)$ *is a bifurcation point at which saddle-nodes of period* m *points of* P_ε *occur.*

Proof: See Exercise 3.41. □

The reader should note that Hypotheses i), ii), iii), and iv) of Theorems 4.5.8 and 4.5.9 are the same for the appropriate Melnikov functions (homoclinic and subharmonic, respectively). Applying Theorem 4.5.7 to the situation, it would appear we can conclude that quadratic homoclinic tangencies are the limit of saddle-node bifurcations of period m points as $m \to \infty$. However, we have not proved this (even though it is true), since, if we have a convergent sequence of functions, i.e., $\{f_n(x)\}$ with $\lim_{n\to\infty} f_n(x) = f(x)$, it does not follow immediately that $\lim_{n\to\infty} f_n'(x) = f'(x)$ (see Rudin [1964], Theorem 7.17). We will leave this as an exercise for the reader (Exercise 4.20), since we will show this result in a more general setting (i.e., in a nonperturbative framework) in Section 4.7. However, first we will consider these results applied to the damped, forced Duffing oscillator.

4.5F Application to the Damped, Forced Duffing Oscillator

Recall from Section 1.2E that the damped, forced Duffing oscillator is given by

$$\begin{aligned} \dot{x} &= y, \\ \dot{y} &= x - x^3 + \varepsilon(\gamma\cos\phi - \delta y), \\ \dot{\phi} &= \omega. \end{aligned} \tag{4.5.110}$$

For $\varepsilon = 0$, (4.5.110) has a pair of homoclinic orbits (see Exercise 1.2.29) given by

$$q_0^{\pm}(t) = (x_0^{\pm}(t), y_0^{\pm}(t)) = \left(\pm\sqrt{2}\operatorname{sech} t, \mp\sqrt{2}\operatorname{sech} t \tanh t\right); \tag{4.5.111}$$

see Figure 4.5.17.

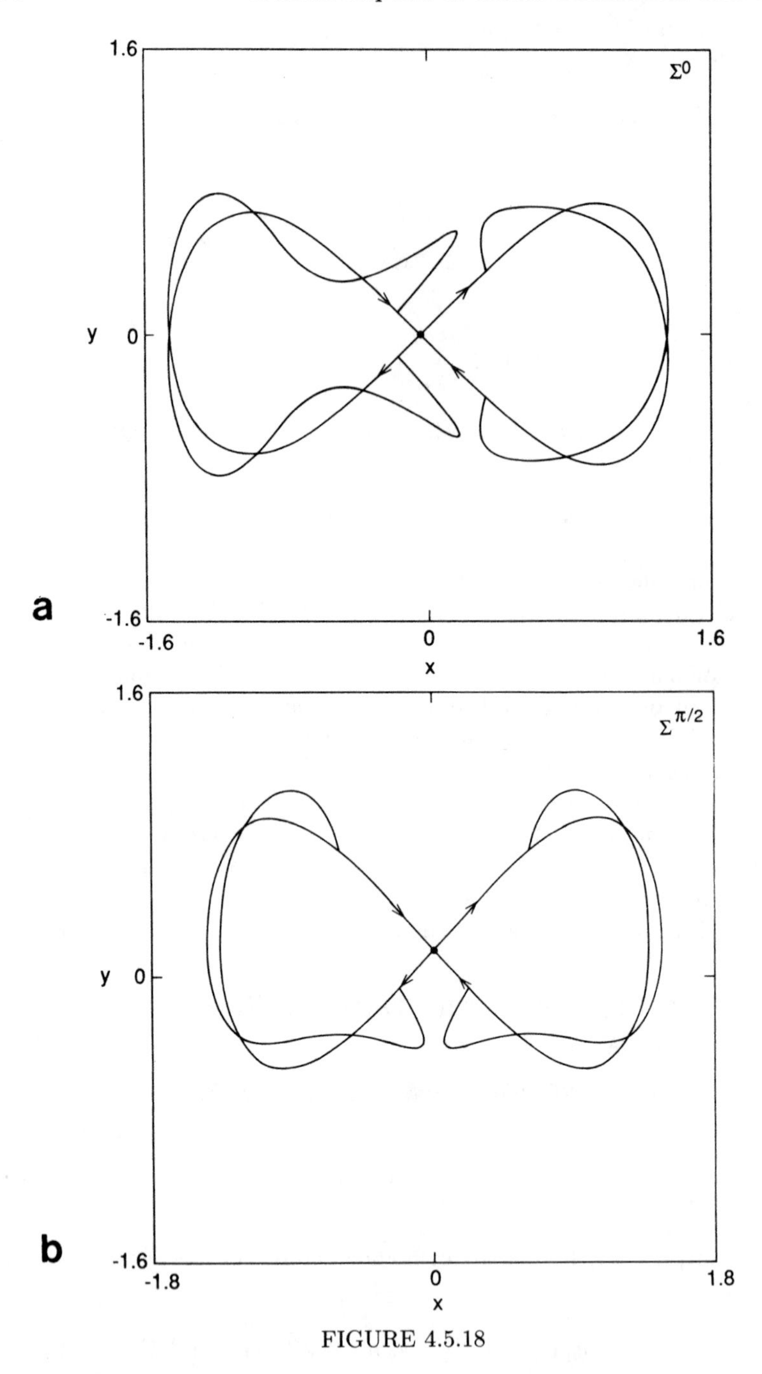

FIGURE 4.5.18

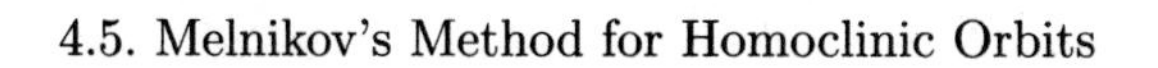

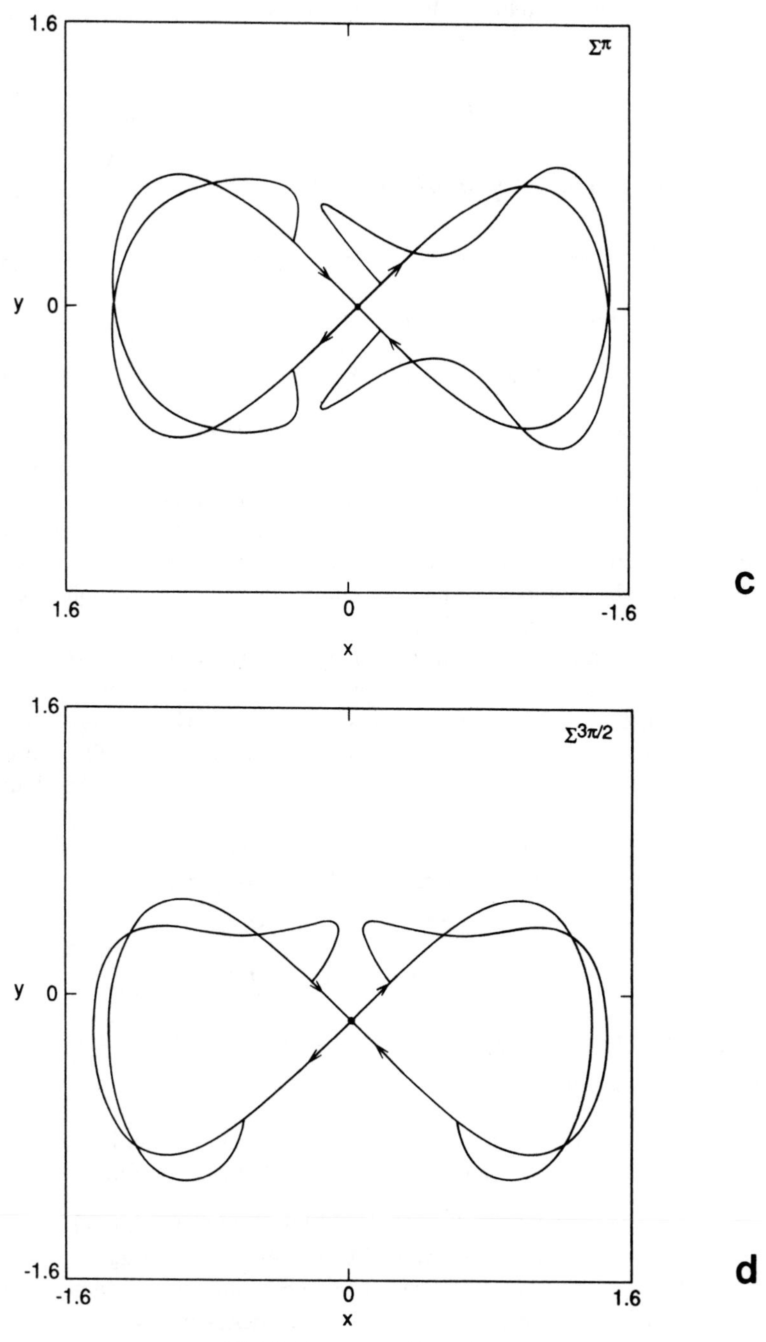

FIGURE 4.5.18. (Cont.)

The homoclinic Melnikov function is given by

$$M^{\pm}(t_0, \phi_0) = \int_{-\infty}^{\infty} [-\delta(y^{\pm}(t))^2 \pm \gamma y^{\pm}(t)\cos(\omega t + \omega t_0 + \phi_0)]dt. \tag{4.5.112}$$

Substituting (4.5.111) into (4.5.112) gives

$$M^{\pm}(t_0, \phi_0) = -\frac{4\delta}{3} \pm \sqrt{2}\gamma\pi\omega \operatorname{sech}\frac{\pi\omega}{2}\sin(\omega t_0 + \phi_0). \tag{4.5.113}$$

Fixing ϕ_0 defines a cross-section

$$\Sigma^{\phi_0} = \{(x, y, \phi) \in \mathbb{R} \times \mathbb{R} \times S^1 \mid \phi = \phi_0\}, \tag{4.5.114}$$

where the Melnikov function describes the splitting of the stable and unstable manifolds of the hyperbolic fixed point defined on the cross-section. Let $P_\varepsilon^{\phi_0}$ denote the Poincaré map of the cross-section Σ^{ϕ_0} defined by the flow generated by (4.5.110) and consider the case $\delta = 0$. Then, using the Melnikov function (4.5.113) and Remark 1 of Section 4.5C, it is easy to verify that the stable and unstable manifolds of the hyperbolic fixed point P_{ϕ_0} intersect as in Figure 4.5.18 for $\phi_0 = 0$, $\pi/2$, π, and $3\pi/2$. Figure 4.5.18 illustrates an important point; namely, that altering the cross-section on which the Poincaré map is defined can change the symmetry properties of the Poincaré map. This can often result in substantial savings in computer time in the numerical computation of Poincaré maps. We will explore these issues in Exercise 4.21 as well as consider how Figure 4.5.18 changes for $\delta \neq 0$.

From (4.5.113) it is easy to see that the condition for the manifolds to intersect in terms of the parameters (δ, ω, γ) is given by

$$\delta < \left(\frac{3\pi\omega}{2\sqrt{2}\operatorname{sech}\frac{\pi\omega}{2}}\right)\gamma. \tag{4.5.115}$$

In Figure 4.5.19 we graph the critical surface $\delta = \left(\frac{3\pi\omega}{2\sqrt{2}\operatorname{sech}(\pi\omega/2)}\right)\gamma$, and we note the following.

1. The condition (4.5.115) for intersection of the manifolds is independent of the particular cross-section Σ^{ϕ_0} (as it should be).

2. If the "right-hand" branches of the stable and unstable manifolds intersect, then the "left-hand" branches also intersect, and vice-versa. However, as seen in Figure 4.5.18, the geometry of intersections of the right-hand branches and left-hand branches may differ.

Thus, (4.5.115) is a criterion for chaos in the damped, forced Duffing oscillator as a function of the parameters (δ, ω, γ).

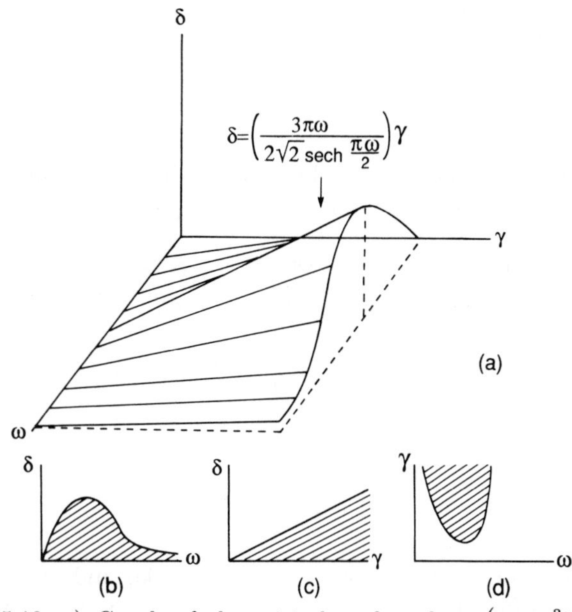

FIGURE 4.5.19. a) Graph of the critical surface $\delta = \left(\frac{3\pi\omega}{2\sqrt{2}\,\mathrm{sech}\,(\pi\omega/2)}\right)\gamma$. b) Cross-section of the critical surface for $\gamma =$ constant. c) Cross-section of the critical surface for $\omega =$ constant. d) Cross-section of the critical surface for $\delta =$ constant.

Now let us return to the periodic orbits inside and outside Γ_{p_0}. The subharmonic Melnikov function for the periodic orbits inside Γ_{p_0} on Σ^0 was given by (see Section 1.2E)

$$\bar{M}_{1,i}^{m/1}(t_0, 0) = -\delta J_1(m, 1) + \gamma J_2(m, 1, \omega) \sin \omega t_0, \tag{4.5.116}$$

where

$$J_1(m, 1) = \frac{2}{3}\left[2(2-k^2)E(k) - 4(1-k^2)K(k)\right]/(2-k^2)^{3/2}$$

and

$$J_2(m, 1) = \sqrt{2}\pi\omega \;\mathrm{sech}\, \frac{\pi m K(\sqrt{1-k^2})}{K(k)}.$$

Recall that $k \in (0, 1)$ is the elliptic modulus, and $K(k)$ and $E(k)$ are complete elliptic integrals of the first and second kind, respectively. The resonance relation for the inner periodic orbits was given by

$$\frac{2\pi m}{\omega} = 2\sqrt{2-k^2}K(k). \tag{4.5.117}$$

Similarly, the subharmonic Melnikov function for the outer periodic orbits was given by (see Exercise 1.2.30)

$$\bar{M}_{1,0}^{m/1} = -\delta \hat{J}_1(m, 1) + \gamma \hat{J}_2(m, 1, \omega) \sin \omega t_0, \tag{4.5.118}$$

where $\hat{J}_1(m,1)$ and $\hat{J}_2(m,1,\omega)$ are computed in Exercise 1.2.30. The resonance relation for the outer periodic orbits was given by

$$\frac{2\pi m}{\omega} = 4\sqrt{2k^2-1}K(k). \tag{4.5.119}$$

From (4.5.116) and (4.5.118) we obtained saddle-node bifurcation curves for the creation of period mT subharmonics given by

$$\text{inner subharmonics} \qquad \gamma = R^m(\omega)\delta, \tag{4.5.120a}$$
$$\text{outer subharmonics} \qquad \gamma = \hat{R}^m(\omega)\delta, \tag{4.5.120b}$$

where

$$R^m(\omega) \equiv \frac{J_1(m,1)}{J_2(m,1,\omega)}$$

and

$$\hat{R}^m(\omega) = \frac{\hat{J}_1(m,1)}{\hat{J}_2(m,1,\omega)}.$$

Figure 1.2.43 showed the lines defined by (4.5.120a) and (4.5.120b) in the $\gamma-\delta$ plane as a function of m (for fixed $\omega=1$); we reproduce that figure here as Figure 4.5.20. Recall that, e.g., for

$$\frac{\gamma}{\delta} > R^m(\omega),$$

$2m$ fixed points of the mth iterate of the Poincaré map existed near the order m resonance level (i.e., near the unperturbed periodic orbit labeled by the k-value such that the resonance relation $\frac{2\pi m}{\omega} = 2\sqrt{2-k^2}K(k)$ holds) with m of the fixed points being saddles and m sinks (for $\delta>0$). Hence, by Theorem 4.5.9, the lines (4.5.120a) and (4.5.120b) define saddle-node bifurcation curves in which the bifurcating periodic orbits may have arbitrarily high period.

Now we want to consider the limit of (4.5.120) as $m\to\infty$. Substituting the respective resonance relations into (4.5.116) and (4.5.118), it is easy to verify (using elliptic function identities that can be found in, e.g., Byrd and Friedman [1971]) that, as $m\to\infty$, $k\to 1$ (by the resonance relations) and both (4.5.120a) and (4.5.120b) converge to

$$\delta = \left(\frac{2\sqrt{2}}{3\pi\omega\operatorname{sech}(\pi\omega/2)}\right)\gamma, \tag{4.5.121}$$

i.e., the homoclinic bifurcation curve. Hence, we have shown that in the damped, forced Duffing oscillator the homoclinic bifurcation curve is the countable limit (from both sides) of saddle-node bifurcation curves to periodic orbits of higher and higher periods. This brings up several interesting points which we will explore further in Section 4.7.

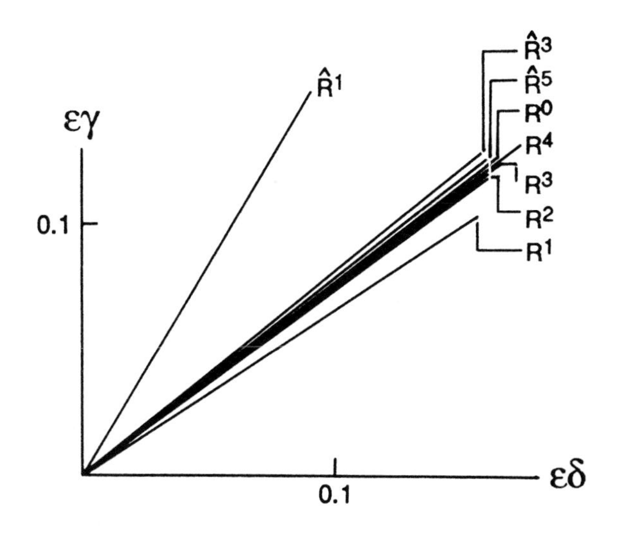

FIGURE 4.5.20.

Finally we remark that Melnikov-type methods have been developed for multi-degree-of-freedom systems and for systems having more general time dependencies. These techniques also deal with orbits homoclinic to invariant sets other than periodic orbits. A complete exposition of this theory can be found in Wiggins [1988].

4.6 Geometry and Dynamics in the Tangle

We have seen that the dynamics near transverse homoclinic points of two-dimensional maps can be very complicated. At the same time, the geometrical structure associated with the stable and unstable manifolds of a hyperbolic point can also be very complicated. In this section we want to explore some aspects of the relationships between dynamics and geometry in the homoclinic (or heteroclinic) tangle.

Consider a $\mathbf{C}^r$ $(r \geq 1)$ diffeomorphism of $\mathbb{R}^2$

$$f: \mathbb{R}^2 \longrightarrow \mathbb{R}^2 \tag{4.6.1}$$

having a hyperbolic periodic point at p_0, i.e., for some integer $k \geq 1$ $f^k(p_0) = p_0$. As described in Section 4.4, without loss of generality we can assume that $k = 1$ by applying our arguments to f^k rather than f. We will explore some of the dynamical consequences of this in the exercises. As an additional technical assumption we suppose that f is orientation preserving, i.e., $\det Df > 0$. If f does not preserve orientation, then we

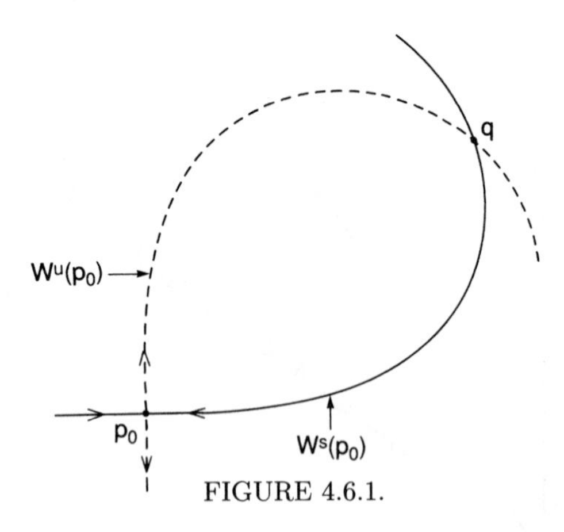

FIGURE 4.6.1.

apply our arguments to f^2, which does preserve orientation. The need for this assumption will become clear later on; however, recall from Exercise 1.2.35 that Poincaré maps arising from vector fields preserve orientation.

Now suppose that the stable and unstable manifolds of p_0, denoted $W^s(p_0)$ and $W^u(p_0)$, respectively, intersect at a point q as shown in Figure 4.6.1. The point q is said to be *homoclinic to* p_0 or simply a *homoclinic point* when the invariant set to which the point asymptotes in positive and negative time is understood. If $W^s(p_0)$ and $W^u(p_0)$ are transversal at q, then q is called a *transverse homoclinic point*. At this point we *do not* assume that $W^s(p_0)$ and $W^u(p_0)$ are transverse at q.

Consider the orbit of q under f

$$\{\ldots, f^{-2}(q), f^{-1}(q), q, f(q), f^2(q), \ldots\}. \tag{4.6.2}$$

Since q lies in both $W^s(p_0)$ and $W^u(p_0)$, and these manifolds are *invariant,* then the infinite number of points in (4.6.2) must lie in *both* $W^s(p_0)$ and $W^u(p_0)$. Now $W^s(p_0)$ and $W^u(p_0)$ are $\mathbf{C}^r$ manifolds; hence, their image under a $\mathbf{C}^r$ diffeomorphism (either f or f^{-1}) must also be $\mathbf{C}^r$. This implies that

$$\bigcup_{n=0}^{k} f^{-n}(W^s_{\text{loc}}(p_0)) \tag{4.6.3a}$$

and

$$\bigcup_{n=0}^{k} f^{n}(W^u_{\text{loc}}(p_0)) \tag{4.6.3b}$$

are $\mathbf{C}^r$ curves, for k as large as we like. Therefore, $W^s(p_0)$ and $W^u(p_0)$ must wind amongst each other intersecting (at least) along the infinite number of points given in (4.6.2), as we illustrate in Figure 4.6.2. This geometrical structure has been called a *homoclinic tangle*, and we now want to develop the concepts necessary to describe it more precisely.

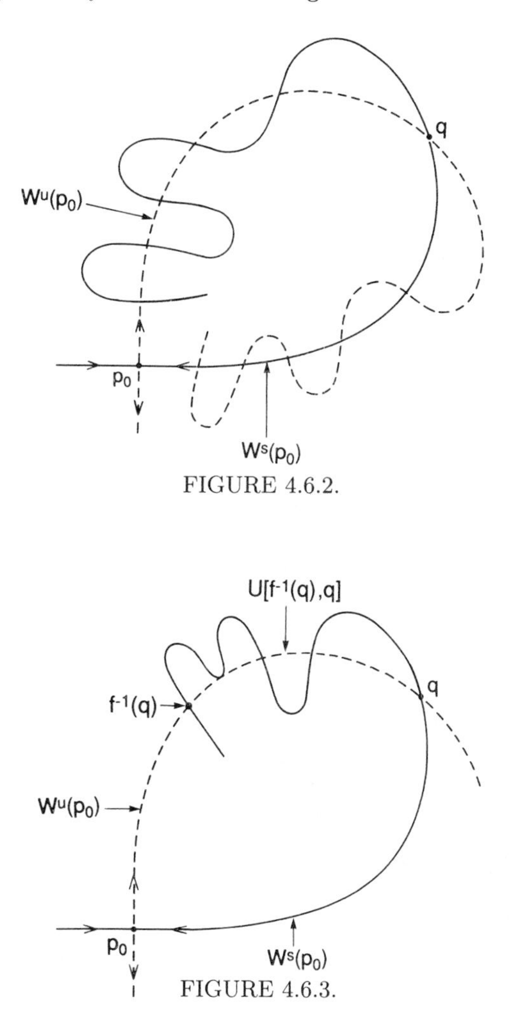

FIGURE 4.6.2.

FIGURE 4.6.3.

4.6A Pips and Lobes

Let $U[f^{-1}(q), q]$ denote the segment of $W^u(p_0)$ with endpoints at $f^{-1}(q)$ and q; see Figure 4.6.3. Then $W^s(p_0)$ intersects $U[f^{-1}(q), q]$ at q and $f^{-1}(q)$ and also at k points in between q and $f^{-1}(q)$ (note: for orientation preserving maps $k \geq 1$, see Exercise 4.29). We illustrate this in Figure 4.6.3 for $k = 3$. Without loss of generality we can assume $k = 1$ as shown in Figure 4.6.4; later we will show how to generalize our results for $k > 1$. We denote the segment of $W^s(p_0)$ with endpoints at q and $f^{-1}(q)$ by $S[f^{-1}(q), q]$.

The homoclinic point q shown in Figure 4.6.4 is a special type of homoclinic point that we call a *primary intersection point* or *pip* (this term is originally due to Easton [1986]).

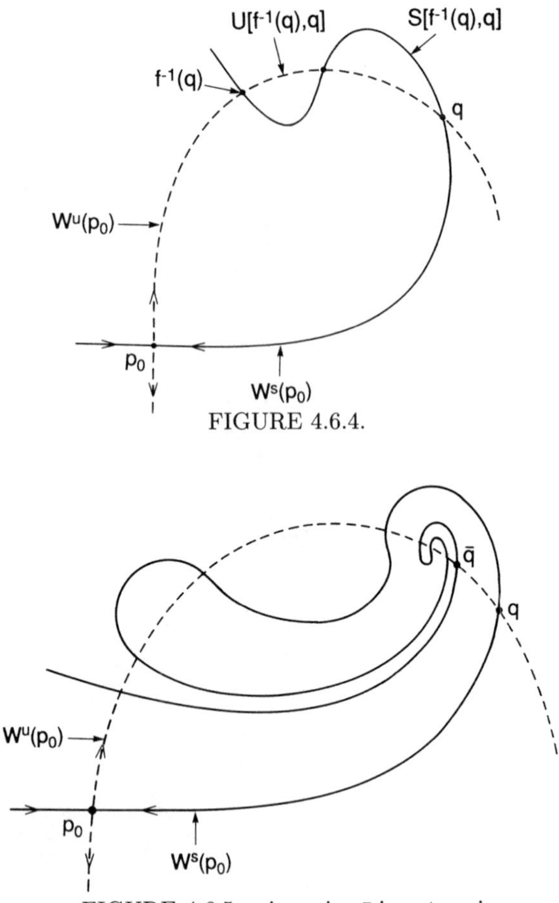

FIGURE 4.6.4.

FIGURE 4.6.5. q is a pip. $\bar{q}$ is not a pip.

DEFINITION 4.6.1 Suppose $q \in W^s(p_0) \cap W^u(p_0)$. Let $S[p_0, q]$ denote the segment of $W^s(p_0)$ from p_0 to q, and $U[p_0, q]$ denote the segment of $W^u(p_0)$ from p_0 to q. Then q is called a *primary intersection point* (or pip) if, *other than* p_0, $S[p_0, q]$ and $U[p_0, q]$ intersect only in the point q.

The pips will play an important role in building up the structure of the homoclinic tangle. In Figure 4.6.5 we show examples of homoclinic points that are pips and homoclinic points that are not pips.

Now along $W^s(p_0)$ (resp. $W^u(p_0)$) we can define an ordering of points as follows. For $q_0, q_1 \in W^s(p_0)$ (resp. $W^u(p_0)$) we say that $q_0 \underset{s}{<} q_1$ (resp. $q_0 \underset{u}{<} q_1$) if q_0 is closer than q_1 to p_0 in arclength measured along $W^s(p_0)$ (resp. $W^u(p_0)$).

The following lemmas are quite useful.

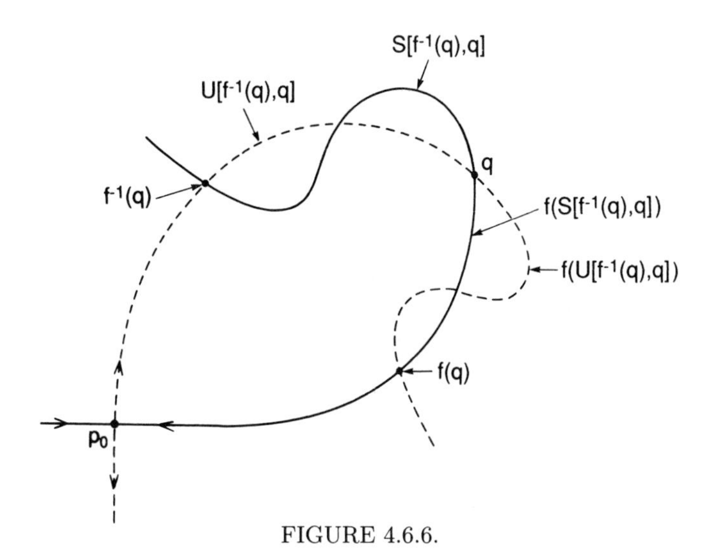

FIGURE 4.6.6.

Lemma 4.6.1 *If* $q_0 \underset{s}{<} q_1$ *(resp.* $q_0 \underset{u}{<} q_1$*), then* $f(q_0) \underset{s}{<} f(q_1)$ *(resp.* $f(q_0) \underset{u}{<} f(q_1)$*).*

Proof: This follows from the fact that f preserves orientation. We leave the details as an exercise for the reader (see Exercise 4.34). □

Lemma 4.6.2 *If* q *is a pip, then* $f^k(q)$ *is a pip for all* k*.*

Proof: If the lemma is true for $k = 1$ and -1, then, by induction, it follows that it is true for all k. The easiest proof is by contradiction. Assume q is a pip and $f(q)$ (or $f^{-1}(q)$) is *not* a pip. In order for this to be true, Lemma 4.6.1 will have to be violated, giving rise to a contradiction. We leave the details as an exercise for the reader (see Exercise 4.35). □

Hence, by Lemmas 4.6.1 and 4.6.2, $f(U[f^{-1}(q), q])$ and $f(S[f^{-1}(q), q])$ appear as in Figure 4.6.6. Iterating $U[f^{-1}(q), q]$ and $S[f^{-1}(q), q]$ further under f and f^{-1} gives the complicated geometrical structure shown in Figure 4.6.7. In Figure 4.6.7 we have very carefully avoided showing pieces of $W^u(p_0)$ intersecting pieces of $W^s(p_0)$ as they accumulate on p_0; we will address this important point shortly. However, first we give an important definition.

DEFINITION 4.6.2 Let q and q_1 be two adjacent pips (i.e., there are no other pips on $U[q, q_1]$ (or, equivalently $S[q, q_1]$)) between q and q_1, and let $U[q, q_1]$ and $S[q, q_1]$ be the segments of $W^u(p_0)$ and $W^s(p_0)$, respectively, with q and q_1 as endpoints. Then we refer to the region bounded by $U[q, q_1]$ and $S[q, q_1]$ as a *lobe;* see Figure 4.6.8. For any lobe L, $\mu(L)$ will denote the area of L.

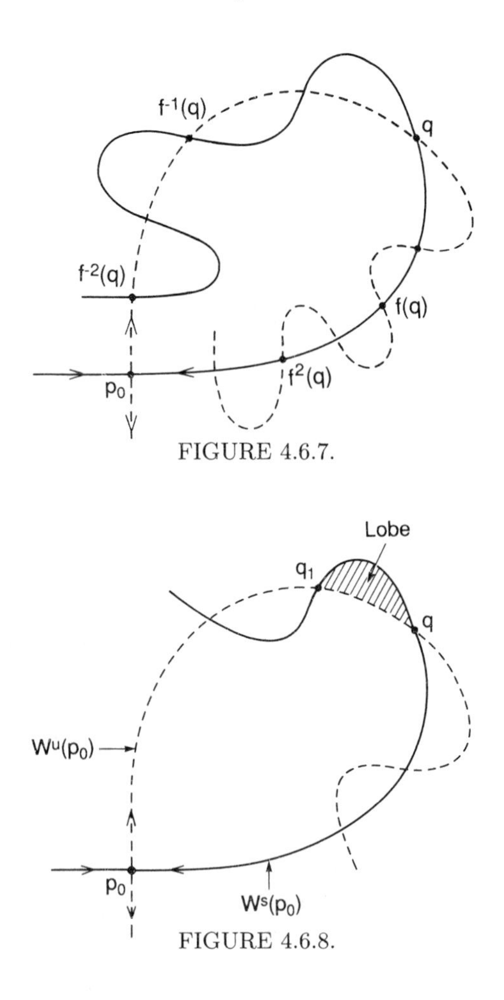

FIGURE 4.6.7.

FIGURE 4.6.8.

From Lemma 4.6.2 and the invariance of $W^s(p_0)$ and $W^u(p_0)$, it follows that, for any lobe L, $f^k(L)$ is also a lobe, for all k. The lobes will play a very important role in our analysis of transport in phase space. However, besides the pips, there are other *secondary intersection points or sips* which complicate matters further.

Consider the lobes labeled L_1 and L_2 in Figure 4.6.9. Then, for positive integers k and n sufficiently large, $f^k(L_1)$ must cut through $f^{-n}(L_2)$ as shown in Figure 4.6.9 (this can be proved rigorously using the lambda lemma). Let $X = f^k(L_1) \cap f^{-n}(L_2)$; then $f^n(X) = f^{n+k}(L_1) \cap L_2$ is contained in L_2 as shown in Figure 4.6.10. Therefore, iterates of lobes will intersect other lobes. We will prove a much stronger result later on; however, for now we want to note the main difference between pips and sips, which is illustrated in Figure 4.6.11. Namely, once a pip enters a neighborhood of the hyperbolic fixed point under iteration by f, it remains

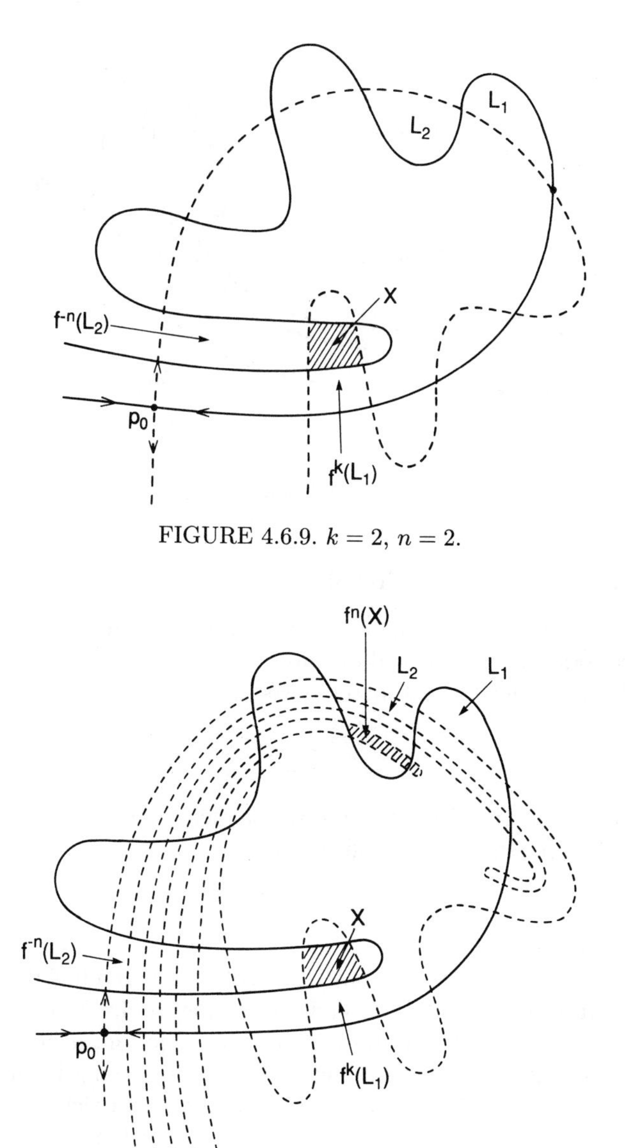

FIGURE 4.6.9. $k = 2$, $n = 2$.

FIGURE 4.6.10. $k = 2$, $n = 2$.

in the neighborhood. Sips, on the other hand, may enter and leave the vicinity of the hyperbolic fixed point many times before finally remaining near the hyperbolic fixed point under all forward (or backward) iterations by f.

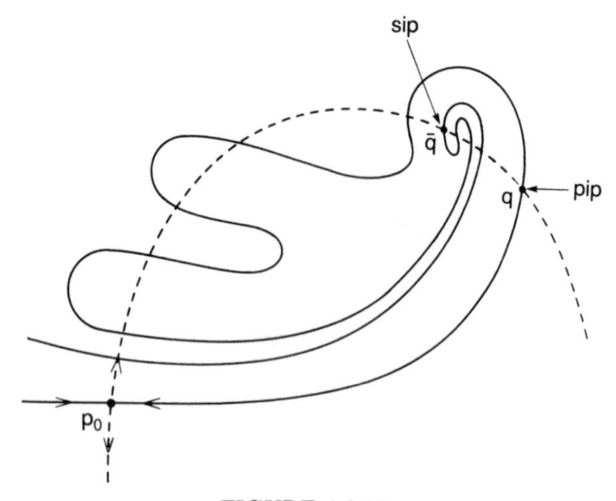

FIGURE 4.6.11.

4.6B Transport in Phase Space

We now want to show how these various definitions can be used. However, first we consider a motivational example.

Consider the following autonomous vector field in the plane

$$\begin{aligned} \dot{x} &= y, \\ \dot{y} &= x - x^2. \end{aligned} \tag{4.6.4}$$

This is the equation of motion for a "particle" moving in a single-well potential with potential energy given by

$$V(x) = -\frac{x^2}{2} + \frac{x^3}{3}. \tag{4.6.5}$$

In Figure 4.6.12a we show the phase space of (4.6.4) and in 4.6.12b a graph of the potential energy (4.6.5). Thus, from Figure 4.6.12 it is clear that (4.6.4) has two qualitatively different types of motion, a bounded, oscillatory type of motion corresponding to a particle moving in the potential well and an unbounded motion corresponding to a particle outside of the potential well. These two types of motion are separated in phase space by the homoclinic orbit connecting the saddle point at the origin to itself, or the *separatrix*. In this system, initial conditions in the well remain in the well forever, and initial conditions outside the well become unbounded.

Now consider the nonautonomous vector field in the plane

$$\begin{aligned} \dot{x} &= y, \\ \dot{y} &= x - x^2 - \delta y + \gamma \cos \omega t. \end{aligned} \tag{4.6.6}$$

Heuristically, this system could be thought of as a particle in a single well potential subjected to an external periodic forcing ($\gamma \cos \omega t$) and damping

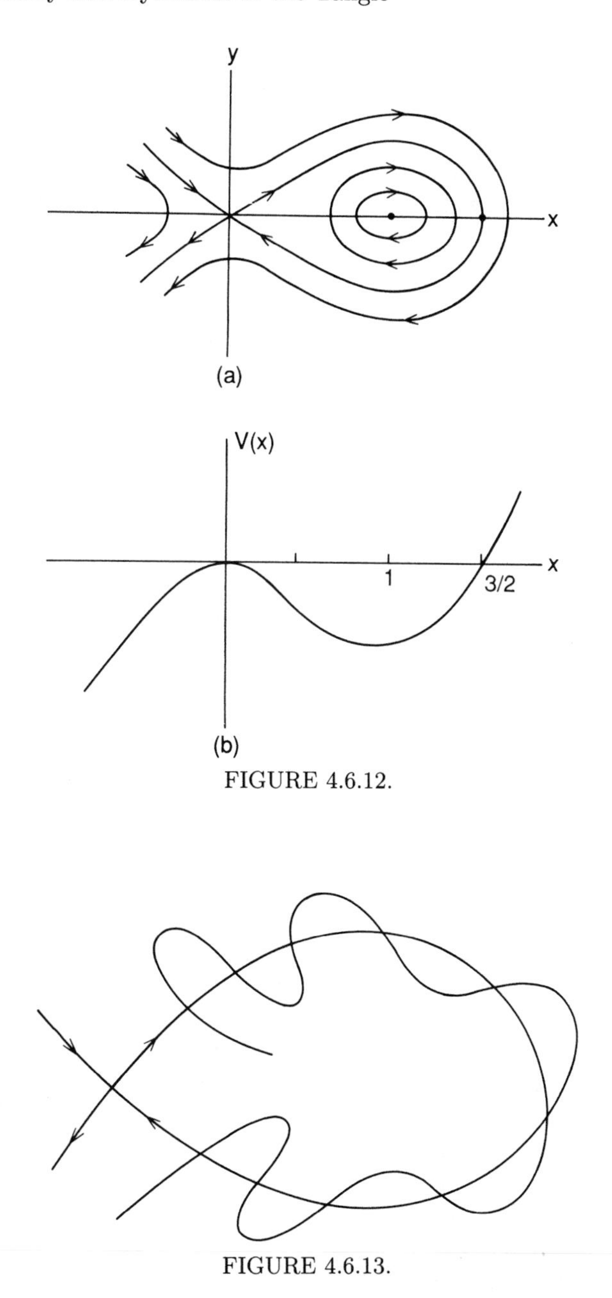

FIGURE 4.6.12.

FIGURE 4.6.13.

$(-\delta y)$. We have proven that for appropriate choices of parameters (i.e., $\gamma - \delta - \omega$) the Poincaré map associated with (4.6.6) has transversal homoclinic orbits, as shown in Figure 4.6.13. Hence, there are vast differences between the dynamics of the integrable vector field (4.6.4) and the vector field (4.6.6). In particular, (4.6.6) has chaotic dynamics but, also, the sep-

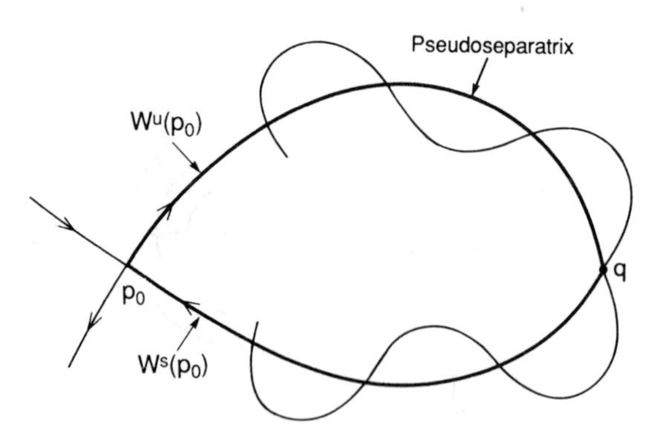

FIGURE 4.6.14.

aratrix in (4.6.4) has been violently deformed in (4.6.6). This means that in (4.6.6) particles with initial conditions in the potential well may escape the well and, similarly, particles with initial conditions outside the potential well may enter the well. The mechanism for describing this is through the dynamics of the lobes. Thus, one speaks of the problem of transport in phase space between regions exhibiting qualitatively different motions. We now want to develop some techniques for dealing with this problem.

The first problem we must deal with involves specifying what we mean by the term "separatrix" in a map such as that shown in Figure 4.6.13. We return to the general setting described at the beginning of this section.

DEFINITION 4.6.3 Choose any pip $q \in W^s(p_0) \cap W^u(p_0)$. Then the region bounded by $U[p_0, q] \cup S[p_0, q]$ is referred to as a *pseudoseparatrix;* see Figure 4.6.14.

Note that if $W^s(p_0) \cap W^u(p_0)$ contains one pip, then it contains an infinity of pips. Therefore, there are infinitely many choices for the pseudoseparatrix. The obvious question thus arises as to which choice should be made. This depends on the context of the *specific* problem under consideration. For example, in the single-well potential problem it is probably most natural to choose the pseudoseparatrix so that it is as close as possible to the separatrix in the associated integrable problem.

Once a pseudoseparatrix is chosen, then the phase space is divided into two disjoint components, labeled R_1 and R_2 in Figure 4.6.15. The problem of transport of phase space that we shall study concerns how initial conditions in R_1 (resp. R_2) may enter R_2 (resp. R_1). We will see that this is completely determined by the geometry and dynamics of the lobes.

i) The Mechanism for Transport: Turnstiles and Lobe Dynamics

We suppose that $S[f^{-1}(q), q]$ intersects $U[f^{-1}(q), q]$ in precisely one pip (besides $f^{-1}(q)$ and q), as shown in Figure 4.6.16 (we will discuss the

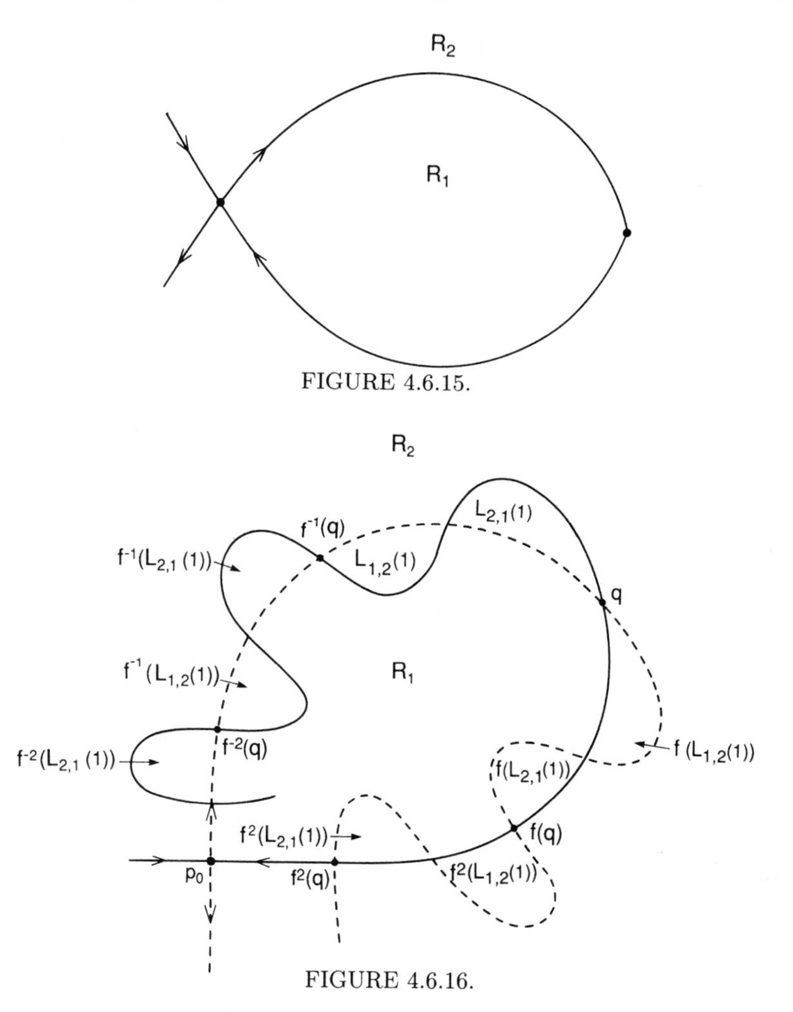

FIGURE 4.6.15.

FIGURE 4.6.16.

generalization to more than one pip later on). Then precisely two lobes are formed, one lying in R_1 and the other lying in R_2, which we label $L_{1,2}(1)$ and $L_{2,1}(1)$, respectively (note: the reason for this somewhat cumbersome notation will become apparent shortly). Now $f(L_{1,2}(1))$ and $f(L_{2,1}(1))$ appear as in Figure 4.6.16. Note that $L_{1,2}(1)$ enters R_2 under iteration by f, and $L_{2,1}(1)$ enters R_1 under iteration by f. This is the mechanism for transport across the pseudoseparatrix and appears to have been discussed explicitly for the first time by Channon and Lebowitz [1980] and Bartlett [1982]. The lobes formed by $U[f^{-1}(q), q] \cap S[f^{-1}(q), q]$ have been called a *turnstile* by MacKay, Meiss, and Percival [1984]. In Figure 4.6.16 we show several forward and backward iterates of the turnstile, and we make the following main observation.

> *Initial conditions in R_i entering R_j on the nth iterate of f must be in $L_{i,j}(1)$ on the $(n-1)$ iterate of f, $i, j = 1, 2$.*

Now we can answer some questions concerning the transport of phase space across the pseudoseparatrix.

Flux Across the Pseudoseparatrix

The area of phase space crossing the pseudoseparatrix from R_1 into R_2 under one iteration of f is given by $\mu(f(L_{1,2}(1)))$. Similarly, the area of phase space crossing the pseudoseparatrix from R_2 into R_1 under one iteration of f is given by $\mu(f(L_{2,1}(1)))$. Thus, the total area of phase space crossing the pseudoseparatrix from R_1 into R_2 (resp. R_2 into R_1) under n iterations of f is $n\mu(f(L_{1,2}(1)))$ (resp. $n\mu(f((L_{2,1}(1)))$.

Let us consider two slightly different questions.

1. What is the area occupied by points that are in R_1 initially (i.e., at $t = 0$) that enter R_2 on the nth iteration by f?
2. What is the area in R_2 at $t = n$ that is occupied by points that were in R_1 at $t = 0$?

The answer to Question 1 is *not* $\mu(f((L_{1,2}(1)))$ and the answer to Question 2 is *not* $n\mu(f((L_{1,2}(1)))$. This is because we are not just interested in arbitrary points crossing the pseudoseparatrix but, rather, in points which have a specified location initially. Thus, with each point in the plane it is important to keep track of whether it was in R_1 or R_2 initially. We can do this by calling points in R_i at $t = 0$ species S_i, $i = 1, 2$. In this terminology, Questions 1 and 2 can be equivalently stated as follows.

1. What is the flux of species S_1 into R_2 on the nth iterate?
2. What is the amount of species S_1 in R_2 after the nth iterate?

Heuristically, we can think of species S_1 as black fluid and species S_2 as white fluid, and we are interested in how the fluids mix amongst each other under the dynamics generated by f.

We first must establish some terminology and definitions. Let

$$L_{i,j}(n)$$

denote the lobe that leaves R_i and enters R_j on the nth iterate and let

$$L_{i,j}^k(n) \equiv L_{i,j}(n) \cap R_k;$$

thus we have

$$f^{n-1}(L_{i,j}(n)) \equiv L_{i,j}(1).$$

The main quantities that we wish to compute are

$$a_{i,j}(n)$$

flux of species S_i into R_j on the nth iterate and

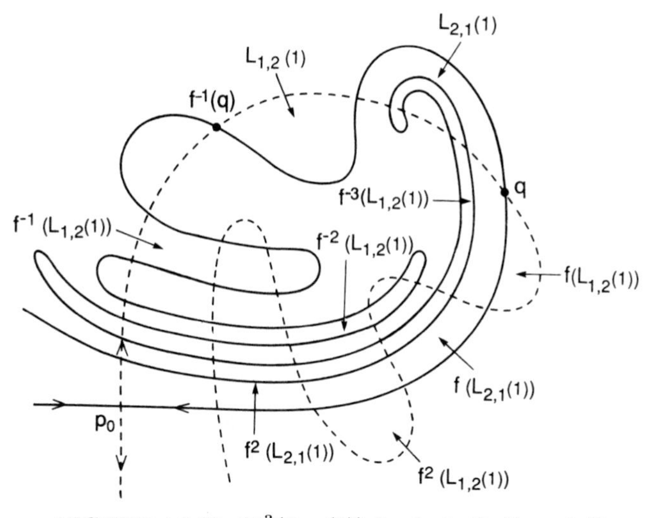

FIGURE 4.6.17. $f^{-3}(L_{1,2}(1))$ lies in both R_1 and R_2.

$$T_{i,j}(n)$$

area occupied by species S_i in region R_j immediately after the nth iterate.

Then it follows from these definitions that

$$a_{1,2}(n) = \mu(f^n(L^1_{1,2}(n)))$$

and

$$T_{1,2}(n) = \sum_{k=1}^{n} \mu(f^n(L^1_{1,2}(k))).$$

Next we want to compute $L^1_{1,2}(k)$. Using the "main observation" noted above, a point in R_1 at $t = 0$ cannot enter R_2 on the kth iterate unless it is in $L_{1,2}(1)$ on the $(k-1)$ iterate. Thus, points in $f^{-k+1}(L_{1,2}(1))$ enter R_2 on the kth iterate. However, $f^{-k+1}(L_{1,2}(1))$ may not contain only species S_1, since $f^{-k+1}(L_{1,2}(1))$ may intersect $f^{-\ell+1}(L_{2,1}(1))$, $\ell = 1, \cdots, k$, and, thus, lie in R_2; see Figure 4.6.17. Hence, we have

$$L^1_{1,2}(k) = f^{-k+1}(L_{1,2}(1)) - \bigcup_{\ell=0}^{k-1} \big(f^{-k+1}(L_{1,2}(1)) \cap f^{-\ell}(L_{2,1}(1))\big).$$

Now, since the sets $f^{-k+1}(L_{1,2}(1)) \cap f^{-\ell}(L_{2,1}(1))$ are disjoint, we have

$$\begin{aligned}\mu(f^n(L^1_{1,2}(k))) &= \mu\big(f^{n-k+1}(L_{1,2}(1))\big)\\ &\quad - \sum_{\ell=0}^{k-1} \mu\big(f^{n-k+1}(L_{1,2}(1)) \cap f^{n-\ell}(L_{2,1}(1))\big).\end{aligned}$$

Substituting this expression into the formulae for $a_{1,2}(n)$ and $T_{1,2}(n)$ gives

$$a_{1,2}(n) = \mu(f(L_{1,2}(1))) - \sum_{k=0}^{n-1} \mu(f(L_{1,2}(1)) \cap f^{n-k}(L_{2,1}(1))) \tag{4.6.7}$$

and

$$T_{1,2}(n) = \sum_{k=1}^{n} \mu(f^{n-k+1}(L_{1,2}(1))) - \sum_{k=1}^{n}\sum_{\ell=0}^{k-1} \mu(f^{n-k+1}(L_{1,2}(1)) \cap f^{n-\ell}(L_{1,2}(1))). \tag{4.6.8}$$

(Note: we leave it as an exercise for the reader to show that $f^{-n+1}(L_{1,2}(1))$ *cannot* intersect $f^{-k+1}(L_{2,1}(1))$, $k = n, n+1, \cdots$.) We make the important remark that (4.6.7) and (4.6.8) have implications for *all* points in R_1 but are obtained entirely by iterating the turnstile. Thus, for $i = 1$, $j = 2$, the answer to Question 1 above is given by (4.6.7) and the answer to Question 2 by (4.6.8). If we think back to the motivational problem of the particle moving in the single well potential subject to external periodic excitation and damping, the quantity

$$\frac{T_{1,2}(n)}{\mu(R_1)} \tag{4.6.9}$$

can be interpreted as the probability that a particle *starting in the well* leaves the well on the nth iterate of the associated Poincaré map. Thus, we can give a statistical description of the motion using the deterministic structures (i.e., the stable and unstable manifolds of p_0) that give rise to the chaotic dynamics.

Let us make several remarks concerning these results.

Remark 1. In general we should have

$$a_{1,2}(n) = \mu(f^n(L_{1,2}^1(n))) - \mu(f^n(L_{2,1}^1(n)))$$

and

$$T_{1,2}(n) = \sum_{k=1}^{n} \left[\mu(f^n(L_{1,2}^1(k))) - \mu(f^n(L_{2,1}^1(k)))\right];$$

however, for the geometry of this particular homoclinic tangle, we have

$$L_{2,1}^1(k) = \emptyset \quad \forall\, k.$$

Remark 2. For area preserving maps, i.e., for any set $A \subset \mathbb{R}^2$, $\mu(f(A)) = \mu(A)$, we have

$$a_{i,j}(n) = T_{i,j}(n) - T_{i,j}(n-1).$$

While this should be obvious, the reader can verify this fact by direct substitution into the definitions.

Remark 3. A general theory of transport through a countable number of regions separated by segments of stable and unstable manifolds of hyperbolic periodic points can be found in Rom-Kedar and Wiggins [1989]. In this section we have dealt with transport between two regions only. When more than two regions are involved, the geometry and dynamics (and notation) become much more complicated; see Exercises 4.39 and 4.40.

We now want to give a bit more geometrical insight into the formulae (4.6.7) and (4.6.8); for this it will be more simple for expository purposes to consider the situation where f is area preserving, i.e.,

$$\det(Df) = 1. \tag{4.6.10}$$

In this case we have

$$\begin{aligned} \mu\big(f(L_{1,2}(1))\big) &= \mu(L_{1,2}(1)), \\ \mu\big(f(L_{1,2}(1)) \cap f^{n-k}(L_{2,1}(1))\big) &= \mu\big(L_{1,2}(1) \cap f^{n-k-1}(L_{2,1}(1))\big), \end{aligned} \tag{4.6.11}$$

and

$$\begin{aligned} \mu\big(f^{n-k+1}(L_{1,2}(1))\big) &= \mu(L_{1,2}(1)), \\ \mu\big(f^{n-k+1}(L_{1,2}(1)) \cap f^{n-\ell}(L_{2,1}(1))\big) &= \mu\big(L_{1,2}(1) \cap f^{k-1-\ell}(L_{2,1}(1))\big). \end{aligned} \tag{4.6.12}$$

Substituting (4.6.11) and (4.6.12) into (4.6.7) and (4.6.8), respectively, and reindexing gives

$$a_{1,2}(n) = \mu(L_{1,2}(1)) - \sum_{k=1}^{n-1} \mu(L_{1,2}(1) \cap f^k(L_{2,1}(1))) \tag{4.6.13}$$

and

$$T_{1,2}(n) = n\mu(L_{1,2}(1)) - \sum_{k=1}^{n-1} (n-k)\mu(L_{1,2}(1) \cap f^k(L_{2,1}(1))), \tag{4.6.14}$$

where we have used the fact that by construction $L_{1,2}(1) \cap L_{2,1}(1) = \emptyset$. In words, the sets $L_{1,2}(1) \cap f^k(L_{2,1}(1))$ are the points that leave R_1 and enter R_2 under one iteration by f subject to the condition that they entered R_1 from R_2 k iterations earlier. In Figure 4.6.18 we show some of these sets.

An asymptotic analysis of the transport formula (4.6.14) allows us to prove an interesting result.

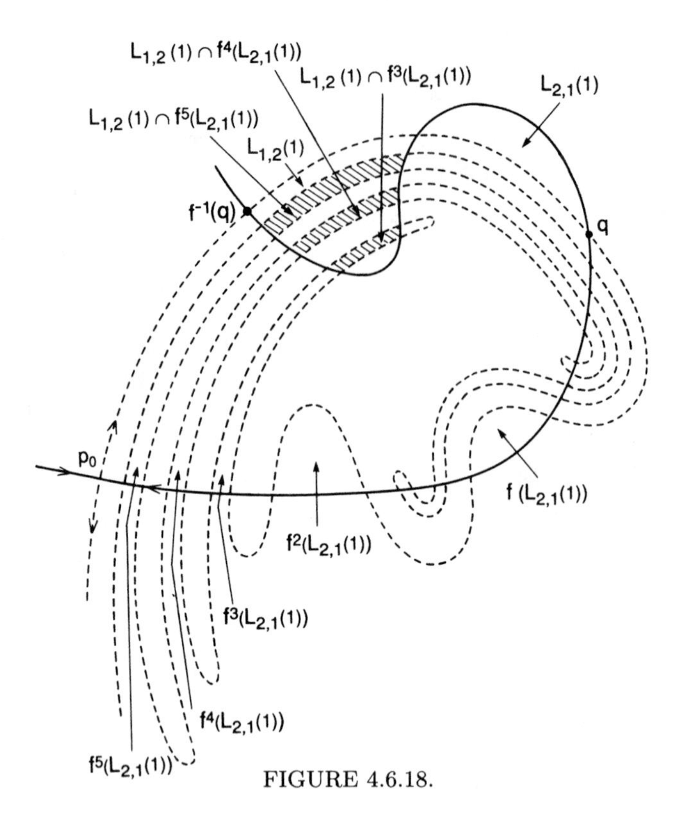

FIGURE 4.6.18.

Proposition 4.6.3

$$\mu(L_{1,2}(1)) = \sum_{k=1}^{\infty} \mu(L_{1,2}(1) \cap f^k(L_{2,1}(1))).$$

Proof: Note that we have

$$T_{1,2}(n) \leq \mu(R_1). \tag{4.6.15}$$

We begin by rearranging (4.6.14) to obtain

$$\begin{aligned} n\Big[\mu(L_{1,2}(1)) - \sum_{k=1}^{n-1} \mu\big(L_{1,2}(1) \cap f^k(L_{2,1}(1))\big)\Big] \\ + \sum_{k=1}^{n-1} k\mu\big(L_{1,2}(1) \cap f^k(L_{2,1}(1))\big) \leq \mu(R_1). \end{aligned} \tag{4.6.16}$$

Note that

$$\mu(L_{1,2}(1)) - \sum_{k=1}^{n-1} \mu\big(L_{1,2}(1) \cap f^k(L_{2,1}(1))\big) > 0 \tag{4.6.17}$$

and

$$\sum_{k=1}^{n-1} k\mu\big(L_{1,2}(1) \cap f^k(L_{2,1}(1))\big) > 0. \tag{4.6.18}$$

Now consider the limit of (4.6.16) as $n \to \infty$. Using (4.6.15) and (4.6.18), if

$$\mu(L_{1,2}(1)) < \sum_{k=1}^{\infty} \mu\big(L_{1,2}(1) \cap f^k(L_{2,1}(1))\big), \tag{4.6.19}$$

then the left-hand side of (4.6.16) becomes unbounded. However, $\mu(R_1)$ is bounded. Hence, we must have

$$\mu(L_{1,2}(1)) = \sum_{k=1}^{\infty} \mu\big(L_{1,2}(1) \cap f^k(L_{2,1}(1))\big). \quad \square \tag{4.6.20}$$

Proposition 4.6.3 says that $L_{1,2}(1)$ is completely filled with images of pieces of $L_{2,1}(1)$. Thus, *all* points that leave R_1 and enter R_2 must have entered R_1 from R_2 at an earlier time, but the time may be arbitrarily large.

We leave this example and consider several technical details that we avoided in our initial discussion.

4.6c Technical Details

Self-Intersecting Turnstiles

Suppose q is the pip chosen to define the pseudoseparatrix. The turnstile is then defined by the intersection of $U[f^{-1}(q), q]$ and $S[f^{-1}(q), q]$, as we showed in Figure 4.6.16, with the two resulting lobes denoted $L_{1,2}(1)$ and $L_{2,1}(1)$. In Figure 4.6.16 we showed $L_{1,2}(1)$ and $L_{2,1}(1)$ each lying entirely on one side of the pseudoseparatrix, i.e., except for the pip between q and $f^{-1}(q)$, $L_{1,2}(1)$ and $L_{2,1}(1)$ do not intersect. However, it may happen that $L_{1,2}(1)$ and $L_{2,1}(1)$ intersect elsewhere, possibly as shown in Figure 4.6.19. In this case we must redefine the two lobes forming the turnstile. Let

$$I = \text{int}(L_{1,2}(1) \cap L_{2,1}(1)), \tag{4.6.21}$$

where "int" denotes the interior of the set.

The two lobes defining the turnstiles are then redefined as

$$\tilde{L}_{1,2}(1) = L_{1,2}(1) - I, \tag{4.6.22}$$

$$\tilde{L}_{2,1}(1) = L_{2,1}(1) - I. \tag{4.6.23}$$

It should be clear that the area crossing the pseudoseparatrix from R_1 (resp. R_2) into R_2 (resp. R_1) in one iteration by f is given by $\mu(f(\tilde{L}_{1,2}(1)))$ (resp. $\mu(f(\tilde{L}_{2,1}(1)))$). Moreover, points leaving R_1 (resp. R_2) and entering

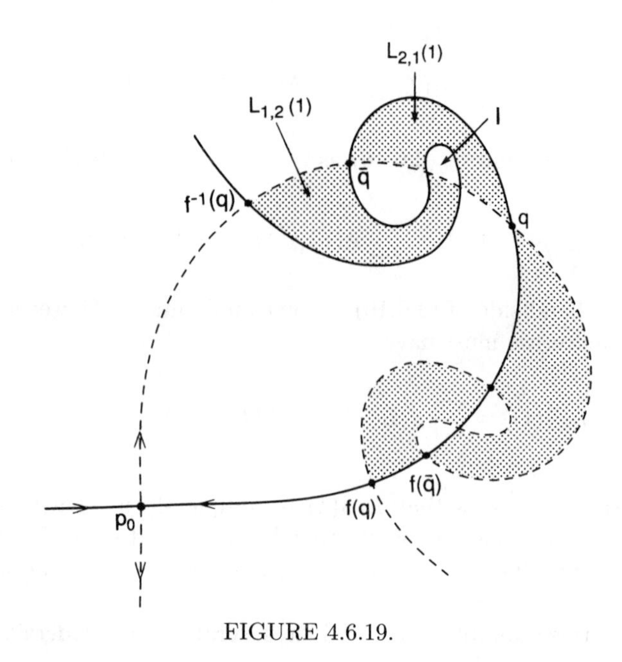

FIGURE 4.6.19.

R_2 (resp. R_1) on the nth iterate must be in $\tilde{L}_{1,2}(1)$ (resp. $\tilde{L}_{2,1}(1)$) on the $(n-1)$ iterate. Hence, the formulae (4.6.7) and (4.6.8) go through with $L_{1,2}(1)$ and $L_{2,1}(1)$ replaced by $\tilde{L}_{1,2}(1)$ and $\tilde{L}_{2,1}(1)$, respectively.

Multi-Lobe Turnstiles

We made the assumption that precisely two lobes were formed by $U[f^{-1}(q), q] \cap S[f^{-1}(q), q]$. However, it may happen that, besides q and $f^{-1}(q)$, $U[f^{-1}(q), q]$ intersects $S[f^{-1}(q), q]$ in k pips, $k > 1$. In Figure 4.6.20 we illustrate a possible case with $k = 2$. The k pips form $k+1$ lobes, which we denote by $L_0, L_1, \cdots, L_k$, with n lobes lying inside R_1 and $(k+1)-n$ lobes lying in R_2 (note: if some of the lobes lie in both R_1 and R_2, then apply Technical Detail 1 above). Suppose the numbering of the lobes has been chosen such that

$$L_0, L_1, \cdots, L_{n-1} \subset R_1$$

and

$$L_n, L_{n+1}, \cdots, L_k \subset R_2;$$

then we define

$$L_{1,2}(1) \equiv L_0 \cup L_1 \cup \cdots \cup L_{n-1} \tag{4.6.24}$$

and

$$L_{2,1}(1) \equiv L_n \cup L_{n+1} \cup \cdots \cup L_k; \tag{4.6.25}$$

see Figure 4.6.20 where $k = 2$, $n = 2$.

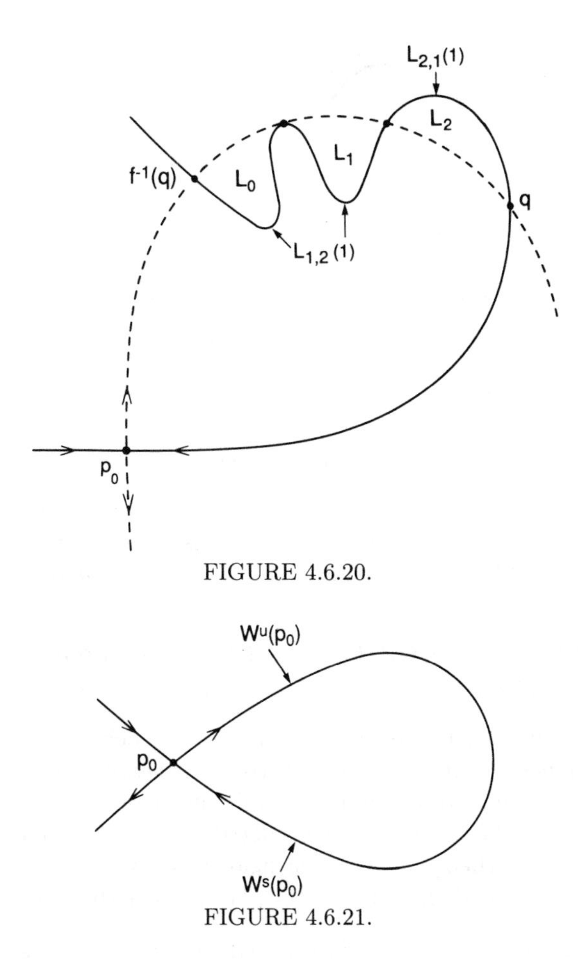

FIGURE 4.6.20.

FIGURE 4.6.21.

It should be clear that the area crossing the pseudoseparatrix from R_1 (resp. R_2) into R_2 (resp. R_1) in one iteration by f is given by $\mu(f(L_{1,2}(1)))$ (resp. $\mu(f(L_{2,1}(1))))$. Moreover, points leaving R_1 (resp. R_2) and entering R_2 (resp. R_1) on the nth iterate must be in $L_{1,2}(1)$ (resp. $L_{2,1}(1)$) on the $(n-1)$ iterate. Hence, formulae (4.6.7) and (4.6.8) hold with $L_{1,2}(1)$ and $L_{2,1}(1)$ defined as in (4.6.24) and (4.6.25), respectively.

Pathological Intersections

We have assumed that $W^s(p_0)$ and $W^u(p_0)$ intersect in a discrete set of points. By the Kupka–Smale theorem (see Palis and de Melo [1982]), this situation is generic. However, suppose $W^s(p_0)$ and $W^u(p_0)$ coincide along a branch, as shown in Figure 4.6.21. In this case no transport across the separatrix occurs; it is exactly like the case of a planar vector field. To consider an even more exotic case, suppose $W^s(p_0)$ and $W^u(p_0)$ intersect along a discrete set of *arcs*, as shown in Figure 4.6.22. In this nongeneric

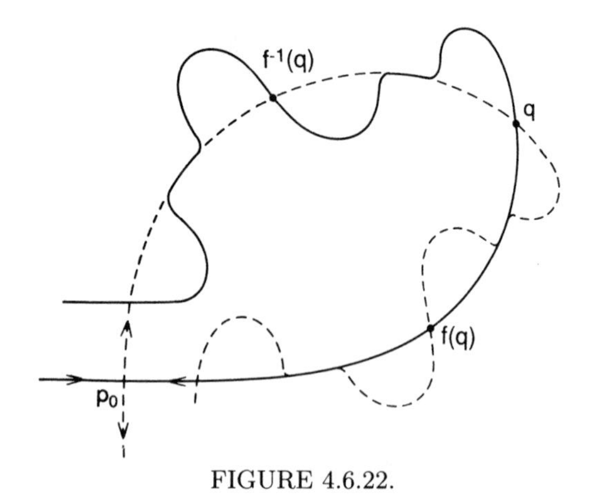

FIGURE 4.6.22.

situation (we know of no examples where this occurs), the theory would need to be modified. Note that this cannot occur for *analytic* maps (why?).

4.6D Application of the Melnikov Theory to Transport

The theory of transport of phase space across homoclinic and heteroclinic tangles developed above is not a perturbation theory. However, the Melnikov theory can be used to compute certain quantities arising in the transport theory when the maps being considered arise as the Poincaré maps of t-periodically perturbed planar vector fields possessing a homoclinic orbit (i.e., the vector fields defined in (4.5.1)). We present three main results.

Proposition 4.6.4 *The zeros of $M(t_0, \phi_0)$ correspond to pips of the Poincaré map defined on the cross-section Σ^{ϕ_0}.*

Proof: This is a consequence of the definition of the points in the stable and unstable manifolds that are detected by the Melnikov function described in Definition 4.5.1. □

Proposition 4.6.5 *Consider the Poincaré map defined on the cross-section Σ^{ϕ_0}. Suppose*

i) $M(\bar{t}_0, \phi_0) = 0$;

ii) $\dfrac{\partial M}{\partial t_0}(\bar{t}_0, \phi_0) \neq 0$;

iii) *for* $t_0 \in [\bar{t}_0, \bar{t}_0 + T)$, $M(t_0, \phi_0)$ *has precisely* n *zeros.*

Then for any pip q of the Poincaré map, $U[f^{-1}(q), q] \cap S[f^{-1}(q), q]$ forms exactly n lobes. If n is even, $n/2$ lobes lie on one side of the pseudoseparatrix *and the remaining $n/2$ lobes lie on the opposite side of the pseudoseparatrix. If n is odd, $(n-1)/2$ lobes lie on one side of the pseudoseparatrix and the remaining $(n+1)/2$ lobes lie on the opposite side of the pseudoseparatrix.*

Proof: This follows from Proposition 4.6.4 and by appealing to the appropriate definitions. We leave the details as an exercise for the reader (see Exercise 4.32). □

Proposition 4.6.6 *Let L be a lobe defined by the pips $q_1 \equiv q_0(-t_{01})+\mathcal{O}(\varepsilon)$ and $q_2 = q_0(-t_{02}) + \mathcal{O}(\varepsilon)$ on the cross-section Σ^{ϕ_0}. Then*

$$\mu(L) = \varepsilon \int_{t_{01}}^{t_{02}} |M(t_0, \phi_0)| dt_0 + \mathcal{O}(\varepsilon^2).$$

Proof: For ε sufficiently small,

$$\mu(L) = \int_{q_1}^{q_2} \ell(s) ds, \tag{4.6.26}$$

where ds represents the unit of arclength along $W^s(\gamma_\varepsilon(t)) \cap \Sigma^{\phi_0}$ and $\ell(s)$ is the perpendicular distance between $W^s(\gamma_\varepsilon(t)) \cap \Sigma^{\phi_0}$ and $W^u(\gamma_\varepsilon(t)) \cap \Sigma^{\phi_0}$. We must take ε sufficiently small so that $\ell(s)$ along $S[q_1, q_2]$ intersects $U[q_1, q_2]$ in a unique point. This follows from the $\mathbf{C}^r$ dependence of the stable and unstable manifolds on ε. We can change the parametrization of (4.6.26) from s to t_0 as follows

$$ds = \frac{ds}{dt_0} dt_0 = \big[\|DH(q_0(-t_0))\| + \mathcal{O}(\varepsilon)\big] dt_0. \tag{4.6.27}$$

Now, by the $\mathbf{C}^r$ dependence of the stable and unstable manifolds on ε, we have (see Figure 4.6.23)

$$\begin{aligned} \ell(s) &= |d(t_0, \phi_0; \varepsilon)| + \mathcal{O}(\varepsilon^2) \\ &= \varepsilon \frac{|M(t_0, \phi_0)|}{\|DH(q_0(-t_0))\|} + \mathcal{O}(\varepsilon^2). \end{aligned} \tag{4.6.28}$$

Substituting (4.6.27) and (4.6.28) into (4.6.26) gives

$$\mu(L) = \varepsilon \int_{t_{01}}^{t_{02}} |M(t_0, \phi_0)| dt_0 + \mathcal{O}(\varepsilon^2). \quad \square \tag{4.6.29}$$

For an alternative proof of Proposition 4.6.6 for the Hamiltonian case, see Kaper et al. [1989].

We end by remarking that the theory of transport in phase space across pseudoseparatrices presented in this section is a special case of a more complete theory developed in Rom-Kedar and Wiggins [1989]. They consider the phase space of a two-dimensional, orientation-preserving diffeomorphism divided up into N disjoint regions, say R_i, $i = 1, \dots, N$, separated by pseudoseparatrices. At $t = 0$ each region is uniformly filled with

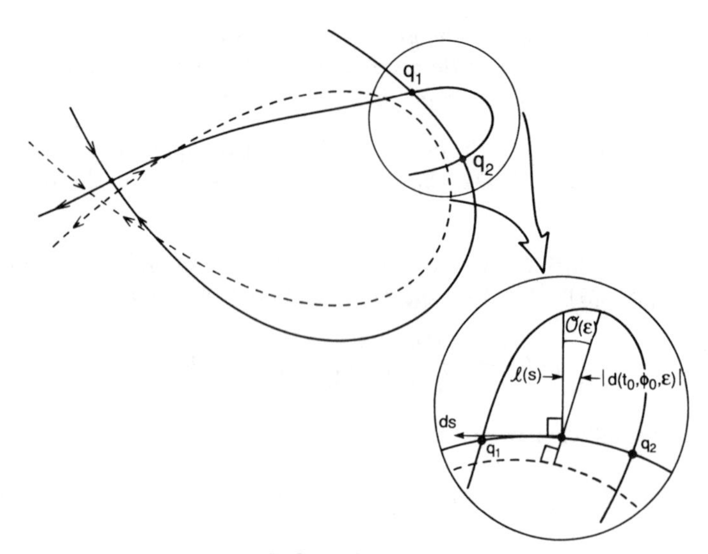

FIGURE 4.6.23.

species S_i. Using only the dynamics of the turnstiles associated with the pseudoseparatrices, they give formulae for the amount of species S_i contained in region R_j at $t = n \geq 0$ for any $i, j = 1, \cdots, N$. We will explore some generalizations along the lines of the example given in this section in Exercises 4.39 and 4.40.

4.7 Homoclinic Bifurcations: Cascades of Period-Doubling and Saddle-Node Bifurcations

In Section 4.4 we described the complex dynamics associated with a transverse homoclinic orbit to a hyperbolic fixed point. In particular, the map possessed a countable infinity of unstable periodic orbits of all periods. We now want to consider the situation of a bifurcation to transverse homoclinic orbits. Specifically, we consider a one-parameter family of diffeomorphisms of the plane having a hyperbolic periodic orbit (which, without loss of generality, we can assume is a fixed point). Referring to the parameter as μ, suppose, for $\mu > \mu_0$, the stable and unstable manifolds of the fixed point do not intersect and, for $\mu < \mu_0$, they intersect transversely (the reader might peek ahead to Figure 4.7.2). Hence, a natural question arises; as we go from $\mu > \mu_0$ (i.e., no horseshoe) to $\mu < \mu_0$ (i.e., many horseshoes), how are all the unstable periodic orbits created? We will see that under certain conditions the creation of the complicated dynamics associated with a transverse homoclinic orbit to a hyperbolic periodic orbit is an infinite sequence (or cascade) of period-doubling and saddle-node bifurcations. The set-up for the analysis will be very similar to that given in Section 4.4 for the proof of Moser's theorem. Specifically, we will analyze a sufficiently large iterate of the map that is defined in a neighborhood (in both phase and parameter space) of a homoclinic point. We begin by stating our assumptions.

We consider a one-parameter family of two-dimensional $\mathbf{C}^r$ ($r \geq 3$) diffeomorphisms

$$z \mapsto f(z;\mu), \qquad z \in \mathbb{R}^2, \quad \mu \in I \subset \mathbb{R}^1, \tag{4.7.1}$$

where I is some interval in $\mathbb{R}^1$. We have the following assumption on the map.

Assumption 1: Existence of a Hyperbolic Fixed Point. For all $\mu \in I$,

$$f(0,\mu) = 0. \tag{4.7.2}$$

Moreover, $z = 0$ is a hyperbolic fixed point with the eigenvalues of $Df(0,\mu)$ given by $\rho(\mu)$, $\lambda(\mu)$, with

$$0 < \rho(\mu) < 1 < \lambda(\mu) < \frac{1}{\rho(\mu)}. \tag{4.7.3}$$

We remark that since (4.7.3) is satisfied for all $\mu \in I$, we will often omit denoting the explicit dependence of the eigenvalues ρ and λ on μ unless it is relevant to the specific argument being discussed. We denote the stable and unstable manifolds of the hyperbolic fixed point by $W^s_\mu(0)$ and $W^u_\mu(0)$, respectively.

Assumption 2: Existence of a Homoclinic Point. At $\mu = 0$, $W^s_0(0)$ and $W^u_0(0)$ intersect.

Assumption 3: Behavior Near the Fixed Point. There exists some neighborhood $\mathcal{N}$ of the origin such that the map takes the form

$$f(x,y;\mu) = (\rho x, \lambda y), \tag{4.7.4}$$

where x and y are local coordinates in $\mathcal{N}$.

Note that Assumption 3 implies that $W^s_\mu(0) \cap \mathcal{N}$ and $W^u_\mu(0) \cap \mathcal{N}$ are given by the local coordinate axes.

Our final assumption will place more specific conditions on the geometry of the intersection of $W^s_\mu(0)$ with $W^u_\mu(0)$ at $\mu = 0$. This is most conveniently done in terms of the local return map in a neighborhood of a homoclinic point which we now derive.

We are interested in the dynamics near a homoclinic point. Therefore, using the same construction as that given in the proof of Moser's theorem in Section 4.4, we will derive a map of a neighborhood of a homoclinic point into itself, which is given by f^N, for some N (large). The construction proceeds as follows. Assumption 3 implies that there exists a point $(0, y_0) \in$

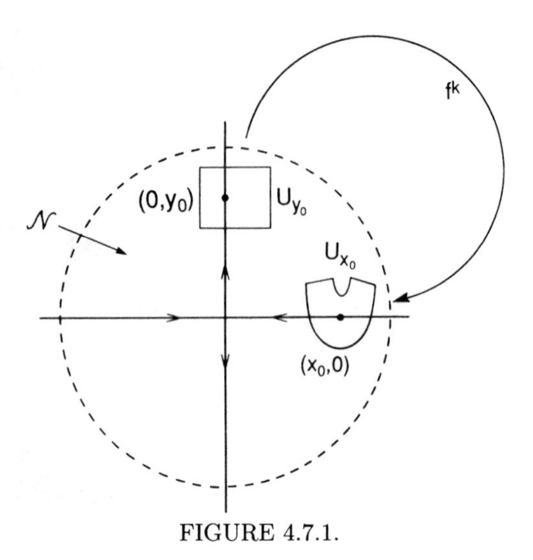

FIGURE 4.7.1.

$W_0^u(0) \cap \mathcal{N}$ and a point $(x_0, 0) \in W_0^s(0) \cap \mathcal{N}$ such that $f^k(0, y_0; 0) = (x_0, 0)$ for some $k \geq 1$. Thus, following the construction in Section 4.4, we can find a neighborhood of $(0, y_0)$, $U_{y_0} \subset \mathcal{N}$, and a neighborhood of $(x_0, 0)$, $U_{x_0} \subset \mathcal{N}$, such that $f^k(U_{y_0}; 0) = U_{x_0}$; see Figure 4.7.1 (for all the details, see the construction in Section 4.4). Note that by continuity, $f^k(\cdot; \mu)$ will be defined in U_{y_0} for μ sufficiently small. We can now state our final assumption.

Assumption 4: Quadratic Homoclinic Tangency at $\mu = 0$. We assume that, in U_{y_0}, $f^k(\cdot; \mu)$ has the form

$$\begin{aligned} f^k : U_{y_0} &\longrightarrow U_{x_0} \\ (x, y) &\longmapsto (x_0 - \beta(y - y_0), \mu + \gamma x + \delta(y - y_0)^2), \end{aligned} \tag{4.7.5}$$

with $\beta, \gamma, \delta > 0$. Hence, for μ near zero, $W_\mu^s(0)$ and $W_\mu^u(0)$ behave as in Figure 4.7.2.

From Lemma 4.4.1 and the arguments given in Section 4.4 it follows that there exists an integer N_0 such that, for all $n \geq N_0$, we can find subsets $U_{x_0}^n \subset U_{x_0}$ such that $f^n(x, y; \mu) = (\rho^n x, \lambda^n y)$ maps $U_{x_0}^n$ into U_{y_0}, i.e.,

$$f^n(U_{x_0}^n; \mu) \subset U_{y_0} \tag{4.7.6}$$

for μ sufficiently small (note: this can be trivially verified due to the fact that we have assumed f is linear in $\mathcal{N}$). Therefore,

$$f^n \circ f^k \equiv f^{n+k} \colon f^{-k}(U_{x_0}^n; \mu) \longrightarrow U_{y_0},$$

$$(x, y) \longmapsto (\rho^n(x_0 - \beta(y - y_0)), \lambda^n(\mu + \gamma x + \delta(y - y_0)^2)), \tag{4.7.7}$$

is well defined for μ sufficiently small (note: in general $n = n(\mu)$).

We now give our first result.

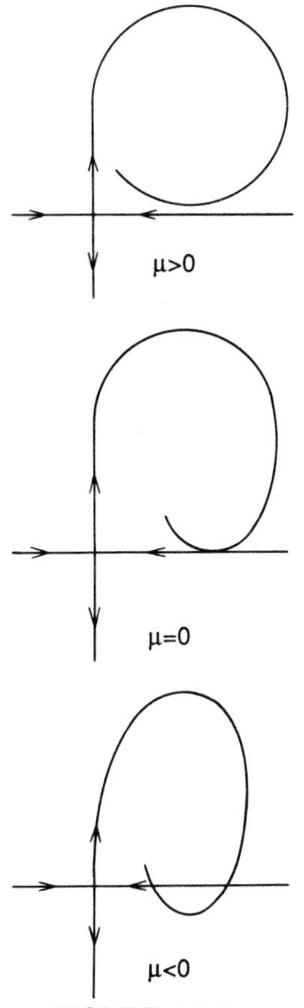

FIGURE 4.7.2.

Theorem 4.7.1 (Gavrilov and Silnikov [1973]) *At $\mu = 0$ there exists an integer N_0 such that, for all $n \geq N_0$, there exists a set $\Lambda_{n+k} \subset \mathcal{N}$, invariant under f^{n+k}, such that $f^{n+k}|_{\Lambda_{n+k}}$ is topologically conjugate to a full shift on two symbols.*

Proof: The proof can be found in Gavrilov and Silnikov [1973] or Guckenheimer and Holmes [1983]. The basic idea is to choose a neighborhood of $(0, y_0)$ so that $f^{n+k}(\cdot;\mu)$ maps it back over itself and Assumptions 1 and 3 of Section 4.3 hold; see Figure 4.7.3. In Exercise 4.43 we outline the steps necessary to prove this theorem. $\square$

The next theorem tells us how the periodic orbits in the sets Λ_{n+k}, $n \geq N_0$, are created as μ decreases through zero.

Theorem 4.7.2 (Gavrilov and Silnikov [1973]) *There exists an inte-*

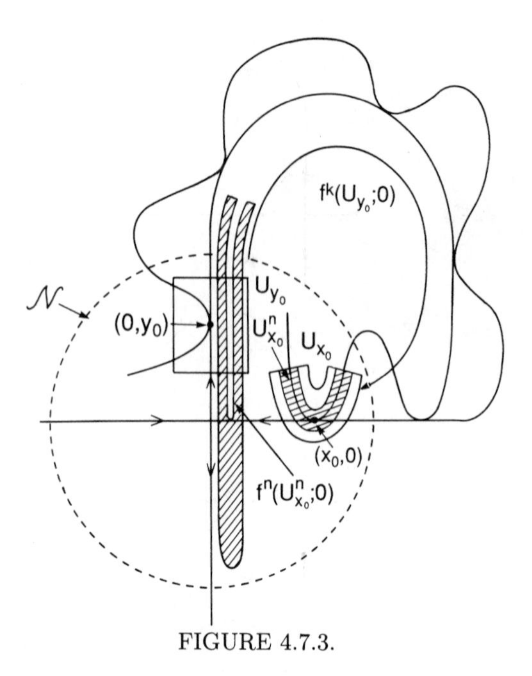

FIGURE 4.7.3.

ger N_0 and infinite sequences of parameter values

$$\{\mu_{SN}^{n+k}\}, \qquad n \geq N_0,$$

$$\{\mu_{PD}^{n+k}\}, \qquad n \geq N_0$$

with $\mu_{SN}^{n+k} > 0$, $\mu_{PD}^{n+k} > 0$ and $\mu_{SN}^{n+k} \underset{n\to\infty}{\longrightarrow} 0$, $\mu_{PD}^{n+k} \underset{n\to\infty}{\longrightarrow} 0$ such that μ_{SN}^{n+k} corresponds to a saddle-node bifurcation value for f^{n+k}, and μ_{PD}^{n+k} corresponds to a period-doubling bifurcation value for f^{n+k}. Moreover,

$$\mu_{SN}^{N_0+k} > \mu_{PD}^{N_0+k} > \mu_{SN}^{N_0+1+k} > \mu_{PD}^{N_0+1+k} > \cdots > \mu_{SN}^{N_0+m+k} > \mu_{PD}^{N_0+m+k} > \cdots \tag{4.7.8}$$

with

$$\mu_{SN}^{n+k} \sim \lambda^{-n} \quad as \ \ n \to \infty \tag{4.7.9a}$$

and

$$\mu_{PD}^{n+k} \sim \lambda^{-n} \quad as \ \ n \to \infty. \tag{4.7.9b}$$

Before proving Theorem 4.7.2 we want to make several remarks.

Remark 1. In the saddle-node bifurcation at $\mu = \mu_{SN}^{n+k}$, the node is actually a sink. The two orbits created are period $n+k$ orbits for f. The sink created in the saddle-node bifurcation subsequently loses stability in a period-doubling bifurcation at $\mu = \mu_{PD}^{n+k}$ (hence, we must have $\mu_{PD}^{n+k} < \mu_{SN}^{n+k}$), resulting in the creation of a period $2(n+k)$ sink for f. This bifurcation scenario will be verified in the course of the proof of Theorem 4.7.2.

Remark 2. Theorem 4.7.2 tells us how the countable infinity of periodic orbits in the horseshoes for $\mu \leq 0$ are created; namely, the periodic orbits are created in saddle-node bifurcations, and the period is increased through period-doubling bifurcations.

Remark 3. Equation (4.7.9) gives us the rate by which the period-doubling and saddle-node bifurcation values accumulate on $\mu = 0$. This rate is *not* universal but depends on the size of the unstable eigenvalue of the hyperbolic fixed point.

Remark 4. Theorems 4.7.1 and 4.7.2 are stated explicitly for dissipative or nonconservative maps, i.e., $\lambda\rho < 1$. However, similar results will hold for area-preserving maps where we have $\lambda\rho = 1$; see Newhouse [1983] and Exercise 4.83.

We now begin the proof of Theorem 4.7.2.

Proof: The proof is constructive. The condition for fixed points of f^{n+k} is given by

$$x = \rho^n x_0 - \beta\rho^n (y - y_0), \tag{4.7.10a}$$

$$y = \mu\lambda^n + \gamma\lambda^n x + \delta\lambda^n (y - y_0)^2, \tag{4.7.10b}$$

where it is important to recall that x_0, y_0, γ, β, δ, and $\rho > 0$. By substituting (4.7.10a) into (4.7.10b) we obtain

$$\begin{aligned} \delta\lambda^n y^2 - (\beta\gamma\lambda^n\rho^n + 2\delta\lambda^n y_0 + 1)y + \delta\lambda^n y_0^2 \\ + \beta\gamma\lambda^n\rho^n y_0 + \gamma\rho^n\lambda^n x_0 + \mu\lambda^n = 0. \end{aligned} \tag{4.7.11}$$

Solving (4.7.11) yields

$$\begin{aligned} y = {} & \frac{\beta\gamma\lambda^n\rho^n + 2\delta\lambda^n y_0 + 1}{2\delta\lambda^n} \\ & \pm \frac{1}{2\delta\lambda^n}\big[((\beta\gamma\lambda^n\rho^n + 2\delta\lambda^n y_0 + 1)^2 \\ & - 4\delta\lambda^n(\delta\lambda^n y_0^2 + \beta\gamma\lambda^n\rho^n y_0 + \gamma\lambda^n\rho^n x_0 + \mu\lambda^n)\big]^{1/2}. \end{aligned} \tag{4.7.12}$$

After some algebra, the expression under the radical in (4.7.12) can be simplified so that (4.7.12) becomes

$$\begin{aligned} y = {} & \frac{\beta\gamma\lambda^n\rho^n + 2\delta\lambda^n y_0 + 1}{2\delta\lambda^n} \pm \frac{1}{2\delta\lambda^n} \\ & \cdot\sqrt{4\delta\lambda^{2n}\Big[\frac{(\beta\gamma\rho^n + \lambda^{-n})^2}{4\delta} + (y_0\lambda^{-n} - \gamma\rho^n x_0) - \mu\Big]}. \end{aligned} \tag{4.7.13}$$

Note that (4.7.13) is a function of n and μ. Thus, (4.7.13) gives the y coordinate of a fixed point which can be substituted into (4.7.10a) to obtain the x coordinate. Note that since (4.7.10a) is linear in x, for a fixed y coordinate there is a unique x coordinate for the fixed point. Thus, in studying the numbers of fixed points and their bifurcations, it suffices to study only (4.7.13).

From (4.7.13) we can easily see that there are *no fixed points* for

$$\mu > \frac{(\beta\gamma\rho^n + \lambda^{-n})^2}{4\delta} + (y_0\lambda^{-n} - \gamma\rho^n x_0) \tag{4.7.14}$$

and *two fixed points* for

$$\mu < \frac{(\beta\gamma\rho^n + \lambda^{-n})^2}{4\delta} + (y_0\lambda^{-n} - \gamma\rho^n x_0). \tag{4.7.15}$$

Therefore,

$$\mu = \frac{(\beta\gamma\rho^n + \lambda^{-n})^2}{4\delta} + (y_0\lambda^{-n} - \gamma\rho^n x_0) \tag{4.7.16}$$

is a *bifurcation value* for f^{n+k}.

Next, we verify that this is a saddle-node bifurcation. This can be shown directly. From (4.7.7), the matrix associated with the linearized map is given by

$$Df^{n+k} = \begin{pmatrix} 0 & -\beta\rho^n \\ \gamma\lambda^n & 2\delta\lambda^n(y - y_0) \end{pmatrix}; \tag{4.7.17}$$

hence we have

$$\det Df^{n+k} = \gamma\beta\rho^n\lambda^n, \tag{4.7.18a}$$

$$\operatorname{tr} Df^{n+k} = 2\delta\lambda^n(y - y_0), \tag{4.7.18b}$$

with eigenvalues, χ_1 and χ_2, of (4.7.17) given by

$$\chi_{1,2} = \frac{\operatorname{tr} Df^{n+k}}{2} \pm \frac{1}{2}\sqrt{(\operatorname{tr} Df^{n+k})^2 - 4\det Df^{n+k}}. \tag{4.7.19}$$

At the bifurcation value (4.7.16) there is only one fixed point where, using (4.7.18a), (4.7.18b), and (4.7.19), the eigenvalues of (4.7.17) are given by

$$\chi_1 = 1, \qquad \chi_2 = \gamma\beta\rho^n\lambda^n. \tag{4.7.20}$$

Note that since $\rho\lambda < 1$, by taking n sufficiently large, χ_2 can be made arbitrarily small.

At this point we want to check the stability of the bifurcating fixed points. From (4.7.18) and using $\rho\lambda < 1$, we see that for n sufficiently large, the eigenvalues of (4.7.17) are approximately given by

$$\chi_1 \approx \operatorname{tr} Df^{n+k}, \tag{4.7.21a}$$

$$\chi_2 \approx 0, \tag{4.7.21b}$$

and by substituting (4.7.13) into (4.7.18b) (and neglecting terms of $\mathcal{O}(\rho^n\lambda^n)$) we have

$$\chi_1 \approx 1 \pm \sqrt{4\delta\lambda^{2n}\left(\frac{(\gamma\beta\rho^n+\lambda^{-n})^2}{4\delta}+(y_0\lambda^{-n}-\gamma x_0\rho^n)-\mu\right)}. \tag{4.7.22}$$

Thus, for the branch of fixed points with y coordinate given by

$$y = \frac{\beta\gamma\lambda^n\rho^n+2\delta\lambda^n y_0+1}{2\delta\lambda^n}+\frac{1}{2\delta\lambda^n} \cdot\sqrt{4\delta\lambda^{2n}\left(\frac{(\beta\gamma\rho^n+\lambda^{-n})^2}{4\delta}+(y_0\lambda^{-n}-\gamma\rho^n x_0)-\mu\right)}, \tag{4.7.23}$$

the eigenvalues associated with the linearized map are, for n sufficiently large, approximately given by

$$\begin{aligned}\chi_1 &\approx 1+\sqrt{4\delta\lambda^{2n}\left(\frac{(\gamma\beta\rho^n+\lambda^{-n})^2}{4\delta}+(y_0\lambda^{-n}-\gamma x_0\rho^n)-\mu\right)},\\ \chi_2 &\approx 0.\end{aligned} \tag{4.7.24}$$

Hence, for

$$\mu < \frac{(\gamma\beta\rho^n+\lambda^{-n})^2}{4\delta}+(y_0\lambda^{-n}-\gamma x_0\rho^n) \tag{4.7.25}$$

(which, from (4.7.16), is the saddle-node bifurcation value), it is easy to see that this fixed point is always a *saddle.* Similarly, for the branch of fixed points given by

$$y = \frac{\beta\gamma\lambda^n\rho^n+2\delta\lambda^n y_0+1}{2\delta\lambda^n} -\frac{1}{2\delta\lambda^n}\sqrt{4\delta\lambda^n\left(\frac{(\beta\gamma\rho^n+\lambda^{-n})^2}{4\delta}+(y_0\lambda^{-n}-\gamma\rho^n x_0)-\mu\right)}, \tag{4.7.26}$$

the eigenvalues associated with the linearized map are, for n sufficiently large, approximately given by

$$\begin{aligned}\chi_1 &\approx 1-\sqrt{4\delta\lambda^{2n}\left(\frac{(\beta\gamma\rho^n+\lambda^{-n})^2}{4\delta}+(y_0\lambda^{-n}-\gamma\rho^n x_0)-\mu\right)},\\ \chi_2 &\approx 0.\end{aligned} \tag{4.7.27}$$

Therefore, for μ "slightly" less than $\frac{(\beta\gamma\rho^n+\lambda^{-n})^2}{4\delta}+(y_0\lambda^{-n}-\gamma\rho^n x_0)$, it is easy to see that this fixed point is a *sink.* However, as μ decreases further, it is possible for χ_1 to decrease through -1 and, consequently, for this branch of fixed points to undergo a period-doubling bifurcation. We now want to study this possibility.

We are considering the branch of fixed points with y coordinate given by

$$y = \frac{\beta\gamma\lambda^n\rho^n + 2\delta\lambda^n y_0 + 1}{2\delta\lambda^n} - \frac{1}{2\delta\lambda^n} \cdot \sqrt{4\delta\lambda^{2n}\left(\frac{(\beta\gamma\rho^n + \lambda^{-n})^2}{4\delta} + (y_0\lambda^{-n} - \gamma\rho^n x_0) - \mu\right)}. \quad (4.7.28)$$

From (4.7.19), the condition for an eigenvalue of Df^{n+k} to be -1 is

$$1 + \det Df^{n+k} = -\operatorname{tr} Df^{n+k}. \quad (4.7.29)$$

Substituting (4.7.18a) and (4.7.18b) into (4.7.29) yields

$$1 + \gamma\beta\rho^n\lambda^n = 2\delta\lambda^n y_0 - 2\delta\lambda^n y, \quad (4.7.30)$$

and substituting (4.7.28) into (4.7.30) yields

$$1+\gamma\beta\rho^n\lambda^n = \sqrt{\delta\lambda^{2n}\left(\frac{(\beta\gamma\rho^n + \lambda^{-n})^2}{4\delta} + (y_0\lambda^{-n} - \gamma\rho^n x_0) - \mu\right)}. \quad (4.7.31)$$

By solving (4.7.31) for μ, we obtain

$$\mu = -\frac{3}{4\delta}(\gamma\beta\rho^n + \lambda^{-n})^2 + (y_0\lambda^{-n} - \gamma\rho^n x_0). \quad (4.7.32)$$

Hence, (4.7.32) is the bifurcation value for the period-doubling bifurcation of the sink created in the saddle-node bifurcation. We leave it as an exercise for the reader to verify that the period-doubling bifurcation is "generic" (see Exercise 4.44).

Let us summarize what we have shown thus far. The map f^{n+k}, for n sufficiently large, undergoes a *saddle-node* bifurcation at

$$\mu_{SN}^{n+k} = \frac{(\beta\gamma\rho^n + \lambda^{-n})^2}{4\delta} + (y_0\lambda^{-n} - \gamma\rho^n x_0). \quad (4.7.33)$$

In this bifurcation two fixed points of f^{n+k} are created, a saddle and a sink. As μ decreases below μ_{SN}^{n+k}, the saddle remains a saddle but the sink undergoes a period-doubling bifurcation at

$$\mu_{PD}^{n+k} = -\frac{3(\beta\gamma\rho^n + \lambda^{-n})^2}{4\delta} + (y_0\lambda^{-n} - \gamma\rho^n x_0). \quad (4.7.34)$$

It is easy to see from (4.7.33) and (4.7.34) that we have

$$\mu_{PD}^{n+k} < \mu_{SN}^{n+k}.$$

Also, for n sufficiently large, we have

$$y_0\lambda^{-n} - \gamma\rho^n x_0 >> (\beta\gamma\rho^n + \lambda^{-n})^2. \tag{4.7.35}$$

Using (4.7.35) along with the fact that

$$(y_0\lambda^{-n} - \gamma\rho^n x_0) = \lambda^{-n}(y_0 - \gamma\rho^n\lambda^n x_0) \tag{4.7.36}$$

with $\rho\lambda < 1$ implies that

$$\mu_{SN}^{n+k} > 0$$

and

$$\mu_{PD}^{n+k} > 0$$

for n sufficiently large. Next we need to show that

$$\mu_{SN}^{n+1+k} < \mu_{PD}^{n+k}.$$

From (4.7.33) and (4.7.34), we have

$$\mu_{SN}^{n+1+k} = \frac{(\beta\gamma\rho^{n+1} - \lambda^{-n-1})^2}{4\delta} + \lambda^{-n-1}(y_0 - \gamma\rho^{n+1}\lambda^{-n-1}x_0) \tag{4.7.37}$$

and

$$\mu_{PD}^{n+k} = -\frac{3(\beta\gamma\rho^n - \lambda^{-n})^2}{4\delta} + \lambda^{-n}(y_0 - \gamma\rho^n\lambda^n x_0). \tag{4.7.38}$$

Using (4.7.35) and (4.7.36), along with the fact that $\lambda > 1$, we can easily see from (4.7.37) and (4.7.38) that, for n sufficiently large, we have

$$\mu_{SN}^{n+1+k} < \mu_{PD}^{n+k}. \tag{4.7.39}$$

Equation (4.7.8) now follows from (4.7.39) by induction. Finally, it follows from (4.7.35), (4.7.33) and (4.7.34) that

$$\mu_{SN}^{n+k} \sim \lambda^{-n} \quad \text{as} \quad n \to \infty$$

and

$$\mu_{PD}^{n+k} \sim \lambda^{-n} \quad \text{as} \quad n \to \infty. \qquad \square$$

The following corollary is an obvious consequence of the proof of Theorem 4.7.2.

Corollary 4.7.3 *Let p_0 denote a point of quadratic tangency of $W_0^s(0)$ and $W_0^u(0)$. Then for all n sufficiently large there is a parameter value, μ^{n+k}, with $\mu^{n+k} \to 0$ as $n \to \infty$, such that f has a periodic sink, p_n, of period $n+k$ with $p_n \to p_0$ as $n \to \infty$.*

The reason for pulling out this corollary from Theorem 4.7.2 is that if we put it together with some deep results of Newhouse concerning the persistence of quadratic homoclinic tangencies, we get a very provocative result. (Note: Newhouse's results do not require our more restrictive Assumptions 3 and 4 given above but, rather, they apply to any two-dimensional diffeomorphism having a *dissipative* hyperbolic periodic point whose stable and unstable manifolds have a quadratic tangency.) We now want to give a brief description of the implications of Newhouse's results in the context of this section.

It should be clear that a specific point of tangency of $W_0^s(0)$ and $W_0^u(0)$ can easily be destroyed by the slightest perturbation. However, Newhouse [1974] has proven the following result.

Theorem 4.7.4 *For* $\varepsilon > 0$, *let* $I_\varepsilon = \{\mu \in I \mid |\mu| < \varepsilon\}$. *Then, for every* $\varepsilon > 0$, *there exists a nontrivial interval* $\hat{I}_\varepsilon \subset I_\varepsilon$ *such that* $\hat{I}_\varepsilon$ *contains a dense set of points at which* $W_\mu^s(0)$ *and* $W_\mu^u(0)$ *have a quadratic homoclinic tangency.*

Proof: This "parametrized" version of Newhouse's theorem is due to Robinson [1983]. □

Heuristically, Theorem 4.7.4 says that if we destroy the quadratic homoclinic tangency at $\mu = 0$ by varying μ slightly, then for a dense set of parameter values containing $\mu = 0$, we have a quadratic homoclinic tangency somewhere else in the homoclinic tangle. Thus, Corollary 4.7.3 can be applied to each of these tangencies so that Corollary 4.7.3 and Theorem 4.7.4 together imply that there are parameter values at which the map has infinitely many periodic attractors which coexist with Smale horseshoe–type dynamics. This phenomenon is at the heart of the difficulties encountered in proving that a two-dimensional map possesses a "strange attractor." We will discuss this in much more detail in Section 4.11.

Let us now comment on the generality of our results. In particular, in Assumption 2 we assumed that our map is linear in a neighborhood, $\mathcal{N}$, of the origin, and in Assumption 3 we assumed that the form of the map defined outside of a neighborhood of the origin is given as in (4.7.5). It follows from the work of Gavrilov and Silnikov [1972], [1973] that our results are not restricted by these assumptions in the sense that if we assume the most general forms for f in $\mathcal{N}$ and for f^k, Theorems 4.7.1 and 4.7.2 are unchanged. We remark that this generally holds for the study of the orbit structure near orbits homoclinic to hyperbolic periodic points. A return map defined near a homoclinic point constructed as the composition of an iterate of the *linearized map* near the origin with low-order terms in the Taylor expansion of an iterate of the map outside of a neighborhood of the origin is sufficient to capture the qualitative dynamics in a sufficiently small neighborhood of the homoclinic point. Let us now briefly describe the set-up considered by Gavrilov and Silnikov.

In local coordinates in $\mathcal{N}$, Gavrilov and Silnikov showed that a general $\mathbf{C}^r$ $(r \geq 3)$ diffeomorphism can be written in the form

$$\begin{pmatrix} x \\ y \end{pmatrix} \longmapsto \begin{pmatrix} \lambda(\mu)x + f(x,y;\mu)x \\ \rho(\mu)y + g(x,y;\mu)y \end{pmatrix}. \tag{4.7.40}$$

For the form of f^k acting outside of $\mathcal{N}$ (but mapping a neighborhood of a homoclinic point on the local unstable manifold in $\mathcal{N}$ to a neighborhood of the local stable manifold in $\mathcal{N}$), they assumed the completely general form

$$f^k\colon \begin{pmatrix} x \\ y \end{pmatrix} \longmapsto \begin{pmatrix} x_0 + F(x, y - y_0; \mu) \\ G(x, y - y_0; \mu) \end{pmatrix}. \tag{4.7.41}$$

The assumption of quadratic homoclinic tangency at $\mu = 0$ requires

$$G_y(0,0,0) = 0, \tag{4.7.42}$$

$$G_{yy}(0,0,0) \neq 0 \tag{4.7.43}$$

(note: since f is a diffeomorphism, then (4.7.42) implies that we must have $G_x(0,0,0) \neq 0$ and $F_y(0,0,0) \neq 0$). Gavrilov and Silnikov then simplify (4.7.41) as follows; letting

$$y - y_0 = \phi(x, \mu) \tag{4.7.44}$$

be the (unique) solution of

$$G_y(x, y - y_0, \mu) = 0 \tag{4.7.45}$$

(which can be solved by the implicit function theorem as a result of (4.7.43)), (4.7.41) can be rewritten as

$$\begin{pmatrix} x \\ y \end{pmatrix} \longmapsto \begin{pmatrix} x_0 + F(x, y - y_0; \mu) \\ E(\mu) + C(x,\mu)x + D(x,\mu)(y - y_0 - \phi(x,\mu))^2 \end{pmatrix}, \tag{4.7.46}$$

where

$$\begin{aligned} E(\mu) &\equiv G(0, \phi(0,\mu), \mu), \quad E(0) = 0, \\ C(0,0) &\equiv c, \\ 2D(0, y_0, 0) &\equiv d. \end{aligned}$$

The reader should note the similarity of (4.7.46) and (4.7.5). The numbers c and d in (4.7.46) describe the geometry of the tangency of the stable and unstable manifolds. Gavrilov and Silnikov show that there are ten cases to consider depending on the signs of λ, ρ, c, and d. (Note: yes, there are 16 possible combinations of signs of these parameters, but Gavrilov and Silnikov show how to reduce the number of possibilities.)

Five of the cases correspond to orientation-preserving maps with the remaining five corresponding to orientation-reversing maps. The example we treated corresponds to one of the five orientation-preserving maps with λ, ρ, $c, d > 0$. The structure of the period-doubling and saddle-node cascades can be different for the remaining cases, and we refer the reader to Gavrilov and Silnikov [1972], [1973] for the details. We close with some final remarks.

Remark 1: The Codimension of the Homoclinic Bifurcation. By the term "homoclinic bifurcation," we mean the creation of transverse homoclinic orbits to a hyperbolic periodic point of a two-dimensional diffeomorphism as parameters are varied. Since the stable and unstable manifolds of the hyperbolic periodic point are codimension one, their transversal intersection occurs stably in a one-parameter family of maps (see Theorem 4.5.8). Now recall the definition of "codimension of a bifurcation" given in Section 3.1D. Roughly speaking, the codimension is the number of parameters in which the corresponding parametrized family is stable under perturbations. We have seen in this section that there is an infinity of saddle-node and period-doubling bifurcation values accumulating on the quadratic homoclinic tangency parameter value. Thus the type of bifurcation is codimension infinity by the standard definitions. In particular, one cannot find a versal deformation for this type of bifurcation satisfying the standard definitions given in Chapter 3.

Remark 2. There have been a number of papers by the Maryland group over the last few years that have shed much light on the creation of horseshoes. In particular, we refer the reader to Yorke and Alligood [1985] and Alligood et al. [1987]. In Tedeschini-Lalli and Yorke [1986] the question of the measure of the set of parameter values for which the map possesses an infinite number of coexisting periodic sinks is addressed. For a "generic" map, the measure of the set is shown to be zero.

4.8 Orbits Homoclinic to Hyperbolic Fixed Points in Three-Dimensional Autonomous Vector Fields

In this section we will study the orbit structure near orbits homoclinic to hyperbolic fixed points of three-dimensional autonomous vector fields. The term "near" refers to both phase space and parameter space. We will see that in some cases Smale horseshoe–type behavior may arise. In parametrized systems the creation of the horseshoes may be accompanied by cascades of period-doubling and saddle-node bifurcations as described in Theorem 4.7.2, or the horseshoes may "explode" into creation at a critical parameter value. We will see that the nature of the orbit structure near the homoclinic orbits depends mainly on two properties of the vector field:

1. the nature of the eigenvalues of the linearized vector field at the fixed point;

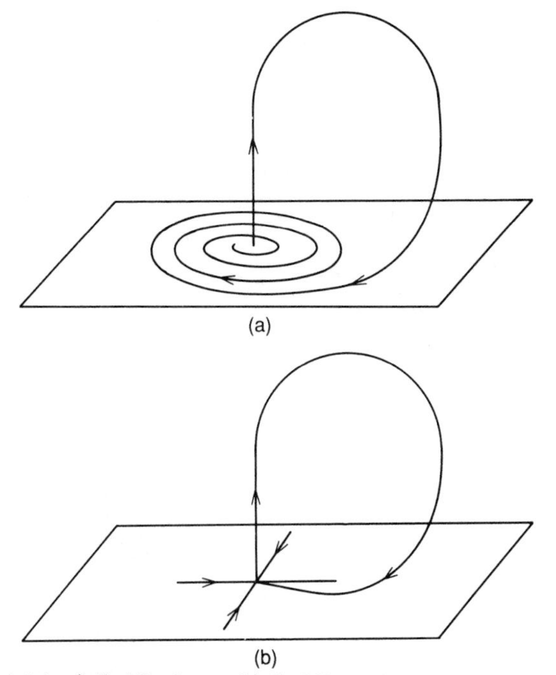

FIGURE 4.8.1. a) Saddle-focus. b) Saddle with purely real eigenvalues.

2. the existence of multiple homoclinic orbits to the same hyperbolic fixed point which could be a consequence of *symmetries* of the vector field.

Regarding Property 1, it should be clear that there are only two possibilities for the three eigenvalues associated with the linearized vector field.

1. *Saddle* λ_1, λ_2, λ_3 real with $\lambda_1, \lambda_2 < 0$, $\lambda_3 > 0$.

2. *Saddle-focus* $\rho \pm i\omega$, λ with $\rho < 0$, $\lambda > 0$.

All other possibilities for hyperbolic fixed points follow from these two via time reversal. We will analyze each situation individually, but first we want to describe the general technique of analysis that will apply to both cases.

The Technique of Analysis

Consider a three-dimensional autonomous $\mathbf{C}^r$ $(r \geq 2)$ vector field having a hyperbolic fixed point at the origin with a two-dimensional stable manifold and a one-dimensional unstable manifold such that a homoclinic orbit connects the origin to itself (i.e., $W^u(0) \cap W^s(0) \neq \emptyset$) (see Figure 4.8.1 for the two possibilities according to the nature of the linearized flow near the origin). The strategy will be to define a two-dimensional cross-section to the vector field near the homoclinic orbit, and to construct a map of the cross-section into itself from the flow generated by the vector field. This is

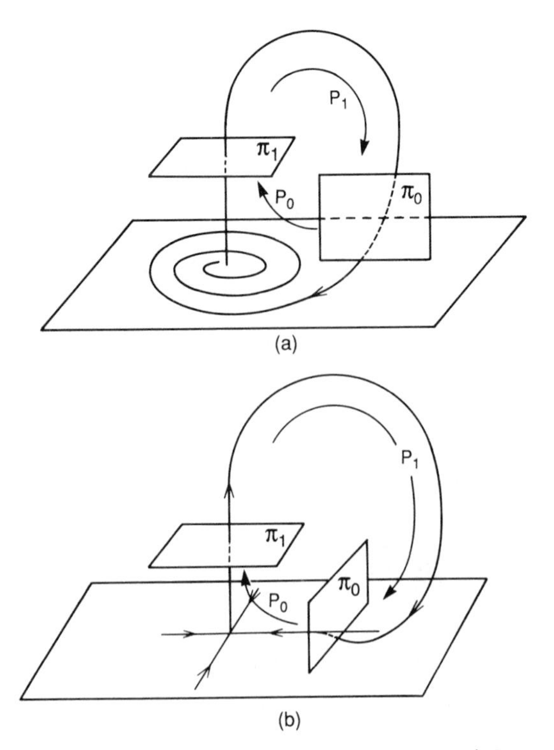

FIGURE 4.8.2. Poincaré map near the homoclinic orbit. a) Saddle-focus. b) Saddle with real eigenvalues.

exactly the same idea that was used in Section 1.2A and was used to prove Moser's theorem in Section 4.4. Let us be more precise.

Consider cross-sections Π_0 and Π_1 transverse to the homoclinic orbit and located in a "sufficiently small neighborhood of the origin" as shown in Figure 4.8.2. We construct a Poincaré map of Π_0 into itself

$$P: \Pi_0 \longrightarrow \Pi_0,$$

which will be the composition of two maps, one constructed from the flow near the origin

$$P_0: \Pi_0 \longrightarrow \Pi_1$$

and the other constructed from the flow defined outside a neighborhood of the origin

$$P_1: \Pi_1 \longrightarrow \Pi_0.$$

Then we have

$$P \equiv P_1 \circ P_0: \Pi_0 \longrightarrow \Pi_0;$$

see Figure 4.8.2. Thus, the entire construction requires four steps.

Step 1: Define Π_0 *and* Π_1. We will do this in each of the cases individually. As is typical, the choice of a cross-section on which to define a Poincaré map requires some knowledge of the geometrical structure of the phase space. A clever choice can often simplify the computations considerably.

Step 2: Construction of P_0. For Π_0 and Π_1 located sufficiently close to the origin, the map of Π_0 into Π_1 is "essentially" given by the flow generated by the linearized vector field. We put "essentially" in quotes, because using the linearized vector field to construct P_0 does introduce an error. However, the error can be made arbitrarily small by taking Π_0 and Π_1 sufficiently small and close to the origin; this is proven in Wiggins [1988]. Moreover, this error is truly negligible in the sense that it does not affect our results. Therefore, we will construct P_0 from the flow generated by the linearized vector field in order to avoid unnecessary technical distractions.

Step 3: Construction of P_1. Let $p_0 \equiv W^u(0) \cap \Pi_0$ and $p_1 \equiv W^u(0) \cap \Pi_1$. Then the time of flight from p_0 to p_1 is finite, since we are outside of a neighborhood of the fixed point. We will assume that, except for the origin, the homoclinic orbit is bounded away from all other possible fixed points of the vector field. Then, by continuity with respect to initial conditions, for Π_0 sufficiently small, the flow generated by the vector field maps Π_0 onto Π_1. This implies that the map P_1 is defined for Π_0 sufficiently small.

Thus, P_0 is defined, but how is it computed? Taylor expanding P_1 about p_0 gives

$$P_1(h) = p_0 + DP_1(p_1)h + \mathcal{O}(|h|^2),$$

where h represents coordinates on Π_1 centered at p_1. Now for Π_1 sufficiently small the $\mathcal{O}(|h|^2)$ term in this expression can be made arbitrarily small. Therefore, for $P_1 \colon \Pi_1 \to \Pi_0$, we will take

$$P_1(h) = p_0 + DP_1(p_1)h.$$

Of course, this approximation to P_1 introduces an error. However, in Wiggins [1988] it is shown that the error is truly negligible in the sense that it does not affect our results.

Step 4: Construction of $P \equiv P_1 \circ P_0$. With P_0 and P_1 defined the construction of P is obvious.

Let us now make some heuristic remarks. Our analysis will give us information on the orbit structure in a sufficiently small neighborhood of the homoclinic orbit. The map P_0 can be constructed exactly from the linearized vector field (since we can "solve" linear, constant coefficient ordinary differential equations). Hence, we can compute how Π_0 is stretched, contracted, and, possibly, folded as it passes near the fixed point. Now P_1 might appear to present a problem, since we cannot even compute $DP_1(p_1)$ without solving for the flow generated by the nonlinear vector field. Fortunately, and perhaps surprisingly, it will turn out that we do not need to know $DP_1(p_1)$ exactly, only that it is *compatible with the geometry of the homoclinic orbit.* This will be made clear in the examples to which we now turn.

4.8A Orbits Homoclinic to a Saddle-Point with Purely Real Eigenvalues

Consider the following

$$\begin{aligned} \dot{x} &= \lambda_1 x + f_1(x, y, z; \mu), \\ \dot{y} &= \lambda_2 y + f_2(x, y, z; \mu), \qquad (x, y, z, \mu) \in \mathbb{R}^1 \times \mathbb{R}^1 \times \mathbb{R}^1 \times \mathbb{R}^1, \\ \dot{z} &= \lambda_3 z + f_3(x, y, z; \mu), \end{aligned} \tag{4.8.1}$$

where the f_i are $\mathbf{C}^2$ and they vanish at $(x, y, z, \mu) = (0, 0, 0, 0)$ and are nonlinear in x, y, and z. Hence, (4.8.1) has a fixed point at the origin with eigenvalues given by λ_1, λ_2, and λ_3. We make the following assumptions.

Assumption 1. $\lambda_1, \lambda_2 < 0$, $\lambda_3 > 0$.

Assumption 2. At $\mu = 0$, (4.8.1) possesses a homoclinic orbit Γ connecting $(x, y, z) = (0, 0, 0)$ to itself. Moreover, we assume that the homoclinic orbit breaks as shown in Figure 4.8.3 for $\mu > 0$ and $\mu < 0$.

The following remarks are now in order.

Remark 1. We assume that the parameter dependence is contained in the f_i and not in the eigenvalues λ_1, λ_2, and λ_3. This is mainly for convenience and does not affect the generality of our results; see Exercise 4.45.

Remark 2. In Figure 4.8.3 we drew the homoclinic orbit entering a neighborhood of the origin along a curve that is tangent to the y-axis at the origin. This assumes that $\lambda_2 > \lambda_1$ and that the system is *generic.* We deal with these issues in Exercise 4.46. Our results will not change for generic systems if $\lambda_1 \geq \lambda_2$; see Exercises 4.48 and 4.49.

We will analyze the orbit structure in a neighborhood of Γ in the standard way by computing a Poincaré map on an appropriately chosen cross-section. We choose two rectangles transverse to the flow, which are defined as follows

$$\begin{aligned} \Pi_0 &= \big\{(x, y, z) \in \mathbb{R}^3 \,|\, |x| \leq \varepsilon, y = \varepsilon, 0 < z \leq \varepsilon\big\}, \\ \Pi_1 &= \big\{(x, y, z) \in \mathbb{R}^3 \,|\, |x| \leq \varepsilon, |y| \leq \varepsilon, z = \varepsilon\big\}, \end{aligned} \tag{4.8.2}$$

for some $\varepsilon > 0$; see Figure 4.8.4.

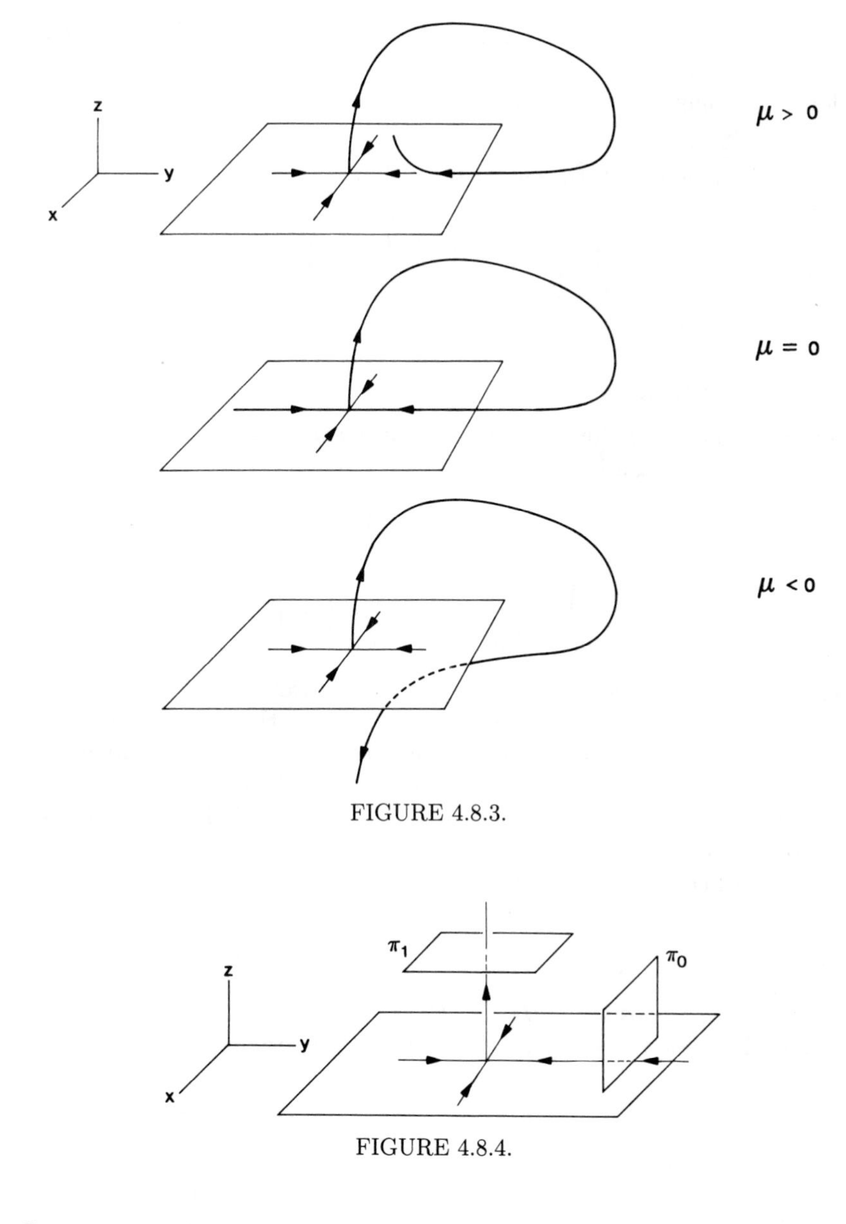

FIGURE 4.8.3.

FIGURE 4.8.4.

Computation of P_0

The flow linearized at the origin is given by

$$\begin{aligned} x(t) &= x_0 e^{\lambda_1 t}, \\ y(t) &= y_0 e^{\lambda_2 t}, \\ z(t) &= z_0 e^{\lambda_3 t}, \end{aligned} \tag{4.8.3}$$

and the time of flight from Π_0 to Π_1 is given by

$$t = \frac{1}{\lambda_3} \log \frac{\varepsilon}{z_0}. \tag{4.8.4}$$

Hence, the map

$$P_0 \colon \Pi_0 \to \Pi_1$$

is given by (leaving off the subscript 0's)

$$\begin{pmatrix} x \\ \varepsilon \\ z \end{pmatrix} \mapsto \begin{pmatrix} x\left(\frac{\varepsilon}{z}\right)^{\lambda_1/\lambda_3} \\ \varepsilon\left(\frac{\varepsilon}{z}\right)^{\lambda_2/\lambda_3} \\ \varepsilon \end{pmatrix}. \tag{4.8.5}$$

Computation of P_1

Following the discussion of Step 3 in our general analysis above, we take as P_1 the following affine map

$$P_1 \colon \Pi_1 \longrightarrow \Pi_0$$

$$\begin{pmatrix} x \\ y \\ \varepsilon \end{pmatrix} \mapsto \begin{pmatrix} 0 \\ \varepsilon \\ 0 \end{pmatrix} + \begin{pmatrix} a & b & 0 \\ 0 & 0 & 0 \\ c & d & 0 \end{pmatrix} \begin{pmatrix} x \\ y \\ 0 \end{pmatrix} + \begin{pmatrix} e\mu \\ 0 \\ f\mu \end{pmatrix}, \tag{4.8.6}$$

where a, b, c, d, e, and f are constants. Note from Figure 4.8.3 that we have $f > 0$, so we may rescale the parameter μ so that $f = 1$. Henceforth, we will assume that this has been done. Let us briefly explain the form of (4.8.6). On Π_0 the y coordinate is fixed at $y = \varepsilon$. This explains why there are only zeros in the middle row of the linear part of (4.8.6). Also, the z coordinate of Π_1 is fixed at $z = \varepsilon$. This explains why there are only zeros in the third column of the matrix in (4.8.6).

The Poincaré Map $P \equiv P_1 \circ P_0$

Forming the composition of P_0 and P_1, we obtain the Poincaré map defined in a neighborhood of the homoclinic orbit having the following form.

$$P \equiv P_1 \circ P_0 \colon \Pi_0 \to \Pi_0,$$

$$\begin{pmatrix} x \\ z \end{pmatrix} \mapsto \begin{pmatrix} ax\left(\frac{\varepsilon}{z}\right)^{\lambda_1/\lambda_3} + b\varepsilon\left(\frac{\varepsilon}{z}\right)^{\lambda_2/\lambda_3} + e\mu \\ cx\left(\frac{\varepsilon}{z}\right)^{\lambda_1/\lambda_3} + d\varepsilon\left(\frac{\varepsilon}{z}\right)^{\lambda_2/\lambda_3} + \mu \end{pmatrix}, \tag{4.8.7}$$

where Π_0 is chosen sufficiently small so that $P_1 \circ P_0$ is defined.

We reiterate that the approximate Poincaré map (4.8.7) is valid for ε sufficiently small and x and z sufficiently small. For ε sufficiently small, the approximation of P_0 by the linearized flow is valid and, for x and z sufficiently small, the approximation of P_1 by the affine map P_1 is valid. Note the ε, x, and z are independent.

Calculation of Fixed Points of P

Now we look for fixed points of the Poincaré map (which will correspond to periodic orbits of (4.8.1). First some notation; let

$$A = a\varepsilon^{\lambda_1/\lambda_3}, \quad B = b\varepsilon^{1+(\lambda_2/\lambda_3)}, \quad C = c\varepsilon^{\lambda_1/\lambda_3}, \quad D = d\varepsilon^{1+(\lambda_2/\lambda_3)}.$$

Then the condition for fixed points of (4.8.7) is

$$x = Axz^{|\lambda_1|/\lambda_3} + Bz^{|\lambda_2|/\lambda_3} + e\mu, \tag{4.8.8a}$$

$$z = Cxz^{|\lambda_1|/\lambda_3} + Dz^{|\lambda_2|/\lambda_3} + \mu. \tag{4.8.8b}$$

Solving (4.8.8a) for x as a function of z gives

$$x = \frac{Bz^{|\lambda_2|/\lambda_3} + e\mu}{1 - Az^{|\lambda_1|/\lambda_3}}. \tag{4.8.9}$$

We will restrict ourselves to a sufficiently small neighborhood of the homoclinic orbit so that z can be taken sufficiently small in order that the denominator of (4.8.9) can be taken to be 1 (see Exercise 4.47). Substituting this expression for x into (4.8.8b) gives the following condition for fixed points of (4.8.7) in terms of z and μ only

$$z - \mu = CBz^{|\lambda_1+\lambda_2|/\lambda_3} + Ce\mu z^{|\lambda_1|/\lambda_3} + Dz^{|\lambda_2|/\lambda_3}. \tag{4.8.10}$$

We will graphically display the solutions of (4.8.10) for μ sufficiently small and near zero by graphing the left-hand side of (4.8.10) and the right-hand side of (4.8.10) and seeking intersections of the curves.

First, we want to examine the slope of the right-hand side of (4.8.10) at $z = 0$. This is given by the following expression

$$\begin{aligned} &\frac{d}{dz}\left(CBz^{|\lambda_1+\lambda_2|/\lambda_3} + Ce\mu z^{|\lambda_1|/\lambda_3} + Dz^{|\lambda_2|/\lambda_3}\right) \\ &\qquad = \frac{|\lambda_1+\lambda_2|}{\lambda_3} CBz^{\frac{|\lambda_1+\lambda_2|}{\lambda_3}-1} + \frac{|\lambda_1|}{\lambda_3} Ce\mu z^{\frac{|\lambda_1|}{\lambda_3}-1} \\ &\qquad + \frac{|\lambda_2|}{\lambda_3} Dz^{\frac{|\lambda_2|}{\lambda_3}-1}. \end{aligned} \tag{4.8.11}$$

We recall that P_1 is invertible so that $ad - bc \neq 0$. This implies that $AD - BC \neq 0$ so that C and D cannot both be zero. Therefore, at $z = 0$, (4.8.11) takes the values

$$\begin{array}{ll} \infty & \text{if } |\lambda_1| < \lambda_3 \text{ or } |\lambda_2| < \lambda_3 \\ 0 & \text{if } |\lambda_1| > \lambda_3 \text{ and } |\lambda_2| > \lambda_3. \end{array}$$

There are four possible cases, two each for both the infinite-slope and zero-slope situations. The differences in these situations depend mainly on global

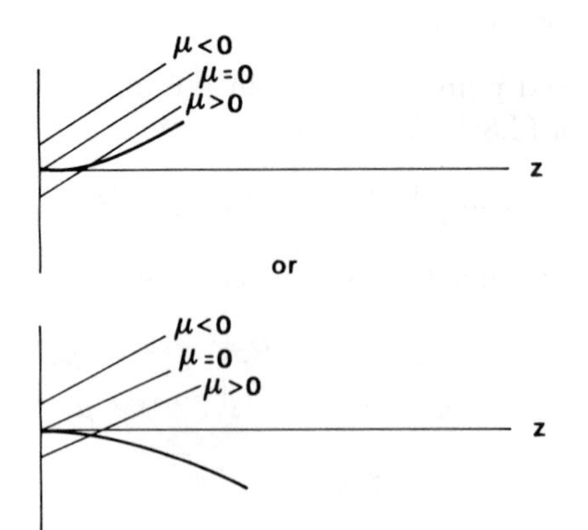

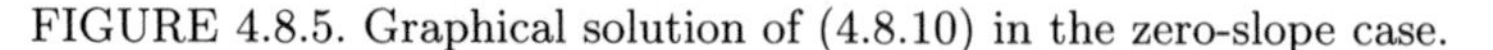

FIGURE 4.8.5. Graphical solution of (4.8.10) in the zero-slope case.

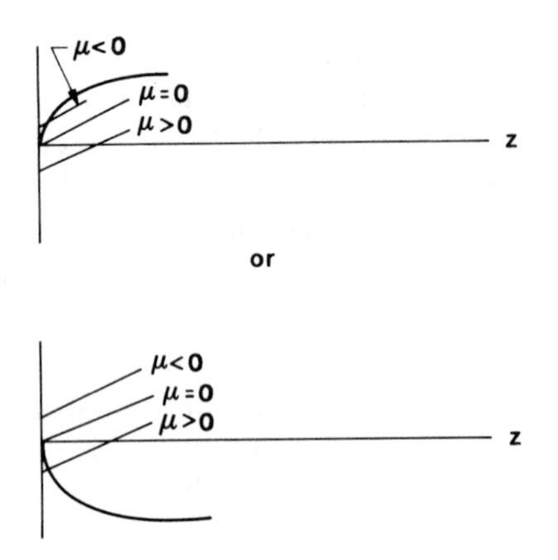

FIGURE 4.8.6. Graphical solution of (4.8.10) in the infinite-slope case.

effects, i.e., the relative signs of A, B, C, D, e, and μ. We will consider this more carefully shortly. Figure 4.8.5 illustrates the graphical solution of (4.8.10) in the zero-slope case. The two-slope cases illustrated in Figure 4.8.5 give the same result, namely, that for $\mu > 0$ a periodic orbit bifurcates from the homoclinic orbit.

In the infinite-slope case the two possible situations are illustrated in Figure 4.8.6. Interestingly, in the infinite-slope case we get two different results; namely, in one case we get a periodic orbit for $\mu < 0$, and in the other case a periodic orbit for $\mu > 0$. So what is going on? As we will shortly see, there is a global effect in this case that our local analysis does not detect. Now we want to explain this global effect.

Let τ be a tube beginning and ending on Π_0 and Π_1, respectively, which contains Γ. Then $\tau \cap W^s(0)$ is a two-dimensional strip which we denote as $\mathcal{R}$. Suppose, *without twisting* $\mathcal{R}$, that we join together the two ends of $\mathcal{R}$.

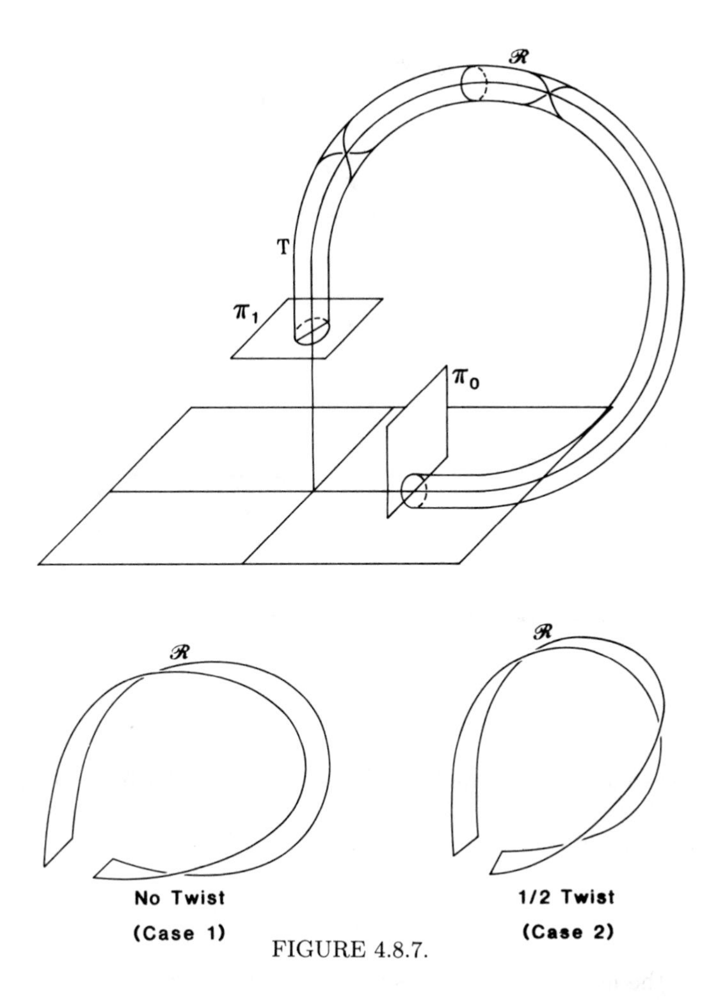

FIGURE 4.8.7.

Then there are two possibilities: 1) $W^s(0)$ experiences an even number of half-twists inside τ, in which case, when the ends of $\mathcal{R}$ are joined together it is homeomorphic to a cylinder, or 2) $W^s(0)$ experiences an odd number of half-twists inside τ, in which case, when the ends of $\mathcal{R}$ are joined together it is homeomorphic to a Mobius strip; see Figure 4.8.7. The reader should verify this experimentally with a strip of paper.

We now want to discuss the dynamical consequences of these two situations. First, consider the rectangle $\mathcal{D} \subset \Pi_0$ shown in Figure 4.8.8, which has its lower horizontal boundary in $W^s(0)$. We want to consider the shape of the image of $\mathcal{D}$ under P_0. From (4.8.5), P_0 is given by

$$\begin{pmatrix} x \\ \varepsilon \\ z \end{pmatrix} \mapsto \begin{pmatrix} x\left(\frac{\varepsilon}{z}\right)^{\lambda_1/\lambda_3} \\ \varepsilon\left(\frac{\varepsilon}{z}\right)^{\lambda_2/\lambda_3} \\ \varepsilon \end{pmatrix} \equiv \begin{pmatrix} x' \\ y' \\ \varepsilon \end{pmatrix}, \tag{4.8.12}$$

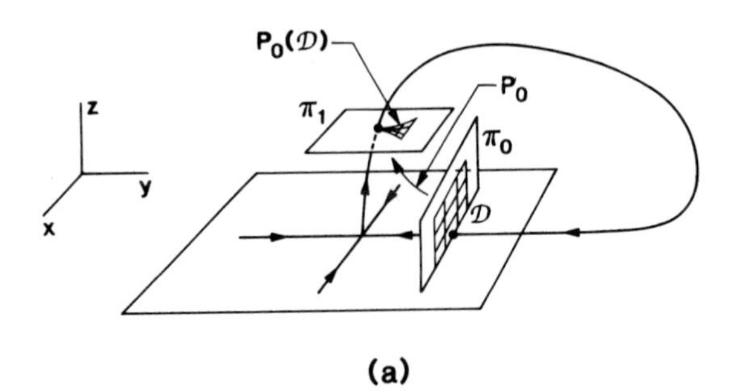

(a)

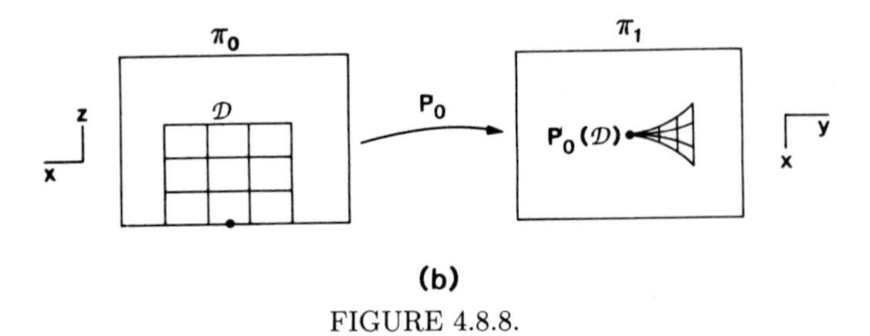

(b)

FIGURE 4.8.8.

where, to avoid confusion, we label the coordinates in Π_1 by x' and y'. Now consider a horizontal line in $\mathcal{D}$, i.e., a line with $z =$ constant. From (4.8.12) we see that this line is mapped to a line in Π_1 given by

$$y' = \varepsilon \left(\frac{\varepsilon}{z}\right)^{\lambda_2/\lambda_3} = \text{constant}. \tag{4.8.13}$$

However, the length of this line is not preserved, since

$$\frac{x'}{x} = \left(\frac{\varepsilon}{z}\right)^{\lambda_1/\lambda_3} \underset{z\to 0}{\longrightarrow} 0 \tag{4.8.14}$$

because $\lambda_1 < 0 < \lambda_3$. Next consider a vertical line in $\mathcal{D}$, i.e., a line with $x =$ constant. From (4.8.12) we see that

$$\frac{y'}{z} = \varepsilon^{1+\frac{\lambda_2}{\lambda_3}} z^{-\frac{\lambda_2}{\lambda_3}-1}. \tag{4.8.15}$$

Hence, for $-\lambda_2 > \lambda_3$, the length of vertical lines is contracted under P_0 as $z \to 0$ and, for $-\lambda_2 < \lambda_3$, the length of vertical lines is expanded under P_0 as $z \to 0$. Now, (4.8.14) implies that a horizontal line in $\mathcal{D}$ on the stable manifold of the origin is contracted to a point (i.e., P_0 is not defined here). From these remarks we thus see that $\mathcal{D}$ is mapped to the "half-bowtie" shape as shown in Figure 4.8.8. Note that for $\lambda_2 > \lambda_1$ the

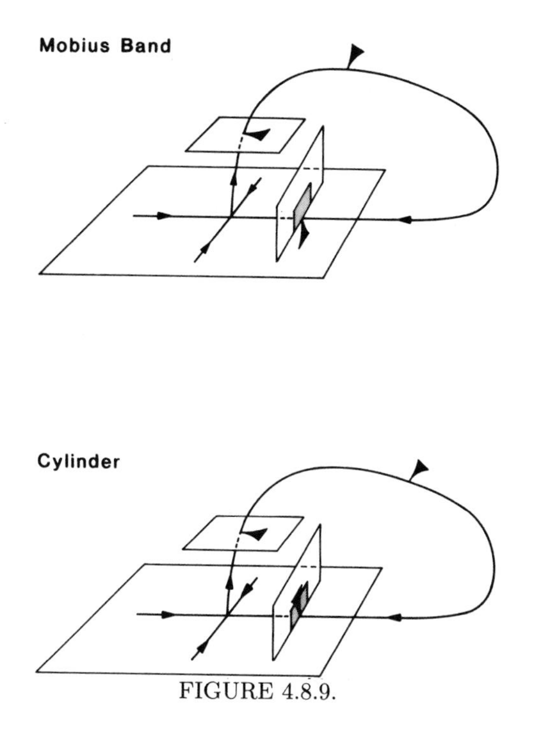

FIGURE 4.8.9.

"vertical" boundary of the "half-bowtie" is tangent to the y-axis at the origin. If $\lambda_2 < \lambda_1$, it would be tangent to the x-axis at the origin; see Exercise 4.48. Since we are assuming $\lambda_2 > \lambda_1$ for the purpose of illustrating the construction and geometry of the maps, we show only the first case.

Under the map P_1 the half-bowtie $P_0(\mathcal{D})$ is mapped back around Γ with the sharp tip of $P_0(\mathcal{D})$ coming back near $\Gamma \cap \Pi_0$. In the case where $\mathcal{R}$ is homeomorphic to a cylinder, $P_0(\mathcal{D})$ twists around an even number of times in its journey around Γ and comes back to Π_0 lying above $W^s(0)$. In the case where $\mathcal{R}$ is homeomorphic to a mobius strip, $P_0(\mathcal{D})$ twists around an odd number of times in its journey around Γ and returns to Π_0 lying below $W^s(0)$; see Figure 4.8.9.

At this point, we will return to the four different cases that arose in locating the bifurcated periodic orbits and see which particular global effect occurs.

Recall from (4.8.10) that the z components of the fixed points were obtained by solving

$$z = CBz^{\frac{|\lambda_1+\lambda_2|}{\lambda_3}} + Ce\mu z^{\frac{|\lambda_1|}{\lambda_3}} + Dz^{\frac{|\lambda_2|}{\lambda_3}} + \mu. \tag{4.8.16}$$

The right-hand side of this equation thus represents the z-component of the first return of a point to Π_0. Then, at $\mu = 0$, the first return will be positive if we have a cylinder (C) and negative if we have a mobius band

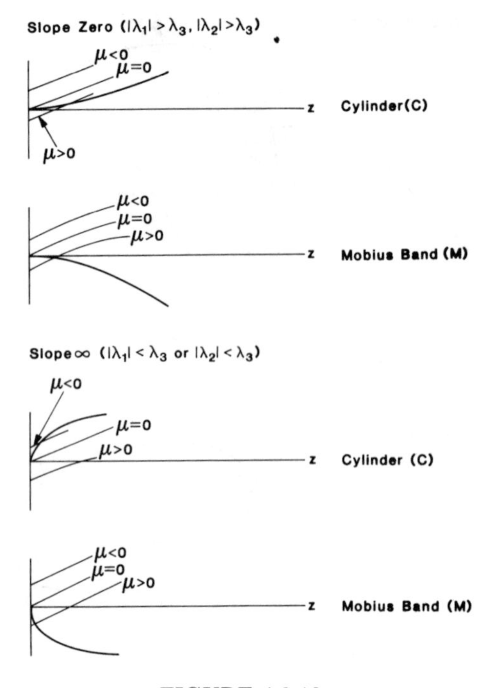

FIGURE 4.8.10.

(M). Using this remark, we can go back to the four cases and label them as in Figure 4.8.10.

We now address the question of stability of the bifurcated periodic orbits.

Stability of the Periodic Orbits

The derivative of (4.8.7) is given by

$$DP = \begin{pmatrix} Az^{\frac{|\lambda_1|}{\lambda_3}} & \frac{|\lambda_1|}{\lambda_3} Axz^{\frac{|\lambda_1|}{\lambda_3}-1} + \frac{|\lambda_2|}{\lambda_3} Bz^{\frac{|\lambda_2|}{\lambda_3}-1} \\ Cz^{\frac{|\lambda_1|}{\lambda_3}} & \frac{|\lambda_1|}{\lambda_3} Cxz^{\frac{|\lambda_1|}{\lambda_3}-1} + \frac{|\lambda_2|}{\lambda_3} Dz^{\frac{|\lambda_2|}{\lambda_3}-1} \end{pmatrix} \tag{4.8.17}$$

Stability is determined by considering the nature of the eigenvalues of (4.8.17). The eigenvalues of DP are given by

$$\gamma_{1,2} = \frac{\operatorname{tr} DP}{2} \pm \frac{1}{2}\sqrt{(\operatorname{tr} DP)^2 - 4\det(DP)}, \tag{4.8.18}$$

where

$$\det DP = \frac{|\lambda_2|}{\lambda_3}(AD - BC)z^{\frac{|\lambda_1+\lambda_2|-\lambda_3}{\lambda_3}},$$

$$\operatorname{tr} DP = Az^{\frac{|\lambda_1|}{\lambda_3}} + \frac{|\lambda_1|}{\lambda_3} Cxz^{\frac{|\lambda_1|}{\lambda_3}-1} + \frac{|\lambda_2|}{\lambda_3} Dz^{\frac{|\lambda_2|}{\lambda_3}-1}. \tag{4.8.19}$$

Substituting equation (4.8.9) for x at a fixed point into the expression for $\operatorname{tr} DP$ gives

$$\operatorname{tr} DP = Az^{\frac{|\lambda_1|}{\lambda_3}} + \frac{|\lambda_1|}{\lambda_3} CBz^{\frac{|\lambda_1+\lambda_2|}{\lambda_3}-1} + \frac{|\lambda_2|}{\lambda_3} Dz^{\frac{|\lambda_2|}{\lambda_3}-1} + \frac{|\lambda_1|}{\lambda_3} Ce\mu z^{\frac{|\lambda_1|}{\lambda_3}-1}. \tag{4.8.20}$$

Let us note the following important facts.
For z sufficiently small

$$\det DP \text{ is } \begin{cases} \text{a) arbitrarily large for } |\lambda_1 + \lambda_2| < \lambda_3; \\ \text{b) arbitrarily small for } |\lambda_1 + \lambda_2| > \lambda_3. \end{cases}$$

$$\operatorname{tr} DP \text{ is } \begin{cases} \text{a) arbitrarily large for } |\lambda_1| < \lambda_3 \text{ or } |\lambda_2| < \lambda_3; \\ \text{b) arbitrarily small for } |\lambda_1| > \lambda_3 \text{ and } |\lambda_2| > \lambda_3. \end{cases}$$

Using these facts along with (4.8.18) and (4.8.19) we can conclude the following.

1. For $|\lambda_1| > \lambda_3$ and $|\lambda_2| > \lambda_3$, both eigenvalues of DP can be made arbitrarily small by taking z sufficiently small.

2. For $|\lambda_1 + \lambda_2| > \lambda_3$ and $|\lambda_1| < \lambda_3$ and/or $|\lambda_2| < \lambda_3$, one eigenvalue can be made arbitrarily small and the other eigenvalue can be made arbitrarily large by taking z sufficiently small.

3. For $|\lambda_1 + \lambda_2| < \lambda_3$, both eigenvalues can be made arbitrarily large by taking z sufficiently small.

We summarize our results in the following theorem.

Theorem 4.8.1 *For $\mu \neq 0$ and sufficiently small, a periodic orbit bifurcates from Γ in (4.8.1). The periodic orbit is a*

i) *sink for $|\lambda_1| > \lambda_3$ and $|\lambda_2| > \lambda_3$;*

ii) *saddle for $|\lambda_1 + \lambda_2| > \lambda_3$, $|\lambda_1| < \lambda_3$, and/or $|\lambda_2| < \lambda_3$;*

iii) *source for $|\lambda_1 + \lambda_2| < \lambda_3$.*

We remark that the construction of the Poincaré map used in the proof of Theorem 4.8.1 was for the case $\lambda_2 > \lambda_1$ (see Figure 4.8.3); however, the same result holds for $\lambda_2 < \lambda_1$ and $\lambda_1 = \lambda_2$. We leave the details to the reader in Exercises 4.48 and 4.49.

Next we consider the case of two homoclinic orbits connecting the saddle type fixed point to itself and show how under certain conditions chaotic dynamics may arise.

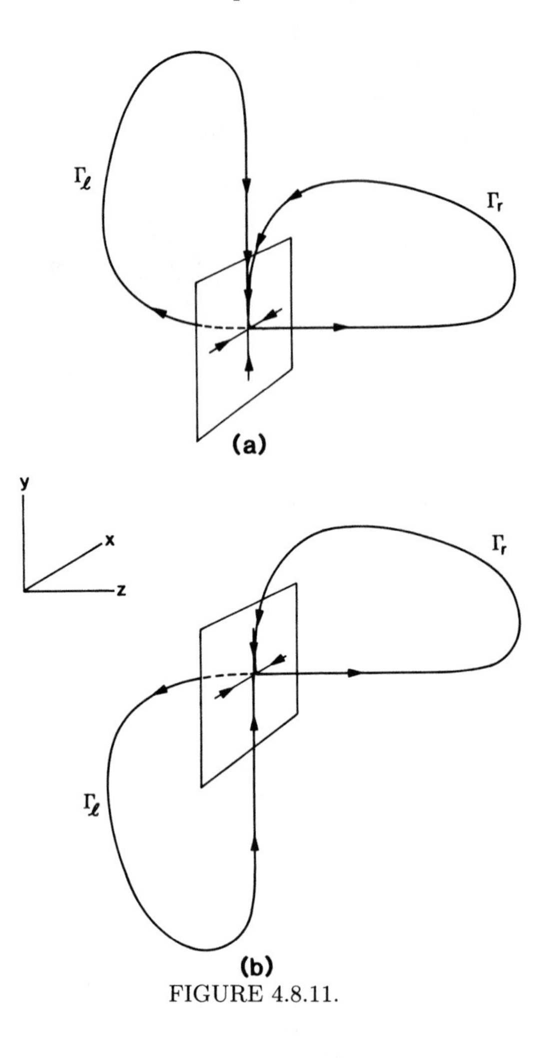

FIGURE 4.8.11.

i) Two Orbits Homoclinic to a Fixed Point Having Real Eigenvalues

We consider the same system as before; however, we now replace Assumption 2 with Assumption 2′ given below.

Assumption 2′. Equation (4.8.1) has a pair of orbits, Γ_r, Γ_ℓ, homoclinic to $(0,0,0)$ at $\mu = 0$, and Γ_r and Γ_ℓ lie in separate branches of the unstable manifold of $(0,0,0)$. There are thus two possible generic pictures illustrated in Figure 4.8.11.

Note that the coordinate axes in Figure 4.8.11 have been rotated with respect to those in Figure 4.8.3. This is merely for artistic convenience. We will consider only the configuration of Case a in Figure 4.8.11; however, the

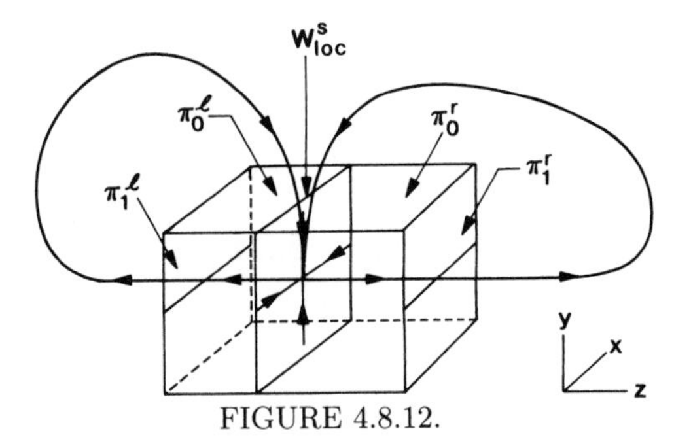

FIGURE 4.8.12.

same analysis (and most of the resulting dynamics) will go through for Case b. Our goal will be to establish that the Poincaré map constructed near the homoclinic orbits contains the chaotic dynamics of the Smale horseshoe or, more specifically, that it contains an invariant Cantor set on which it is homeomorphic to the full shift on two symbols (see Section 4.2).

We begin by constructing the local cross-sections to the vector field near the origin. We define

$$\begin{aligned}
\Pi_0^r &= \{(x,y,z) \in \mathbb{R}^3 \mid y=\varepsilon,\ |x| \le \varepsilon,\ 0 < z \le \varepsilon\},\\
\Pi_0^\ell &= \{(x,y,z) \in \mathbb{R}^3 \mid y=\varepsilon,\ |x| \le \varepsilon,\ -\varepsilon \le z < 0\},\\
\Pi_1^r &= \{(x,y,z) \in \mathbb{R}^3 \mid z=\varepsilon,\ |x| \le \varepsilon,\ 0 < y \le \varepsilon\},\\
\Pi_1^\ell &= \{(x,y,z) \in \mathbb{R}^3 \mid z=-\varepsilon,\ |x| \le \varepsilon,\ 0 < y \le \varepsilon\},
\end{aligned} \tag{4.8.21}$$

for $\varepsilon > 0$ and small; see Figure 4.8.12 for an illustration of the geometry near the origin.

Now recall the global twisting of the stable manifold of the origin. We want to consider the effect of this in our construction of the Poincaré map. Let τ_r (resp. τ_ℓ) be a tube beginning and ending on Π_1^r (resp. Π_1^ℓ) and Π_0^r (resp. Π_0^ℓ) which contains Γ_r (resp. Γ_ℓ) (see Figure 4.8.7). Then $\tau_r \cap W^s(0)$ (resp. $\tau_\ell \cap W^s(0)$) is a two-dimensional strip, which we denote as $\mathcal{R}_r$ (resp. $\mathcal{R}_\ell$). If we join together the two ends of $\mathcal{R}_r$ (resp. $\mathcal{R}_\ell$) *without twisting* $\mathcal{R}_r$ (resp. $\mathcal{R}_\ell$), then $\mathcal{R}_r$ (resp. $\mathcal{R}_\ell$) is homeomorphic to either a cylinder or a mobius strip (see Figure 4.8.7). Thus, this global effect gives rise to three distinct possibilities.

1. $\mathcal{R}_r$ and $\mathcal{R}_\ell$ are homeomorphic to cylinders.

2. $\mathcal{R}_r$ is homeomorphic to a cylinder and $\mathcal{R}_\ell$ is homeomorphic to a mobius strip.

3. $\mathcal{R}_r$ and $\mathcal{R}_\ell$ are homeomorphic to mobius strips.

These three cases manifest themselves in the Poincaré map as shown in Figure 4.8.13.

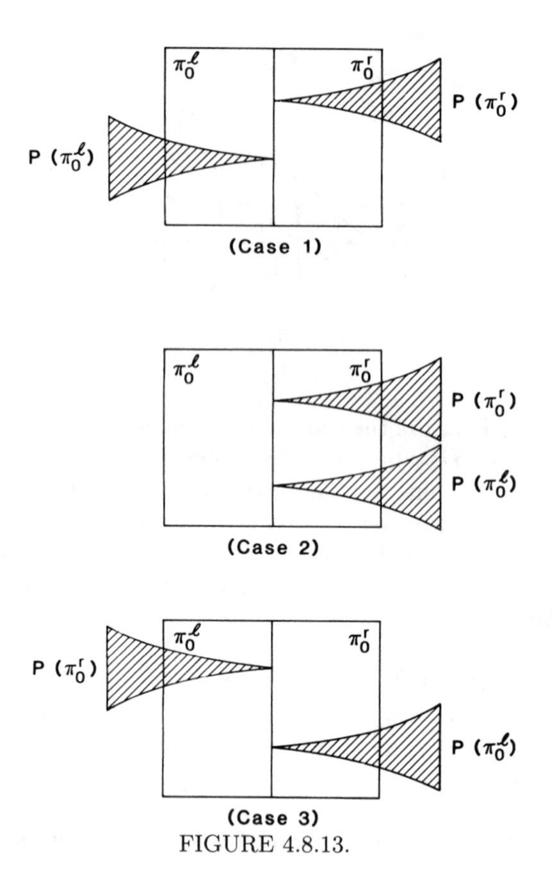

FIGURE 4.8.13.

We now want to motivate how we might expect a horseshoe to arise in these situations. Consider Case 1. Suppose we vary the parameter μ so that the homoclinic orbits break, with the result that the images of Π_0^r and Π_0^ℓ move in the manner shown in Figure 4.8.14. The question of whether or not we would expect such behavior in a one-parameter family of three-dimensional vector fields will be addressed shortly.

From Figure 4.8.14 one can begin to see how we might get horseshoe-like dynamics in this system. We can choose μ_h-horizontal strips in Π_0^r and Π_0^ℓ, which are mapped over themselves in μ_v-vertical strips as μ is varied, as shown in Figure 4.8.15. The conditions on the relative magnitudes of the eigenvalues at the fixed point will insure the appropriate stretching and contracting directions. Note that no horseshoe behavior is possible at $\mu = 0$.

Of course, many things need to be justified in Figure 4.8.15, namely, the stretching and contraction rates and also that the little "half-bowties" behave correctly as the homoclinic orbits are broken. However, rather than go through the three cases individually, we will settle for studying a specific

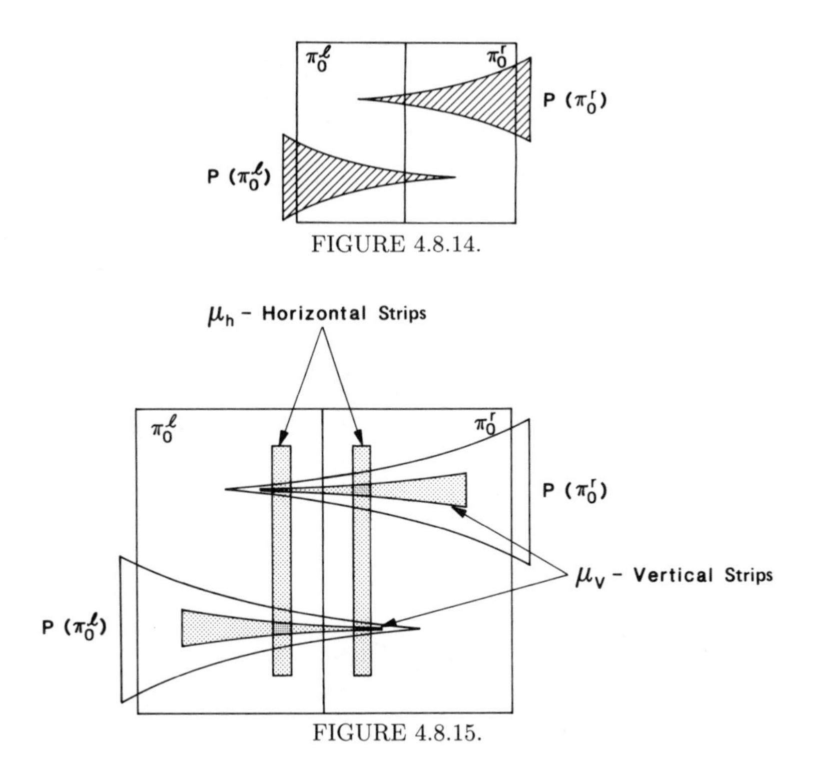

FIGURE 4.8.14.

FIGURE 4.8.15.

example and refer the reader to Afraimovich, Bykov, and Silnikov [1983] for detailed discussions of the general case. However, first we want to discuss the role of parameters.

In a three-dimensional vector field one would expect that varying a parameter would result in the destruction of a particular homoclinic orbit. In the case of two homoclinic orbits we cannot expect that the behavior of both homoclinic orbits can be controlled by a single parameter resulting in the behavior shown in Figure 4.8.13. For this we would need two parameters where each parameter could be thought of as "controlling" a particular homoclinic orbit. In the language of bifurcation theory this is a global codimension-two bifurcation problem. However, if the vector field contains a symmetry, e.g., (4.8.1) is invariant under the change of coordinates $(x, y, z) \to (-x, y, -z)$, which represents a $180°$ rotation about the y axis, then the existence of one homoclinic orbit necessitates the existence of another so that one parameter controls both. For simplicity, we will treat the symmetric case and refer the reader to Afraimovich, Bykov, and Silnikov [1983] for a discussion of the nonsymmetric cases. The symmetric case is of historical interest, since this is precisely the situation that arises in the much-studied Lorenz equations; see Sparrow [1982].

The case we will consider is characterized by the following properties.

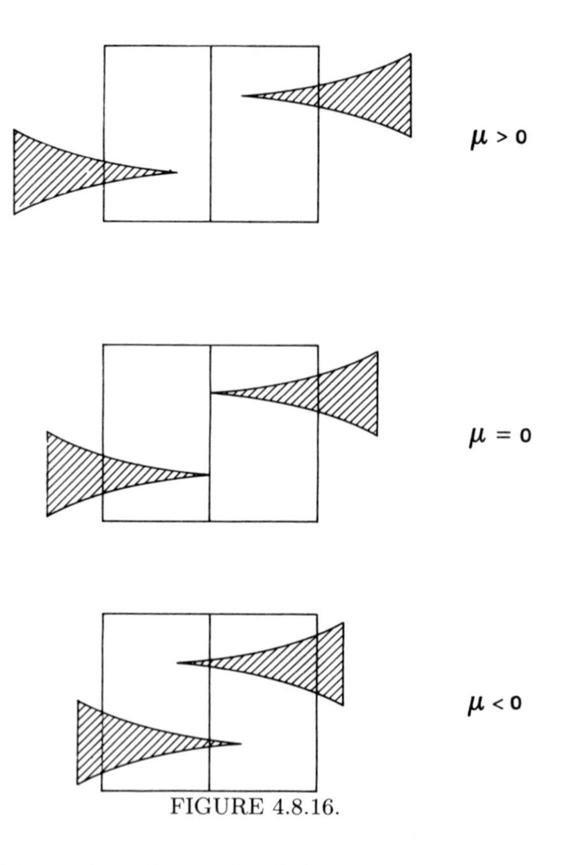

FIGURE 4.8.16.

Assumption 1′. $0 < -\lambda_2 < \lambda_3 < -\lambda_1$, $d \neq 0$.

Assumption 2′. Equation (4.8.1) is invariant under the coordinate transformation $(x, y, z) \to (-x, y, -z)$, and the homoclinic orbits break for μ near zero in the manner shown in Figure 4.8.16.

Assumption A1′ insures that the Poincaré map has a strongly contracting direction and a strongly expanding direction (recall from (4.8.6) that d is an entry in the matrix defining P_1, and $d \neq 0$ is a generic condition). The reader should recall the discussion of Figure 4.8.8, which explains the geometry behind these statements.

Now, the Poincaré map P of $\Pi_0^r \cup \Pi_0^\ell$ into $\Pi_0^r \cup \Pi_0^\ell$ consists of two parts

$$P_r \colon \Pi_0^r \to \Pi_0^r \cup \Pi_0^\ell, \tag{4.8.22}$$

with P_r given by (4.8.7), and

$$P_\ell \colon \Pi_0^\ell \to \Pi_0^r \cup \Pi_0^\ell, \tag{4.8.23}$$

where by the symmetry we have

$$P_\ell(x, z; \mu) = -P_r(-x, -z; \mu). \tag{4.8.24}$$

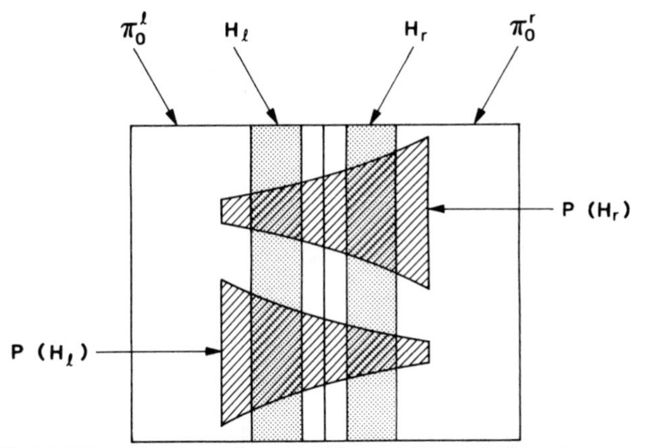

FIGURE 4.8.17. μ_h-horizontal strips H_r and H_ℓ and their image under P.

Our goal is to show that, for $\mu < 0$, P contains an invariant Cantor set on which it is topologically conjugate to the full shift on two symbols. This is done in the following theorem.

Theorem 4.8.2 *There exists $\mu_0 < 0$ such that, for $\mu_0 < \mu < 0$, P possesses an invariant Cantor set on which it is topologically conjugate to the full shift on two symbols.*

Proof: The method behind the proof of this theorem is the same as that used in the proof of Moser's theorem (see Section 4.4). In $\Pi_0^r \cup \Pi_0^\ell$ we locate two disjoint μ_h-horizontal strips that are mapped over themselves in two μ_v-vertical strips so that Assumptions 1 and 3 of Section 4.3 hold.

We choose $\mu < 0$ fixed. Then we choose two μ_h-horizontal strips, one in Π_0^r and one in Π_0^ℓ, where the "horizontal" coordinate is the z axis. We choose the horizontal sides of the strips to be parallel to the x-axis so that $\mu_h = 0$. Then, under the Poincaré map P defined in (4.8.22) and (4.8.23), since $\lambda_3 < -\lambda_1$ and μ is fixed, we can choose the two μ_h-horizontal strips sufficiently close to $W^s(0)$ so that the image of each μ_h-horizontal strip intersects both horizontal boundaries of each of the μ_h-horizontal strips as shown in Figure 4.8.17. Thus, it follows that Assumption 1 holds.

Next we must verify that Assumption 3 holds. This follows from a calculation very similar to that which we performed to show that Assumption 3 holds in the proof of Moser's theorem in Section 4.4. It uses the fact that $-\lambda_2 < \lambda_3 < -\lambda_1$ and $d \neq 0$. We leave the details as an exercise for the reader. □

We left out many of the details of the proof of Theorem 4.8.2. In Exercise 4.52 we outline how one would complete the missing details.

The dynamical consequences of Theorem 4.8.2 are stunning. For $\mu \geq 0$, there is nothing spectacular associated with the dynamics near the (broken) homoclinic orbits. However, for $\mu < 0$, the horseshoes and their attendant chaotic dynamics appear seemingly out of nowhere. This particular type of global bifurcation has been called a *homoclinic explosion.*

Observations and Additional References

We have barely scratched the surface of the possible dynamics associated with orbits homoclinic to a fixed point having real eigenvalues in a third-order ordinary differential equation. There are several issues which deserve a more thorough investigation.

1. *Two Homoclinic Orbits without Symmetry.* See Afraimovich, Bykov, and Silnikov [1983] for the references therein.

2. *The Existence of Strange Attractors.* Horseshoes are chaotic invariant sets, yet all the orbits in the horseshoes are unstable of saddle type. Nevertheless, it should be clear that horseshoes may exhibit a striking effect on the dynamics of any system. In particular, they are often the chaotic heart of numerically observed strange attractors. For work on the "strange attractor problem" associated with orbits homoclinic to fixed points having real eigenvalues in a third-order ordinary differential equation, see Afraimovich, Bykov, and Silnikov [1983]. Most of the work done on such systems has been in the context of the Lorenz equations. References for Lorenz attractors include Sparrow [1982], Guckenheimer and Williams [1980], and Williams [1980]. Recently, some breakthroughs have been made in proving the existence of strange attractors in such equations by Rychlik [1987] and Robinson [1988].

3. *Bifurcations Creating the Horseshoe.* In the homoclinic explosion an infinite number of periodic orbits of all possible periods are created. The question arises concerning precisely how these periodic orbits were created and how they are related to each other. This question also has relevance to the strange attractor problem.

   In recent years Birman, Williams, and Holmes have been using the *knot type* of a periodic orbit as a bifurcation invariant in order to understand the appearance, disappearance, and interrelation of periodic orbits in third-order ordinary differential equations. Roughly speaking, a periodic orbit in three dimensions can be thought of as a knotted closed loop. As system parameters are varied, the periodic orbit may never intersect itself due to uniqueness of solutions. Hence, the knot type of a periodic orbit cannot change as parameters are varied. The knot type is therefore a bifurcation invariant as well as a key tool for developing a classification scheme for periodic orbits. For references, see Birman and Williams [1983a,b], Holmes [1986a], [1987], and Holmes and Williams [1985].

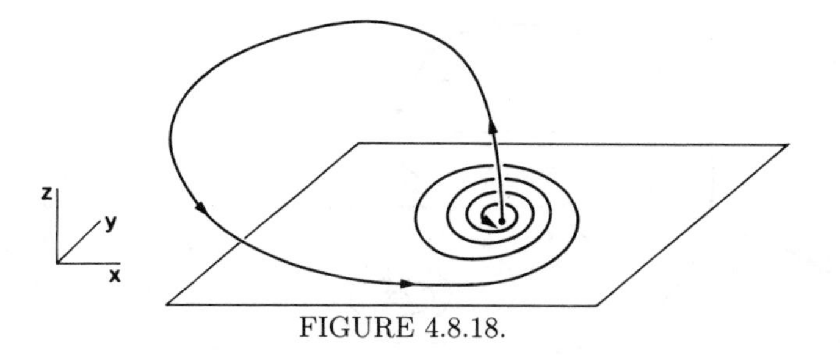

FIGURE 4.8.18.

4.8B Orbits Homoclinic to a Saddle-Focus

We now consider the dynamics near an orbit homoclinic to a fixed point of saddle-focus type in a third-order ordinary differential equation. This has become known as the *Silnikov phenomenon,* since it was first studied by Silnikov [1965].

We consider an equation of the following form

$$\begin{aligned} \dot{x} &= \rho x - \omega y + P(x,y,z), \\ \dot{y} &= \omega x + \rho y + Q(x,y,z), \\ \dot{z} &= \lambda z + R(x,y,z), \end{aligned} \tag{4.8.25}$$

where P, Q, R are $\mathbf{C}^2$ and $\mathcal{O}(2)$ at the origin. It should be clear that $(0,0,0)$ is a fixed point and that the eigenvalues of (4.8.25) linearized about $(0,0,0)$ are given by $\rho \pm i\omega$, λ (note that there are no parameters in this problem at the moment; we will consider bifurcations of (4.8.25) later). We make the following assumptions on the system (4.8.25).

Assumption 1. Equation (4.8.25) possesses a homoclinic orbit Γ connecting $(0,0,0)$ to itself.

Assumption 2. $\lambda > -\rho > 0$.

Thus, $(0,0,0)$ possesses a two-dimensional stable manifold and a one-dimensional unstable manifold which intersect nontransversely; see Figure 4.8.18.

In order to determine the nature of the orbit structure near Γ, we construct a Poincaré map defined near Γ in the manner described at the beginning of this section.

Computation of P_0

Let Π_0 be a rectangle lying in the $x-z$ plane, and let Π_1 be a rectangle parallel to the $x-y$ plane at $z=\varepsilon$; see Figure 4.8.19. As opposed to the case of purely real eigenvalues, Π_0 will require a more detailed description. However, in order to do this we need to better understand the dynamics of the flow near the origin.

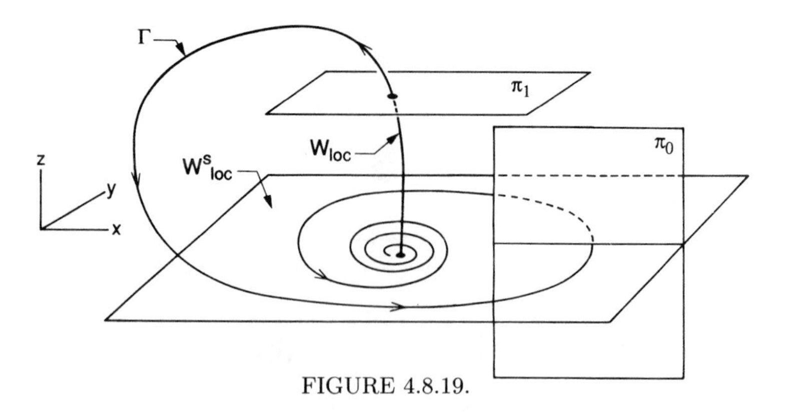

FIGURE 4.8.19.

The flow generated by (4.8.25) linearized about the origin is given by

$$\begin{aligned} x(t) &= e^{\rho t}(x_0 \cos \omega t - y_0 \sin \omega t), \\ y(t) &= e^{\rho t}(x_0 \sin \omega t + y_0 \cos \omega t), \\ z(t) &= z_0 e^{\lambda t}. \end{aligned} \tag{4.8.26}$$

The time of flight for points starting on Π_0 to reach Π_1 is found by solving

$$\varepsilon = z_0 e^{\lambda t} \tag{4.8.27}$$

or

$$t = \frac{1}{\lambda} \log \frac{\varepsilon}{z_0}. \tag{4.8.28}$$

Thus, P_0 is given by (omitting the subscript 0's)

$$P_0 \colon \Pi_0 \to \Pi_1,$$

$$\begin{pmatrix} x \\ 0 \\ z \end{pmatrix} \mapsto \begin{pmatrix} x\left(\frac{\varepsilon}{z}\right)^{\rho/\lambda} \cos\left(\frac{\omega}{\lambda} \log \frac{\varepsilon}{z}\right) \\ x\left(\frac{\varepsilon}{z}\right)^{\rho/\lambda} \sin\left(\frac{\omega}{\lambda} \log \frac{\varepsilon}{z}\right) \\ \varepsilon \end{pmatrix}. \tag{4.8.29}$$

We now consider Π_0 more carefully. For Π_0 arbitrarily chosen it is possible for points on Π_0 to intersect Π_0 many times before reaching Π_1. In this case, P_0 would not map Π_0 diffeomorphically onto $P_0(\Pi_0)$. We want to avoid this situation, since the conditions for a map to possess the dynamics of the shift map described in Section 4.3 are given for diffeomorphisms. According to (4.8.26), it takes time $t = 2\pi/\omega$ for a point starting in the $x - z$ plane with $x > 0$ to return to the $x - z$ plane with $x > 0$. Now let $x = \varepsilon$, $0 < z \leq \varepsilon$ be the right-hand boundary of Π_0. Then if we choose $x = \varepsilon e^{2\pi\rho/\omega}$, $0 < z \leq \varepsilon$ to be the left-hand boundary of Π_0, no point starting in the interior of Π_0 returns to Π_0 before reaching Π_1. We take this as the definition of Π_0:

$$\Pi_0 = \left\{(x, y, z) \in \mathbb{R}^3 \,|\, y = 0,\ \varepsilon e^{2\pi\rho/\omega} \leq x \leq \varepsilon,\ 0 < z \leq \varepsilon\right\}. \tag{4.8.30}$$

Π_1 is chosen large enough to contain $P_0(\Pi_0)$ in its interior.

We now want to describe the geometry of $P_0(\Pi_0)$. Π_1 is coordinatized by x and y, which we will label as x', y' to avoid confusion with the coordinates of Π_0. Then, from (4.8.29), we have

$$(x', y') = \left(x \left(\frac{\varepsilon}{z}\right)^{\rho/\lambda} \cos\left(\frac{\omega}{\lambda} \log \frac{\varepsilon}{z}\right), x \left(\frac{\varepsilon}{z}\right)^{\rho/\lambda} \sin\left(\frac{\omega}{\lambda} \log \frac{\varepsilon}{z}\right)\right). \quad (4.8.31)$$

Polar coordinates on Π_1 give a clearer picture of the geometry. Let

$$r = \sqrt{x'^2 + y'^2}, \qquad \frac{y'}{x'} = \tan\theta.$$

Then (4.8.31) becomes

$$(r, \theta) = \left(x \left(\frac{\varepsilon}{z}\right)^{\rho/\lambda}, \frac{\omega}{\lambda} \log \frac{\varepsilon}{z}\right). \quad (4.8.32)$$

Now consider a vertical line in Π_0, i.e., a line with $x =$ constant. By (4.8.32) it gets mapped into a logarithmic spiral. A horizontal line in Π_0, i.e., a line with $z =$ constant, gets mapped onto a radial line emanating from $(0, 0, \varepsilon)$. Consider the rectangles

$$R_k = \left\{(x, y, z) \in \mathbb{R}^3 \mid y = 0, \ \varepsilon e^{\frac{2\pi\rho}{\omega}} \le x \le \varepsilon, \ \varepsilon e^{\frac{-2\pi(k+1)\lambda}{\omega}} \le z \le \varepsilon e^{\frac{-2\pi k\lambda}{\omega}}\right\}. \quad (4.8.33)$$

Then we have

$$\Pi_0 = \bigcup_{k=0}^{\infty} R_k.$$

We study the geometry of the image of a rectangle R_k by determining the behavior of its horizontal and vertical boundaries under P_0. We denote these four line segments as

$$\begin{aligned}
h^u &= \left\{(x, y, z) \in \mathbb{R}^3 \mid y = 0, \ z = \varepsilon e^{\frac{-2\pi k\lambda}{\omega}}, \varepsilon e^{\frac{2\pi\rho}{\omega}} \le x \le \varepsilon\right\},\\
h^\ell &= \left\{(x, y, z) \in \mathbb{R}^3 \mid y = 0, \ z = \varepsilon e^{\frac{-2\pi(k+1)\lambda}{\omega}}, \varepsilon e^{\frac{2\pi\rho}{\omega}} \le x \le \varepsilon\right\},\\
v^r &= \left\{(x, y, z) \in \mathbb{R}^3 \mid y = 0, \ x = \varepsilon, \varepsilon e^{\frac{-2\pi(k+1)\lambda}{\omega}} \le z \le \varepsilon e^{\frac{-2\pi k\lambda}{\omega}}\right\},\\
v^\ell &= \left\{(x, y, z) \in \mathbb{R}^3 \mid y = 0, \ x = \varepsilon e^{\frac{2\pi\rho}{\omega}}, \varepsilon e^{\frac{-2\pi(k+1)\lambda}{\omega}} \le z \le \varepsilon e^{\frac{-2\pi k\lambda}{\omega}}\right\};
\end{aligned} \quad (4.8.34)$$

see Figure 4.8.20. The images of these line segments under P_0 are given by

$$\begin{aligned}
P_0(h^u) &= \left\{(r, \theta, z) \in \mathbb{R}^3 \mid z = \varepsilon, \theta = 2\pi k, \varepsilon e^{\frac{2\pi(k+1)\rho}{\omega}} \le r \le \varepsilon e^{\frac{2\pi k\rho}{\omega}}\right\},\\
P_0(h^\ell) &= \left\{(r, \theta, z) \in \mathbb{R}^3 \mid z = \varepsilon, \theta = 2\pi(k+1), \varepsilon e^{\frac{2\pi(k+2)\rho}{\omega}} \le r \le \varepsilon e^{\frac{2\pi(k+1)\rho}{\omega}}\right\},\\
P_0(v^r) &= \left\{(r, \theta, z) \in \mathbb{R}^3 \mid z = \varepsilon, 2\pi k \le \theta \le 2\pi(k+1), r(\theta) = \varepsilon e^{\frac{\rho\theta}{\omega}}\right\},\\
P_0(v^\ell) &= \left\{(r, \theta, z) \in \mathbb{R}^3 \mid z = \varepsilon, 2\pi k \le \theta \le 2\pi(k+1), r(\theta) = \varepsilon e^{\frac{\rho(2\pi+\theta)}{\omega}}\right\},
\end{aligned} \quad (4.8.35)$$

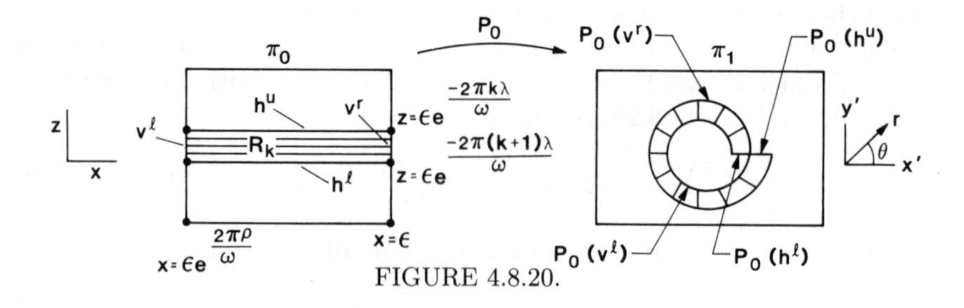

FIGURE 4.8.20.

so that $P_0(R_k)$ appears as in Figure (4.8.20).

The geometry of Figure 4.8.20 should give a strong indication that horseshoes may arise in this system.

Computation of P_1

From the discussion at the beginning of this section, we approximate P_1 by an affine map as follows

$$P_1 \colon \Pi_1 \to \Pi_0,$$

$$\begin{pmatrix} x \\ y \\ \varepsilon \end{pmatrix} \mapsto \begin{pmatrix} a & b & 0 \\ 0 & 0 & 0 \\ c & d & 0 \end{pmatrix} \begin{pmatrix} x \\ y \\ 0 \end{pmatrix} + \begin{pmatrix} \overline{x} \\ 0 \\ 0 \end{pmatrix}, \tag{4.8.36}$$

where $(\overline{x}, 0, 0) \equiv \Gamma \cap \Pi_0$ is the intersection of the homoclinic orbit with Π_0 (note: by our choice of Π_0, Γ intersects Γ_0 only once). We remark that the structure of the 3×3 matrix in (4.8.36) comes from the fact that the coordinates of Π_1 are x and y with $z = \varepsilon =$ constant and the coordinates of Π_0 are x and z with $y = 0$.

The Poincaré Map $P \equiv P_1 \circ P_0$

Composing (4.8.29) and (4.8.36) gives

$$P \equiv P_1 \circ P_0 \colon \Pi_0 \to \Pi_0,$$

$$\begin{pmatrix} x \\ z \end{pmatrix} \mapsto \begin{pmatrix} x \left(\frac{\varepsilon}{z}\right)^{\rho/\lambda} \left[a \cos\left(\frac{\omega}{\lambda} \log \frac{\varepsilon}{z}\right) + b \sin\left(\frac{\omega}{\lambda} \log \frac{\varepsilon}{z}\right)\right] + \overline{x} \\ \\ x \left(\frac{\varepsilon}{z}\right)^{\rho/\lambda} \left[c \cos\left(\frac{\omega}{\lambda} \log \frac{\varepsilon}{z}\right) + d \sin\left(\frac{\omega}{\lambda} \log \frac{\varepsilon}{z}\right)\right] \end{pmatrix}, \tag{4.8.37}$$

where Π_0 is chosen sufficiently small (by taking ε small). Thus, $P(\Pi_0)$ appears as in Figure 4.8.21.

Our goal is to show that P contains an invariant Cantor set on which it is topologically conjugate to a full shift on (at least) two symbols.

Consider the rectangle R_k shown in Figure 4.8.22. In order to verify the proper behavior of horizontal and vertical strips in R_k, it will be necessary

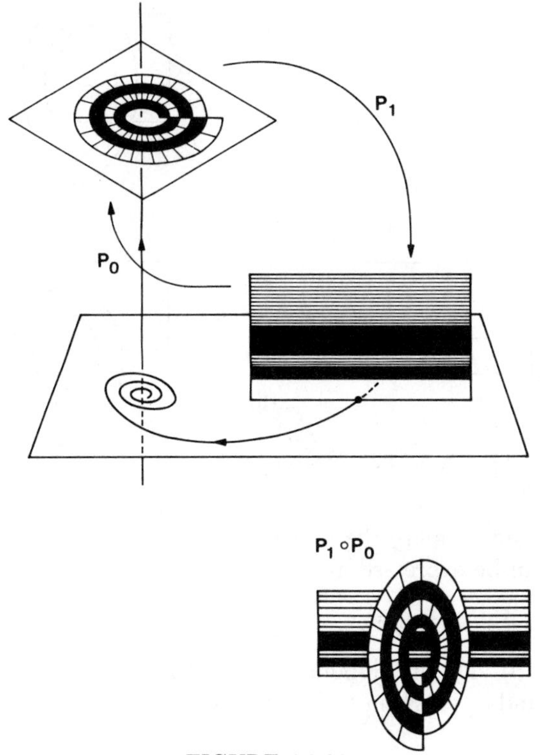

FIGURE 4.8.21.

to verify that the inner and outer boundaries of $P(R_k)$ both intersect the upper boundary of R_k as shown in Figure 4.8.22 or, in other words, the upper horizontal boundary of R_k intersects (at least) two points of the inner boundary of $P(R_k)$. Additionally, it will be useful to know how many rectangles above R_k $P(R_k)$ also intersects in this manner. We have the following lemma.

Lemma 4.8.3 *Consider R_k for fixed k sufficiently large. Then the inner boundary of $P(R_k)$ intersects the upper horizontal boundary of R_i in (at least) two points for $i \geq k/\alpha$ where $1 \leq \alpha < -\lambda/\rho$. Moreover, the preimage of the vertical boundaries of $P(R_k) \cap R_i$ is contained in the vertical boundary of R_k.*

Proof: The z coordinate of the upper horizontal boundary of R_i is given by

$$\overline{z} = \varepsilon e^{\frac{-2\pi i \lambda}{\omega}}, \tag{4.8.38}$$

and the point on the inner boundary of $P_0(R_k)$ closest to $(0, 0, \varepsilon)$ is given by

$$r_{\min} = \varepsilon e^{\frac{4\pi\rho}{\omega}} e^{\frac{2\pi k\rho}{\omega}}. \tag{4.8.39}$$

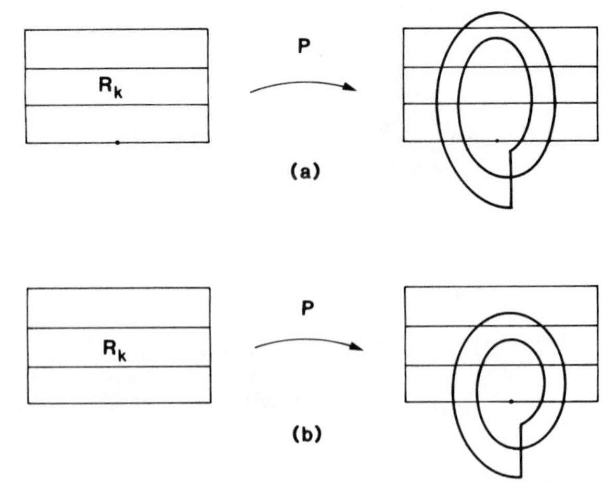

FIGURE 4.8.22. Two possibilities for $P(R_k) \cap R_k$.

Since P_1 is an affine map, the bound on the inner boundary of $P(R_k) = P_1 \circ P_0(R_k)$ can be expressed as

$$\overline{r}_{\min} = K\varepsilon e^{\frac{4\pi\rho}{\omega}} e^{\frac{2\pi k\rho}{\omega}} \tag{4.8.40}$$

for some $K > 0$. The inner boundary of $P(R_k)$ will intersect the upper horizontal boundary of R_i in (at least) two points provided

$$\frac{\overline{r}_{\min}}{\overline{z}} > 1. \tag{4.8.41}$$

Using (4.8.38) and (4.8.40), we compute this ratio explicitly and find

$$\frac{\overline{r}_{\min}}{\overline{z}} = Ke^{\frac{4\pi\rho}{\omega}} e^{\frac{2\pi}{\omega}(k\rho+i\lambda)}. \tag{4.8.42}$$

Because $Ke^{4\pi\rho/\omega}$ is a fixed constant, the size of (4.8.42) is controlled by the $e^{(2\pi/\omega)(k\rho+i\lambda)}$ term. In order to make (4.8.42) larger than one, it is sufficient that $k\rho+i\lambda$ is taken sufficiently large. By Assumption 2 we have $\lambda+\rho > 0$, so for $i \geq k/\alpha$, $1 \leq \alpha < -\lambda/\rho$, $k\rho + i\lambda$ is positive, and for k sufficiently large, (4.8.42) is larger than one.

We now describe the behavior of the vertical boundaries of R_k. Recall Figure 4.8.22a. Under P_0 the vertical boundaries of R_k map to the inner and outer boundaries of an annulus-like object. P_1 is an invertible affine map; hence, the inner and outer boundaries of $P_0(R_k)$ correspond to the inner and outer boundaries of $P(R_k) = P_1 \circ P_0(R_k)$. Therefore, the preimage of the vertical boundary of $P(R_k) \cap R_i$ is contained in the vertical boundary of R_k. □

Lemma 4.8.3 points out the necessity of Assumption 2, since, if we had instead $-\rho > \lambda > 0$, then the image of R_k would fall below R_k for k sufficiently large, as shown in Figure 4.8.22b.

We now can state our main theorem.

Theorem 4.8.4 *For k sufficiently large, R_k contains an invariant Cantor set, Λ_k, on which the Poincaré map P is topologically conjugate to a full shift on two symbols.*

Proof: The proof is very similar to the proof of both Moser's theorem (see Section 4.4) and Theorem 4.8.2. In R_k we must find two disjoint μ_h-horizontal strips that are mapped over themselves in μ_v-vertical strips on which Assumptions 1 and 3 of Section 4.3 hold.

The fact that such μ_h-horizontal strips can be found on which Assumption 1 holds follows from Lemma 4.8.3. The fact that Assumption 3 holds follows from a calculation similar to that given in Moser's theorem and Theorem 4.8.2. In Exercise 4.54 we outline how one fills in the details of this proof. □

We make several remarks.

Remark 1. The dynamics of P are often described by the phrase "P has a countable infinity of horseshoes."

Remark 2. Note from Lemma 4.8.3 that the horseshoes in the different R_k can interact. This would lead to different ways of setting up the symbolic dynamics. We will deal with this issue in Exercise 4.55.

Remark 3. If one were to break the homoclinic orbit with a perturbation, then only a finite number of the Λ_k would survive. We deal with this issue in Exercise 4.56.

i) The Bifurcation Analysis of Glendinning and Sparrow

Now that we have seen how complicated the orbit structure is in the neighborhood of an orbit homoclinic to a fixed point of saddle-focus type, we want to get an understanding of how this situation occurs as the homoclinic orbit is created. In this regard, the analysis given by Glendinning and Sparrow [1984] is insightful.

Suppose that the homoclinic orbit in (4.8.25) depends on a scalar parameter μ in the manner shown in Figure 4.8.23. We construct a parameter-dependent Poincaré map in the same manner as when we discussed the case of a fixed point with all real eigenvalues. This map is given by

$$\begin{pmatrix} x \\ z \end{pmatrix} \mapsto \begin{pmatrix} x\left(\frac{\varepsilon}{z}\right)^{\rho/\lambda}\left[a\cos\frac{\omega}{\lambda}\log\frac{\varepsilon}{z} + b\sin\frac{\omega}{\lambda}\log\frac{\varepsilon}{z}\right] + e\mu + \overline{x} \\ \\ x\left(\frac{\varepsilon}{z}\right)^{\rho/\lambda}\left[c\cos\frac{\omega}{\lambda}\log\frac{\varepsilon}{z} + d\sin\frac{\omega}{\lambda}\log\frac{\varepsilon}{z}\right] + f\mu \end{pmatrix}, \tag{4.8.43}$$

where, from Figure 4.8.23, we have $f > 0$. We have already seen that this map possesses a countable infinity of horseshoes at $\mu = 0$, and we

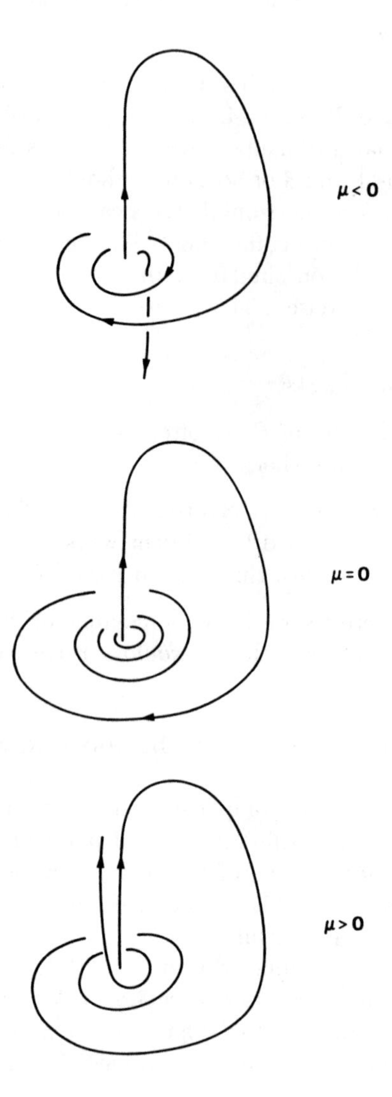

FIGURE 4.8.23.

know that each horseshoe contains periodic orbits of all periods. To study how the horseshoes are formed in this situation as the homoclinic orbit is formed is a difficult (and unsolved) problem. We will tackle a more modest problem which will still give us a good idea about some things that are happening; namely, we will study the fixed points of the above map. Recall that the fixed points correspond to periodic orbits which pass through a neighborhood of the origin *once* before closing up. First we put the map in a form which will be easier to work with. The map can be written in the form

$$\begin{pmatrix} x \\ z \end{pmatrix} \mapsto \begin{pmatrix} x\left(\frac{\varepsilon}{z}\right)^{\rho/\lambda} p\cos\left(\frac{\omega}{\lambda}\log\frac{\varepsilon}{z}+\phi_1\right)+e\mu+\overline{x} \\ x\left(\frac{\varepsilon}{z}\right)^{\rho/\lambda} q\cos\left(\frac{\omega}{\lambda}\log\frac{\varepsilon}{z}+\phi_2\right)+\mu \end{pmatrix}, \tag{4.8.44}$$

where we have rescaled μ so that $f=1$ (note that f must be positive).

Now let

$$-\delta=\frac{\rho}{\lambda},\quad \alpha=p\varepsilon^{-\delta},\quad \beta=q\varepsilon^{-\delta},$$

$$\xi=-\frac{\omega}{\lambda},\quad \Phi_1=\frac{\omega}{\lambda}\log\varepsilon+\phi_1,\quad \Phi_2=\frac{\omega}{\lambda}\log\varepsilon+\phi_2.$$

Then the map takes the form

$$\begin{pmatrix} x \\ z \end{pmatrix} \mapsto \begin{pmatrix} \alpha x z^{\delta}\cos(\xi\log z+\Phi_1)+e\mu+\overline{x} \\ \beta x z^{\delta}\cos(\xi\log z+\Phi_2)+\mu \end{pmatrix}. \tag{4.8.45}$$

Now we will study the fixed points of this map and their stability and bifurcations.

Fixed Points

The fixed points are found by solving

$$x=\alpha x z^{\delta}\cos(\xi\log z+\Phi_1)+e\mu+\overline{x}, \tag{4.8.46a}$$
$$z=\beta x z^{\delta}\cos(\xi\log z+\Phi_2)+\mu. \tag{4.8.46b}$$

Solving (4.8.46a) for x as a function of z gives

$$x=\frac{e\mu+\overline{x}}{1-\alpha z^{\delta}\cos(\xi\log z+\Phi_1)}. \tag{4.8.47}$$

Substituting (4.8.47) into (4.8.46b) gives

$$(z-\mu)(1-\alpha z^{\delta}\cos(\xi\log z+\Phi_1))=(e\mu+\overline{x})\beta z^{\delta}\cos(\xi\log z+\Phi_2). \tag{4.8.48}$$

Solving (4.8.48) gives us the z-component of the fixed point; substituting this into (4.8.46a) gives us the x-component of the fixed point. In order to

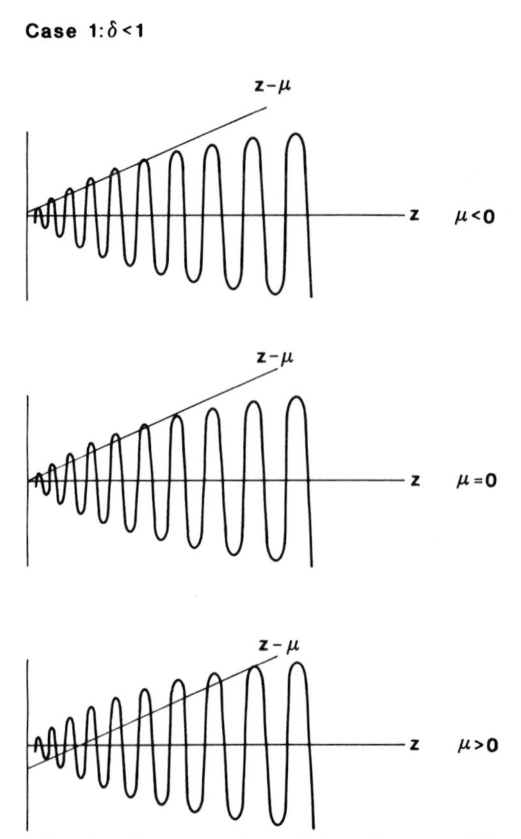

FIGURE 4.8.24. Graphical construction of the solutions of (4.8.50) for $\delta < 1$.

obtain an idea about the solutions of (4.8.48) we will assume that z is so small that

$$1 - \alpha z^{\delta} \cos(\xi \log z + \Phi_1) \sim 1; \tag{4.8.49}$$

then the equation of the z component of the fixed point will be

$$(z - \mu) = (e\mu + \overline{x})\beta z^{\delta} \cos(\xi \log z + \Phi_2). \tag{4.8.50}$$

We solve (4.8.50) graphically by drawing the graph of both the right- and left-hand sides of (4.8.50) and looking for points of intersection. There are various cases shown in Figure 4.8.24.

$\underline{\delta < 1}$. In the the case $\delta < 1$, we have

$\mu < 0$ finite number of fixed points;

$\mu = 0$ countable infinity of fixed points;

$\mu > 0$ finite number of fixed points.

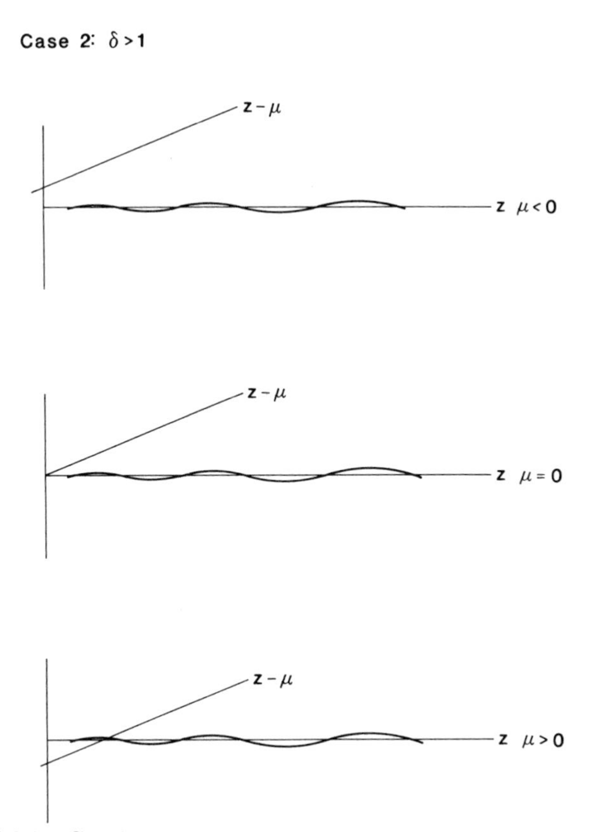

FIGURE 4.8.25. Graphical construction of the solutions of (4.8.50) for $\delta > 1$.

$\underline{\delta > 1}$. The next case is $\delta > 1$, i.e., Assumption 2 does not hold. We show the results in Figure 4.8.25. In the case $\delta > 1$, we have

$\mu \leq 0$ there are no fixed points except the one at $z = \mu = 0$ (i.e., the homoclinic orbit).

$\mu > 0$ For $z > 0$, there is one fixed point for each μ. This can be seen as follows: the slope of the wiggly curve is of order $z^{\delta-1}$, which is small for z small, since $\delta > 1$. Thus, the $z - \mu$ line intersects it only once.

Again, the fixed points which we have found correspond to periodic orbits of the parametrized version of (4.8.25) which pass *once* through a neighborhood of zero before closing up. Our knowledge of these fixed points allows us to draw the following bifurcation diagrams in Figure 4.8.26.

The $\delta > 1$ diagram should be clear; however, the $\delta < 1$ diagram may be confusing. The wiggly curve in the diagram above represents periodic orbits. It should be clear from Figure 4.8.26 that periodic orbits are born in pairs, and the one with the lower z value has the higher period (since it

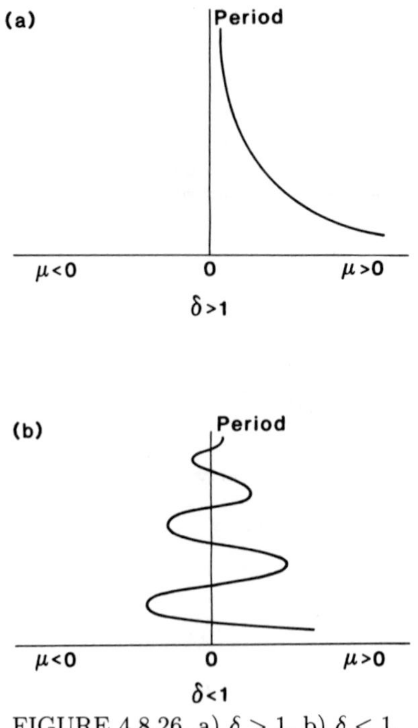

FIGURE 4.8.26. a) $\delta > 1$. b) $\delta < 1$.

passes closer to the fixed point). We will worry more about the structure of this curve as we proceed.

Stability of the Fixed Points

The Jacobian of the map is given by

$$\begin{pmatrix} A & C \\ D & B \end{pmatrix},$$

where

$$\begin{aligned} A &= \alpha z^{\delta} \cos(\xi \log z + \Phi_1), \\ B &= \beta x z^{\delta-1}[\delta \cos(\xi \log z + \Phi_2) - \xi \sin(\xi \log z + \Phi_2)], \\ C &= \alpha x z^{\delta-1}[\delta \cos(\xi \log z + \Phi_1) - \xi \sin(\xi \log z + \Phi_1)], \\ D &= \beta z^{\delta} \cos(\xi \log z + \Phi_2). \end{aligned} \tag{4.8.51}$$

The eigenvalues of the matrix are given by

$$\lambda_{1,2} = \frac{1}{2}\left\{(A+B) \pm \sqrt{(A+B)^2 - 4(AB - CD)}\right\}. \tag{4.8.52}$$

$\underline{\delta > 1}$. For $\delta > 1$, it should be clear that the eigenvalues will be small if z is small (since both z^{δ} and $z^{\delta-1}$ are small). Hence, for $\delta > 1$, the one periodic orbit existing for $\mu > 0$ is stable for μ small, and the homoclinic orbit at $\mu = 0$ is an attractor.

The case $\delta < 1$ is more complicated.

<u>$\delta < 1$</u>. First notice that the determinant of the matrix given by $AB - CD$ only contains terms of order $z^{2\delta-1}$, so the map will be

area-contracting	$\frac{1}{2} < \delta < 1,$
area-expanding	$0 < \delta < \frac{1}{2},$

for z sufficiently small.

We would thus expect different results in these two different δ ranges.

Now recall that the wiggly curve whose intersection with $z - \mu$ gave the fixed points was given by

$$(e\mu + \overline{x})\beta z^{\delta} \cos(\xi \log z + \Phi_2).$$

Thus, from (4.8.51), a fixed point corresponding to a maximum of this curve corresponds to $B = 0$, and a fixed point corresponding to a zero crossing of this curve corresponds to $D = 0$. We now want to look at the stability of fixed points satisfying these conditions.

<u>$D = 0$</u>. In this case $\lambda_1 = A$, $\lambda_2 = B$. Thus, for z small, λ_1 is small and λ_2 is always large; hence, the fixed point is a saddle. Note in particular that, for $\mu = 0$, D is very close to zero; hence all periodic orbits will be saddles as expected.

<u>$B = 0$</u>. The eigenvalues are given by

$$\lambda_{1,2} = \frac{1}{2}\left[A \pm \sqrt{A^2 + 4CD}\right],$$

and both eigenvalues will have large or small modulus depending on whether CD is large or small, since

$$A^2 \sim z^{2\delta} \text{ can be neglected compared with } CD \sim z^{2\delta-1},$$
$$A \sim z^{\delta} \text{ can be neglected compared with } \sqrt{CD} \sim z^{\delta-(1/2)}.$$

Whether or not CD is small depends on whether $0 < \delta < \frac{1}{2}$ or $\frac{1}{2} < \delta < 1$. Hence, we have

stable fixed points for	$\frac{1}{2} < \delta < 1.$
unstable fixed points for	$0 < \delta < \frac{1}{2}.$

Now we want to put everything together for other z values (i.e., for z such that $B, D \neq 0$).

Consider Figure 4.8.27, which is a blow-up of Figure 4.8.24 for various parameter values. In this figure the intersection of the two curves gives us the z coordinate of the fixed points.

Now we describe what happens at each parameter value shown in Figure 4.8.27.

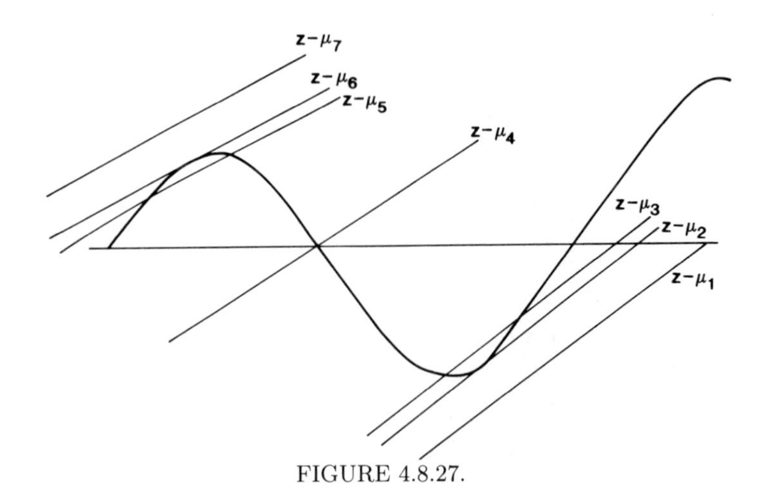

FIGURE 4.8.27.

$\mu = \mu_6$ At this point we have a tangency, and we know that a saddle-node pair will be born in a saddle-node bifurcation.

$\mu = \mu_5$ At this point we have two fixed points; the one with the lower z value has the larger period. Also, the one at the maximum of the curve has $B = 0$; therefore, it is stable for $\mu > \frac{1}{2}$, unstable for $\delta < \frac{1}{2}$. The other fixed point is a saddle.

$\mu = \mu_4$ At this point the stable (unstable) fixed point has become a saddle since $D = 0$. Therefore, it must have changed its stability type via a period-doubling bifurcation.

$\mu = \mu_3$ At this point $B = 0$ again; therefore, the saddle has become either purely stable or unstable again. This must have occurred via a reverse period-doubling bifurcation.

$\mu = \mu_2$ A saddle-node bifurcation occurs.

Hence, we finally arrive at Figure 4.8.28.

Next we want to get an idea of the size of the "wiggles" in Figure 4.8.28 because, if the wiggles are small, that implies that the one-loop periodic orbits are only visible for a narrow range of parameters. If the wiggles are large, we might expect there to be a greater likelihood of observing the periodic orbits.

Let us denote the parameter values at which the tangent to the curve in Figure 4.8.28 is vertical by

$$\mu_i, \mu_{i+1}, \cdots, \mu_{i+n}, \cdots \to 0, \tag{4.8.53}$$

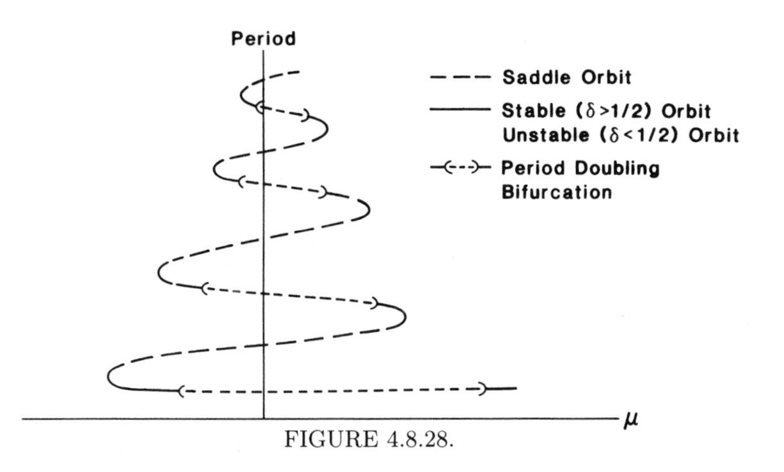

FIGURE 4.8.28.

where the μ_i alternate in sign. Now recall that the z component of the fixed point was given by the solutions to the equations

$$z - \mu = (e\mu + \overline{x})\beta z^\delta \cos(\xi \log z + \Phi_2). \tag{4.8.54}$$

Thus, we have

$$z_i - \mu_i = (e\mu_i + \overline{x})\beta z_i^\delta \cos(\xi \log z_i + \Phi_2), \tag{4.8.55}$$

$$z_{i+1} - \mu_{i+1} = (e\mu_{i+1} + \overline{x})\beta z_{i+1}^\delta \cos(\xi \log z_{i+1} + \Phi_2). \tag{4.8.56}$$

From (4.8.55) and (4.8.56), we obtain

$$\mu_i = \frac{z_i - \overline{x}\beta z_i^\delta \cos(\xi \log z_i + \Phi_2)}{1 + e\beta z_i^\delta \cos(\xi \log z_i + \Phi_2)}, \tag{4.8.57}$$

$$\mu_{i+1} = \frac{z_{i+1} - \overline{x}\beta z_{i+1}^\delta \cos(\xi \log z_{i+1} + \Phi_2)}{1 + e\beta z_{i+1}^\delta \cos(\xi \log z_{i+1} + \Phi_2)}. \tag{4.8.58}$$

Now note that we have

$$\xi \log z_{i+1} - \xi \log z_i \approx \pi \Rightarrow \frac{z_{i+1}}{z_i} \approx \exp \frac{\pi}{\xi}, \tag{4.8.59}$$

and we assume that $z << 1$ so that

$$1 + e\beta z_{i(i+1)}^\delta \cos(\xi \log z_{i(i+1)} + \Phi_2) \sim 1. \tag{4.8.60}$$

Finally, we obtain

$$\frac{\mu_{i+1}}{\mu_i} = \frac{z_{i+1} + [\overline{x}\beta \cos(\xi \log z_i + \Phi_2)] z_{i+1}^\delta}{z_i - [\overline{x}\beta \cos(\xi \log z_i + \Phi_2)] z_i^\delta}. \tag{4.8.61}$$

Now, in the limit as $z \to 0$, (4.8.61) becomes

$$\frac{\mu_{i+1}}{\mu_i} \approx -\left(\frac{z_{i+1}}{z_i}\right)^\delta \approx -\exp\left(\frac{\pi\delta}{\xi}\right). \tag{4.8.62}$$

Recall that $\delta = -\rho/\lambda$, $\xi = -\omega/\lambda$. We thus obtain

$$\lim_{i\to\infty} \frac{\mu_{i+1}}{\mu_i} = -\exp\frac{\rho\pi}{\omega}. \tag{4.8.63}$$

This quantity governs the size of the oscillations we see in Figure 4.8.28.

Subsidiary Homoclinic Orbits

Now we will show that, as we break our original homoclinic orbit (the *principal* homoclinic orbit), other homoclinic orbits of a different nature arise, and the Silnikov picture is repeated for these new homoclinic orbits. This phenomenon was first noted by Hastings [1982], Evans et al. [1982], Gaspard [1983], and Glendinning and Sparrow [1984]. We follow the argument of Gaspard.

When we break the homoclinic orbit, the unstable manifold intersects Π_0 at the point $(e\mu + \overline{x}, \mu)$. Thus, if $\mu > 0$, this point can be used as an initial condition for our map. Now, if the z component of the image of this point is zero, we will have found a new homoclinic orbit which passes once through a neighborhood of the origin before falling back into the origin. This condition is given by

$$0 = \beta(e\mu + \overline{x})\mu^\delta \cos(\xi \log \mu + \Phi_2) + \mu \tag{4.8.64}$$

or

$$-\mu = \beta(e\mu + \overline{x})\mu^\delta \cos(\xi \log \mu + \Phi_2). \tag{4.8.65}$$

We find the solutions for this graphically for $\delta > 1$ and $\delta < 1$ in the same manner as we investigated the equations for the fixed points; see Figure 4.8.29.

Thus, for $\delta > 1$, the only homoclinic orbit is the principal homoclinic orbit which exists at $\mu = 0$.

For $\delta < 1$, we get a countable infinity of μ values

$$\mu_i, \mu_{i+1}, \cdots, \mu_{i+n}, \cdots \to 0, \tag{4.8.66}$$

for which these *subsidiary* or *double-pulse* homoclinic orbits exist, as shown in Figure 4.8.30. Note for each of these homoclinic orbits, we can reconstruct our original Silnikov picture of a countable infinity of horseshoes. For a reference dealing with double-pulse homoclinic orbits for the case of real eigenvalues, see Yanagida [1987].

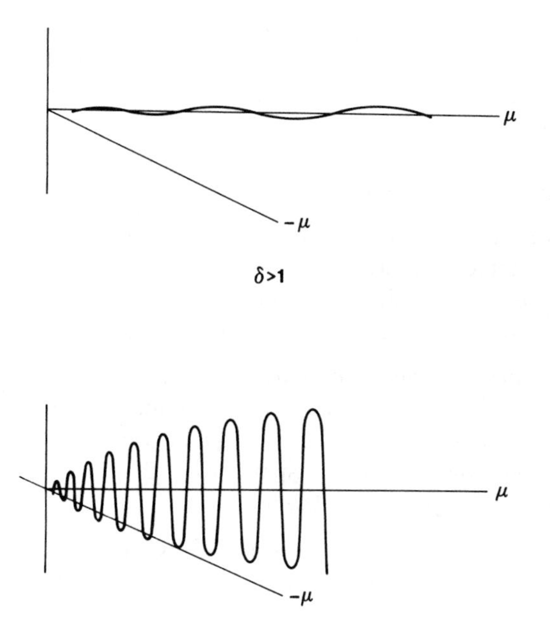

FIGURE 4.8.29. Graphical construction of the solutions of (4.8.65).

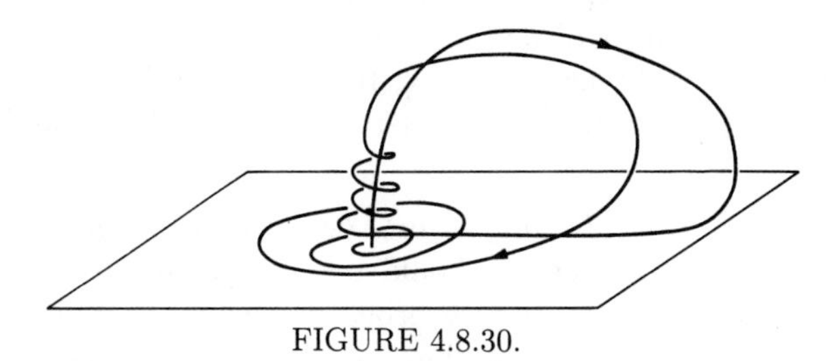

FIGURE 4.8.30.

Observations and General Remarks

Remark 1: Comparison Between the Saddle with Real Eigenvalues and the Saddle-Focus. Before leaving three dimensions, we want to reemphasize the main differences between the two cases studied.

Real Eigenvalues. In order to have horseshoes it was necessary to start with two homoclinic orbits. Even so, there were no horseshoes near the homoclinic orbits until the homoclinic orbits were broken such as might happen by varying a parameter. It was necessary to know the global twisting of orbits around the homoclinic orbits in order to determine how the horseshoe was formed.

Complex Eigenvalues. One homoclinic orbit is sufficient for a countable infinity of horseshoes whose existence does not require first breaking the homoclinic connection. Knowledge of global twisting around the homoclinic orbit is unnecessary, since the spiralling associated with the imaginary part of the eigenvalues tends to "smear" trajectories uniformly around the homoclinic orbit.

While an extensive amount of work exists concerning Silnikov's phenomenon, there are still some open problems.

Remark 2: Strange Attractors. Silnikov-type attractors have not attracted the great amount of attention that has been given to Lorenz attractors. The topology of the spiralling associated with the imaginary parts of the eigenvalues makes the Silnikov problem more difficult.

Remark 3: Creation of the Horseshoes and Bifurcation Analysis. We have given part of the bifurcation analysis of Glendinning and Sparrow [1984]. Their paper also contains some interesting numerical work and conjectures. The reader should also consult Gaspard, Kapral, and Nicolis [1984]. Knot theory has not been applied to this problem.

Remark 4: Nonhyperbolic Fixed Points. There appear to be little or no results concerning orbits homoclinic to nonhyperbolic fixed points in three dimensions.

Remark 5: Applications. The Silnikov phenomenon arises in a variety of applications. See, for example, Arneodo, Coullet, and Tresser [1981a,b], [1985], Arneodo, Coullet, Spiegel, and Tresser [1985], Arneodo, Coullet, and Spiegel [1982], Gaspard and Nicolis [1983], Hastings [1982], Pikovskii, Rabinovich, and Trakhtengerts [1979], Rabinovich [1978], Rabinovich and Fabrikant [1979], Roux, Rossi, Bachelart, and Vidal [1981], and Vyshkind and Rabinovich [1976].

Remark 6: The General Technique for Analyzing the Orbit Structure Near Homoclinic Orbits. The reader should note the similarities between the analyses of the orbit structure near orbits homoclinic to 1) hyperbolic periodic points of two-dimensional diffeomorphisms (or, equivalently, hyperbolic periodic orbits of autonomous, three-dimensional vector fields); 2) hyperbolic fixed points of three-dimensional autonomous vector fields having purely real eigenvalues; and 3) hyperbolic fixed points of three-dimensional autonomous vector fields having a pair of complex conjugate eigenvalues. In all three cases a return map was constructed in a neighborhood of a point along the homoclinic orbit that consisted of the composition of two maps. One map described the dynamics near the hyperbolic invariant set (i.e., periodic orbit or fixed point) and the other map described the dynamics near the homoclinic orbit outside of a neighborhood of the hyperbolic invariant set. Due to the fact that the fixed point is hyperbolic, the first map is well approximated from the linearization of the dynamical system (either map or vector field) about the hyperbolic invariant set (either periodic orbit or fixed point). The second map, if we restrict ourselves to a sufficiently small neighborhood of the homoclinic orbit, is well approximated by an affine map. In all cases, in order to show that chaotic invariant sets (more precisely, an invarant Cantor set on which the dynamics are topologically conjugate to a full shift on N symbols) are present near the homoclinic orbit, we find N μ_h-horizontal strips that are mapped over themselves in μ_v-vertical strips so that Assumptions 1 and 3 of Section 4.3 hold. The nature of the map near the hyperbolic invariant set is responsible for the verification of Assumption 3 (i.e., expansion and contraction in the appropriate directions). The nature of the map outside of the hyperbolic invariant set is responsible for the verification of Assumption 1. However, note that in the case of transversal orbits homoclinic to hyperbolic periodic orbits of two-dimensional diffeomorphisms, we did not need any special assumptions on the eigenvalues of the periodic orbit or the nature of the intersection of the stable and unstable manifolds (hyperbolicity and transversality were sufficient). This is very different from the situation of orbits homoclinic to hyperbolic fixed points of three-dimensional autonomous vector fields, where special assumptions on the relative magnitudes of the eigenvalues at the fixed point and the nature of the (nontransversal) homoclinic orbit(s) were necessary.

Remark 7. For generalizations to orbits homoclinic to hyperbolic fixed points of n-dimensional ($n \geq 4$) autonomous vector fields, the reader should consult Wiggins [1988].

Remark 8. For generalizations to heteroclinic cycles connecting hyperbolic fixed points of autonomous vector fields, the reader should consult Wiggins [1988].

4.9 Global Bifurcations Arising from Local Codimension-Two Bifurcations

In Section 3.1E we studied the bifurcation of a fixed point of a vector field in the situation where the matrix associated with the linearization of the vector field at the bifurcation point had two zero eigenvalues, and in Section 3.1F we studied the situation where the matrix had a zero and a pure imaginary pair of eigenvalues (with any remaining eigenvalues having nonzero real part). In both cases we saw that dynamical phenomena arose which could not be explained by any local bifurcation analysis, and in this section we want to attempt to complete the analysis. In the case of the double-zero eigenvalue we will succeed completely. In the case of the zero-pure imaginary pair we will only achieve partial success. We begin with the double-zero eigenvalue.

4.9A The Double-Zero Eigenvalue

Recall from Section 3.1E that the truncated normal form associated with this bifurcation is given by

$$\begin{aligned}\dot{x} &= y,\\ \dot{y} &= \mu_1 + \mu_2 y + x^2 + bxy, \qquad b = \pm 1.\end{aligned} \tag{4.9.1}$$

Equation (4.9.1) applies to generic vector fields, i.e., there are no symmetries and we treat the case $b = +1$.

Recall that (4.9.1) has no periodic orbits for $\mu_1 > 0$ and, for $\mu_1 < 0$, periodic orbits are created in a Poincaré-Andronov-Hopf bifurcation. Therefore, there must be some other bifurcation occurring which accounts for the destruction of the periodic orbits as μ_1 increases through zero. In Section 3.1E we gave some heuristic arguments as to why this should be a *homoclinic or saddle-connection* bifurcation, and now we want to prove this.

We begin by rescaling the dependent variables and parameters of (4.9.1) as follows

$$x = \varepsilon^2 u, \quad y = \varepsilon^3 v, \quad \mu_1 = -\varepsilon^4, \quad \mu_2 = \varepsilon^2 \nu_2 \ (\varepsilon > 0), \tag{4.9.2}$$

and we rescale the independent variable time as follows

$$t \longrightarrow \frac{t}{\varepsilon},$$

so that (4.9.1) becomes

$$\begin{aligned}\dot{u} &= v,\\ \dot{v} &= -1 + u^2 + \varepsilon(\nu_2 v + uv).\end{aligned} \tag{4.9.3}$$

Notice that, in the original variables, we are interested in $\mu_1 < 0$ and that our rescaling allows us to interpret our results in this parameter regime. (Note: the reader should be somewhat irritated that we have simply pulled this particular rescaling "out of the air." For now, we will proceed with the analysis, but at the end of this section we will discuss "why it works.")

The single most important characteristic of this rescaling is that, for $\varepsilon = 0$, the rescaled equations (4.9.3) become a completely integrable Hamiltonian system with Hamiltonian function given by

$$H(u, v) = \frac{v^2}{2} + u - \frac{u^3}{3}. \tag{4.9.4}$$

Melnikov's method can then be used to perform a global analysis that includes the effects of the higher order terms of the normal form. The phase space of this completely integrable Hamiltonian system is shown in Figure 4.9.1. Thus, the vector field

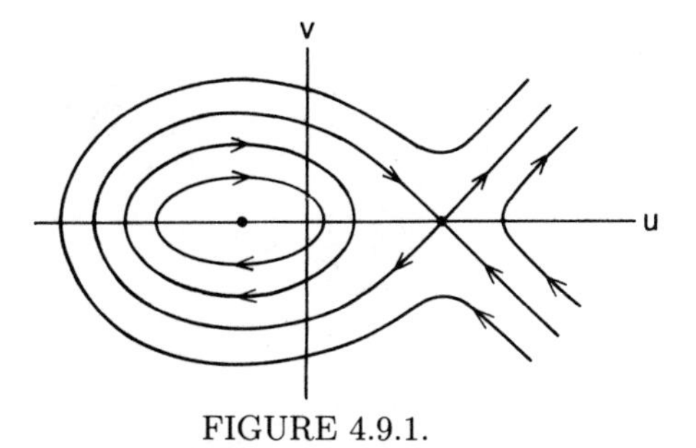

FIGURE 4.9.1.

$$\begin{aligned}\dot{u} &= v, \\ \dot{v} &= -1 + u^2,\end{aligned} \tag{4.9.5}$$

has a hyperbolic fixed point at

$$(u, v) = (1, 0),$$

an elliptic fixed point at

$$(u, v) = (-1, 0),$$

and a one-parameter family of periodic orbits surrounding the elliptic fixed point. We denote the latter by

$$(u^\alpha(t), v^\alpha(t)), \qquad \alpha \in [-1, 0) \tag{4.9.6}$$

with period T^α where

$$(u^{-1}(t), v^{-1}(t)) = (-1, 0) \tag{4.9.7}$$

and

$$\begin{aligned}\lim_{\alpha \to 0} (u^\alpha(t), v^\alpha(t)) &= (u_0(t), v_0(t)) \\ &= \left(1 - 3\,\mathrm{sech}^2 \frac{t}{\sqrt{2}}, 3\sqrt{2}\,\mathrm{sech}^2 \frac{t}{\sqrt{2}} \tanh \frac{t}{\sqrt{2}}\right)\end{aligned} \tag{4.9.8}$$

is a homoclinic orbit connecting the hyperbolic fixed point to itself.

The Melnikov theory can now be used to determine the effect of the $\mathcal{O}(\varepsilon)$ part of (4.9.3) on this integrable structure. The homoclinic Melnikov function is given by

$$M(\nu_2) = \int_{-\infty}^{\infty} v_0(t)\big[\nu_2 v_0(t) + u_0(t) v_0(t)\big]\,dt. \tag{4.9.9}$$

Using the expression for $u_0(t)$ and $v_0(t)$ given in (4.9.8), (4.9.9) becomes

$$M(\nu_2) = 7\nu_2 - 5$$

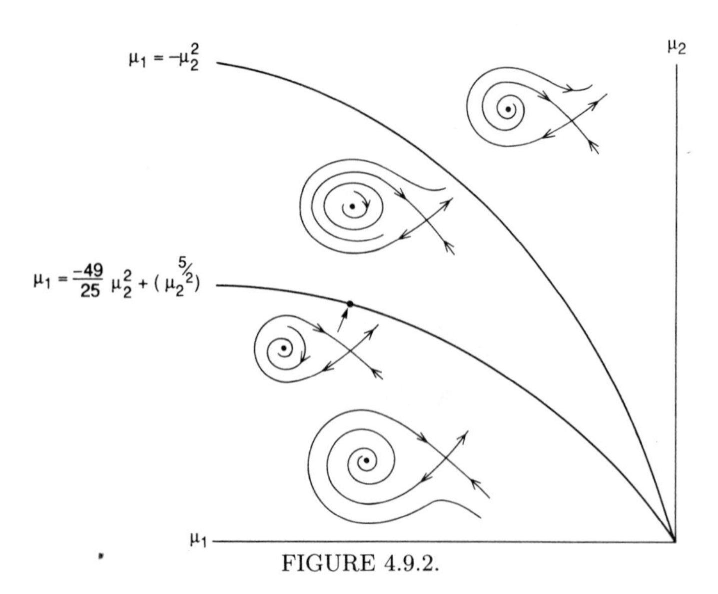

FIGURE 4.9.2.

or

$$M(\nu_2) = 0 \Rightarrow \nu_2 = \frac{5}{7}; \tag{4.9.10}$$

hence, a bifurcation curve on which the stable and unstable manifolds of the hyperbolic fixed point coincide is given by

$$\nu_2 = \frac{5}{7} + \mathcal{O}(\varepsilon). \tag{4.9.11}$$

We now want to translate (4.9.11) back into our original parameter values. Using (4.9.2) and (4.9.11), we obtain

$$\mu_1 = -\left(\frac{49}{25}\right)\mu_2^2 + \mathcal{O}(\mu_2^{5/2}) \tag{4.9.12}$$

for the homoclinic bifurcation curve. Note that we have

$$M > 0 \qquad \text{for } \mu_1 > -\frac{49}{25}\mu_2^2, \tag{4.9.13a}$$

and

$$M < 0 \qquad \text{for } \mu_1 < -\frac{49}{25}\mu_2^2, \tag{4.9.13b}$$

which give us the relative orientations of the stable and unstable manifolds of the hyperbolic fixed points that we show in Figure 4.9.2.

Thus, we have shown the existence of a global mechanism for periodic orbits to be created and destroyed. Can we now actually claim that the periodic orbit created in the Poincaré–Andronov–Hopf bifurcation is the one that is destroyed in the homoclinic bifurcation? No, we cannot, because it is possible for nonlocal saddle-node bifurcations of periodic orbits to occur in the parameter region between the Poincaré–Andronov–Hopf and homoclinic bifurcation curves. We can check whether or not such bifurcations occur by computing the Melnikov function for the periodic orbits. This is given by

$$M(\alpha;\nu_2)=\nu_2\int_0^{T^\alpha}(v^\alpha(t))^2dt+\int_0^{T^\alpha}u^\alpha(t)(v^\alpha(t))^2dt. \tag{4.9.14}$$

The condition $M(\alpha;\nu_2)=0$ implies the existence of a periodic orbit for (4.9.3). Using (4.9.14), this is equivalent to

$$\nu_2=\frac{\int_0^{T^\alpha}u^\alpha(t)(v^\alpha(t))^2dt}{\int_0^{T^\alpha}(v^\alpha(t))^2dt}\equiv f(\alpha). \tag{4.9.15}$$

Now, if $f(\alpha)$ is a monotone function on $[-1,0]$, we can conclude that (4.9.3) has a unique periodic orbit created in a Poincaré–Andronov–Hopf bifurcation and destroyed in a homoclinic bifurcation. It turns out that $f(\alpha)$ is indeed monotone. This can be verified by computing the expressions for $(u^\alpha(t),v^\alpha(t))$ directly in terms of elliptic functions and then analytically evaluating $f(\alpha)$ from (4.9.15). Once this is done, monotonicity properties of $f(\alpha)$ may be studied. This is a fairly messy (but, in principle, straightforward) calculation that we leave as an exercise for the interested reader. We note that this result was obtained by Bogdanov [1975], Takens [1974], and Carr [1981]. Therefore, the complete bifurcation diagram for (4.9.1) with $b=+1$ is as shown in Figure 4.9.3 and we end with the following remarks.

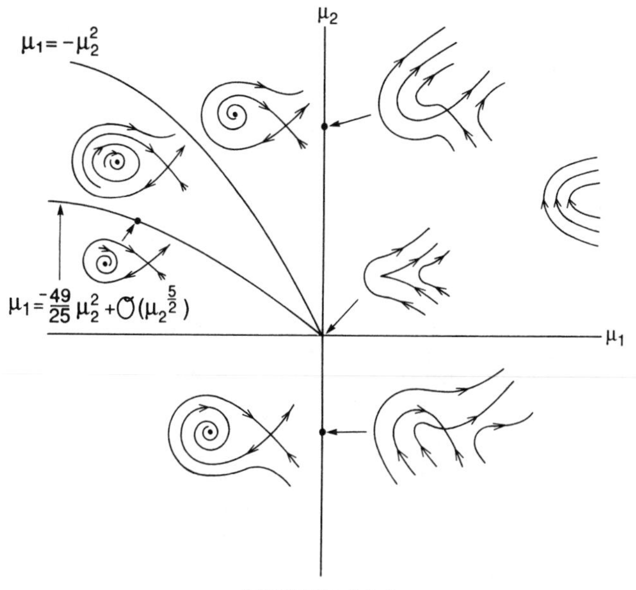

FIGURE 4.9.3.

Remark 1. The local bifurcation analysis of (4.9.1) did not require any smallness restrictions on μ_1 and μ_2. The global bifurcation analysis does require μ_1 and μ_2 to be "sufficiently small."

Remark 2. It is now fairly easy to show that restoring the higher order terms in the normal form (4.9.1) does not qualitatively affect the bifurcation diagram. We will outline the necessary arguments in Exercise 4.61.

4.9B A Zero and a Pure Imaginary Pair of Eigenvalues

The normal form associated with this bifurcation is three-dimensional. However, the symmetry in the linear part associated with the pure imaginary eigenvalues enabled us to decouple one of the coordinates from the remaining two so that we could begin our analysis using phase plane techniques. In some sense, the dynamics in this phase plane can be viewed as an approximation to a Poincaré map of the full three-dimensional normal form. This is also a result of the symmetry of the linear part. Our analysis will proceed in the following steps.

Step 1. Analyze global bifurcations of the associate truncated, two-dimensional normal form.

Step 2. Interpret in terms of the truncated, three-dimensional normal form.

Step 3. Discuss the effects of higher order terms in the normal form.

Step 1. Recall from Section 3.1F that the associated two-dimensional normal form of interest is given by

$$\begin{aligned} \dot{r} &= \mu_1 r + arz, \\ \dot{z} &= \mu_2 + br^2 - z^2, \end{aligned} \tag{4.9.16}$$

where $a \neq 0$ and $b = \pm 1$. There were essentially only four distinct cases to study, and only two admitted the possibility of global bifurcations (for r, z small). They were denoted by

Case IIa,b	$a < 0, b = +1,$
Case III	$a > 0, b = -1.$

In Case IIa,b we are interested in the dynamics near the μ_2-axis for $\mu_2 < 0$. In Case III we are interested in the dynamics near the μ_2-axis for $\mu_2 > 0$. In both cases the normal form was integrable on the μ_2-axis (with the appropriate sign of μ_2); thus, we would expect higher order terms in the normal form to drastically affect the dynamics in this parameter regime. Our strategy will be to include the cubic terms in the normal form and to introduce a scaling of the variables so that we obtain a perturbed Hamiltonian system. Then a Melnikov-type analysis can be used to determine the number of periodic orbits and possible homoclinic bifurcations.

From (3.1.170), restoring the cubic terms to (4.9.16) gives

$$\begin{aligned} \dot{r} &= \mu_1 r + arz + (cr^3 + dr^2 z), \\ \dot{z} &= \mu_2 + br^2 - z^2 + (er^2 z + f z^3). \end{aligned} \tag{4.9.17}$$

In Guckenheimer and Holmes [1983] it is shown that coordinate changes can be introduced so that all cubic terms except for z^3 in (4.9.17) can be eliminated (see Exercise 4.63 where we outline this procedure). Hence, without loss of generality, we can analyze the following normal form

$$\begin{aligned} \dot{r} &= \mu_1 r + arz, \\ \dot{z} &= \mu_2 + br^2 - z^2 + f z^3. \end{aligned} \tag{4.9.18}$$

We next rescale the dependent variables and parameters as follows

$$r = \varepsilon u, \quad z = \varepsilon v, \quad \mu_1 = \varepsilon^2 \nu_1, \quad \mu_2 = \varepsilon^2 \nu_2, \tag{4.9.19}$$

and we rescale time as follows

$$t \longrightarrow \varepsilon t,$$

so that (4.9.17) becomes

$$\begin{aligned} \dot{u} &= auv + \varepsilon \nu_1 u, \\ \dot{v} &= \nu_2 + bu^2 - v^2 + \varepsilon f v^3. \end{aligned} \tag{4.9.20}$$

At $\varepsilon = 0$, the vector field has the first integral (for $a \neq 1$)

$$F(u,v) = \frac{a}{2} u^{2/a} \left[\nu_2 + \frac{b}{1+a} u^2 - v^2 \right]. \tag{4.9.21}$$

Unfortunately, it is not Hamiltonian, but can be made Hamiltonian (Guckenheimer and Holmes [1983]) by multiplying the right-hand side of (4.9.20) by the integrating factor $u^{(2/a)-1}$ to obtain

$$\begin{aligned} \dot{u} &= au^{2/a} v + \varepsilon \nu_1 u^{2/a}, \\ \dot{v} &= -bu^{(2/a)-1} + bu^{(2/a)+1} - u^{(2/a)-1} v^2 + \varepsilon f u^{(2/a)-1} v^3, \end{aligned} \tag{4.9.22}$$

where we let $\nu_2 = \mp 1$ when $b = \pm 1$ since, for Case IIa,b, we are interested in $\mu_2 < 0$ and, for Case III, we are interested in $\mu_2 > 0$. For $\varepsilon = 0$, (4.9.22) is Hamiltonian with Hamiltonian function

$$H(u,v) = \frac{1}{2} u^{2/a} v^2 + \frac{ab}{2} u^{2/a} - \frac{ab}{2(a+1)} u^{(2/a)+2}, \qquad a+1 \neq 0$$

or

$$H(u,v) = -\frac{1}{2} u^{-2} v^2 - \frac{b}{2} u^{-2} - b \log u, \qquad a+1 = 0. \tag{4.9.23}$$

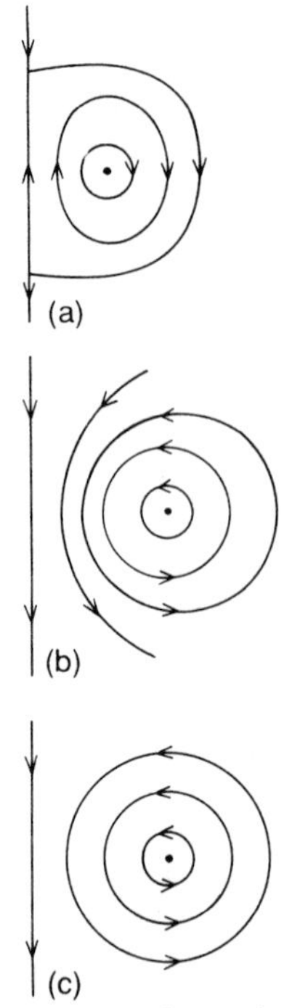

FIGURE 4.9.4. Integrable structure of (4.9.22) for $\varepsilon = 0$. a) Case III; b) Case IIa,b, $-1 < a < 0$; c) Case IIa,b, $a \leq -1$.

In Figure 4.9.4 we show the level sets of the Hamiltonian (i.e., the orbits of (4.9.22) for $\varepsilon = 0$) for the relevant cases. We see from this figure that, for Case IIa,b, the integrable Hamiltonian system has a one-parameter family of periodic orbits surrounding an elliptic fixed point with the orbits becoming unbounded in amplitude. In Case III the integrable Hamiltonian system has a one-parameter family of periodic orbits surrounding an elliptic fixed point that limit on a heteroclinic cycle. In both cases we denote the one-parameter family of orbits by

$$(u^\alpha(t), v^\alpha(t)), \qquad \alpha \in [-1, 0),$$

with period T^α, where $(u^{-1}(t), v^{-1}(t))$ is an elliptic fixed point in both Case IIa,b and Case III, and $\lim_{\alpha \to 0}(u^\alpha(t), v^\alpha(t))$ is an unbounded periodic orbit in Case IIa,b and a heteroclinic cycle in Case III.

The Melnikov functions are given by

$$M(\alpha; \nu_1)$$
$$= af \int_0^{T^\alpha} (u^\alpha(t))^{(4/a)-1}(v^\alpha(t))^4 dt$$

$$- \nu_1 \int_0^{T^\alpha} \left[b(u^\alpha(t))^{(4/a)+1} - b(u^\alpha(t))^{(4/a)-1} + (u^\alpha(t))^{(4/a)-1}(v^\alpha(t))^2 \right] dt. \tag{4.9.24}$$

Therefore, $M(\alpha; \nu_1) = 0$ is equivalent to

$$\nu_1 = \frac{af \int_0^{T^\alpha} (u^\alpha(t))^{(4/a)-1}(v^\alpha(t))^4 dt}{\int_0^{T^\alpha} [b(u^\alpha(t))^{(4/a)+1} - b(u^\alpha(t))^{(4/a)-1} + (u^\alpha(t))^{(4/a)-1}(v^\alpha(t))^2] dt} \equiv f(\alpha). \tag{4.9.25}$$

We are interested in

Case IIa,b	$a < 0, b = 1, f \neq 0, \nu_1 < 0,$
Case III	$a > 0, b = -1, f \neq 0, \nu_1 > 0.$

Thus, if $f(\alpha)$ is a monotone function of α (for a, b, and f fixed as above), then (4.9.22) has a unique periodic orbit which is born in a Poincaré-Andronov-Hopf bifurcation (we consider stability in Exercise 4.64) and grows monotonically in amplitude in Case IIa,b and disappears in a heteroclinic bifurcation in Case III. However, proof that (4.9.25) is monotonic in α is a formidable and difficult problem, since the integrals cannot be evaluated explicitly in terms of elementary integrals. Fortunately, it has recently been proven by Zoladek [1984], [1987] (see also Carr et al. [1985], van Gils [1985] and Chow et al. [1989]) that (4.9.25) is indeed monotone in α. The techniques in these papers for proving monotonicity involve complicated estimates that we will not go into here. In Figure 4.9.5 we show an example of a possible bifurcation for Case III with $a = 2$, $f < 0$.

Step 2. We now turn to Step 2, interpreting the dynamics of the two-dimensional vector field in terms of the three-dimensional dynamics. The truncated, three-dimensional normal form is given by

$$\begin{aligned} \dot{r} &= \mu_1 r + arz, \\ \dot{z} &= \mu_2 + br^2 - z^2 + fz^3, \\ \dot{\theta} &= \omega + \cdots . \end{aligned} \tag{4.9.26}$$

Thus, the r and z components of (4.9.26) are independent of θ, and the discussion in Section 3.1F still holds. Namely, it is the case that the periodic orbits become invariant two tori in the full three-dimensional phase space and, in Case III, the heteroclinic cycle becomes such that the two-dimensional stable manifold and one-dimensional unstable manifold of the hyperbolic fixed point with $z < 0$ coincides with the two-dimensional unstable manifold and one-dimensional stable manifold of the hyperbolic fixed point with $z > 0$ creating the invariant sphere (with invariant axis), as shown in Figure 4.9.6. Both situations are radically altered by the addition of the higher order terms in the normal form and we now turn to a discussion of this situation in Step 3.

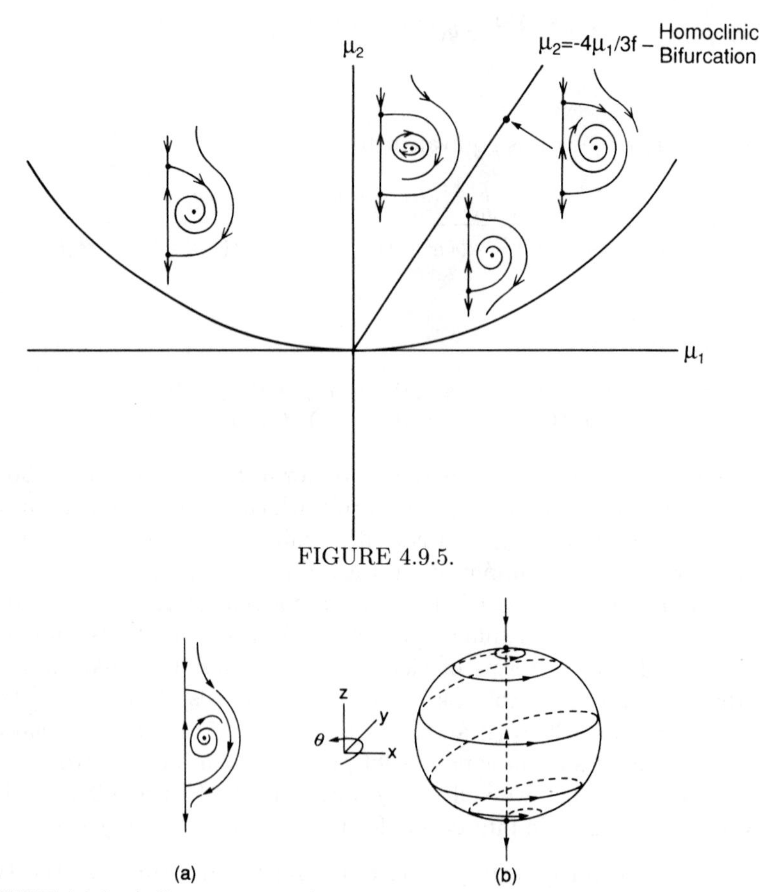

FIGURE 4.9.5.

FIGURE 4.9.6. a) Cross-section of the heteroclinic cycle for the truncated normal form. b) Heteroclinic cycle for the truncated normal form.

Step 3. The problem of how the invariant two-tori of the truncated normal form in Cases IIa,b and III and the heteroclinic cycle in Case III are affected by the higher order terms of the normal form is difficult, and the full story is not yet known. We will briefly summarize the main issues and known results.

Invariant Two-Tori

Concerning the invariant two-tori of the truncated normal form in Cases IIa,b and Case III there are two main questions that arise.

1. Do the two-tori persist when the effects of the higher order terms of the normal form are considered?

2. In the truncated normal form the orbits on the invariant two-tori are either periodic or quasiperiodic, densely covering the torus (see Section 1.2E). Do invariant two-tori having quasiperiodic flow persist when the effects of the higher order terms of the normal form are considered?

In answering the first question techniques from the persistence theory of normally hyperbolic invariant manifolds are used (see, e.g., Fenichel [1971], [1979], [1977] and Hirsch, Pugh, and Shub [1977]). The application of these techniques is not straightforward, since the strength of the normal hyperbolicity depends on the bifurcation parameter. Results showing that invariant two-tori persist have been obtained by Iooss and Langford [1980] and Scheurle and Marsden [1984].

In answering the second question small divisor problems arise which necessitate the use of KAM-type techniques (see, e.g., Siegel and Moser [1971]). This is very much beyond the scope of this book. However, we mention that some results along these lines have been obtained by Scheurle and Marsden [1984]. Their results imply that on a Cantor set of parameter values having positive Lebesgue measure one has invariant two-tori having quasiperiodic flow. In Exercise 3.47 we consider some simple examples that highlight some of the main issues associated with this second question.

Heteroclinic Cycle

In Case III the truncated normal form has two saddle-type fixed points on the z axis; p_1 having a two-dimensional stable manifold and a one-dimensional unstable manifold and p_2 having a two-dimensional unstable manifold and a one-dimensional stable manifold with $W^s(p_1)$ coinciding with $W^u(p_2)$ to form the invariant sphere as shown in Figure 4.9.6. The axis of the sphere is formed from the coincidence of a branch of $W^u(p_1)$ with a branch of $W^s(p_2)$.

We expect the effects of the higher order terms of the normal form to drastically alter this situation. Generically, we would expect the one-dimensional unstable manifold of p_1 and the one-dimensional stable manifold of p_2 to *not* intersect in the three-dimensional phase space. Similarly, generically we would expect the two-dimensional stable manifold of p_2 and the two-dimensional unstable manifold of p_1 to intersect along one-dimensional orbits in three-dimensional phase space. We illustrate these two solutions in Figure 4.9.7.

Now it may happen that when this degenerate structure is broken the branch of $W^u(p_1)$ inside the sphere falls into $W^s(p_1)$, or, similarly, the branch of $W^s(p_2)$ inside the sphere falls into $W^u(p_2)$; see Figure 4.9.8. If this happens, then from Section 4.8B the reader should realize that it is possible for a return map defined near one or the other of these homoclinic orbits to possess a countable infinity of Smale horseshoes, i.e., Silnikov's phenomenon may occur. Moreover, from Section 4.8B,i) a countable in-

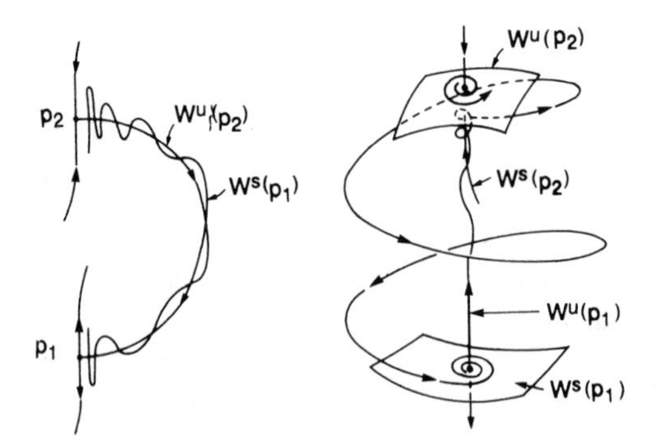

FIGURE 4.9.7. a) Cross-section of the manifolds for the full normal form. b) Homoclinic orbit for the full normal form.

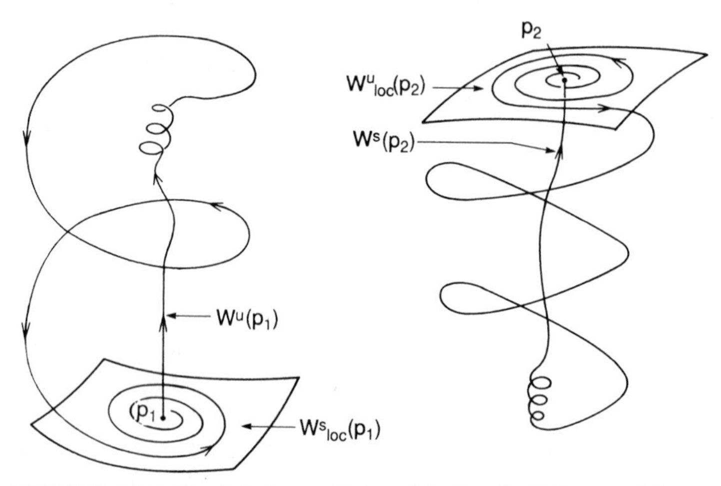

FIGURE 4.9.8. Possible homoclinic orbits for the full normal form.

finity of period-doubling and saddle-node bifurcation values would accumulate on the parameter value at which the homoclinic orbit was formed. Thus, this particular local codimension-two bifurcation would actually be codimension-infinity when all of the dynamical effects are included (see the comments at the end of Section 4.7).

The fact that this local codimension-two bifurcation point can exhibit Silnikov's phenomenon in its versal deformation has been proved by Broer and Vegter [1984]. They prove that the full normal form possesses orbits homoclinic to a hyperbolic fixed point. Their result is very delicate in the sense that, due to the rotational symmetry of the linear part of the normal form, the symmetry is preserved to all orders in the normal form (i.e., the r and z components of the normal form are independent of θ). Thus, the homoclinic orbits are a result of exponentially small terms that are not picked up in the Taylor expansion and subsequent normal form transformation of the vector field.

The Hamiltonian-Dissipative Decomposition

The key aspect that enabled the preceeding analyses to go through was the fact that one could find an "inspired" rescaling that transformed the normal form into a perturbation of an integrable Hamiltonian system with the higher order terms in the normal form part of the perturbation (note: the perturbation need *not* also be Hamiltonian). Once this has been accomplished, a wealth of techniques for the global analysis of nonlinear dynamical systems can then be employed; for example, Melnikov theory, perturbation theory for normally hyperbolic invariant manifolds, and KAM theory. However, the real question is, "in a given problem, how can we find the rescalings which turn the problem into a perturbation of a completely integrable Hamiltonian system?"

The rescalings in this section are due to the cleverness of Takens [1974], Bogdanov [1975], Guckenheimer and Holmes [1983], Kopell and Howard [1975], and Iooss and Langford [1980]. However, in recent years there has been an effort to understand the structure of the normal form that leads to a splitting of the normal form into a Hamiltonian part and a dissipative part so that the appropriate rescalings can be generated by some computational procedure. We refer the reader to Lewis and Marsden [1989] and Olver and Shakiban [1988].

4.10 Liapunov Exponents

Consider the $\mathbf{C}^r$ $(r \geq 1)$ vector field

$$\dot{x} = f(x), \qquad x \in \mathbb{R}^n. \tag{4.10.1}$$

Let $x(t)$ be a trajectory of (4.10.1) satisfying $x(0) = x_0$. We want to describe the orbit structure of (4.10.1) near $x(t)$. In particular, we want to know the geometry associated with the attraction and/or repulsion of orbits of (4.10.1) relative to $x(t)$. For this it is natural to first consider the orbit structure of the linearization of (4.10.1) about $x(t)$ given by

$$\dot{\xi} = Df(x(t))\xi, \qquad \xi \in \mathbb{R}^n. \tag{4.10.2}$$

Let $X(t)$ be the fundamental solution matrix of (4.10.2) and let e be a vector in $\mathbb{R}^n$. Then the *coefficient of expansion in the direction e along the orbit through* x_0 is defined to be

$$\lambda_t(x_0, e) \equiv \frac{\|X(t)e\|}{\|e\|}, \tag{4.10.3}$$

where $\|\cdot\| = \sqrt{\langle \cdot, \cdot \rangle}$ with $\langle \cdot, \cdot \rangle$ denoting the standard scalar product on $\mathbb{R}^n$. Note that $\lambda_t(x_0, e)$ is a time-dependent quantity that also depends on a particular orbit of (4.10.1) and a particular direction along this orbit. The *Liapunov characteristic exponent* (or just Liapunov exponent) *in the direction* e *along the orbit through* x_0 is defined to be

$$\chi(x_0, e) \equiv \overline{\lim_{t\to\infty}} \frac{1}{t} \log \lambda_t(x_0, e). \tag{4.10.4}$$

This definition brings up two points that we need to address.

1. Equation (4.10.4) is an asymptotic quantity. Therefore, in order for it to make sense, we must at least know that $x(t)$ exists for all $t > 0$. This will be true if the phase space is a compact, boundaryless manifold or if x_0 lies in a positively invariant region.

2. For computational purposes it would be advantageous if we could replace $\overline{\lim}$ by lim. We deal with this question next.

Let $\{e_1, \cdots, e_n\}$ be any orthonormal basis in $\mathbb{R}^n$; then we can form the quantity $\sum_{i=1}^n \chi(x_0, e_i)$. An orthonormal basis of $\mathbb{R}^n$ is called a *normal basis* if $\sum_{i=1}^n \chi(x_0, e_i)$ attains its minimum on the basis. Liapunov [1966] gave an explicit construction of such bases (therefore they exist). The fundamental solution matrix $X(t)$ is called *regular as* $t \to \infty$ if 1) $\lim_{t\to\infty} \frac{1}{t} \log|\det X(t)|$ exists and is finite and 2) for each normal basis $\{e_1, \cdots, e_n\}$

$$\sum_{i=1}^n \chi(x_0, e_i) = \lim_{t\to\infty} \frac{1}{t} \log|\det X(t)|.$$

Now we can state the main theorem due to Liapunov [1966].

Theorem 4.10.1 *If $X(t)$ is regular as $t \to \infty$, then*

$$\chi(x_0, e) = \lim_{t\to\infty} \frac{1}{t} \log \lambda_t(x_0, e) \tag{4.10.5}$$

exists and is finite for any vector $e \in \mathbb{R}^n$.

Proof: See Liapunov [1966] or Oseledec [1968]. □

We remark that the fundamental solution matrix, $X(t)$, of (4.10.2) is associated with a particular trajectory, $x(t)$, of (4.10.1). Thus, if we considered a different trajectory, $\tilde{x}(t)$, of (4.10.1), then the fundamental solution matrix associated with the vector field linearized about $\tilde{x}(t)$ may (and most probably will) have different properties. In particular, it may not be regular as $t \to \infty$, even though $X(t)$ may be regular as $t \to \infty$.

Let us now consider a few examples.

EXAMPLE 4.10.1 Consider the linear, scalar vector field

$$\dot{x} = ax, \qquad x \in \mathbb{R}^1, \tag{4.10.6}$$

where a is a constant. Equation (4.10.6) has three orbits, $x = 0$, $x > 0$, and $x < 0$, but the fundamental solution matrix associated with each orbit is given by

$$X(t) = e^{at}. \tag{4.10.7}$$

Thus, using (4.10.7) and (4.10.5), we see that each orbit of (4.10.6) has only one Liapunov exponent, and the Liapunov exponent of each orbit is a. Thus, if $a > 0$, trajectories of (4.10.6) separate exponentially as $t \to \infty$.

EXAMPLE 4.10.2 Consider a planar, Hamiltonian system in a region of phase space where the vector field is given in action-angle variables (see Section 1.2D,ii)) as follows

$$\begin{aligned} \dot{I} &= 0, \\ \dot{\theta} &= \Omega(I), \end{aligned} \qquad (I, \theta) \in \mathbb{R}^+ \times S^1. \tag{4.10.8}$$

Then, a trajectory of (4.10.8) is given by

$$\begin{aligned} I &= \text{constant}, \\ \theta(t) &= \Omega(I)t + \theta_0. \end{aligned} \tag{4.10.9}$$

Linearizing (4.10.8) about (4.10.9) gives

$$\begin{pmatrix} \dot{\xi}_1 \\ \dot{\xi}_2 \end{pmatrix} = \begin{pmatrix} 0 & 0 \\ \frac{\partial \Omega}{\partial I}(I) & 0 \end{pmatrix} \begin{pmatrix} \xi_1 \\ \xi_2 \end{pmatrix}. \tag{4.10.10}$$

The fundamental solution matrix of (4.10.10) is easily computed and found to be

$$X(t) = \begin{pmatrix} C & C \\ C\frac{\partial \Omega}{\partial I}(I)t & 0 \end{pmatrix}, \tag{4.10.11}$$

where C is a constant. Letting $\delta\theta$ represent a vector tangent to a trajectory and δI a vector normal to a trajectory using (4.10.11) and (4.10.5), we easily obtain

$$\begin{aligned} \chi(I, \delta I) &= 0, \\ \chi(I, \delta\theta) &= 0, \end{aligned}$$

for any I labeling a trajectory of (4.10.8) defined by (4.10.9).

EXAMPLE 4.10.3 Consider the vector field

$$\begin{aligned} \dot{x} &= x - x^3, \\ \dot{y} &= -y. \end{aligned} \tag{4.10.12}$$

In Example 1.1.15 we saw that (4.10.12) has a saddle at $(x, y) = (0, 0)$ and sinks at $(\pm 1, 0)$. Moreover, the closed interval $[-1, 1]$ on the x-axis is an attracting set; see Figure 4.10.1. We want to compute the Liapunov exponents associated with orbits in this attracting set. The attracting set

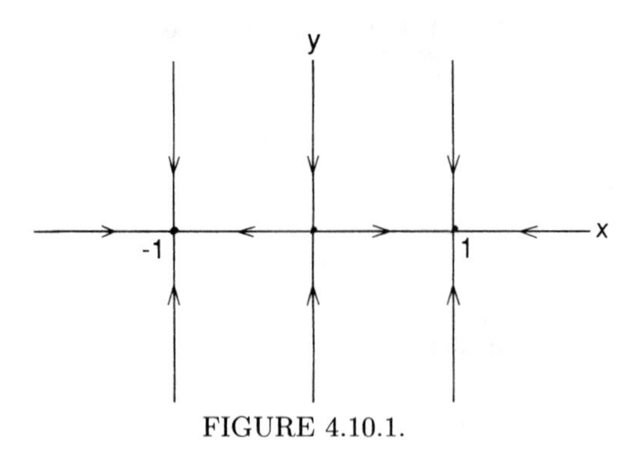

FIGURE 4.10.1.

$[-1, 1]$ contains five orbits, the fixed points $(x, y) = (0, 0)$, $(\pm 1, 0)$ and the open intervals $(-1, 0)$ and $(0, 1)$. Each orbit has two Liapunov exponents. We will compute the Liapunov exponents of each orbit individually. We let $\delta x \equiv (1, 0)$ denote a tangent vector in the x direction and $\delta y \equiv (0, 1)$ denote a tangent vector in the y direction. It should be clear that at each point of the attracting set δx and δy are a basis of $\mathbb{R}^2$. The Liapunov exponents for the three fixed points are trivial to obtain and we merely state the results.

$$(0,0) \qquad \begin{aligned} \chi((0,0), \delta x) &= +1, \\ \chi((0,0), \delta y) &= -1. \end{aligned}$$

$$(-1,0) \qquad \begin{aligned} \chi((-1,0), \delta x) &= -2, \\ \chi((-1,0), \delta y) &= -1. \end{aligned}$$

$$(+1,0) \qquad \begin{aligned} \chi((1,0), \delta x) &= -2, \\ \chi((1,0), \delta y) &= -1. \end{aligned}$$

The Liapunov exponents for the orbits $0 < x < 1$, $y = 0$, and $-1 < x < 0$, $y = 0$, require a little more work. Note that (4.10.12) is unchanged under the coordinate transformation $x \to -x$. Therefore, the Liapunov exponents for the orbit $0 < x < 1$, $y = 0$, are the same as for the orbit $-1 < x < 0$, $y = 0$.

Linearizing (4.10.12) about $(x(t), 0)$ gives

$$\begin{pmatrix} \dot{\xi}_1 \\ \dot{\xi}_2 \end{pmatrix} = \begin{pmatrix} 1 - 3x^2(t) & 0 \\ 0 & -1 \end{pmatrix} \begin{pmatrix} \xi_1 \\ \xi_2 \end{pmatrix}, \tag{4.10.13}$$

where $x(t)$ is a trajectory in $0 < x < 1$, $y = 0$. Integrating the x-component of (4.10.12) gives

$$x^2(t) = \frac{e^{2t}}{e^{2t} + 1}. \tag{4.10.14}$$

Substituting (4.10.14) into (4.10.13), we obtain the fundamental solution matrix

$$X(t) = \begin{pmatrix} e^{2t}(1+e^{-2t})^{-3/2} & 0 \\ 0 & e^{-t} \end{pmatrix}. \tag{4.10.15}$$

Using (4.10.5) and (4.10.15), we obtain the Liapunov exponents

$$0 < x < 1,\ y = 0 \qquad \begin{aligned} &\chi((0 < x < 1, y = 0), \delta x) = -2, \\ &\chi((0 < x < 1, y = 0), \delta y) = -1, \end{aligned}$$

$$-1 < x < 0,\ y = 0 \qquad \begin{aligned} &\chi((-1 < x < 0, y = 0), \delta x) = -2, \\ &\chi((-1 < x < 0, y = 0), \delta y) = -1. \end{aligned}$$

We end this section with some final remarks.

Remark 1. The Liapunov exponent in the direction e along the orbit through x_0 is unchanged if the initial condition of the trajectory along the orbit is varied. In particular, for any $t_1 \in \mathbb{R}$, let $x(t_1) \equiv x_1$; then $\chi(x_0, e) = \chi(x_1, e)$. This should be intuitively clear from the fact that the exponents are limits as $t \to \infty$; see Exercise 4.70.

Remark 2. We can view the Liapunov exponents of a given orbit as the long time average of the real parts of the eigenvalues of the fundamental solution matrix associated with the linearization of the vector field about the orbit. Therefore, they give us information concerning local expansion and contraction of phase space only and nothing about twisting and folding.

Remark 3. In general, Liapunov exponents are *not* continuous functions of the orbits. This can be seen from Example 4.10.3.

Remark 4. It should be clear that if the orbit is a fixed point or periodic orbit, then the Liapunov exponents are, in the first case, the real parts of the eigenvalues associated with the matrix of the vector field linearized about the fixed point and, in the second case, the real parts of the Floquet exponents. Thus, in some sense the theory of Liapunov exponents is a generalization of linear stability theory for arbitrary orbits (but see Goldhirsch et al. [1987]).

This brings up an interesting point. Associated with the linear eigenspaces associated with vector fields linearized about fixed points and periodic orbits are manifolds invariant under the full nonlinear dynamics where orbits have the same asymptotic behavior as in the linearized system. We refer to these as the stable and unstable manifolds. Might an *arbitrary* orbit possess stable and unstable manifolds having dimension equal to the number of negative and positive Liapunov exponents, respectively, associated with the orbit? The answer to this question is yes, and it has been proved by Pesin [1976], [1977], but see also Sacker and Sell [1974], [1976a], [1976b], [1978], and [1980].

We have only hinted in this section at the various properties of Liapunov exponents. For more information the reader should consult Liapunov [1966], Bylov et al. [1966], and Oseledec [1968]. Benettin et al. [1980a], [1980b] give an algorithm for computing all the Liapunov exponents of an orbit; see also the interesting paper of Goldhirsch et al. [1987]. We develop many additional properties of Liapunov exponents in the exercises.

4.11 Chaos and Strange Attractors

In this final section we want to examine what is meant by the term "chaos" as applied to deterministic dynamical systems as well as the notion of a "strange attractor." We will begin by giving several definitions of properties that should be characteristic of "chaos" and then consider several examples that possess one, several, or all of these properties.

We consider $\mathbf{C}^r$ $(r \geq 1)$ autonomous vector fields and maps on $\mathbb{R}^n$ denoted as follows

$$\text{vector field} \qquad \dot{x} = f(x), \tag{4.11.1a}$$

$$\text{map} \qquad x \mapsto g(x). \tag{4.11.1b}$$

We denote the flow generated by (4.11.1a) by $\phi(t,x)$ and we assume that it exists for all $t > 0$. We assume that $\Lambda \subset \mathbb{R}^n$ is a compact set invariant under $\phi(t,x)$ (resp. $g(x)$), i.e., $\phi(t,\Lambda) \subset \Lambda$ for all $t \in \mathbb{R}$ (resp. $g^n(\Lambda) \subset \Lambda$ for all $n \in \mathbb{Z}$, except that if g is not invertible, we must take $n \geq 0$). Then we have the following definitions.

DEFINITION 4.11.1 The flow $\phi(t,x)$ (resp. $g(x)$) is said to have *sensitive dependence on initial conditions on* Λ if there exists $\varepsilon > 0$ such that, for any $x \in \Lambda$ and any neighborhood U of x, there exists $y \in U$ and $t > 0$ (resp. $n > 0$) such that $|\phi(t,x) - \phi(t,y)| > \varepsilon$ (resp. $|g^n(x) - g^n(y)| > \varepsilon$).

Roughly speaking, Definition 4.11.1 says that for any point $x \in \Lambda$, there is (at least) one point arbitrarily close to Λ that diverges from x. Some authors require the rate of divergence to be exponential; for reasons to be explained later, we will not do so. As we will see in the examples, taken by itself sensitive dependence on initial conditions is a fairly common feature in many dynamical systems.

DEFINITION 4.11.2. Λ is said to be *chaotic* if

1. $\phi(t,x)$ (resp. $g(x)$) has sensitive dependence on initial conditions on Λ.

2. $\phi(t,x)$ (resp. $g(x)$) is topologically transitive on Λ.

Some authors (e.g., Devaney [1986]) add an additional requirement to Definition 4.11.2.

3. The periodic orbits of $\phi(t, x)$ (resp. $g(x)$) are dense in Λ.

We will not explicitly include point 3 as part of the definition of a chaotic invariant set, but we will examine its importance and relationship to "chaos."

We will now consider several examples that exhibit the properties described in these two definitions.

EXAMPLE 4.11.1 Consider the following vector field on $\mathbb{R}^1$

$$\dot{x} = ax, \qquad x \in \mathbb{R}^1, \tag{4.11.2}$$

with $a > 0$. The flow generated by (4.11.2) is given by

$$\phi(t, x) = e^{at} x. \tag{4.11.3}$$

From (4.11.3) we conclude the following.

1. $\phi(t, x)$ has no periodic orbits.

2. $\phi(t, x)$ is topologically transitive on the *noncompact* sets $(0, \infty)$ and $(-\infty, 0)$.

3. $\phi(t, x)$ has sensitive dependence on initial conditions on $\mathbb{R}^1$ since for any $x_0, x_1 \in \mathbb{R}^1$, with $x_0 \neq x_1$,

$$|\phi(t, x_0) - \phi(t, x_1)| = e^{at}|x_0 - x_1|.$$

Hence, the distance between any two points grows (exponentially) in time.

EXAMPLE 4.11.2 Consider the vector field

$$\begin{aligned} \dot{r} &= \sin\frac{\pi}{r}, \\ \dot{\theta} &= r, \end{aligned} \qquad (r, \theta) \in \mathbb{R}^+ \times S^1. \tag{4.11.4}$$

The flow generated by (4.11.4) has a countable infinity of periodic orbits given by

$$(r(t), \theta(t)) = \left(\frac{1}{n}, \frac{t}{n} + \theta_0\right), \qquad n = 1, 2, 3, \cdots. \tag{4.11.5}$$

It is easy to verify that the periodic orbits are stable for n even and unstable for n odd. Hence, in a compact region of the phase space, (4.11.4) has a countable infinity of unstable periodic orbits. We leave it as an exercise for the reader to verify that (4.11.4) may exhibit sensitive dependence on initial conditions in (open) annuli bounded by adjacent stable periodic orbits (see Exercise 4.72). However, (4.11.4) is only topologically transitive in the (open) annuli bounded by adjacent stable and unstable periodic orbits.

EXAMPLE 4.11.3 Consider the vector field on the two torus, $T^2 \equiv S^1 \times S^1$

$$\begin{aligned} \dot{\theta}_1 &= \omega_1, \\ \dot{\theta}_2 &= \omega_2, \end{aligned} \qquad (\theta_1, \theta_2) \in T^2, \tag{4.11.6}$$

with

$$\frac{\omega_1}{\omega_2} = \text{irrational}. \tag{4.11.7}$$

Then, it follows from Section 1.2A, Example 1.2.3, that the flow generated by (4.11.6) is topologically transitive on T^2. From (4.11.7) it is easy to see that the flow generated by (4.11.6) has no periodic orbits. We leave it as an exercise for the reader (see Exercise 4.73) to show that on T^2 the flow generated by (4.11.6) does *not* have sensitive dependence on initial conditions.

EXAMPLE 4.11.4 Consider the following integrable twist map

$$\begin{pmatrix} I \\ \theta \end{pmatrix} \mapsto \begin{pmatrix} I \\ 2\pi\Omega(I) + \theta \end{pmatrix} \equiv \begin{pmatrix} f_1(I,\theta) \\ f_2(I,\theta) \end{pmatrix}, \qquad (I,\theta) \in \mathbb{R}^+ \times S^1, \tag{4.11.8}$$

and

$$\frac{\partial \Omega}{\partial I}(I) \neq 0 \qquad \text{(twist condition)}. \tag{4.11.9}$$

The nth iterate of (4.11.8) is easily calculated and is given by

$$\begin{pmatrix} I \\ \theta \end{pmatrix} \mapsto \begin{pmatrix} I \\ 2\pi n\Omega(I) + \theta \end{pmatrix} \equiv \begin{pmatrix} f_1^n(I,\theta) \\ f_2^n(I,\theta) \end{pmatrix}. \tag{4.11.10}$$

The simple form of (4.11.8) and (4.11.10) enable us to easily verify the following.

1. Equation (4.11.8) is *not* topologically transitive, since all orbits remain on invariant circles.

2. The periodic orbits of (4.11.8) are dense in the phase space. This uses the twist condition.

3. Equation (4.11.8) has sensitive dependence on initial conditions due to the twist condition. This can be seen as follows. For (I_0, θ_0), $(I_1, \theta_1) \in \mathbb{R}^+ \times S^1$, with $I_0 \neq I_1$, we have

$$\begin{aligned} &|(f_1^n(I_0,\theta_0) - f_1^n(I_1,\theta_1)), (f_2^n(I_0,\theta_0) - f_2^n(I_1,\theta_1))| \\ &= |((I_0 - I_1), (2\pi n(\Omega(I_0) - \Omega(I_1)) + (\theta_0 - \theta_1))|. \end{aligned} \tag{4.11.11}$$

Therefore, from (4.11.9), $\Omega(I_0) - \Omega(I_1) \neq 0$. Thus, we see from (4.11.11) that as n increases, the θ components of nearby points drift apart. However, the rate of separation is *not* exponential.

EXAMPLE 4.11.5 Consider

$$\sigma: \Sigma^N \longrightarrow \Sigma^N, \tag{4.11.12}$$

where Σ^N is the space of bi-infinite sequences of N symbols and σ is the shift map as described in Section 4.2. Then we have proven the following.

1. Σ^N is compact and invariant (see Proposition 4.2.4).

2. σ is topologically transitive, i.e., σ has an orbit that is dense in Σ^N (see Proposition 4.2.7).

3. σ has sensitive dependence on initial conditions (see Section 4.1E).

4. σ has a countable infinity of periodic orbits (see Proposition 4.2.7) that are dense in Σ^N (see Exercise 4.74).

Thus, Σ^N is a chaotic, compact invariant set for σ. From Sections 4.3, 4.4, and 4.8 we know that two-dimensional maps and three-dimensional autonomous vector fields may possess compact invariant sets on which the dynamics are topologically conjugate to (4.11.12). In all of these cases, homoclinic (and possibly heteroclinic) orbits are the underlying mechanism that gives rise to such behavior. This is important, because this knowledge enables us to develop techniques (e.g., Melnikov's method) that predict (in terms of the system parameters) when chaotic dynamics occur in specific dynamical systems.

We make the following remarks concerning these examples.

Remark 1. Example 4.11.1 illustrates why we require chaotic invariant sets to be compact.

Remark 2. Example 4.11.2 illustrates why having an infinite number of unstable periodic orbits in a compact, invariant region of phase space is not by itself a sufficient condition for chaotic dynamics.

Remark 3. Example 4.11.3 describes a vector field having a compact invariant set (which is actually, the entire phase space) on which the dynamics are topologically transitive, but it does not have sensitive dependence on initial conditions.

Remark 4. Example 4.11.4 describes a two-dimensional integrable map that has sensitive dependence on initial conditions and the periodic orbits are dense in the phase space, but it is not topologically transitive.

Thus, taken together, Examples 4.11.1 through 4.11.4 show the importance of Definition 4.11.2 being satisfied completely. Example 4.11.5 shows how chaotic dynamics can arise in many dynamical systems. However, it does not address the question of observability.

DEFINITION 4.11.3 Suppose $\mathcal{A} \subset \mathbb{R}^n$ is an attractor. Then $\mathcal{A}$ is called a *strange attractor* if it is chaotic.

Hence, if we want to prove that a dynamical system has a strange attractor we might proceed as follows.

Step 1. Find a trapping region, $\mathcal{M}$, in the phase space (see Definition 1.1.13).

Step 2. Show that $\mathcal{M}$ contains a chaotic invariant set Λ. In practice, this means showing that inside $\mathcal{M}$ is a homoclinic orbit (or heteroclinic cycle) which has associated with it an invariant Cantor set on which the dynamics are topologically conjugate to a full shift on N symbols (recall Sections 4.4 and 4.8).

Step 3. Then, from Definition 1.1.26,

$$\bigcap_{t>0} \phi(t, \mathcal{M}) \quad \left(\text{resp.} \bigcap_{n>0} g^n(\mathcal{M}) \right) \equiv \mathcal{A} \tag{4.11.13}$$

is an attracting set. Moreover, $\Lambda \subset \mathcal{A}$ (see Exercise 4.75) so that $\mathcal{A}$ contains a mechanism that gives rise to sensitive dependence on initial conditions; in order to conclude that $\mathcal{A}$ is a strange attractor we need only demonstrate the following.

1. The sensitive dependence on initial conditions on Λ extends to $\mathcal{A}$;
2. $\mathcal{A}$ is topologically transitive.

Hence, in just three steps we can show that a dynamical system possesses a strange attractor. In this book we have developed techniques and seen examples of how to carry out Steps 1 and 2. However, the third step is the killer, namely, showing that $\mathcal{A}$ is topologically transitive. This is because a single, *stable* orbit in $\mathcal{A}$ will destroy topological transitivity and, in Section 4.7, we have seen that periodic sinks are always associated with quadratic homoclinic tangencies. Moreover, as a result of Newhouse's work, at least for two-dimensional dissipative maps, these homoclinic tangencies are persistent in the sense that if we destroy a particular tangency we will create another elsewhere in the homoclinic tangle. This is largely the reason why there is yet to be an analytical proof of the existence of a strange attractor for the periodically forced, damped Duffing oscillator, despite an enormous amount of numerical evidence (see Greenspan and Holmes [1984]).

At present, there exist rigorous results concerning strange attractors (by our Definition 4.11.3) in the following areas.

1. *One-Dimensional Non-Invertible Maps.* For maps such as

$$x \mapsto \mu x(1 - x)$$

or

$$x \mapsto x^2 - \mu$$

with μ a parameter there now exists a fairly complete existence theory for strange attractors. The reader should consult Jakobsen [1981], Misiurewicz [1981], Johnson [1987], and Guckenheimer and Johnson [1989].

2. *Hyperbolic Attractors of Two-Dimensional Maps.* Plykin [1974], Nemytskii and Stepanov [1989] and Newhouse [1980] have constructed examples of hyperbolic attracting sets which satisfy Definition 4.11.2. These examples are somewhat artificial in the sense that one would not expect them to arise in Poincaré maps of ordinary differential equations that arise in typical applications.

3. *Lorenz-Like Systems.* The topology associated with the Lorenz equations (see Sparrow [1982]) avoids many of the problems associated with Newhouse sinks. Consequently, in the past few years there has been much progress in proving that the Lorenz equations (along with slightly modified versions of the Lorenz equations) possess a strange attractor. The reader is referred to Sinai and Vul [1981], Afraimovich, Bykov, and Silnikov [1983], Rychlik [1987], and Robinson [1988].

4. *The Hénon Map.* The Hénon map is defined to be

$$\begin{aligned} x &\mapsto y, \\ y &\mapsto -\varepsilon x + \mu - y^2, \end{aligned} \tag{4.11.14}$$

where ε and μ are parameters. Over the past ten years a large amount of numerical evidence has suggested the existence of a strange attractor in this map. Recently, Benedicks and Carleson [1988] have proven that, for ε small, (4.11.14) does indeed possess a strange attractor.

Thus, the "strange attractor problem" is still far from being solved. In particular, there is a need for thoroughly studied examples in higher dimensions, for vector fields in dimensions larger than three and for maps of dimension larger than two. A variety of examples of systems undergoing (rigorously proven) chaotic behavior in high dimensions can be found in Wiggins [1988]; however, the attractive nature of the chaos in these examples has not been studied.

If a dynamical system possesses a chaotic invariant set, then an obvious question arises: namely, how is the chaos manifested in terms of "random" or "unpredictable" behavior of the system? The answer to this question depends on the geometry of the construction of the chaotic invariant set

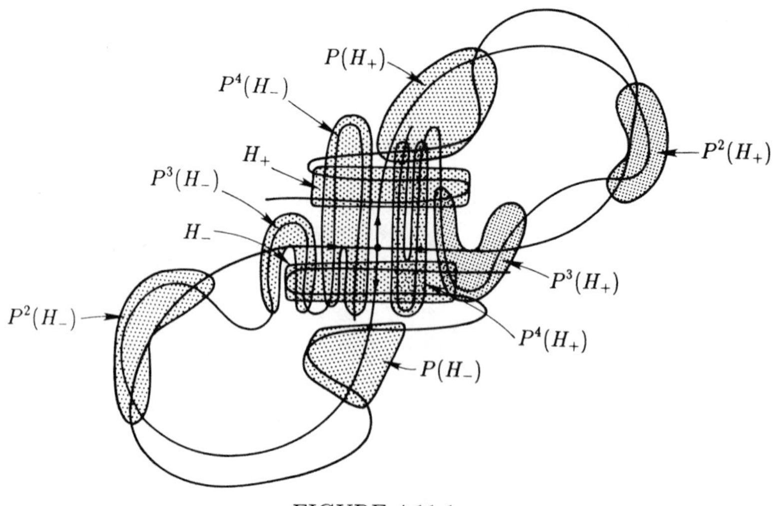

FIGURE 4.11.1.

and, thus, the answer varies from problem to problem (as we should expect). Let us consider an example, our old friend the periodically forced, damped Duffing oscillator.

This system is given by

$$\begin{aligned} \dot{x} &= y, \\ \dot{y} &= x - x^3 + \varepsilon(-\delta y + \gamma \cos \omega t). \end{aligned} \tag{4.11.15}$$

We know from Section 4.5F that, for ε sufficiently small and $\delta < \left(\frac{3\pi\omega}{2\sqrt{2}\,\mathrm{sech}(\pi\omega/2)}\right)\gamma$, the Poincaré map associated with (4.11.15) possesses transverse homoclinic orbits to a hyperbolic fixed point. Theorem 4.4.2 implies that (4.11.15) has chaotic dynamics. However, we want to interpret this chaos specifically in terms of the dynamics of (4.11.15). We will do this by constructing the chaotic invariant set geometrically and describing the associated symbolic dynamics geometrically. Our discussion will be heuristic, but, at this stage, the reader should easily be able to supply the necessary rigor (see Exercise 4.81).

Consider the two "horizontal" strips labeled H_+ and H_- in Figure 4.11.1. Under the Poincaré map, denoted P, H_+ and H_- are mapped over themselves in iterates as shown heuristically in Figure 4.11.1. It should be clear that the horizontal (resp. vertical) boundaries of H_+ and H_- map to horizontal (resp. vertical) boundaries of $P^4(H_+)$ and $P^4(H_-)$, respectively. Thus, one can show that Assumptions 1 and 3 of Section 4.3 hold (see Exercise 4.80). Therefore, $H_+ \cup H_-$ contains an invariant Cantor set, Λ, on which the dynamics are topologically conjugate to a full shift on two symbols. A point starting in $\Lambda \cap H_+$ makes a circuit around the right-hand homoclinic tangle before returning to Λ. A point starting in $\Lambda \cap H_-$ makes a circuit around the left-hand homoclinic tangle before returning to Λ. A symbol sequence such as

$$(\cdots + + + - - - + \cdot - + - - \cdots)$$

thus corresponds to an initial condition starting in H_-, going to H_+ under P^4 (hence making a circuit around the left-hand homoclinic tangle), and then going back to H_- under P^4 (hence making a circuit around the right-hand homoclinic tangle), etc. The geometrical meaning of "chaos" should be clear for this system, but the reader should do Exercise 4.80.

We end this section with some final remarks.

Remark 1. The dynamics of the full shift on N symbols best describes what we mean by the term "chaos" as applied to deterministic dynamical systems. The system is purely deterministic; however, the dynamics are such that our inability to *precisely* specify the initial conditions results in behavior that appears random or unpredictable.

Remark 2. We did not include in our definition of the chaotic invariant set (Definition 4.11.2) the requirement of density of periodic points. If the chaotic invariant set is hyperbolic, then, by the shadowing lemma (see, e.g., Shub [1987]), it follows immediately that the periodic points are dense. Moreover, Grebogi et al. [1985] have obtained numerical evidence for the existence of chaotic attractors in maps of the N-torus which have the property that orbits in the attractor densely cover the N-torus.

Remark 3. In our definition of sensitive dependence on initial conditions (Definition 4.11.1) we did not require the separation rate to be exponential. This is because it appears now that the strange attractors observed in numerical experiments of typical dynamical systems arising in applications will not, in general, be hyperbolic. Hence, one should expect parts of the attractor to exhibit nonexponential contraction or expansion rates. This is an area where new analytical techniques need to be developed.

Remark 4. Positive Liapunov exponents have been a standard criterion for deciding when a dynamical system is "chaotic" over the past few years. Examples 4.10.1 and 4.10.3 show that this criterion should be interpreted with caution; see also Exercise 4.79.

Exercises

4.1 From Section 4.3 recall

$$\mathcal{L}_\Lambda^u = \{\text{continuous line bundles over } \Lambda \text{ contained in } \mathcal{S}_\Lambda^u\}$$

with typical "points" in $\mathcal{L}_\Lambda^u$ denoted by

$$\mathcal{L}_\Lambda^u(\alpha(z_0)) \equiv \bigcup_{z_0 \in \Lambda} L_{\alpha(z_0)}^u,$$

$$\mathcal{L}_\Lambda^u(\beta(z_0)) \equiv \bigcup_{z_0 \in \Lambda} L_{\beta(z_0)}^u,$$

where

$$L_{\alpha(z_0)}^u = \{(\xi_{z_0}, \eta_{z_0}) \in \mathbb{R}^2 \mid \xi_{z_0} = \alpha(z_0)\eta_{z_0}\},$$
$$L_{\beta(z_0)}^u = \{(\xi_{z_0}, \eta_{z_0}) \in \mathbb{R}^2 \mid \xi_{z_0} = \beta(z_0)\eta_{z_0}\}.$$

We defined a metric on $\mathcal{L}_\Lambda^u$ as follows

$$\|\mathcal{L}_\Lambda^u(\alpha(z_0)) - \mathcal{L}_\Lambda^u(\beta(z_0))\| = \sup_{z_0 \in \Lambda} |\alpha(z_0) - \beta(z_0)|. \tag{E4.1}$$

a) Prove that (E4.1) is indeed a metric.

b) Prove that $\mathcal{L}_\Lambda^u$ is a complete metric space with the metric (E4.1).

c) Recall the unstable invariant line bundle constructed in Theorem 4.3.6 that we denoted

$$E^u = \bigcup_{z_0 \in \Lambda} E_{z_0}^u.$$

For $\zeta_{z_0} = (\xi_{z_0}, \eta_{z_0}) \in E_{z_0}^u$ prove that

$$|Df^{-1}(z_0)\zeta_{z_0}| < \lambda|\zeta_{z_0}|$$

where $0 < \lambda < 1$.

4.2 Prove Theorem 4.3.7. *Hints:* 1) show that the map F defined on $\mathcal{L}_\Lambda^u$ in (4.3.53) can be extended to a map on line bundles over $\Lambda_{-\infty}$ (the μ_v-vertical curves) and not just Λ.

2) Next, let the graph of $x = v(y)$ be a μ_v-vertical curve in $\Lambda_{-\infty}$ and let $z_0 = (x_0, y_0)$ be a point on that curve. Let T_{z_0} denote the set of lines $\xi = \alpha(z_0)\eta$ with

$$\alpha(z_0) = \lim_{n\to\infty} \frac{v(y_n) - v(y_n')}{y_n - y_n'},$$

where $y_n \neq y'_n$ are two sequences approaching y_0 for which this limit exists. Show that $|\alpha(z_0)| \leq \mu_v$ and that the set of $\alpha(z_0)$, z_0 fixed, satisfying this is closed.

3) Let

$$\omega(T_{z_0}) = \max \alpha(z_0) - \min \alpha(z_0) \leq 2\mu_v,$$

where the maximum and minimum is taken over the set defined above. Show that *if* $\omega(T_{z_0}) = 0$, then the curve has a derivative at z_0 and that, since the two sequences, y_n and y'_n, were arbitrary, the derivative is continuous.

4) Finally, we will be through if, from Step 3, we show that $\omega(T_{z_0}) = 0$. This is done as follows. First show that $F(T_{z_0}) = T_{z_0}$, $z_0 \in \Lambda_{-\infty}$, by using the mean value theorem. Next use the contraction property to show that $\omega(T_{z_0}) = \omega(F(T_{z_0})) \leq \frac{1}{2}\omega(T_{z_0})$ and from this conclude that $\omega(T_{z_0}) = 0$. Does it follow that $E^u_{z_0}$ agrees with the tangent to this $\mathbf{C}^1$ curve?

If you need help see Moser [1973].

4.3 Consider the invariant set Λ constructed in Theorem 4.3.5. Using Theorems 4.3.6, 4.3.7, and 4.3.8 describe in detail the stable and unstable manifolds of Λ.

4.4 *Horseshoes are Structurally Stable.* Suppose that a map $f: D \to \mathbb{R}^2$ satisfies the hypothesis of Theorem 4.3.5. Then it possesses an invariant Cantor set Λ. Show that, for ε sufficiently small, the map $f + \varepsilon g$ (with g $\mathbf{C}^r$, $r \geq 1$, on D) also possesses an invariant Cantor set Λ_ε. Moreover, show that Λ_ε can be constructed so that $(f + \varepsilon g)\big|_{\Lambda_\varepsilon}$ is topologically conjugate to $f\big|_\Lambda$.

4.5 In Section 4.3 we considered a map f defined on the unit square D. Suppose that D was an arbitrary closed set in $\mathbb{R}^2$. What modifications would be necessary in order for Theorems 4.3.3, 4.3.5, 4.3.6, and 4.3.7 to hold?

4.6 Consider the map

$$\begin{aligned} \xi &\mapsto \lambda\xi + g_1(\xi, \eta), \\ \eta &\mapsto \mu\eta + g_2(\xi, \eta), \end{aligned} \qquad (\xi, \eta) \in U \subset \mathbb{R}^2, \tag{E4.2}$$

defined in (4.4.1). Under the transformation given in (4.4.3)

$$(x, y) = (\xi - h^u(\eta), \eta - h^s(\xi)),$$

with

$$\begin{aligned} W^s_{\text{loc}}(0) &= \text{graph} \;\; h^s(\xi), \\ W^u_{\text{loc}}(0) &= \text{graph} \;\; h^u(\eta). \end{aligned}$$

Show that (E4.2) takes the form

$$\begin{aligned} x &\mapsto \lambda x + f_1(x,y), \\ y &\mapsto \mu y + f_2(x,y), \end{aligned}$$

with

$$\begin{aligned} f_1(0,y) &= 0, \\ f_2(x,0) &= 0. \end{aligned}$$

What are the specific forms of f_1 and f_2 in terms of g_1 and g_2?

4.7 Consider the region V shown in Figure 4.4.1. Show that V can be chosen so that $f^{k_0}(V)$ and $f^{-k_1}(V_1)$ appear as in Figure 4.4.1. In particular, show that for some positive integers $k_0, k_1 > 0$, both $f^{k_0}(V)$ and $f^{k_1}(V)$ lie in the first quadrant with their sides coinciding with the particular pieces of $W^s(0)$ and $W^u(0)$ as shown in Figure 4.4.1.

4.8 Following the outline given after Theorem 4.4.3, prove the Smale–Birkhoff homoclinic theorem.

4.9 Discuss the dynamical similarities and differences between Moser's theorem (Theorem 4.4.2) and the Smale–Birkhoff homoclinic theorem (Theorem 4.4.3). In particular, how do orbits differ in the invariant sets constructed in each theorem?

4.10 Suppose $f: \mathbb{R}^2 \to \mathbb{R}^2$ is $\mathbf{C}^r$ ($r \geq 1$), having a hyperbolic fixed point at p_0 whose stable and unstable manifolds intersect transversely as shown in Figure E4.1a.

We are interested in the dynamics near p_0. Suppose that in local coordinates (x,y) near p_0 the linearization of f has the form

$$Df(p_0): \begin{pmatrix} x \\ y \end{pmatrix} \mapsto \begin{pmatrix} \lambda & 0 \\ 0 & \mu \end{pmatrix} \begin{pmatrix} x \\ y \end{pmatrix}, \qquad \begin{aligned} \lambda &< 1, \\ \mu &> 1, \end{aligned}$$

so that orbits near p_0 appear as in Figure E4.1b. However, we know that the stable and unstable manifolds near p_0 oscillate infinitely often as shown in Figure 4.1a.

Are Figures E4.1a and E4.1b contradictory? If not, show how orbits in Figure E4.1b are manifested in Figure E4.1a.

4.11 Consider a $\mathbf{C}^r$ ($r \geq 1$) diffeomorphism

$$f: \mathbb{R}^2 \to \mathbb{R}^2$$

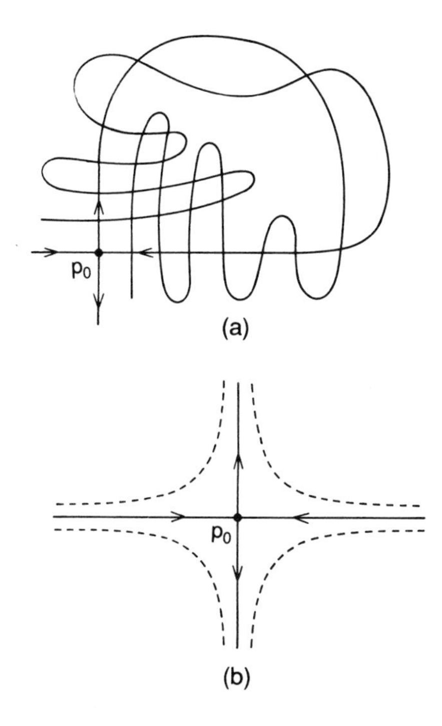

FIGURE E4.1.

having hyperbolic fixed points at p_0 and p_1, respectively. Suppose $q \in W^s(p_0) \cap W^u(p_1)$; then q is called a *heteroclinic* point, and if $W^s(p_0)$ intersects $W^u(p_1)$ transversely at q, q is called a *transverse heteroclinic point.* In order to be more descriptive, sometimes q is referred to as being *heteroclinic to p_0 and p_1*; see Figure E4.2a.

a) Does the existence of a transverse heteroclinic point imply the existence of a Cantor set on which some iterate of f is topologically conjugate to a full shift on N ($N \geq 2$) symbols?

b) Additionally, suppose that $W^u(p_0)$ intersects $W^s(p_1)$ transversely to form a *heteroclinic cycle* as shown in Figure E4.2b. Show that in this case one can find an invariant Cantor set on which some iterate of f is topologically conjugate to a full shift on N ($N \geq 2$) symbols. (*Hint:* mimic the proof of Theorem 4.4.2.)

c) Suppose that a branch of $W^u(p_0)$ coincides with a branch of $W^s(p_0)$, yet $W^s(p_0)$ intersects $W^u(p_1)$ transversely at q; see Figure E4.2c. Does it follow that you can find an invariant Cantor set on which some iterate of f is topologically conjugate to a full shift on N ($N \geq 2$) symbols?

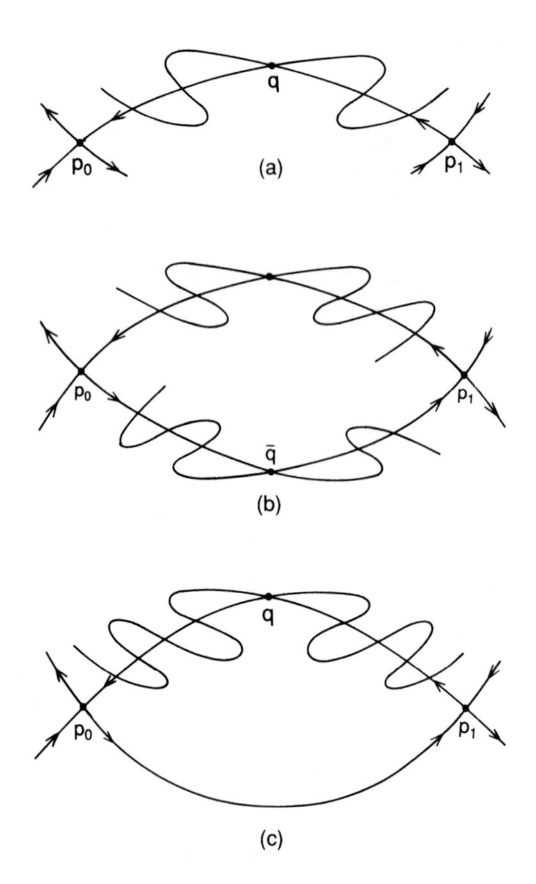

FIGURE E4.2.

4.12 Consider the parametrization of the homoclinic manifold given in (4.5.9). Show that the map

$$(t_0, \phi_0) \mapsto (q_0(-t_0), \phi_0), \qquad (t_0, \phi_0) \in \mathbb{R}^1 \times S^1,$$

is $\mathbf{C}^r$, one-to-one, and onto.

4.13 Recall Proposition 4.5.1. Prove that, for ε sufficiently small, $\gamma_\varepsilon(t)$ persists as a periodic orbit of period $T = 2\pi/\omega$ having the same stability type as $\gamma(t)$. For information concerning the persistence of the local stable and unstable manifolds, see Fenichel [1971] or Wiggins [1988].

4.14 Suppose that Γ_γ intersects Π_p transversely at some $p = (q_0(-t_0), \phi_0)$. Show that, for ε sufficiently small, $W^s(\gamma_\varepsilon(t))$ and $W^u(\gamma_\varepsilon(t))$ each intersect Π_p transversely at a distance $\mathcal{O}(\varepsilon)$ from p.

4.15 Recall Definition 4.5.1. Show that the points $p^s_{\varepsilon,\bar{i}}$ and $p^u_{\varepsilon,\bar{i}}$ "closest" to $\gamma_\varepsilon(t)$ in the sense of Definition 4.5.1 are unique. (*Hint:* study the proof of Lemma 4.5.2.)

4.16 Recall the set-up for the proof of Lemma 4.5.2. Choose

$$(q_0^s, \phi_0) \in W^s_{\text{loc}}(\gamma(t)) \cap \mathcal{N}(\varepsilon_0)$$

and

$$(q_\varepsilon^s, \phi_0) \in W^s_{\text{loc}}(\gamma_\varepsilon(t)) \cap \mathcal{N}(\varepsilon_0),$$

with the trajectories

$$(q_0^s(t), \phi(t)) \in W^s(\gamma(t))$$

and

$$(q_\varepsilon^s(t), \phi(t)) \in W^s(\gamma_\varepsilon(t))$$

satisfying

$$(q_0^s(0), \phi(0)) = (q_0^s, \phi_0)$$

and

$$(q_\varepsilon^s(0), \phi(0)) = (q_\varepsilon^s, \phi_0).$$

Prove that

$$\left|(q_\varepsilon^s(t), \phi(t)) - (q_0^s(t), \phi(t))\right| = \mathcal{O}(\varepsilon_0)$$

for $0 < t < \infty$.

4.17 Suppose $q_\varepsilon^s(t) \in W^s(\gamma_\varepsilon(t))$ is a solution of (4.5.6). Then show that

$$\left.\frac{\partial q_\varepsilon^s(t)}{\partial \varepsilon}\right|_{\varepsilon=0} \equiv q_1^s(t)$$

is bounded in t as $t \to \infty$. (*Hint:* as $t \to \infty$, $q_1^s(t)$ should behave as $\left.\frac{\partial \gamma_\varepsilon(t)}{\partial \varepsilon}\right|_{\varepsilon=0}$.

Does the same result hold for solutions in $W^u(\gamma_\varepsilon(t))$ as $t \to -\infty$?

4.18 Recall Theorem 4.5.6. Show that if $M(t_0, \phi_0) \neq 0$ for all $(t_0, \phi_0) \in \mathbb{R}^1 \times S^1$, then $W^s(\gamma_\varepsilon(t)) \cap W^u(\gamma_\varepsilon(t)) = \emptyset$. (*Hint:* study the proof of Lemma 4.5.2.)

4.19 Prove Theorem 4.5.7. Explain why we must take $n = 1$.

4.20 Recall Theorems 4.5.7, 4.5.8, and 4.5.9. Prove that a bifurcation value corresponding to a quadratic homoclinic tangency is the limit of a sequence of parameter values corresponding to subharmonic saddle-node bifurcations to successively higher and higher periods.

4.21 Consider the Poincaré maps associated with the damped, periodically forced Duffing equation on the cross-sections Σ^0, $\Sigma^{\pi/2}$, Σ^π, and $\Sigma^{3\pi/2}$ shown in Figure 4.5.18 for $\delta = 0$. Describe in detail how the geometry of the stable and unstable manifolds change on each cross-section for $\delta \neq 0$. (*Hint:* use the Melnikov function.)

4.22 *Melnikov's Method for Autonomous Perturbations*

Suppose we consider the $\mathbf{C}^r$ $(r \geq 2)$ vector field

$$\begin{aligned} \dot{x} &= \frac{\partial H}{\partial y}(x,y) + \varepsilon g_1(x,y;\mu,\varepsilon), \\ \dot{y} &= -\frac{\partial H}{\partial x}(x,y) + \varepsilon g_2(x,y;\mu,\varepsilon), \end{aligned} \qquad (x,y,\mu) \in \mathbb{R}^3,$$

or

$$\dot{q} = JDH(q) + \varepsilon g(q;\mu,\varepsilon), \tag{E4.3}$$

where

$$\begin{aligned} q &\equiv (x,y), \\ DH &= \begin{pmatrix} \frac{\partial H}{\partial x} \\ \frac{\partial H}{\partial y} \end{pmatrix}, \\ J &= \begin{pmatrix} 0 & 1 \\ -1 & 0 \end{pmatrix}, \\ g &= (g_1, g_2), \end{aligned}$$

with ε small and μ regarded as a parameter.

Suppose that the unperturbed system (i.e., (E4.3) with $\varepsilon = 0$) satisfies Assumptions 1 and 2 of Section 4.5A.

a) Show that the function

$$\overline{M}_1(\alpha;\mu) = \int_0^{T^\alpha} (DH \cdot g)(q^\alpha(t);\mu)dt$$

can be used to prove the existence of periodic orbits in the perturbed systems.

(*Hint:* proceed along the lines of the case for nonautonomous perturbations and derive a Poincaré map. In this case the Poincaré map will be one-dimensional. For the domain of the Poincaré map, choose a radial line emanating from the fixed point of center type. Use action-angle variables and argue that the first return time to the Poincaré section is $T^\alpha + \mathcal{O}(\varepsilon)$. Show that neglecting the $\mathcal{O}(\varepsilon)$ term in the first return time only results in an error of $\mathcal{O}(\varepsilon^2)$ in the Poincaré map.)

b) In the autonomous case, what is the analog of the nondegeneracy condition

$$\frac{\partial \Omega}{\partial I} \frac{\partial M_1^{m/n}(I, \theta)}{\partial \theta} \neq 0$$

in the autonomous case? (Recall the nondegeneracy condition was necessary in order to use the implicit function theorem to argue that the higher order terms in the Poincaré map could be neglected.)

c) Show that if the perturbation is Hamiltonian, then we have

$$\overline{M}_1(\alpha; \mu) = 0.$$

Can you explain this result geometrically in terms of the level sets of the Hamiltonian?

d) Discuss the geometrical meaning of

$$M(\mu) = \int_{-\infty}^{\infty} (DH \cdot g)(q_0(t), \mu) dt.$$

Compare $M(\mu)$ with $M(t_0, \phi_0)$ derived in Section 4.5.

4.23 Recall Exercise 1.2.19 which considered the librational motion of an arbitrarily shaped satellite in a planar, elliptical orbit. The equations of motion were given by

$$\begin{aligned} \psi'' + 3K_i \sin\psi \cos\psi = \varepsilon[&2\mu \sin\theta(\psi' + 1) \\ &+ 3\mu K_i \sin\psi \cos\psi \cos\theta] + \mathcal{O}(\varepsilon^2). \end{aligned}$$

Use Melnikov's method to study orbits homoclinic to hyperbolic periodic orbits for $\varepsilon \neq 0$. Describe the physical manifestation of any chaotic dynamics that arise in this problem.

4.24 Recall our discussion of the *driven Morse oscillator* given in Exercise 1.2.20. The equations were given by

$$\begin{aligned} \dot{x} &= y, \\ \dot{y} &= -\mu\left(e^{-x} - e^{-2x}\right) + \varepsilon\gamma\cos\omega t. \end{aligned} \tag{E4.4}$$

a) Show that for $\varepsilon = 0$, $(x, y) = (\infty, 0)$ is a nonhyperbolic fixed point of (E4.4) that is connected to itself by a homoclinic orbit.

We would like to apply Melnikov's theory to (E4.4) in order to see if (E4.4) has horseshoes; however, the fixed point having the homoclinic orbit is nonhyperbolic. Therefore, the theory developed in Section 4.5 does not immediately apply. Schecter [1987a], [1987b] has extended Melnikov's method so that it applies to nonhyperbolic fixed points. However, we will not develop his technique.

Instead, we introduce the following transformation of variables

$$x = -2\log u, \qquad y = v, \tag{E4.5}$$

and reparametrize time as follows

$$\frac{ds}{dt} = -\frac{u}{2}.$$

b) Rewrite (E4.4) in these new variables and show that the resulting equation, for $\varepsilon = 0$, has a *hyperbolic* fixed point at the origin that is connected to itself by a homoclinic orbit. Apply Melnikov's method in order to study homoclinic orbits for $\varepsilon \neq 0$.

c) Describe the resulting chaotic dynamics in both the $x - y$ and $u - v$ coordinate systems.

This problem was originally solved by Bruhn [1989]. The transformation (E4.5) is known as a "McGehee transformation" in honor of Richard McGehee, who first cooked it up in order to study a degenerate fixed point at infinity in a celestial mechanics problem (see McGehee [1974]). Such singularities frequently arise in mechanics and coordinate transformations such as (E4.5) can greatly facilitate the analysis. An excellent introduction to such problems can be found in Devaney [1982].

4.25 Recall Exercise 1.2.33. Consider the following two-degree-of-freedom Hamiltonian system

$$\begin{aligned} \dot{\phi} &= v, \\ \dot{v} &= -\sin\phi + \varepsilon(x-\phi), \\ \dot{x} &= y, \\ \dot{y} &= -\omega^2 x - \varepsilon(x-\phi), \end{aligned} \qquad (\phi, v, x, y) \in S^1 \times \mathbb{R}^1 \times \mathbb{R}^1 \times \mathbb{R}^1, \tag{E4.6}$$

with Hamiltonian

$$H^\varepsilon(\phi, v, x, y) = \frac{v^2}{2} - \cos\phi + \frac{y^2}{2} + \frac{\omega^2 x}{2} + \frac{\varepsilon}{2}(x-\phi)^2.$$

Use the *method of reduction* developed in Exercise 1.2.32 and the homoclinic Melnikov method developed in Section 4.5 to study orbits homoclinic to hyperbolic periodic orbits in (E4.6). How do the homoclinic orbits change as one moves from level set to level set of the Hamiltonian? Also, what does it mean for a periodic orbit to be hyperbolic in a two-degree-of-freedom Hamiltonian system (*hint:* one "direction" is irrelevant).

4.26 Consider the $\mathbf{C}^r$ ($r \geq 2$) vector field

$$\begin{aligned} \dot{x} &= f_1(x,y) + \varepsilon g_1(x,y,t;\varepsilon), \\ \dot{y} &= f_2(x,y) + \varepsilon g_2(x,y,t;\varepsilon), \end{aligned} \qquad (x,y) \in \mathbb{R}^2,$$

or

$$\dot{q} = f(q) + \varepsilon g(q, t; \varepsilon), \tag{E4.7}$$

where

$$\begin{aligned} q &\equiv (x, y), \\ f &\equiv (f_1, f_2), \\ g &\equiv (g_1, g_2), \end{aligned}$$

with ε small and $g(q, t; \varepsilon)$ periodic in t with period $T = 2\pi/\omega$.

Assumption: For $\varepsilon = 0$, (E4.7) has a hyperbolic fixed point at p_0 that is connected to itself by a homoclinic orbit, $q_0(t)$, i.e., $\lim_{t\to\pm\infty} q_0(t) = p_0$.

a) Derive a measure of the distance between the stable and unstable manifolds of the hyperbolic periodic orbit that persists in (E4.7) for ε sufficiently small. (*Hint:* follow the steps in Section 4.5 as closely as possible. If you need help, see Melnikov [1963].)

b) Using the parametrization of the unperturbed homoclinic orbit in terms of $(t_0, \phi_0) \in \mathbb{R}^1 \times S^1$ as defined in (4.5.9), is the Melnikov function obtained in part a) periodic in both t_0 and ϕ_0? Explain fully the reasons behind your answer.

4.27 How is the Melnikov theory modified if in the unperturbed vector field we instead had two hyperbolic fixed points, p_1 and p_2, connected by a heteroclinic orbit, $q_0(t)$, i.e., $\lim_{t\to\infty} q_0(t) = p_1$ and $\lim_{t\to-\infty} q_0(t) = p_2$? (*Hint:* follow the development of the homoclinic Melnikov theory in Section 4.5. You should arrive at the same formula for the distance between the stable and unstable manifolds; however, the geometrical interpretation will be different.)

4.28 Consider the vector field

$$\begin{aligned} \dot{\theta} &= \varepsilon v, \\ \dot{v} &= -\varepsilon \sin\theta + \varepsilon^2 \gamma \cos\omega t, \end{aligned} \qquad (\theta, v) \in S^1 \times \mathbb{R}^1, \quad \varepsilon \text{ small.}$$

Apply Melnikov's method to show that the Poincaré map associated with this equation has transverse homoclinic orbits. What problems arise? Can any conclusions be drawn for homoclinic orbits arising in the following vector field?

$$\begin{aligned} \dot{\theta} &= \varepsilon v, \\ \dot{v} &= -\varepsilon \sin\theta + \varepsilon^2(-\delta v + \gamma \cos\omega t). \end{aligned}$$

What implications do these examples have for applying Melnikov's method to vector fields

$$\dot{x} = \varepsilon f(x, t), \qquad f - T\text{-periodic in } t, \quad x \in \mathbb{R}^2,$$

that are transformed into

$$\dot{y} = \overline{f}(y) + \varepsilon g(y,t), \qquad y \in \mathbb{R}^2, \quad \overline{f}(y) = \frac{1}{T}\int_0^T f(y,t)dt,$$

by the method of averaging?

We refer the reader to Holmes, Marsden, and Scheurle [1988] for more examples of problems of this type along with some rigorous results.

4.29 Recall the discussion at the beginning of Section 4.6A. Let k denote the number of intersection points of $W^s(p_0)$ with $U[f^{-1}(q), q]$ *between the points* q *and* $f^{-1}(q)$.

a) Show that if f is orientation-preserving, then we must have $k \geq 1$.

b) What further restrictions can be put on k if f is orientation-preserving *and* all of the intersection points are transversal?

c) What restrictions can be put on k if f is orientation- *and* area-preserving?

4.30 In our discussion of flux across the pseudoseparatrix in Section 4.6B we stated that $f^{-n+1}(L_{1,2}(1))$ may intersect $f^{-k+1}(L_{2,1}(1))$ for $k = 1, \cdots, n-1$. Why can't $f^{-n+1}(L_{1,2}(1))$ intersect $f^{-k+1}(L_{2,1}(1))$ for $k \geq n$? (*Hint:* look ahead to Section 4.6C under self-intersecting turnstiles.)

4.31 Prove Proposition 4.6.4.

4.32 Prove Proposition 4.6.5.

4.33 Let f denote the Poincaré map derived from a planar, linearly damped, periodically forced oscillator, i.e., a vector field of the form

$$\begin{aligned} \dot{x} &= y, \\ \dot{y} &= -\delta y + f(x,t), \end{aligned} \qquad (x,y) \in \mathbb{R}^2,$$

where $\delta > 0$ and $f(x,t)$ is periodic in t with period $T > 0$. Show that for any set $A \in \mathbb{R}^2$

$$\mu(f(A)) = \delta\mu(A),$$

where $\mu(\cdot)$ denotes the area of a set.

4.34 Prove Lemma 4.6.1.

4.35 Prove Lemma 4.6.2.

4.36 Consider the vector field discussed in Section 4.6B

$$\begin{aligned} \dot{x} &= y, \\ \dot{y} &= x - x^3 - \varepsilon\delta y + \varepsilon\gamma\cos\omega t, \end{aligned}$$

where δ and γ have been scaled so that we can apply the Melnikov method.

a) Compute the homoclinic Melnikov function and describe the surface in $\gamma-\delta-\omega$ space where the bifurcation to homoclinic orbits occurs.

b) Describe how the lobe structure varies with $\gamma-\delta-\omega$ and, hence, how escape from the potential well depends on the parameters.

4.37 Suppose that the general map f discussed in Section 4.6 satisfies

$$\mu(f(A)) = \delta\mu(A)$$

for any set $A \subset \mathbb{R}^2$ with some positive constant. Then study the limit of Equation (4.6.8) as $n \to \infty$. Compare the result you obtained to that in Proposition 4.6.3.

Using the result obtained in Exercise 4.32 as well as Proposition 4.6.3, compare transport in and out of the potential well in the vector field

$$\begin{aligned} \dot{x} &= y, \\ \dot{y} &= x - x^2 - \varepsilon\delta y + \varepsilon\gamma\cos\omega t \end{aligned}$$

for $\delta = 0$ and $\delta > 0$.

4.38 Consider the damped, periodically forced Duffing oscillator

$$\begin{aligned} \dot{x} &= y, \\ \dot{y} &= x - x^3 + \varepsilon(-\delta y + \gamma\cos\omega t). \end{aligned} \tag{E4.8}$$

Suppose the parameters are chosen such that (E4.8) possesses transverse homoclinic orbits to a hyperbolic fixed point. Choose any lobe in the homoclinic tangle. Since (E4.8) is dissipative for $\delta > 0$ one would expect the lobe area to shrink in size under iteration by the Poincaré map. Check if this is so by computing the lobe area using the Melnikov function (see Section 4.6D). Explain your result.

4.39 Consider the periodically forced, undamped pendulum

$$\begin{aligned} \dot{\theta} &= v, \\ \dot{v} &= -\sin\theta + \gamma\sin\omega t, \end{aligned} \qquad (\theta, v) \in S^1 \times \mathbb{R}^1, \tag{E4.9}$$

with $\gamma, \omega > 0$. For $\gamma = 0$ the phase space of (E4.9) appears as in Figure E4.3a. In particular, $(\theta, v) = (\pi, 0)$ is a hyperbolic fixed point of (E4.9) having a pair of homoclinic orbits. This situation is most conveniently illustrated by depicting the phase space on the plane and identifying the points whose θ coordinates differ by an integer multiple of 2π; see Figure E4.3b. The homoclinic orbits divide the phase space into three disjoint regions denoted R_1, R_2, and R_3 in Figure E4.3b. The goal of this problem is to study transport between these regions for $\gamma \neq 0$.

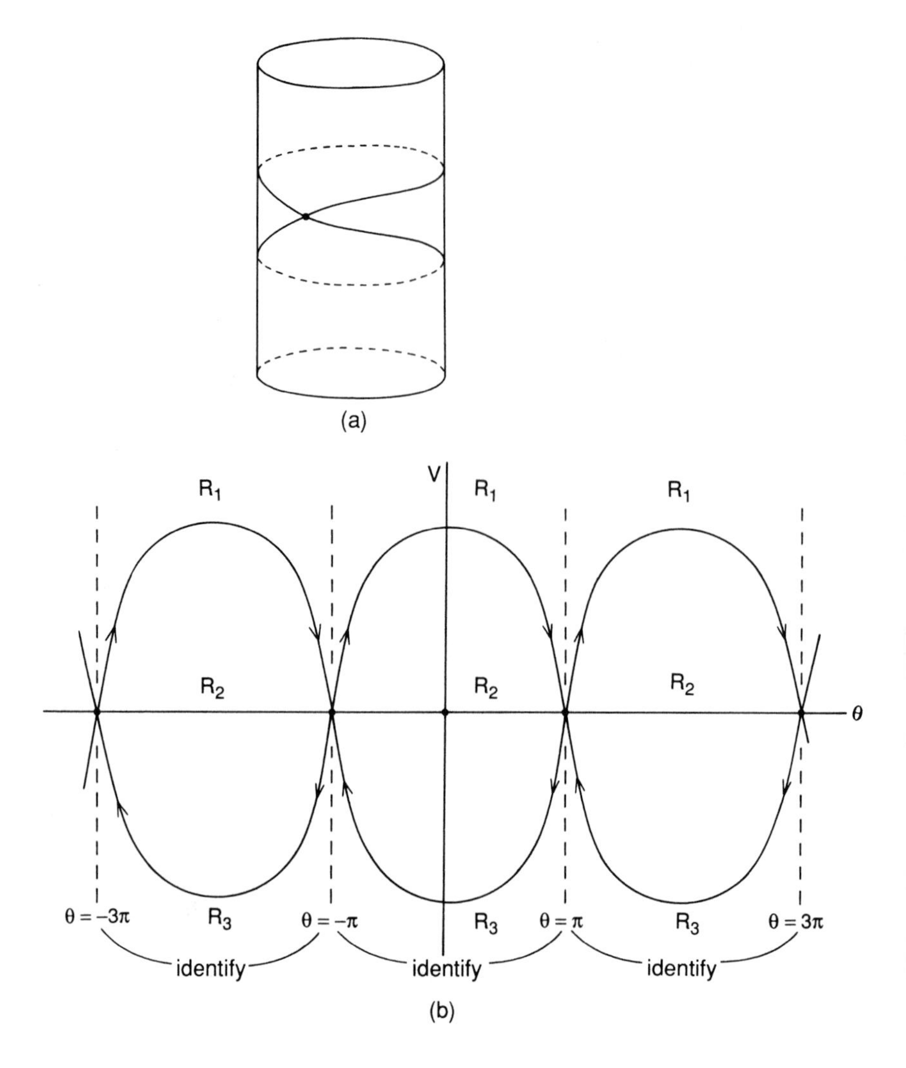

FIGURE E4.3.

For $\gamma \neq 0$ consider the "suspended" system

$$\begin{aligned} \dot{\theta} &= v, \\ \dot{v} &= -\sin\theta + \gamma \sin\phi, \qquad (\theta, v, \phi) \in S^1 \times \mathbb{R}^1 \times S^1, \\ \dot{\phi} &= \omega, \end{aligned}$$

with the usual Poincaré map defined on the cross-section Σ^{ϕ_0}.

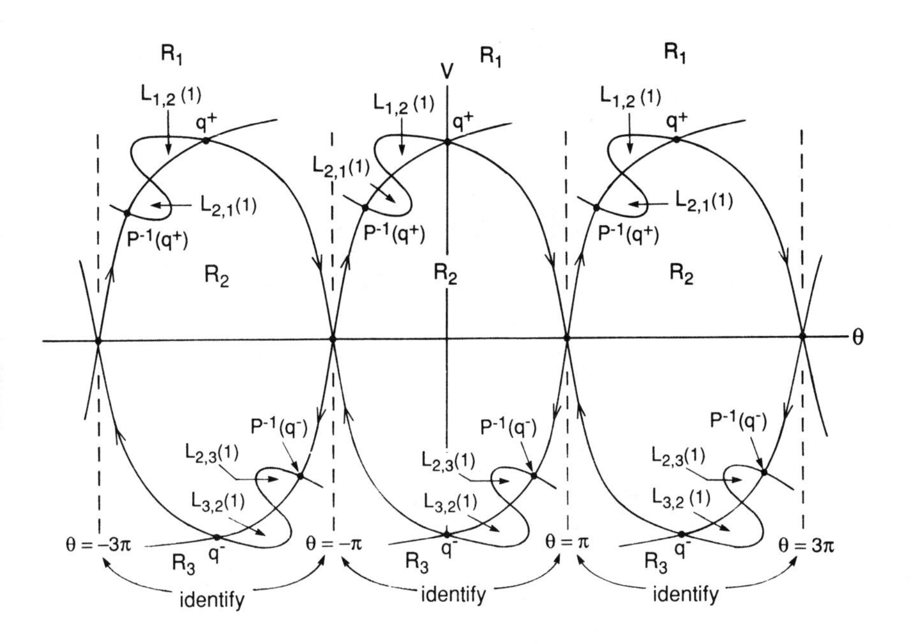

FIGURE E4.4.

a) Show that, for γ sufficiently small, the Poincaré map on the cross-section Σ^0, denoted P, has pips at $(\theta, v) = (0, v^+) \equiv q^+$ and $(\theta, v) = (0, v^-) \equiv q^-$, with $v^+ = -v^- > 0$, and that there are precisely two lobes between q^+ and $P^{-1}(q^+)$ (denoted $L_{1,2}(1)$ and $L_{2,1}(1)$, respectively) and precisely two lobes between q^- and $P^{-1}(q^-)$ (denoted $L_{2,3}(1)$ and $L_{3,2}(1)$, respectively); see Figure E4.4. (*Hint:* use Melnikov's method.)

b) Show that P is invariant under the coordinate transformation

$$(\theta, v) \mapsto (-\theta, v)$$

plus time reversal.

c) Does the symmetry in part b) imply that there must exist two pips on $\theta = 0$?

d) Prove that the following conservation laws hold.

Conservation of Species

$$\sum_{j=1}^{3} (T_{i,j}(n) - T_{i,j}(n-1)) = 0, \qquad i = 1, 2, 3.$$

Conservation of Area

$$\sum_{i=1}^{3} (T_{i,j}(n) - T_{i,j}(n-1)) = 0, \qquad j = 1, 2, 3.$$

e) Show that part b) implies

$$\begin{aligned} T_{1,3}(n) &= T_{3,1}(n), \\ T_{2,1}(n) &= T_{2,3}(n), \\ T_{1,2}(n) &= T_{3,2}(n). \end{aligned}$$

f) If we view $T_{i,j}(n) - T_{i,j}(n-1)$, $i, j = 1, 2, 3$, as nine unknown quantities, show that the conservation laws from part d) and the relationships from part e) can be used to form eight independent equations for these nine unknowns. From this conclude that knowledge of *one* of the $T_{i,j}(n)$ allows us to determine the remaining eight.

g) Show that

$$\begin{aligned} T_{3,1}(n) = \sum_{m=1}^{n-1} (n-m)\{ & \mu(L_{2,1}(1) \cap P^m(L_{3,2}(1))) \\ & - \mu(L_{2,1}(1) \cap P^m(L_{2,3}(1))) \\ & - \mu(L_{1,2}(1) \cap P^m(L_{3,2}(1))) \\ & + \mu(L_{1,2}(1) \cap P^m(L_{2,3}(1)))\}. \end{aligned} \tag{E4.10}$$

(*Hint:* $P^m(L_{2,3}(1))$ may intersect $L_{3,2}(1)$, $P^m(L_{2,1}(1))$ may intersect $L_{1,2}(1)$, and $P^m(L_{3,2}(1))$ may intersect $L_{2,1}(1)$. These possible intersections of turnstile lobes provide the "route" for all of the terms shown in (E4.10). See Rom-Kedar and Wiggins [1989] for additional help.)

h) Discuss the numerical computation of (E4.10). Compare this procedure with a Monte Carlo calculation of the transport of phase space between the three regions.

i) Consider the damped, periodically forced pendulum

$$\begin{aligned} \dot{\theta} &= v, \\ \dot{v} &= -\sin\theta - \delta v + \gamma \sin \omega t, \end{aligned} \qquad \delta > 0.$$

Construct two pseudoseparatrices that separate the phase space into three disjoint regions. Assume that δ and γ are sufficiently small so that the Melnikov theory can be applied. Are there pips on $\theta = 0$?

j) Show that the conservation laws from part d) now become

$$\sum_{j=1}^{3} (T_{i,j}(n) - T_{i,j}(n-1)) = 0, \qquad i = 1, 2, 3,$$

$$\sum_{i=1}^{3} T_{i,j}(n) = \delta \sum_{i=1}^{3} T_{i,j}(n-1), \qquad j = 1, 2, 3.$$

k) Do the relations in part e) still hold?

l) Derive $T_{3,1}(n)$.

We remark that we have organized this exercise around the forced pendulum. However, little of our problem used the explicit form of the equations (with the exception of the Melnikov theory used to establish the existence and location of the pips as well as the number of lobes in a turnstile; however, this could have been carried out numerically). It was the form of the geometry of the manifolds that was most important.

The geometry of the manifolds associated with the forced pendulum are common to many examples. In some sense, it is a normal form for the 1:1 resonance in forced oscillators and two-dimensional maps such as the standard map (see Lichtenberg and Lieberman [1982]).

4.40 Consider the periodically forced undamped Duffing oscillator

$$\begin{aligned} \dot{x} &= y, \\ \dot{y} &= x - x^3 + \gamma \cos \phi, \qquad (x, y) \in \mathbb{R}^2, \\ \dot{\phi} &= \omega, \end{aligned} \tag{E4.11}$$

with $\gamma > 0$. Denote the Poincaré map associated with (E4.11) on the cross-section Σ^0 by P.

a) Show that for γ sufficiently small (E4.11) has pips $(x^+, 0) \equiv q^+$ and $(x^-, 0) \equiv q^-$ with $x^+, -x^- > 0$. Use these two pips to define pseudoseparatrices that divide the plane into three disjoint regions denoted R_1, R_2 and R_3; see Figure E4.5. Show that there are precisely two lobes between q^+ and $P^{-1}(q^+)$ (denoted $L_{1,2}(1)$ and $L_{2,1}(1)$, respectively) and two lobes between q^- and $P^{-1}(q^-)$ (denoted $L_{3,2}(1)$ and $L_{2,3}(1)$, respectively). (*Hint:* use Melnikov's method.)

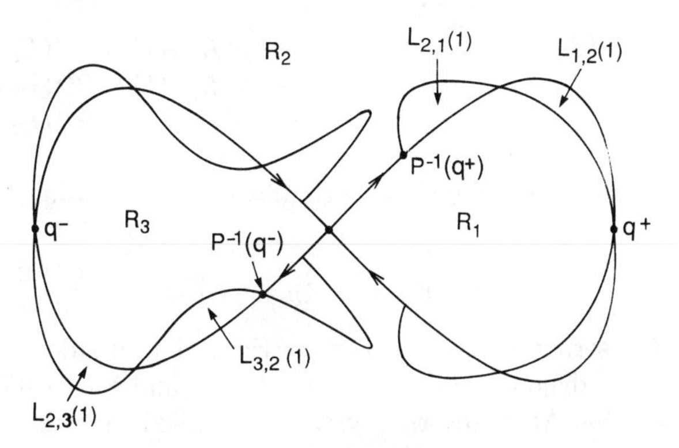

FIGURE E4.5.

The goal of this problem is to study transport between these three regions. Recall the motivational example discussed in Section 4.6B – a particle moving in a single-well potential subject to periodic excitation and damping. This exercise could be viewed as the study of a particle moving in a two-well (symmetric) potential subject to periodic excitation. The transport problem is concerned with the particle jumping from well to well, escape from either well, or capture into either well.

b) Show that P is invariant under the coordinate transformation

$$(x, y) \to (x, -y)$$

plus time reversal.

c) Does the symmetry in part b) imply that there must exist two pips on the x-axis?

d) Prove that the following conservation laws hold.

Conservation of Species

$$\sum_{j=1}^{3} (T_{i,j}(n) - T_{i,j}(n-1)) = 0, \qquad i = 1, 2, 3,$$

Conservation of Area

$$\sum_{i=1}^{e} (T_{i,j}(n) - T_{i,j}(n-1)) = 0, \qquad j = 1, 2, 3.$$

e) Show that $T_{1,3}(n) = T_{3,1}(n)$.

f) Show that

$$\begin{aligned} T_{3,1}(n) = \sum_{m=1}^{n-1} (n-m)\{ & \mu(L_{2,1}(1) \cap P^m(L_{3,2}(1))) \\ & - \mu(L_{2,1}(1) \cap P^m(L_{2,3}(1))) \\ & - \mu(L_{1,2}(1) \cap P^m(L_{3,2}(1))) \\ & + \mu(L_{1,2}(1) \cap P^m(L_{2,3}(1)))\}. \end{aligned}$$

g) Consider the damped, periodically forced Duffing oscillator

$$\begin{aligned} \dot{x} &= y, \\ \dot{y} &= x - x^3 - \delta y + \gamma \cos \omega t, \end{aligned} \qquad \delta > 0.$$

Construct two pseudoseparatrices that divide the plane into three disjoint regions. Assume that γ and δ are sufficiently small so that Melnikov theory can be applied. Are there pips on the x-axis?

h) Show that the conservation laws from part d) now become

$$\sum_{j=1}^{3}(T_{i,j}(n) - T_{i,j}(n-1)) = 0, \qquad i = 1, 2, 3,$$

$$\sum_{i=1}^{3} T_{i,j}(n) = \delta \sum_{i=1}^{3} T_{i,j}(n-1), \qquad i = 1, 2, 3.$$

i) Is it still true that $T_{1,3}(n) = T_{3,1}(n)$?

j) Derive $T_{3,1}(n)$.

4.41 In Exercises 4.39 and 4.40 the two branches of the stable and unstable manifolds of a hyperbolic fixed point were used to divide the phase space into three disjoint regions. However, the underlying phase space was very different in the two exercises. In Exercise 4.39 it was a cylinder and in Exercise 4.40 it was the plane. Discuss the similarities and differences in the transport of phase space between the three regions for these two different situations.

4.42 Recall Exercise 1.2.21. The equations for fluid particle motions in this flow are given by

$$\begin{aligned} \dot{x}_1 &= \frac{\partial \psi_0}{\partial x_2}(x_1, x_2) + \varepsilon \frac{\partial \psi_1}{\partial x_2}(x_1, x_2, t), \\ \dot{x}_2 &= -\frac{\partial \psi_0}{\partial x_1}(x_1, x_2) - \varepsilon \frac{\partial \psi_1}{\partial x_2}(x_1, x_2, t), \end{aligned} \tag{E4.12}$$

where

$$\begin{aligned} \psi_0(x_1, x_2) &= -x_2 + R\cos x_1 \sin x_2, \\ \psi_1(x_1, x_2, t) &= \frac{\gamma}{2}\left[\left(1 - \frac{2}{\omega}\right)\cos(x_1 + \omega_1 t + \theta)\right. \\ &\quad + \left.\left(1 + \frac{2}{\omega}\right)\cos(x_1 - \omega t - \theta)\right] \sin x_2. \end{aligned}$$

For $\varepsilon = 0$, $R > 1$, the streamlines of (E4.12) appear as in Figure E4.6. Note the two hyperbolic fixed points on the x_1-axis, denoted p_+ and p_-, respectively. The fixed points are connected by a pair of heteroclinic orbits denoted Γ_0 and Γ_u. This heteroclinic cycle forms a region of trapped fluid that is shaded in Figure E4.6.

a) For $\varepsilon \neq 0$ show that Γ_0 persists and results in a barrier which fluid cannot cross.

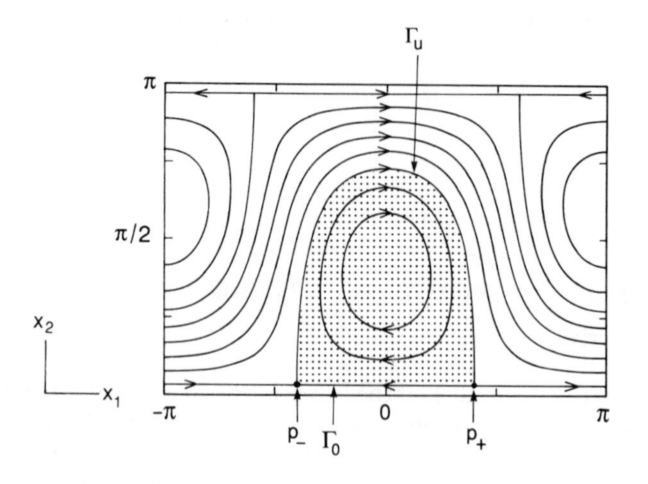

FIGURE E4.6.

b) Show that for $\varepsilon \neq 0$, Γ_u breaks up giving rise to transverse heteroclinic orbits. This provides a mechanism for fluid to mix between the two regions in the time-dependent fluid flow.

c) Show that a horseshoe exists in the heteroclinic tangle for $\varepsilon \neq 0$. (*Hint:* see Rom-Kedar et al. [1989].) Hence, chaotic fluid particle trajectories exist.

d) The discussion of lobes, pips, and transport in Section 4.6 was developed solely for homoclinic orbits. However, show that the same definitions and transport mechanism apply for heteroclinic orbits (see Rom-Kedar and Wiggins [1989]).

e) Use the results from Section 4.6 to study transport between the two regions that were bounded by $\Gamma_0 \cup \Gamma_u \cup \{p_+\} \cup \{p_-\}$ for $\varepsilon = 0$. In particular

   1. Define pseudoseparatrices between the regions in the associated Poincaré map.
   2. Construct the turnstile.
   3. Discuss flux between the two regions.
   4. If the two regions are defined as R_1 and R_2, compute $T_{1,2}(n)$ and compare it to Equation (4.6.14). Does it follow that $T_{1,2}(n) = T_{2,1}(n)$?

4.43 Prove Theorem 4.7.1. (*Hints:* the idea is to find a region S_n in U_{y_0} such that Assumptions 1 and 3 of Section 4.3 hold for f^{n+k}. 1) Let $S_n = \{(x, y) \mid |y - y_0| \leq \varepsilon,\ 0 \leq x \leq \nu^n\}$ where $\rho < \nu < \frac{1}{\lambda}$. The idea is to show that $f^{n+k}(S_n; 0)$ intersects S_n in two μ_v-vertical strips. Then the pre-image of these two μ_v-vertical strips will be μ_h-horizontal strips with proper boundary behavior and $0 \leq \mu_h \mu_v < 1$. This can be accomplished in several steps. First show that under $f^k(\cdot;0)$, vertical lines in S_n (i.e., lines with $x = c =$ constant) map to parabolas in U_{x_0} given by the graph of $y = \gamma c + \frac{\delta}{\beta^2}(x - x_0)^2$. 2) Show that for n sufficiently large, the x-components of points in $f^{n+k}(\cdot;\mu)$ are smaller than ν^n. 3) Finally, show that for $\varepsilon = \varepsilon(n) \sim (y_0 \lambda^{-n}/\delta)^{1/2}$, $f^{n+k}(S_n;0)$ cuts through the top and bottom horizontal boundaries of S_n. From these three facts you should be able to find μ_h-horizontal and μ_v-vertical strips so that Assumption 1 is satisfied.

The proof that Assumption 3 holds is very similar to the same step carried out in Theorem 4.4.2. Use the fact that

$$Df^{n+k}(x,y;0) = \begin{pmatrix} 0 & -\beta\rho^n \\ \gamma\lambda^n & 2\delta\lambda^n(y - y_0) \end{pmatrix}$$

and that, for $|y - y_0| \sim (y_0\lambda^{-n}/\delta)^{1/2}$ and n sufficiently large, this Jacobian is essentially

$$Df^{n+k}(x,y;0) \sim \begin{pmatrix} 0 & 0 \\ \gamma\lambda^n & 2(y_0\delta\lambda^n)^{1/2} \end{pmatrix}.$$

(Drawing figures in each case should help.)

4.44 Recall the proof of Theorem 4.7.2. In showing that the sink created in the saddle-node bifurcation of f^{n+k} subsequently underwent a period-doubling bifurcation at

$$\mu = -\frac{3}{4\delta}(\gamma\beta\rho^n + \lambda^{-n})^2 + (y_0\lambda^{-n} - \gamma\rho^n x_0),$$

we only showed that the map had an eigenvalue of -1 at this parameter value. Examine the nonlinear terms (possibly do a center manifold reduction) to show that this period-doubling bifurcation is indeed nondegenerate or "generic."

4.45 Recall the discussion in 4.8A and specifically the vector field (4.8.1). Would there be any qualitative changes in the results of these sections if the eigenvalues depended on the parameter μ? How would this situation best be handled?

4.46 Show that *generically* for $\lambda_2 > \lambda_1$, the homoclinic orbit in (4.8.1) for $\mu = 0$ is tangent to the y axis in the $x - y$ plane at the origin.

Construct a *nongeneric* example where this does not occur and explain why your example is not generic. (*Hint:* find an appropriate symmetry.)

4.47 In Equation (4.8.9) we took

$$\frac{1}{1 - Az^{|\lambda_1|/\lambda_3}} \sim 1 \qquad \text{as} \quad z \to 0.$$

Show that if instead we take

$$\frac{1}{1 - Az^{|\lambda_1|/\lambda_3}} = 1 + Az^{|\lambda_1|/\lambda_3} + \cdots$$

our results are not affected for z sufficiently small.

4.48 From Section 4.8A, consider the case $\lambda_2 < \lambda_1$. Show that in this case the homoclinic orbit of (4.8.1) is tangent to the x-axis at the origin in the $x - y$ plane. Construct a Poincaré map near the homoclinic orbit following Section 4.8A and show that Theorem 4.8.1 still holds. In describing the "half-bowtie" shape of $P_0(\Pi_0)$ in Π_1, compare it with the case $\lambda_1 < \lambda_2$.

4.49 From Section 4.8A, consider the case $\lambda_2 = \lambda_1$. Describe the geometry of the return of the homoclinic orbit to the origin. Construct a Poincaré map following Section 4.8A and show that Theorem 4.8.1 still holds. In describing the "half-bowtie" shape of $P_0(\Pi_0)$ in Π_1, compare it with the cases $\lambda_1 > \lambda_2$ and $\lambda_2 < \lambda_1$.

4.50 Argue that if (4.8.1) possesses only one homoclinic orbit, then the Poincaré map defined near the homoclinic orbit cannot possess an invariant Cantor set on which it is topologically conjugate to a full shift on N $(N \geq 2)$ symbols.

4.51 Recall the discussion in Section 4.8A, i). In Assumption $1'$, discuss the necessity and geometry behind the requirement $d \neq 0$.

4.52 Work out all of the details in the proof of Theorem 4.8.2. (*Hint:* mimic the proof of Moser's theorem in Section 4.4.)

4.53 Recall the discussion in Section 4.8A, i). Suppose we instead chose configuration b) in Figure 4.8.11 with Assumption $1'$ and Assumption $2'$ still holding. Is Theorem 4.8.2 still valid in this case? If so, what modifications must be made in order to carry out the proof?

4.54 Work out all of the details in the proof of Theorem 4.8.4. (*Hint:* mimic the proof of Moser's theorem in Section 4.4.)

4.55 In Theorem 4.8.4 we proved the existence of an invariant Cantor set $\Lambda_k \subset R_k$ such that the Poincaré map restricted to Λ_k was topologically conjugate to a full shift on two symbols. This was true for all k sufficiently large. Show that Theorem 4.8.4 can be modified (in particular, the choice of μ_h-horizontal and μ_v-vertical strips) so that

Π_0 contains an invariant Cantor set on which the Poincaré map is topologically conjugate to a full shift on N symbols with N arbitrarily large. (*Hint:* use Lemma 4.8.3 and see Wiggins [1988] if you need help.)

Is there a difference in the dynamics of this invariant set and the dynamics in $\bigcup_{k \geq k_0} \Lambda_k$ constructed in Theorem 4.8.4?

4.56 Recall the construction in Theorem 4.8.4. Suppose the Poincaré map P is perturbed (as might occur in a one-parameter family). Show that, for sufficiently small perturbations, an infinite number of the Λ_k are destroyed yet a finite number survive. Does this contradict the fact that horseshoes are structurally stable? (See Exercise 4.4 and Wiggins [1988].)

4.57 Recall the discussion in Section 4.8B. Suppose (4.8.25) is invariant under the coordinate transformation

$$(x, y, z) \to (-x, -y, -z)$$

with Assumptions 1 and 3 still holding.

a) Show that (4.8.25) must possess two orbits homoclinic to the origin. Draw the two homoclinic orbits in the phase space. Denote the homoclinic orbits by Γ_0 and Γ_1.

b) Construct a Poincaré map in a neighborhood of $\Gamma_0 \cup \Gamma_1 \cup \{(0,0,0)\}$ and show that the map has an invariant Cantor set on which the dynamics are topologically conjugate to a full shift on two symbols.

c) If we denote the two symbols by 0 and 1, show that the motion in phase space is such that a '0' corresponds to a trajectory following close to Γ_0 and a '1' corresponds to a trajectory following close to Γ_1. Hence, give a geometrical description of the manifestation of chaos in phase space. (See Wiggins [1988] if you need help.)

4.58 Suppose we reverse the direction of time in (4.8.25), i.e., we let

$$t \to -t,$$

with Assumptions 1 and 2 still holding. Describe the dynamics near the homoclinic orbit.

4.59 Consider the vector field (4.8.25). Suppose that Assumption 1 holds but Assumption 2 is replaced by the following.

Assumption 2′. $-\rho > \lambda > 0$.

Describe the dynamics near the homoclinic orbit. (See Holmes [1980] for help.)

4.60 Show that the function $f(\alpha)$ defined in (4.9.15) is monotone.

4.61 Show that the two-parameter family

$$\begin{aligned} \dot{x} &= y, \\ \dot{y} &= \mu_1 + \mu_2 y + x^2 + bxy, \end{aligned} \qquad b = \pm 1, \tag{E4.13}$$

is a versal deformation of a fixed point of a planar vector field at which the matrix associated with the linearization has the form

$$\begin{pmatrix} 0 & 1 \\ 0 & 0 \end{pmatrix}.$$

(*Hint:* the idea is to show that the neglected higher order terms in the normal form do not introduce any qualitatively new dynamics in the sense that the bifurcation diagram in Figure 4.9.3 is unchanged. Begin by considering the fixed points and local bifurcations and show that these are qualitatively unchanged. Next, consider the global behavior, i.e., homoclinic orbits and "large amplitude" periodic orbits. The Melnikov theory can be used here.

Once all of these results are established, does it then follow that (E4.13) is a versal deformation?)

4.62 Recall Exercise 3.32, the double-zero eigenvalue with the symmetry $(x, y) \to (-x, -y)$. The normal form was given by

$$\begin{aligned} \dot{x} &= y, \\ \dot{y} &= \mu_1 x + \mu_2 y + cx^3 - x^2 y, \end{aligned} \qquad c = \pm 1. \tag{E4.14}$$

In this exercise we want to analyze possible global behavior that might occur.

a) For $c = +1$, using the rescaling

$$x = \varepsilon u, \quad y = \varepsilon^2 v, \quad \mu_1 = -\varepsilon^2, \quad \mu_2 = \varepsilon^2 \nu_2$$

and $t \to \frac{t}{\varepsilon}$, show that (E4.14) becomes

$$\begin{aligned} \dot{u} &= v, \\ \dot{v} &= -u + u^3 + \varepsilon(\nu_2 v - u^2 v). \end{aligned} \tag{E4.15}$$

b) Show that (E4.15) is Hamiltonian for $\varepsilon = 0$ and draw the phase portrait.

c) Use the Melnikov theory to show that (E4.15) has a heteroclinic connection on

$$\mu_2 = -\frac{\mu_1}{5} + \cdots. \tag{E4.16}$$

What is the form of the higher order terms in (E4.16) (i.e., $\mathcal{O}(\mu_1^\alpha)$, where α is some number)? This is important for determining the behavior of the bifurcation curves at the origin.

d) Show that (E4.15) has a unique periodic orbit for $\mu_1 < 0$ between $\mu_2 = 0$ and $\mu_2 = -\frac{\mu_1}{5} + \cdots$.

e) Draw the complete bifurcation diagram for (E4.14) with $c = +1$. Is (E4.14) a versal deformation for $c = +1$?

f) For $c = -1$, using the rescaling

$$x = \varepsilon u, \quad y = \varepsilon^2 v, \quad \mu_1 = \varepsilon^2, \quad \mu_2 = \varepsilon^2 \nu_2,$$

and $t \to \frac{t}{\varepsilon}$, show that (E4.14) becomes

$$\begin{aligned} \dot{u} &= v, \\ \dot{v} &= u - u^3 + \varepsilon(\nu_2 v - u^2 v). \end{aligned} \tag{E4.17}$$

g) Show that (E4.17) is Hamiltonian for $\varepsilon = 0$ and draw the phase portrait.

h) Using the Melnikov theory, show that (E4.17) undergoes a homoclinic bifurcation on

$$\mu_2 = \frac{4}{5}\mu_1 + \cdots, \tag{E4.18}$$

and a saddle-node bifurcation of periodic orbits on

$$\mu_2 = c\mu_1 + \cdots, \tag{E4.19}$$

with $c \approx .752$. Hence, for $\mu_1 > 0$, between $\mu_2 = \mu_1$ and $\mu_2 = \frac{4}{5}\mu_1 + \cdots$, (E4.14) has three periodic orbits for $c = -1$. What are their stabilities? Between $\mu_2 = \frac{4}{5}\mu_1 + \cdots$ and $\mu_2 = c\mu_1 + \cdots$ (E4.14) has two periodic orbits for $c = -1$. What are their stabilities? Below $\mu_2 = c\mu_1 + \cdots$ there are no periodic orbits for $c = -1$.

In (E4.18) and(E4.19) what is the form of the higher order terms (i.e., $\mathcal{O}(\mu_1^\alpha)$ where α is some number).

1. Draw the complete bifurcation diagram for (E4.14) with $c = -1$. Is (E4.14) a versal deformation for $c = -1$?

(*Hint:* the Melnikov theory for autonomous systems is developed in Exercise 4.22 and the Melnikov theory for heteroclinic orbits is developed in Exercise 4.27.)

4.63 Show that all the cubic terms, *except* z^3, can be eliminated from (4.9.17) so that it takes the form of (4.9.18). (*Hint:* this result is due to J. Guckenheimer.)

Consider the following coordinate transformation

$$\begin{aligned} s &= r(1 + gz), \\ w &= z + hr^2 + iz^2, \\ \tau &= (1 + jz)^{-1} t, \end{aligned}$$

where g, h, i, j are unspecified constants, they will be chosen to make the equations simpler. In these new coordinates (4.9.17) becomes

$$\begin{aligned}\frac{ds}{d\tau} &= \mu_1 s + asw + (c + bg - ah)s^3 + (d - g - ai + aj)sw^2 \\ &\quad + R_s(s, w, \mu_1, \mu_2), \\ \frac{dw}{d\tau} &= \mu_2 + bs^2 - w^2 + (e - 2bg + 2(a+1)h + 2bi + bj)s^2 w \\ &\quad + (f - j)w^3 + R_w(s, w, \mu_1, \mu_2),\end{aligned}$$

where the remainder terms are $\mathcal{O}(4)$ in s, w, μ_1, μ_2. We will ignore R_s and R_w and choose g, h, i, j so as to make the cubic terms as simple as possible. If we think of the cubic terms as being in a vector space spanned by

$$\begin{pmatrix} s^3 \\ 0 \end{pmatrix}, \begin{pmatrix} sw^2 \\ 0 \end{pmatrix}, \begin{pmatrix} 0 \\ s^2 w \end{pmatrix}, \begin{pmatrix} 0 \\ w^3 \end{pmatrix},$$

the problem of annihilating the cubic terms reduces to solving the linear problem

$$Ax = \theta,$$

where

$$x = \begin{pmatrix} g \\ h \\ i \\ j \end{pmatrix}, \qquad \theta = \begin{pmatrix} -c \\ -d \\ -e \\ -f \end{pmatrix},$$

$$A = \begin{pmatrix} b & -a & 0 & 0 \\ -1 & 0 & -a & a \\ -2b & 2a+2 & 2b & b \\ 0 & 0 & 0 & -1 \end{pmatrix}.$$

It is not hard to show that A has rank 3 and that we can eliminate all cubic terms except $\begin{pmatrix} 0 \\ w^3 \end{pmatrix}$. Thus, since our transformation did not change our equation at $\mathcal{O}(2)$ and below, we can write our normal form in the r, z coordinates as

$$\begin{aligned}\dot{r} &= \mu_1 r + arz, \\ \dot{z} &= \mu_2 + br^2 - z^2 + fz^3\end{aligned}$$

(where f can take on all values).

4.64 Once the cubic terms have been restored, study the Poincaré–Andronov–Hopf bifurcation in Case IIa,b and Case III.

4.65 Verify that the homoclinic bifurcation shown in Figure 4.9.5 occurs for $a = 2$, $f < 0$.

4.66 Suppose that $W^u(p_1)$ falls into $W^s(p_1)$ as shown in Figure 4.9.8. What are the conditions on the normal form in order that the hypotheses of Theorem 4.8.4 hold.

Suppose $W^s(p_2)$ falls into $W^u(p_2)$ as shown in Figure 4.9.8. Can horseshoes also occur in this case? (*Hint:* see Exercise 4.68, b).) What conditions must the normal form satisfy?

4.67 Consider a $\mathbf{C}^r$ (r as large as necessary) autonomous vector field in $\mathbb{R}^3$. We denote coordinates in $\mathbb{R}^3$ by $x - y - z$. Suppose the vector field has two hyperbolic fixed points, p_1 and p_2, respectively, in the $x - y$ plane.

Local Assumption: The vector field linearized at p_1 has the form

$$\begin{aligned} \dot{x} &= \lambda_1 x, \\ \dot{y} &= \lambda_2 y, \\ \dot{z} &= \lambda_3 z, \end{aligned}$$

with $\lambda_1 > 0$, $\lambda_3 < \lambda_2 < 0$.

The vector field linearized at p_2 has the form

$$\begin{aligned} \dot{x} &= \rho x - \omega y, \\ \dot{y} &= \omega x + \rho y, \\ \dot{z} &= \lambda z, \end{aligned}$$

with $\rho < 0$, $\lambda > 0$, $\omega \neq 0$.

Global Assumption: p_1 and p_2 are connected by a heteroclinic orbit, denoted Γ_{12} that lies in the $x - y$ plane.

p_2 and p_1 are connected by a heteroclinic orbit, Γ_{21}, that lies outside the plane; see Figure E4.7.

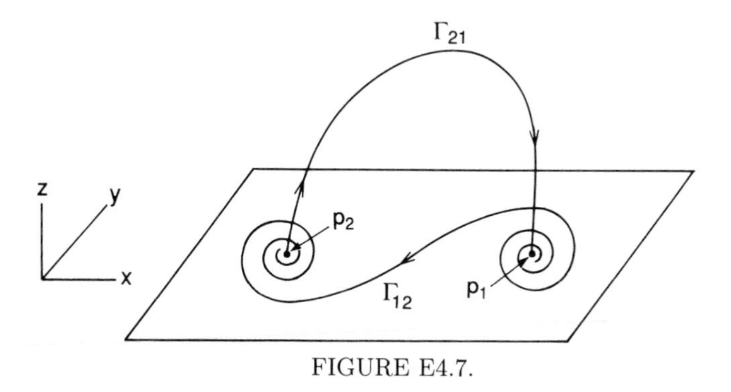

FIGURE E4.7.

Thus, $\Gamma_{12} \cup \Gamma_{21} \cup \{p_1\} \cup \{p_2\}$ form a heteroclinic cycle. The goal is to construct a Poincaré map, P, in a neighborhood of the heteroclinic cycle and prove the following theorem.

Theorem E4.1. (Tresser [1984]) *P possesses a countable number of horseshoes provided*

$$\frac{\rho\lambda_2}{\lambda\lambda_1} < 1.$$

(*Hint:* P will be constructed from the composition of four maps. Define cross sections Π_{01}, Π_{11}, Π_{12}, and Π_{02} appropriately; see Figure E4.7. $\Pi_{0,1}$ and Π_{11} should be sufficiently close to p_1, and Π_{12} and Π_{02} should be chosen sufficiently close to p_2. If the coordinates on these cross-sections are chosen appropriately, then we can derive maps as follows

$$P_{01}\colon \Pi_{01} \to \Pi_1,$$

$$\begin{pmatrix} x_1 \\ \varepsilon \\ z_1 \end{pmatrix} \mapsto \begin{pmatrix} \varepsilon \\ \varepsilon\left(\frac{\varepsilon}{x_1}\right)^{\lambda_2/\lambda_1} \\ z_1\left(\frac{\varepsilon}{x_1}\right)^{\lambda_3/\lambda_1} \end{pmatrix},$$

$$P_{02}\colon \Pi_{02} \to \Pi_{12},$$

$$\begin{pmatrix} x_2 \\ 0 \\ z_2 \end{pmatrix} \mapsto \begin{pmatrix} x_2\left(\frac{\varepsilon}{z_2}\right)^{\rho/\lambda} \cos\frac{\omega}{\lambda}\log\frac{\varepsilon}{z_2} \\ x_2\left(\frac{\varepsilon}{z_2}\right)^{\rho/\lambda} \sin\frac{\omega}{\lambda}\log\frac{\varepsilon}{z_2} \\ \varepsilon \end{pmatrix},$$

$$P_{12}\colon \Pi_{12} \to \Pi_{01},$$

$$\begin{pmatrix} x_2 \\ y_2 \\ \varepsilon \end{pmatrix} \mapsto \begin{pmatrix} 0 \\ 0 \\ \varepsilon \end{pmatrix} + \begin{pmatrix} a_2 & b_2 & 0 \\ c_2 & d_2 & 0 \\ 0 & 0 & 0 \end{pmatrix}\begin{pmatrix} x_2 \\ y_2 \\ 0 \end{pmatrix},$$

$$P_{11}\colon \Pi_{11} \to \Pi_{02},$$

$$\begin{pmatrix} \varepsilon \\ y_1 \\ z_2 \end{pmatrix} \mapsto \begin{pmatrix} \varepsilon \\ 0 \\ 0 \end{pmatrix} + \begin{pmatrix} 0 & 0 & 0 \\ a & a_1 & b_1 \\ 0 & c_1 & d_1 \end{pmatrix}\begin{pmatrix} 0 \\ y_1 \\ z_1 \end{pmatrix},$$

where $x_1 - y_1 - z_1$ denote coordinates near p_1, and $x_2 - y_2 - z_2$ denote coordinates near p_2. These maps are approximations (see the discussion at the beginning of Section 4.8); discuss their validity and specify all steps in their derivation.

Then the Poincaré map near the heteroclinic cycle is defined as

$$P \equiv P_{11} \circ P_{01} \circ P_{12} \circ P_{02}\colon \Pi_{02} \to \Pi_{02}.$$

The rest of the proof is very much the same as Theorem 4.8.4. (See Wiggins [1988] for additional help.)

4.68 Consider a $\mathbf{C}^r$ (r as large as necessary) autonomous vector field in $\mathbb{R}^3$. We denote coordinates in $\mathbb{R}^3$ by $x - y - z$. Suppose the vector field has two hyperbolic fixed points, denoted p_1 and p_2, respectively, which lie in the $x - y$ plane.

Local Assumption: The vector field linearized at p_1 has the form

$$\begin{aligned}\dot{x} &= \rho_1 x - \omega_1 y,\\ \dot{y} &= \omega_1 x + \rho_1 y,\\ \dot{z} &= \lambda_1 z,\end{aligned}$$

with

$$\lambda_1 > 0, \ \rho_1 < 0 \qquad \text{and} \qquad \omega_1 \neq 0.$$

The vector field linearized at p_2 has the form

$$\begin{aligned}\dot{x} &= \rho_2 x - \omega_2 y,\\ \dot{y} &= \omega_2 x + \rho_2 y,\\ \dot{z} &= \lambda_2 z,\end{aligned}$$

with $\lambda_2 < 0$, $\rho_2 > 0$, and $\omega_2 \neq 0$.

Global Assumption: There exists a trajectory Γ_{12} in the $x - y$ plane connecting p_1 to p_2.

There exists a trajectory Γ_{21} connecting p_2 to p_1.

See Figure E4.8 for an illustration of the geometry. Γ_{12} and Γ_{21} are examples of *heteroclinic orbits,* i.e., an orbit that is biasymptotic to two different fixed points. $\Gamma_{12} \cup \Gamma_{21} \cup \{p_1\} \cup \{p_2\}$ is said to form a *heteroclinic cycle.*

a) Define a Poincaré map in the neighborhood of the heteroclinic cycle and determine if there are conditions on the eigenvalues (i.e., ρ, ρ, λ, and λ) such that the map possesses an invariant Cantor set on which the dynamics are topologically conjugate to a full shift on N ($N \geq 2$) symbols.

b) Consider the case

$$\begin{aligned}|\rho_2| &= |\rho_1|,\\ |\lambda_2| &= |\lambda_1|,\\ |\omega_2| &= |\omega_1|.\end{aligned}$$

Can horseshoes exist in this case? What is the relevance of this case to Case III of the *truncated* (hence symmetric) three-dimensional normal form discussed in Section 4.9B?

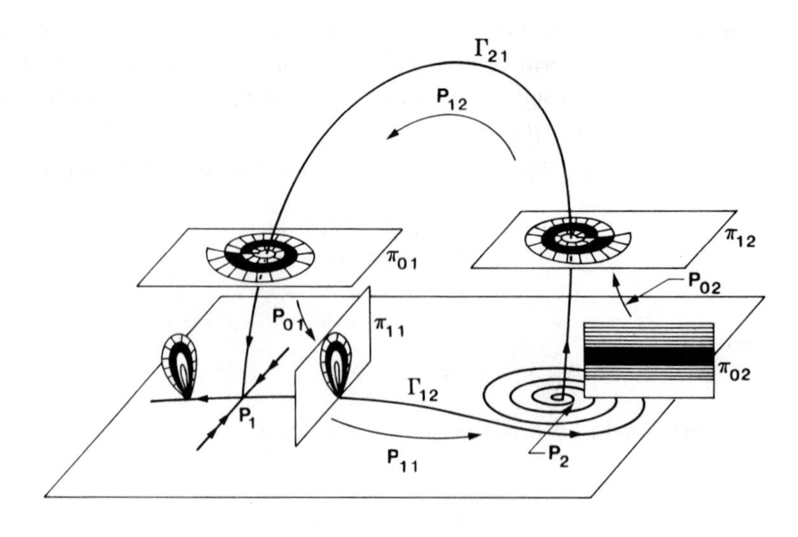

FIGURE E4.8.

4.69 Recall the motivational example for deriving Poincaré maps near homoclinic orbits described in Section 1.2A. We now describe the analog of that example for heteroclinic orbits.

Consider a $\mathbf{C}^r$ (r as large as necessary) two-parameter family of vector fields in the plane having a hyperbolic fixed points p_1 and p_2, respectively.

Local Assumption: The vector field linearized at p_1 is given by

$$\begin{aligned} \dot{x}_1 &= \alpha_1 x_1, \\ \dot{y}_1 &= \beta_1 y_1, \end{aligned} \qquad \alpha_1 < 0,\ \beta_1 > 0,$$

and the vector field linearized at p_2 is given by

$$\begin{aligned} \dot{x}_2 &= \alpha_2 x_2, \\ \dot{y}_2 &= \beta_2 y_2, \end{aligned} \qquad \alpha_2 > 0,\ \beta_2 < 0,$$

where α_i, β_i, $i = 1, 2$ are constants.

Global Assumption. p_1 and p_2 are connected by a heteroclinic cycle. We denote the heteroclinic orbit going from p_1 to p_2 in positive time by Γ_{12} and the heteroclinic orbit going from p_2 to p_1 in positive time by Γ_{21}.

The heteroclinic cycle depends on the parameters as follows. For $\mu \equiv (\mu_1, \mu_2)$, let $\mathcal{N}$ be a neighborhood of zero in the μ_1, μ_2 plane; then we assume the following.

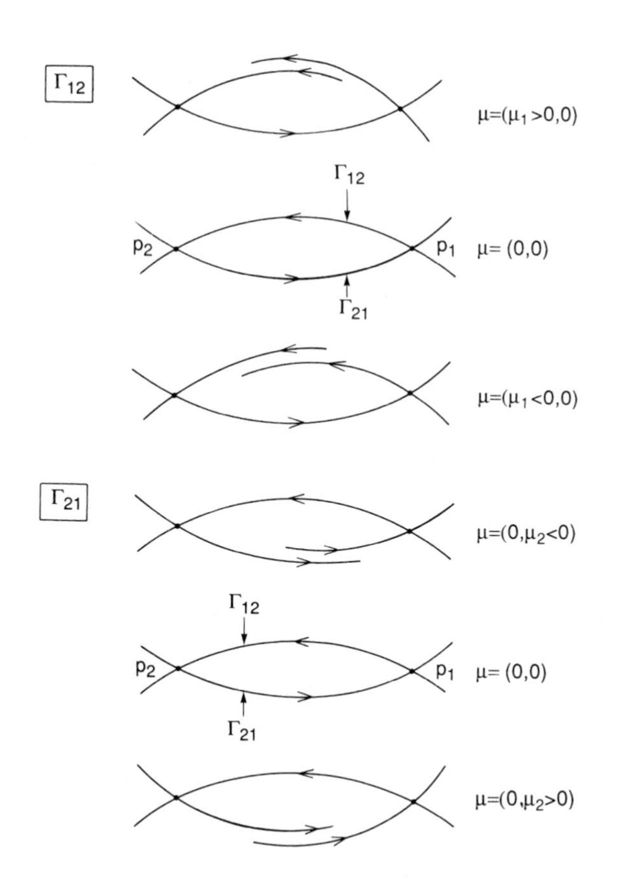

FIGURE E4.9.

1) Γ_{12} exists for all $\mu \in \{(\mu_1, \mu_2) \mid \mu_1 = 0\} \cap \mathcal{N} \equiv \mathcal{N}_1$.
2) Γ_{21} exists for all $\mu \in \{(\mu_1, \mu_2) \mid \mu_2 = 0\} \cap \mathcal{N} \equiv \mathcal{N}_2$.

Furthermore, we assume that Γ_{12} and Γ_{21} break "transversely" as shown in Figure E4.9.

Construct a two-parameter family of Poincaré maps near the heteroclinic cycle and study the bifurcation and stability of periodic orbits. (*Hint:* apply the ideas from Section 1.2A, Case 3 and Exercise 4.66.)

4.70 Recall the discussion in Section 4.10. Show that

$$\chi(x_0, e) = \chi(x(T), e)$$

for any finite T.

4.71 Can a $\mathbf{C}^r$ map or flow depend on initial conditions in a $\mathbf{C}^r$ manner and also exhibit sensitive dependence on initial conditions? Explain.

4.72 Recall Example 4.11.2. Do all or only some orbits in the open annuli bounded by adjacent stable periodic orbits exhibit sensitive dependence on initial conditions?

4.73 Recall Example 4.11.3. Show that the flow generated by (4.11.6) is topologically transitive on T^2.

4.74 Show that for the dynamical system

$$\sigma \colon \Sigma^N \to \Sigma^N$$

the periodic orbits are dense in Σ^N.

4.75 Let

$$x \mapsto g(x), \qquad x \in \mathbb{R}^n,$$

be a $\mathbf{C}^r$ $(r \geq 1)$ map. Suppose $\mathcal{M} \subset \mathbb{R}^n$ is a trapping region with $\Lambda \subset \mathcal{M}$ a chaotic invariant set. Then defining

$$\mathcal{A} \equiv \bigcap_{n>0} g^n(\mathcal{M}),$$

show that

$$\Lambda \subset \mathcal{A}.$$

4.76 Consider a planar, Hamiltonian vector field. Must all Liapunov exponents of *every* orbit be zero?

4.77 Show that any trajectory of a vector field must have at least one zero Liapunov exponent. (*Hint:* consider the direction tangent to the orbit.)

4.78 Very often one hears the phrase,

> *A dynamical system is chaotic if it has one positive Liapunov exponent.*

Discuss what this phrase means in light of the discussion in Sections 4.10 and 4.11. Consider both dissipative and nondissipative systems.

4.79 In any realistic example, the Liapunov exponents of an orbit must be calculated numerically. In this case, you can see that a problem may arise. Namely, a Liapunov exponent is a number obtained in the limit $t \to \infty$, and, in practice, one can only compute for a finite amount of time.

Recall Example 4.10.3. Consider an initial condition

$$(x, y) = (\varepsilon, 0),$$

and the direction

$$\delta x = (1, 0).$$

Let

$$\chi_t(x_0, e) = \frac{1}{t} \log \lambda_t(x_0, e),$$

and compute

$$\chi_t((\varepsilon, 0), \delta x)$$

for this example. It should follow from the discussion in the example that for some T we have

$$\chi_t((\varepsilon, 0), \delta x) \leq 0, \qquad t \in [T, \infty).$$

Let $T_0(\varepsilon)$ be the value of t such that

$$\chi_{T_0(\varepsilon)}((\varepsilon, 0), \delta x) = 0.$$

a) Compute $T_0(\varepsilon)$ and graph it as a function of ε.

b) What can you conclude from this example concerning the numerical computation of Liapunov exponents?

4.80 Recall the discussion of chaos in the phase space of the damped, periodically forced Duffing oscillator at the end of Section 4.11. The goal of this exercise is to make the heuristic arguments given in that discussion rigorous.

a) Draw the homoclinic tangle correctly for the Poincaré map on a given cross-section (say Σ^0). You may want to use a computer.

b) Find candidates for two μ_h-horizontal strips, denoted H_0 and H_1, which map over themselves in μ_v-vertical strips under some iterate of the Poincaré map so that Assumptions 1 and 3 of Section 4.3 are satisfied. Choose the horizontal strips so that the relationship between the motion in phase space and the dynamics on the invariant set is as described in Section 4.11.

c) Describe the relationship between the number of iterates needed to form the chaotic invariant set and the parameters $\gamma - \delta - \omega$ (see Holmes and Marsden [1982] for help).

4.81 *The Attracting Set of the Damped, Periodically Forced Duffing Oscillator.* Consider the damped, periodically forced Duffing oscillator

$$\begin{aligned} \dot{x} &= y, \\ \dot{y} &= x - x^3 + \varepsilon(-\delta y + \gamma \cos \omega t), \end{aligned} \tag{E4.20}$$

with ε small and $\delta, \gamma, \omega > 0$. Choose

$$\gamma \in \left(\hat{R}^M(\omega)\delta, \hat{R}^{M-2}(\omega)\delta\right).$$

Recall from 1.2E, i) that this choice of parameters implies that the outer resonance band of order M (odd) is excited but the further out resonance bands of order $M-2$ are not excited. Let P denote the Poincaré map associated with (E4.20) and let p denote the period M saddle point on the order M resonance band. Then prove the following result due to Holmes and Whitley [1984].

Theorem E4.2. (Holmes and Whitley) $A \equiv \bigcap_{n \geq 0} P^n(D) = \overline{W^u(p)}$, *where D is a trapping region containing the order M resonance band.*

(*Hints:* From 1.2E, ii), we know that both components intersect the boundary of D only once. This separates D into two disjoint regions, S and T, as labeled in Figure E4.10 (obviously this picture is very idealized, i.e., we are ignoring many things; however, it contains the essence of what we want to describe). In the following we will make our arguments with the $M(M+2)$ iterate of the Poincaré-map, denoted $\tilde{P} \equiv P^{M(M+2)}$; thus, p and q will be fixed points for this map, where q denotes the saddle on the $M+2$ resonance band.

Now from Section 1.2E,ii) both components of $W^s(q)$ intersect $W^i(p)$ transversely. We will consider two specific points on these intersections which we will label x, y with the ordering p, x, y on $W^u(p)$ away from p. Iterating these points under $\tilde{P}^{-1}$ we obtain the points x', y' with the ordering p, x', y', x, y along $W^u(p)$ away from p as shown in Figure E4.11. The arcs $\widehat{xx'}$ of $W^s(q)$ and $W^u(p)$ bound a closed region R, which we show shaded in Figure E4.11.

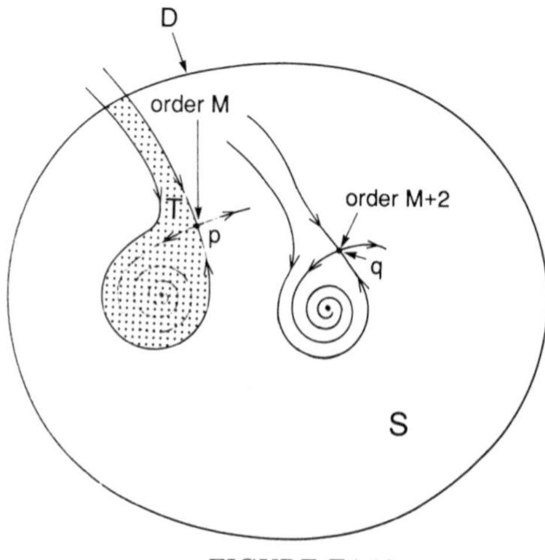

FIGURE E4.10.

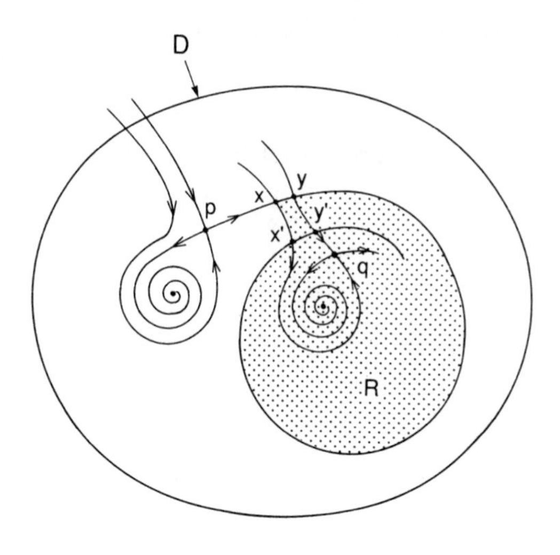

FIGURE E4.11.

Now that we have described the set-up, we give a brief outline of the steps needed in order to complete the proof.

1) Show $\overline{W^u(p)} \subset A$ (easy).
2) Show $A \subset \overline{W^u(p)}$.

This is demonstrated as follows.

a) Argue that all points on the Mth resonance level which are not trapped on it pass through to the next lower level. (This has already been established through Melnikov theory.)
b) Show that $\overline{R} \equiv \bigcap_{n \geq 0} P^n(R) \subset W^u(p)$.
c) Argue that all points eventually fall into an $\overline{R}$ constructed on some lower resonance level.

Discuss in detail why $A \subset \overline{W^u(p)}$ follows from a), b), and c). Also, why is it sufficient to consider only one (fixed) saddle-sink pair on the order M and order $M+2$ resonance bands?

4.82 Often one hears the phrase,

> *For diffeomorphisms of dimension two and larger and for vector fields of dimension three and larger, homoclinic orbits produce chaos.*

Is this statement generally true? Give a complete discussion with examples.

4.83 Show that Theorem 4.7.2 holds for area-preserving maps. (*Hint:* use the implicit function theorem proof given in Tedeschini-Lalli and Yorke [1986].)

4.84 Does the result of Theorem 4.7.1 hold for area-preserving maps?

Bibliography

Abraham, R.H. and Marsden, J.E. [1978]. *Foundations of Mechanics.* Benjamin/Cummings: Menlo Park, CA.

Abraham, R.H., Marsden, J.E., and Ratiu, T. [1988]. *Manifolds, Tensor Analysis, and Applications.* Springer-Verlag: New York, Heidelberg, Berlin.

Abraham, R.H. and Shaw, C.D. [1984]. *Dynamics—The Geometry of Behavior, Part Three: Global Behavior.* Aerial Press, Inc.: Santa Cruz.

Afraimovich, V.S., Bykov, V.V., and Silnikov, L.P. [1983]. On structurally unstable attracting limit sets of Lorenz attractor type. *Trans. Moscow Math. Soc.* **2**, 153–216.

Alekseev, V.M. [1968a]. Quasirandom dynamical systems, I. *Math. USSR-Sb.* **5**, 73–128.

Alekseev, V.M. [1968b]. Quasirandom dynamical systems, II. *Math. USSR-Sb.* **6**, 505–560.

Alekseev, V.M. [1969]. Quasirandom dynamical systems, III. *Math. USSR-Sb.* **7**, 1–43.

Alligood, K.T., Yorke, E.D., and Yorke, J.A. [1987]. Why period-doubling cascades occur in periodic orbit creation followed by stability shedding. *Physica* **28D**, 197–205.

Andronov, A.A. [1929]. Application of Poincaré's theorem on "bifurcation points" and "change in stability" to simple autooscillatory systems. *C.R. Acad. Sci. Paris* **189** (15), 559–561.

Andronov, A.A., Leontovich, E.A., Gordon, I.I., and Maier, A.G. [1971]. *Theory of Bifurcations of Dynamic Systems on a Plane.* Israel Program of Scientific Translations: Jerusalem.

Andronov, A.A. and Pontryagin, L. [1937]. Systèmes grossiers. *Dokl. Akad. Nauk. SSSR* **14**, 247–251.

Arnéodo, A., Coullet, P., and Tresser, C. [1981a]. A possible new mechanism for the onset of turbulence. *Phys. Lett.* **81A**, 197–201.

Arnéodo, A., Coullet, P., and Tresser, C. [1981b]. Possible new strange attractors with spiral structure. *Comm. Math. Phys.* **79**, 573–579.

Arnéodo, A., Coullet, P., and Tresser, C. [1982]. Oscillators with chaotic behavior: An illustration of a theorem by Shil'nikov. *J. Statist. Phys.* **27**, 171–182.

Arnéodo, A., Coullet, P., and Spiegel, E. [1982]. Chaos in a finite macroscopic system. *Phys. Lett.* **92A**, 369–373.

Arnéodo, A., Coullet, P., Spiegel, E., and Tresser, C. [1985]. Asymptotic chaos. *Physica* **14D**, 327–347.

Arnold, V.I. [1972]. Lectures on bifurcations in versal families. *Russian Math. Surveys* **27**, 54–123.

Arnold, V.I. [1973]. *Ordinary Differential Equations.* M.I.T. Press: Cambridge, MA.

Arnold, V.I. [1977]. Loss of stability of self oscillations close to resonances and versal deformations of equivariant vector fields. *Functional Anal. Appl.* **11**(2), 1–10.

Arnold, V.I. [1978]. *Mathematical Methods of Classical Mechanics.* Springer-Verlag: New York, Heidelberg, Berlin.

Arnold, V.I. [1983]. *Geometrical Methods in the Theory of Ordinary Differential Equations.* Springer-Verlag: New York, Heidelberg, Berlin.

Aubry, S. [1983a]. The twist map, the extended Frenkel–Kontorova model and the devil's staircase. *Physica* **7D**, 240–258.

Aubry, S. [1983b]. Devil's staircase and order without periodicity in classical condensed matter. *J. Physique* **44**, 147–162.

Baer, S.M., Erneux, T., and Rinzel, J. [1989]. The slow passage through a Hopf bifurcation: Delay, memory effects, and resonance. *SIAM J. Appl. Math.* **49**, 55–71.

Baider, A. [1989]. Unique normal forms for vector fields and Hamiltonians. *J. Differential Equations* **78**, 33–52.

Baider, A. and Churchill, R.C. [1988]. Uniqueness and non-uniqueness of normal forms for vector fields. *Proc. Roy. Soc. Edinburgh Sect. A* **108**, 27–33.

Bartlett, J.H. [1982]. Limits of stability for an area-preserving polynomial mapping. *Celestial Mech.* **28**, 295–317.

Benedicks, M. and Carleson, L. [1988]. The Dynamics of the Hénon Map. preprint.

Benettin, G., Galgani, L., Giorgilli, A., and Strelcyn, J.-M. [1980a]. Lyapunov characteristic exponents for smooth dynamical systems and for Hamiltonian systems; a method for computing all of them, Part 1: Theory. *Meccanica* **15**, 9–20.

Benettin, G., Galgani, L., Giorgilli, A., and Strelcyn, J.-M. [1980b]. Lyapunov characteristic exponents for smooth dynamical systems and for

Hamiltonian systems; a method for computing all of them, Part II: Numerical application. *Meccanica* **15**, 21–30.

Birkhoff, G.D. [1927]. *Dynamical Systems.* A.M.S. Coll. Publications, vol. 9, reprinted 1966. American Mathematical Society: Providence.

Birkhoff, G.D. [1935]. Nouvelles Recherches sur les systèmes dynamiques. *Mem. Point. Acad. Sci. Novi. Lyncaei* **1**, 85–216.

Birman, J.S. and Williams, R.F. [1983a]. Knotted periodic orbits in dynamical systems I: Lorenz's equations. *Topology* **22**, 47–82.

Birman, J.S. and Williams, R.F. [1983b]. Knotted periodic orbits in dynamical systems II: Knot holders for fibred knots. *Contemp. Math.* **20**, 1–60.

Bogdanov, R.I. [1975]. Versal deformations of a singular point on the plane in the case of zero eigenvalues. *Functional Anal. Appl.* **9**(2), 144–145.

Bost, J. [1986]. Tores invariants des systèmes dynamiques Hamiltoniens. *Astérisque* 133–134. 113–157.

Bowen, R. [1970]. Markov partitions for Axiom A diffeomorphisms. *Amer. J. Math.* **92**, 725–747.

Bowen, R. [1978]. *On Axiom A Diffeomorphisms.* CBMS Regional Conference Series in Mathematics, vol. 35. A.M.S. Publications: Providence.

Boyce, W.E. and DiPrima, R.C. [1977]. *Elementary Differential Equations and Boundary Value Problems.* Wiley: New York.

Broer, H.W. and Vegter, G. [1984]. Subordinate Sil'nikov bifurcations near some singularities of vector fields having low codimension. *Ergodic Theory and Dynamical Systems* **4**, 509–525.

Bruhn, B. [1989]. Homoclinic bifurcations in simple parametrically driven systems. *Ann. Physik* **46**, 367–375.

Bryuno, A.D. [1989]. *Local Methods in Nonlinear Differential Equations. Part I. The Local Method of Nonlinear Analysis of Differential Equations. Part II. The Sets of Analyticity of a Normalizing Transformation.* Springer-Verlag: New York, Heidelberg, Berlin.

Bylov, B.F., Vinograd, R.E., Grobman, D.M., and Nemyckii, V.V. [1966]. *Theory of Liapunov Characteristic Numbers.* Moscow (Russian).

Byrd, P.F. and Friedman, M.D. [1971]. *Handbook of Elliptic Integrals for Scientists and Engineers.* Springer-Verlag: New York, Heidelberg, Berlin.

Carr, J. [1981]. *Applications of Center Manifold Theory.* Springer-Verlag: New York, Heidelberg, Berlin.

Carr, J., Chow, S.-N., and Hale, J.K. [1985]. Abelian integrals and bifurcation theory. *J. Differential Equations* **59**, 413–436.

Celletti, A. and Chierchia, L. [1988]. Construction of analytic KAM surfaces and effective stability bounds. *Comm. Math. Phys.* **118**, 119–161.

Channon, S.R. and Lebowitz, J.L. [1980]. Numerical experiments in stochasticity and homoclinic oscillation. *Ann. New York Acad. Sci.* **357**, 108–118.

Chillingworth, D.R.J. [1976]. *Differentiable Topology with a View to Applications.* Pitman: London.

Chorin, A.J. and Marsden, J.E. [1979]. *A Mathematical Introduction to Fluid Mechanics.* Springer-Verlag: New York, Heidelberg, Berlin.

Chow, S.-N. and Hale, J.K. [1982]. *Methods of Bifurcation Theory.* Springer-Verlag: New York, Heidelberg, Berlin.

Chow, S.-N., Li, C., and Wang, D. [1989]. Uniqueness of periodic orbits of some vector fields with codimension two singularities. *J. Differential Equations* **77**, 231–253.

Conley, C. [1978]. *Isolated Invariant Sets and the Morse Index.* CBMS Regional Conference Series in Mathematics, vol. 38. American Mathematical Society: Providence.

Coullet, P. and Spiegel, E.A. [1983]. Amplitude equations for systems with competing instabilities. *SIAM J. Appl. Math.* **43**, 774–819.

Cushman, R. and Sanders, J.A. [1986]. Nilpotent normal forms and representation theory of $sl(2, R)$. In *Multi-Parameter Bifurcation Theory,* M. Golubitsky and J. Guckenheimer (eds.) Contemporary Mathematics, vol. 56. American Mathematical Society, Providence.

Devaney, R.L. [1982]. Blowing up singularities in classical mechanical systems. *Amer. Math. Monthly* **89**(8), 535–552.

Devaney, R.L. [1986]. *An Introduction to Chaotic Dynamical Systems.* Benjamin/Cummings: Menlo Park, CA.

Dubrovin, B.A., Fomenko, A.T., and Novikov, S.P. [1984]. *Modern Geometry—Methods and Applications, Part I. The Geometry of Surfaces, Transformation Groups, and Fields.* Springer-Verlag: New York, Heidelberg, Berlin.

Dubrovin, B.A., Fomenko, A.T., and Novikov, S.P. [1985]. *Modern Geometry—Methods and Applications, Part II. The Geometry and Topology of Manifolds.* Springer-Verlag: New York, Heidelberg, Berlin.

Dugundji, J. [1966]. *Topology.* Allyn and Bacon: Boston.

Easton, R.W. [1986]. Trellises formed by stable and unstable manifolds in the plane. *Trans. Amer. Math. Soc.* **294**, 714–732.

Elphick, C., Tirapegui, E., Brachet, M.E., Coullet, P., and Iooss, G. [1987]. A simple global characterization for normal forms of singular vector fields. *Physica* **29D**, 95–127.

Erneux, T. and Mandel, P. [1986]. Imperfect bifurcation with a slowly varying control parameter. *SIAM J. Appl. Math.* **46**, 1–16.

Evans, J.W., Fenichel, N., and Feroe, J.A. [1982]. Double impulse solutions in nerve axon equations. *SIAM J. Appl. Math.* **42**(2), 219–234.

Fenichel, N. [1971]. Persistence and smoothness of invariant manifolds for flows. *Indiana Univ. Math. J.* **21**, 193–225.

Fenichel, N. [1974]. Asymptotic stability with rate conditions. *Indiana Univ. Math. J.* **23**, 1109–1137.

Fenichel, N. [1977]. Asymptotic stability with rate conditions, II. *Indiana Univ. Math. J.* **26**, 81–93.

Fenichel, N. [1979]. Geometric singular perturbation theory for ordinary differential equations. *J. Differential Equations* **31**, 53–98.

Franks, J.M. [1982]. *Homology and Dynamical Systems.* CBMS Regional Conference Series in Mathematics, vol. 49. A.M.S. Publications: Providence.

Galin, D.M. [1982]. Versal deformations of linear Hamiltonian systems. *Amer. Math. Soc. Trans.* **118**, 1–12.

Gambaudo, J.M. [1985]. Perturbation of a Hopf bifurcation by external time-periodic forcing. *J. Differential Equations* **57**, 172–199.

Gantmacher, F.R. [1977]. *Theory of Matrices,* vol. 1. Chelsea: New York.

Gantmacher, F.R. [1989]. *Theory of Matrices,* vol. 2. Chelsea: New York.

Gaspard, P. [1983]. Generation of a countable set of homoclinic flows through bifurcation. *Phys. Lett.* **97A**, 1–4.

Gaspard, P., Kapral, R., and Nicolis, G. [1984]. Bifurcation phenomena near homoclinic systems: A two parameter analysis. *J. Statist. Phys.* **35**, 697–727.

Gaspard, P. and Nicolis, G. [1983]. What can we learn from homoclinic orbits in chaotic systems? *J. Statist. Phys.* **31**, 499–518.

Gavrilov, N.K. and Silnikov, L.P. [1972]. On three dimensional dynamical systems close to systems with a structurally unstable homoclinic curve, I. *Math. USSR-Sb.* **17**, 467–485.

Gavrilov, N.K. and Silnikov, L.P. [1973]. On three dimensional dynamical systems close to systems with a structurally unstable homoclinic curve, II. *Math. USSR-Sb.* **19**, 139–156.

Gibson, C.G. [1979]. *Singular Points of Smooth Mappings.* Pitman: London.

Glendinning, P. and Sparrow, C. [1984]. Local and global behavior near homoclinic orbits. *J. Statist. Phys.* **35**, 645–696.

Goggin, M.E. and Milonni, P.W. [1988]. Driven Morse oscillator: Classical chaos, quantum theory, and photodissociation. *Phys. Rev. A* **37**, 796–806.

Goldhirsch, I., Sulem, P.-L., and Orszag, S.A. [1987]. Stability and Lyapunov stability of dynamical systems: A differential approach and a numerical method. *Physica* **27D**, 311–337.

Goldstein, H. [1980]. *Classical Mechanics,* 2nd ed. Addison-Wesley: Reading, MA.

Golubitsky, M. and Guillemin, V. [1973]. *Stable Mappings and Their Singularities.* Springer-Verlag: New York, Heidelberg, Berlin.

Golubitsky, M. and Schaeffer, D.G. [1985]. *Singularities and Groups in Bifurcation Theory,* vol. 1. Springer-Verlag: New York, Heidelberg, Berlin.

Golubitsky, M. and Stewart, I. [1987]. Generic bifurcation of Hamiltonian systems with symmetry. *Physica* **24D**, 391–405.

Golubitsky, M., Stewart, I., and Schaeffer, D.G. [1988]. *Singularities and Groups in Bifurcation Theory,* vol. 2. Springer-Verlag: New York, Heidelberg, Berlin.

Grebenikov, E.A. and Ryabov, Yu.A. [1983]. *Constructive Methods in the Analysis of Nonlinear Systems.* Mir: Moscow.

Grebogi, C., Oh, E., and Yorke, J.A. [1985]. Attractors on an N-torus: Quasiperiodicity versus chaos. *Physica* **15D**, 354–373.

Greenspan, B.D. and Holmes, P.J. [1983]. Homoclinic orbits, subharmonics, and global bifurcations in forced oscillations. In *Nonlinear Dynamics and Turbulence,* G. Barenblatt, G. Iooss, and D.D. Joseph (eds.), pp. 172–214. Pitman: London.

Greenspan, B.D. and Holmes, P.J. [1984]. Repeated resonance and homoclinic bifurcation in a periodically forced family of oscillators. *SIAM J. Math. Anal.* **15**, 69–97.

Grobman, D.M. [1959]. Homeomorphisms of systems of differential equations. *Dokl. Akad. Nauk SSSR* **128**, 880.

Guckenheimer, J. [1981]. On a codimension two bifurcation. In *Dynamical Systems and Turbulence,* D.A. Rand and L.S. Young (eds.), pp. 99–142. Springer Lecture Notes in Mathematics, vol. 898. Springer-Verlag: New York, Heidelberg, Berlin.

Guckenheimer, J. and Holmes, P.J. [1983]. *Nonlinear Oscillations, Dynamical Systems, and Bifurcations of Vector Fields.* Springer-Verlag: New York, Heidelberg, Berlin.

Guckenheimer, J. and Johnson, S. [1989]. Distortion of S-unimodal maps, preprint.

Guckenheimer, J. and Williams, R.F. [1980]. Structural stability of the Lorenz attractor. *Publ. Math. IHES* **50**, 73–100.

Haberman, R. [1979]. Slowly varying jump and transition phenomena associated with algebraic bifurcation problems. *SIAM J. Appl. Math.* **37**, 69–105.

Hadamard, J. [1898]. Les surfaces à curbures opposés et leurs lignes géodesiques. *Journ. de Math.* **5**, 27–73.

Hale, J. [1980]. *Ordinary Differential Equations.* Robert E. Krieger Publishing Co., Inc.: Malabar, Florida.

Hale, J.K. and Lin, X.-B. [1986]. Symbolic dynamics and nonlinear semiflows. *Ann. Mat. Pura Appl.* **144**(4), 224–259.

Hartman, P. [1960]. A lemma in the theory of structural stability of differential equations. *Proc. Amer. Math. Soc.* **11**, 610–620.

Hassard, B.D., Kazarinoff, N.D., and Wan, Y.-H. [1980]. *Theory and Applications of the Hopf Bifurcation.* Cambridge University Press: Cambridge.

Hastings, S. [1982]. Single and multiple pulse waves for the Fitzhugh–Nagumo equations. *SIAM J. Appl. Math.* **42**, 247–260.

Hausdorff, [1962]. *Set Theory.* Chelsea: New York.

Henry, D. [1981]. *Geometric Theory of Semilinear Parabolic Equations.* Springer Lecture Notes in Mathematics, vol. 840. Springer-Verlag: New York, Heidelberg, Berlin.

Herman, M.R. [1988]. Existence et non existence de Tores Invariants par des diffeomorphismes symplectiques, preprint.

Hirsch, M.W. [1976]. *Differential Topology.* Springer-Verlag: New York, Heidelberg, Berlin.

Hirsch, M.W., Pugh, C.C., and Shub, M. [1977]. *Invariant Manifolds.* Springer Lecture Notes in Mathematics, vol. 583. Springer-Verlag: New York, Heidelberg, Berlin.

Hirsch, M.W. and Smale, S. [1974]. *Differential Equations, Dynamical Systems, and Linear Algebra.* Academic Press: New York.

Holmes, C. and Holmes, P.J. [1981]. Second order averaging and bifurcations to subharmonics in Duffing's equation. *J. Sound and Vibration* **78**(4), 161–174.

Holmes, C.A. and Wood, D. [1985]. Studies of a complex Duffing equation in nonlinear waves on plane Poiseuille flow, preprint, Imperial College, London.

Holmes, P.J. [1980]. A strange family of three-dimensional vector fields near a degenerate singularity. *J. Differential Equations* **37**, 382–404.

Holmes, P.J. [1985]. Dynamics of a nonlinear oscillator with feedback control I: Local analysis. *Trans. ASME J. Dyn. Sys. Control* **107**, 159–165.

Holmes, P.J. [1986a]. Knotted periodic orbits in suspensions of Smale's horseshoe: Period multiplying and cabled knots. *Physica* **21D**, 7–41.

Holmes, P.J. [1986b]. Spatial structure of time-periodic solutions of the Ginzburg–Landau equation. *Physica* **23D**, 84–90.

Holmes, P.J. [1987]. Knotted periodic orbits in suspensions of annulus maps. *Proc. Roy. Soc. London Ser. A* **411**, 351–378.

Holmes, P.J. and Marsden, J.E. [1982]. Horseshoes in perturbation of Hamiltonian systems with two degrees of freedom. *Comm. Math. Phys.* **82**, 523–544.

Holmes, P.J., Marsden, J.E., and Scheurle, J. [1988]. Exponentially small splitting of separatrices with applications to KAM theory and degenerate bifurcations. Contemporary Mathematics, vol. 81, pp. 213–243. American Mathematical Society: Providence.

Holmes, P.J. and Moon, F.C. [1983]. Strange attractors in nonlinear mechanics. *Trans. ASME J. Appl. Mech.* **50**, 1021–1032.

Holmes, P.J. and Rand, D.A. [1978]. Bifurcations of the forced van der Pol oscillator. *Quart. Appl. Math.* **35**, 495–509.

Holmes, P.J. and Whitley, D.C. [1984]. On the attracting set for Duffing's equation I: Analytical methods for small force and damping. In *Partial Differential Equations and Dynamical Systems,* W. Fitzgibbon III (ed.), pp. 211–240. Pitman: London.

Holmes, P.J. and Williams, R.F. [1985]. Knotted periodic orbits in suspensions of Smale's horseshoe: Torus knots and bifurcation sequences. *Arch. Rational Mech. Anal.* **90**, 115–194.

Hopf, E. [1942]. Abzweigung einer periodischen Lösung von einer stationären Lösung eines Differentialsystems. *Ber. Math. Phys. Sächsische Akademie der Wissenschaften Leipzig* **94**, 1–22 (see also the English translation in Marsden and McCracken [1976]).

Iooss, G. [1979]. *Bifurcation of Maps and Applications.* North Holland: Amsterdam.

Iooss, G. and Langford, W.F. [1980]. Conjectures on the routes to turbulence via bifurcation. In *Nonlinear Dynamics,* R.H.G. Helleman (ed.), pp. 489–505. New York Academy of Sciences: New York City, NY.

Jakobson, M.V. [1981]. Absolutely continuous invariant measures for one-parameter families of one-dimensional maps. *Comm. Math. Phys.* **81**, 39–88.

Johnson, R.A. [1986]. Exponential dichotomy, rotation number, and linear differential operators with bounded coefficients. *J. Differential Equations* **61**, 54–78.

Johnson, R.A. [1987]. m-Functions and Floquet exponents for linear differential systems. *Ann. Mat. Pura Appl. (4)* vol. CXLVII, 211–248.

Johnson, S. [1987]. Singular measures without restrictive intervals. *Comm. Math. Phys.* **110**, 185–190.

Kaper, T.J., Kovačič, G., and Wiggins, S. [1989]. Melnikov functions, action, and lobe area in Hamiltonian systems, preprint, California Institute of Technology, Pasadena, CA.

Kaplan, B.Z. and Kottick, D. [1983]. Use of a three-phase oscillator model for the compact representation of synchronomous generators. *IEEE Trans. Magn.,* vol. MAG-19, 1480–1486.

Kaplan, B.Z. and Kottick, D. [1985]. A compact representation of synchronous motors and unregulated synchronous generators. *IEEE Trans. Magn.,* vol. MAG-21, 2657–2663.

Kaplan, B.Z. and Kottick, D. [1987]. Employment of three-phase compact oscillator models for representing comprehensively two synchronous generator systems. *Elect. Mach. Power Systems* **12**, 363–375.

Kaplan, B.Z. and Yardeni, D. [1989]. Possible chaotic phenomenon in a three-phase oscillator. *IEEE Trans. Circuits and Systems* **36**(8), 1148–1151.

Kato, T. [1980]. *Perturbation Theory for Linear Operators,* Corrected 2nd Ed. Springer-Verlag: New York, Heidelberg, Berlin.

Katok, A. and Bernstein, D. [1987]. Birkhoff periodic orbits for small perturbations of completely integrable Hamiltonian systems with convex Hamiltonians. *Invent. Math.* **88**, 225–241.

Kelley, A. [1967]. The stable, center-stable, center, center-unstable, unstable manifolds. An appendix in *Transversal Mappings and Flows,* R. Abraham and J. Robbin. Benjamin: New York.

Kevorkian, J. and Cole, J.D. [1981]. *Perturbation Methods in Applied Mathematics.* Springer-Verlag: New York, Heidelberg, Berlin.

Kocak, H. [1984]. Normal forms and versal deformations of linear Hamiltonian systems. *J. Differential Equations* **51**, 359–407.

Kopell, N. and Howard, L.N. [1975]. Bifurcations and trajectories joining critical points. *Adv. in Math.* **18**, 306–358.

Kummer, M. [1971]. How to avoid "secular" terms in classical and quantum mechanics. *Nuovo Cimento B,* 123–148.

Landau, L.D. and Lifschitz, E.M. [1976]. *Mechanics.* Pergamon: Oxford.

Landman, M.J. [1987]. Solutions of the Ginzburg–Landau equation of interest in shear flow transition. *Stud. Appl. Math.* **76**(3), 187–238.

Langford, W.F. [1979]. Periodic and steady mode interactions lead to tori. *SIAM J. Appl. Math.* **37**(1), 22–48.

Langford, W.F. [1985]. A review of interactions of Hopf and steady-state bifurcations. In *Nonlinear Dynamics and Turbulence,* G. Barenblatt, G. Iooss, and D.D. Joseph (eds.), pp. 215–237. Pitman: London.

LaSalle, J.P. and Lefschetz, S. [1961]. *Stability by Liapunov's Direct Method.* Academic Press: New York.

Lebovitz, N.R. and Schaar, R.J. [1975]. Exchange of stabilities in autonomous systems. *Stud. Appl. Math.* **54**, 229–260.

Lebovitz, N.R. and Schaar, R.J. [1977]. Exchange of stabilities in autonomous systems, II. Vertical bifurcations. *Stud. Appl. Math.* **56**, 1–50.

Lerman, L.M. and Silnikov, L.P. [1989]. Homoclinic structures in infinite-dimensional systems. *Siberian Math. J.* **29**(3), 408–417.

Levinson, N. [1949]. A second order differential equation with singular solutions. *Ann. Math.* **50**, 127–153.

Lewis, D. and Marsden, J. [1989]. A Hamiltonian-dissipative decomposition of normal forms of vector fields, preprint, University of California-Berkeley.

Liapunov, A.M. [1966]. *Stability of Motion.* Academic Press: New York.

Lichtenberg, A.J. and Lieberman, M.A. [1982]. *Regular and Stochastic Motion.* Springer-Verlag: New York, Heidelberg, Berlin.

de la Llave, R. and Rana, D. [1988]. Accurate strategies for small divisor problems, preprint.

Lochak, P. and Meunier, C. [1988]. *Multiphase Averaging for Classical Systems.* Springer-Verlag: New York, Heidelberg, Berlin.

McGehee, R. [1974]. Triple collision in the collinear three body problem. *Invent. Math.* **27**, 191–227.

MacKay, R.S. [1988]. A criterion for non-existence of invariant tori for Hamiltonian systems. *Physica D* (to appear).

MacKay, R.S., Meiss, J.D., and Percival, I.C. [1984]. Transport in Hamiltonian systems. *Physica* **13D**, 55–81.

MacKay, R.S., Meiss, J.D., and Stark, J. [1989]. Converse KAM theory for symplectic twist maps, preprint.

MacKay, R.S. and Percival, I.C. [1985]. Converse KAM: Theory and practice. *Comm. Math. Phys.* **98**, 469–512.

Mandel, P. and Erneux, T. [1987]. The slow passage through a steady bifurcation: Delay and memory effects. *J. Statist. Phys.* **48**, 1059–1070.

Marsden, J.E. and McCracken, M. [1976]. *The Hopf Bifurcation and Its Applications.* Springer-Verlag: New York, Heidelberg, Berlin.

Mather, J. [1982]. Existence of quasi-periodic orbits for twist maps of the annulus, *Topology* **21**(4), 457–467.

Mather, J. [1984]. Non-existence of invariant circles. *Ergodic Theory Dynamical Systems* **4**, 301–311.

Mather, J. [1986]. A criterion for the non-existence of invariant circles. *Publ. Math. IHES* **63**, 153–204.

Melnikov, V.K. [1963]. On the stability of the center for time periodic perturbations. *Trans. Moscow Math. Soc.* **12**, 1–57.

Meyer, K.R. [1986]. Counter-examples in dynamical systems via normal form theory. *SIAM Rev.* **28**, 41–51.

Milnor, J. [1985]. On the concept of attractor. *Comm. Math. Phys.* **99**, 177–195.

Misiurewicz, M. [1981]. The structure of mapping of an interval with zero entropy. *Publ. Math. IHES* **53**, 5–16.

Mitropol'skii, Y.A. [1965]. *Problems of the Asymptotic Theory of Nonstationary Vibrations.* Israel Program for Scientific Translations.

Modi, V.S. and Brereton, R.C. [1969]. Periodic solutions associated with the gravity–gradient-oriented system: Part I. Analytical and numerical determination. *AIAA J.* **7**, 1217–1225.

Morozov, A.D. [1976]. A complete qualitative investigation of Duffing's equation. *Differential Equations* **12**, 164–174.

Morozov, A.D. and Silnikov, L.P. [1984]. On nonconservative periodic systems close to two-dimensional Hamiltonian. *PMM USSR* **47**, 327–334.

Morse, M. and Hedlund, G.A. [1938]. Symbolic dynamics. *Amer. J. Math.* **60**, 815–866.

Moser, J. [1966a]. A rapidly convergent iteration method and non-linear partial differential equations, I. *Ann. Scuola Norm. Sup. Pisa Cl. Sci.* **20**(2), 265–315.

Moser, J. [1966b]. A rapidly convergent iteration method and non-linear partial differential equations, II. *Ann. Scuola Norm. Sup. Pisa Cl. Sci.* **20**(3), 499–535.

Moser, J. [1968]. Lectures on Hamiltonian systems. *Mem. Amer. Math. Soc.* **81**, American Mathematical Society: Providence.

Moser, J. [1973]. *Stable and Random Motions in Dynamical Systems.* Princeton University Press: Princeton.

Moses, E. and Steinberg, V. [1988]. Mass transport in propagating patterns of convection. *Phys. Rev. Lett.* **60**(20), 2030–2033.

Murdock, J. and Robinson, C. [1980]. Qualitative dynamics from asymptotic expansions: Local theory. *J. Differential Equations* **36**, 425–441.

Naimark, J. [1959]. On some cases of periodic motions depending on parameters. *Dokl. Akad. Nauk. SSSR* **129**, 736–739.

Nayfeh, A.H. and Mook, D.T. [1979]. *Nonlinear Oscillations.* John Wiley: New York.

Nehorošev, N.N. [1972]. Action-angle variables and their generalizations. *Trans. Moscow Math. Soc.* **26**, 180–198.

Neishtadt, A.I. [1987]. Persistence of stability loss for dynamical bifurcations, I. *Differential Equations* **23**, 1385–1391.

Neishtadt, A.I. [1988]. Persistence of stability loss for dynamical bifurcations, II. *Differential Equations* **24**, 171–176.

Nemytskii, V.V. and Stepanov, V.V. [1989]. *Qualitative Theory of Differential Equations.* Dover: New York.

Newell, A.C. [1985]. *Solitons in Mathematics and Physics.* CBMS-NSF Regional Conference Series in Applied Mathematics, vol. 48, SIAM: Philadelphia.

Newhouse, S.E. [1974]. Diffeomorphisms with infinitely many sinks. *Topology* **13**, 9–18.

Newhouse, S.E. [1979]. The abundance of wild hyperbolic sets and non-smooth stable sets for diffeomorphisms. *Publ. Math. IHES* **50**, 101–151.

Newhouse, S.E. [1980]. Lectures on dynamical systems. In *Dynamical Systems.* C.I.M.E. Lectures, Bressanone, Italy, June 1978, pp. 1–114. Birkhauser: Boston.

Newhouse, S.E. [1983]. Generic properties of conservative systems. In *Chaotic Behavior of Deterministic Systems.* Les Houches 1981, G. Iooss, R.H.G. Helleman, and R. Stora (eds.). North-Holland: Amsterdam, New York.

Newhouse, S. and Palis, J. [1973]. Bifurcations of Morse–Smale dynamical systems. In *Dynamical Systems,* M.M. Peixoto (ed.). Academic Press: New York, London.

Newton, P.K. and Sirovich, L. [1986a]. Instabilities of the Ginzburg–Landau equation: Periodic solutions. *Quart. Appl. Math.* **44**(1), 49–58.

Newton, P.K. and Sirovich, L. [1986b]. Instabilities of the Ginzburg–Landau equation: Part II, secondary bifurcation. *Quart. Appl. Math.* **44**(2), 367–374.

Nitecki, Z. [1971]. *Differentiable Dynamics.* M.I.T. Press: Cambridge.

Ottino, J.M. [1989]. *The Kinematics of Mixing: Stretching, Chaos, and Transport.* Cambridge University Press: Cambridge.

Olver, P.J. [1986]. *Applications of Lie Groups to Differential Equations.* Springer-Verlag: New York, Heidelberg, Berlin.

Olver, P.J. and Shakiban, C. [1988]. Dissipative decomposition of ordinary differential equations. *Proc. Roy. Soc. Edinburgh Sect. A* **109**, 297–317.

Oseledec, V.I. [1968]. A multiplicative ergodic theorem. Liapunov characteristic numbers for dynamical systems. *Trans. Moscow Math. Soc.* **19**, 197–231.

Palis, J. and deMelo, W. [1982]. *Geometric Theory of Dynamical Systems: An Introduction.* Springer-Verlag: New York, Heidelberg, Berlin.

Peixoto, M.M. [1962]. Structural stability on two-dimensional manifolds. *Topology* **1**, 101–120.

Percival, I.C. [1979]. Variational principles for invariant tori and cantori. In *Nonlinear Dynamics and the Beam-Beam Interaction,* M. Month and J.C. Herrera (eds.), *Am. Inst. of Phys. Conf. Proc.* **57**, 302–310.

Percival, I. and Richards, D. [1982]. *Introduction to Dynamics.* Cambridge University Press: Cambridge.

Pesin, Ja.B. [1976]. Families of invariant manifolds corresponding to nonzero characteristic exponents. *Math. USSR-Izv.* **10**(6), 1261–1305.

Pesin, Ja.B. [1977]. Characteristic Lyapunov exponents and smooth ergodic theory. *Russian Math. Surveys* **32**(4), 55–114.

Pikovskii, A.S., Rabinovich, M.I., and Trakhtengerts, V.Yu. [1979]. Onset of stochasticity in decay confinement of parametric instability. *Soviet Phys. JETP* **47**, 715–719.

Pliss, V.A. [1964]. The reduction principle in the theory of stability of motion. *Soviet Math.* **5**, 247–250.

Plykin, R. [1974]. Sources and sinks for A-diffeomorphisms. *Math. USSR-Sb.* **23**, 233–253.

Poincaré, H. [1892]. *Les Méthodes Nouvelles de la Mécanique Céleste,* vol. I. Gauthier-Villars: Paris.

Poincaré, H. [1899]. *Les Méthodes Nouvelles de la Mécanique Céleste,* 3 vols. Gauthier-Villars: Paris.

Poincaré, H. [1929]. Sur les propriétés des fonctions définies par les équations aux différences partielles. *Oeuvres,* Gauthier-Villars: Paris, pp. XCIX–CX.

Rabinovich, M.I. [1978]. Stochastic self-oscillations and turbulence. *Soviet Phys. Uspekhi* **21**, 443–469.

Rabinovich, M.I. and Fabrikant, A.L. [1979]. Stochastic self-oscillation of waves in non-equilibrium media. *Soviet Phys. JETP* **50**, 311–323.

Rand, R.H. and Armbruster, D. [1987]. *Perturbation Methods, Bifurcation Theory and Computer Algebra.* Springer-Verlag: New York, Heidelberg, Berlin.

Robinson, C. [1983]. Bifurcation to infinitely many sinks. *Comm. Math. Phys.* **90**, 433–459.

Robinson, C. [1988]. Bifurcation to a transitive attractor of Lorenz type, preprint.

Rom-Kedar, V., Leonard, A., and Wiggins, S. [1989]. An analytical study of transport, mixing, and chaos in an unsteady vortical flow. *J. Fluid Mech.* (to appear).

Rom-Kedar, V. and Wiggins, S. [1989]. Transport in two-dimensional maps. *Arch. Rational Mech. Anal.* (in press).

Roux, J.C., Rossi, A., Bachelart, S., and Vidal, C. [1981]. Experimental observations of complex dynamical behavior during a chemical reaction. *Physica* **2D**, 395–403.

Rudin, W. [1964]. *Principles of Mathematical Analysis.* McGraw-Hill: New York.

Ruelle, D. [1973]. Bifurcations in the presence of a symmetry group. *Arch. Rational Mech. Anal.* **51**, 136–152.

Ruelle, D. [1981]. Small random perturbations of dynamical systems and the definition of attractors. *Comm. Math. Phys.* **82**, 137–151.

Rychlik, M. [1987]. Lorenz attractors through Silnikov type bifurcations. Part I, preprint.

Sacker, R.S. [1965]. On invariant surfaces and bifurcations of periodic solutions of ordinary differential equations. *Comm. Pure Appl. Math.* **18**, 717–732.

Sacker, R.J. and Sell, G.R. [1974]. Existence of dichotomies and invariant splittings for linear differential systems I. *J. Differential Equations* **15**, 429–458.

Sacker, R.J. and Sell, G.R. [1976a]. Existence of dichotomies and invariant splittings for linear differential systems II. *J. Differential Equations* **22**, 478–496.

Sacker, R.J. and Sell, G.R. [1976b]. Existence of dichotomies and invariant splittings for linear differential systems III. *J. Differential Equations* **22**, 497–522.

Sacker, R.J. and Sell, G.R. [1978]. A spectral theory for linear differential systems. *J. Differential Equations* **27**, 320–358.

Sacker, R.J. and Sell, G.R. [1980]. The spectrum of an invariant submanifold. *J. Differential Equations* **37**, 135–160.

Sanders, J.A. and Verhulst, F. [1985]. *Averaging Methods in Nonlinear Dynamical Systems.* Springer-Verlag: New York, Heidelberg, Berlin.

Schecter, S. [1985]. Persistent unstable equilibria and closed orbits of a singularly perturbed system. *J. Differential Equations* **60**, 131–141.

Schecter, S. [1987a]. The saddle-node separatrix-loop bifurcation. *SIAM J. Math. Anal.* **18**, 1142–1156.

Schecter, S. [1987b]. Melnikov's method of a saddle-node and the dynamics of the forced Josephson junction. *SIAM J. Math. Anal.* **18**, 1699–1715.

Schecter, S. [1988]. Stable manifolds in the method of averaging. *Trans. Amer. Math. Soc.* **308**, 159–176.

Scheurle, J. and Marsden, J.E. [1984]. Bifurcation to quasi-periodic tori in the interaction of steady state and Hopf bifurcations. *SIAM J. Math. Anal.* **15**(6), 1055–1074.

Schwartz, A.J. [1963]. A generalization of a Poincaré–Bendixson theorem to closed two-dimensional manifolds. *Amer. J. Math.* **85**, 453–458; errata, ibid. **85**, 753.

Sell, G.R. [1971]. *Topological Dynamics and Differential Equations.* Van Nostrand-Reinhold: London.

Sell, G.R. [1978]. The structure of a flow in the vicinity of an almost periodic motion. *J. Differential Equations* **27**, 359–393.

Shub, M. [1987]. *Global Stability of Dynamical Systems.* Springer-Verlag: New York, Heidelberg, Berlin.

Siegel, C.L. [1941]. On the Integrals of Canonical Systems. *Ann. Math.* **42**, 806–822.

Siegel, C.L. and Moser, J.K. [1971]. *Lectures on Celestial Mechanics.* Springer-Verlag: New York, Heidelberg, Berlin.

Sijbrand, J. [1985]. Properties of center manifolds. *Trans. Amer. Math. Soc.* **289**, 431–469.

Silnikov, L.P. [1965]. A Case of the Existence of a Denumerable Set of Periodic Motions. *Sov. Math. Dokl.* **6**, 163–166.

Sinai, J.G. and Vul, E. [1981]. Hyperbolicity conditions for the Lorenz model. *Physica* **2D**, 3–7.

Smale, S. [1963]. Diffeomorphisms with many periodic points. In *Differential and Combinatorial Topology,* S.S. Cairns (ed.), pp. 63–80. Princeton University Press: Princeton.

Smale, S. [1966]. Structurally stable systems are not dense. *Amer. J. Math.* **88**, 491–496.

Smale, S. [1967]. Differentiable dynamical systems. *Bull. Amer. Math. Soc.* **73**, 747–817.

Smale, S. [1980]. *The Mathematics of Time: Essays on Dynamical Systems, Economic Processes and Related Topics.* Springer-Verlag: New York, Heidelberg, Berlin.

Smoller, J. [1983]. *Shock Waves and Reaction-Diffusion Equations.* Springer-Verlag: New York, Heidelberg, Berlin.

Šošitaĭšvili, A.N. [1975]. Bifurcations of topological type of a vector field near a singular point. *Trudy Sem. Petrovsk.* **1**, 279–309.

Sparrow, C. [1982]. *The Lorenz Equations.* Springer-Verlag: New York, Heidelberg, Berlin.

Stark, J. [1988]. An exhaustive criterion for the non-existence of invariant circles for area-preserving twist maps. *Comm. Math. Phys.* **117**, 177–189.

Sternberg, S. [1957]. On local C^n contractions of the real line. *Duke Math. J.* **24**, 97–102.

Sternberg, S. [1957]. Local contractions and a theorem of Poincaré. *Amer. J. Math.* **79**, 809–824.

Sternberg, S. [1958]. On the structure of local homeomorphisms of Euclidean n-space, II. *Amer. J. Math.* **80**, 623–631.

Takens, F. [1974]. Singularities of vector fields. *Publ. Math. IHES* **43**, 47–100.

Takens, F. [1979]. Forced oscillations and bifurcations. *Comm. Math. Inst. Rijksuniv. Utrecht* **3**, 1–59.

Tedeschini-Lalli, L. and Yorke, J.A. [1986]. How often do simple dynamical processes have infinitely many coexisting sinks? *Comm. Math. Phys.* **106**, 635–657.

Tresser, C. [1984]. About some theorems by L.P. Silnikov. *Ann. Inst. H. Poincaré* **40**, 440–461.

van Gils, S.A. [1985]. A note on "Abelian integrals and bifurcation theory." *J. Differential Equations* **59**, 437–441.

van der Meer, J.-C. [1985]. *The Hamiltonian Hopf Bifurcation.* Springer Lecture Notes in Mathematics, vol. 1160. Springer-Verlag: New York, Heidelberg, Berlin.

Vyshkind, S.Ya. and Rabinovich, M.I. [1976]. The phase stochastization mechanism and the structure of wave turbulence in dissipative media. *Soviet Phys. JETP* **44**, 292–299.

Weiss, J.B. and Knobloch, E. [1989]. Mass transport and mixing by modulated traveling waves. *Phys. Rev. A* (to appear).

Wiggins, S. [1988]. *Global Bifurcations and Chaos: Analytical Methods.* Springer-Verlag: New York, Heidelberg, Berlin.

Wiggins, S. and Holmes, P.J. [1987a]. Periodic orbits in slowly varying oscillators. *SIAM J. Math. Anal.* **18**, 542–611.

Wiggins, S. and Holmes, P.J. [1987b]. Homoclinic orbits in slowly varying oscillators. *SIAM J. Math. Anal.* **18**, 612–629. (See also 1988, *SIAM J. Math. Anal.* **19**, 1254–1255, errata.)

Williams, R.F. [1980]. Structure of Lorenz attractors. *Publ. Math. IHES* **80**, 59–72.

Yanagida, E. [1987]. Branching of double pulse solutions from single pulse solutions in nerve axon equations. *J. Differential Equations* **66**, 243–262.

Yorke, J.A. and Alligood, K.T. [1985]. Period doubling cascades of attractors: A prerequisite for horseshoes. *Comm. Math. Phys.* **101**, 305–321.

Zoladek, H. [1984]. On the versality of symmetric vector fields in the plane. *Math. USSR-Sb.* **48**, 463–492.

Zoladek, H. [1987]. Bifurcations of certain family of planar vector fields tangent to axes. *Differential Equations* **67**, 1–55.

Index